JACARANDA SCIENCE QUEST 8

AUSTRALIAN CURRICULUM | FOURTH EDITION

JACARANDA SCIENCE QUEST 8

AUSTRALIAN CURRICULUM | FOURTH EDITION

GRAEME LOFTS

MERRIN J. EVERGREEN

CONTRIBUTING AUTHORS

Sarah Beamish | Simone Meakin | Billie Murray | Adele Norton | Robert Stokes

REVIEWED BY

Courtney Rubie, Wiradjuri woman

Fourth edition published 2023 by
John Wiley & Sons Australia, Ltd
Level 4, 600 Bourke Street, Melbourne, Vic 3000

First edition published 2011
Second edition published 2015
Third edition published 2018

Typeset in 10.5/13 pt TimesLT Std

ISBN: 978 1 394 15119 6

Front cover images: © LEE WOODGATE / IKON IMAGES / SCIENCE PHOTO LIBRARY

Illustrated by various artists, diacriTech and Wiley Composition Services

Typeset in India by diacriTech

A catalogue record for this book is available from the National Library of Australia

Printed in Singapore
M WEP311141 210924

The Publishers of this series acknowledge and pay their respects to Aboriginal Peoples and Torres Strait Islander Peoples as the traditional custodians of the land on which this resource was produced.

This suite of resources may include references to (including names, images, footage or voices of) people of Aboriginal and/or Torres Strait Islander heritage who are deceased. These images and references have been included to help Australian students from all cultural backgrounds develop a better understanding of Aboriginal and Torres Strait Islander Peoples' history, culture and lived experience.

It is strongly recommended that teachers examine resources on topics related to Aboriginal and/or Torres Strait Islander Cultures and Peoples to assess their suitability for their own specific class and school context. It is also recommended that teachers know and follow the guidelines laid down by the relevant educational authorities and local Elders or community advisors regarding content about all First Nations Peoples.

All activities in this resource have been written with the safety of both teacher and student in mind. Some, however, involve physical activity or the use of equipment or tools. **All due care should be taken when performing such activities**. To the maximum extent permitted by law, the author and publisher disclaim all responsibility and liability for any injury or loss that may be sustained when completing activities described in this resource.

The Publisher acknowledges ongoing discussions related to gender-based population data. At the time of publishing, there was insufficient data available to allow for the meaningful analysis of trends and patterns to broaden our discussion of demographics beyond male and female gender identification.

CONTENTS

About this resource

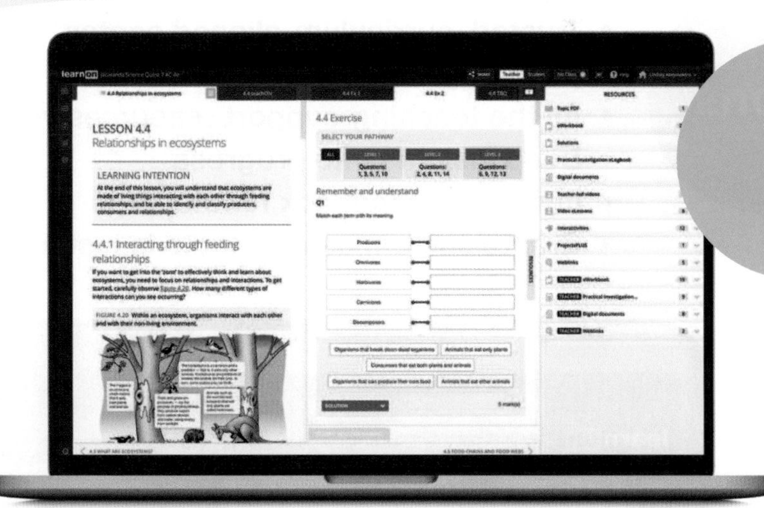

NEW FOR

AUSTRALIAN CURRICULUM V9.0

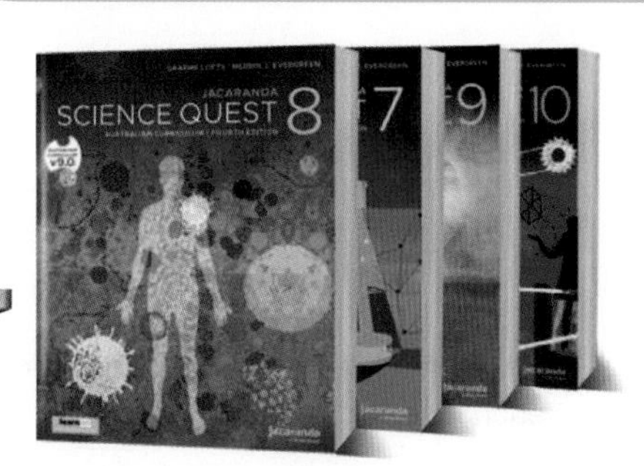

JACARANDA

SCIENCE QUEST 8 AUSTRALIAN CURRICULUM

FOURTH EDITION

Developed by teachers for students

Tried, tested and trusted. Every lesson in the new *Jacaranda Science Quest* series has been carefully designed to support teachers and help students succeed by sparking their curiosity about the world around them.

Because both what and how students learn matter

Learning is personal

Whether students need a challenge or a helping hand, you'll find what you need to create engaging lessons.

Whether in class or at home, students can access carefully scaffolded and sequenced lessons to support in-depth Science Inquiry Skills development and step students through scientific inquiry with engaging interactive content and practical investigations. Automatically marked, differentiated question sets are all supported by detailed solutions — so students can get unstuck and progress!

Learning is effortful

Learning happens when students push themselves. With learnON, Australia's most powerful online learning platform, students can challenge themselves, build confidence and ultimately achieve success.

Learning is rewarding

Through real-time results data, students can track and monitor their own progress and easily identify areas of strength and weakness.

And for teachers, Learning Analytics provide valuable insights to support student growth and drive informed intervention strategies.

Learn online with Australia's most

Everything you need for each of your lessons in one simple view

- **Trusted, curriculum-aligned content**
- **Engaging, rich multimedia**
- **All the teaching-support resources you need**
- **Deep insights into progress**
- **Immediate feedback for students**
- **Create custom assignments in just a few clicks.**

Practical teaching advice and ideas for each lesson provided in teachON

Reading content and rich media including embedded videos and interactivities

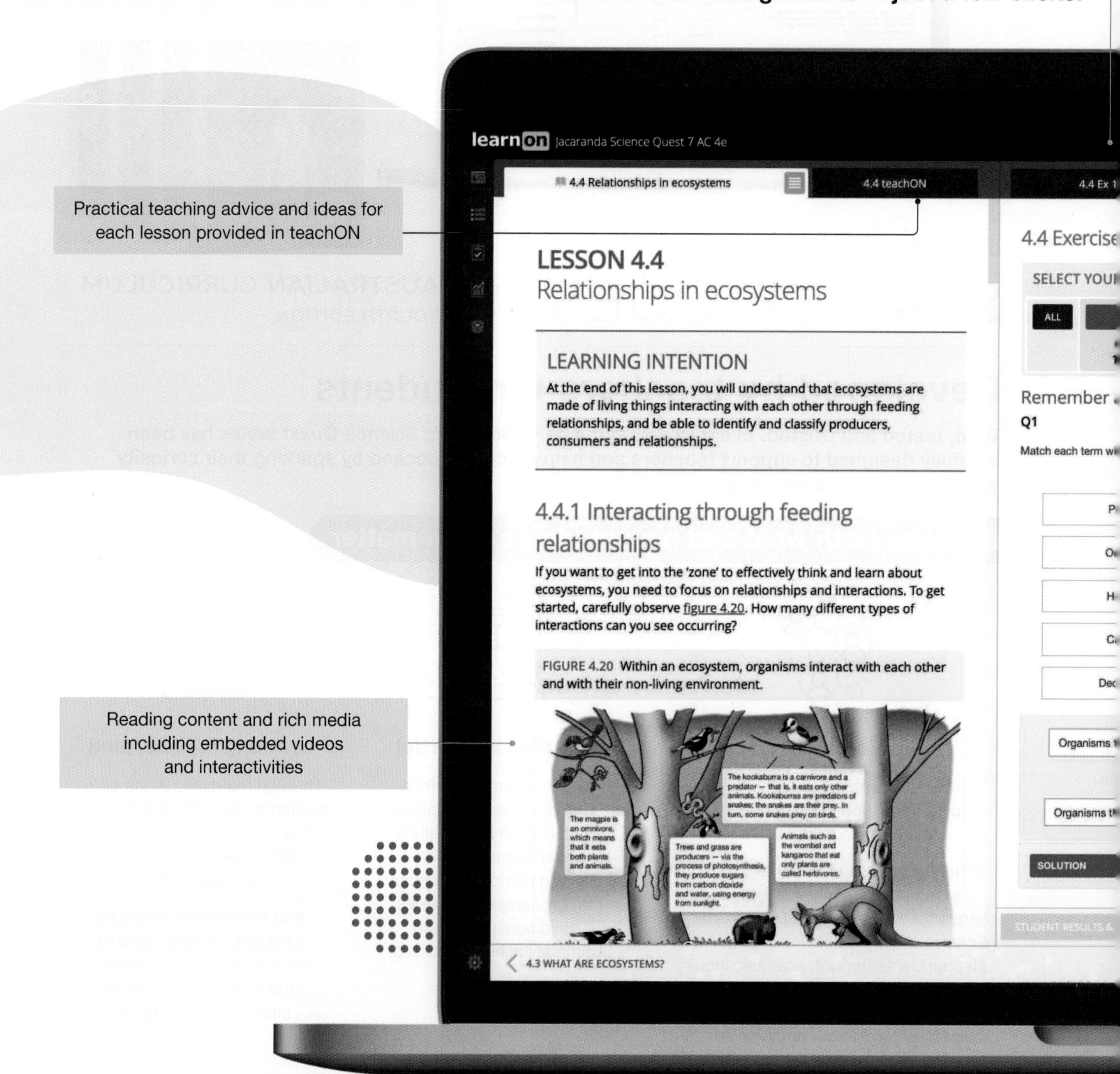

powerful learning tool, learnON

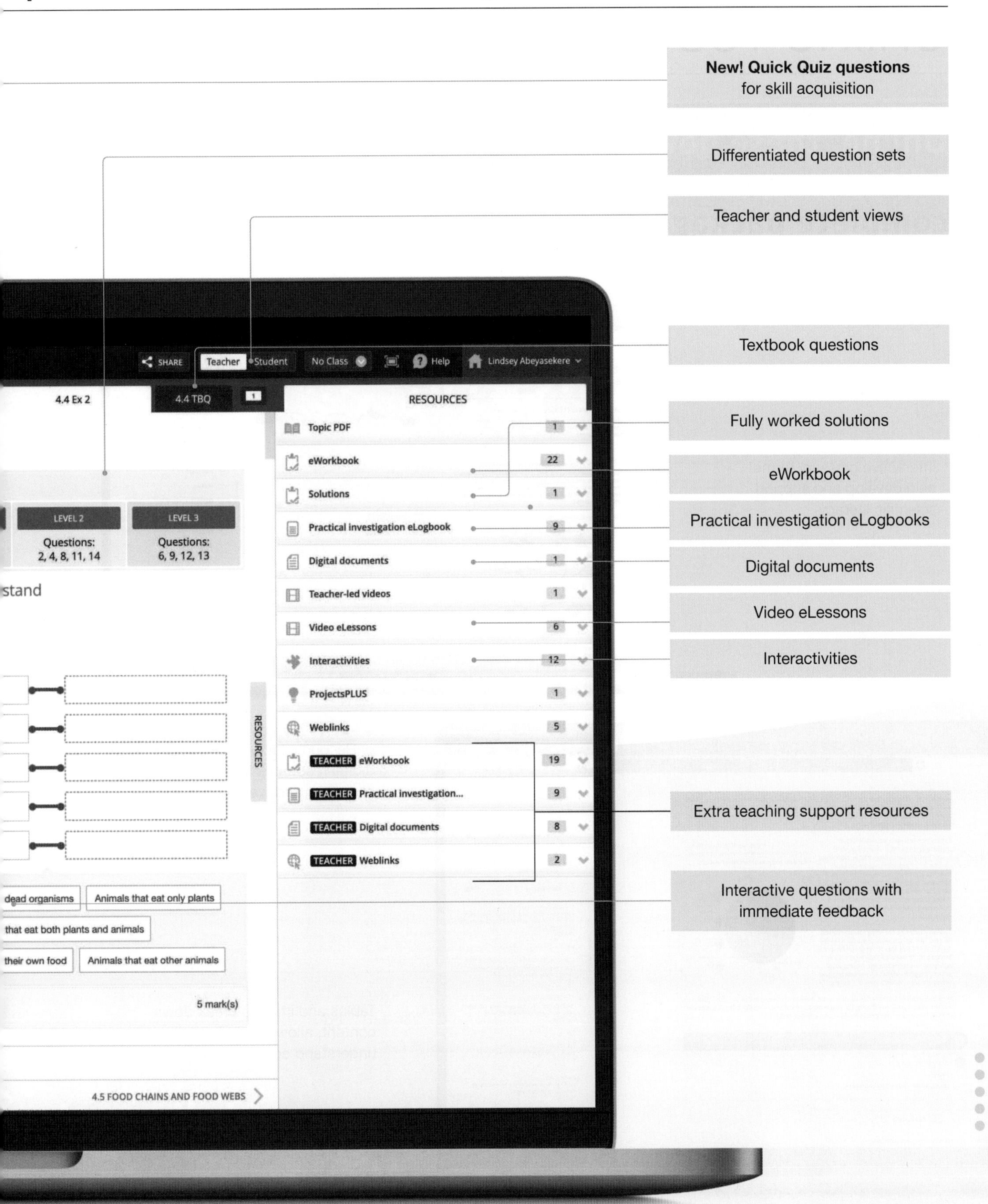

Get the most from your online resources

Online, these new editions are the **complete package**

Trusted Jacaranda theory, plus tools to support teaching and make learning more engaging, personalised and visible.

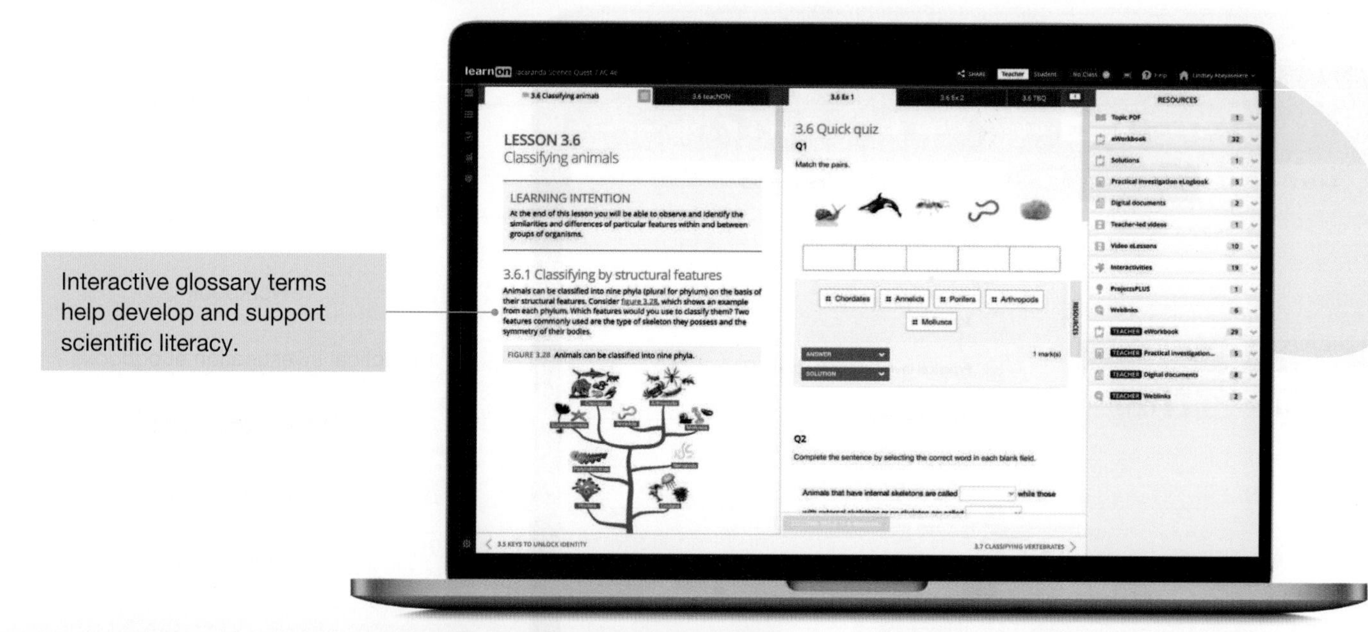

Interactive glossary terms help develop and support scientific literacy.

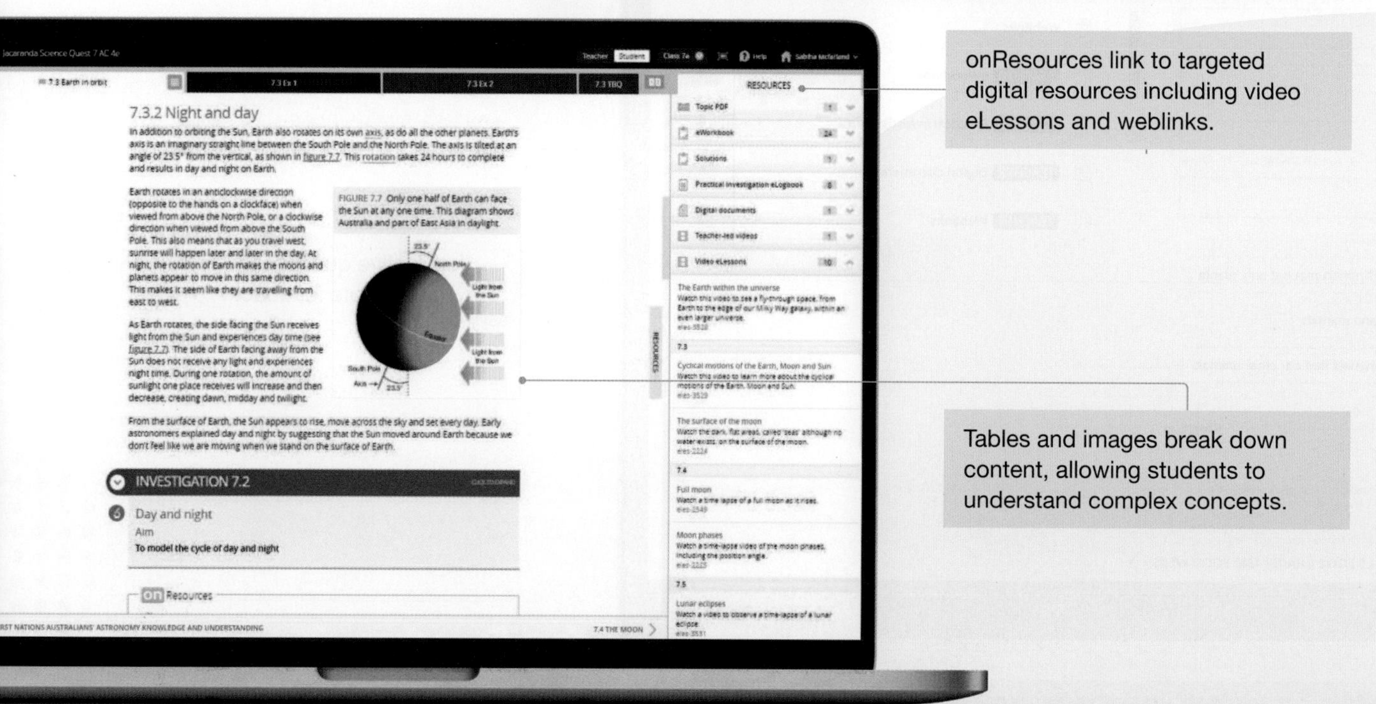

onResources link to targeted digital resources including video eLessons and weblinks.

Tables and images break down content, allowing students to understand complex concepts.

rand new! Quick
uiz questions for skill
cquisition in every
esson.

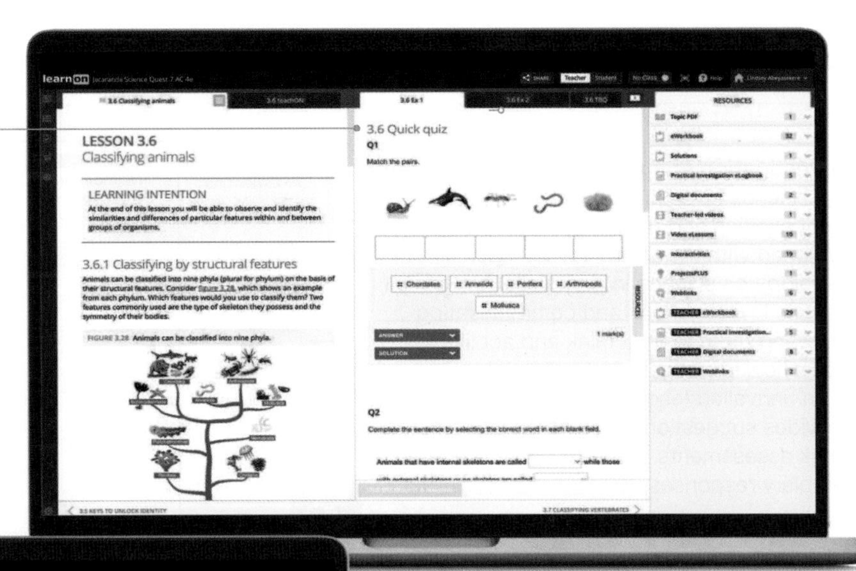

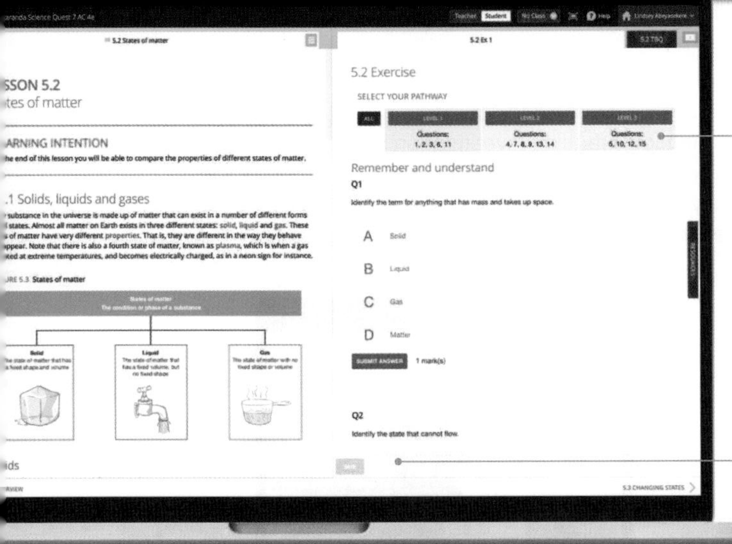

Three differentiated question sets, with immediate feedback in every lesson, enable students to challenge themselves at their own level.

Instant reports give students visibility into progress and performance.

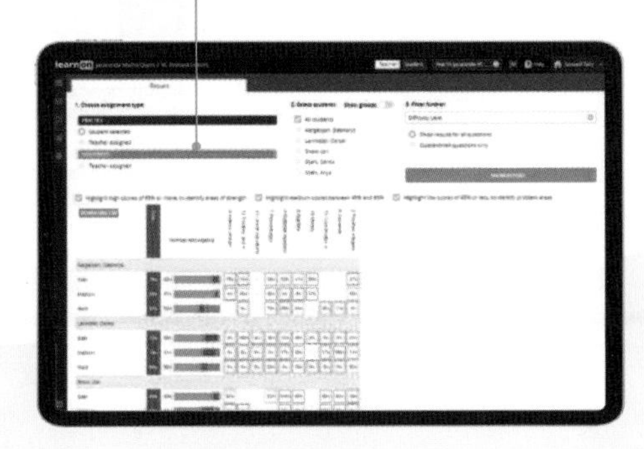

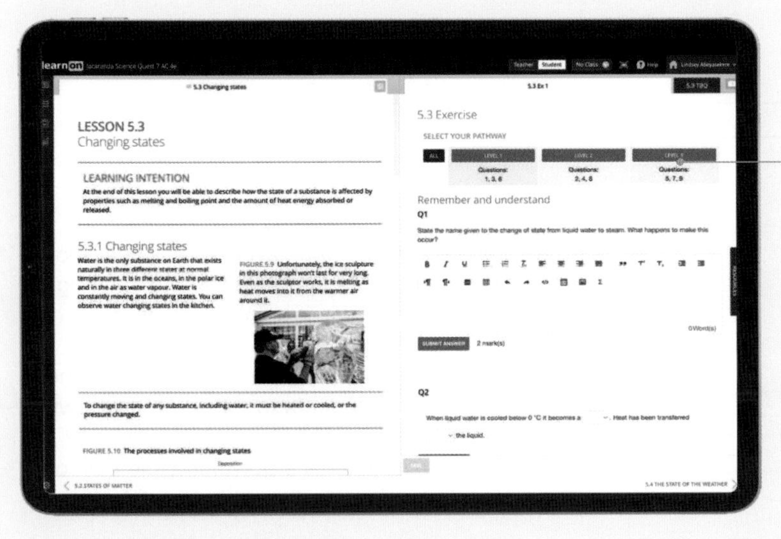

Every question has immediate, corrective feedback to help students overcome misconceptions as they occur and get unstuck as they study independently — in class and at home.

Practical Investigation eLogbook

The **practical investigation eLogbook** ignites curiosity through science investigation work, with an extensive range of exciting and meaningful practical investigations. Aligned with the scientific method, students can develop rich science inquiry skills in conducting scientific investigations and communicating their findings, allowing them to truly think and act like scientists! The practical investigation eLogbook is supported with an unrivalled teacher and laboratory guide, which provides suggestions for differentiation and alteration, risk assessments, expected practical results and exemplary responses.

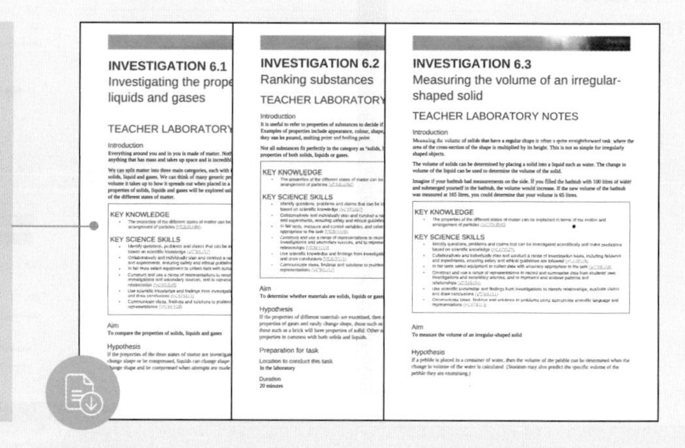

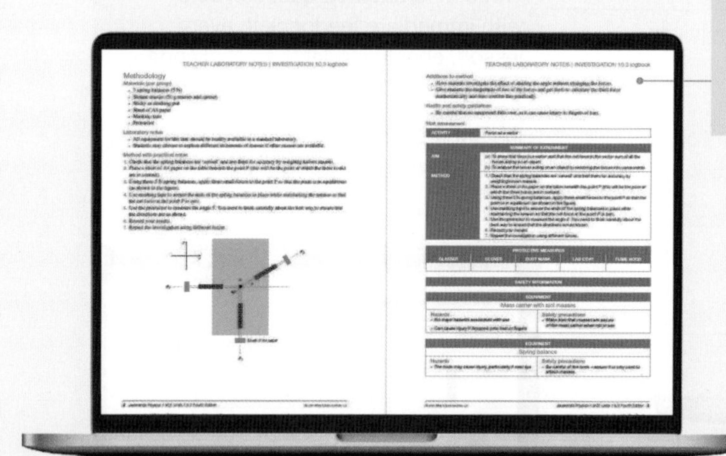

Enhanced practical investigation support includes practical investigation videos and an eLogbook with fully customisable practical investigations — including teacher advice and risk assessments.

eWorkbook

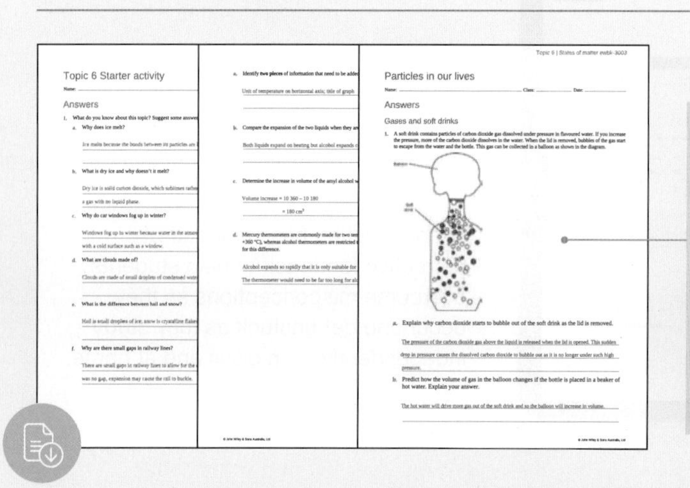

The **eWorkbook** is the perfect companion to the series, adding another layer of individualised learning opportunities for students, and catering for multiple entry and exit points in student learning. The eWorkbook also features fun and engaging activities for students of all abilities and offers a space for students to reflect on their own learning. The new eWorkbook and eWorkbook solutions are available as a downloadable PDF or a customisable Word document in learnON.

A wealth of teacher resources

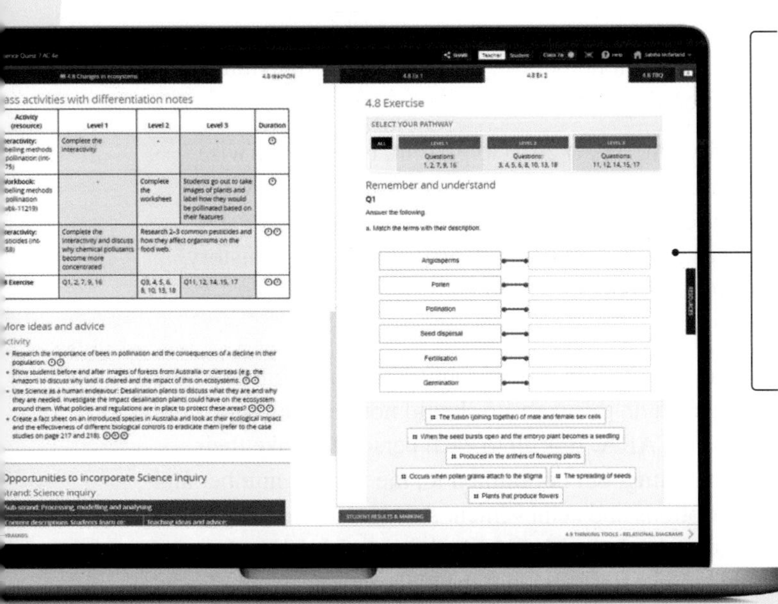

Enhanced teaching-support resources for every lesson, including:

- work programs and curriculum grids
- practical teaching advice
- three levels of differentiated teaching programs
- quarantined topic tests (with solutions)

Customise and assign

An inbuilt testmaker enables you to create custom assignments and tests from the complete bank of thousands of questions for immediate, spaced and mixed practice.

Reports and results

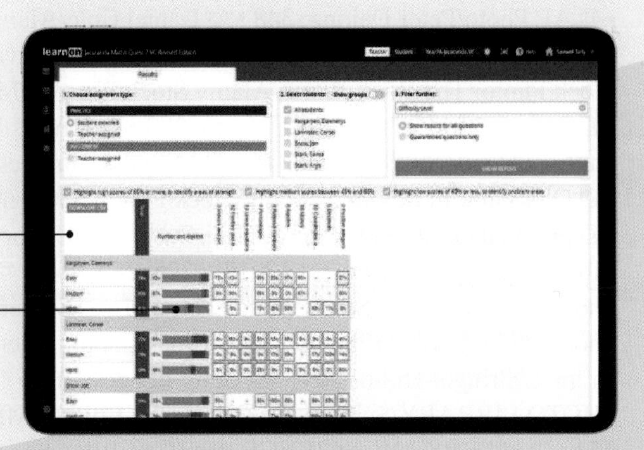

Data analytics and instant reports provide data-driven insights into progress and performance within each lesson and across the entire course.

Show students (and their parents or carers) their own assessment data in fine detail. You can filter their results to identify areas of strength and weakness.

Acknowledgements

The authors and publisher would like to thank the following copyright holders, organisations and individuals for their assistance and for permission to reproduce copyright material in this book.

Images

• © wollertz/Adobe Stock Photos: **392** • © monticello/Adobe Stock: **521** • © juananbarros/Adobe Stock Photos: **15** • © Archivist/Adobe Stock Photos: **260** • © zsv3207/Adobe Stock Photos: **29** • © Orlando Florin Rosu/Adobe Stock Photos: **429** • © Steve/Adobe Stock Photos: **496** • © microscience/Adobe Stock Photos: **172** • © vkilikov/Adobe Stock Photos: **552** • © wowinside/Adobe Stock Photos: **343** • © Sergey Nivens/Adobe Stock Photos: **139** • © CA Irene Lorenz/Adobe Stock Photos: **465** • © antikwar1/Adobe Stock Photos: **211** • © fizkes/Adobe Stock Photos: **22** • © josehidalgo87/Adobe Stock Photos: **227** • © Soumyadip/Adobe Stock Photos: **552** • © Виктория Попова/Adobe Stock: **282** • © fotofabrika/Adobe Stock Photos: TimurD/Adobe Stock Photos, TADDEUS/Adobe Stock Photos, **271** • © Chris/Adobe Stock Photos: **488** • © magicmine/Adobe Stock Photos: **150** • © oliver de la haye/Adobe Stock Photos: **467** • © Thomas/Adobe Stock Photo: **28** • © Alex Photo/Adobe Stock Photos: **22** • © Luis/Adobe Stock Photos: **135** • © BillionPhotos.com/Adobe Stock Photos: **112** • © playwalker/Adobe Stock Photos: **501** • © POWER AND SYRED/SCIENCE PHOTO LIBRARY: **199** • © By Unknown author - hp.ujf.cas.cz (uploader=–Kuebi 18:28, 10 April 2007 (UTC)), Public Domain, https://commons.wikimedia.org/w/index.php?curid=20426111, **50** • © Pascal Goetgheluck/Science Source: **122** • © Viaframe/Stone/Getty Images: **181** • © Source: Mooneydriver/iStockphot: **327** • © calcassa/Science Photo Library: **452** • © GARY HINCKS/SCIENCE PHOTO LIBRARY: **479** • © Hemera/Hemera/Getty Images: **402** • © AP Photo/Peter Dejong: **348** • © Daniel Cole/Alamy Stock Photo: **82** • © GRANGER - Historical Picture Archvie/Alamy Stock Photo: **459** • © Greg Balfour Evans/Alamy Stock Photo: **499** • © Historic Images/Alamy Stock Photo: The Book Worm/Alamy Stock Photo, **274** • © Imaginechina Limited/Alamy Stock Photo: **348** • © Janet Horton/Alamy Stock Photo: **283** • © Juniors Bildarchiv GmbH/Alamy Stock Photo: **379** • © mark phillips/Alamy Stock Photo: **346** • © Milleflore Images - Holidays Events/Alamy Stock Photo: **282** • © Oleksandr Prykhodko/Alamy Stock Photo: **520** • © picturelibrary/Alamy Stock Photo: **183** • © pOrbital.com/Shutterstock: **275** • © Science History Images/Alamy Stock Photo: **276** • © Science Photo Library/Alamy Stock Photo: **12, 129, 275, 283, 469** • © Science Photo Library/Alamy Stock Photo: Science Photo Library/Alamy Stock Photo, **274, 494** • © Steven May/Alamy Stock Photo: **267** • © UPI/Alamy Stock Photo: **492** • © WorldPhotos/Alamy Stock Photo: Science History Images/Alamy Stock Photo, **275** • © Hulton Archive/Stringer/Hulton Archive/Getty Images: **263** • © PaulMaguire/iStock/Getty Images: **421** • © SCIENCE PHOTO LIBRARY/Science Photo Library: **179** • © MIKKEL JUUL JENSEN/SCIENCE PHOTO LIBRARY: **454** • © MAREK MIS/SCIENCE PHOTO LIBRARY: **63** • © Shannon Morris/Newspix: **51** • © Science Photo

Library/Getty Images: **169** • © STEVE GSCHMEISSNER/SCIENCE PHOTO LIBRARY/Science Photo Library/Getty Images: **192** • © AstroStar/Shutterstock: **1** • © Coprid/Shutterstock: **17** • © Everett Historica/Shutterstock: **332** • © Keith Wheatley/Shutterstock: **336** • © Morphart Creation/Shutterstock: **453** • © Aldona Griskeviciene/Shutterstock: **111** • © Aleksandr Pobedimskiy/Shutterstock: **402** • © Andrew Burgess/Shutterstock: **308** • © Andrew Mayovskyy/Shutterstock: **375** • © Antoine2K/Shutterstock: **267** • © AridOcean/Shutterstock: **511** • © ArtOfPhotos/Shutterstock: **380** • © Artography/Shutterstock: **405** • © AuntSpray/Shutterstock: **423** • © Barbol/Shutterstock: SIRIKWAN DOKUTA/Shutterstock, **71** • © boscorelli/Shutterstock: **522** • © ChameleonsEye/Shutterstock: **529** • © Chu KyungMin/Shutterstock: **192** • © cpaulfell/Shutterstock: By Carlofnetcong - Own work, CC BY-SA 4.0, https://commons.wikimedia.org/w/index.php?curid=47899375, **351** • © Damsea/Shutterstock: **88** • © Darrenp/Shutterstock: **498** • © Darryl Brooks/Shutterstock: **500** • © dotshock/Shutterstock: **548** • © Dr Morley Read/Shutterstock: **477** • © ESB Professional/Shutterstock: **549** • © Faizal Ramli/Shutterstock: **499** • © Federico Rostagno/Shutterstock: **408** • © fen deneyim/Shutterstock: **521** • © fran_kie/Shutterstock: **18** • © Frans Delian/Shutterstock: **500** • © Georgios Kollidas/Shutterstock: **48, 263** • © ggw/Shutterstock: **307** • © gresei/Shutterstock: **314** • © hpphtns/Shutterstock: **551** • © ian woolcock/Shutterstock: **495** • © Infomages - Sam Lee/Shutterstock: **201** • © James Steidl/Shutterstock: **380** • © Jason Patrick Ross/Shutterstock: **493** • © Javani LLC/Shutterstock: **377** • © JurateBuiviene/Shutterstock: **360** • © Kaspars Grinvalds/Shutterstock: **524** • © Kevin Case/Shutterstock: **64** • © Kirill Chernyshev/Shutterstock: **390** • © kmls/Shutterstock: **515** • © Kzenon/Shutterstock: **24** • © mansong suttakarn/Shutterstock: **220** • © Mark Nazh/Shutterstock: **5** • © Matthijs Wetterauw/Shutterstock: **472** • © Merlin74/Shutterstock: **425** • © Michael Rosskothen/Shutterstock: **518** • © michal812/Shutterstock: **393, 405** • © MRS.Siwaporn/Shutterstock: **315** • © mTaira/Shutterstock: **486** • © netsuthep/Shutterstock: **428** • © Oleg Mikhaylov/Shutterstock: **520** • © Olivier Le Moal/Shutterstock: **556** • © panitanphoto/Shutterstock: **20** • © Roschetzky Photography/Shutterstock: **543** • © Rubi Rodriguez Martinez/Shutterstock: **490** • © schankz/Shutterstock: **523** • © Barbol/Shutterstock: **87** • © sirtravelalot/Shutterstock: **448** • © Sleepyhobbit/Shutterstock: **522** • © solarseven/Shutterstock: **521** • © SpeedKingz/Shutterstock: **13** • © Steven Tritton/Shutterstock: **338** • © Trofimov Denis/Shutterstock: **523** • © Tyler Boyes/Shutterstock: **380, 393** • © urbanbuzz/Shutterstock: **312** • © Valdis Skudre/Shutterstock: **95** • © Vasily Smirnov/Shutterstock: **333** • © Vibe Images/Shutterstock: **390** • © Yes058/Shutterstock: Montree/Adobe Stock Photos, **391** • © tkyszk/Shutterstock: **525** • © SCIENCE PHOTO LIBRARY: **180** • © Getty Images/Universal Images Gr: **386** • © LIBRARY OF CONGRESS/SCIENCE PHOTO LIBRARY: **180** • © LIBRARY OF CONGRESS/SCIENCE PHOTO LIBRARY: **459** • © Caspar Benson/fStop/Alamy Stock Photo: **331** • © Stephen Barnes/Animals/Alamy Stock Photo: **228** • © Sueddeutsche Zeitung Photo/Alamy Stock Photo: **48** • © OMIKRON/SCIENCE PHOTO LIBRARY: **180** • © QA INTERNATIONAL/SCIENCE SOURCE/SCIENCE PHOTO LIBRARY: **491** • © active nerve cells: Bundles of collagen fibers in a connective tissue stained with a silver method. The collagen fibers show a typical wavy appearance, Light micrograph showing a simple columnar epithelium. Human gallbladder. H&E stain, WMM - Striated cardiac myocytes showing yellow pigment lipofuscin granules near the nuclei. The myocytes are joined by intercalated disks. Light micrograph. H&E stain, LM of human blood smear showing red & white cells, **234** • © Adwo/Shutterstock: **482** • © Amanda Padbury: **351** • © Andreas Bjerkeholt/Shutterstock: **271** • © Andrei Shumskiy/Shutterstock: **258** • © Apaterson/Shutterstock: **424** • © Apollofoto/Shutterstock: **355** • © Baciu/Shutterstock: **423** • © Be.sign/Adobe Stock: GVS/Adobe Stock, **543** • © BillionPhotos.com/Adobe Stock Photos: Cookie Studio/Adobe Stock Photos, Cookie Studio/Adobe Stock Photos, carballo/Adobe Stock Photos, **303** • © Brett M. Lewis: **74** • © BrunoWeltmann/Shutterstock: **297** • © Budimir Jevtic/Shutterstock: **361** • © By Dr Graham Beards - Own work, CC BY-SA 3.0, https://commons.wikimedia.org/w/index.php?curid=20521666, **177** • © By Islander61 - Own work, CC BY-SA 4.0, https://commons.wikimedia.org/w/index.php?curid=95028362, **337** • © By KarlaPanchuk - Own work, CC BY-SA 4.0, https://commons.wikimedia.org/w/index.php?curid=46200993, **381** • © By Leonardo da Vinci - http://www.ntv.ru/novosti/151536/, Public Domain, https://commons.wikimedia.org/w/index.php?curid=9931161, **133** • © By Leonardo da Vinci, Public Domain, https://commons.wikimedia.org/w/index.php?curid=42138639, **133** • © By Mulletsrokk - Own work, CC BY-SA 3.0, https://commons.wikimedia.org/w/index.php?curid=10270500, **123** • © By Mx. Granger - Own work, CC0, https://commons.wikimedia.org/w/index.php?curid=35170896, www.sandatlas.org/Shutterstock, **408** • © By Rita Greer - The original is an oil painting on board by Rita Greer, history painter, 2009. This was digitized by Rita and sent via email to the

Department of Engineering Science, Oxford University, where it was subsequently uploaded to Wikimedia., FAL, https://commons.wikimedia.org/w/index.php?curid=7667280, Chronicle/Alamy Stock Photo, ggw/Adobe Stock Photos, **67** • © By Zatelmar - Own work, CC BY-SA 3.0, https://commons.wikimedia.org/w/index.php?curid=2777766, Lebendkulturen.de/Shutterstock, M I (Spike) Walker/Alamy Stock Photo, **72** • © calcassa/Adobe Stock Photos: **452** • © Commonwealth of Australia (Geoscience Australia) 2020: **396** • © constantincornel/Adobe Stock Photos: **106** • © Corepics VOF/Shutterstock: **409** • © Courtesy of the University of Gothenburg: **184** • © Courtesy: Taronga Zoo: **57** • © Cre8tive Images/Shutterstock: DNY59/E+/Getty Images, **316** • © Creativa Images/Shutterstock: **163** • © Creative Commons: **395** • © Daniel Cole/Alamy Stock Photo: **349** • © dc222/Adobe Stock Photos: **130** • © desdemona72/Shutterstock: **9** • © Designua/Shutterstock: **200** • © Dimarion/Shutterstock: **94** • © Dinoton/Shutterstock: **376** • © Dominik Hladik/Shutterstock: **311** • © Don Hammond/Design Pics/Corbis: **447** • © donsimon/Shutterstock: **302** • © Downunderphoto/Adobe Stock Photos: **424** • © DR P. MARAZZI/Science Photo Library (SPL): Proxima Studio/Shutterstock, **347** • © Dr. Amy Engevik: **74** • © Dr. Grigorii Timin and Prof. Michel C. Milinkovitch: **74** • © Dr. Olivier Leroux: **74** • © Eacham Historical Society: **416, 417** • © Ethan Daniels/Shutterstock: **211** • © Ewa Studio/Shutterstock: **38** • © gali estrange/Shutterstock: **46** • © GARY HINCKS/SCIENCE PHOTO LIBRARY: **458, 487** • © ggw1962/Shutterstock: **317** • © Gino Santa Maria/Shutterstock: **368** • © Gordon Saunders/Shutterstock: **340, 370** • © Greg Rakozy/Unsplash: **255** • © HEARing CRC: **137** • © Heritage Image Partnership Ltd/Alamy Stock Photo: **134** • © ianlusung/Shutterstock: GoodStudio/Shutterstock, siridhata/Shutterstock, ghrzuzudu/Shutterstock, **90** miles/Shutterstock, lady-luck/Shutterstock, Abstract man 24/Shutterstock, tynyuk/Shutterstock, Hafiz1902 Studio/Shutterstock, Glinskaja Olga/Shutterstock, **90** miles/Shutterstock, Sailor Johnny/Shutterstock, **527** • © ilikestudio/Shutterstock: **230** • © Illustration: Peter Trusler: Australia's Megafauna; ©Australian Postal Corporation 2008, **432** • © Image courtesy of the National Archives of Australia. A6135: **K28**/12/84/16 - People - Painter on Sydney Harbour Bridge, **1984, 335** • © Image Point Fr/Shutterstock: **200** • © Image Point Fr/Shutterstock: ruigsantos/Shutterstock, **184** • © imagedb.com/Shutterstock: **315** • © Ivan Cholakov/Shutterstock: **265** • © Jakub Krechowicz/Shutterstock: **134** • © James St. John/Flickr: **291** • © James Steidl/Shutterstock: **309, 517** • © janaka Dharmasena/Adobe Stock: **134** • © John Crux Photography/Moment Open/Getty Images: **444** • © John Wiley & Sons Australia/Photo by Coo-ee Picture Library: **281** • © John Wiley & Sons Australia/Photo by Coo-ee Picture Library: **271** • © Kate Kunz/Corbis/Getty Images: **8** • © Keith Corrigan/Alamy Stock Photo: **132** • © Kenishirotie/Shutterstock: ktsdesign/Shutterstock, ViktoriyaFivko/Shutterstock, **528** • © khuruzero/Shutterstock: **174** • © Lebendkulturen.de/Shutterstock: Lebendkulturen.de/Shutterstock, Picture Partners/Shutterstock, **87** • © Lightspring/Shutterstock: **356** • © Lindsey Moore/Shutterstock: **369** • © Lisa/Adobe Stock Photos: **208** • © Lou-Foto/Alamy Stock Photo: **498** • © Maksym Gorpenyuk/Shutterstock: **516** • © marcel/Adobe Stock Photos: **270** • © Matthijs Wetterauw/Shutterstock: **403** • © Maya Kruchankova/Shutterstock: **369** • © MDPI: By Mike Renlund - Seismic Footings, CC BY 2.0, https://commons.wikimedia.org/w/index.php?curid=12560348, **499** • © Melissa Brandes/Shutterstock: **211** • © michal812/Shutterstock: **397** • © molekuul.be/Shutterstock: **350** • © Monty Rakusen/Alamy Stock Photo: **32** • © Morne de Klerk/Newspix: **52** • © NASA: **52, 341, 350** • © Neale Taylor: **329** • © Neveshkin Nikolay/Shutterstock: **180** • © NigelSpiers/Shutterstock: **505** • © NOAA/SCIENCE PHOTO LIBRARY: **458** • © NOAA/USGS: **466** • © Ody_Stocker/Shutterstock: Standard Studio/Shutterstock, Nevada31/Shutterstock, **95** • © Oleg Mikhaylov/Shutterstock: wellphoto/Shutterstock, **517** • © Olga Danylenko/Shutterstock: **473** • © Olivier Le Queinec/Shutterstock: buruhtan/Shutterstock, ppart/Shutterstock, **544** • © Opas Chotiphantawanon/Shutterstock: **224** • © OrangeDeer/Adobe Stock Photos: molekuul_be/Shutterstock, **256** • © Pascal Warnant: **3** • © Pascale Warnant: **19, 45, 46** • © Patricia Hofmeester/Shutterstock: **39** • © Petr Bukal/Shutterstock: Arnont.tp/Shutterstock, **345** • © Pictorial Press Ltd/Alamy Stock Photo: **136** • © Radu Razvan/Shutterstock: **314** • © Rattiya Thongdumhyu/Shutterstock: **85** • © Richard Cisar-Wright/Newspix: **52** • © Rob kemp/Shutterstock: **391** • © Robert A. Smith/Australian Opal Centre: **422** • © Robyn Ohehir: **203** • © Roy LANGSTAFF/Alamy Stock Photo: Mirka Moksha/Shutterstock, Phil Hill Geo Pics/Alamy Stock Photo, **408** • © s-ts/Shutterstock: **538** • © SARIN KUNTHONG/Shutterstock: **339** • © Science History Images/Alamy Stock Photo: Rattiya Thongdumhyu/Shutterstock, **80** • © Science Photo Library/Alamy Stock Photo: **7** • © SGM/Shutterstock: **171** • © shawlin/Adobe Stock Photos: **449** • © shipfactory/Shutterstock: **27** • © Shutterstock: **297** • © sonsam/Shutterstock: **401** • © sonsam/Shutterstock: Orlando_Stocker/Shutterstock, Nicola Renna/Shutterstock, **6** • © Source: By Robertharrison95, CC BY-SA 3.0, https://commons.wikimedia.org/w/index.php?curid=106021583, **277** • © Source: After Johnson © Commonwealth of Australia Geoscience Australia 2020: **512** • © Source: Based on data from Department of Climate Change, Energy, the Environment and Water. 2022. Waste and Resource Recovery Data Hub - National waste data viewer, **362** • © Source: By Charles Janet, CC BY-SA 4.0, https://en.wikipedia.org/wiki/Alternative_periodic_tables, **276** • © Source: By DePiep - Own work, CC BY-SA 3.0, https://commons.wikimedia.org/w/index.php?curid=27766488, **276** • © Source: Community Safety – Earthquake by Geoscience Australia which is © Commonwealth of Australia and is provided under a Creative Commons Attribution 4.0 International Licence and is subject to the disclaimer of warranties in section 5 of that licence.: **485, 497** • © Source: Cooperative Institute for Meteorological Satellite Studies, University of Wisconsin-Madison, **456** • © Source: Future Directions for the National Earthquake Hazard Map - Scientific Figure on ResearchGate by Geoscience Australia which is © Commonwealth of Australia and is provided under a Creative Commons Attribution 4.0 International Licence and is subject to the disclaimer of warranties in section 5 of that licence.: **502** • © Source: MAPgraphics Pty Ltd: **455** • © Source: NDBC DART® Program Public Domain: **484** • © Source: Surachit: 2007. Oceanic spreading, Wikimedia Commons. Licensed under CC BY-SA 3.0., **463** • © Source: Vladimir Melnik/Shutterstock: **163** • © Source: ©Commonwealth of Australia Geoscience Australia 2020: **490** • © Source:By Rezmason - Own work, CC BY-SA 4.0, https://commons.wikimedia.org/w/index.php?curid=54218895, **277** • © SpeedKingz/Shutterstock: **2** • © Stephanie Frey/Shutterstock: **416** • © STEVE GSCHMEISSNER/SCIENCE PHOTO LIBRARY: **122** • © STEVE GSCHMEISSNER/SCIENCE PHOTO LIBRARY: POWER AND SYRED/SCIENCE PHOTO LIBRARY, SUSUMU NISHINAGA/SCIENCE PHOTO LIBRARY, **65** • © sumikophoto/Shutterstock: **403** • © sumstock/Shutterstock: **320** • © The Coeliac Society of Australia: The Coeliac Society of Australia, **165** • © The Nobel Foundation: **332** • © The University of Western Australia: **164** • © Tim Masters/Shutterstock: **5** • © Trevor Chriss/Alamy Stock Photo: **344** • © Tyler Boyes/Shutterstock: **391** • © Tyler Boyes/Shutterstock: Tyler Boyes/Shutterstock, Tyler Boyes/Shutterstock, **411** • © udaix/Shutterstock: **328** • © Val Thoermer/Shutterstock: **321** • © VectorMine/Adobe Stock Photos: **229** • © VectorMine/Shutterstock: **451** • © VectorMine/Adobe Stock Photos: **112** • © Wajarri Yamaji Aboriginal Corporation/Image courtesy of State Library of Western Australia: Wajarri Yamaji Aboriginal Corporation/Image courtesy of State Library of Western Australia, **418** • © wavebreakmedia/Shutterstock: **309** • © WavebreakMediaMicro/Adobe Stock Photos: Taras Grebinets/Adobe Stock Photos, **8** • © wawritto/Shutterstock: **14** • © WEHI. 1G Royal Pde Parkville 3052 VIC Australia. www.wehi.edu.au: **165** • © Wiley art: **137** • © www.sandatlas.org/Shutterstock: **392** • © XXLPhoto/Shutterstock: **402** • © Ari N/Shutterstock: **358** • © Excentro/Shutterstock: **359** • © frantic/Alamy Stock Photo: **525** • © Mega Pixel/Shutterstock: **357** • © nikkytok/Shutterstock: **358** • © Oleksiy Mark/Shutterstock: **358** • © urbanbuzz/Shutterstock: **358** • © Yuri Samsonov/Shutterstock: **357** • © zhu difeng/Shutterstock: **525** • © Source: Redraw from Rock ID Lab - VISTA HEIGHTS 8TH GRADE SCIENCE (weebly.com): **413** • © Anton Starikov/Shutterstock: **17** • © JAKKS Pacific, Inc.: **69** • © Emre Terim/Shutterstock: **109** • © Sebastian Kaulitzki/Shutterstock: **142** • © Jose Luis Calvo/Shutterstock: **142, 234** • © Science History Images/Alamy Stock Photo: **142, 234** • © Everett Histrorical/Shutterstock: **332** • © agnormark/Adobe Stock Photos: **332** • © Don B. Stevenson/Alamy Stock Photo: **336** • © Auscape International Pty Ltd/Alamy Stock Photo: **396** • © By Constant Aimé Marie Cap (1842–1915) before - https://image.invaluable.com/housePhotos/Bernaerts/92/603692/H0233-L114049833.jpg, Public Domain, https://commons.wikimedia.org/w/index.php?curid=86125206: **453** • © AlienCat/Adobe Stock Photos: **208** • © Evan Lorne/Shutterstock: **357** • © Natursports/Shutterstock: **524**

Text

• © Source: Geoscience Australia, Earthquake, 2021. Commonwealth of Australia. Licensed under CC BY 4.0.: **485** • © Source: Mount Gambier Volcanic Complex State Heritage Area, 1992. Department for Environment and Water. Licensed under CC BY 3.0.: **494** • © Source: What drives the plates?, Earth Learning Idea. Retrieved from: https://www.earthlearningidea.com/PDF/217_Slab_pull.pdf.: **463**

Every effort has been made to trace the ownership of copyright material. Information that will enable the publisher to rectify any error or omission in subsequent reprints will be welcome. In such cases, please contact the Permissions Section of John Wiley & Sons Australia, Ltd.

1 Investigating science

LESSON SEQUENCE

SCIENCE INQUIRY AND INVESTIGATIONS

Science inquiry is a central component of the Science curriculum. Investigations, supported by a **Practical investigation eLogbook** and **teacher-led videos**, are included in this topic to provide opportunities to build Science inquiry skills through undertaking investigations and communicating findings.

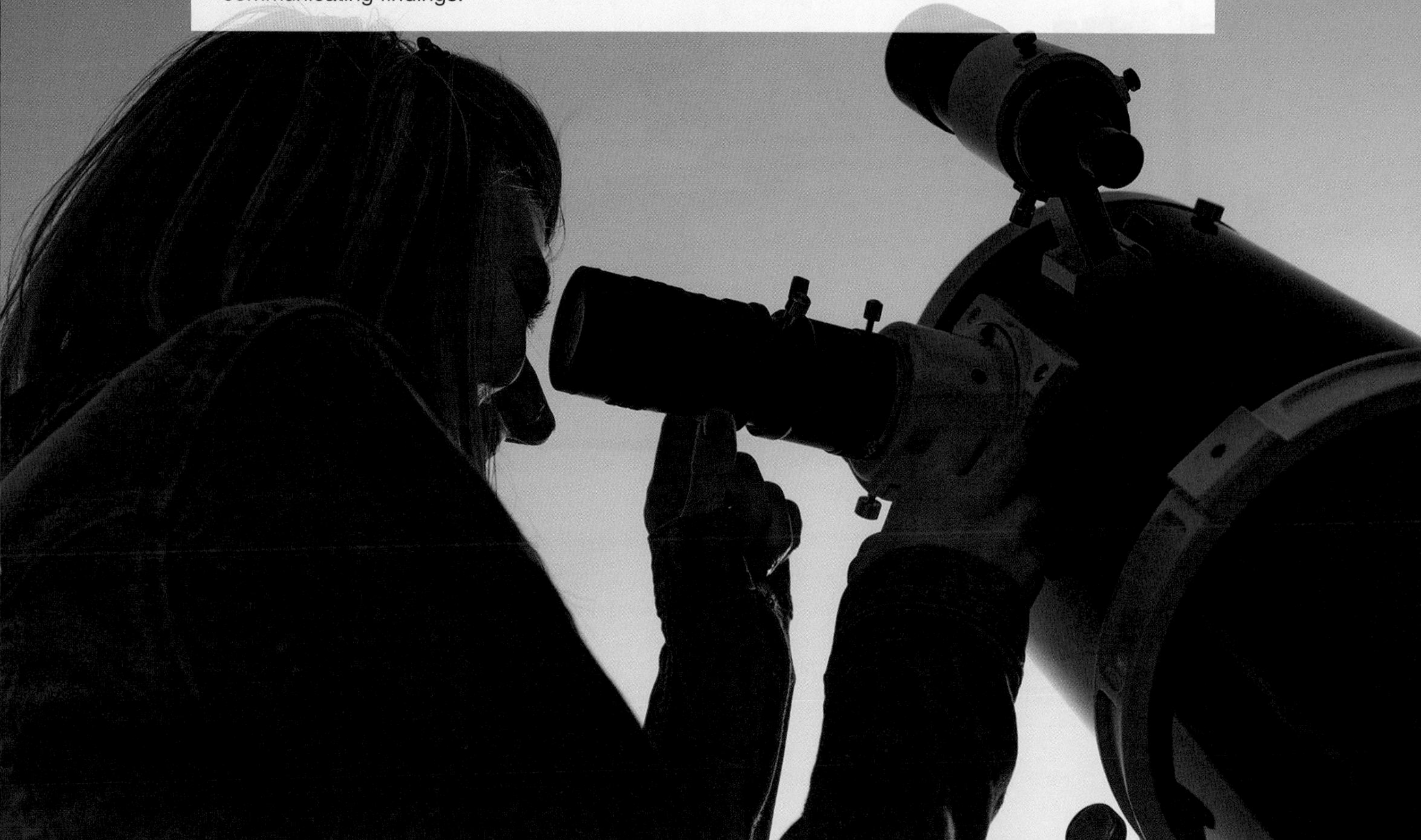

LESSON
1.1 Overview

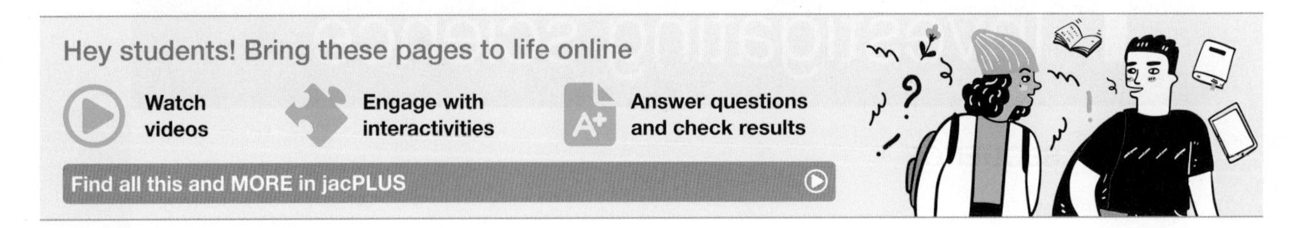

1.1.1 Introduction

Although science is a body of knowledge, it is also a way of solving problems and finding the answers to questions. Scientific knowledge is always growing and changing because scientists design and perform new investigations.

Observing, measuring, constructing tables, drawing graphs and forming conclusions are just some of the skills used in conducting scientific investigations.

FIGURE 1.1 Scientific investigation drives scientific progress.

1.1.2 Think about science

1. What is the connection between a pendulum, a playground swing and a metronome?
2. Why are graphs useful to scientists?
3. How do you start your own scientific investigation?
4. What do the letters CSIRO stand for?

1.1.3 Science inquiry

Researching the CSIRO

1. What is the CSIRO?
2. The CSIRO's website describes some of the research done by CSIRO scientists. Australia has a significant role in scientific research; we are world leaders in many different research areas. Read the information provided for one area of research that the CSIRO is involved with and summarise this research in point form.
3. Form groups of three. Explain to the other two students the area of research you have just read about. Try doing this without referring to your notes.

 Resources

 **Weblink** CSIRO

elog-2115

INVESTIGATION 1.1

Milk now or later?

You have just finished making yourself a cup of coffee when the phone rings. For your coffee to stay as warm as possible, should you add the milk now or after you have finished talking on the phone? Does your answer depend on the length of the phone call?

Aim

To compare the rate of cooling of hot coffee with and without the addition of milk

Materials

- kettle
- 2 identical cups
- instant coffee
- milk
- 2 thermometers or a data logger with 2 temperature probes
- 2 measuring cylinders

A data logger can be used for this investigation.

Method

1. Your teacher will assign a particular length of 'phone call' time to each group of students.
2. Make a hypothesis about this investigation.
3. Heat some water in a kettle and use it to make two cups of instant coffee. Use the same type of cup and the same amount of hot water and coffee powder.
4. Place a thermometer or temperature probe in each cup of coffee. If you are using a data logger, set it to collect results for at least 10 minutes.
5. Add 40 mL of milk to one of the cups.
6. If you are using thermometers, record the temperature of the coffee in both cups every 30 seconds.
7. After your phone call time has passed, add 40 mL of milk to the second cup.
8. Continue measuring the temperature in both cups every 30 seconds until 10 minutes has passed since you added the milk to the second cup.

Results

1. If you used thermometers, record your results in a table.

TABLE Results of investigation 1.1

Time (minutes)	Temperature (°C)	
	Milk added at time 0	Milk added after 'phone call'
0.0		
0.5		
1.0		
1.5		

2. Plot line graphs of your results on the same set of axes. Put time on the horizontal axis and temperature on the vertical axis.
3. If you used a data logger, a graph is plotted automatically. If necessary, adjust the settings so that the graph shows the temperatures measured by both probes on the same set of axes. Put the graph into the results section of your experiment report.
4. Was you hypothesis correct? Explain you results.

Discussion

1. Does hot coffee cool faster than warm coffee? How can you tell from your graph?
2. Did the two lines on the graph cross at any stage? What does this indicate?
3. Does the length of the 'phone call' affect the results? Compare your graphs with those of other groups.
4. Why was it important to put exactly the same amount of water in both cups and to use the same type of cup?
5. What are the advantages and disadvantages of using a data logger for this experiment?
6. Explain how this experiment could be improved.
7. How could this experiment lead to further experiments?

Conclusion

Summarise the findings for this investigaiton.

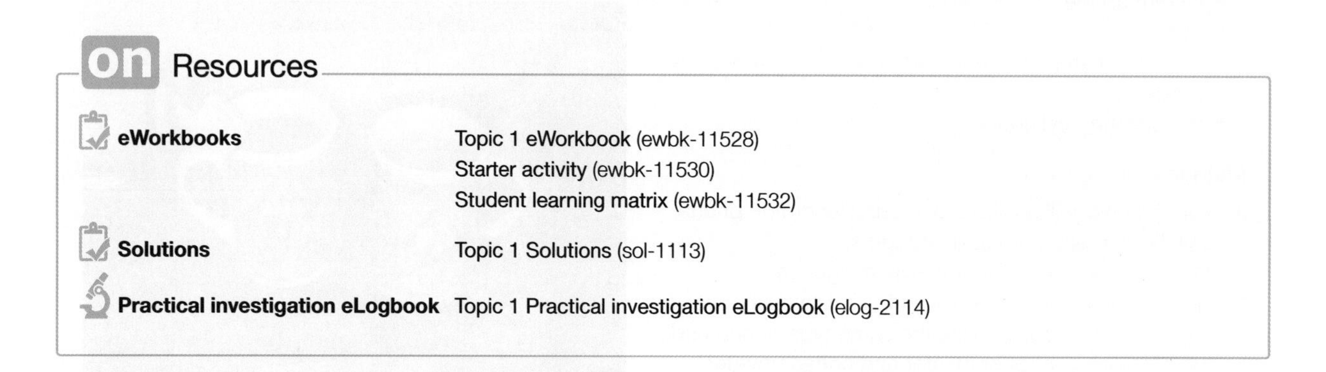

on Resources

eWorkbooks Topic 1 eWorkbook (ewbk-11528)
Starter activity (ewbk-11530)
Student learning matrix (ewbk-11532)

Solutions Topic 1 Solutions (sol-1113)

Practical investigation eLogbook Topic 1 Practical investigation eLogbook (elog-2114)

LESSON
1.2 Investigating skills

LEARNING INTENTION

At the end of this lesson you will understand that you need to take into consideration all aspects of fair testing, available equipment and safe investigation when planning investigations.

1.2.1 Safety rules

Conducting scientific investigations in a laboratory can be exciting, but accidents can happen if experiments are not carried out carefully. There are general rules that must be followed for your own safety and the safety of others, but there may be more for your school laboratory.

filter funnel a funnel used with filter paper to separate solids from liquids

beaker a container for mixing or heating substances

test tube a thin glass container for holding, heating or mixing small amounts of substances

measuring cylinder a cylinder used to measure volumes of liquids accurately

Investigating safely

- Use a **filter funnel** when pouring from a bottle or container without a lip.
- Never put wooden test-tube holders near a flame.
- Always turn the tap on before putting a **beaker**, **test tube** or **measuring cylinder** under the stream of water.
- Remember that most objects get very hot when exposed to heat or a naked flame.
- Do not use tongs to lift or move beakers.

ALWAYS . . .

- follow the teacher's instructions
- wear safety glasses and a laboratory coat or apron, and tie back long hair when mixing or heating substances
- point test tubes away from your eyes and away from your fellow students
- push chairs in and keep walkways clear
- inform your teacher if you break equipment, spill chemicals, or cut or burn yourself
- wait until hot equipment has cooled before putting it away
- clean your workspace — do not leave any equipment on the bench
- dispose of waste as instructed by your teacher
- wash your hands thoroughly after handling any substances in the laboratory.

FIGURE 1.2 It is important to use appropriate safety equipment in a laboratory.

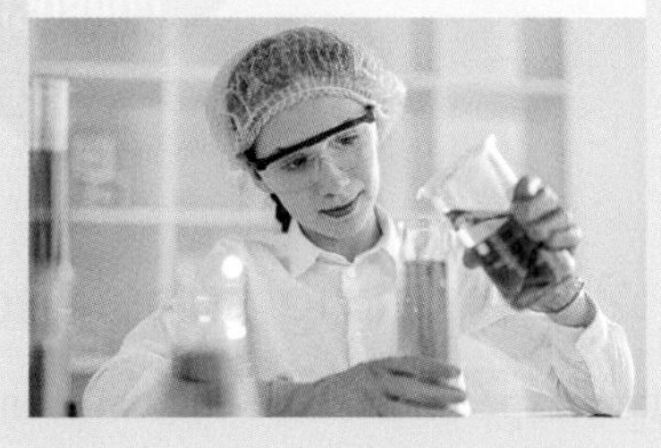

NEVER . . .

- enter the laboratory without your teacher's permission
- run or push in the laboratory
- eat or drink in the laboratory
- smell or taste chemicals unless your teacher says it is OK. When you do need to smell substances, fan the odour to your nose with your hand.
- leave an experiment unattended
- conduct your own experiments without the teacher's approval
- put solid materials down the sink
- pour hazardous chemicals down the sink (check with your teacher)
- put hot objects or broken glass in the bin.

FIGURE 1.3 Experiments should not be left unattended.

1.2.2 Working with dangerous chemicals

Your teacher will tell you how to handle the chemicals in each experiment. At times, you may come across warning labels on the substances you use.

TABLE 1.1 Different hazardous symbols and their meaning

	Always wear gloves and **safety glasses** when using chemicals with this symbol. Corrosive substances can cause severe damage to skin and eyes. Acid is an example of a **corrosive** substance.
	These substances are easily set on fire so keep them away from flames. Methylated spirits is **flammable**.
	Chemicals with this label can cause death or serious injury if swallowed or breathed in. They are also dangerous when touched without gloves because they can be absorbed by the skin. Mercury is a toxic substance.

FIGURE 1.4 Examples of warning labels

1.2.3 Heating substances

Many experiments that you will conduct in the laboratory require heating. In school laboratories, heating is usually done with a Bunsen burner, outlined in figure 1.5.

The barrel, shown in figure 1.5a, is the part of the burner from where the flame comes out. Never touch the barrel with your bare hand as it is very hot long after use. The base of the burner is the safest part to touch, as it is designed to not get hot.

The air hole controls the amount of oxygen that can mix with the gas in the burner. This air hole can be partially or completely open by turning it. The collar is a metal ring that can be adjusted to alter the amount of oxygen that can enter the burner. Always light the Bunsen burner with the air holes completely closed. The gas inlet is where the gas enters the burner and mixes with the oxygen.

The flames of a Bunsen burner vary depending on their temperature and use, as seen in figure 1.5b. The bright yellow, visible flame is known as the safety flame. This flame will be least hot due to less oxygen in the gas. The light blue, less visible flame is known as the heating or roaring flame. This flame is used when heating something. Always tie hair back and wear safety glasses and a laboratory coat or apron when using a Bunsen burner.

safety glasses plastic glasses used to protect the eyes during experiments

corrosive describes a chemical that wears away the surface of substances, especially metals

flammable describes substances, such as methylated spirits, that burn easily

int-8166

ewbk-11577

FIGURE 1.5 a. The components of a Bunsen burner **b.** The yellow visible flame is known as the safety flame and is less hot than the blue flame.

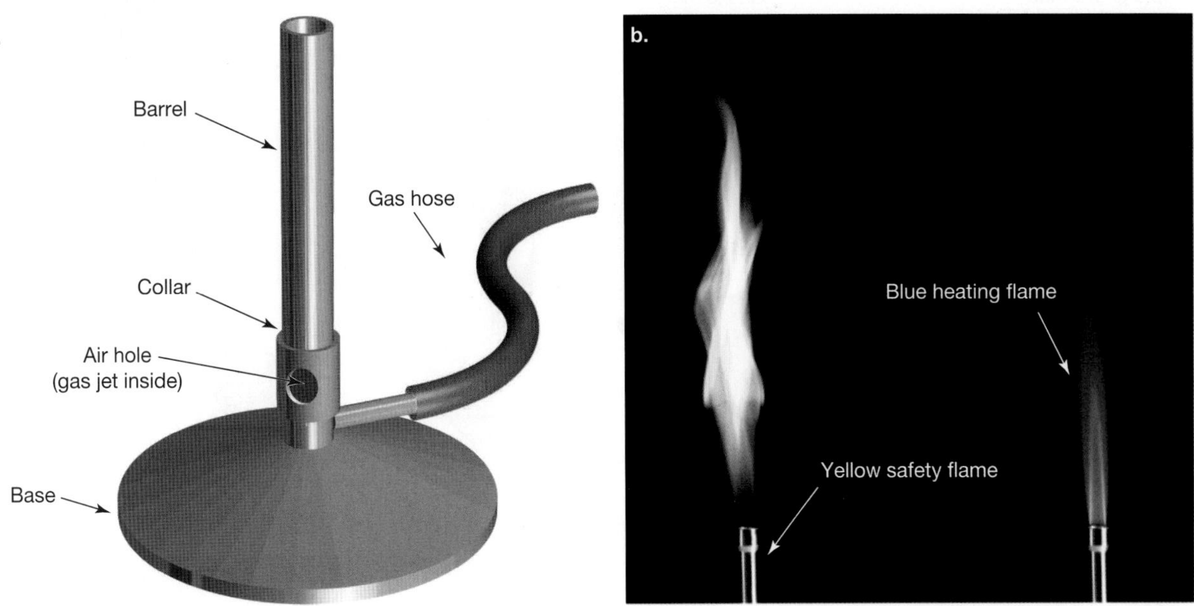

eles-2360

A GUIDE TO USING THE BUNSEN BURNER

1. Place the Bunsen burner on a heatproof mat.
2. Check that the gas tap is in the 'off' position.
3. Connect the rubber hose to the gas tap.
4. Close the air hole of the Bunsen burner collar.
5. Light a match and hold it a few centimetres above the barrel.
6. Turn on the gas tap and a yellow flame will appear.
7. Adjust the flame by moving the collar until the air hole is open and a blue flame appears.
8. Remember to close the collar to return the flame to yellow when the Bunsen burner is not in use.

FIGURE 1.6 A brief flowchart showing the main steps of using a Bunsen burner

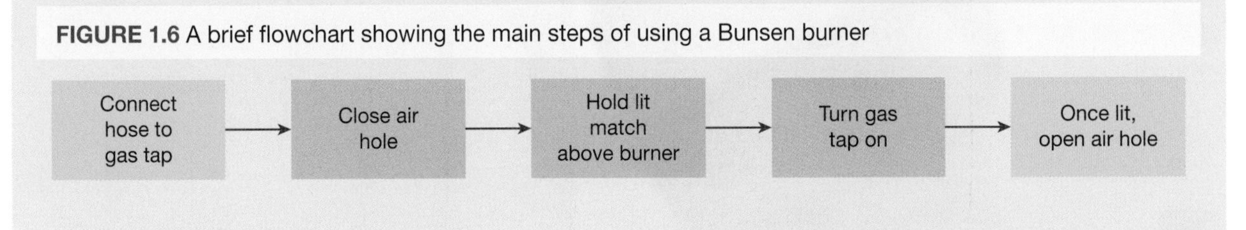

Heating containers

There are a number of pieces of equipment you will need in the laboratory when using a Bunsen burner for heating; these are shown in figure 1.7. Beakers and evaporating dishes can be placed straight onto a gauze mat for heating. Never look directly into a container while it is being heated. Wait until the equipment has cooled properly before handling it. Never heat an empty container as it is likely to shatter or melt.

Heating a test tube

Tripods and gauze mats are not used when heating test tubes. Hold the test tube with a test-tube holder. Keep the base of the test tube above the flame. Make sure that the test tube points away from you and other students.

FIGURE 1.7 How to heat containers and test tubes over a Bunsen burner

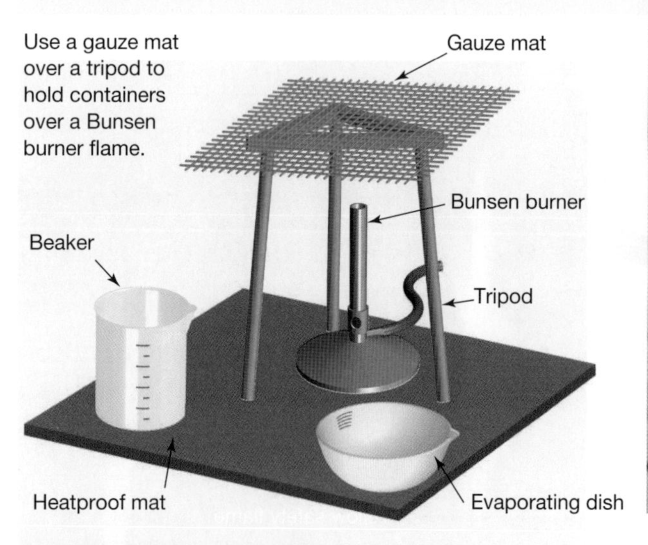

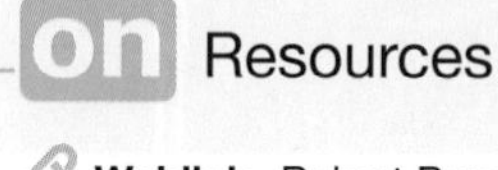

Resources

Weblink Robert Bunsen

1.2.4 Glassware

Pouring a liquid into a test tube

Pour liquids carefully into the test tube from a beaker or measuring cylinder. Use a filter funnel when pouring from bottles or containers without a lip.

FIGURE 1.8 **a.** Pouring a liquid into a test tube **b.** Using a filter funnel to transfer liquid from one container to another

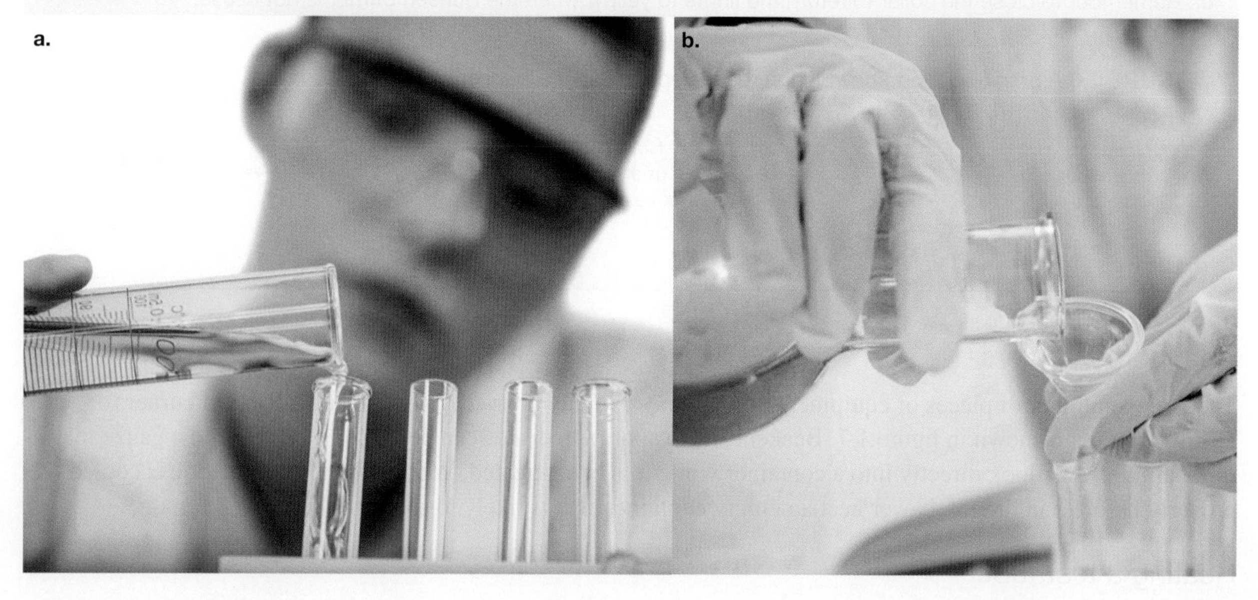

Shaking a test tube

There are two ways to shake substances in a test tube (figure 1.9).

Method 1

Hold the top of the test tube and gently move its base in a sideways direction. This method is good to use with non-hazardous substances that do not need to be shaken vigorously. This is the method you will use most of the time.

Method 2

Use a stopper when a substance needs to be mixed by shaking vigorously. Place an appropriately sized stopper into the mouth of the test tube. With your thumb over the stopper and your hand securely around the test tube, shake the test tube with an up and down motion. Shake a test tube in this way only if instructed to do so by the teacher.

FIGURE 1.9 Two correct ways to shake a test tube

1.2.5 Using electricity safely

Electrical equipment in the science laboratory should be used with great care, just as it should be in the home or workplace.

Never:

- place heavy electrical appliances near the edge of a bench or table
- allow water near electrical cords, plugs or power points
- place objects other than the correct electrical plug into a power point
- use appliances with damaged cords or exposed wires.

1.2.6 Precise measurements

Precision is the degree to which repeated measurements produce the same result. It tells us how close a series of measurements are to each other, as can be seen in figure 1.10. If there is a large variation in the results, the precision is low. If the results are all very similar, and only vary by a very small amount, then the precision is high.

The degree of precision of the measurements taken in an experiment depends on the instruments that have been used. If you want to measure the length of your classroom, you could use a trundle wheel with marks every 10 cm, or you could use a tape measure marked in millimetres. The tape measure would provide the most precise measurement. Similarly, to measure 100 mL of water, you could use a measuring cylinder that is graduated in millilitres or you could use a measuring cup that is marked every 100 mL. The measuring cylinder would provide a more precise measurement than the cup. A set of scales that measures mass to two decimal places is more precise than one that measures mass to one decimal place.

FIGURE 1.10 Precision and accuracy can be visualised on a target.

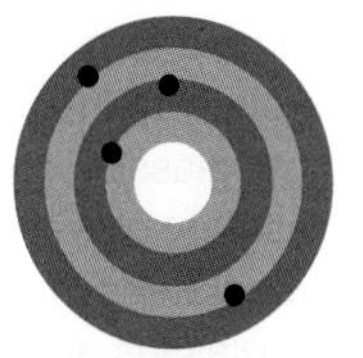

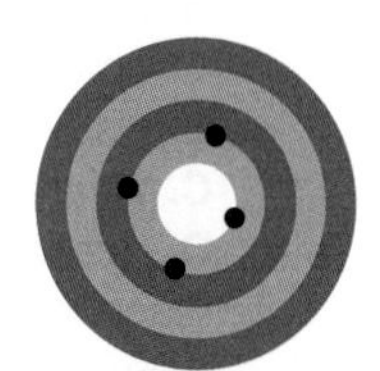

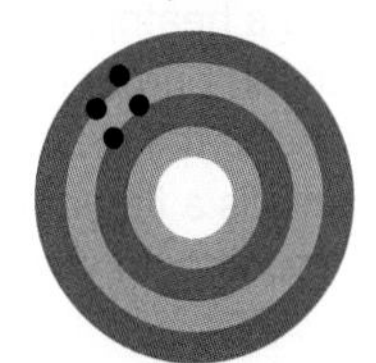

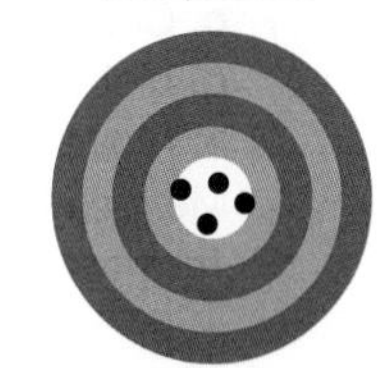

1.2.7 Accurate results

Accuracy is different to precision. Accuracy refers to how close an experimental measurement is to a known value. Sometimes results that are not precise can still be accurate, if the average of them is close to the actual value. This can be seen in figure 1.10.

precision how close multiple measurements of the same investigation are to each other

accuracy how close an experimental measurement is to a known value

A small measuring cylinder can provide a reasonably precise measurement of a volume of water but, if it is not read at eye level, the measurement may not be accurate. A set of bathroom scales might display a reading with two decimal places but, if you use it on carpeted floor, it may not provide an accurate measurement of your mass if it is designed to be used on a hard floor. To ensure that your results are accurate you should use measuring instruments correctly, and in some instances it may be necessary to **calibrate** the instruments. To calibrate a set of scales, for example, you could place an object that has a mass of exactly 100.00 g on the scale and adjust the scale until it reads exactly 100.00 g.

calibrate to check or adjust a measuring instrument to ensure accurate measurements

1.2 Activities

learnon

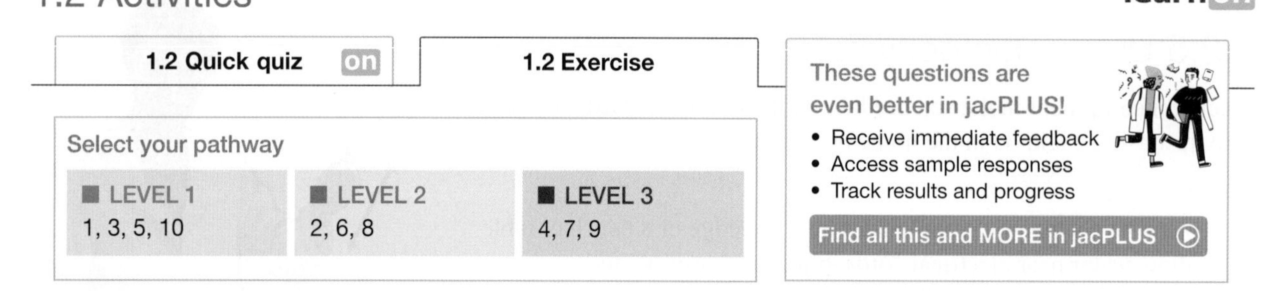

Remember and understand

1. MC How should a substance in a test tube be shaken if you are not instructed to shake it vigorously?
   A. Hold the top of the test tube firmly and gently move the base of the test tube in a sideways direction.
   B. Hold the test tube in the middle and move the test tube up and down.
   C. Place a stopper on the test tube so the substance does not fall out and shake hard.
   D. Place a stopper on the test tube and tip the test tube up and down.
2. MC Which TWO of the following are dangers in the science laboratory?
   A. Using an electronic balance to measure the mass of a substance before cleaning up some spilled water on the bench next to it
   B. Listening carefully to instructions
   C. Wearing safety glasses while entering the laboratory
   D. Not wearing safety glasses while heating a liquid in a beaker
3. Place the following steps in order to show how to heat substances in a test tube safely.
   a. Ensure safety glasses and laboratory coats are worn.
   b. Hold the test tube over the flame with the open end pointing away from people's faces.
   c. Check there is a heatproof mat below the Bunsen burner.
   d. Light the Bunsen burner correctly.
   e. Move the base of the test tube gently in and out of the blue flame to heat the substance.
4. Methylated spirits is a flammable liquid. Explain what this means.
5. True or false? You should always wear gloves when working with corrosive and/or toxic substances.

Apply and analyse

6. MC Which of the following is a safety precaution that must be followed in a scientific laboratory?
   A. Dispose of solid materials down the sink.
   B. Make sure that the gas tap is in the 'on' position when turning on the Bunsen burner.
   C. Use tongs to lift and move beakers.
   D. Long hair should be tied back when heating or mixing substances in the laboratory.

7. **MC** Why should a test tube be standing in a test-tube rack when you are pouring a liquid into it?
   A. A test-tube rack is more stable, so spills are less likely to occur.
   B. A test-tube rack is more stable, so spills are more likely to occur.
   C. So the test tube can be labelled
   D. So the test tube does not burn you
8. Three students weighed a standard mass of 10.0 g using the same balance. They all returned a result of 12.4 g. Based on these results, describe the balance used in terms of its accuracy and its precision.

Evaluate and create

9. Create a six-step flowchart to illustrate the correct method for lighting a Bunsen burner.
10. Which one safety rule do you feel is the most important when you are mixing two liquids and heating them? Create a poster to illustrate the rule.

Fully worked solutions and sample responses are available in your digital formats.

LESSON
1.3 SkillBuilder — Measuring and reading scales

LEARNING INTENTION

At the end of this lesson you will be able to read and record measurements accurately.

online only

Why do we need to measure and read scales?

When conducting experiments, it is critical that measurements and data are recorded accurately. Whether measuring volume or temperature, or interpreting alternate scales, it is important that they are recorded accurately.

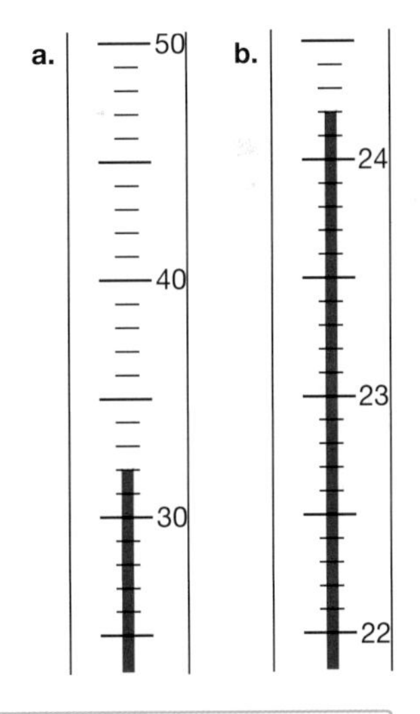

Go online to access:
- **Tell me:** an overview of the skill and its application in science
- **Show me:** a video and a step-by-step process to explain the skill
- **Let me do it:** an interactivity, question set and Skillbuilder activity for you to practice and consolidate your understanding of the skill

on Resources

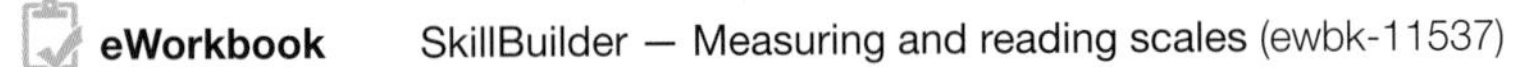
eWorkbook SkillBuilder — Measuring and reading scales (ewbk-11537)

Video eLesson SkillBuilder — Measuring and reading scales (eles-4153)

Interactivity SkillBuilder — Measuring and reading scales (int-0201)

LESSON
1.4 SkillBuilder — Using a Bunsen burner

LEARNING INTENTION

At the end of this lesson you will understand how to operate a Bunsen burner safely.

Why use a Bunsen burner?

Many experiments in the laboratory require heating using a Bunsen burner. A Bunsen burner provides heat when a mixture of air and gas is lit. Bunsen burners heat objects or liquids with a naked flame, and therefore there are precautions that must be taken to ensure the safe usage of a Bunsen burner.

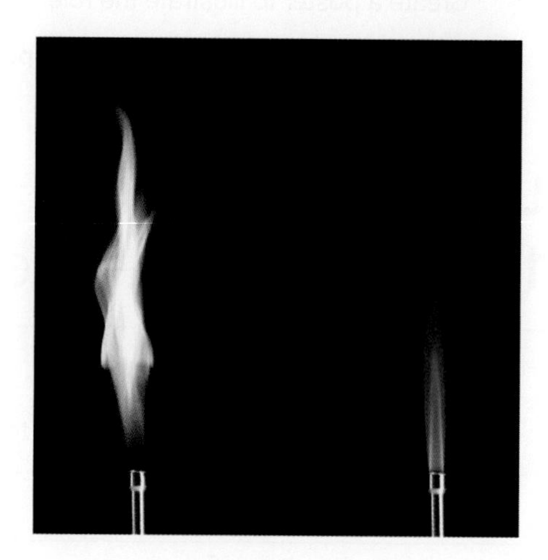

Go online to access:
- **Tell me:** an overview of the skill and its application in science
- **Show me:** a video and a step-by-step process to explain the skill
- **Let me do it:** an interactivity, question set and Skillbuilder activity for you to practice and consolidate your understanding of the skill

on Resources

eWorkbook	SkillBuilder — Using a Bunsen burner (ewbk-11539)
Video eLesson	SkillBuilder — Using a Bunsen burner (eles-4154)
Interactivity	SkillBuilder — Using a Bunsen burner (int-8088)

LESSON
1.5 Planning your own investigation

LEARNING INTENTION

At the end of this lesson you will know why it is important to have a valid aim and hypothesis.

1.5.1 The scientific method

As a science student you are required to undertake scientific investigations. These investigations will not only help you understand scientific concepts — they can be a lot of fun! Scientists around the world all follow what is known as the **scientific method** (figure 1.12). This allows scientists to examine each other's work and build on the scientific knowledge gained. An important aspect of science is being able to reproduce someone else's experiment. The more evidence a scientist has about a theory, the more accepted the theory will be.

scientific method a systematic and logical process of investigation to test hypotheses and answer questions based on data or experimental observations

The skills you will develop in conducting scientific investigations include the following:

- Questioning and predicting
- Planning and conducting
- Recording and processing
- Analysing and evaluating
- Communicating scientifically.

Whenever you take a trip away from home, you need to plan ahead and have some idea of where you are going.

You need to know how you are going to get there, what you need to pack and have some idea of what you are going to do when you get there.

It's the same with an experimental investigation. Planning ahead increases your chances of success.

FIGURE 1.11 Discuss your ideas with others.

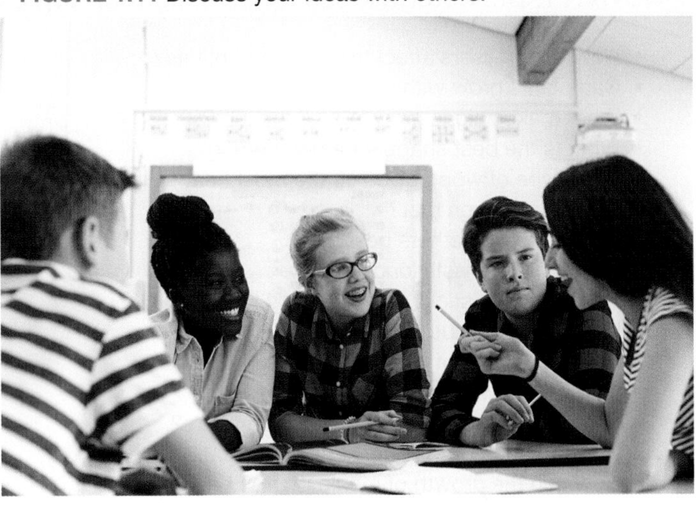

FIGURE 1.12 The scientific method

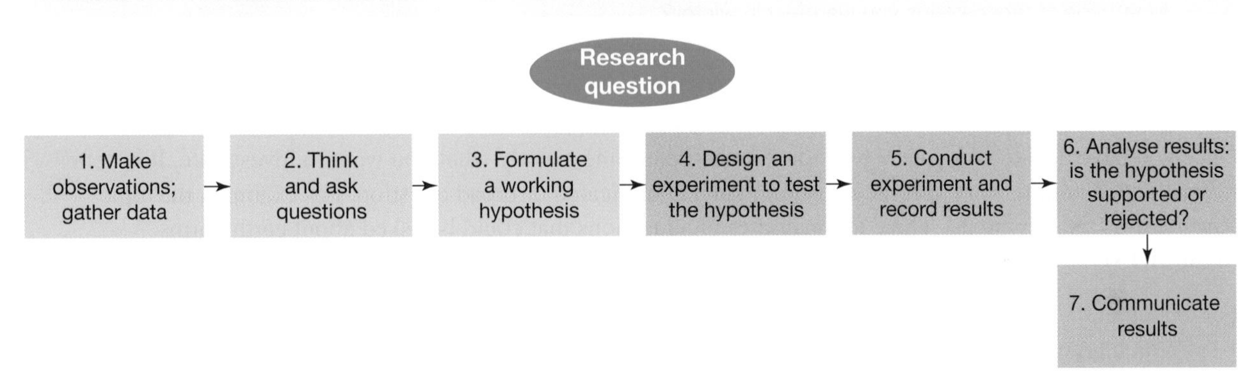

Finding a topic

The first step in the scientific method is to develop a **research question**. You can think of this as finding a topic.

research question a question that is the focus of the investigation, which the experiment aims to answer

Your investigation is much more likely to be of high quality if you choose a topic that you will enjoy working on. These steps might help you choose a good topic:

1. Start by searching for a general area of interest. List your hobbies and other interests.
2. Do you have a friend or relative who might be able to help you in a scientific investigation? Write down the topic areas in which you could get help.
3. Discuss the possible research topics you have already written down with a group of fellow students. Listen carefully to their ideas. They might help you to decide on your own topic. Write down your ideas.
4. Have a look through the list of ideas shown in the following activity box. Even if none of the suggested topics appeal to you, they may help you to think of other ideas. For example, 'How strong is sticky tape?' could lead you to consider topics such as the strength of glass, wood, paper, plastics or some other material. Brainstorm possible topics with your friends and make your own list of suggested investigations.
5. Search online or in a library for resources about the topic areas that you have already written down. You might also find journals or magazines that include articles about these topic areas. Use reliable websites and do not rely on just one source.

ACTIVITY: Brainstorm ideas for investigation topics

- How do fertilisers affect the growth of plants?
- Can plants grow without soil?
- What makes algae grow in an aquarium?
- What is the best shape for a boomerang?
- What type of wood gives off the most heat while burning?
- What makes iron rust?
- Which paint weathers best?
- Which battery lasts longest?
- How strong is sticky tape?
- Which type of glue is best?
- How much weight can a plastic bag hold?
- Which food wrap keeps food freshest?
- How effective are pre-wash stain removers?
- Which fabrics burn faster?
- How can the growth of mould on fruit be slowed down?
- Which concrete mixture is strongest?
- What type of fishing line is the strongest?
- Does the thickness of a rubber band affect how far it stretches?
- What type of paper aeroplane flies furthest?
- What is the best recipe for soap bubble mixture?
- Do tall people jump higher and further than short people?
- What type of fabric keeps you warmest in winter?

Defining the question

Once you have decided on your topic, you need to determine exactly what you want to investigate. It is better to start with a simple, very specific question than a complicated or broad question. For example, the topic 'earthworms' is very broad. There are many simple questions that could be asked about earthworms (figure 1.13).

For example:

- Which type of soil do earthworms prefer?
- How much do earthworms eat?
- Do earthworms prefer meat or vegetables?
- How fast does a population of earthworms grow?

Your question needs to be realistic. In defining the question, you need to consider whether:

- you can obtain the background information that you need
- the equipment that you need is available
- the investigation can be completed in the time you have available
- the question is safe to investigate.

FIGURE 1.13 There are many problems relating to earthworms that could be investigated.

1.5.2 Writing an aim

Your investigation should have a clear and realistic **aim**. Your aim should be very specific. The aim of an investigation is its purpose, or the reason for doing it. Some examples of aims are:

- to find out how the weight and shape of paper aeroplanes affects how far they fly
- to compare the effect of different fertilisers on the growth of pea plants
- to investigate whether different coloured lights affect the growth of algae in an aquarium
- to investigate how exposing iron to salty water affects how quickly it rusts.

'To investigate if the weight of paper planes makes them fly better' is not a suitable aim because 'fly better' has not been defined. 'Fly better' could mean fly further, fly in a straighter line or stay in the air longer. A better aim would be 'To investigate how the weight of paper planes affects their flight distance and time in the air'.

When you have decided what your aim is, make sure that it is realistic. You should be able to answer 'yes' to each of the following questions.

- Is my aim simple and clear?
- Will I be able to get the background information that I need?
- Is the equipment I need for my experiments available or can it be made?
- Is the question a safe one to investigate?

If you answer 'no' to any of these questions you need to rethink your aim.

1.5.3 Forming a hypothesis

A **hypothesis** is a reasoned prediction about the outcome of an experiment; it must be able to be supported by evidence. Your hypothesis should relate to your aim and should be **testable** and **falsifiable** with an experimental investigation. A hypothesis is usually written in an 'IF ... THEN ...' format (e.g. 'IF the balloon is more inflated, THEN it will fly for longer when let go'). The results of your investigation will either support (agree with) or not support (disagree with or refute) the hypothesis. It is not possible to prove conclusively that a hypothesis is correct.

FIGURE 1.14 Will nylon be better than cotton? Your own experience might help you form a hypothesis.

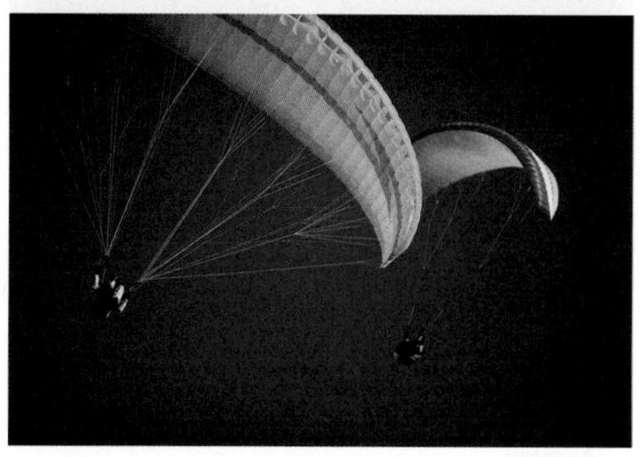

When scientists make a hypothesis, they usually carry out a number of experiments to test it. Sometimes, a number of teams of scientists test the same hypothesis with slightly different experiments. Even if the results of each experiment agree with the hypothesis, the scientists could never say that the hypothesis is proven to be correct. They would say that each experiment has provided further evidence to support the hypothesis.

Your hypothesis should be based on what you know about the topic or what you have already observed. For example, if you are trying to design the best parachute for a toy, you should read about parachutes before writing your hypothesis. You might also recall that when you are walking in the rain, a cotton T-shirt soaks up a lot of water and becomes heavy, whereas a nylon jacket does not soak up water. As a result, your hypothesis might be: 'If either nylon or cotton are used for the parachutes of two identical toys, then the toy with a nylon parachute will take longer to reach the ground because the fabric is more tightly woven'.

A statement that cannot be tested with a scientific experiment is not a suitable hypothesis. You will explore hypotheses in more depth in lessons 1.6 SkillBuilder — Writing an aim and forming a hypothesis, 1.8 Controlling variables and 1.9 SkillBuilder — Controlled, dependent and independent variables.

aim a statement outlining the purpose of an investigation

hypothesis a suggested, testable explanation for observations or experimental results; it acts as a prediction for the investigation

testable able to be supported or proven false through the use of observations and investigation

falsifiable can be proven false

TABLE 1.2 Simple examples of results that can support or disprove a hypothesis

Observation	Hypothesis	Test of hypothesis	Result	Conclusion
The phone is not charging.	The charger is faulty.	Use a different charger.	The phone still does not charge.	Hypothesis rejected
The car won't start.	The battery is flat.	Replace the battery.	The car still won't start.	Hypothesis rejected
The measurement on the scales is far higher than expected.	The scales were not set to zero before use.	Reset the scales back to zero.	The reading is closer to the expected value.	Hypothesis supported
The water did not freeze at 0 °C.	There was a contaminant in the water.	Freeze water that is pure (distilled).	The water froze at 0 °C.	Hypothesis supported

1.5.4 Ethical considerations

Some research problems may not be appropriate to investigate due to ethical considerations. Ethics have to do with what is considered to be right and wrong. Different groups in society have varying opinions about certain types of research. While many people accept that testing medicines on animals is necessary, others feel very strongly that no research should be carried out on animals. Scientists involved in medical research are often required to have their research proposals reviewed by an ethics committee. The potential benefits of the research are taken into consideration. In your investigations, you should not do research that has the potential to cause stress or harm to people or animals. Your research should not be upsetting to people. This is particularly relevant if your research involves a survey.

1.5.5 Working in groups

Working in groups has many advantages. You can divide up a task to get more work done in a short period of time. Each group member brings along their interests, expertise and skills and, if these are used effectively, the quality of the work produced will be increased. When doing practical work, each team member can have a different role so that the task can be carried out efficiently.

Group work also has some pitfalls. Resentment builds up when the work is not divided up fairly. Group members might have different ideas about the best way to carry out the project and waste time trying to come to an agreement. An effective way to avoid some of these pitfalls is to assign each group member a role at the start of the project. Think about each group member's skills and decide on roles accordingly.

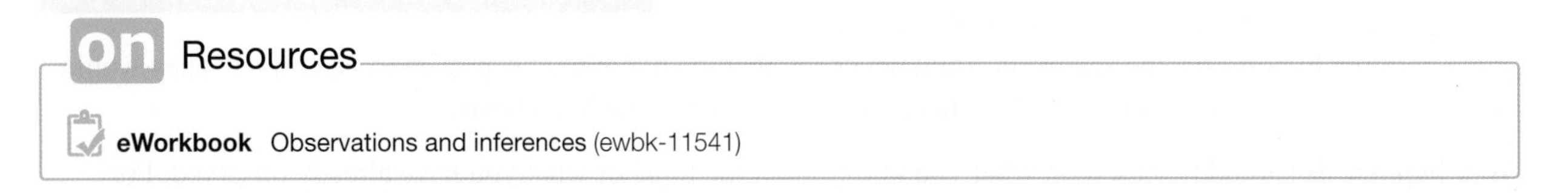

on Resources

eWorkbook Observations and inferences (ewbk-11541)

1.5 Activities

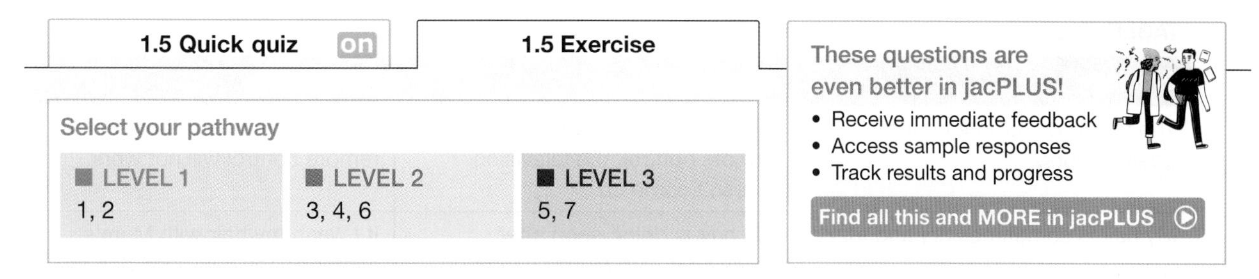

Remember and understand

1. MC Which four of the following questions should you ask about your aim before it is final?
   A. Is my aim simple and clear?
   B. Will I be able to get the background information that I need?
   C. Is my aim interesting enough?
   D. Is the equipment I need for my experiment available or can it be made?
   E. Is my aim too simple?
   F. Is the question a safe one to investigate?
   G. Is my aim too long?
2. MC What does the term hypothesis mean?
   A. A question you test with an experiment
   B. A sensible guess about the outcome of an experiment based on what you know is testable
   C. An observation you make during the experiment that may support the aim
   D. Research you undertake before commencing an experiment
3. How can you test a hypothesis?

Apply and analyse

4. MC Which of the following is a more suitable aim for an investigation about glue rather than 'To find out which glue is best'?
   A. To find out which brand of glue is the cheapest when gluing paper
   B. To find out which brand of glue takes the longest time to dry
   C. To find out which brand of glue is the strongest when gluing paper
   D. To find out which brand of glue is colourless when dry

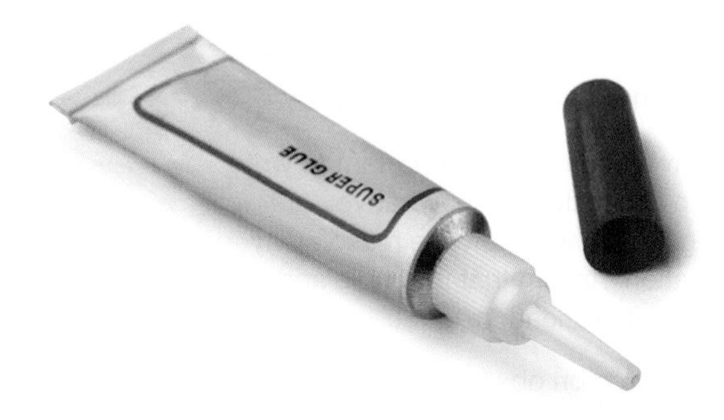

5. MC Which four of the following would be the most suitable hypotheses?
   A. White chocolate tastes better than dark chocolate.
   B. Washing powder X removes tomato sauce stains faster than washing powder Y.
   C. Plants grow faster under red light than under green light.
   D. Sagittarians are nicer people than Leos.
   E. Playing video games increases the muscle strength in your thumbs.
   F. Science teachers perform better in IQ tests than English teachers.
   G. Science teachers are more interesting people than English teachers.

6. **SIS** Consider the following table.

TABLE Examples of how problems and observations can lead to hypotheses

Problem	Observation	Hypothesis
The television remote control doesn't work.	If I press the 'on' button on the remote control, the television doesn't come on.	If the batteries are flat then the remote control will not work.
My hair is sometimes dry and frizzy.	My hair is driest soon after washing it with Mum's shampoo.	If I wash my hair with Mum's shampoo my hair becomes dry.
No parrots come to our bird feeder.	There is bread in the bird feeder, and magpies and miner birds feed there.	If I feed birds bread then parrots will not eat it.

Describe how you could test each of the three hypotheses.

Evaluate and create

7. **SIS** Write an outline of an investigation you could do to test each of the three hypotheses from the table in question 6.

Fully worked solutions and sample responses are available in your digital formats.

LESSON
1.6 SkillBuilder — Writing an aim and forming a hypothesis

LEARNING INTENTION

At the end of this lesson you will be able to write aims and hypotheses.

online only

Why do we need to write aims and hypotheses?

When you conduct a scientific investigation, it is important to write an aim and a hypothesis. An aim is a statement of what you are trying to find out in your investigation. It is simply the reason why you are conducting the investigation. An aim that is simple and clear will allow you to focus on the investigation.

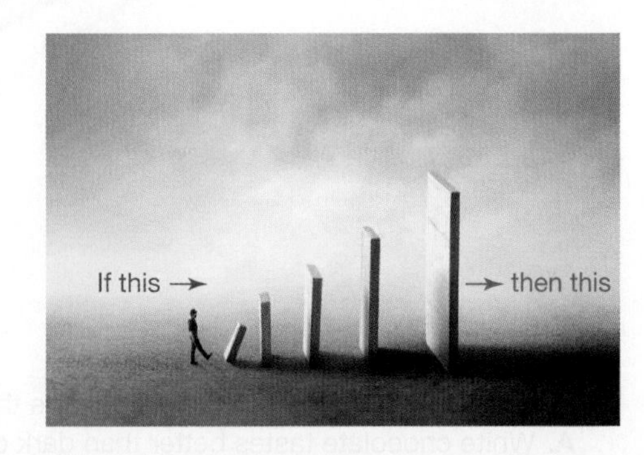

A hypothesis is an idea which is based on observation that may explain a phenomenon and it must be able to be tested. It should be related to your aim and it is a statement, not a question. A hypothesis cannot be proven correct, but the results of your experiment will either support your hypothesis or not support your hypothesis.

Go online to access:

- **Tell me:** an overview of the skill and its application in science
- **Show me:** a video and a step-by-step process to explain the skill
- **Let me do it:** an interactivity, question set and Skillbuilder activity for you to practice and consolidate your understanding of the skill

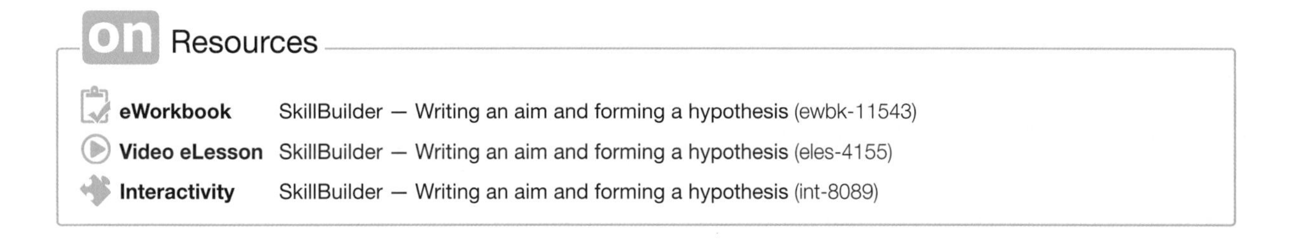

LESSON
1.7 Record keeping and research

LEARNING INTENTION

At the end of this lesson you will know how to conduct effective research and how to keep good records of your investigation.

1.7.1 Background research

Scientists do experiments to test hypotheses, which are based on observations as well as the previous discoveries of other scientists.

Before designing their experiments, scientists do background research, which usually includes reading reports written by other scientists. Scientists also need to keep records of all their observations and any changes they make to the design of their experiments. When you conduct your own research investigation, you will probably be asked to do this by keeping a logbook.

FIGURE 1.15 Part of a blog site used by a researcher to share the results of her investigations into acid–base indicators

1.7.2 What is a logbook?

A **logbook** is a document in which you keep a record of all the work you do towards an investigation. Each entry should be dated like a diary.

A logbook can be written by hand on paper, on a computer, or it can even be written in an app or as a website. A blog is a website that has dated entries so it can be used as a logbook. It has the added advantage that you can invite other people, such as your friends, parents and teachers, to look at your work and post comments. You should check with your teacher on the format required for your logbook.

logbook a complete record of an investigation from the time a search for a topic is started

In your logbook, you might include the following items.

Information to include in your logbook

- A timeline or other evidence of planning your time
- Notes about conversations you had with teachers, friends, parents or experts and how these conversations affected your project. Make sure you record each person's details so you can acknowledge their contribution in your report.
- Background information from research you did. Include all the details you need for your **bibliography**.
- A plan or rough outline of the method you will use for your experiment(s)
- All of your results of all your experiments (these may be presented roughly at this stage)
- Notes about any problems you encountered during your project and how you dealt with these
- Information on any changes you made to your original plan
- First drafts of your reports, including your thoughts about your conclusions

1.7.3 Researching your topic

Before you start your own experiments, you should find out more about your topic.

As well as increasing your general knowledge of the topic, you need to find out whether others have investigated your problem. Information already available about your topic might help you to design your experiments. It might also help you to explain your results.

Make notes on your topic as you find information. You may be able to include some relevant background information in your report.

FIGURE 1.16 Make notes on your topic.

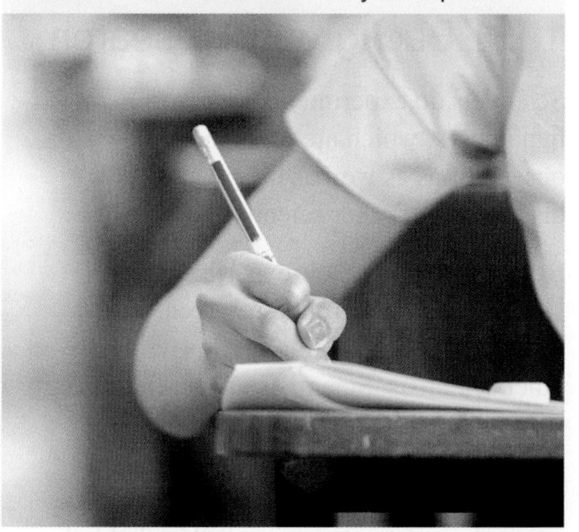

How to use information

Make notes on information that is relevant to your research topic. Think about what you really need to know. You need information that will help you to:

- plan your experiments
- understand your results later on
- show in your report how your research relates to everyday life or why your research is important.

You will need to keep an accurate list in your logbook of the steps you have taken and the resources you have used.

The internet

The internet provides a wealth of information on almost every topic imaginable. Use a search engine such as Google or Bing. The success of your search will depend on a thoughtful choice of keywords. Don't just look at the first things that appear but scroll through the first page looking at the sources of the findings. Try to find reliable websites that are specific to your search words.

bibliography a list of references and sources at the end of a scientific report

1.7.4 Reliable information

Most research is now carried out over the internet, but a great deal of information on the internet is unreliable. Reliable information is information that can be trusted. Imagine that your friend emailed you a link to a website that said that brushing your teeth with peppermint-flavoured toothpaste causes your tongue to turn green over time. How could you check that the information was reliable?

- Check the URL (uniform resource locator). This is the address you type in to access the page. The ending of the URL can be useful for assessing the reliability of a web page. A URL with an .edu ending is usually from an educational institution such as a school or university. The ending .gov is used for government websites while an .org ending usually indicates that the web page is associated with a nonprofit organisation.
- Look for information about the author of the web page, the organisation associated with the web page and the date on which the information was last updated. A web page that provides no such information is less likely to be reliable. If an author's name is provided, what are their qualifications? Do they have expertise in the area they are commenting on? Are they likely to show bias? For websites associated with an organisation, is the organisation likely to benefit from a particular viewpoint about an issue?
- Check the information against other sources. Experimental results are considered reliable if the experiment, when repeated a number of times, consistently produces similar results. Similarly, the reliability of information from secondary sources can be assessed by checking it against other sources.

Using the library

Another good place to start is the school library. There are several different types of information sources in the library, including those in the following list. Ask a librarian for help.

Nonfiction books

Use the subject index catalogue to learn where to find books with information about your topic. Your library catalogue is most likely to be stored in a computer database. You might need to ask the librarian to help you use the catalogue at first. It is a good idea to browse through the contents list of science textbooks. Your topic may appear.

Reference books

These include encyclopedias, atlases and yearbooks. The index of a good encyclopedia is a great place to start looking for information.

Journals and magazines

There are quite a few scientific journals that are suitable for use by school students. They provide up-to-date information. Your library may have an index for journals, such as 'Guidelines', which you can use to find articles on your topic. You may, however, need to browse. Some journals to look for are: *New Scientist, Ecos, Australasian Science, Habitat, Popular Science, Choice* and *Double Helix.*

Information file

Many school libraries keep collections of digital files of newspaper articles on topics of interest. Ask your school librarian if you don't know how to access these resources.

Audiovisual resources

The library may have slides, videos and audio CDs that can be used or borrowed. These resources can be located using the subject index catalogue. Your librarian may also recommend some podcasts.

Beyond the library

Information on your topic may also be available from the following sources.

FIGURE 1.17 Look for information beyond the library.

Your science teacher

This may seem obvious, but many people do not even think to ask. Your science teacher may also be able to direct you to other sources of information.

Government departments and agencies

Federal, state and local government departments and agencies may be able to provide you with information or advice on your topic. Try searching government web pages, which usually list contact details. A polite email to the appropriate department or agency is the best way to ask for help.

Industry

Information on some topics can be obtained from certain industries. For example, if you were testing glues for strength or batteries to find which ones last longest, the manufacturers might have useful information. Use the internet to find contact details. A polite email is often the best way to ask for help.

Relatives or friends

Perhaps you or a relative know somebody who works in your area of interest. Let your friends and relatives know about your intended research.

In your logbook, complete a checklist like the one in the following highlighted box to see if you have thoroughly searched sources of information.

Logbook checklist for collecting information

- [ ] The internet

School library:

- [ ] Nonfiction books
- [ ] Reference books
- [ ] Journals and magazines
- [ ] Information files
- [ ] Audiovisual resources

Beyond the library:

- [ ] Your science teacher
- [ ] Government departments and agencies
- [ ] Industry
- [ ] Relatives or friends
- [ ] Other sources

FIGURE 1.18 In your logbook, keep an accurate list of resources that you have used.

Resources

Weblink How to evaluate sources for reliability

1.7 Activities

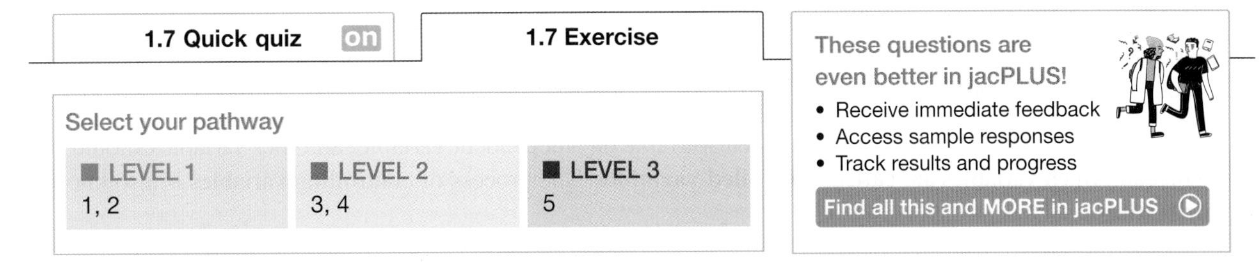

Remember and understand

1. MC Why is a logbook a bit like a diary?
   A. All entries are dated.
   B. Nothing is allowed to be drawn in your logbook.
   C. A logbook is always written on paper.
   D. Both are written as if events will happen in the future.
2. What does the term 'blog' mean?

Apply and analyse

3. You can research information about science topics in science textbooks and on the internet.
   a. Explain why you would not find the results of scientific research that was done last month in a science textbook.
   b. Outline some advantages and disadvantages of using the internet as a source of information.
4. List the resources that you could use to research your investigation topic:
   a. in your school library
   b. outside the school library.

Evaluate and create

5. Imagine you are a scientist. Assess the advantages and disadvantages of maintaining a blog rather than keeping a logbook in your office.

Fully worked solutions and sample responses are available in your digital formats.

LESSON
1.8 Controlling variables

LEARNING INTENTION

At the end of this lesson you will be able to describe the differences between controlled, dependent and independent variables.

1.8.1 Understanding variables

In order to complete a successful investigation, you need to make sure that your experiments are well designed.

Once you've decided exactly what you are going to investigate, you need to be aware of:

- which variables need to be controlled and which variables can be changed
- whether a control is necessary
- what observations and measurements you will make and what equipment you will need to make them
- the importance of repeating experiments (replication) to make your results more reliable
- how you will record and analyse your **data**.

data information collected that can be used for studying or analysing

A poorly designed investigation is likely to produce a conclusion that is not **valid**.

A **variable** is an observation or measurement that can change during an experiment. You should change only one variable at a time in an experiment. It is important that you understand and identify the different types of variables in your experiment.

When you are testing the effect of an independent variable on a dependent variable, all other variables should be kept constant. Such variables are called **controlled variables**. The process of controlling variables is also known as **fair testing**.

Dependent and independent variables

When was the last time you were on a swing? A playground swing is simply a large **pendulum**. A pendulum is a suspended object that is free to swing to and fro. Each complete swing is called an **oscillation**. The time taken for one complete oscillation of a pendulum is called its **period**. Pendulums are used mainly as measuring instruments. Their most well-known use is in clocks, such as grandfather clocks.

FIGURE 1.19 A playground swing is simply a large pendulum.

To answer questions scientifically, we need to perform a controlled investigation, which must also be reliable. Investigation 1.2 examines a swinging pendulum, where the variables are controlled. To ensure reliability, the measurements in the investigation need to be accurate, repeated and averaged.

There are several factors that affect the period of a pendulum. They include:

- the length of the pendulum
- the total mass that is swinging
- the height from which the pendulum is released.

- **Independent variable**: The variable that you deliberately change during an experiment
- **Qualitative data**: The variable that is being affected by the independent variable — that is, the variable you are measuring.

The variable that you are measuring (in this case, the period of the pendulum) is the dependent variable. The variable that you are investigating is the independent variable. In investigation 1.2, you will investigate two independent variables: the mass of the pendulum and the length of the pendulum. However, it is important that we only investigate one variable at a time. This allows a fair test.

Fair testing

Scientific investigations must be fair tests. In the first part of investigation 1.2, the independent variable is the mass of the pendulum. All variables other than the independent variable must be controlled; that is, they must be kept the same. If they were not, you could not tell which variable was affecting the period of the pendulum. You might find it helpful when designing your own investigations to use a table similar to table 1.3 to identify all the variables.

valid sound or true; a valid conclusion can be supported by other scientific investigations

variables quantities or conditions in an experiment that can change

controlled variables the conditions that must be kept the same throughout an experiment

fair test a test that changes only one variable and controls all other variables when attempting to answer a scientific question

pendulum an object swinging on the end of a string, chain or rod

oscillation one complete swing of a pendulum

period the time taken for one oscillation of a pendulum

independent variable the variable that the scientist chooses to change to observe its effect on another variable

dependent variable a variable that is expected to change when the independent variable is changed; the dependent variable is observed or measured during the experiment

TABLE 1.3 Experiment: How does mass affect the period of a pendulum?

Independent variable	• The mass of the pendulum
Dependent variable	• The period of the pendulum
Controlled variables	• The length of the pendulum • The angle of release • The method of release

INVESTIGATION 1.2

elog-2117

tlvd-10723

The period of a pendulum

Aim

To investigate the effects of mass and length on the period of a pendulum

Materials

- length of string (at least 80 cm long)
- set of slotted masses
- retort stand with bosshead
- pair of scissors
- a 1 m ruler
- stopwatch or clock with a second hand
- protractor
- rod

Method

Part 1: The effect of mass

1. Write a hypothesis for this part of the investigation.
2. Set up your pendulum so it can swing freely. Start with the largest possible length and the smallest weight.
3. Copy the table in the results section into your logbook, and record the mass and the length of the pendulum in it. The length should be measured from the top of the pendulum to the bottom of the swinging mass, as shown in the figure.
4. Using a protractor, pull the mass aside so that the angle of release is about 20°. Take note of the height from which the mass is released so that this angle of release is used throughout the experiment.
5. Release the pendulum. Measure the time taken for 10 complete swings of the pendulum. Repeat your measurement at least twice to find the average time for 10 swings.
6. Repeat this procedure for three larger masses, completing the table as you go.

Pendulum setup

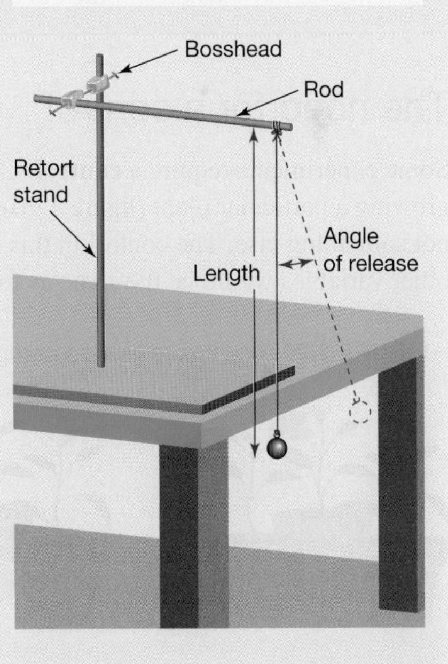

Part 2: The effect of length

7. Write a hypothesis for this part of the investigation.
8. Construct a table like table 1.3 to identify all of the variables that need to be considered for an investigation of the effect of length on the period of a pendulum.
9. Construct a second table in which to record your measurements. Remember that this time you'll be testing four different lengths without changing the mass. Use the same procedure as you did in part 1 for measuring the period.

Results

Part 1: The effect of mass

1. Record the length of your pendulum.
   Length of pendulum = ______ cm
   Angle of release = 20°

2. Record all the measurements in your table and calculate the average time taken for one complete swing (the period).

TABLE Results of investigation 1.2, part 1

Time taken for 10 complete swings (seconds)					
Mass (grams)	**Trial 1**	**Trial 2**	**Trial 3**	**Average**	**Period (seconds)**

Part 2: The effect of length

3. Record your measurements for part 2 of the investigation in a table.
4. Draw a line graph to show how the period of the pendulum is affected by its length. Remember to add a heading to your graph.

Period (s): 0.0, 0.5, 1.0, 1.5, 2.0
Length (cm): 0, 10, 20, 30, 40, 50, 60, 70, 80

Discussion

1. How does the mass of the pendulum affect its period?
2. How does the length of the pendulum affect its period?
3. The period of most standard clock pendulums is one second. Use your graph to predict the length of a standard clock pendulum.
4. Explain why it is a good idea to measure the time for ten swings rather than just one.
5. Suggest one aspect that could be done next time to improve this investigation.

Conclusion

Summarise the findings for Part 1 and Part 2 of this investigation.

The need for a control

Some experiments require a **control**. Consider an experiment investigating which brand of fertiliser is best for growing a particular plant (figure 1.20). A control is needed to ensure that the result is due to the fertilisers and not something else. The control in this experiment would be a pot of plants to which no fertiliser is added. All other variables would be the same as for the other pots in the experiment.

FIGURE 1.20 A control is used to compare the difference in growth to a plant with no fertiliser.

Valid experiments

A valid experiment measures what it actually sets out to measure. If your aim was to find out whether watering plants with seawater affects their growth rate, then comparing the number of radish seeds that germinate after one week when watered with tap water or seawater would not be a valid method. This is because it does not actually measure growth rate — it tests the effect of seawater on seed germination.

control an experimental set-up in which the independent variable is not applied; a control is used to ensure that the result is due to the variable and nothing else

Repeatable and reliable experiments

Replication is the repeating of an experiment to make sure you have collected **reliable data**. In the case of the fertiliser experiment shown in figure 1.20, a more reliable result could be obtained by setting up two, three or four pots for each brand of fertiliser, or having a number of seedlings in each pot. The results are checked for consistency, and an average result for each brand or the control could then be calculated.

A reliable experiment provides consistent results when repeated, even if it is repeated on different days and under slightly different conditions — for example, in a different room or with a different researcher collecting the data. Replication increases the reliability of an experiment. This can involve simply doing the same experiment a few times, or having different groups repeat the same experiment and pooling the data gathered by each group when writing the report.

reliable data data that is able to be replicated in different circumstances but the same conditions

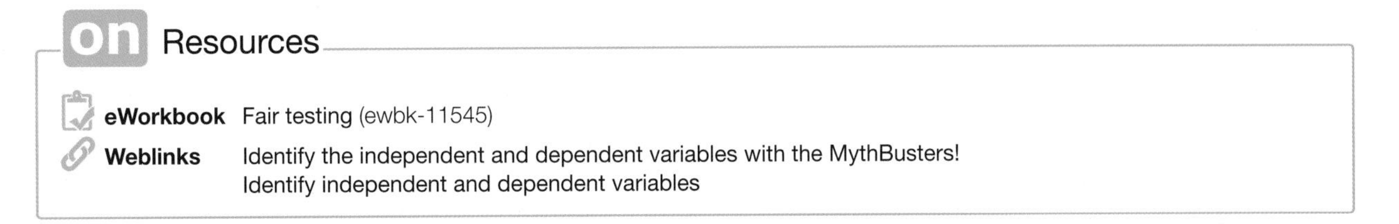

on Resources

eWorkbook Fair testing (ewbk-11545)

Weblinks Identify the independent and dependent variables with the MythBusters!
Identify independent and dependent variables

1.8 Activities

learnon

1.8 Quick quiz on | **1.8 Exercise**

Select your pathway

LEVEL 1	LEVEL 2	LEVEL 3
1, 5	2, 4, 6	3, 7

These questions are even better in jacPLUS!
- Receive immediate feedback
- Access sample responses
- Track results and progress

Find all this and MORE in jacPLUS

Remember and understand

1. What is a variable?
2. Complete the sentences to describe the difference between a dependent variable and an independent variable.
   During an experiment, the researcher purposely alters the ____________ variable. The ____________ variable is then an outcome or value that results from the change in the ______________ variable.

Apply and analyse

3. Why is it important to control variables in a scientific investigation?
4. A metronome is an upside-down pendulum. A metronome's period is changed by moving the sliding mass up or down. To make the period of the metronome longer, should you move the sliding mass up or down?
5. Identify the independent and dependent variables in:
   a. part 1 of investigation 1.2
   b. part 2 of investigation 1.2.
6. In investigation 1.2 you conducted three trials for each measurement and calculated an average. List two reasons for the repetition.

A metronome

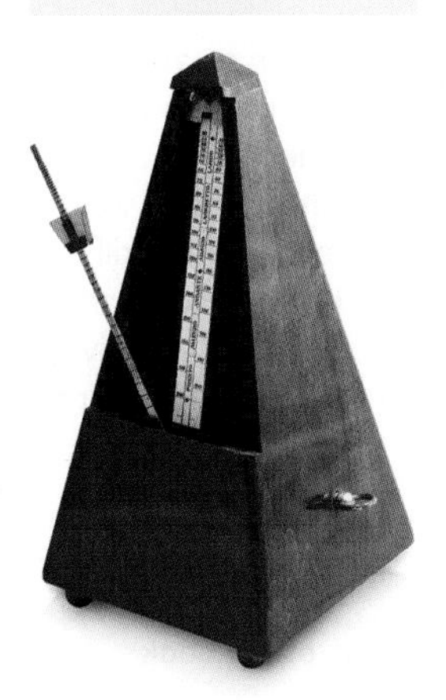

Evaluate and create

7. You are to conduct an investigation to answer the question 'Does the angle of release affect the period of a pendulum?'
   a. Identify the independent and dependent variables for this investigation.
   b. Using the variables defined in part **a**, write a hypothesis.
   c. Write a method on how to investigate and test your hypothesis.
   d. Perform your investigation using your method in part **c** and write a brief report on your findings. In your conclusion, state clearly whether your results supported your hypothesis.

Fully worked solutions and sample responses are available in your digital formats.

LESSON 1.9 SkillBuilder — Controlled, dependent and independent variables

LEARNING INTENTION

At the end of this lesson you will be able to identify independent, dependent and controlled variables.

online only

What is the difference between controlled, dependent and independent variables?

In order to answer a question scientifically, a controlled investigation needs to be performed. In a controlled investigation every variable except the one being tested is held constant, which stops the results being affected by an uncontrolled factor. The variable that you are investigating is called the independent variable. The variable that you are measuring is called the dependent variable.

Go online to access:

- **Tell me:** an overview of the skill and its application in science
- **Show me:** a video and a step-by-step process to explain the skill
- **Let me do it:** an interactivity, question set and Skillbuilder activity for you to practice and consolidate your understanding of the skill

on Resources

eWorkbook	SkillBuilder — Controlled, dependent and independent variables (ewbk-11547)
Video eLesson	SkillBuilder — Controlled, dependent and independent variables (eles-4156)
Interactivity	SkillBuilder — Controlled, dependent and independent variables (int-8090)

LESSON
1.10 Scientific reports

LEARNING INTENTION

At the end of this lesson you will understand how to write a scientific report.

1.10.1 Getting approval

Almost all scientists need the approval of their employer before they commence an investigation. As a student, you should not commence an investigation until your plan has been approved by your teacher.

1. Title

Choose a title in the form of a question — you may decide to change it before your work is completed.

2. The aim or problem

Briefly state what you intend to investigate or the question that you intend to answer.

Aim: To study the behaviour of slaters

Problem: What makes algae grow in an aquarium?

3. Hypothesis

Using the information that you know or have discovered while deciding on your aim and question, make an educated guess about the answer to your problem or what you expect to find out. It is important to be creative and objective, and to use logical reasoning when devising a hypothesis and testing it.

4. Outline of experiment

Explain how you intend to test your hypothesis, and briefly outline the experiments you intend to conduct.

5. Equipment

List any equipment you need for your experiments.

6. Resources

List the sources of information that you have used or intend to use. This list should include library resources, organisations and people.

1.10.2 Performing your experiments

Once your teacher has approved your plan, you may begin your experiments. Detail how you conducted your experiments in your logbook. All observations and measurements should be recorded. Use tables where possible to record your data. Use graphs to display your data.

Some information about using tables, graphs and data loggers is provided in lessons 1.11–1.15.

Where appropriate, measurements should be repeated and an average value determined. All measurements — not just the averages — should be recorded in your logbook.

Photographs should be taken if appropriate.

FIGURE 1.21 All observations and measurements should be recorded.

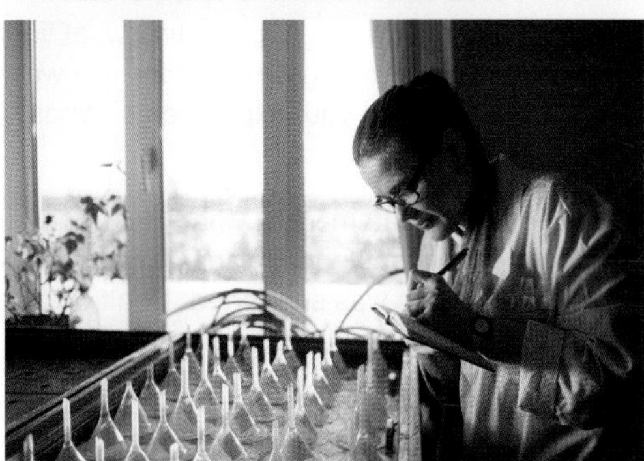

You might need to change your experiments if you get results you don't expect. If things go wrong, record what happened. Knowing what went wrong allows you to improve your experiment and technique. Any major changes should be checked with your teacher.

1.10.3 Writing your report

Check with your teacher about what is required — teachers might not want all the following sections, or might want a poster or some other format. You can begin writing your report as soon as you have planned your investigation, but it cannot be completed until your observations are complete. Your report should be typed or neatly written on A4 paper. It should begin with a table of contents, and the pages should be numbered. Your report should include the following headings (unless they are not applicable to your investigation).

discussion a detailed area of a scientific report that explains the results and how they link back to the relevant concepts; it also includes suggestions for improvements to the experiment

Scientific report structure

Abstract

Briefly summarise your experiments and your main conclusions. Even though this appears at the beginning of your report, it is best not to write it until after you have completed the rest of your report.

Introduction

Present all relevant background information. Include a statement of the problem that you are investigating, saying why it is relevant or important. You could also explain why you became interested in the topic.

Aim

State the purpose of your investigation – that is, what you are trying to find out.

Hypothesis

Using the knowledge you already have about your topic, make a guess about what you will find out by doing your investigation.

Materials and method

Describe in detail how you carried out your experiments. Begin with a list of the equipment used and include photographs of your equipment if appropriate. The description of the method must be detailed enough to allow somebody else to repeat your experiments. It should also convince the reader that the variables in your investigation are well controlled. Labelled diagrams can be used to make your description clear. Using a step-by-step outline makes your method easier to follow.

Results

Observations and measurements (data) are presented in this section. Wherever possible, present data as a table so that they are easy to read. Graphs can be used to help you and the reader interpret data. Each table and graph should have a title. Ensure that you use the most appropriate type of graph for your data (see sections 1.11.2 and 1.11.3).

Discussion

In this **discussion**, begin by stating what your results indicate about the answer to your question. Explain how your results might be useful. Outline any weaknesses in your design or difficulties in measuring here. Explain how you could improve your experiments. What further experiments are suggested by your results?

Conclusion

This is a brief statement of what you found out and may link with the final paragraph of your 'Discussion'. It is a good idea to read your aim again before you write your conclusion. Your conclusion should also state whether your hypothesis was supported. Don't be disappointed if it is not supported. Some scientists deliberately set out to reject hypotheses!

Bibliography

Make a list of books and other printed or audiovisual material to which you have referred. The list should include enough detail to allow the source of information to be easily found by the reader. Arrange the sources in alphabetical order.

The way a resource is listed depends on whether it is a book, magazine (or journal) or website. For each resource, list the following information in the order shown:

- Author(s), if known (book, magazine or website)
- Title of book or article, or name of website
- Volume number or issue (magazine)
- URL (website) and the date you accessed it
- Publisher (book or magazine), if not in the title
- Place of publication, if given (book)
- Year of publication (book, magazine or website)
- Chapter or pages used (book).

Some examples of different sources are as follows:

- Taylor, N., Stubbs A., Stokes, R. (2020) *Jacaranda Chemistry 2 VCE Units 3 & 4*. 2nd edition. Milton: John Wiley & Sons.
- Gregg, J. (2014) 'How Smart are Dolphins?' *Focus Science and Technology,* Issue 264, February 2014, BBC, pages 52–57.
- *Bridge Building and Safe Design*, John Daly, Safe Design Australia, last updated October 2018, www.safedesignaustralia.com.au/bridge-building-safe-design/, accessed 25 May 2020.

Acknowledgements

List the people and organisations who gave you help or advice. You should state how each person or organisation assisted you.

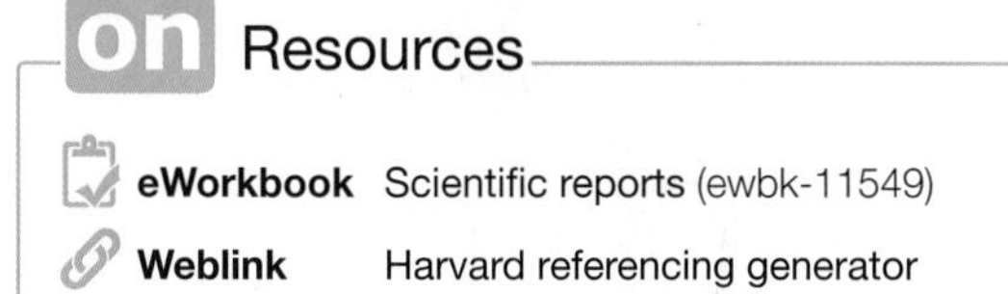

on Resources

eWorkbook Scientific reports (ewbk-11549)

Weblink Harvard referencing generator

1.10 Activities

1.10 Quick quiz	1.10 Exercise

These questions are even better in jacPLUS!
- Receive immediate feedback
- Access sample responses
- Track results and progress

Find all this and MORE in jacPLUS

Select your pathway

LEVEL 1	LEVEL 2	LEVEL 3
1	2	3, 4

Remember and understand

1. Complete the table by identifying which section of an investigation report the content should be located in.

Content	Investigation report section
The purpose of the experiment	
A brief summary of your investigation and findings	
A table showing all the measurements you recorded	
A list of the books and other resources you used to find information for your project	
A diagram of the equipment you used	
A statement that relates the results back to the aim and outlines what your results show	

Apply and analyse

2. When scientists write up their investigations for publication in a scientific journal, the abstract is one of the most important parts of the report. Explain why the abstract is usually read by many more people than the full report.
3. Explain why it is important for scientists to publish their investigations in scientific journals and to read the reports written by other scientists.

Evaluate and create

4. There have been instances where scientists have faked their results or committed other types of scientific misconduct.
   a. Enter the words 'scientific misconduct' in a search engine to find examples of such instances.
   b. Why do you think that some scientists might be tempted to fake or fabricate their results?
   c. Explain why cases of scientific misconduct are damaging to all scientists.
   d. What do you think might happen to scientists who are found to have faked their results?

Fully worked solutions and sample responses are available in your digital formats.

LESSON
1.11 Presenting your data

LEARNING INTENTION

At the end of this lesson you will be able to use different types of diagrammatic, graphical and physical representations of data, and consider their strengths and limitations.

1.11.1 Presenting your data

Observations and measurements obtained from an investigation are called data. Data can be qualitative or quantitative.

- **Qualitative data** is expressed in words. It is also known as categorical data — you can think of this data falling into categories. It is descriptive and can be easily observed but not measured. There are two types of qualitative data:
  - Ordinal data can be ordered or ranked. This could be levels (1st, 2nd, 3rd . . .) or opinions (strongly agree, agree, disagree, strongly disagree).
  - Nominal data cannot be organised in a logical sequence. This could include colours or brand names.
- **Quantitative data** (or numerical data) can be precisely measured and have values that are expressed in numbers. There are two types of quantitative data:
  - Continuous data can take any numerical value, such as the temperature of a substance.
  - Discrete data can only take on set values that can be counted, such as the number of students with green eyes.

qualitative data categorical data that examines the quality of something (e.g. colour or gender) rather than a measurement or quantity

quantitative data numerical data that examines the quantity of something (e.g. length or time)

Having collected the data, it is important to present them clearly in a way that another person reading or studying them can understand. Tables and graphs are a great way to organise data.

1.11.2 Using tables

Many different types of data can be collected in scientific experiments. Data is often presented in tables.

Tables

Tables can be used to record data to help separate and organise your information. All tables should:

- have a heading
- display the data clearly, with the independent variable in the first column and the dependent variable in later columns
- include units in the column headings and not with every data point
- be designed to be easy to read.

An example of a simple table is shown in table 1.4; it includes all the features you need to remember when constructing a table.

Always include a title for your table.

TABLE 1.4 Temperature of the Earth at different depths

Depth (km)	Temperature (°C)
0	15
1	44
2	73
3	102
4	130
5	158
6	187
7	215
8	242

Include the measurement units in the headings.

The column headings show clearly what has been measured.

Use a ruler to draw lines for rows, columns and borders.

Enter the data in the body of the table. Do not include units in this part of the table.

You may need to construct more complex tables, such as table 1.5, to present your research project results.

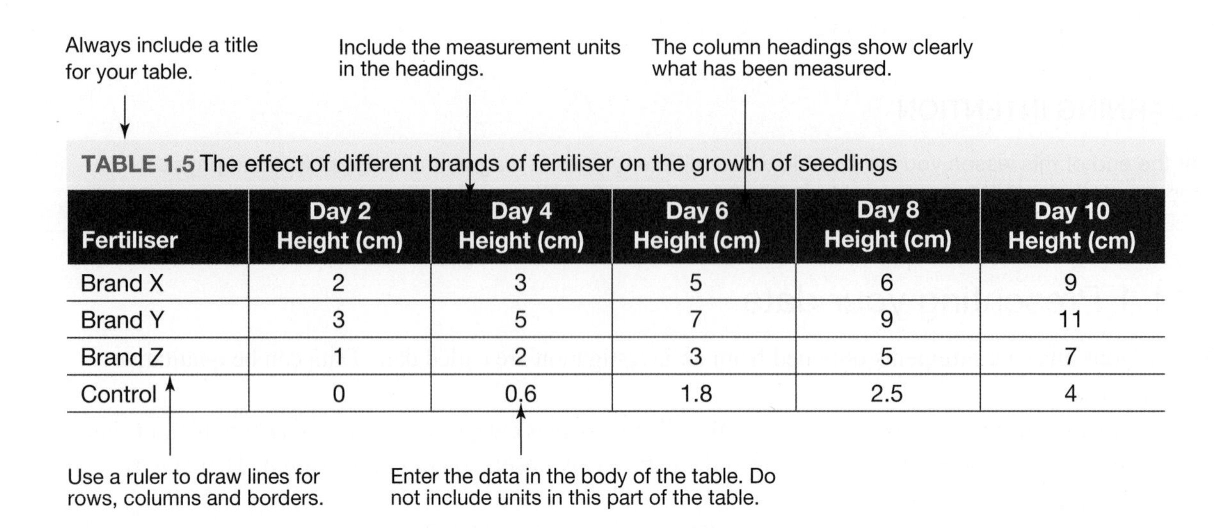

TABLE 1.5 The effect of different brands of fertiliser on the growth of seedlings

Fertiliser	Day 2 Height (cm)	Day 4 Height (cm)	Day 6 Height (cm)	Day 8 Height (cm)	Day 10 Height (cm)
Brand X	2	3	5	6	9
Brand Y	3	5	7	9	11
Brand Z	1	2	3	5	7
Control	0	0.6	1.8	2.5	4

1.11.3 Using graphs

Organising data as a graph is a widely recognised way to make a clear presentation. Graphs make it easier to read and interpret information, find trends and draw conclusions. Just like tables, graphs should always have a heading.

The most common types of graphs are as follows:

- *Pie charts and divided bar charts.* These are used to show frequencies or portions of a whole. This includes percentages or fractions.
- *Bar/column graphs*: These are used when one piece of data is qualitative and the other is quantitative. The bars are separated from each other. The horizontal axis has no scale because it simply shows categories. The vertical axis has a scale showing the units of measurements.
- *Histograms*: These are a special type of bar graph that show continuous categories, and are often used when examining frequency. The bars are not separated.
- *Scatterplots*: These require both sets of data to be numerical (or quantitative). Each dot represents one observation. A scatterplot can easily show trends between data sets, and correlations can be seen.
- *Line graphs*: These are scatterplots with the dots joined. The dots are usually joined using a straight line, but sometimes the line is curved. They are used for continuous data.

FIGURE 1.22 A pie chart

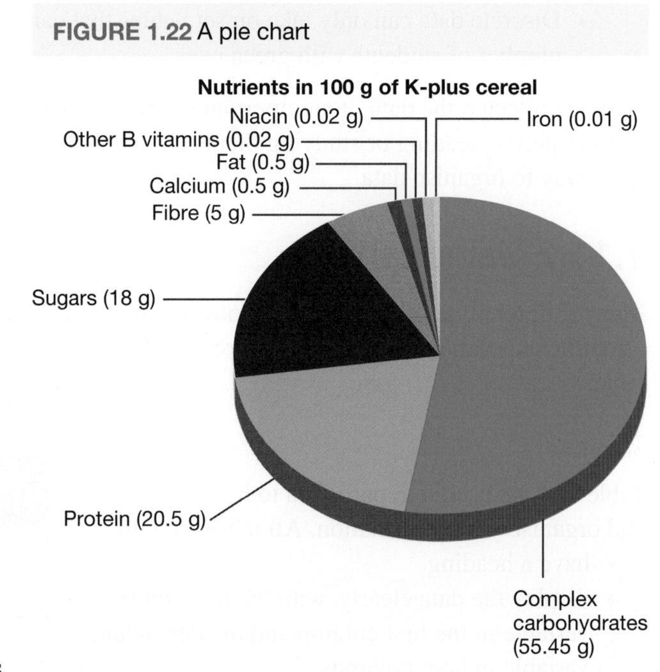

Pie charts (or sector graphs)

A pie chart (also known as a sector graph) is a circle divided into sections that represent parts of the whole. This type of graph may be used when the data can be added as parts of a whole. The example in figure 1.22 shows the food types, vitamins and minerals that make up the nutrients in a breakfast cereal.

Divided bar graphs

Divided bar graphs are also used to represent parts of a whole. However, the data are represented as a long rectangle, rather than a circle, divided into sections. The example in figure 1.23 shows the type of footwear worn to school today by male and female students.

FIGURE 1.23 A divided bar graph

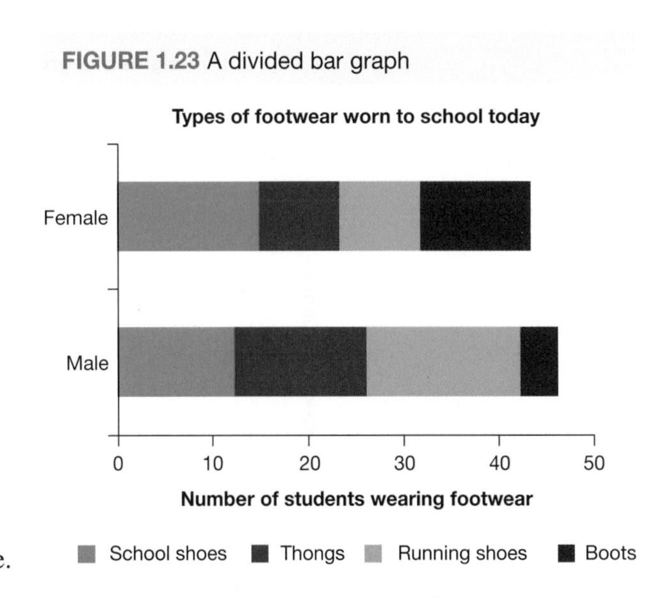

Column graphs and bar graphs

A column graph (sometimes called a bar graph) has two axes and uses rectangles (columns or bars) to represent each piece of data. The height or length of the rectangles represents the values in the data. The width of the rectangles is kept constant. This type of graph can be used when the data cannot be connected and are therefore not continuous — that is, when one piece of data is qualitative and the other is quantitative.

Figure 1.24 shows data on the average height to which different balls bounced during an experiment. Each column represents a different type of ball.

Figure 1.25 shows the lengths of different metal bars when heated. Each bar represents a different metal.

FIGURE 1.24 A column graph

FIGURE 1.25 A bar graph

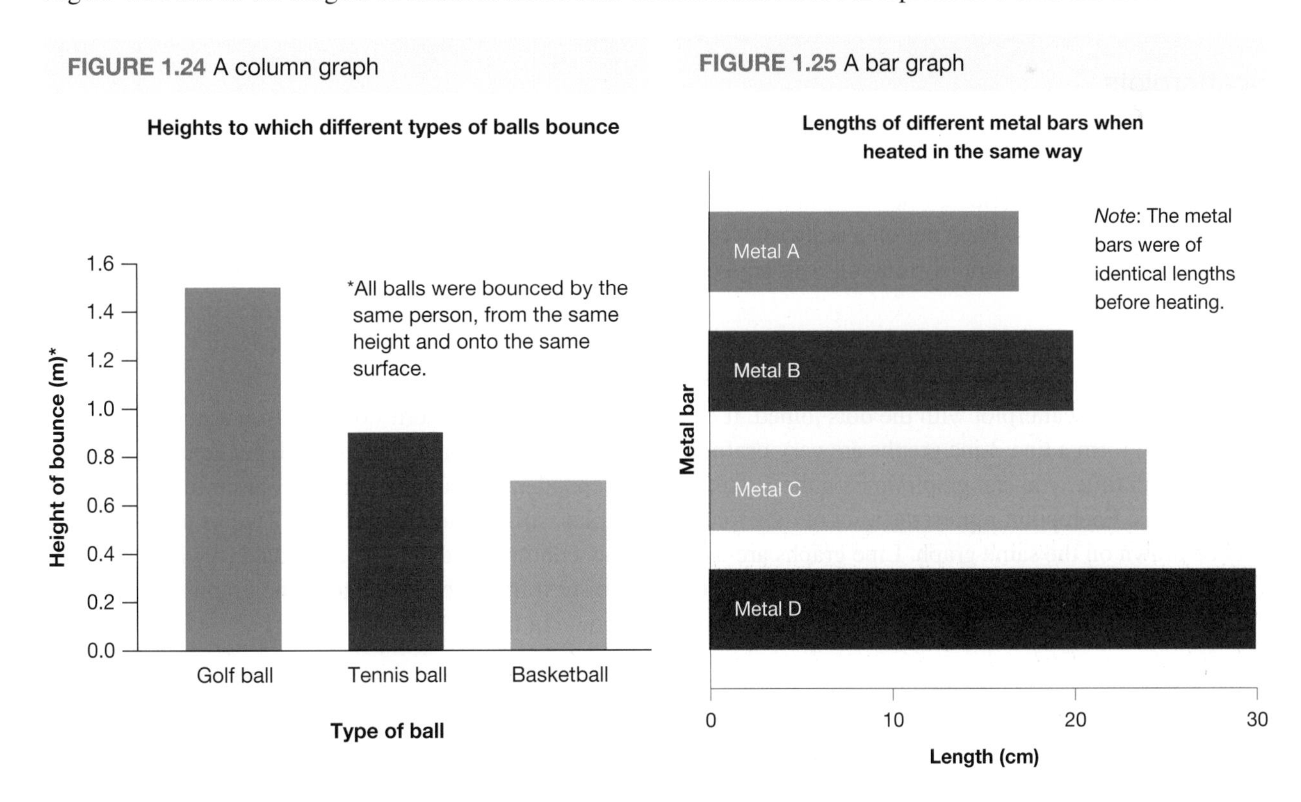

Histograms

Histograms are similar to column graphs except the columns touch because the data are continuous. They are often used to present the results of surveys. In figure 1.26, each column represents the number of students of a particular height.

FIGURE 1.26 A histogram

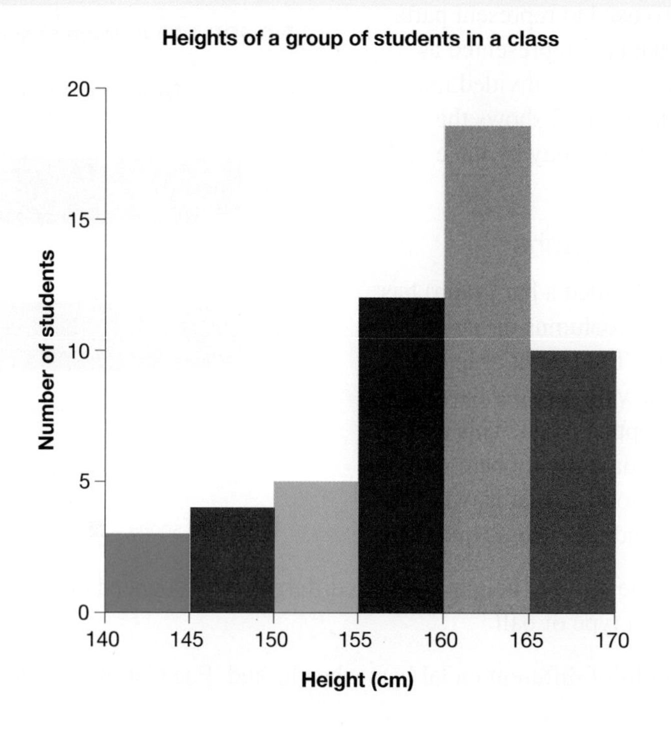

Scatterplots

Scatterplots require both sets of data to be numerical. A scatterplot has two axes: a horizontal axis and a vertical axis. The horizontal axis is known as the *x*-axis, and the vertical axis is known as the *y*-axis. You would normally graph the independent variable (the one you changed) on the *x*-axis, and the dependent variable (the one you measured) on the *y*-axis. Each dot on a scatterplot represents one observation. A scatterplot can easily show trends between data sets, and correlations can be seen. A **line of best fit** may be added to how the overall trend in the data.

Line graphs

A line graph is a scatterplot with the dots joined. It displays information as a series of points on a graph that are joined to form a line. Line graphs are very useful to show change over time. When the dependent variable changes with time, you can graph time on the *x*-axis and the dependent variable (such as height or temperature) on the *y*-axis. Each point represents a set of data for two variables, such as height and time. Two or more lines may be drawn on the same graph. Line graphs are used to show continuous data — that is, data in which the values follow on from each other. As with scatterplots, a line of best fit can be added to a line graph to represent the overall trend in the data. The features of line graphs are shown in figure 1.27.

line of best fit a smooth curve or line that passes as close as possible to all plotted points on a graph

FIGURE 1.27 Features of a line graph

3. Setting up and labelling the axes

Graphs represent a relationship between two variables.

Usually the independent variable is plotted on the horizontal *x*-axis and the dependent variable on the vertical *y*-axis.

After deciding on the variable for each axis, you must clearly label each axis with the variable and its units. The units are written in brackets after the name of the variable.

2. Title

Tell the reader what the graph is about! The title describes the results of the investigation or the relationship between variables.

1. Grid

Graphs should always be drawn on grid paper so values are accurately placed. Drawing freehand on lined or plain paper is not accurate enough for most graphs.

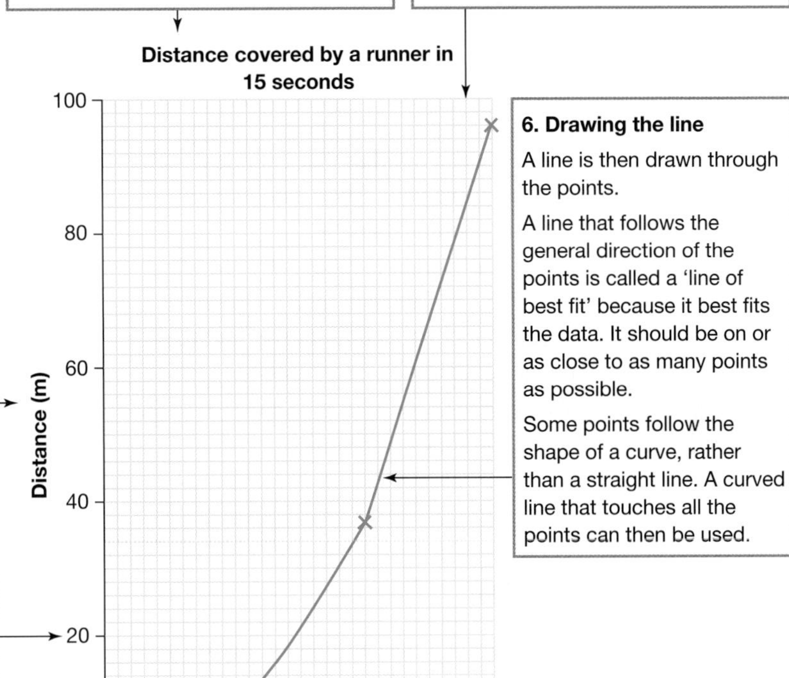

6. Drawing the line

A line is then drawn through the points.

A line that follows the general direction of the points is called a 'line of best fit' because it best fits the data. It should be on or as close to as many points as possible.

Some points follow the shape of a curve, rather than a straight line. A curved line that touches all the points can then be used.

4. Setting up the scales

Each axis should be marked into units that cover the entire range of the measurement. For example, if the distance ranges from 0 m to 96 m, then 0 m and 100 m could be the lowest and highest values on the vertical scale. The distance between the top and bottom values is then broken up into equal divisions and marked. The horizontal axis must also have its own range of values and uniform scale (which does not have to be the same scale as the vertical axis). The most important points about the scales are:

- they must show the entire range of measurements
- they must be uniform; that is, show equal divisions for equal increases in value.

5. Putting in the values

A point is made for each pair of values from the data table (the meeting point of two imaginary lines from each axis). The points should be clearly visible. Only include a point for (0, 0) if you have the data for this point.

TABLE 1.6 Data table

Distance (m)	Time (s)
0	4
8	5
37	10
96	15

SAMPLE PROBLEM 1: Choosing types of graphs

Identify the type of graph that would be most appropriate to display the following data:

a. Data from Melbourne Zoo showing how the mass of a baby elephant has increased over time

b. The mass of each elephant at Melbourne Zoo

c. The proportion of visitors using various modes of transport to travel to Melbourne Zoo.

THINK	WRITE
a. The mass of one elephant is a number that changes over time, so it is quantitative data. Mass can take any numerical value, so it is continuous data.	Mass is continuous data, so a line graph would be the best choice.
b. We compare the mass of different elephants by showing the name of each elephant and its mass at a set point in time. The name of each elephant is qualitative, and the mass of each elephant is quantitative (continuous).	As we have both qualitative and quantitative data, a bar or column graph would be the best choice.
c. The proportion of visitors using various modes of transport shows fractions or percentages of a whole.	As the data is showing the proportion of people using different modes of transport, the best choice would be a pie chart or divided bar chart.

1.11.4 Interpolation

Line graphs can be used to estimate measurements that were not actually made in an investigation. Table 1.7 shows the results of an experiment in which a student measured how many spoons of sugar dissolved in a cup of tea at various temperatures. Once a line of best fit has been drawn, we can determine the mass of sugar that would dissolve at temperatures within the range that were investigated. Predicting values from within the range of the experiment like this is called interpolation.

TABLE 1.7 Amount of sugar that dissolves in one cup of tea at different temperatures

Temperature (°C)	Mass of sugar dissolved (g)
0	4
20	30
40	60
60	98
80	120
100	160

The student did not measure how much sugar dissolved at 50 °C, but we can work this out by interpolation. First, we need to plot the data collected in the experiment. Then we read off the graph the amount of sugar that would dissolve at 50 °C (shown by dotted line 1 in the graph in figure 1.28). The same procedure can be used to work out the water temperature that would be needed to dissolve 130 g of sugar in one cup of tea. This is shown by dotted line 2.

FIGURE 1.28 Using a line graph for interpolation

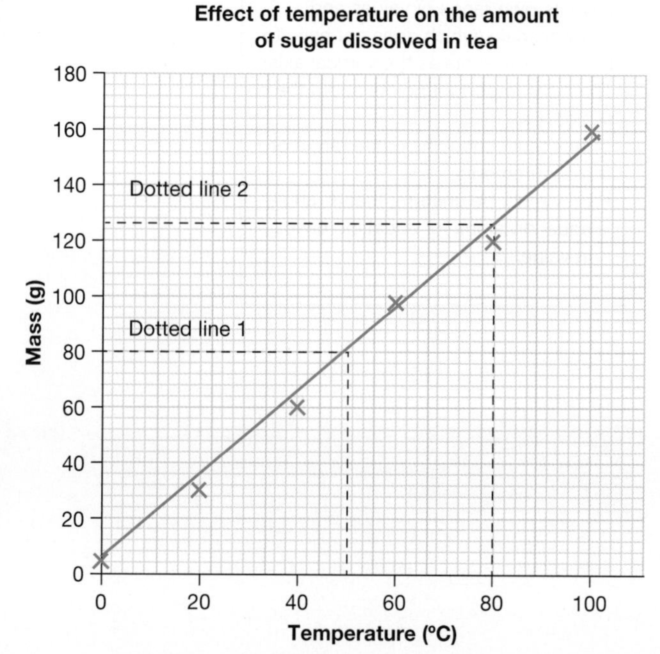

1.11.5 Extrapolation

In many cases it is also possible to assume that the two variables will hold the same relationship beyond the values that have been plotted. By extending a line of best fit past the range tested we can make predictions on data outside of the range tested. This is called extrapolation. Consider table 1.8, which shows the results obtained when different masses were attached to a spring and the increase in length of the spring was measured.

TABLE 1.8 Amount a spring stretched when various masses were attached

Mass attached to the spring (kg)	Length by which spring stretched (cm)
0.0	0
0.5	8
1.0	16
1.6	26
?	32

If you want to predict the mass needed to stretch the spring by 32 cm, you need to plot the data on a graph and extrapolate the value.

The data in table 1.8 are plotted on the graph (figure 1.29). Values have been plotted up to a mass of 1.6 kg and an increase in length of 26 cm. The line on the graph has been projected onwards (as the dotted lines show). This extrapolation shows that a mass of 2 kg will stretch the spring 32 cm. It should be noted that, unlike interpolation, extrapolation is not a very reliable technique. If possible, always try to test values directly.

FIGURE 1.29 Using a line graph for extrapolation

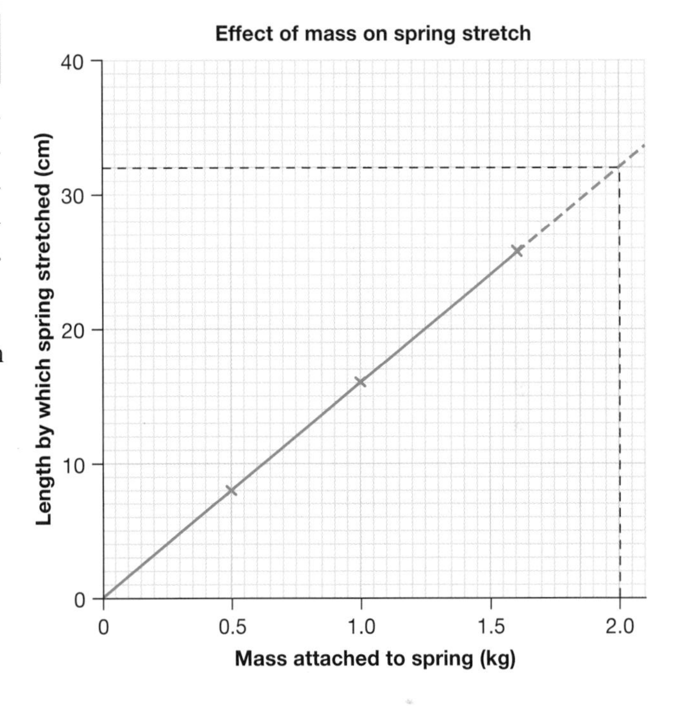

SAMPLE PROBLEM 2: Drawing a line graph

A student conducted an experiment to see how temperature affected the amount of sugar that would dissolve in a cup of tea. Each cup contained the same volume of tea, and the sugar was stirred in at an equal rate for each cup. The results obtained are shown in the following table.

TABLE Amount of sugar that dissolved in one cup of tea

Temperature (°C)	Mass of sugar dissolved (g)
0	4
20	30
40	60
60	98
80	120
100	160

Graph the data in the table.

THINK	WRITE
1. Set up the grid.	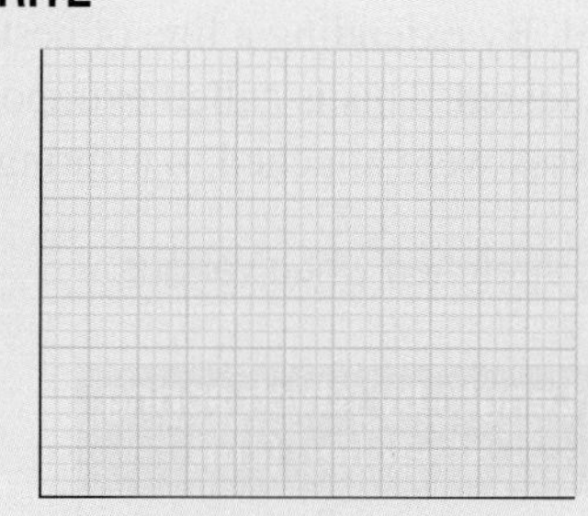
2. Give the graph a title.	Effect of temperature on the amount of sugar dissolved in tea 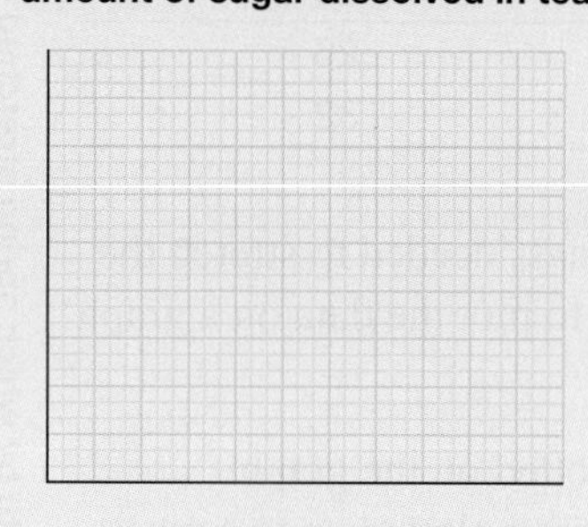
3. Set up the axes and label them.	Effect of temperature on the amount of sugar dissolved in tea 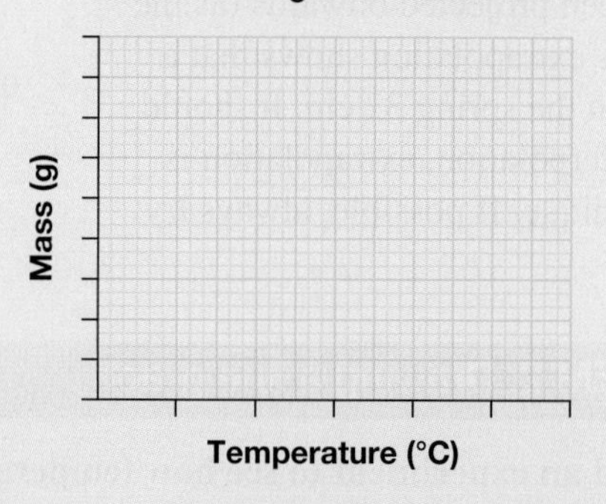
4. Place the scales on the axes.	Effect of temperature on the amount of sugar dissolved in tea 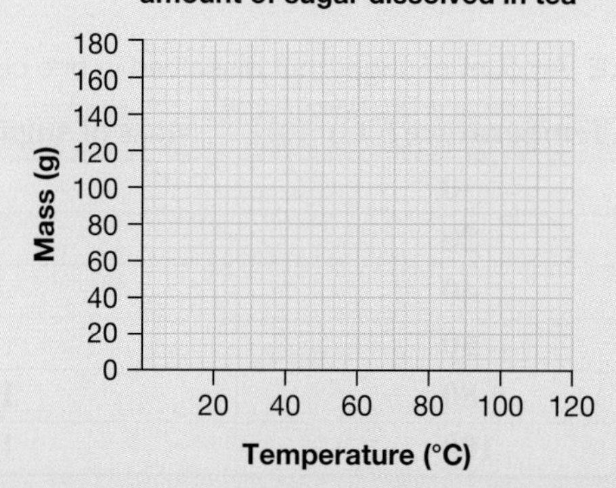

5. Plot each pair of values as a point marked with an x. Make sure that each point is clearly visible. Don't forget to plot (0, 4) because you have the data for this point.

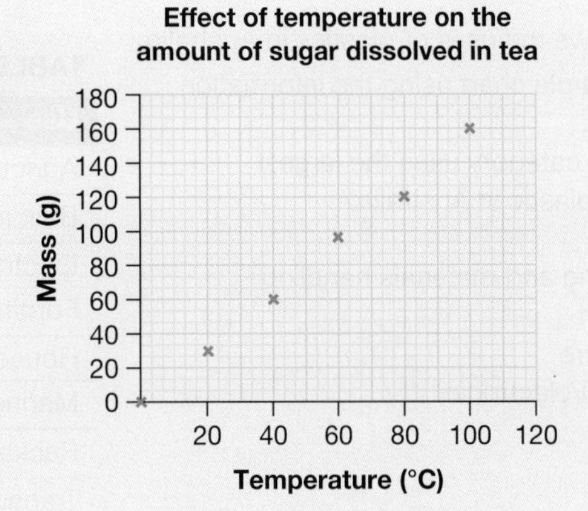

6. Draw a line of best fit; that is, a line drawn in between the points so that some points are on the line, some are below it and some are above.

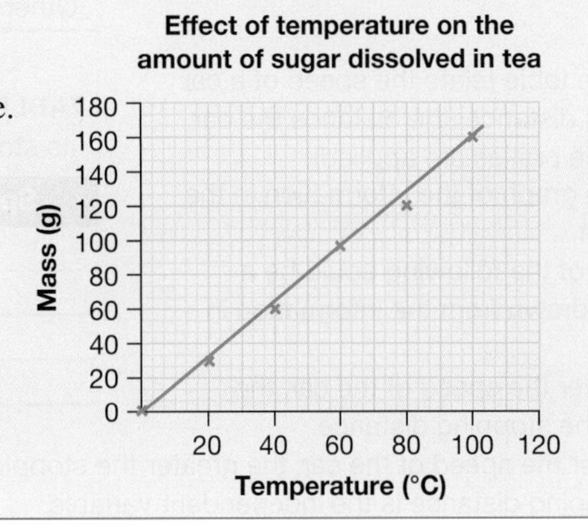

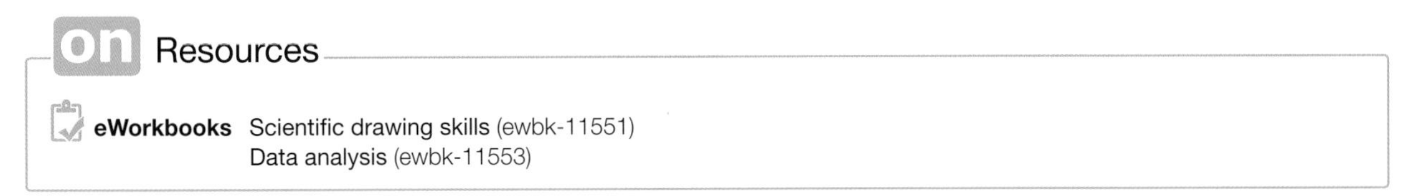

1.11 Activities

learnon

1.11 Quick quiz on	1.11 Exercise

Select your pathway

LEVEL 1	LEVEL 2	LEVEL 3
1, 3	2, 5	4, 6

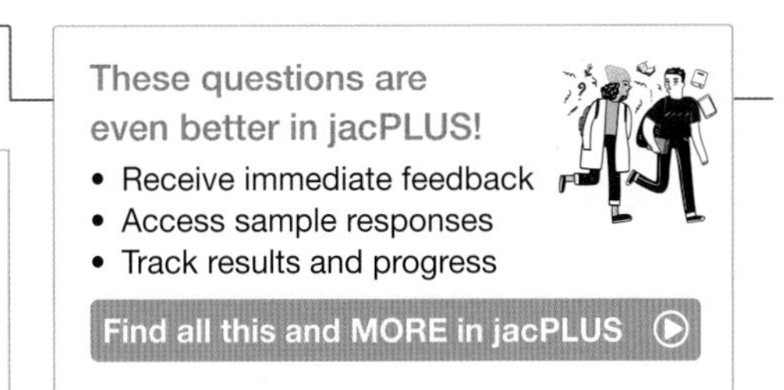

Remember and understand

1. Define the term *interpolation*.

Apply and analyse

2. Would you choose a column graph or a histogram to best represent a set of continuous, numerical data? Explain your reasoning.

Evaluate and create

3. The table shows the uses of plastics in Australia.
   a. Construct a pie chart using the information in the table.
   b. MC Which category uses the largest amount of plastic in Australia?
      A. Building
      B. Packaging and materials handling
      C. Transport
      D. Agriculture
      E. Electrical/electronic

TABLE Uses of plastics in Australia

Use	Percentage (%)
Agriculture	4.0
Building	24.0
Electrical/electronic	8.0
Furniture and bedding	8.0
Housewares	4.0
Marine, toys and leisure	2.0
Packaging and materials handling	31.0
Transport	5.0
Others	14.0

4. The data in the table relate the speed of a car to its stopping distance (the distance the car travels after the brakes are applied).
   a. Construct a graph of the information in the table shown.
   b. MC Which of the following could be a conclusion drawn from the information in the graph?
      A. The slower the speed of the car, the greater the stopping distance.
      B. The faster the speed of the car, the greater the stopping distance.
      C. The stopping distance is the independent variable.
      D. The faster the speed of the car, the smaller the stopping distance.

TABLE Relationship between the speed of a car and its stopping distance

Speed of car (m/s)	Stopping distance (m)
10	12
20	36
30	72
40	120

5. The boiling point of water changes with air pressure. For example, water does not boil at 100 °C at the top of Mount Everest, where the air pressure is less than the pressure at sea level. The following data show the boiling point of water at various air pressure values.
   a. Construct a graph of the information in the table.
   b. Describe the shape of the graph.
   c. What is the pressure of the atmosphere at sea level when water boils at 100 °C?
   d. MC Would it take a longer or shorter time to boil water at the top of Mount Everest compared with at sea level?
      A. It would take less time because air pressure is lower. The lower the air pressure, the lower the boiling point of water.
      B. It would take more time because air pressure is lower. The lower the air pressure, the higher the boiling point of water.
      C. It would take more time because air pressure is higher. The higher the air pressure, the higher the boiling point of water.
      D. It would take less time because air pressure is higher. The higher the air pressure, the lower the boiling point of water.

TABLE Boiling point of water at different air pressures

Air pressure (kPa)	Boiling point of water (°C)
1	20
7	40
21	60
45	80
101	100
200	120

6. The graph shows the increase in mass of a growing pondweed.
   a. What was the mass of the plant after three weeks of growth?
   b. How long did it take for the plant to weigh 250 g?
   c. MC What would be a reasonable prediction for the mass of the plant after six weeks of growth?
      A. 500 g
      B. 300 g
      C. 1 kg
      D. 100 g
   d. Can you be sure that your prediction for part **c** is accurate? Suggest why it may not be accurate.
   e. True or false? Extrapolations are more reliable than interpolations. Justify your response.

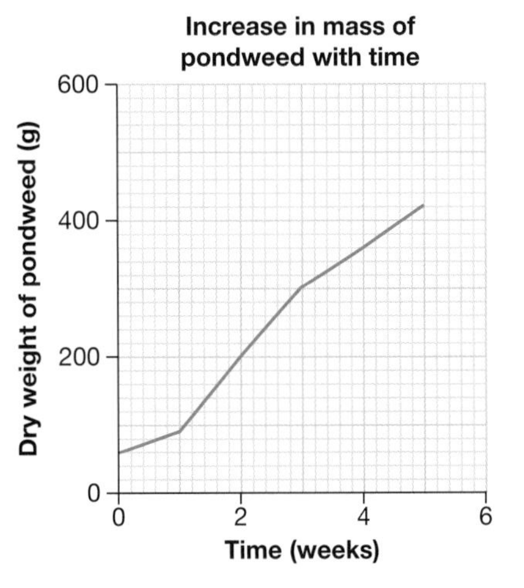

Fully worked solutions and sample responses are available in your digital formats.

LESSON
1.12 SkillBuilder — Constructing a pie chart

LEARNING INTENTION

At the end of this lesson you will be able to construct pie charts.

online only

What is a pie chart?

A pie chart, or pie graph, is a graph in which slices or segments represent the size of different parts that make up the whole. The size of the segments is easily seen and can be compared. Pie graphs give us an overall impression of data.

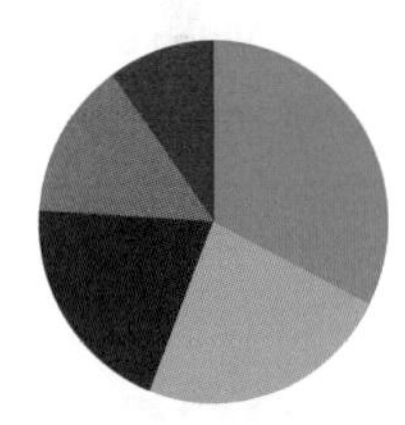
Pie chart parts of a whole

Go online to access:
- **Tell me:** an overview of the skill and its application in science
- **Show me:** a video and a step-by-step process to explain the skill
- **Let me do it:** an interactivity, question set and Skillbuilder activity for you to practice and consolidate your understanding of the skill

on Resources

eWorkbook	Skillbuilder — Constructing a pie graph (ewbk-11555)
Video eLesson	Skillbuilder — Constructing a pie graph (eles-1632)
Interactivity	Skillbuilder — Constructing a pie graph (int-3128)

LESSON
1.13 SkillBuilder — Creating a simple column or bar graph

LEARNING INTENTION

At the end of this lesson you will be able to construct simple column or bar graphs.

online only

What is a column or bar graph?

Column graphs show information or data in columns. In a bar graph the bars are drawn horizontally and in column graphs they are drawn vertically. They can be hand drawn or constructed using computer spreadsheets.

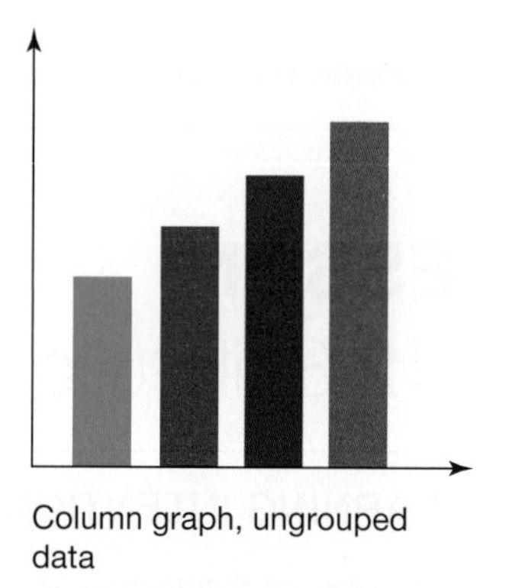

Column graph, ungrouped data

Go online to access:

- **Tell me:** an overview of the skill and its application in science
- **Show me:** a video and a step-by-step process to explain the skill
- **Let me do it:** an interactivity, question set and Skillbuilder activity for you to practice and consolidate your understanding of the skill

on Resources

eWorkbook	Skillbuilder — Creating a simple column or bar graph (ewbk-11557)
Video eLesson	Skillbuilder — Creating a simple column or bar graph (eles-1639)
Interactivity	Skillbuilder — Creating a simple column graph (int-3135)

LESSON
1.14 SkillBuilder — Drawing a line graph

LEARNING INTENTION

At the end of this lesson you will be able to construct line graphs.

online only

What is a line graph?

A line graph displays information as a series of points on a graph that are joined to form a line. Line graphs are very useful to show change over time. They can show a single set of data, or they can show multiple sets, which enables us to compare similarities and differences between two sets of data at a glance.

Go online to access:

- **Tell me:** an overview of the skill and its application in science
- **Show me:** a video and a step-by-step process to explain the skill
- **Let me do it:** an interactivity, question set and Skillbuilder activity for you to practice and consolidate your understanding of the skill

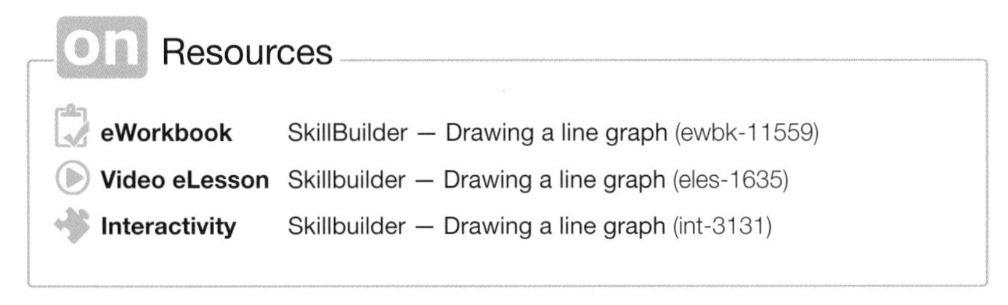
on Resources

eWorkbook	SkillBuilder — Drawing a line graph (ewbk-11559)
Video eLesson	Skillbuilder — Drawing a line graph (eles-1635)
Interactivity	Skillbuilder — Drawing a line graph (int-3131)

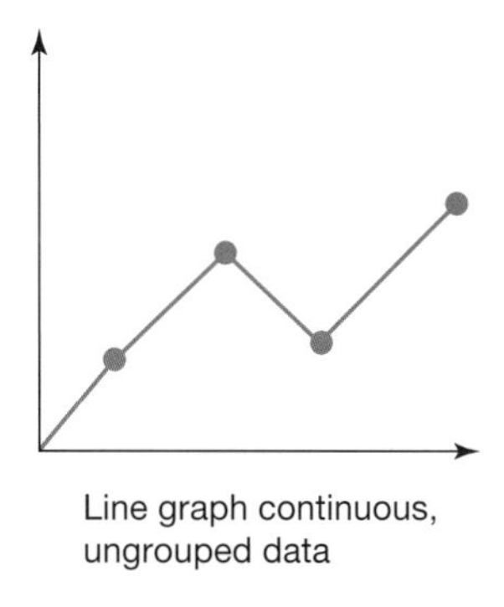

Line graph continuous, ungrouped data

LESSON
1.15 Using data loggers

LEARNING INTENTION

At the end of this lesson you will be able to identify different types of data loggers that can monitor investigations and produce data.

1.15.1 What is a data logger?

A data logger is a type of scientific recording instrument. Many devices record data — for example, fitness monitors and smart watches. A data logger collects and stores measurements that are called data. It has to be attached to a measuring instrument called a **sensor**. The sensor does the measuring and sends the measurements to the data logger.

The real advantage of working with a data logger is that it can store thousands of individual measurements. The measurements can be taken in quick succession or over a long period of time, and the data logger can be programmed to do this automatically. This is why scientists often use data loggers in their work.

Data loggers also tend to be portable and battery-powered, and can therefore be used for applications, such as remote weather monitoring and car crash testing. You may have been in a car that has driven over two closely placed rubber strips on the road — these strips are connected to a data logger used to count traffic.

Of course, to be useful, the stored measurements must be easy to access. That is why the data logger is also attached to either a computer or a graphics calculator. The computer or calculator takes the data and, using special software that comes with the data logger, shows the data as a table, a graph or both.

FIGURE 1.30 Some data loggers have their own touch screen and work like mini computers.

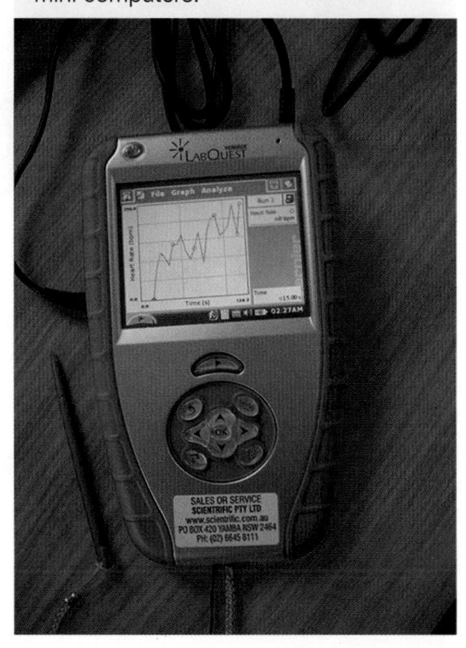

sensor a device connected to an instrument, such as a data logger, that measures and sends information

Other uses for data loggers

Data loggers can be used for just about any experiment in which measurements are taken. All that is needed is the appropriate sensor to be plugged in. It is even possible to plug in several sensors to take different measurements at the same time.

Some of the many different sensors that are available include:

- temperature sensors capable of measuring up to several hundred degrees Celsius
- light intensity sensors
- sound-wave sensors (microphones)
- motion sensors
- magnetic field sensors
- acceleration sensors
- force sensors
- electric current and voltage sensors
- humidity sensors
- blood pressure sensors
- heart rate sensors.

One type of sensor that isn't necessary is a time sensor (stopwatch) because the data logger has its own inbuilt clock that is very accurate. In fact, one of the most useful things about data loggers is their ability to collect measurements at very small and precise time intervals, even as many as a thousand measurements in one second!

FIGURE 1.31 More basic data loggers require the use of a computer to analyse the results.

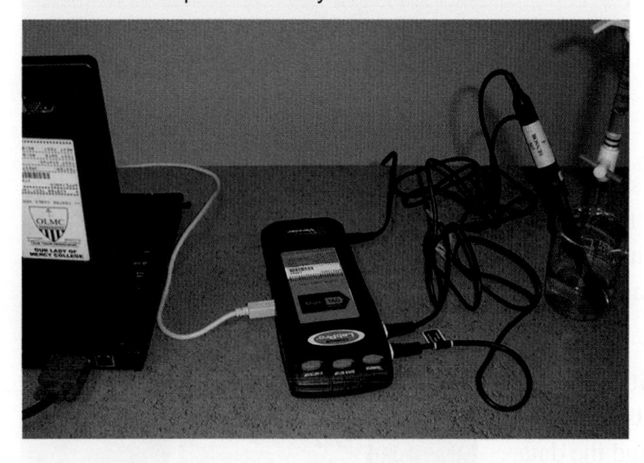

FIGURE 1.32 A data logger for measuring blood pressure

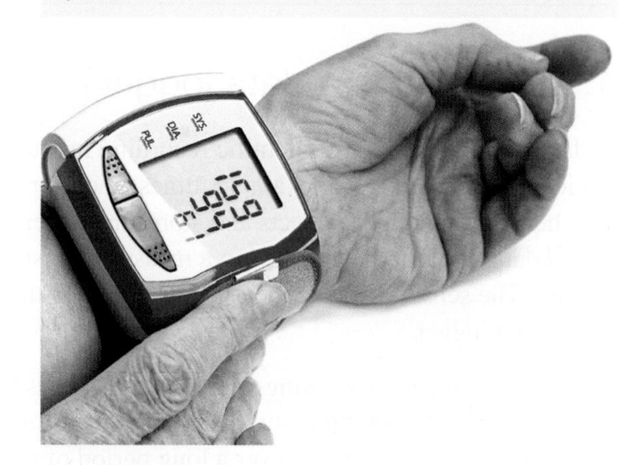

1.15.2 Data loggers in temperature measurement

In investigation 1.1 in section 1.1.3, the measuring instrument you used was a thermometer. You looked at the thermometer every 30 seconds and observed the temperature, which you wrote down in a table. You then made a line graph of temperature against time. If you had used a data logger with a temperature sensor instead of the thermometer, it could have taken the temperature every second and sent it to a computer that automatically tabulated the temperature data and graphed it as well.

1.15 Activities

1.15 Quick quiz	1.15 Exercise

Select your pathway

LEVEL 1	LEVEL 2	LEVEL 3
1	2	3

These questions are even better in jacPLUS!
- Receive immediate feedback
- Access sample responses
- Track results and progress

Find all this and MORE in jacPLUS

Remember and understand

1. Match the word on the left of the table to the correct statement on the right.

Word	Meaning
a. Sensor	**A.** You may need to download the data from the data logger to one of these
b. Data logger	**B.** Piece of information
c. Computer	**C.** These are plugged into the data logger and take the measurements.
d. Data logger software	**D.** Allows you to input data into the data logger or computer by touching it with your finger or a stylus
e. Touch screen	**E.** Allows you to process the data collected by the data logger
f. Data	**F.** Collects and stores data from sensors connected to it

Apply and analyse

2. Sensors are devices that take the measurements that the data logger collects. Outline scientific investigations that could use data collected by sensors that measure:
   a. electric current **b.** heart rate **c.** motion **d.** sound waves **e.** light intensity.

Evaluate and create

3. The graph shows data collected by a data logger for an experiment in which water was heated to boiling point in a beaker. A temperature sensor was used to take the measurements.

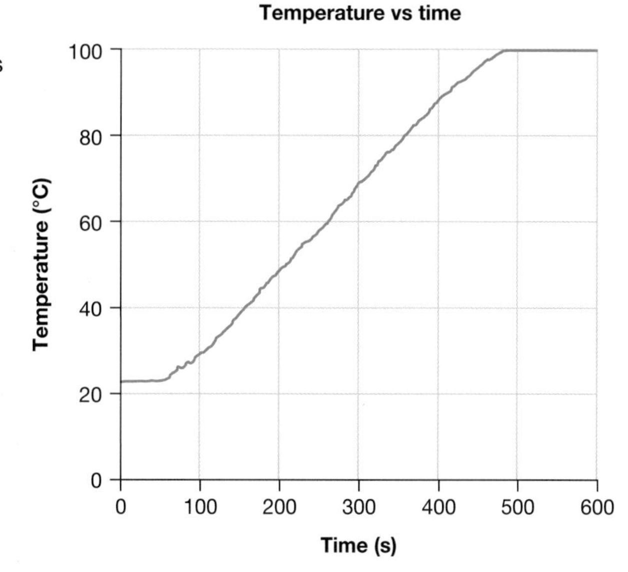

   a. How long did the whole experiment take?
   b. Approximately how long after the experiment began did the heating of water begin?
   c. What was the temperature of the water when heating began?
   d. What was the temperature of the water when heating finished?
   e. Approximately how long after the experiment began did the water begin to boil?
   f. Between 100 and 400 seconds, at what rate (in degrees per second) did the water temperature rise?
   g. The water continued to be heated even when its temperature reached boiling point; yet its temperature did not rise beyond 100 °C. What has happened to all the energy that was being put into the water if it isn't causing the water temperature to rise? (*Hint:* Think about what happens to water while it is boiling.)

Fully worked solutions and sample responses are available in your digital formats.

LESSON
1.16 Famous scientists

LEARNING INTENTION

At the end of this lesson you will be able to describe the discoveries and achievements of four famous scientists.

1.16.1 Famous science investigations

Einstein, Newton, Curie and Pasteur. These are all names that are familiar to us because their contributions to science changed the world. Some facts about their lives and some insights into their discoveries are discussed in this lesson.

Resources

 **Video eLesson** Career spotlight: scientist (eles-0766)

SCIENCE AS A HUMAN ENDEAVOUR: Sir Isaac Newton

Sir Isaac Newton (1642–1727) is probably most well known for his laws of gravitation, which explain the motion of the planets around the sun. According to some historians, his ideas about gravity arose after an apple fell on his head. We will probably never know if this is true.

Isaac Newton was sent to Cambridge University at the age of 18. When the university closed down in 1665 as a result of the Great Plague, young Isaac went home for two years. There he developed his laws of gravitation and his three laws of motion. During his life, he also made discoveries about the behaviour of light and invented a whole new branch of mathematics, called calculus. Much of the scientific knowledge that has been acquired since the seventeenth century is built upon Newton's discoveries during that amazing two-year period.

FIGURE 1.33 Sir Isaac Newton

Resources

 **Video eLesson** Isaac Newton (eles-1771)

SCIENCE AS A HUMAN ENDEAVOUR: Albert Einstein

Albert Einstein (1879–1955) is most well known for his theory of relativity (there are actually two theories of relativity) and the equation $E = mc^2$, which describes how mass can be converted into energy.

Albert Einstein was certainly a slow starter. Although he was fascinated by mathematics, Einstein performed badly at school and left at the age of 15. He returned later and trained as a teacher in Switzerland. Einstein often failed to attend lectures and passed university exams by studying the notes of his classmates.

Einstein's first job was as a junior clerk in a patent office. His work was not demanding and he spent a lot of time doing 'thought' experiments.

FIGURE 1.34 Einstein's first wife, Mileva, was a mathematician. He discussed many of his new ideas with her.

At the age of 26, Einstein began to publish his ideas. These ideas altered our view of the nature of the universe by changing existing laws and discovering new ones.

Einstein explained the photoelectric effect, in which light energy is transformed into electrical energy, and received the Nobel Prize in Physics in 1921 for this.

Einstein's theories of relativity were so different from earlier theories that they were not believed or understood by most scientists. His theory of special relativity explains the behaviour of objects that travel at speeds close to the speed of light. His theory of general relativity explains the effect of gravity on light and predicts that time 'slows down' in the presence of large gravitational forces. These theories provide useful clues about the development and future of the universe.

Einstein's theories suggested that mass could be converted into energy. This idea led to the development of the atomic bomb and nuclear power. Einstein, who was Jewish, fled Germany in 1933 to live and work in the United States. He was an active opponent of nuclear weapons and was involved in the peace movement long before atomic bombs destroyed Hiroshima and Nagasaki at the end of World War II.

SCIENCE AS A HUMAN ENDEAVOUR: Louis Pasteur

Louis Pasteur (1822–1895) proved that infectious diseases were caused by microbes. His ideas became known as 'germ theory'. He also developed several vaccines that made people immune to diseases such as rabies and smallpox. In doing this he has been responsible for saving the lives of millions of people and countless animals.

FIGURE 1.35 One of Pasteur's experiments

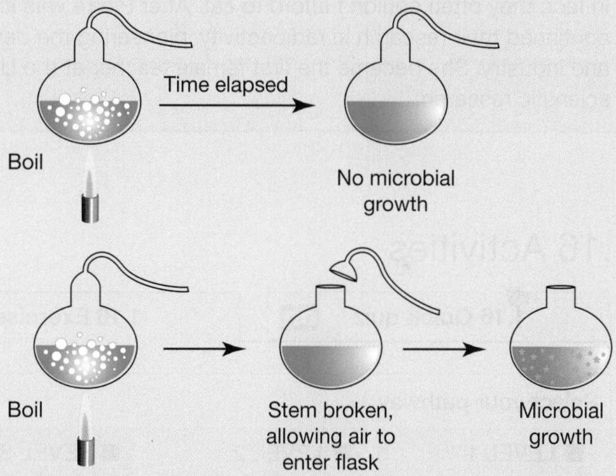

Pasteur began his scientific career in physics and chemistry, but became interested in microbes when he was using light to investigate the differences between chemicals in living and non-living things.

Pasteur's next challenge was to rescue the French wine industry. Wine (and beer) became sour very quickly and this was beginning to have an impact on the French economy, which relied heavily on the export of wine. Pasteur showed that the souring was caused by acids produced by the action of bacteria in the wine. Pasteur invented a process that rapidly heated some of the ingredients of the wine. The rapid heating killed most of the offending microbes without altering the flavour of the wine. The process, known as pasteurisation, was later adapted to slow down the souring of milk.

DISCUSSION

Louis Pasteur conducted many of his experiments on animals. Many of the experiments would now be considered cruel; however, the experiments saved many human and animal lives.

Present the arguments for and against the use of animals in such experiments.

Were the animal experiments justified? Write a brief statement supporting your opinion.

Resources

Interactivity Pasteur's experiment (int-3420)

SCIENCE AS A HUMAN ENDEAVOUR: Marie Curie

Marie Curie (1867–1934) became the first scientist to win two Nobel Prizes when she was awarded the Nobel Prize in Chemistry in 1911 for her discovery of two new elements: polonium and radium. Radium was used in the treatment of cancer until cheaper and safer radioactive materials were developed. Marie Curie's first Nobel Prize, for the study of radioactivity, was shared with her husband, Pierre, and fellow scientist Antoine-Henri Becquerel in 1903.

FIGURE 1.36 Marie Curie with husband Pierre in her laboratory

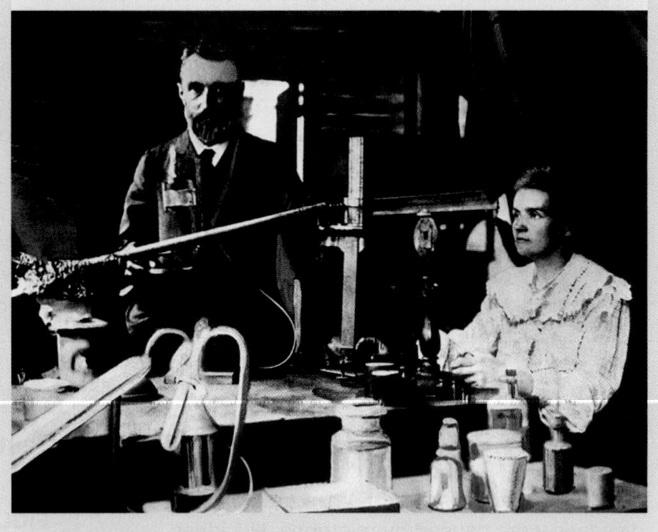

As a child, Marie Sklodowska (her birth name) wanted to study science. However, girls were forbidden to attend university in her native country of Poland. She worked as a private tutor for three years to earn enough money to study at the University of Paris, where she met her future husband, Pierre. They were very poor and spent most of their money on laboratory equipment, leaving very little money for food; in fact, they often couldn't afford to eat. After Pierre was knocked down and killed by a speeding wagon, Marie continued their research in radioactivity, pioneering the development of radioactive materials for use in medicine and industry. She became the first female teacher at the University of Paris and worked hard to raise money for scientific research.

1.16 Activities

learnon

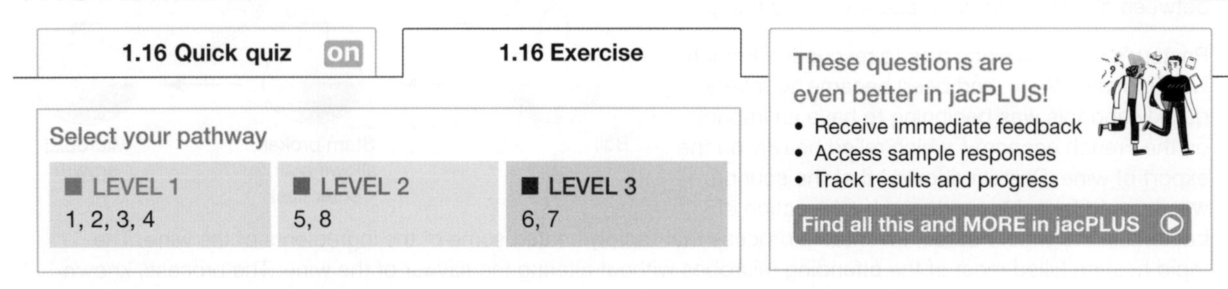

Remember and understand

1. **MC** Louis Pasteur worked in many fields of science during his career. Which of the following did he achieve?
   A. The discovery that infectious diseases are caused by microbes
   B. The development of the theories of gravity and motion
   C. The isolation of two new elements: radium and polonium
   D. The development of the theory of relativity
2. **MC** Sir Isaac Newton worked in many fields of science during his career. Which of the following did he achieve?
   A. The discovery that infectious diseases are caused by microbes
   B. The development of the theories of gravity and motion
   C. The isolation of two new elements: radium and polonium
   D. The development of the theory of relativity

3. MC Marie Curie was the first scientist to win two Nobel prizes. Which of the following did she achieve?
   A. The discovery that infectious diseases are caused by microbes
   B. The development of the theories of gravity and motion
   C. The isolation of two new elements: radium and polonium
   D. The development of the theory of relativity
4. MC Albert Einstein worked in many fields of science during his career. Which of the following did he achieve?
   A. The discovery that infectious diseases are caused by microbes
   B. The development of the theories of gravity and motion
   C. The isolation of two new elements: radium and polonium
   D. The development of the theory of relativity

Apply and analyse

5. Make a quick list of your 'Top 3' scientists of all time. For each one, answer the following questions.
   a. What impact does their work have on your life?
   b. Did they just happen to be in the 'right place at the right time'?
   c. Did they work under adverse conditions?
   d. Did their work save lives?
   e. Did their work have any destructive influence?
   f. What other special qualities make them great?
6. Is it fair to select the single 'greatest' scientist of all time? Explain your answer.

Evaluate and create

7. Imagine that you are one of the three scientists you chose as the greatest scientists of all time. Write a short speech (3–5 minutes) about your life and work, and deliver it to your class. Illustrate your speech with models, diagrams or photographs.
8. Write a biography similar to the four presented in this lesson about one of the following scientists:
   - Michael Faraday (1791–1867)
   - Charles Darwin (1809–1882)
   - Lise Meitner (1878–1968)
   - Barbara McClintock (1902–1992)
   - Peter Doherty (1940–)
   - Stephen Hawking (1942–2018).

Peter Doherty (1940–): Veterinarian and immunologist

Fully worked solutions and sample responses are available in your digital formats.

LESSON
1.17 Project — An inspiration for the future

Scenario

The Florey Medal was established in 1998 by the Australian Institute of Policy and Science in honour of the Australian Nobel Prize-winning scientist Sir Howard Florey, who developed penicillin. It is awarded biennially to an Australian biomedical researcher for significant achievements in biomedical science and human health advancement.

In a similar spirit, the Australian Academy of Science (AAS) hopes to establish an award for outstanding science students. The AAS wishes to name the medal after an Australian scientist who provides the greatest inspiration for young people considering a future career in science. After months of consultation, they have narrowed the choices down to the following:

- David Unaipon (1872–1967): Inventor
- Fred Hollows (1929–1993): Ophthalmologist
- Andrew Thomas (1951–): Astronaut
- Fiona Wood (1958–): Plastic surgeon and burns specialist
- Ian Frazer (1953–): Immunologist
- Graeme Clark (1935–): Otolaryngeal surgeon and engineer

FIGURE 1.37 Andrew Thomas (1951–): Astronaut and engineer

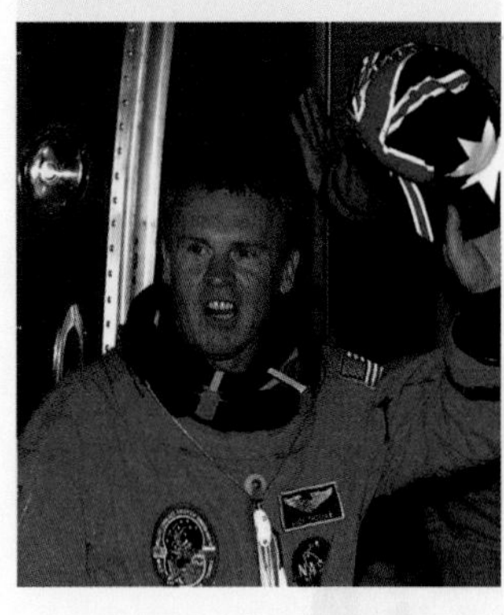

FIGURE 1.38 Fiona Wood (1958–): Plastic surgeon and burns specialist

FIGURE 1.39 Graeme Clark (1935–): Otolaryngeal surgeon and engineer

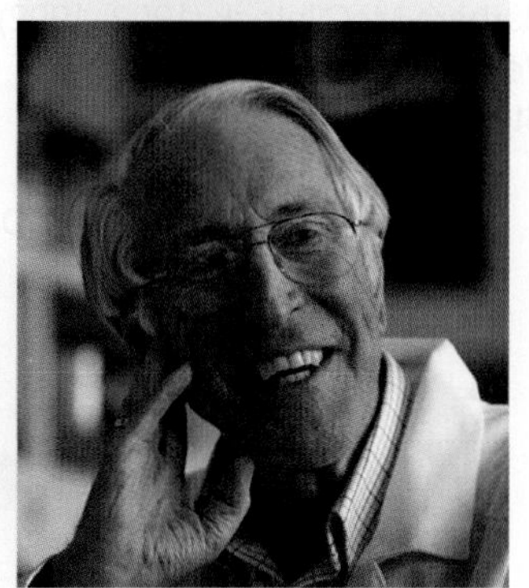

Your task

You will create an 8- to 10-minute podcast in the format of an interviewer discussing with a number of different people which of these scientists would be the best choice to name the AAS medal after. The interviewees (played by group members) should be people who would be likely to have an interest or stake in the award. Examples could include a member of the AAS medal panel, the Minister for Industry, the head of a university science department or science education department, a high school science teacher or even a high school science student. Each interviewee should have their own preference as to which scientist should be selected and at least four scientists should be discussed during the interview.

Resources

ProjectsPLUS An inspiration for the future (pro-0071)

LESSON
1.18 Review

Hey students! Now that it's time to revise this topic, go online to:

Review your results

Watch teacher-led videos

Practise questions with immediate feedback

Find all this and MORE in jacPLUS

Access your topic review eWorkbooks

Resources

Topic review Level 1	Topic review Level 2	Topic review Level 3
ewbk-11561	ewbk-11563	ewbk-11565

1.18.1 Summary

Investigating skills

- There are many materials and procedures in a science laboratory that can be dangerous. To ensure safety, always follow the teacher's instructions and wear appropriate safety clothing.
- Use filter funnels when pouring substances into test tubes from containers without lips.
- Look up the warnings that are associated with the chemicals you are using and adhere to the relevant safety instructions.

Planning your own investigation

- When planning your own investigation consider your interests and think of a question that you might like to investigate.
- All investigations should start with an aim that outlines the purpose of the investigation. Aims should start with 'to find out', 'to compare', 'to investigate' or some other statement of what you intend to investigate.
- A hypothesis is a sensible guess about the outcome of an experiment. When forming your own hypotheses, do some research and use logic to make an educated guess.
- Hypotheses must be a statement that can be disproven. If your hypothesis cannot be disproven there is no point in completing the investigation!
- Ethical considerations must be taken into account when deciding on an investigation. An investigation should not harm or be upsetting to people or animals.

Record keeping and research

- A logbook can be used to keep track of your investigations.
- Each entry in your logbook should be dated like a diary.
- When researching information about the topic you are investigating, use multiple sources such as books, websites, journals and scientific magazines.
- Check that the information you find is reliable. You can do this by researching the author's qualifications or legitimacy of the book/website.

Controlling variables

- In an investigation there will be multiple variables.
- The variable that you measure is called the dependent variable.
- The variable that you purposefully alter to record the effect it has on the dependent variable is called the independent variable.
- All other variables should be kept constant. These variables are called controlled variables.
- It is very important that all variables that you are not purposefully altering are kept constant so that you can be sure that the change in the dependent variable is caused by the change in the independent variable.

Scientific reports

- Scientific reports must contain certain sections so that they can be read and understood by people worldwide.
- Abstracts are brief descriptions of the investigation, including the main conclusions.
- Introductions present all relevant background information that needs to be understood before reading the rest of the report.
- Materials are a list of the materials used in the investigation.
- The method is a detailed description of the steps taken throughout the investigation. It should be clear enough that another person could replicate the investigation by following the steps.
- Results state the observations and data obtained throughout the investigation.
- The discussion is where the results are discussed and put into context. The meaning behind the results should be explained in this section.
- The conclusion is a brief statement of what was learned in the investigation.
- The bibliography is a list of the resources used in the investigation.

Presenting your data

- The data obtained in an investigation can be difficult to read and understand when it is written as numbers on a page.
- To more easily understand the data collected it should be presented visually in graphs and tables.
- Pie charts are a useful way of representing percentages of a whole. The segments of a pie chart represent the percentage of the whole that it accounts for.
- Column and bar graphs are used to compare categories.
- Histograms are used to compare continuous data.
- Interpolation is the prediction of a value from a line of best fit using a value that is within the range tested.
- Extrapolation is the prediction of a value from a line of best fit using a value that is outside the range tested. Extrapolation is not a very reliable method of predicting data.

Using data loggers

- Data loggers are instruments that record data digitally. This reduces the human error in their measurements making them a reliable source of data.
- Some data loggers are also able to plot the data they record to provide accurate graphs automatically.

Famous scientists

- Sir Isaac Newton (1642–1727) was an incredibly influential scientist and mathematician who developed the theory of gravitation that is still used to this day.
- Albert Einstein (1879–1955) is known for developing the theory of relativity, which describes the relationship between energy and matter through the equation $E = mc^2$.
- Louis Pasteur (1822–1895) identified that infectious diseases are caused by microbes, developing a 'germ theory', which he then used to develop vaccines for many deadly diseases. His work saved the lives of millions of people and animals.
- Marie Curie (1867–1934) was the first scientist to win two Nobel Prizes. The first was for her groundbreaking work on radioactivity and the second for the discovery of two new elements: radium and polonium.

1.18.2 Key terms

accuracy how close an experimental measurement is to a known value
aim a statement outlining the purpose of an investigation
beaker a container for mixing or heating substances
bibliography a list of references and sources at the end of a scientific report
calibrate to check or adjust a measuring instrument to ensure accurate measurements

control an experimental set-up in which the independent variable is not applied; a control is used to ensure that the result is due to the variable and nothing else
controlled variables the conditions that must be kept the same throughout an experiment
corrosive describes a chemical that wears away the surface of substances, especially metals
data information collected that can be used for studying or analysing
dependent variable a variable that is expected to change when the independent variable is changed; the dependent variable is observed or measured during the experiment
discussion a detailed area of a scientific report that explains the results and how they link back to the relevant concepts; it also includes suggestions for improvements to the experiment
fair test a test that changes only one variable and controls all other variables when attempting to answer a scientific question
falsifiable can be proven false
filter funnel a funnel used with filter paper to separate solids from liquids
flammable describes substances, such as methylated spirits, that burn easily
hypothesis a suggested, testable explanation for observations or experimental results; it acts as a prediction for the investigation
independent variable the variable that the scientist chooses to change to observe its effect on another variable
line of best fit a smooth curve or line that passes as close as possible to all plotted points on a graph
logbook a complete record of an investigation from the time a search for a topic is started
measuring cylinder a cylinder used to measure volumes of liquids accurately
oscillation one complete swing of a pendulum
pendulum an object swinging on the end of a string, chain or rod
period the time taken for one oscillation of a pendulum
precision how close multiple measurements of the same investigation are to each other
qualitative data categorical data that examines the quality of something (e.g. colour or gender) rather than a measurement or quantity
quantitative data numerical data that examines the quantity of something (e.g. length or time)
reliable data data that is able to be replicated in different circumstances but the same conditions
research question a question that is the focus of the investigation, which the experiment aims to answer
safety glasses plastic glasses used to protect the eyes during experiments
scientific method a systematic and logical process of investigation to test hypotheses and answer questions based on data or experimental observations
sensor a device connected to an instrument, such as a data logger, that measures and sends information
testable able to be supported or proven false through the use of observations and investigation
test tube a thin glass container for holding, heating or mixing small amounts of substances
toxic describes chemicals that are dangerous to touch, inhale or swallow
valid sound or true; a valid conclusion can be supported by other scientific investigations
variables quantities or conditions in an experiment that can change

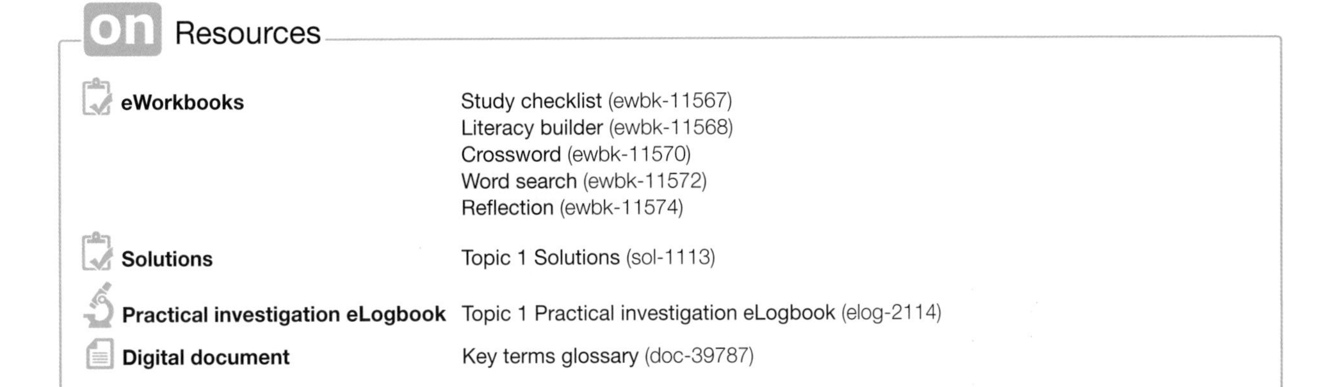

on Resources

eWorkbooks Study checklist (ewbk-11567)
Literacy builder (ewbk-11568)
Crossword (ewbk-11570)
Word search (ewbk-11572)
Reflection (ewbk-11574)

Solutions Topic 1 Solutions (sol-1113)

Practical investigation eLogbook Topic 1 Practical investigation eLogbook (elog-2114)

Digital document Key terms glossary (doc-39787)

1.18 Activities

learnon

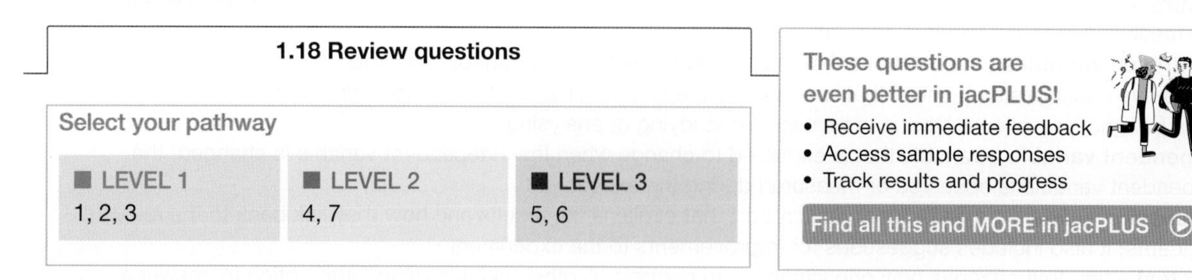

Remember and understand

1. The affinity diagram shown organises some of the ideas used by scientists into four groups. Each category name is a single word and represents an important part of scientific investigations. However, the category names have been jumbled up. What are the correct categories for groups A, B, C and D?

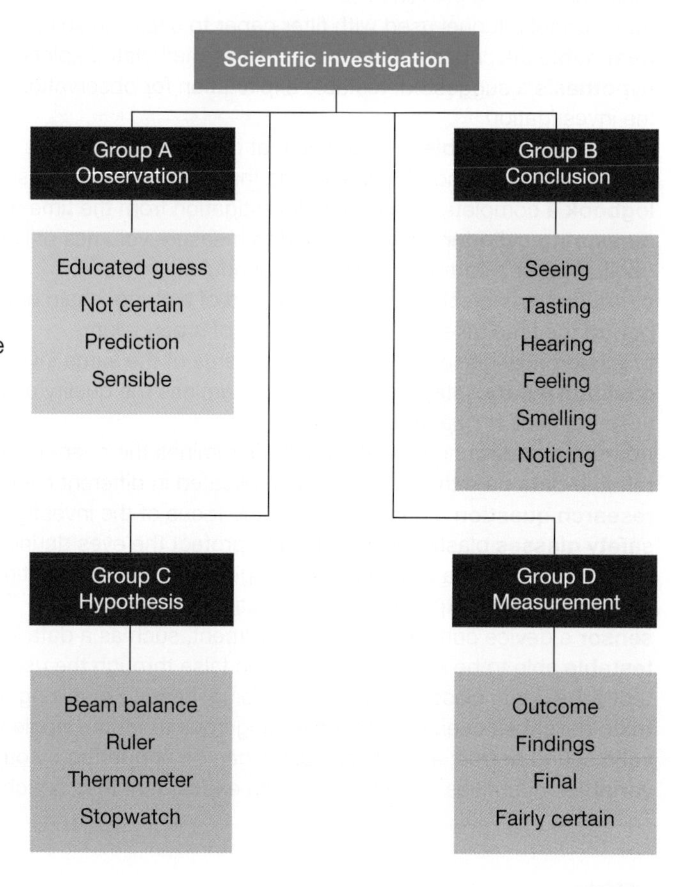

Apply and analyse

2. Bahir was sick of being bitten by mosquitoes. He counted several bites each evening when he sat outside to have dinner. He had heard that burning a citronella candle was a good way to keep mosquitoes away. Design an experiment to test Bahir's idea. List the independent and dependent variables, and the controlled variables needed to make this a fair test. Suggest a control for your experiment.

Evaluate and create

3. Four students each measured the temperature in the same classroom using a thermometer. Their results are shown in the table.

TABLE Temperature as measured by each of four students in the same classroom

Student	Temperature (°C)
1	23.5
2	24.0
3	25.0
4	22.0

a. Construct a bar graph of these results.
b. Propose some possible reasons for the differences between measurements.

4. Jane and Greg decided to test how quickly water would boil when using either the yellow flame or the blue flame of a Bunsen burner. They set up identical experiments, except that Jane used a blue flame and Greg used a yellow flame. Their results are shown in the following graph.
   a. Suggest a title for the graph.
   b. How long did it take for Jane's water to boil?
   c. What was the temperature of Greg's water when Jane's water boiled?
   d. In your own words, explain how you worked out the answers to parts **b** and **c**.
   e. Jane removed her beaker and Greg quickly placed his beaker over Jane's Bunsen burner. Assuming that the temperature of Greg's beaker did not drop while swapping Bunsen burners, predict the time at which his water would boil. Using your own words, explain how you predicted this.
   f. Is this a valid method of investigation?
5. Singalia and Sallyana are two red panda cubs that were born at Sydney's Taronga Zoo. The table shows their masses during their first 22 weeks. The photograph shows one of the cubs being weighed.

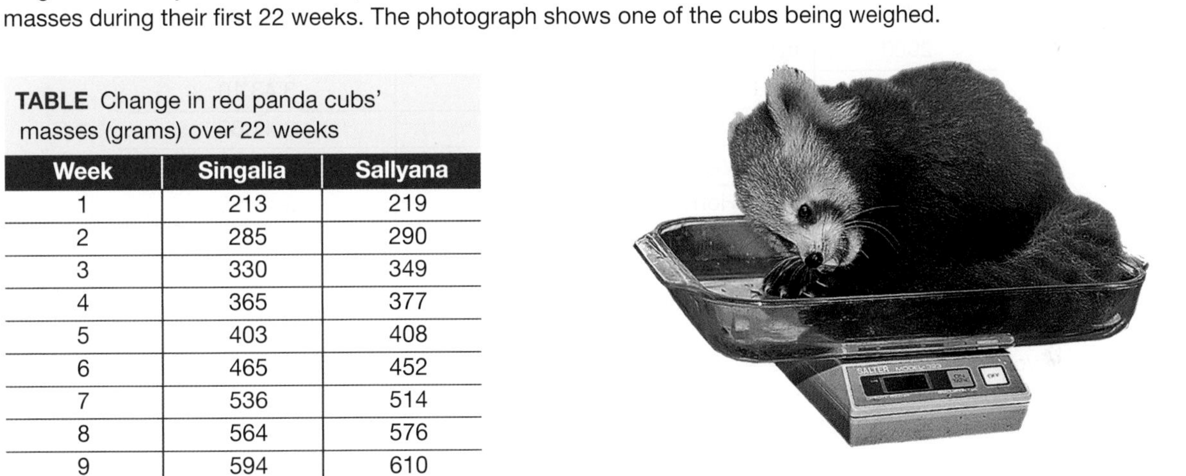

TABLE Change in red panda cubs' masses (grams) over 22 weeks

Week	Singalia	Sallyana
1	213	219
2	285	290
3	330	349
4	365	377
5	403	408
6	465	452
7	536	514
8	564	576
9	594	610
10	650	637
11	703	680
12	714	740
13	814	796
14	872	812
15	956	806
16	1111	786
17	1043	890
18	1130	1000
19	1163	1083
20	1182	1162
21	1225	1218
22	1335	1270

   a. Graph both sets of data onto a grid. Use different symbols for the points for each panda and label each line with the panda's name. You may have to extend the vertical axis to fit in the scale for the pandas' masses (or convert the masses to kilograms and plot in kilograms).
   b. Describe the growth of each of the panda cubs. How do they compare with each other?

c. How long did it take the cubs to double their mass measured in week 1?
d. Did the pandas grow at the same rate during the 22 weeks?
e. Which were the fastest and slowest growth periods for each panda?
f. What age was each of the cubs when it reached 1 kg?
g. At what age would you predict each cub to reach 1.5 kg? Explain how you made your prediction. What assumption did you make to answer the question?

6. The following table shows the winning times for the men's 400 m freestyle swimming event. The data are from various Olympic Games from 1896 to 2016.

TABLE Olympic Games winning times for the men's 400 m freestyle swimming

Year	Name, country	Time (min:s)
1896	Paul Neumann, Austria	8:12.60
1908	Henry Taylor, Great Britain	5:36.80
1920	Norman Ross, USA	5:26.80
1932	Buster Crabbe, USA	4:48.40
1948	Bill Smith, USA	4:41.00
1960	Murray Rose, Australia	4:18.30
1972	Bradford Cooper, Australia	4:00.27
1984	George DiCarlo, USA	3:51.23
1996	Danyon Loader, New Zealand	3:47.97
2000	Ian Thorpe, Australia	3:40.59
2004	Ian Thorpe, Australia	3:43.10
2008	Tae-Hwan Park, Korea	3:41.86
2012	Sun Yang, China	3:40.14
2016	Mack Horton, Australia	3:41:55

a. Are data available for each Olympics every four years?
b. Construct a line graph of the times for the men's 400 m freestyle over these years. Take into account your answer to part **a**.
c. Use your graph to estimate the winning time for this event in the 1956 Melbourne Olympic Games. Is this an example of interpolation or extrapolation?
d. Discuss how the winning times have changed over the 120-year period.
e. Suggest some reasons for the change in winning times.
f. Discuss how you believe the winning times for the men's 400 m freestyle might change over the next 40 years.

7. Create a storyboard that tells the story of the main events in the life of one of these famous scientists.
a. Albert Einstein   b. Sir Isaac Newton   c. Marie Curie   d. Louis Pasteur

Fully worked solutions and sample responses are available in your digital formats.

Online Resources

Below is a full list of **rich resources** available online for this this topic. These resources are designed to bring ideas to life, to promote deep and lasting learning and to support the different learning needs of each individual.

1.1 Overview

eWorkbooks
- Topic 1 eWorkbook (ewbk-11528)
- Starter activity (ewbk-11530)
- Student learning matrix (ewbk-11532)

Solutions
- Topic 1 Solutions (sol-1113)

Practical investigation eLogbooks
- Topic 1 Practical investigation eLogbook (elog-2114)
- Investigation 1.1: Milk now or later? (elog-2115)

Weblink
- CSIRO

1.2 Investigating skills

eWorkbooks
- Safety in the laboratory (ewbk-11533)
- Safety rules (ewbk-11535)

Video eLesson
- How to light a Bunsen burner (eles-2360)

Weblink
- Robert Bunsen

1.3 SkillBuilder — Measuring and reading scales

eWorkbook
- Skillbuilder — Measuring and reading scales (ewbk-11537)

Video eLesson
- Skillbuilder — Measuring and reading scales (eles-4153)

Interactivity
- SkillBuilder — Measuring and reading scales (int-0201)

1.4 SkillBuilder — Using a Bunsen burner

eWorkbook
- Skillbuilder — Using a Bunsen burner (ewbk-11539)

Video eLesson
- Skillbuilder — Using a Bunsen burner (eles-4154)

Interactivity
- Skillbuilder — Using a Bunsen burner (int-8088)

1.5 Planning your own investigation

eWorkbook
- Observations and inferences (ewbk-11541)

1.6 SkillBuilder — Writing an aim and forming a hypothesis

eWorkbook
- SkillBuilder — Writing an aim and forming a hypothesis (ewbk-11543)

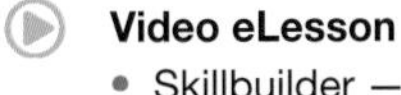

Video eLesson
- Skillbuilder — Writing an aim and forming a hypothesis (eles-4155)

Interactivity
- Skillbuilder — Writing an aim and forming a hypothesis (int-8089)

1.7 Record keeping and research

Weblink
- How to evaluate sources for reliability

1.8 Controlling variables

eWorkbook
- Fair testing (ewbk-11545)

Practical investigation eLogbook
- Investigation 1.2: The period of a pendulum (elog-2117)

Teacher-led video
- Investigation 1.2 The period of a pendulum (tlvd-10723)

Weblinks
- Identify the independent and dependent variables with the MythBusters!
- Identify independent and dependent variables

1.9 SkillBuilder — Controlled, dependent and independent variables

eWorkbook
- Skillbuilder — Controlled, dependent and independent variables (ewbk-11547)

Video eLesson
- Skillbuilder — Controlled, dependent and independent variables (eles-4156)

Interactivity
- Skillbuilder — Controlled, dependent and independent variables (int-8090)

1.10 Scientific reports

eWorkbook
- Scientific reports (ewbk-11549)

Weblink
- Harvard referencing generator

1.11 Presenting your data

eWorkbooks
- Scientific drawing skills (ewbk-11551)
- Data analysis (ewbk-11553)

1.12 SkillBuilder — Constructing a pie chart

eWorkbook
- Skillbuilder — Constructing a pie graph (ewbk-11555)

Video eLesson
- Skillbuilder — Constructing a pie graph (eles-1632)

Interactivity
- Skillbuilder — Constructing a pie graph (int-3128)

1.13 SkillBuilder — Creating a simple column or bar graph

eWorkbook
- Skillbuilder — Creating a simple column or bar graph (ewbk-11557)

Video eLesson
- Skillbuilder — Creating a simple column or bar graph (eles-1639)

Interactivity
- Skillbuilder — Creating a simple column graph (int-3135)

1.14 SkillBuilder — Drawing a line graph

eWorkbook
- Skillbuilder — Drawing a line graph (ewbk-11559)

Video eLesson
- Skillbuilder — Drawing a line graph (eles-1635)

Interactivity
- Skillbuilder — Drawing a line graph (int-3131)

1.16 Famous scientists

Video eLessons
- Career spotlight: scientist (eles-0766)
- Isaac Newton (eles-1771)

Interactivity
- Pasteur's experiment (int-3420)

1.17 Project — An inspiration for the future

ProjectsPLUS
- An inspiration for the future (pro-0071)

1.18 Review

eWorkbooks
- Topic review Level 1 (ewbk-11561)
- Topic review Level 2 (ewbk- 11563)
- Topic review Level 3 (ewbk-11565)
- Study checklist (ewbk-11567)
- Literacy builder (ewbk-11568)
- Crossword (ewbk-11570)
- Word search (ewbk-11572)
- Reflection (ewbk-11574)

Digital document
- Key terms glossary (doc-39787)

To access these online resources, log on to **www.jacplus.com.au**

2 Language of learning

LESSON SEQUENCE

online only

Online Resources

Below is a full list of **rich resources** available online for this topic. These resources are designed to bring ideas to life, to promote deep and lasting learning and to support the different learning needs of each individual.

2.1 Overview

eWorkbooks
- Topic 2 eWorkbook (ewbk-11771)
- Starter activity (ewbk-11773)
- Student learning matrix (ewbk-11775)

Solutions
- Topic 2 Solutions (sol-1114)

2.2 Problem solving with thinking hats

eWorkbook
- Thinking keys (ewbk-11776)

2.4 At first glance

Video eLesson
- This car workshop owner is showing good listening skills when talking with a customer (eles-2563)

2.5 Coded communication

Video eLesson
- What nonverbal communication is this woman showing? (eles-2564)

Weblink
- Non-verbal communication

2.9 Total recall?

Video eLesson
- Neural activity in the brain (eles-2565)

Weblink
- Memory

2.11 Review

eWorkbooks
- Topic review Level 1 (ewbk-11778)
- Topic review Level 2 (ewbk-11780)
- Topic review Level 3 (ewbk-11782)
- Study checklist (ewbk-11784)
- Literacy builder (ewbk-11785)
- Crossword (ewbk-11787)
- Word search (ewbk-11789)
- Personal learning (ewbk-11791)
- Reflecting on individual and group activities (ewbk-11793)
- Reflection (ewbk-11795)

Digital document
- Key terms glossary (doc-39879)

To access these online resources, log on to **www.jacplus.com.au**

3 Cells — the basic units of life

CONTENT DESCRIPTION

Recognise cells as the basic units of living things, compare plant and animal cells, and describe the functions of specialised cell structures and organelles (AC9S8U01)

LESSON SEQUENCE

SCIENCE INQUIRY AND INVESTIGATIONS

Science inquiry is a central component of the Science curriculum. Investigations, supported by a **Practical investigation eLogbook** and **teacher-led videos**, are included in this topic to provide opportunities to build Science inquiry skills through undertaking investigations and communicating findings.

LESSON
3.1 Overview

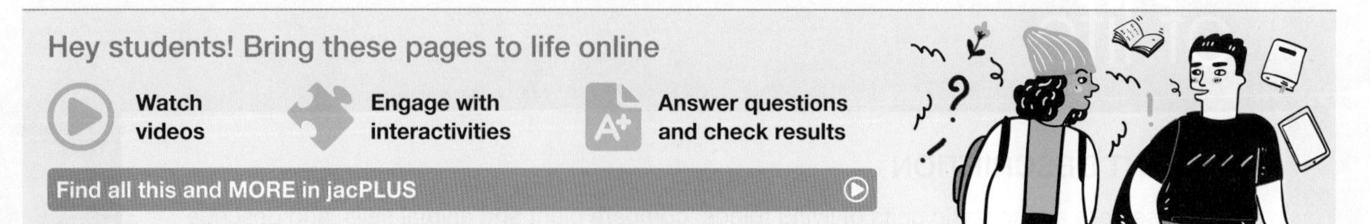

3.1.1 Introduction

Cells are the basic units of all living things. The first cell appeared on Earth about 3.6 billion years ago. It is believed that bacteria living near hydrothermal vents in our oceans may share some similar features to these ancestral cells.

FIGURE 3.1 *Sequoiadendron giganteum.* The oldest tree of this species was over 3200 years, making it one of the oldest living organisms. They grow to an average height of 50–85 metres, with trunk diameters ranging 6–8 metres.

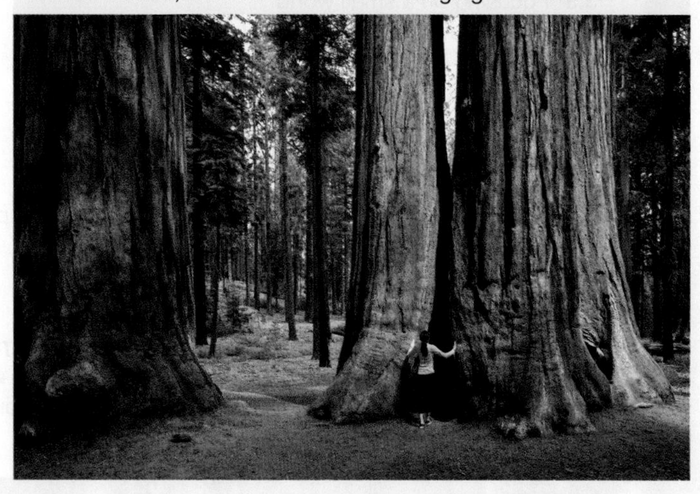

On Earth today, there are many different types of organisms. Some of these organisms are made up of a single cell (unicellular) like those of bacteria mentioned earlier, while others are made up of many cells (multicellular). The evolutionary twists and turns taken from these single-celled organisms to the largest ever land animals, the sauropod dinosaurs, or to the tallest tree, the giant sequoia (*Sequoiadendron giganteum*), or to the largest colony of fungus, the humongous fungus (*Armillaria ostoyae*) spanning 8.9 square kilometres in Oregon, have been plentiful.

The cells that make up organisms differ not just in their number, but also in their size, shape and contents. The cell that makes up one type of unicellular organism may be different from that of another type of unicellular organism. The cells that make up multicellular organisms are also different. The structure of different types of cells and how they are organised within multicellular organisms are well suited to their specific tasks within the organism. No matter the difference, the features of the cell(s) that make up organisms all share the ultimate goal — to keep the organism alive.

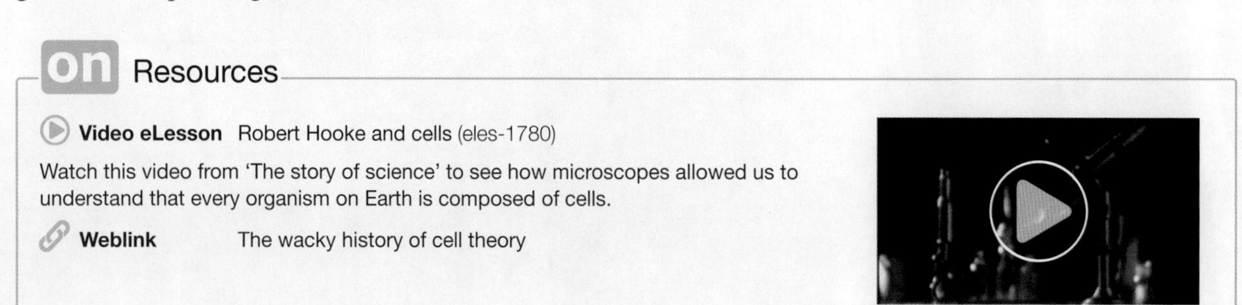

3.1.2 Think about cells

1. How can you make small things look bigger?
2. Which are bigger: animal cells or bacteria?
3. Why are beaches tested for the presence of the bacteria *E. coli*?
4. How does a cell become a clone?
5. Why don't all cells look the same?

3.1.3 Science inquiry

Who am I?

Microscopes are responsible for opening a whole new world to us. They have allowed us to see beyond our own vision. The more developed these microscopes become, the more detail and wonder we are able to observe — but often, rather than answering our questions, they provide us with many more.

The three photos in figure 3.2 show parts of different animals. They were taken with a scanning electron microscope that has a large depth of field and a higher resolution. This means specimens can be magnified to look even bigger, allowing us to observe them with greater detail.

FIGURE 3.2 Different animal parts taken with a scanning electron microscope

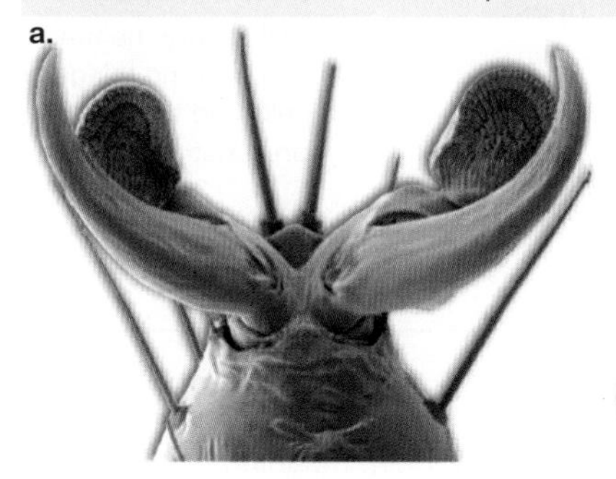

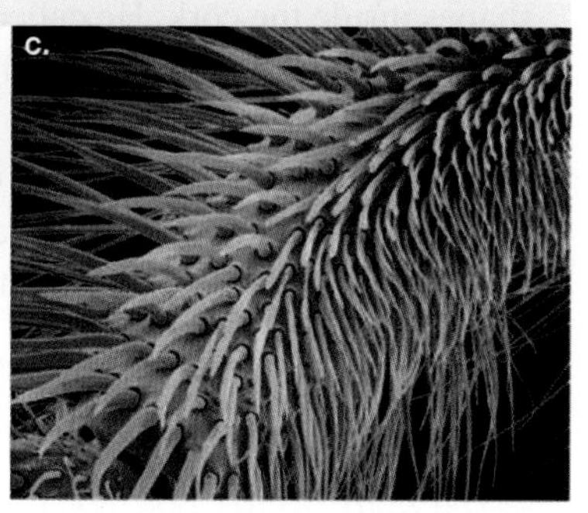

Observe, think and share

1. Look carefully at the photos of each animal part in figure 3.2, and think about:
   a. what they could be
   b. what the purpose of the part may be
   c. what animal they may belong to.
2. Talk through your suggestions with a partner, adding all of the details that you have both observed onto a sheet of paper.
3. Two of these photos show parts of one type of animal, and the other image is from a different animal. Does that information change the way that you look at the details? Which animal do you think two of the parts belong to? Brainstorm to decide which animal the other part could belong to.
4. Suggest other sorts of information that may be helpful in determining which animals these parts belong to and what they are used for.

Resources

eWorkbooks	Topic 3 eWorkbook (ewbk-11848) Starter activity (ewbk-11850) Student learning matrix (ewbk-11852)
 **Solutions**	Topic 3 Solutions (sol-1115)
 **Practical investigation eLogbook**	Topic 3 Practical investigation eLogbook (elog-2123)

LESSON
3.2 A whole new world

LEARNING INTENTION

At the end of this lesson you will understand that our knowledge and understanding of cells has improved as a result of continued scientific investigation, human inventions and technological advancement. This work by scientists led to the creation of scientific theories such as the cell theory.

Science as a human endeavour

3.2.1 The discovery of cells

A whole new world was discovered just over 400 years ago when an English inventor and scientist used magnifying lenses to observe the basic units of which all living things are made. This led to a new way of thinking about living things that required a new scientific language, new classifications and new inventions.

In the seventeenth century, Robert Hooke looked at thin slices of cork under a **microscope** that he had made himself from lenses. He observed small, box-like shapes inside the cork. He called the little boxes that he saw **cells**. Microscopes opened up a whole new world that had never been seen before.

Using microscopes to carefully observe different organisms showed that they were all made up of cells. Observations also showed that many of these cells shared common features, such as the presence of a structure called the **nucleus**.

microscope an instrument used for viewing small objects

cell the smallest unit of life; cells are the building blocks of living things and can be many different shapes and sizes

nucleus a roundish structure inside a cell that acts as its control centre

cell theory the theory that states that all living things are made up of cells and that all cells come from pre-existing cells

WHAT DOES IT MEAN?

The word 'microscope' comes from the Greek words *micrós*, meaning 'small', and *skopein*, meaning 'to view'.

int-3392

FIGURE 3.3 Timeline showing the development of the microscope and **cell theory**

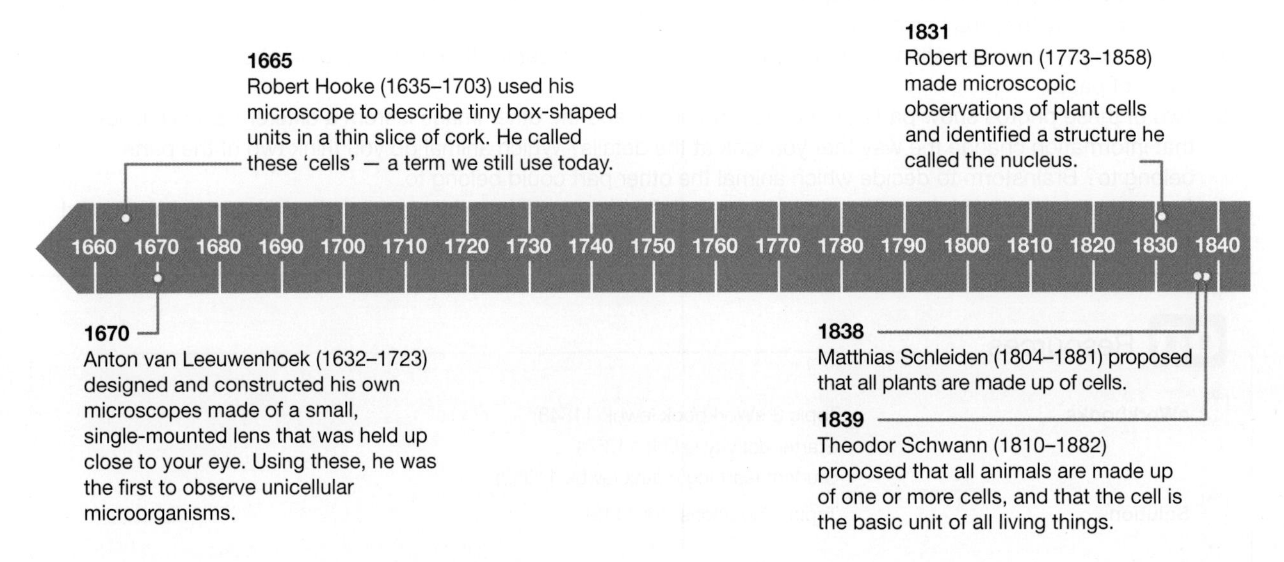

FIGURE 3.4 a. Robert Hooke b. Hooke created a microscope out of lenses c. Cells that make up cork. Do you think the image Hooke observed was as clear as this?

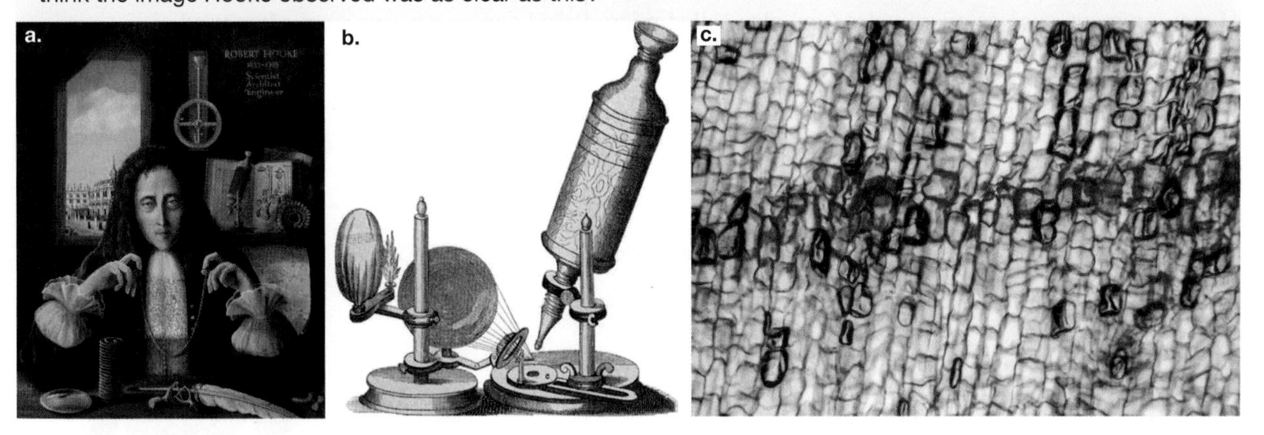

As the magnification provided by microscopes increased, it was seen that although cells shared similar basic structures, there could also be differences between them. Groups of organisms could be made up of cells that differed from the cells of other groups. Some organisms were made up of a single cell (unicellular), whereas others were made up of many cells (multicellular). Different types of cells were also observed within an individual multicellular organism.

Cell theory

1. All organisms are composed of cells.
2. The cell is the basic unit of structure and organisation in organisms.
3. All cells come from pre-existing cells.

3.2.2 Little, littler, littlest ...

With the development of instruments such as microscopes, scientists needed to find words to describe some of the tiniest lengths and time scales in nature. They wanted some simple names to describe, for example, a billionth of a billionth of a metre.

FIGURE 3.3 *(continued)*

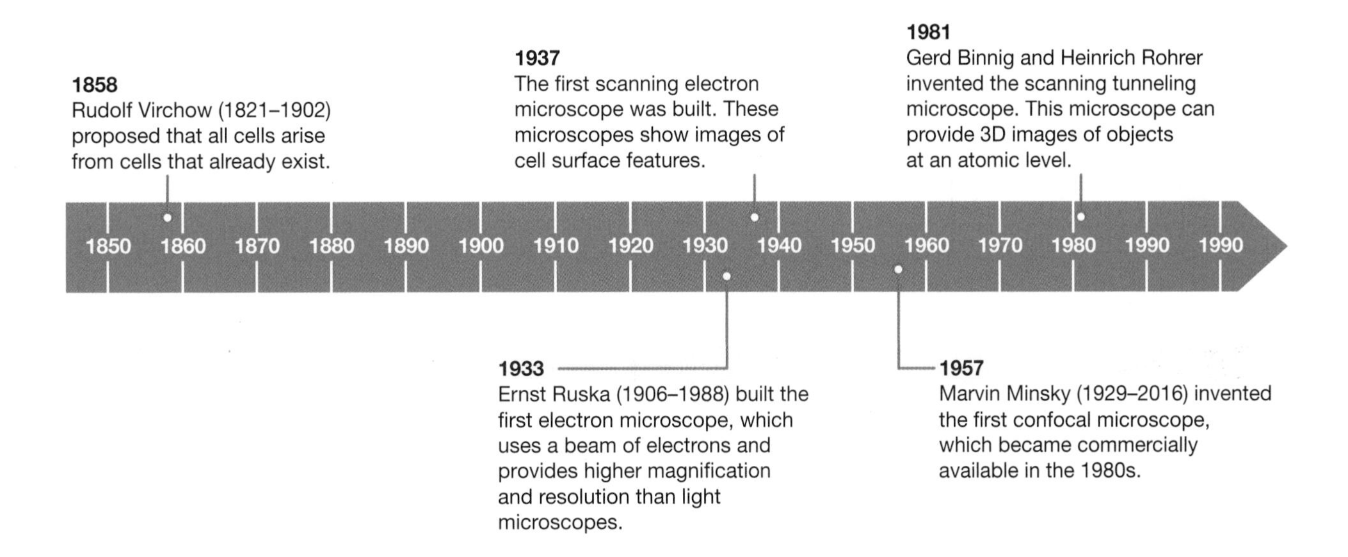

In the microscopic world, there is often a need to describe things in much smaller terms than the units of measurement that you already know, such as metre, centimetre and millimetre. In describing cells, other units of measurement, such as micrometre (μm, also called micron) and nanometre (nm), are often used.

FIGURE 3.5 Different units of measurement are used to describe different sizes.

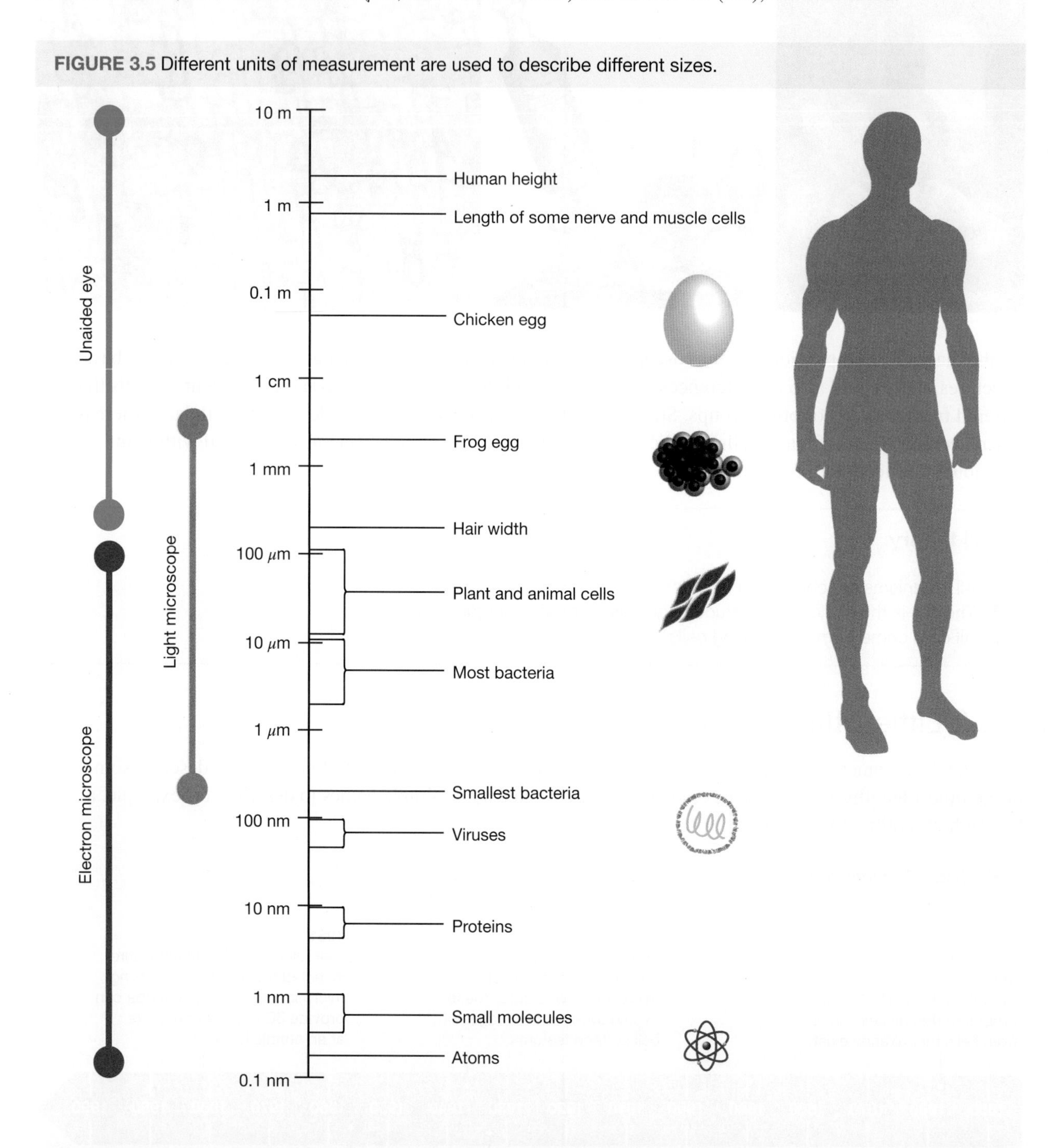

TABLE 3.1 Units of measurement that are often used to describe cells and molecules

Unit	Symbol	Number of units in 1 m
Millimetres	mm	1000
Micrometres	μm	1 000 000
Nanometres	nm	1 000 000 000

CASE STUDY: Scientists who study cells

There are many different types of scientists who study cells. Examples include bacteriologists, cell biologists, clinical microbiologists, cytologists, electron microscopists, genetic scientists and medical microbiologists.

SAMPLE PROBLEM: Evidence-based conclusions

Sam is a bacteriologist who studies *Mycoplasma genitalium*, the smallest known bacteria. In a talk to students, Sam is asked if she uses a microscope to see these small bacteria or whether she can see them without using a microscope.
Do you think Sam would need a microscope to see these bacteria? Use evidence to support your conclusion.

THINK

- The smallest bacteria are between 200 and 300 nm in size.
- The naked eye can only see things that are greater than 100 μm (or 100 000 nm).
- Since the bacteria are much smaller than 100 μm, Sam must use a microscope to see them.
- Sam would need to use an electron microscope.

WRITE

Sam would need to use an electron microscope as the smallest bacteria are much too small to be seen with the naked eye. The smallest bacteria are less than 300 nm in size and the naked eye can only see things that are bigger than 100 000 nm (100 μm).

3.2.3 Present day

A new generation of three-dimensional microscopes are being developed that provide even further details of objects. Superfast electron microscopes enable scientists to capture movement of atoms, and newly invented portable microscopes are becoming important field tools in research and diagnosis of diseases.

FIGURE 3.6 a. A portable microscope for spotting and tracking disease **b.** Looking like a grotesque eyeball, this handheld microscope magnifies your specimens to 200 times their normal size.

a.

b.

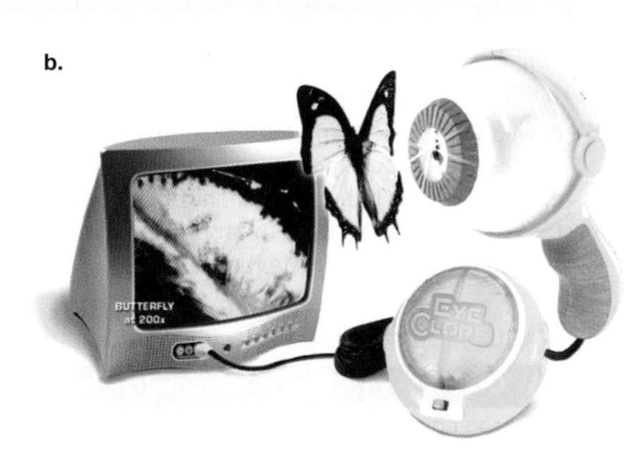

Resources

eWorkbook History of the light microscope (ewbk-11853)

Video eLesson Historic bacteriologists Van Leeuwenhoek, Pasteur and Koch (eles-2026)

3.2 Activities

learnon

3.2 Quick quiz on	3.2 Exercise

Select your pathway

■ LEVEL 1	■ LEVEL 2	■ LEVEL 3
1, 2	3, 5	4, 6

Remember and understand

1. Match the scientist with their cell discovery contribution in the table provided.

Scientist	Cell discovery contribution
a. Anton van Leeuwenhoek	**A.** Built the first electron microscope
b. Robert Hooke	**B.** Proposed that all plants are made up of cells
c. Robert Brown	**C.** Proposed that all animals are made up of cells
d. Matthias Schleiden	**D.** Designed and constructed microscopes, and was the first to observe unicellular microscopic organisms
e. Theodor Schwann	**E.** Proposed that all cells arise from cells that already exist
f. Rudolf Virchow	**F.** Used the term 'cell' to describe the tiny, box-like units in cork
g. Ernst Ruska	**G.** Used the term 'nucleus' to describe a structure found in plant cells

2. Identify:
   a. a feature that all living things have in common
   b. two units often used to describe cells.

Apply and analyse

3. Use the timeline in figure 3.3 to answer the following questions.
   a. In which year did Hooke use the term 'cells' to describe his observation of cork slices?
   b. In which year did Ruska build the electron microscope?
   c. How many years were between:
      i. Hooke first using the term 'cells' and Ruska building the first electron microscope
      ii. Leeuwenhoek's first observation of unicellular microscopic organisms and Schwann's suggestion that all animals are made up of cells?
   d. Credit for developing the cell theory that 'all living things are made up of cells and that cells come from pre-existing cells', is usually attributed to three scientists. Who are they?
   e. Suggest how the development of the microscope has contributed to our understanding of cell structure.
   f. Suggest possible uses for portable microscopes.
4. Use figure 3.5 to complete the table.

TABLE Size of different objects in different units

Object	Size in nanometres (nm)	Size in micrometres (μm)	Size in millimetres (mm)
Frog egg			
Hair (width)			
Plant cell		100	
Bacteria		10	
Protein	10		

Evaluate and create

5. **SIS** A student is presented with two slides with different cells on each. The teacher asks the student to identify which cell is a bacterium and which cell belongs to a plant. They have access to a light microscope and the images are as follows.

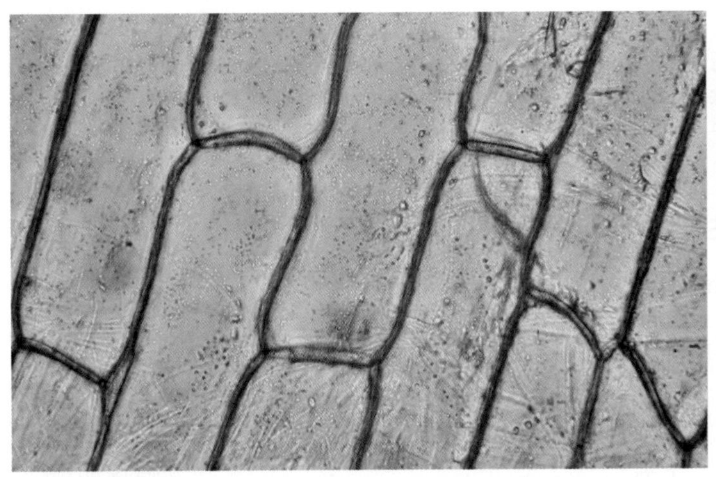

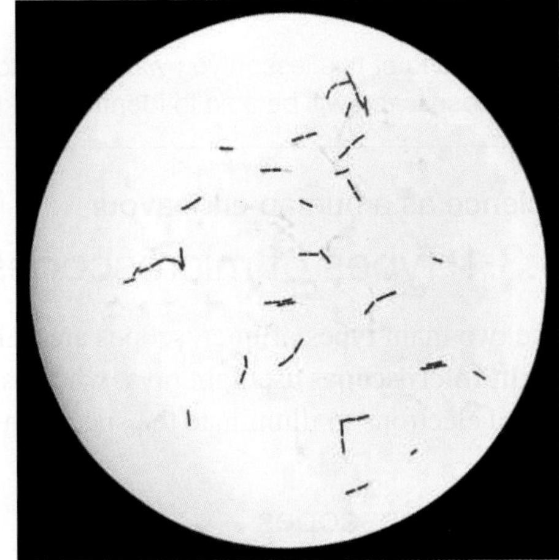

   a. Which feature could be used to identify the different cells?
   b. Which cell do you think is the plant cell? Explain why you reached this conclusion.
   c. What data should be collected as evidence to support your conclusion for part **b**?

6. **SIS** To determine the average size of an elephant skin cell, two methods were suggested. These are outlined as follows:
   - **Method 1** required ten Asian and ten African elephants to have one skin sample taken and measured to create an average using a light microscope.
   - **Method 2** required one Asian and one African elephant to have ten skin samples taken and all of these were used to create the average. These would be measured using a range of light and electron microscopes.

   a. In a table like the one provided, compare and contrast these two methods in terms of accuracy, reproducibility, repeatability, validity and fairness (fair test).

TABLE Comparison of methods to determine cell size

	Method 1	Method 2
Accuracy		
Reproducibility		
Repeatability		
Validity		
Fairness (fair test)		

   b. Create your own experimental design to determine the effect of elephant type on the average size of elephant skin cells. Remember to include the following:
      i. What is the independent variable; that is, what will change?
      ii. What are the controlled variables; that is, what will stay the same?
      iii. What is the dependent variable; that is, what will be measured?
      iv. How will the test be repeatable and reproducible?

Fully worked solutions and sample responses are available in your digital formats.

LESSON
3.3 Focusing on a small world

LEARNING INTENTION

At the end of this lesson you will be able to identify different types of microscopes and understand their specific purposes. You will be able to identify the key features of microscopes and describe how they work.

Science as a human endeavour

3.3.1 Types of microscopes

The two main types of microscopes are light microscopes and electron microscopes. **Light microscopes** use light rays, whereas **electron microscopes** use small particles called electrons to illuminate the specimen being viewed.

light microscope an instrument used for viewing very small objects; a light microscope can magnify things up to 1500 times

electron microscope an instrument used for viewing very small objects; an electron microscope is much more powerful than a light microscope and can magnify things up to a million times

monocular microscope a microscope with a single eyepiece through which the specimen is seen using only one eye

binocular microscope a microscope with two eyepieces through which the specimen is seen using both eyes

stereo microscope a type of binocular microscope through which the detail of larger specimens can be observed

Light microscopes

You may have light microscopes at your school. These may be either **monocular microscopes** (using one eye) or **binocular microscopes** (using two eyes). It is important that the specimen you observe is very thin, so that the light can pass through it. However, one type of binocular microscope, a **stereo microscope**, allows you to see the detail of much larger specimens. Stereo microscopes can be used to observe various objects, including living organisms or parts of them.

Electron microscopes

Transmission electron microscopes (TEMs) show the internal structures of cells, whereas scanning electron microscopes (SEMs) show images of the surface features of the specimen. New electron microscope technologies are being developed, such as superfast electron microscopy, which enables scientists to capture the movement of electrons, and a variety of three-dimensional microscopes that have exciting research and medical applications.

FIGURE 3.7 What's in your water? These images show zooplankton viewed through a scanning electron microscope. **a.** Chaetognath **b.** Daphnia **c.** A rotifer

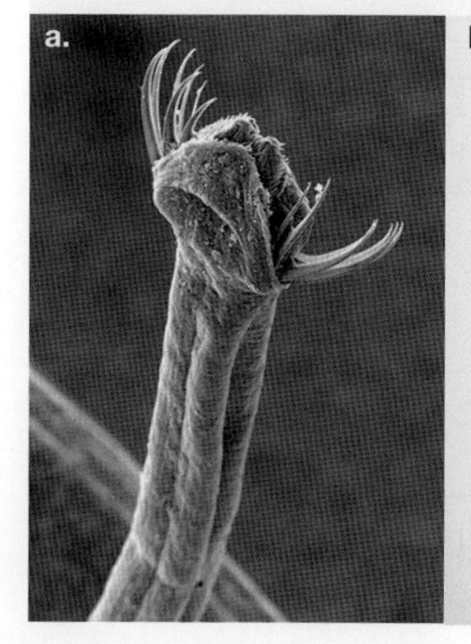

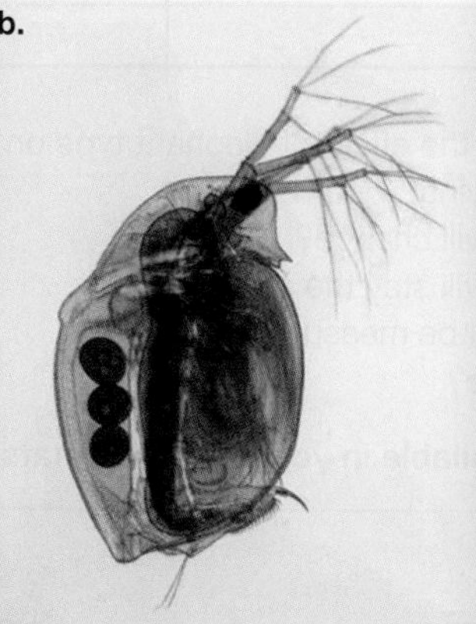

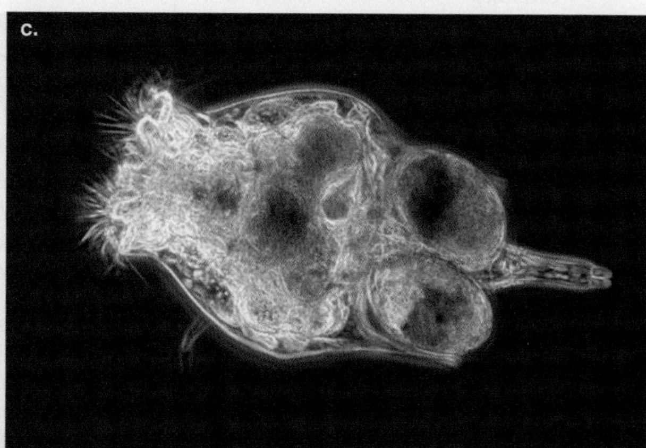

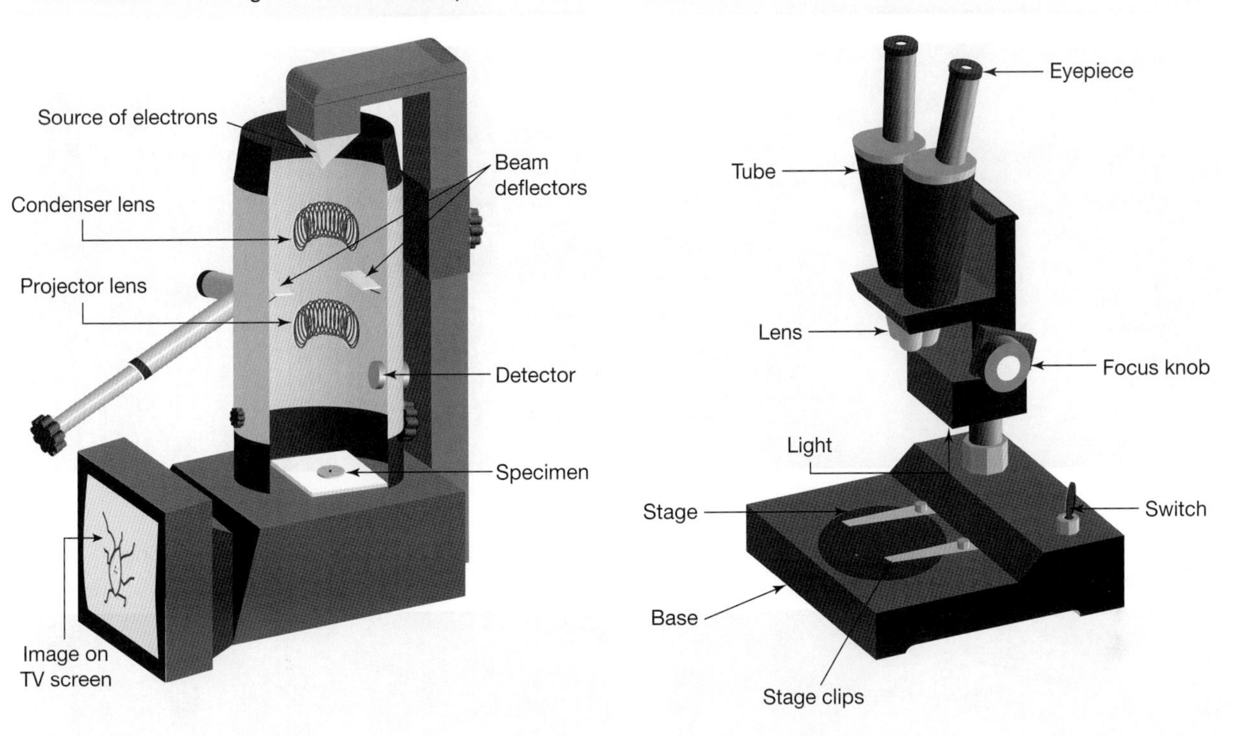

FIGURE 3.8 Scanning electron microscope

FIGURE 3.9 Stereo light microscope

Comparing microscopes

TABLE 3.2 Some comparisons between light microscopes and electron microscopes

Type of microscope	Magnification (how many times bigger)	Resolution (how much detail can be seen)	Advantage(s)	Disadvantage(s)	Examples of detail that can be seen
Light microscope	Up to ×2000	Up to about 500 times better than the human eye	Samples prepared quickly; coloured stains can be used; living cells can be viewed	Limited visible detail	Shapes of cells; some structures inside cells, e.g. nucleus and chloroplasts
Electron microscope	Up to ×2 000 000	Up to about 5 million times better than the human eye	High magnification and resolution	Only dead sections can be viewed; specimen preparation is difficult; very expensive	All parts of cells; viruses

3.3.2 Award-winning images

Microscopes are used not just to observe images of organisms, but also in many other areas of science. Some microscope images win awards recognising not just expertise but also creativity. For example, the Nikon Small World Photomicrography Competition invites photographers and scientists to submit images of all things visible under a microscope. The following figures show examples of some of the 2022 winners.

FIGURE 3.10 The 2022 first place image of the embryonic hand of a Madagascar giant day gecko

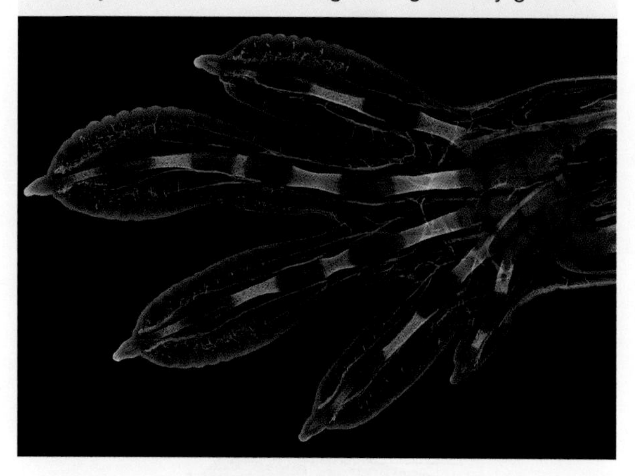

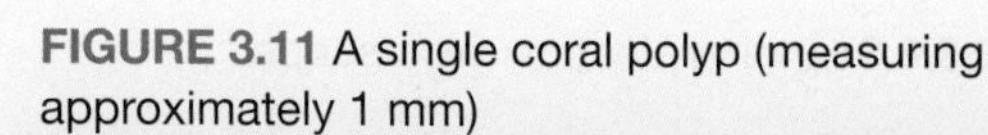

FIGURE 3.11 A single coral polyp (measuring approximately 1 mm)

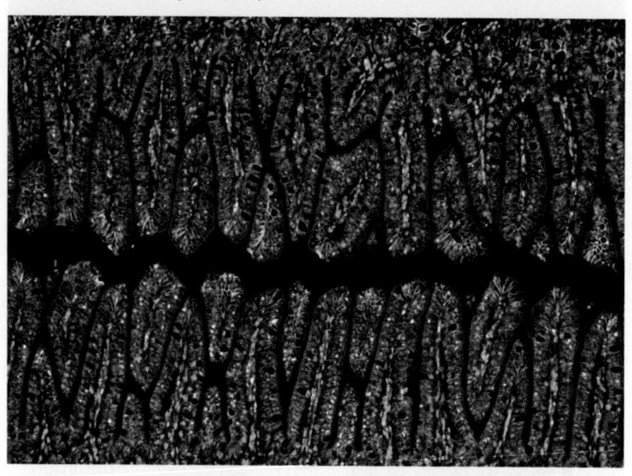

FIGURE 3.12 A cross-section of a white asparagus shoot tip

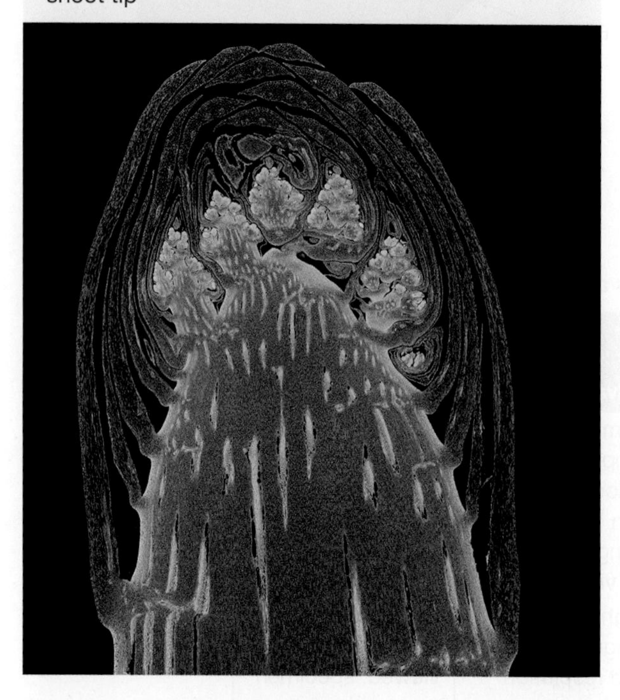

FIGURE 3.13 Villi found in the small intestines

3.3.3 Magnification

The two lenses that determine the **magnification** of your microscope are the eyepiece lens and the objective lens. Each lens has a number on it that signifies its magnification. Multiplying the eyepiece number by the objective lens number will give you the magnification of the microscope. For example:

- eyepiece lens (ocular) magnification = ×10
- objective lens magnification = ×40
- total magnification of the microscope = ×400.

magnification the number of times the image of an object has been enlarged using a lens or lens system; for example, a magnification of two means the object has been enlarged to twice its actual size

int-5702

FIGURE 3.14 As the field of view gets smaller, the magnification gets larger.

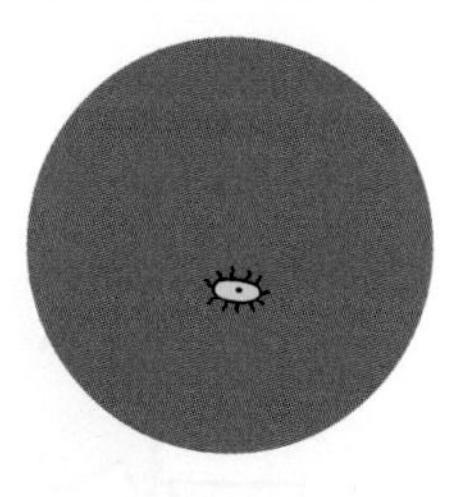

Field of view 4 mm
(4000 μm)
Magnification ×40

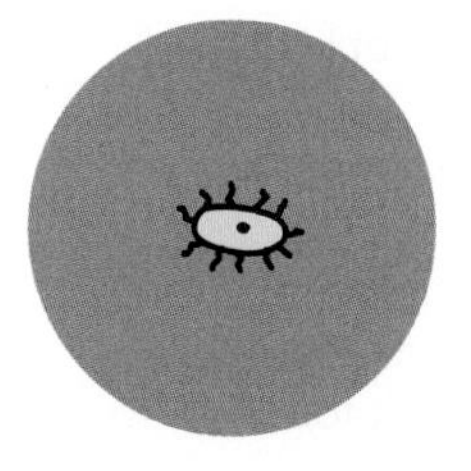

Field of view 1.6 mm
(1600 μm)
Magnification ×100

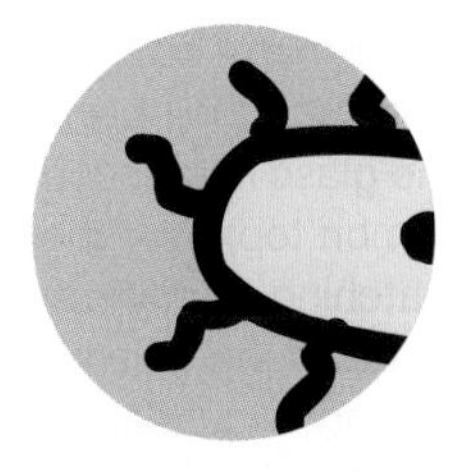

Field of view 0.4 mm
(400 μm)
Magnification ×400

int-3390

FIGURE 3.15 The monocular light microscope

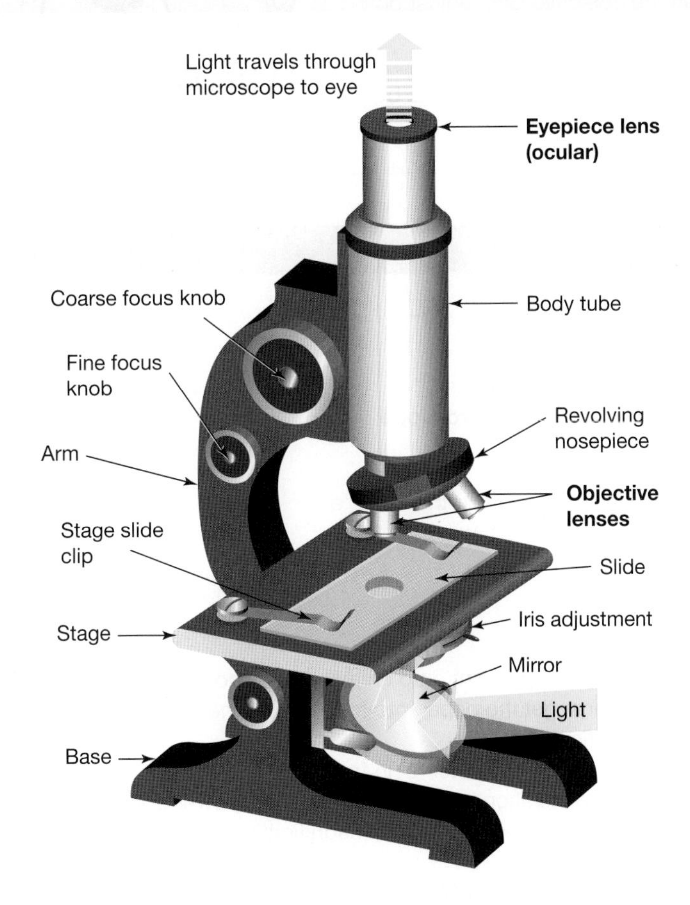

Important points to remember when using a microscope

1. When lifting the microscope, put one hand on the arm of the microscope and one hand under its base.
2. The microscope should be used on a flat surface with the arm closest to you, and not too close to the edge.
3. Take care that the light intensity is not too high, or it might damage your eye.
4. Always begin with the shortest (lowest-power) objective lens.
5. When you have finished using the microscope, return the shortest objective lens into position.
6. Remove the slide, and ensure that the stage is clean.
7. Make sure that when your microscope is not in use, it is always clean and carefully put away.

Using a microscope

1. Adjust your mirror so the appropriate amount of light passes through the hole in the stage.
2. Place the glass microscope slide (with a single hair specimen on top) onto the stage.
3. While watching from the side, use the coarse focus knob to adjust the objective lens or stage, until they are just apart, or just above the slide. Moving it too close may shatter the slide.
4. While looking through the eyepiece lens, carefully turn the coarse focus knob to lower the stage, moving the stage and objective lens apart until the specimen is seen clearly.
5. Carefully use the fine focus knob so that you can see the details of your specimen as clearly as possible. Each time you change the objective lens, adjust only with the fine focus knob.
6. Sketch what you see.
7. Suggest by how many times your specimen has been magnified.

FIGURE 3.16 How to focus your microscope — and how not to!

elog-2125

INVESTIGATION 3.1

Getting into focus with an 'e'

Aim

To practise focusing a monocular light microscope

Materials

- 1 cm square piece of newsprint containing the letter 'e'
- monocular light microscope
- microscope slide
- clear sticky tape

Method

1. Carefully stick the 1 cm square of newsprint onto a clean microscope slide using sticky tape.
2. Using the microscope directions, get the paper into focus using the coarse focus knob and the lowest-power objective lens (smallest magnification).
3. Carefully move the slide until you have a letter 'e' in focus.
4. Change to a higher level of magnification by rotating to a higher-power objective lens.
5. Draw a sketch of what you see under ×100 or ×400 magnification in the results section. Remember not to shade the image, outline only.
6. Record how many 'e's would fit across the field of view. Use this to estimate the size of the 'e' by dividing the size of the field of view by the number of 'e's that would fit across the field of view.
7. Move the slide towards you. Record which direction the 'e' moves.
8. Move the slide to the left. Record which direction the 'e' moves.

Results

1. In which direction did the paper under the microscope move when you moved the slide (a) towards you and (b) to the left?
2. What does the letter 'e' look like under the microscope? Draw a pencil sketch of what you see.
3. Record the magnification that you use and estimate how much of the viewed area is covered by the letter 'e' at this magnification.

Discussion

1. Did changing the magnification change the amount of detail you could see on the 'e'?

2. What occurred when the slide was moved (a) towards you and (b) to the left? Can you suggest a reason for this? (*Hint*: Look at how a light microscope works.)
3. Suggest what the letters 'P' and 'R' would look like under the microscope.
4. Propose a research question that you could explore using a light microscope, and describe how you could investigate it.

Conclusion

Summarise the findings for this investigation.

INVESTIGATION 3.2

Can you tell the difference?

Aim

To view different specimens using a monocular light microscope in order to identify similarities and differences

Materials

- monocular light microscope
- microscope slides
- clear sticky tape
- hair strands (from different individuals)
- spatula
- selection of white powders and crystals (e.g. flour, salt, sugar, baking soda)
- different brands or types of spices and leaf tea
- fibres (e.g. cotton, linen, silk, wool, nylon)

Method

Fibres

1. Remove a fibre strand from one of the materials and stick it on a clean microscope slide using sticky tape.
2. Using the microscope directions, get the fibre into focus using the coarse focus knob and the lowest-power objective lens (smallest magnification). Sketch what you see.
3. Change to a higher level of magnification by rotating to a higher-power objective lens.
4. Draw a sketch of what you see with each magnification in the results section. Remember not to shade the image, outline only.
5. Repeat steps 1–4 for fibres from other materials.

Powders

6. Using a spatula, stick a tiny amount of flour on a clean microscope slide with sticky tape (make sure it is a very thin layer).
7. Using the microscope directions, get the powder into focus using the coarse focus knob and the lowest-power objective lens (smallest magnification). Sketch what you see.
8. Change to a higher level of magnification by rotating to a higher-power objective lens.
9. Draw a sketch of what you see with each magnification in the results section. Remember not to shade the image, outline only.
10. Repeat steps 6–9 for other powders and/or spices and leaf tea.

Results

Sketch the specimens viewed. Include detailed descriptions, the magnification used and an estimate of size next to your diagrams.

Discussion

1. Did changing the magnification change the amount of detail you could see on the different specimens?
2. Identify ways in which the different specimens were similar and ways in which they were different.

Conclusion

Summarise the findings for this investigation.

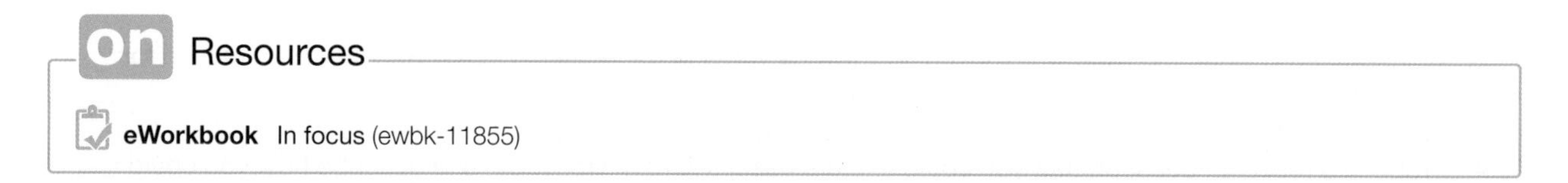

3.3 Activities

learnon

3.3 Quick quiz | 3.3 Exercise

Select your pathway

■ LEVEL 1	■ LEVEL 2	■ LEVEL 3
1, 2, 3, 10, 12, 13	4, 7, 8, 9, 14	5, 6, 11, 15

These questions are even better in jacPLUS!
- Receive immediate feedback
- Access sample responses
- Track results and progress

Find all this and MORE in jacPLUS

Remember and understand

1. **a.** Identify whether the following statements are true or false.

Statement	True or false?
i. A light microscope can produce a greater magnification than an electron microscope.	
ii. Only dead sections can be viewed on an electron microscope.	
iii. Viruses can be viewed on a light microscope.	
iv. Resolution refers to how many times bigger a specimen is, whereas magnification refers to how much detail you can see.	
v. More detail can be seen in thicker specimens when using a monocular light microscope.	

   b. Justify any false responses.
2. Suggest why it is important not to have the light intensity setting too high on a light microscope.
3. Explain the importance of watching from the side of the microscope while using the coarse focus knob.
4. As the field of view of your microscope gets smaller, what happens to the magnification?
5. When you are looking down the microscope, what happens when you move the microscope slide
   a. to the left
   b. to the right
   c. towards you
   d. away from you?

Apply and analyse

6. Use figure 3.14 to answer the following questions. (*Note*: 1000 μm = 1 mm)
   a. Estimate the length of the specimen shown in the diagram at ×40, ×100 and ×400 magnification.
   b. Describe the differences in your observations of the three different magnifications.
7. Create Venn diagrams to distinguish between:
   a. a monocular microscope and a stereo microscope
   b. a light microscope and an electron microscope
   c. a transmission electron microscope and a scanning electron microscope
   d. resolution and magnification
   e. field of view and magnification.
8. If a specimen is 1 mm in length, how big will it appear if it is magnified ×100?
9. If a specimen takes up the entire field of view at ×100, how much of it will be seen at ×400?
10. Sketch a line diagram or take a photo of your microscope and label as many of its parts as you can, using figure 3.15.

11. Copy and complete the table provided.

Ocular lens (eyepiece)	Objective lens	Magnification
×5	×5	×25
×5	×10	
×10		×100
	×40	×400

12. Match the part of the microscope with its function.

Part	Function
a. Objective lens	**A.** Where the slide is placed
b. Slide	**B.** Thin piece of glass where the specimen is placed
c. Stage clip slide	**C.** Magnifies the image
d. Iris adjustment	**D.** Allows large adjustments to the distance between the stage and objective lens, which helps bring images into focus
e. Coarse focus knob	**E.** Adjusts the amount of light reaching the eyepiece
f. Stage	**F.** Allows small adjustments to the distance between the stage and the objective lens, which helps bring the image into closer focus
g. Fine focus knob	**G.** Holds the slide in place

Evaluate and create

13. Design and make a poster that shows either how a microscope should be used or what happens when you use it the wrong way.
14. **SIS** This graph shows how the field of view changes with magnification under a light microscope.

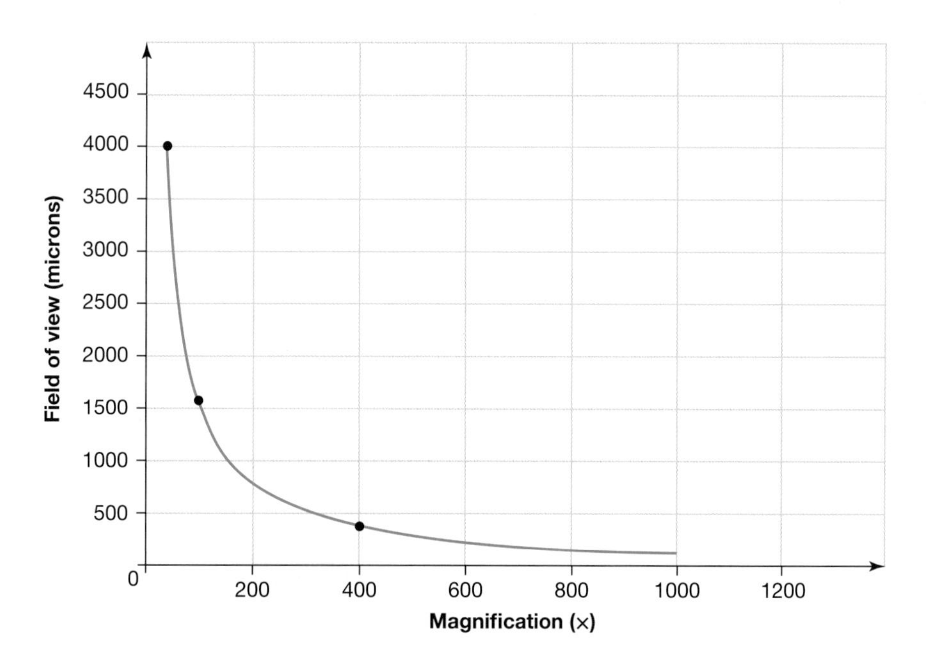

a. The graph is missing a title. How would you title this graph?
b. Some light microscopes have an additional objective lens called an oil immersion lens. This can magnify ×100.
  i. Assuming the eyepiece is ×10, what would the total magnification be if you were using this lens?
  ii. Use the graph to determine how big the field of view would be.
c. The oil immersion lens increases resolution. What does the term *resolution* mean?

15. **SIS** Determine which of the following plant cell images, a or b, is taken with a light microscope, and which is taken with an electron microscope. Use evidence from the images to support your conclusion.

a.

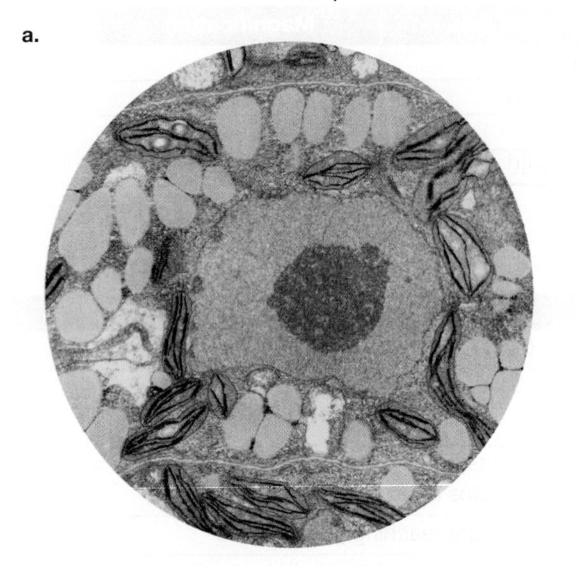

b.

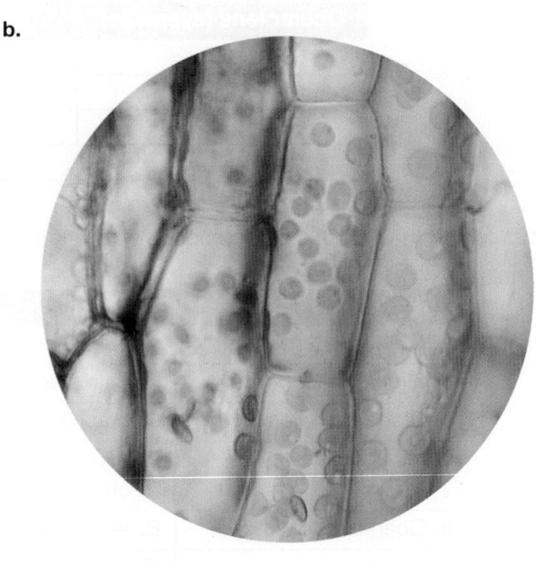

Fully worked solutions and sample responses are available in your digital formats.

LESSON
3.4 Form and function — cell make-up

LEARNING INTENTION

At the end of this lesson you will recognise that organisms consist of a single cell or multiple cells. You will also identify key organelles within cells and describe their function.

3.4.1 Similar, but different

Cells are the building blocks that make up all living things. Organisms may be made up of one cell (**unicellular**) or many cells (**multicellular**). These cells contain small structures called **organelles** that have particular jobs within the cell and function together to keep the organism alive.

unicellular made up of only one cell
multicellular made up of many cells
organelle any specialised structure in a cell that performs a specific function

Cells can be categorised on the basis of the presence and absence of particular organelles and other structural differences. Organisms can be classified by the different types of cells they are made up of.

How big is small?

The size of cells may vary between organisms and within a multicellular organism. Most cells are too small to be seen without a microscope. Cells need to be very small because they have to be able to quickly take in substances they need and remove wastes and other substances (figure 3.17). The bigger a cell is, the longer this process would take.

FIGURE 3.17 The most commonly used unit is the micrometre (μm).

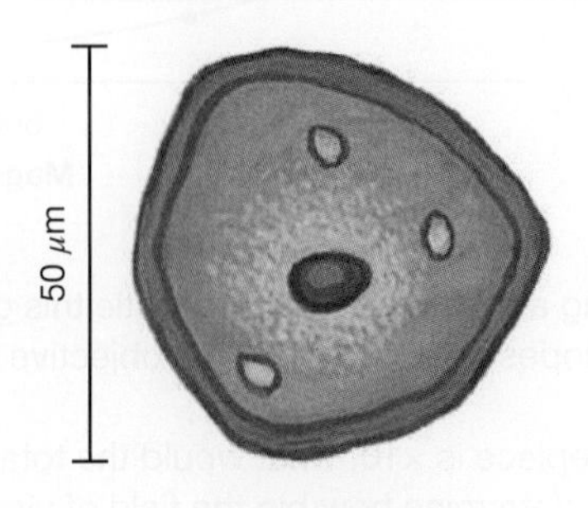

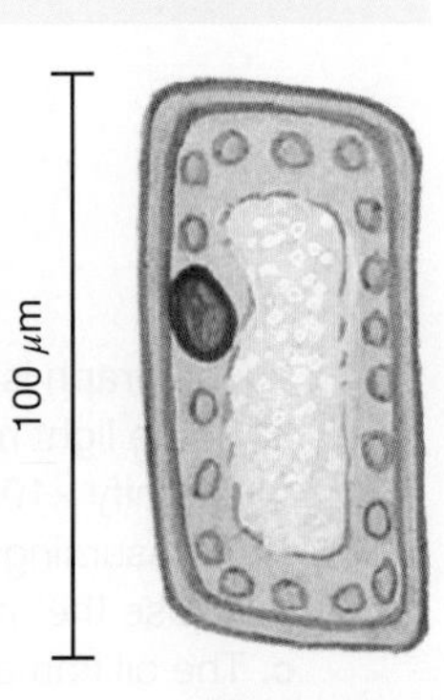

Very small units of measurement are used to describe the size of cells. The most commonly used unit is the **micrometre** (μm). One micrometre equals one millionth (1/1 000 000) of a metre or one thousandth (1/1000) of a millimetre. Check out your ruler to get an idea of how small this is! Most cells are in the range of 1 μm (bacteria) to 100 μm (plant cells).

Advances in technology are creating an increased need for the use of the **nanometre** (nm) as a unit. One nanometre equals 1 billionth (1/1 000 000 000) of a metre. Investigating the organelles within cells and the molecules they react with requires this level of measurement.

Nanotechnology is a rapidly developing field that includes studying and investigating cells at this 'nano level'. While it requires lots of creative, exciting and futuristic 'what if' thinking, it also involves an understanding of the basics of information and ideas that are currently known.

3.4.2 Have it or not?

Prokaryotes such as bacteria were the first type of organism to appear on Earth. The key difference between prokaryotes and all other kingdoms is that members of this group do not contain a nucleus or other membrane-bound organelles. The word prokaryote comes from the Greek terms *pro*, meaning 'before', and *karyon*, meaning 'nut, kernel or fruit stone', referring to the cell nucleus.

FIGURE 3.18 Prokaryotes and eukaryotes

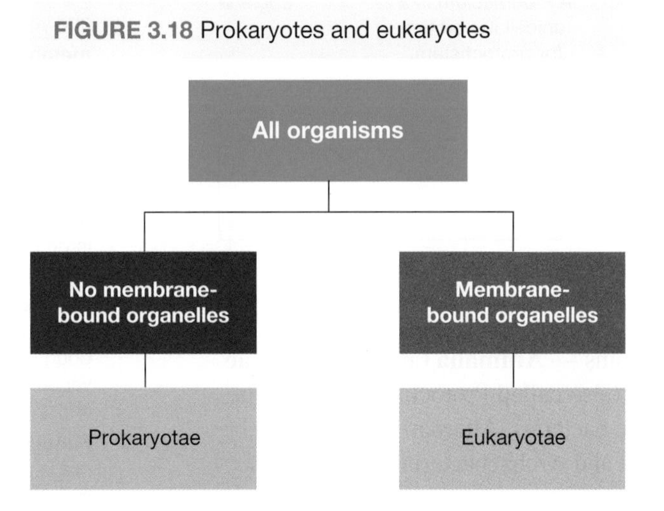

micrometre one millionth of a metre

nanometre one billionth of a metre

nanotechnology a science and technology that focuses on manipulating the structure of matter at an atomic and molecular level

prokaryote any cell or organism without a membrane-bound nucleus (e.g. bacteria)

eukaryote any cell or organism with a membrane-bound nucleus (e.g. plants, animals, fungi and protists)

Eukaryotic organisms made up of eukaryotic cells appeared on Earth billions of years later. As *eu* is the Greek term meaning 'good' or 'true', **eukaryote** can be translated as 'true nucleus'. Members of the kingdoms Animalia, Plantae, Fungi and Protista are eukaryotes, and are made up of cells containing a nucleus and other membrane-bound organelles.

FIGURE 3.19 Eukaryotic cells a. contain a nucleus and membrane-bound organelles, whereas b. prokaryotic cells do not.

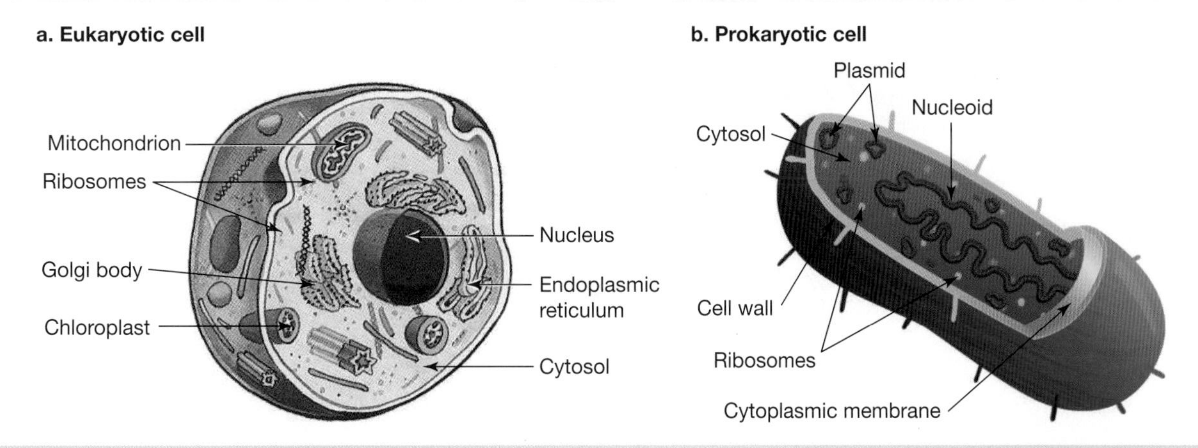

3.4.3 What do we share?

What most cells have in common is that they are made up of a **cell membrane** containing a fluid called **cytosol** and small structures called **ribosomes**. The collective term used to describe the cytosol and all the organelles suspended within it is **cytoplasm**. The hundreds of chemical reactions essential for life that occur within the cytoplasm are referred to as the cell's **metabolism**. The ribosomes are where proteins such as enzymes, which regulate the many chemical reactions important to life, are made. The cell membrane regulates the movement of substances into and out of the cell. This enables the delivery of nutrients and substances essential for reactions, and the removal of wastes.

FIGURE 3.20 Different sizes

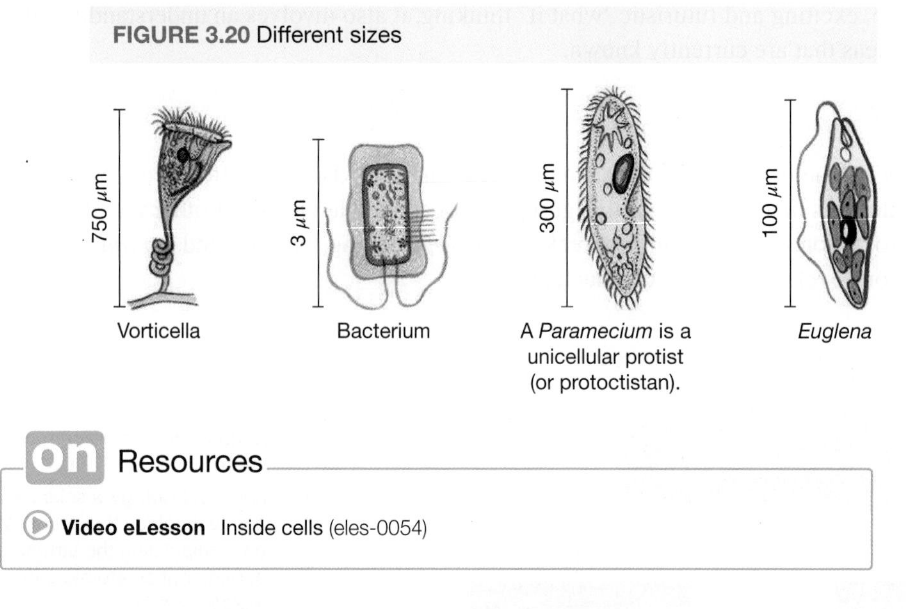

cell membrane the structure that encloses the contents of a cell and allows the movement of some materials in and out

cytosol the fluid found inside cells

ribosomes small structures within a cell in which proteins such as enzymes are made

cytoplasm the jelly-like material inside a cell; it contains many organelles, such as the nucleus and vacuoles

metabolism the chemical reactions occurring within an organism that enable the organism to use energy and grow and repair cells

Animalia the kingdom of organisms that have cells with a membrane-bound nucleus, but no cell wall, large vacuole or chloroplasts (e.g. animals)

Plantae the kingdom of organisms that have cells with a membrane-bound nucleus, cell wall, large vacuole and chloroplasts (e.g. plants)

Fungi the kingdom of organisms made up of cells that possess a membrane-bound nucleus and cell wall, but no chloroplasts (e.g. mushrooms); some fungi can help to decompose dead and decaying matter

Protista the kingdom of organisms made up of cells that possess a membrane-bound nucleus but vary in other features and do not fit into other groups (e.g. protozoans); also called Protoctista

on Resources

Video eLesson Inside cells (eles-0054)

3.4.4 Six kingdoms

Living things can be divided into six kingdoms — **Animalia** (animals), **Plantae** (plants), **Fungi** (e.g. mushrooms), **Protista** (also called Protoctista), Eubacteria (bacteria) and Archaebacteria (extremophile bacteria). Alternatively, they can be divided into five kingdoms, with Eubacteria and Archaebacteria categorised as prokaryotes.

FIGURE 3.21 The six kingdoms of living things

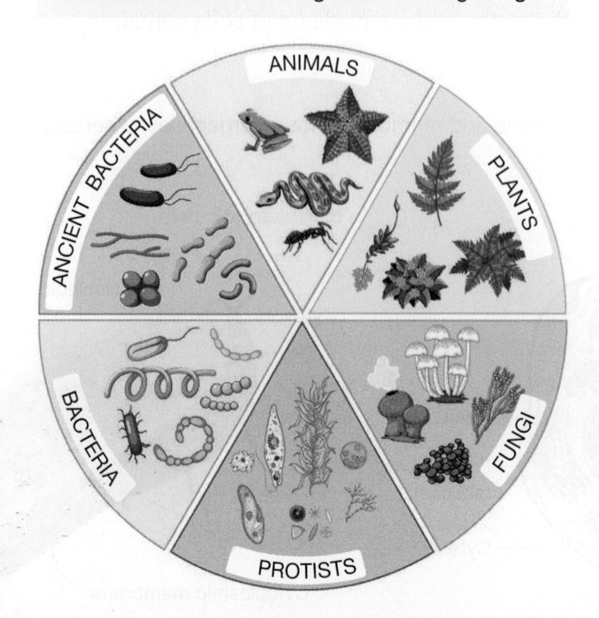

FIGURE 3.22 A key characteristic used to classify organisms into kingdoms is the structure of their cells.

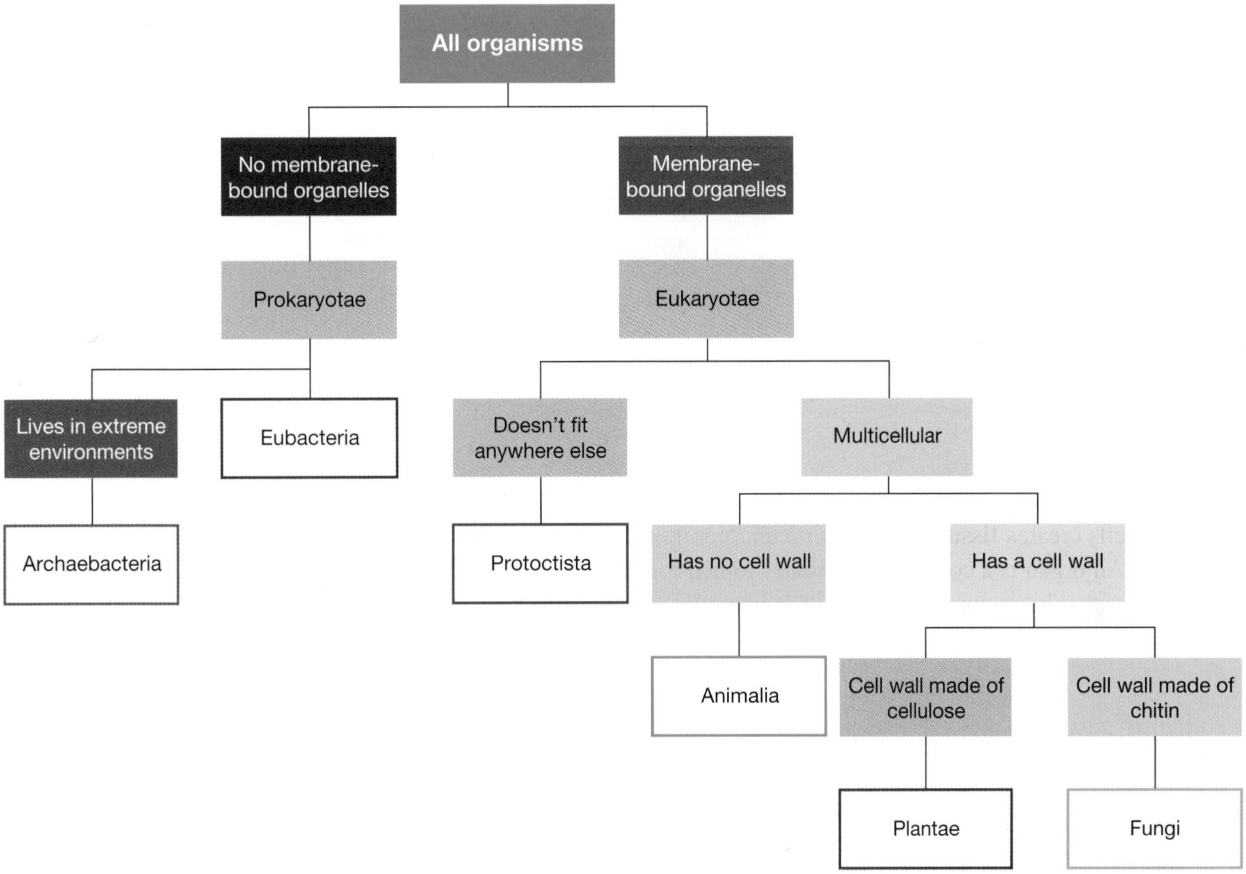

All organisms

No membrane-bound organelles

Membrane-bound organelles

Prokaryotae

Eukaryotae

Lives in extreme environments

Eubacteria

Doesn't fit anywhere else

Multicellular

Archaebacteria

Protoctista

Has no cell wall

Has a cell wall

Animalia

Cell wall made of cellulose

Cell wall made of chitin

Plantae

Fungi

WHAT DOES IT MEAN?

The prefix *uni-* comes from the Latin term meaning 'one'. The prefix *multi-* comes from the Latin term meaning 'many'.

3.4.5 Designed for optimum function

Regardless as to whether an organism is unicellular or multicellular, all cells have a job to do.

In unicellular cells such as bacteria and some protists such as amoeba, *Euglena* or *Paramecium*, the cell is designed to meet all of the organism's needs — gaining nutrients, expelling wastes and reproducing by producing an exact replica of the parent cell.

What nutrients do you think cells need? What waste products do you think cells produce? Why is it important that these waste products are expelled?

FIGURE 3.23 Unicellular organisms

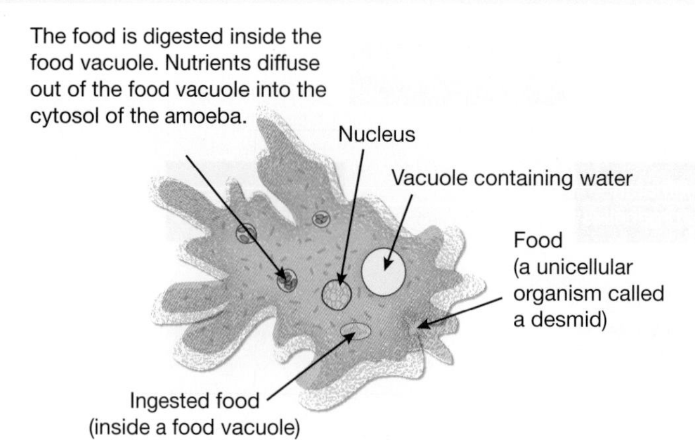

Multicellular organisms have certain cells that carry out certain functions. A group of these cells creates **tissue** that has particular components that specialise it for its function. An organ is a collection of tissues joined together. These organs are part of a system that has a specialised role; for example, the digestive system or the skeletal system.

tissue a group of cells of similar structure that perform a specific function

int-3393

FIGURE 3.24 Different types of cells have particular structures enabling specialised tasks.

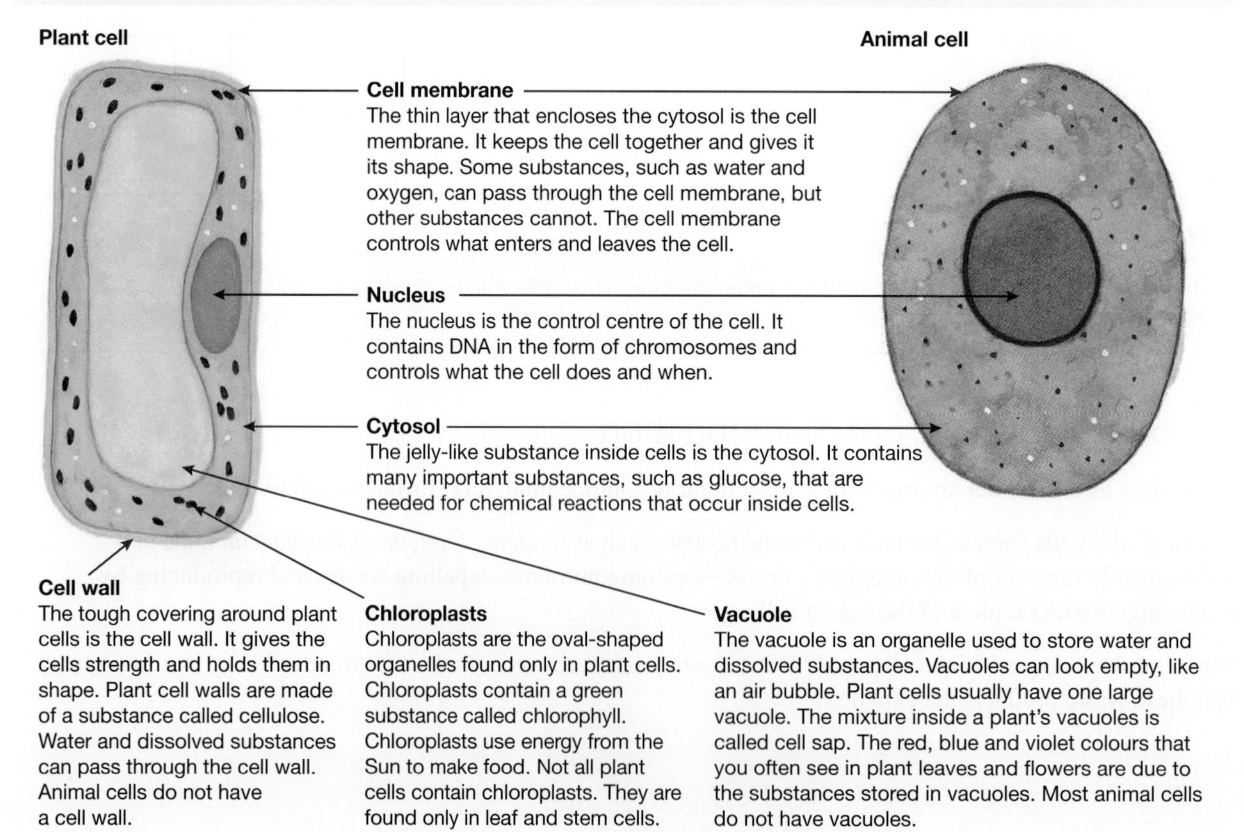

Microfactories

Mitochondria and **chloroplasts** are examples of membrane-bound organelles found in eukaryotic cells. While all eukaryotic cells contain mitochondria, because they are all involved in **cellular respiration**, only those involved in **photosynthesis** (such as those in plant leaves) contain chloroplasts. Chloroplasts contain the green pigment **chlorophyll**. This pigment is used to trap light energy so that it can be converted into chemical energy and used by the cells.

mitochondria small, rod-shaped organelles that are involved in the process of cellular respiration, which results in the conversion of energy into a form that the cells can use

chloroplasts oval-shaped organelles that are involved in the process of photosynthesis, which results in the conversion of light energy into chemical energy

cellular respiration a series of chemical reactions in which the chemical energy in molecules such as glucose is transferred into ATP molecules, which is a form of energy that the cells can use

photosynthesis a series of chemical reactions that occur within chloroplasts in which the light energy is converted into chemical energy; the process also requires carbon dioxide and water, and produces oxygen and sugars, which the plant can use as 'food'

chlorophyll the green-coloured chemical in plants, located in chloroplasts, that absorbs light energy so that it can be used in the process of photosynthesis

EXTENSION: The endosymbiotic theory

The endosymbiotic theory suggests that mitochondria and chloroplasts were once prokaryotic organisms. This theory proposes that, at some time in the past, these organisms were engulfed by another cell and over time they evolved to depend on each other.

FIGURE 3.25 The origin of the eukaryotic cell? Some scientists also suggest that our nucleus may have come from a giant viral ancestor.

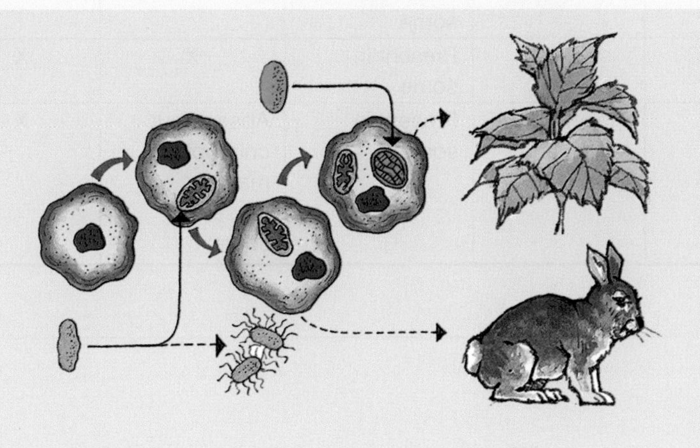

External structures

Specialised external structures such as cilia and flagella may also be a feature of a cell to assist with its movement in the organism.

FIGURE 3.26 The structure and motion of cilia and flagella

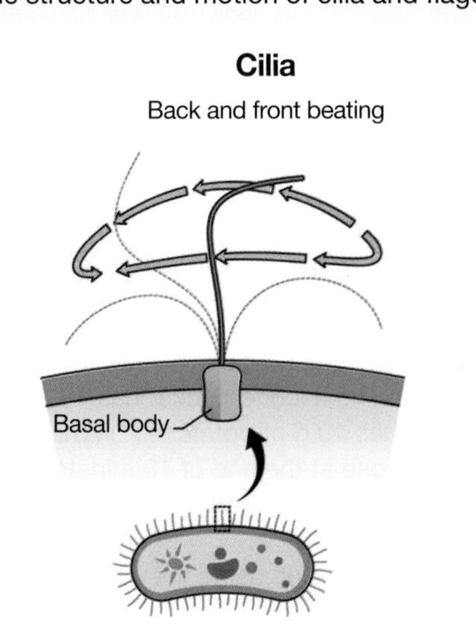

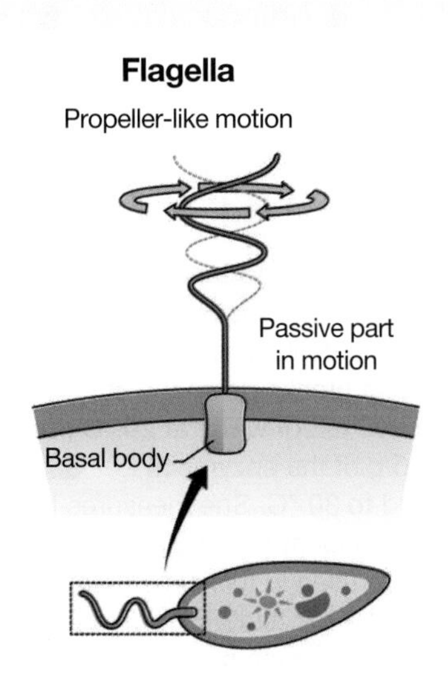

3.4.6 Some differences in the basic cell design in the six kingdoms

TABLE 3.3 Differences in the basic cell design in the six kingdoms

	Kingdom					
Characteristic	**Animalia** (animals; e.g. lizards, fish, spiders, earthworms, sponges)	**Fungi** (e.g. yeasts, moulds, mushrooms, toadstools)	**Plantae** (plants; e.g. ferns, mosses, conifers, flowering plants)	**Protista** (e.g. algae, protozoans)	**Eubacteria** (bacteria and cyanobacteria)	**Archaebacteria**
Number of cells	Multicellular	Usually multicellular but some unicellular	Most multicellular	Unicellular or multicellular	Unicellular	Unicellular
Nucleus	✓	✓	✓	✓	X	X
Cell wall	X	✓	✓	Present in some	✓	✓
Large vacuole	X	X	✓	Present in some	X	X
Chloroplasts	X	X	Present in leaves and stem cells	Present in some	Absent (but chlorophyll may be present in some)	X

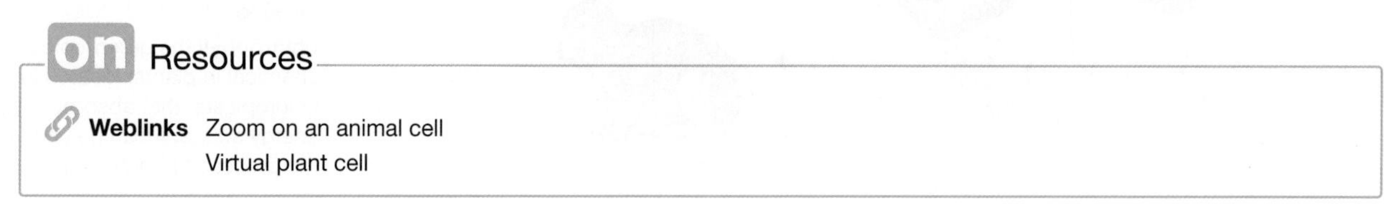

3.4 Activities

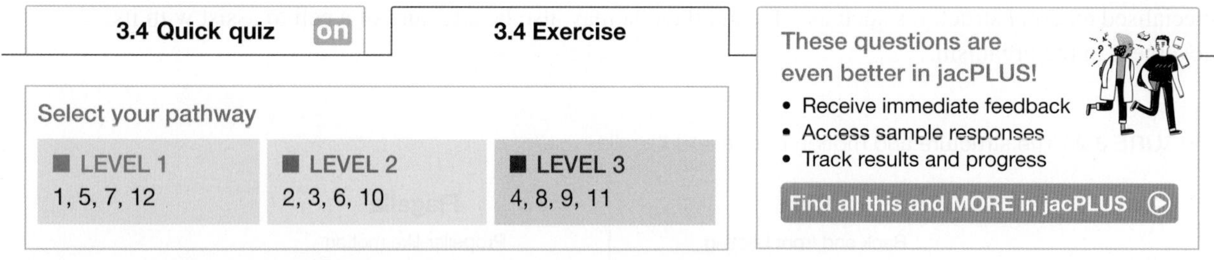

Remember and understand

1. State why the nucleus is important to the cell.
2. Identify where enzymes are made in a cell and state why they are important.
3. **SIS** Emily is trying to determine if increasing the temperature of an enzyme will increase the reaction rate. She uses an enzyme that breaks down the starch in potato into sugar. She sets up three beakers. The first has 10 g of potato in 200 mL of water at 20 °C (room temperature). The second beaker has 10 g of potato in 200 mL of water and 5 g of the enzyme at 20 °C. The third beaker has 10 g of potato in 200 mL of water and 5 g of the enzyme heated to 30 °C. She measures the amount of glucose at the end of 10 minutes.
   a. Identify the independent and dependent variables.
   b. Is a control group used in this experiment? If so, which beaker is it?
   c. Is this experiment a fair test? Provide an explanation for your decision using the definition of fair test.
4. List two organisms from each of the kingdoms.

Use table 3.3 to answer the following questions.

5. Complete the table by identifying which kingdoms relate to the characteristics.

Characteristics	Kingdom
Do not have a cell wall, large vacuole or chloroplasts	
Have a cell wall, large vacuole and chloroplasts	
Have a cell wall, but no large vacuole or chloroplasts	
Has a cell wall and cell membrane but no nucleus	

Apply and analyse

6. **SIS** Review the image given, and answer the following questions.
   a. What features can you identify in the cell?
   b. Hypothesise which kingdom this cell belongs to.
   c. Can a hypothesis be incorrect? Explain your answer.

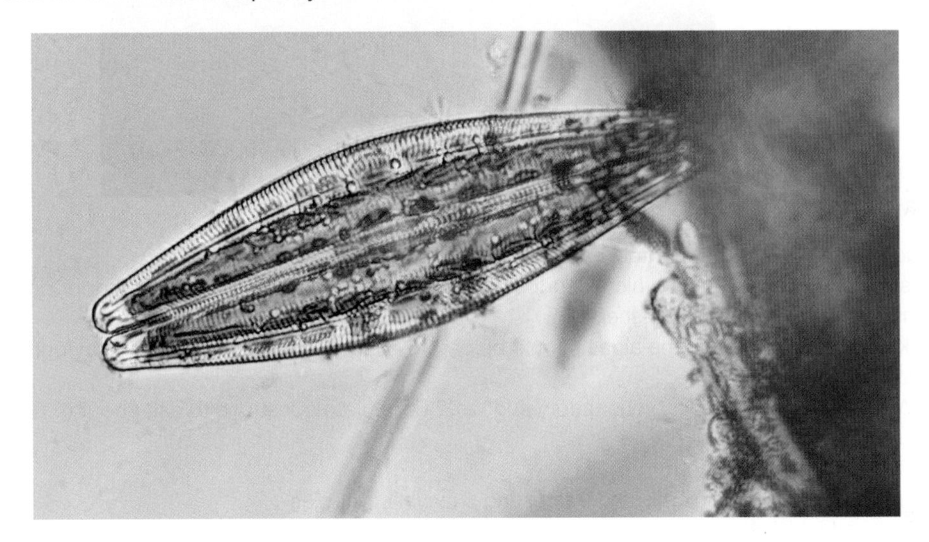

Evaluate and create

7. Make a labelled 2D or 3D model of a cell from one of the kingdoms and explain how the representation models the cell. Use materials available at home, such as drink bottles, egg cartons, cottonwool, wool, cotton or dry foods.
8. **SIS** Algae have many different forms. Broadly, they can be broken into three groups:
   - Green algae
   - Brown algae
   - Red algae.

   Green algae are a group of 9000–12 000 species. They all have a central vacuole and chloroplasts, and some forms have flagella, making them motile (able to move). They can be unicellular, form colonies or be multicellular.

a. Desmid — a unicellular green algae **b.** *Volvox* — a colony formed of unicellular algae **c.** Sea lettuce — a multicellular green algae

a.

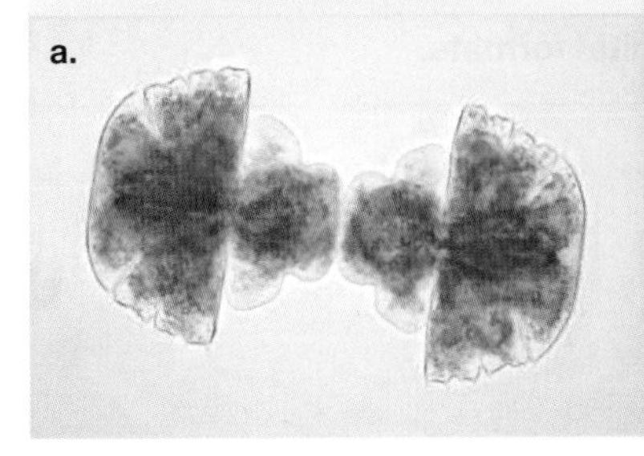

b.

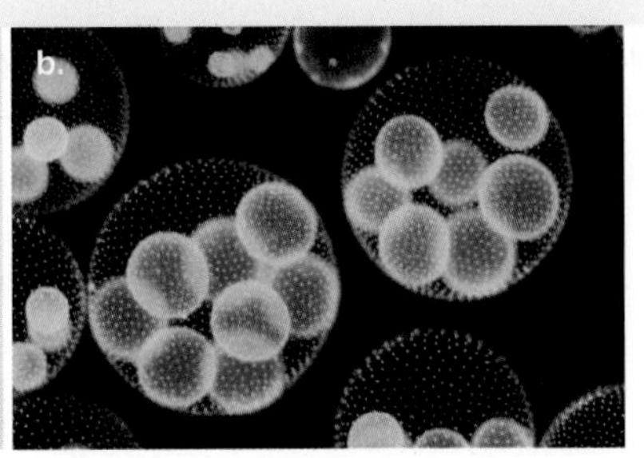

c.

Brown algae are a large group of multicellular algae including many seaweeds. They have chloroplasts surrounded by four membranes (compared to the usual two). Some also possess flagella.
Red algae are a group of approximately 7000 species that have a lifecycle much like that of a fungus. They have no flagella and so they are non-motile. They are mainly multicellular but there are some that are unicellular. They contain chloroplasts.

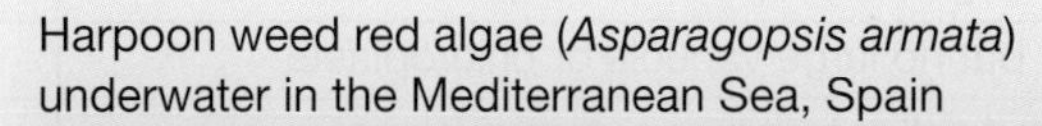
Harpoon weed red algae (*Asparagopsis armata*) underwater in the Mediterranean Sea, Spain

All algae have cell walls that contain cellulose.
There is much discussion over which kingdom the algae should belong to and whether or not the three divisions should be grouped together.
Which kingdom do you think the algae should belong to? Justify your response using evidence from the information given.

9. **SIS** What does the endosymbiotic theory suggest? Formulate questions to ask about it. Research and report on your questions.
10. **SIS** Research and report on:
    a. examples of prokaryotic cells and interesting survival strategies
    b. mitochondrial DNA and haplogroups.
11. **SIS** Research two of the organelles or cells listed. Create a play, and construct puppet models for your characters. Present your play to the class.
    - Nucleus
    - Mitochondrion
    - Chloroplast
    - Prokaryotic cell
    - Protistan cell
    - Animal cell
    - Plant cell
12. Investigate the different types of cells and create your own picture book about them using the following steps.
    a. Construct a matrix table (see lesson 9.5) to show the differences between the cells of the different kingdoms.
    b. Construct a storyboard for a picture book about them.
    c. Create the picture book.

Fully worked solutions and sample responses are available in your digital formats.

LESSON
3.5 Zooming in on life

LEARNING INTENTION

At the end of this lesson you will understand how to prepare a specimen for viewing with a microscope, including the use of dyes to highlight specific cell features. You will also be able to record images using scientific drawing techniques.

3.5.1 Sketching what you see under the microscope

Some points to remember

1. Use a sharp pencil.
2. Draw only the lines that you see (no shading or colouring).
3. Your diagrams should take up about a third to half of a page each.
4. Record the magnification next to each diagram.
5. State the name of the specimen and the date of observation.
6. A written description is also often of considerable value.
7. When you are viewing many cells at one time, it is often useful to select and draw only two or three representative cells for each observation.

FIGURE 3.27 An example of a sketch of a microscope specimen

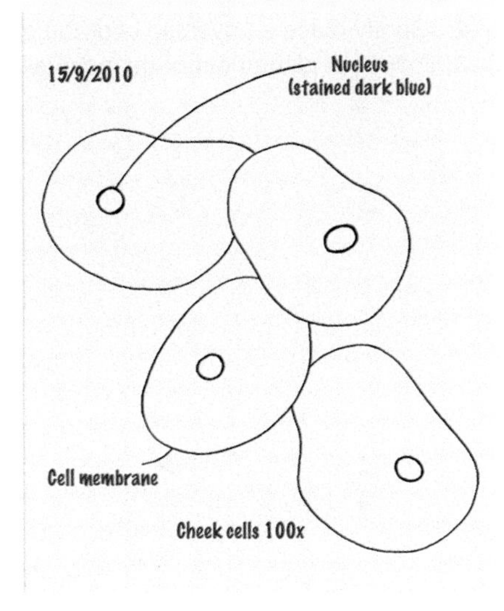

3.5.2 Preparing a specimen

Light microscopes function by allowing light to pass through the specimen to reach your eye. If the specimen is too thick, the object cannot be seen as clearly or may not be seen at all.

Careful peeling, scraping, slicing or squashing techniques can be used to obtain thin specimens of the object to be studied.

Staining a specimen

Many objects are colourless when viewed under the microscope, so specimens are often stained to make them easier to see. Methylene blue, iodine and eosin are some examples of commonly used stains.

Each stain reacts with different chemicals in the specimen. For example, iodine stains starch a blue-black colour.

Take care when using these stains, because they can stain you as well!

elog-2129

INVESTIGATION 3.3

Preparing a wet mount

Aim

To prepare a wet mount and observe microorganisms on a microscope slide

Materials

- light microscope
- coverslips
- pipette
- toothpick
- pond water
- microscope slides (well slides work best for this)
- culture of living microscopic organisms: *Paramecium*, amoeba, rotifers, *Euglena*

Method

1. Use the pipette to put a drop of pond water or microbe culture on a clean microscope slide.

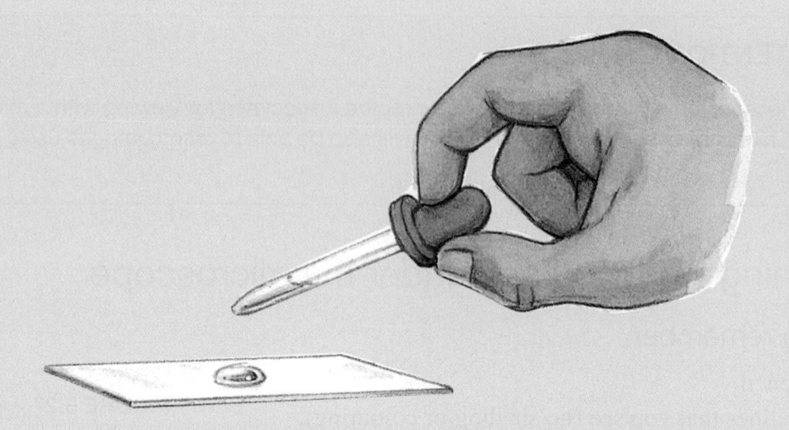

2. Gently place a coverslip over the drop of water by putting one edge down first. Use a toothpick as shown.
3. Incorrect placement of the coverslip can result in air bubbles.

4. Use a microscope to observe the slide.

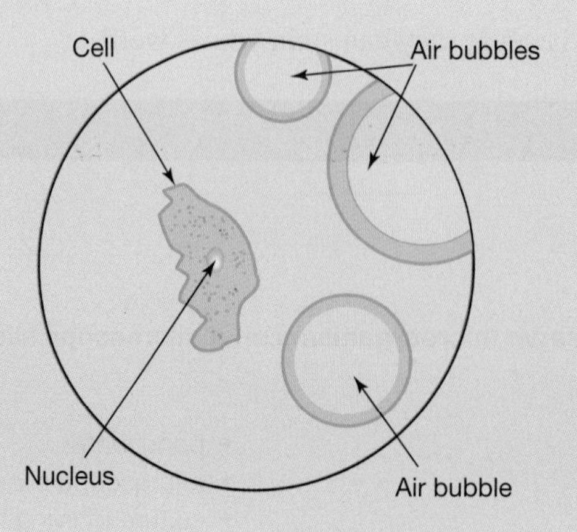

5. Once you have recorded your results, remove the coverslip, rinse and dry the slide, and then prepare a new slide specimen and repeat the steps above.

Results

Draw detailed sketches of what you see. Remember to include a title, the magnifications used and as many observations about the specimen as you can.

Discussion

1. Construct a matrix to show the similarities and differences between the specimens.
2. Suggest reasons for these differences.
3. Use resources on the internet to identify your specimens.
4. Which kingdoms do you think each specimen may belong to? Provide reasons for your classification.
5. Identify two structures you observed in the investigation and find out more about their function (that is, what their 'job' is).
6. You have been observing living specimens. Identify advantages and disadvantages of using living rather than dead specimens or prepared slides.

Conclusion

Summarise the findings for this investigation.

elog-2131

tlvd-10736

INVESTIGATION 3.4

Preparing stained wet mounts

Aim

To prepare, stain and observe a specimen on a microscope slide

Materials

- light microscope
- pipette
- blotting paper
- toothpick
- scalpel
- forceps or tweezers
- microscope slides and coverslips
- water, methylene blue, iodine
- onion, ripe and unripe banana, celery stick, potato

CAUTION

The scalpel has a very sharp blade. Handle it with care.

Method

1. Use the pipette to put a drop of water on a clean microscope slide.

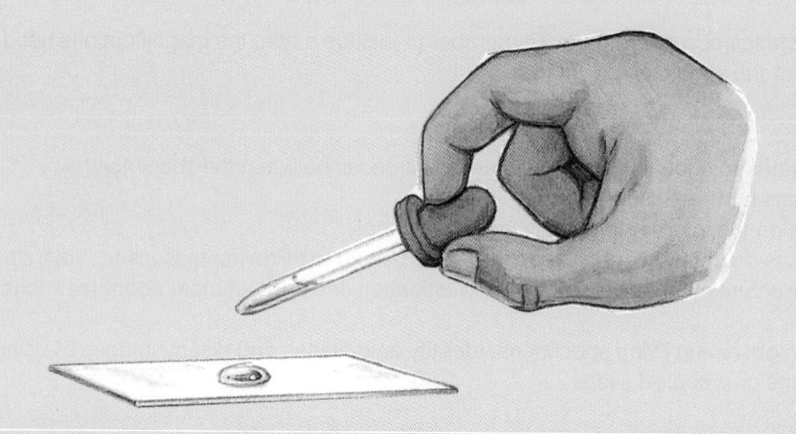

2. Use a scalpel and forceps to peel a small piece of the very thin, almost transparent onion skin from the inside surface of an onion. You may also be able to snap the piece of onion and peel a thin layer of skin off.

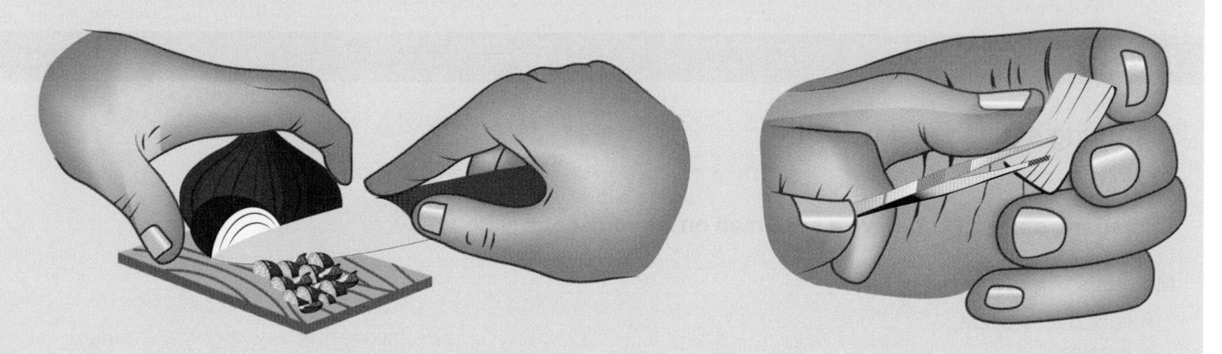

3. Use the forceps to put the thin piece of the onion skin into the drop of water on the microscope slide.

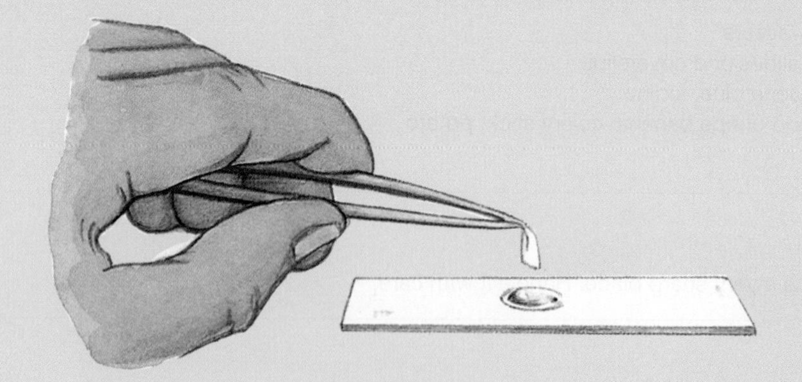

4. Gently place a coverslip over the drop of water containing the onion skin by putting one edge down first. Use a toothpick as in investigation 3.3 to avoid air bubbles. Use blotting paper to soak up any excess water outside the coverslip.

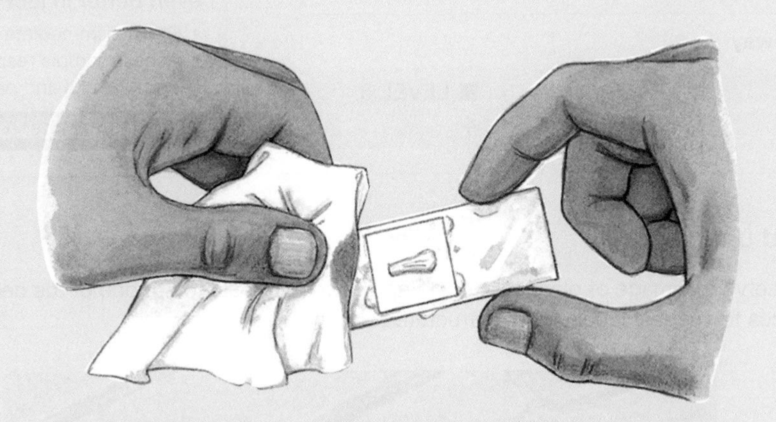

5. Use a microscope to observe the slide; first use low power and then increase the magnification.
6. Prepare another slide of onion skin, except this time add a drop of methylene blue instead of water to the slide. Make sure that you carefully blot excess stain from the slide after you add the coverslip.
7. Observe this stained onion specimen; first use low power, then view at a higher magnification.

Once you have recorded your results:

8. Remove the coverslip, and rinse and dry the slide.
9. Use the previously outlined steps to prepare the following slides:
   - Celery epidermis (outer layer of the celery stem) with and without methylene blue stain
   - Squashed ripe and unripe banana with and without iodine.
   - A very thin slice of potato with iodine.

Results

Draw detailed sketches of what you see. Remember to include a title, the magnifications used and as many comments as you can. Label any parts that you can identify.

Discussion

1. Compare the cells of the stained onion epidermis and the celery epidermis (or potato). Identify their similarities and differences. Suggest reasons for the differences.
2. If a ripe and unripe banana was used, identify their similarities and differences. Suggest reasons for the differences.

Methylene blue is used to stain the nucleus so that it is easier to see. Iodine changes from yellow-brown to a dark blue when it combines with starch.

3. Explain why stains are used. Include reasons for using methylene blue and iodine that relate to your observations in this investigation.
4. Investigate the functions of the structures observed in your stained specimens. Suggest how features of these structures assist their function.
5. Identify strengths, limitations and improvements related to this investigation.

Conclusion

Summarise the findings for this investigation.

Resources

eWorkbook Preparing a stained wet mount (ewbk-11857)

3.5 Activities

learnon

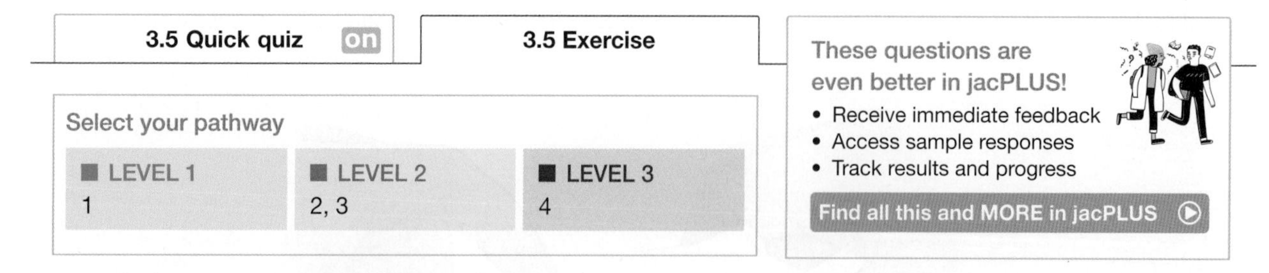

Remember and understand

1. **a.** Carefully observe the image of plant cells given and construct a sketch of one of the cells.
   b. Use references to suggest labels for the structures shown in your sketch.

Apply and analyse

2. **SIS** Carefully observe the student sketches shown. For each diagram, list what is wrong with it and suggest how it could be improved.

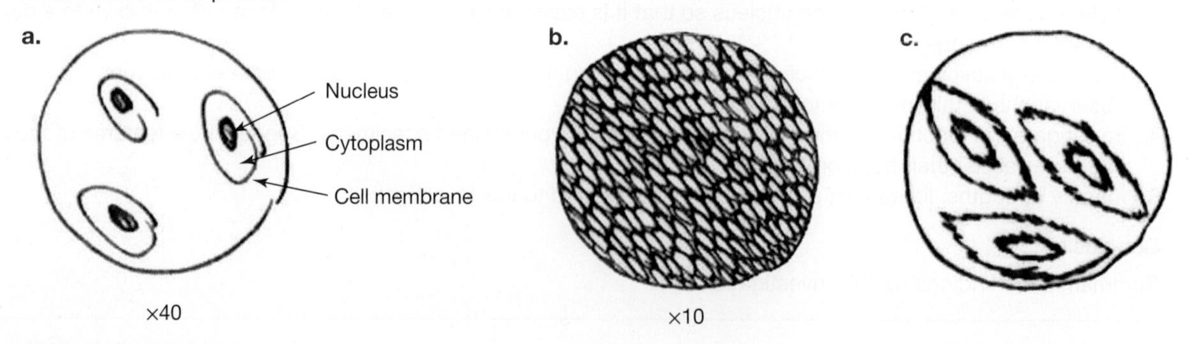

Evaluate and create

3. **SIS** Design a poster that shows others how to prepare a variety of specimens to be viewed under a microscope.

4. SIS Methylene blue has the following warning symbols on the Safety Data Sheet that need to be considered when it is used.

a. What do these symbols mean?
b. What precautions should be taken when using methylene blue?

Fully worked solutions and sample responses are available in your digital formats.

LESSON
3.6 Focus on animal cells

LEARNING INTENTION

At the end of this lesson you will be able to describe how cell shape and size (form) enables the functions of animal cells.

3.6.1 In all shapes and sizes

Cells within an organism may differ in their shape and size. This difference may be due to the particular jobs or functions that the cells carry out within the organism. The human body is made up of more than 20 different types of cells, with each type suited to a particular function.

Nerve cells develop long, thin fibres that quickly carry messages from one cell to another. Cells lining the trachea have hair-like cilia that move fluid and dust particles out of the lungs. Muscle cells contain fibres that contract and relax, and the human sperm cell has a tail, or flagellum, that helps it swim to the egg cell.

Cells can also differ in the organelles that they contain within them. Muscle cells, for example, contain many more mitochondria than other types of cells due to their high energy requirements. Red bloods cells also differ from many other types of cells because, as they mature, they lose their nucleus. This makes more room available for them to carry more oxygen throughout your body.

EXTENSION: Facts about human cells

Did you know these facts about human cells?

- Hair and nails are made of dead cells, and because they are not fed by blood or nerves you can cut them without it hurting.
- A human baby grows from one cell to 2000 million cells in just nine months.
- Red blood cells live for one to four months and each cell travels around your body up to 172 000 times.
- Some of the nerve cells in the human body can be one metre long. But that's small compared with the nerve cells in a giraffe's neck. They are two to three metres long!

int-3395

FIGURE 3.28 The human body is made up of more than 20 different types of cells, with each type suited to a particular function.

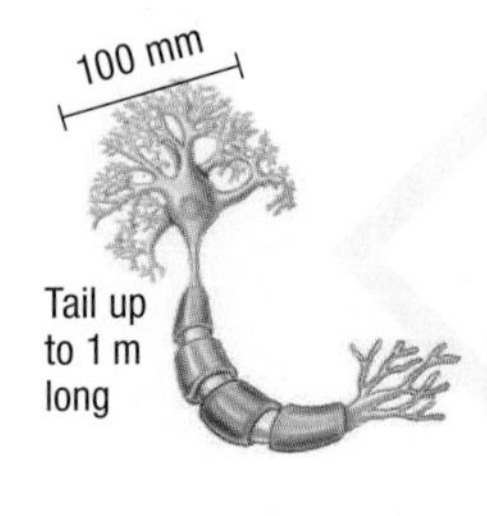

Nerve cells
Nerve cells are very long and have a star shape at one end. The long shape of nerve cells helps them detect and send electrical messages through the body at the speed of a Formula 1 racing car. There are nerve cells all over your body. They allow you to detect touch, smell, taste, sound, light and pain.

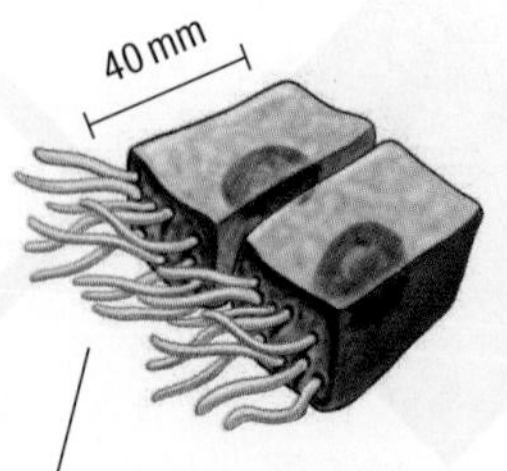

Lung epithelial cells
The cells that line your nose, windpipe and lungs are a type of lining cell. They have hair-like tips called cilia. These cells help protect you by stopping dust and fluid from getting down your windpipe. The cilia can also move these substances away from your lungs. You remove some of these unwanted substances whenever you sneeze, cough or blow your nose.

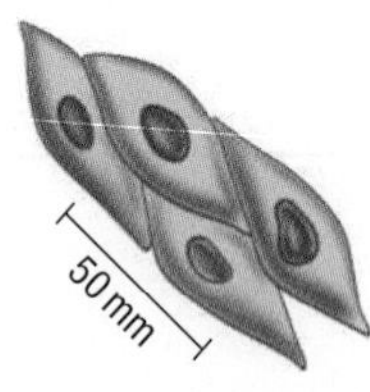

Muscle cells
Muscle cells are long and elastic. Long thin cells can slide further over each other to allow you to move. There are different types of muscle cells. The walls of your blood vessels and parts of your digestive system have 'smooth muscle' cells. The muscles that are joined to your bones are called 'skeletal muscles'. Skeletal muscles work in pairs — one muscle contracts (shortens) and pulls the bone in one direction while the other muscle relaxes.

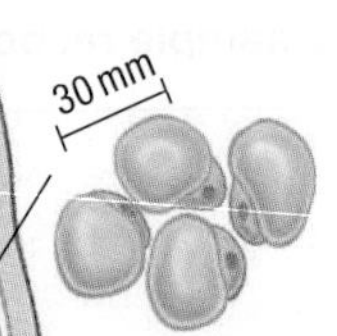

Adipose tissue cells
Some cells store fat. Fat stores a lot of energy for cells to use later. Round shapes are good for holding a lot of material in a small space. Fat cells are mostly found underneath your skin, especially in the chest, waist and buttocks.

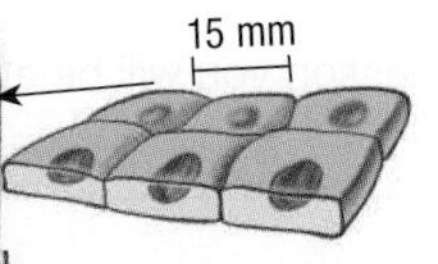

Skin cells
Special cells line the outside surfaces of your body. These are the cells that form your skin. These cells have a flattened shape so they can better cover and protect your body.

Red blood cells
Red blood cells carry oxygen around the body. Their small size allows them to move easily through blood vessels. The nucleus in a red blood cell dies soon after the cell is made. Without a nucleus, red blood cells live for only a few weeks. The body keeps making new blood cells to replace those that have died. Red blood cells are made in bone marrow at the rate of 17 million cells per minute! This is why most people can donate some of their blood to the Red Cross without harm. White blood cells, which are larger than red blood cells, are also made in the bone marrow. Their job is to rid the body of disease-causing organisms and foreign material.

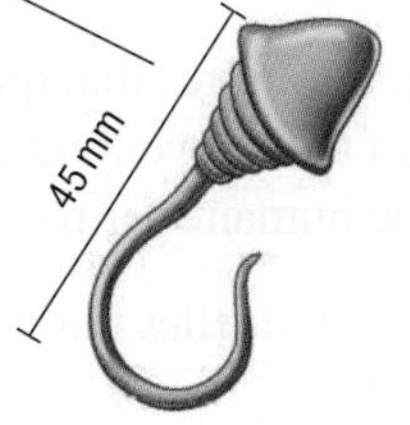

Sperm cells
Sperm cells have long tails that help them swim towards egg cells. Only males have sperm cells.

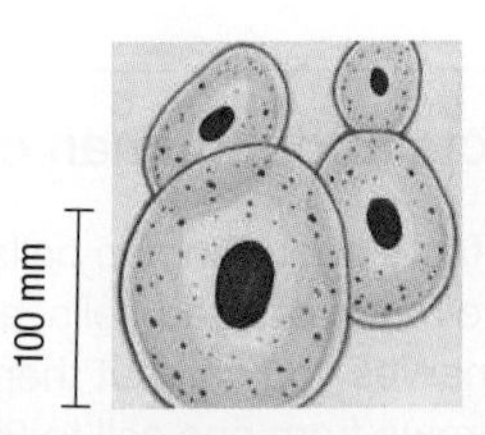

Egg cells
Egg cells are some of the largest cells in a human body. Their large round shape helps them store plenty of food. Only females have egg cells. When a sperm cell moves into an egg cell, the egg cell is fertilised.

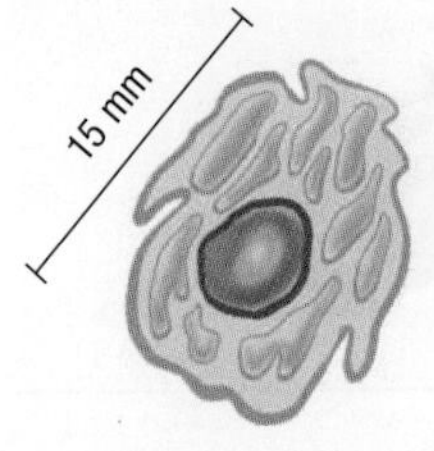

Bone cells
Minerals such as calcium surround your bone cells. The minerals help make bone cells hard and strong. Bone cells need to be hard so that they can keep you upright.

elog-2133

INVESTIGATION 3.5

Animal cells — what's the difference?

Aim

To observe the features of different types of human (animal) cells

Materials

- light microscope
- prepared animal slides: blood cells, muscle cells, cheek cells, nerve cells

Method

Use a microscope to observe the prepared slides.

Results

Record detailed diagrams of your observations. Next to your diagrams, include details of the (a) source of the specimen, (b) type of specimen, (c) magnification used, as well as (d) a detailed description of the specimen.

Discussion

1. Were all of the animal cells you observed the same size? Explain.
2. Did all of the cells observed contain a nucleus? Explain.
3. Identify features that all of the observed animal cells shared.
4. Identify differences between the features of the cells observed.
5. Suggest reasons for the differences between the cells.
6. Compare your cells with those in figure 3.28.
   a. Do your sketched diagrams match the structures shown in the figure? Explain.
   b. Read through the text related to the functions of the different types of cells. Do these match those you suggested in question **5**? Explain.

Conclusion

Summarise the findings for this investigation.

Extension

What are the similarities and differences between human blood cells compared to other animals such a frogs?

3.6 Activities

learn**on**

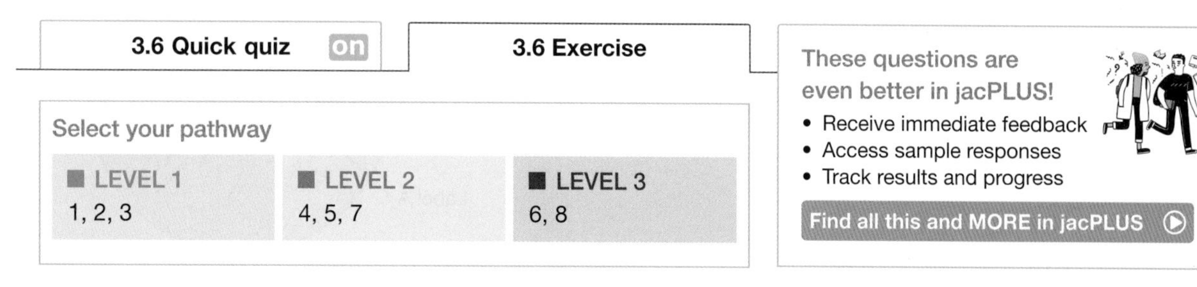

Remember and understand

1. Match the types of cells with their descriptions.

Type of cell	Description
a. Muscle cell	**A.** Has a long tail that helps it to swim towards the egg cell
b. Skin cell	**B.** Long, thin elastic cell that contracts and relaxes
c. Red blood cell	**C.** Flat cell that lines the outside surface of your body
d. Nerve cell	**D.** Very tiny cell that lacks a nucleus when mature, and carries oxygen
e. Sperm cell	**E.** Very long cell, star-shaped at one end; detects and sends messages

2. Identify which features most animal cells have in common. Suggest reasons why.
3. Describe some ways in which cells may differ.
4. Suggest why the cells in a multicellular organism are not all the same. Give examples in your answer.
5. Distinguish between:
   a. skin cells and sperm cells
   b. red blood cells and nerve cells
   c. adipose tissue cells and muscle cells.

Apply and analyse

6. a. SIS Summarise the information from figure 3.28 into a table with the headings: 'Type of cell', 'Function', 'Shape' and 'Size'.
   b. Using this information, determine the average size of an animal cell.
   c. Use a column graph to plot the sizes of the different types of animal cells.
   d. Identify which animal cells are 'above average' in size and which are 'below average'. Suggest reasons for the differences.
   e. Comment on the differences in other features between the cells.
7. Heart cells have large numbers of mitochondria. Explain, with reference to the function of the heart cell, why this is the case.

Evaluate and create

8. SIS Using your own research and the information in figure 3.28, construct a 'peep through' learning wheel that shows the structure and function of the different types of animal cells. Instructions for making a 'peep through' learning wheel are given in figure 3.29.
   a. On an A4 piece of white paper or card, draw two circles: one with a 'tab' (wheel 2) and one without (wheel 1).
   b. Cut out the two rectangular box areas as shown on wheel 1.
   c. Draw in the large and the small rectangles as shown on wheel 2.
   d. Write the animal cell types in the small boxes on wheel 2. Sketch matching diagrams of examples of these cell types in the corresponding large box opposite.
   e. Attach the two wheels, with wheel 1 on top, using a paper fastener.
   f. Rotate your wheel to view examples of types of animal cells.

FIGURE 3.29 How to make a 'peep through' learning wheel

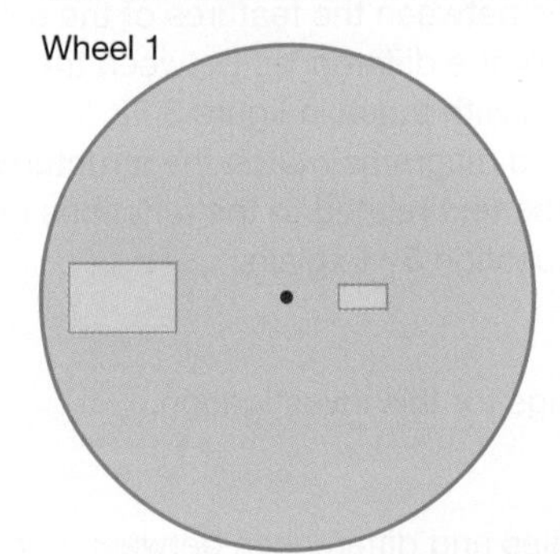

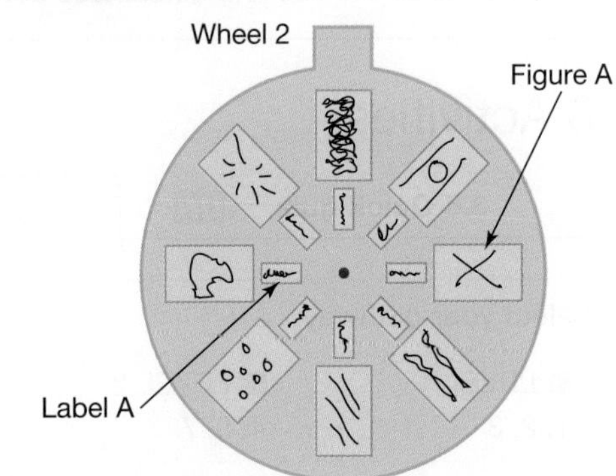

Fully worked solutions and sample responses are available in your digital formats.

LESSON
3.7 Focus on plant cells

LEARNING INTENTION

At the end of this lesson you will be able to identify and describe cells that are specialised in plants.

3.7.1 Have or have not

Like animal cells, plant cells have cytoplasm, a membrane and a nucleus. Unlike animal cells, plant cells have a cellulose cell wall and a large central **vacuole** filled with cell sap. Often plant cells also contain chloroplasts in their leaves, which enable them to make their own food in a process called photosynthesis. In this process, carbon dioxide (from the air) and water (from the roots) move into the chloroplast, leading to the production of glucose (used by the plant) and the release of oxygen.

Carbon dioxide + water → glucose + oxygen

vacuoles sacs within a cell used to store food and wastes; plant cells usually have one large vacuole, while animal cells have several small vacuoles or none at all

guard cells cells on either side of a stoma that work together to control the opening and closing of the stoma

stomata openings mainly on the lower surface of leaves; these pores are opened and closed by guard cells; singular = stoma

FIGURE 3.30 The process of photosynthesis in the leaves of plants

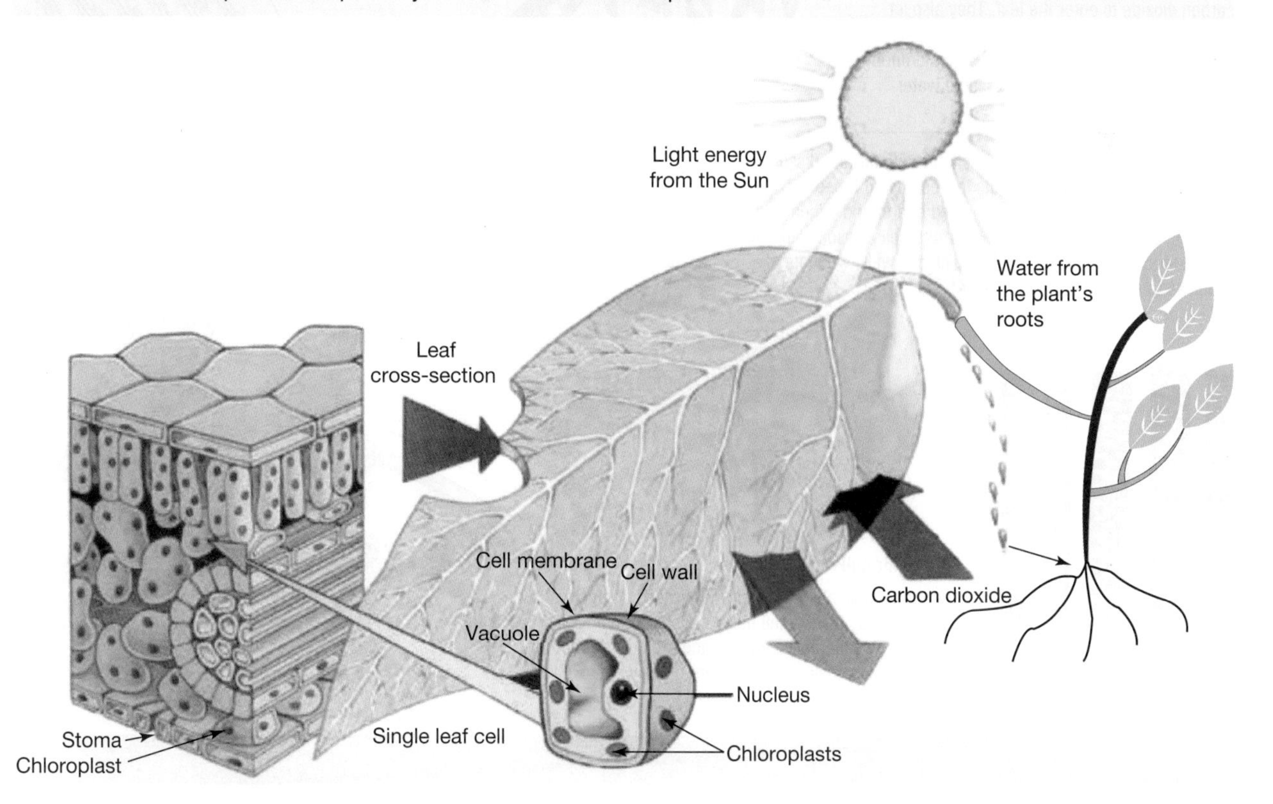

On the surfaces of leaves, there are pairs of special cells called **guard cells**, which surround tiny pores called **stomata**. The guard cells can change shape, opening or closing the stomata. Special cells on the roots extend into microscopic hairs that penetrate between soil particles. The hairs provide a large surface area through which water may be absorbed from the soil.

WHAT DOES IT MEAN?

The word 'xylem' comes from the Greek word *xulan*, meaning 'wood'. The word 'phloem' comes from the Greek word *phloos*, meaning 'bark'.

int-3396

FIGURE 3.31 Some of the types of cells found in plants

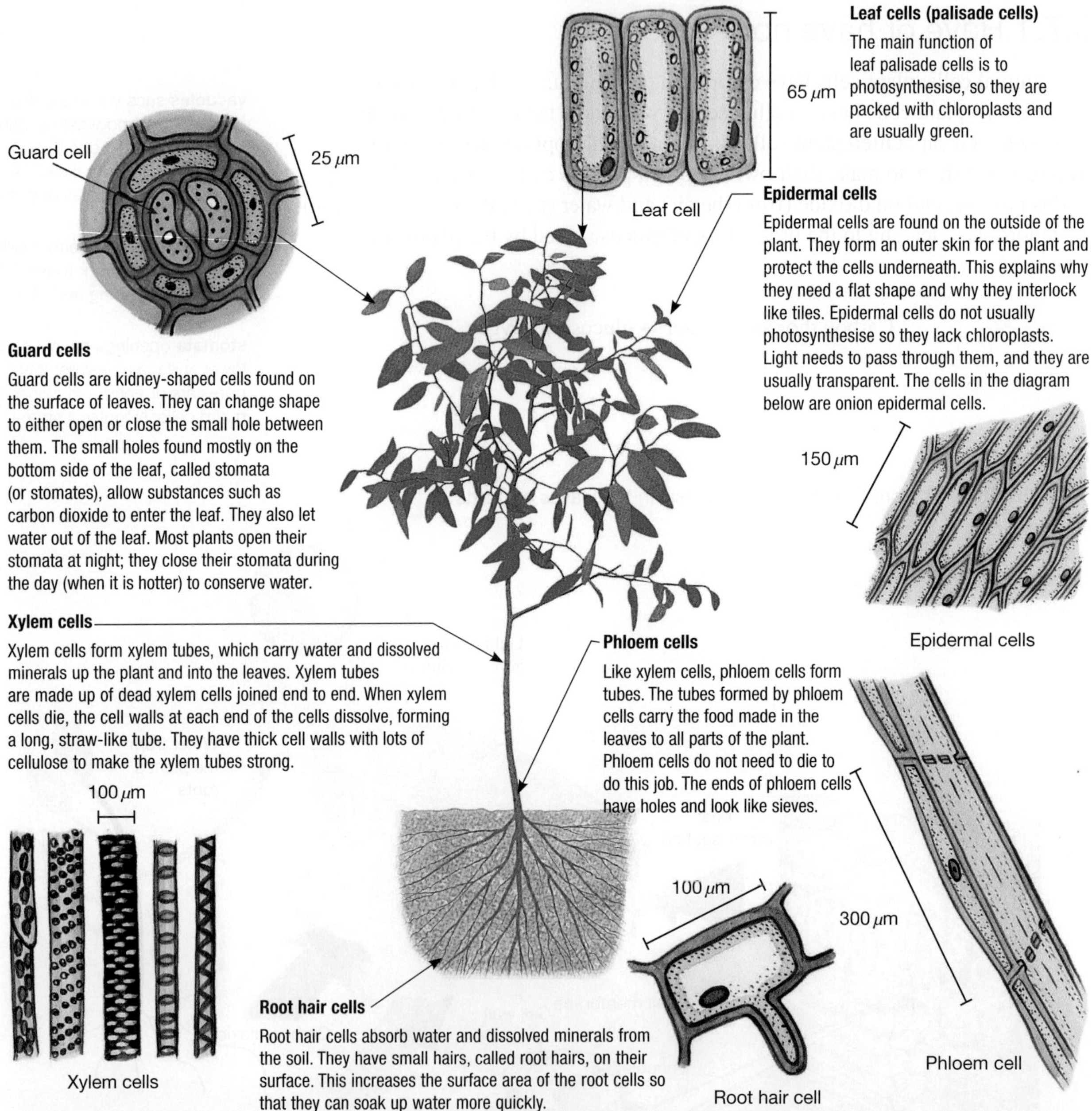

elog-2135

INVESTIGATION 3.6

Plant cells in view

Aim

To observe the features of different types of plant cells

Materials

- light microscope
- prepared plant slides: leaf epidermal cells, root hair cells, stomata/guard cells

Method

Use a microscope to observe the prepared slides.

Results

Record detailed diagrams of your observations. Next to your diagrams, include details of the (a) source of the specimen, (b) type of specimen, (c) magnification used, as well as (d) a detailed description of the specimen.

Discussion

1. Were all of the plant cells the same size? Explain.
2. Did all of the cells observed contain a nucleus? Explain.
3. Identify features that all of the observed plant cells shared.
4. Identify differences between the features of the cells.
5. Suggest reasons for the differences between the cells.
6. Compare your cells with those in figure 3.31.
   a. Do your sketched diagrams match the structures shown in the figure? Explain.
   b. Read through the text related to the functions of the different types of cells. Do these match your answer to question 5? Explain.

Conclusion

Summarise the findings for this investigation.

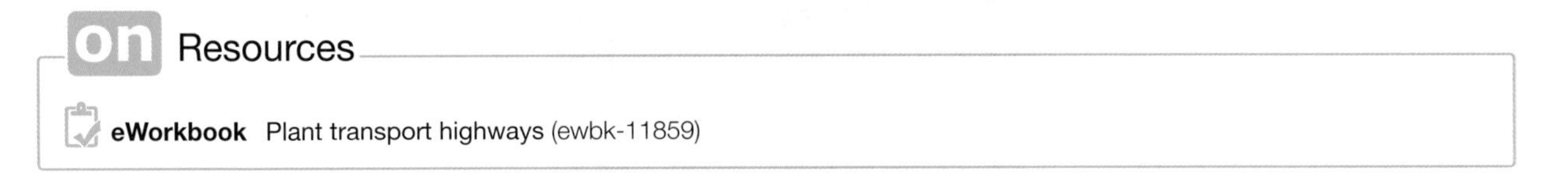

3.7 Activities

learnon

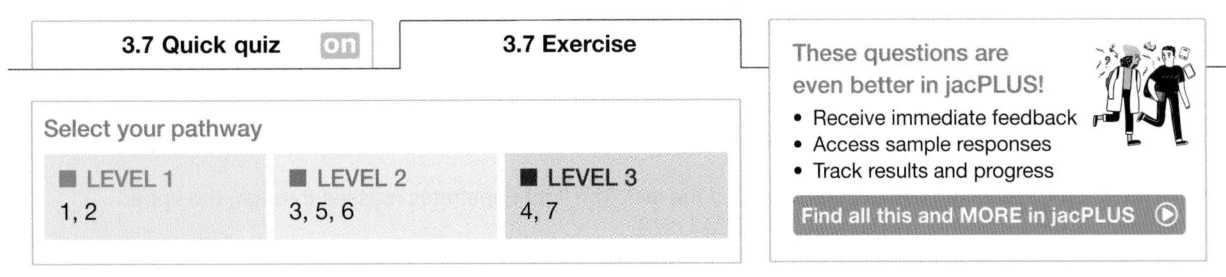

Remember and understand

1. Match the types of cells with their descriptions.

Type of cell	Description
a. Guard cells	A. Sieve-like cells that form tubes that carry food made in the leaves to other parts of the plant
b. Phloem cells	B. Cells with small hairs that increase their surface area so that they can absorb more water
c. Xylem cells	C. Thick-walled cells that carry water up the plant
d. Root hair cells	D. Kidney-shaped cells that can change shape to either open or close the small hole between them, which allows gas exchange between the plant and its environment

2. Describe some ways in which plant cells may differ.
3. Distinguish between:
   a. palisade cells and guard cells
   b. xylem cells and phloem cells
   c. epidermal cells and root hair cells.

Apply and analyse

4. a. **SIS** Summarise the information in figure 3.31 into a table with the headings: 'Type of cell', 'Function', 'Shape' and 'Size'.
   b. Using this information, determine the average size of a plant cell.
   c. Use a bar graph to plot the sizes of the different types of plant cells.
   d. Identify which plant cells are 'above average' in size and which are 'below average'. Suggest reasons for the differences.
   e. Comment on the differences in other features between plant cells.

Evaluate and create

5. Construct a model of a pair of guard cells, using balloons.
6. Using your own research and the information in figure 3.31, construct a 'peep through' learning wheel that shows the structure and function of the different types of plant cells. Instructions for making a 'peep through' learning wheel are given in 3.6 Exercise.
7. **SIS** Consider the following diagram of a leaf.

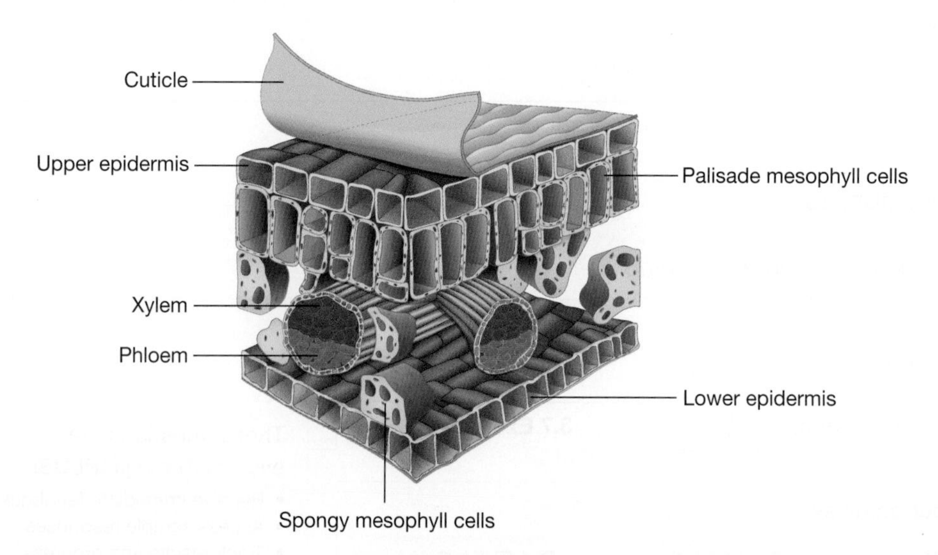

   In most leaves, sunlight hits the upper side of the leaf. The light penetrates (passes through) the upper epidermis before hitting the palisade mesophyll cells.
   a. The mesophyll cells all have chloroplasts. What is the function of chloroplasts?
   b. Suggest a reason the palisade mesophyll cells have more chloroplasts than the spongy mesophyll cells.
   c. During photosynthesis, water is taken up by the roots and combined with carbon dioxide in the air. Glucose is made and stored for future use and oxygen is released. What is the role of the large air spaces between the spongy mesophyll cells?
   d. Explain why there are more stomata on the under surface of the leaf.

Fully worked solutions and sample responses are available in your digital formats.

LESSON
3.8 Plant cells — holding, carrying and guarding

LEARNING INTENTION

At the end of this lesson you will be able to describe the specialised transport systems in plants.

3.8.1 Sweet transport — phloem

As in animal cells, plant cells can work together for a variety of functions to meet their survival needs. Plants have their own transport systems, which consist of many thin tubes made up of different types of cells. Other types of plant cells are involved in water regulation and exchange of important gases, such as oxygen and carbon dioxide, with their environment.

FIGURE 3.32 a. Phloem helps transport and distribute nutrients around the plant. **b.** The structure of phloem allows these nutrients to be transported upwards as well as downwards.

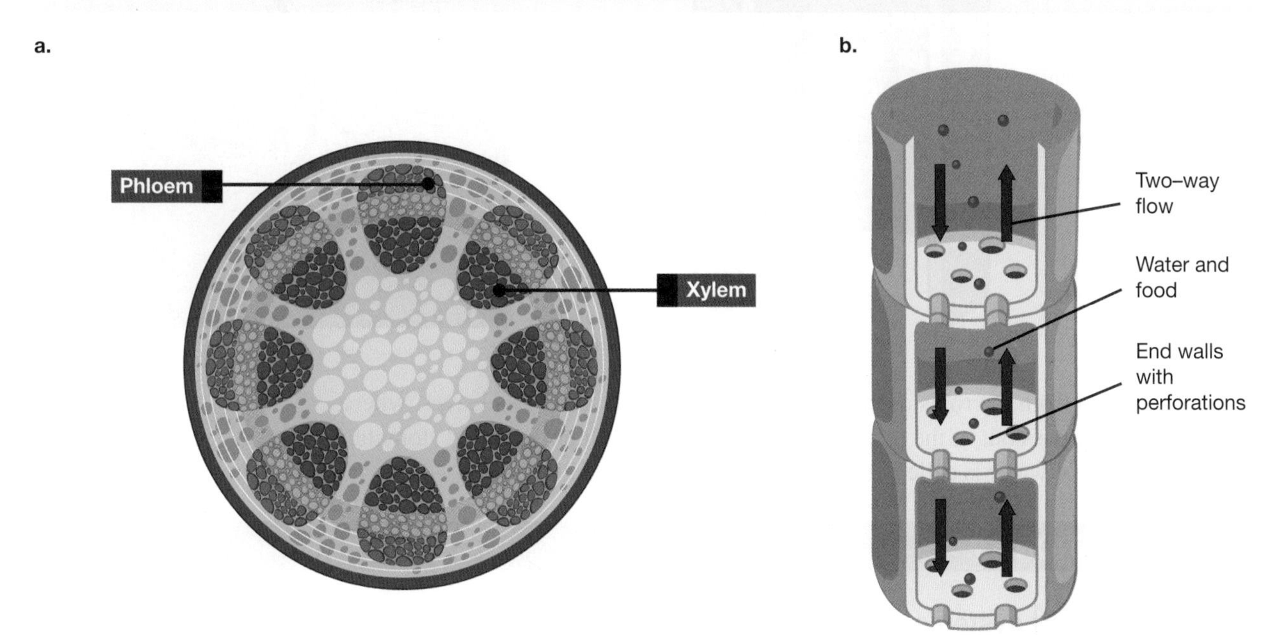

Using the process of photosynthesis, plants make sugar in their leaves. The **phloem** is a system of thin-walled tubes (that are found in the outer part of the stem) that carries this sugar (in the form of glucose or sucrose) from the leaves to other parts of the plant. Phloem consists of living cells called sieve tubes and companion cells. The transport of the sugar solution up and down the plant is called **translocation**.

3.8.2 Water pipes — xylem

Flowering plants also have tubes with strong, thick walls that carry water and minerals up from the roots through the stem to the leaves. These are called **xylem vessels** and are located towards the centre of the stem (figure 3.32a). These tubes are formed from the empty remains of dead cells, the walls of which are strengthened with a woody substance called **lignin**. The xylem is therefore a 'dead' one-way street (figure 3.33), rather than a 'living' two-way highway like phloem.

phloem a type of tissue that transports sugars made in the leaves to other parts of a plant

translocation the process in which sugars and amino acids are transported within a plant by phloem tissue

xylem vessels pipelines for the flow of water up plants, made up of the remains of dead xylem cells fitted end to end with the joining walls broken down; lignin in the cell walls gives them strength

lignin a hard substance in the walls of dead xylem cells that make up the tubes carrying water up plant stems; lignin forms up to 30 per cent of the wood of trees

FIGURE 3.33 The flow of water and minerals is only one way in xylem, away from the roots.

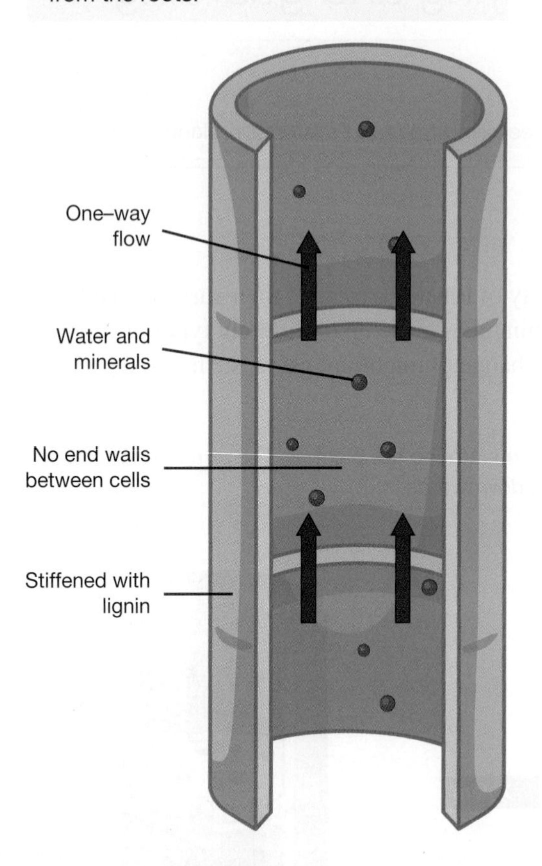

FIGURE 3.34 The movement of water from roots to leaves is known as the transpiration stream.

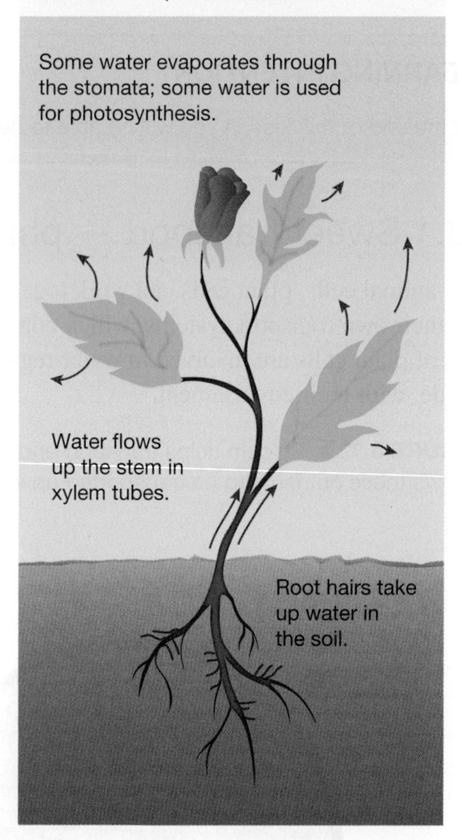

Water moves up from the roots of the plant, through its stem and to its leaves, where some water may pass out of the plant as water vapour through pores called stomata. This movement of water is called the **transpiration stream** (figure 3.34).

transpiration stream the movement of water through a plant as a result of loss of water from the leaves

elog-2137

INVESTIGATION 3.7

Stem transport systems

Aim

To identify xylem cells in celery

Materials

- celery stick (stem and leaves)
- knife
- two 250 mL beakers
- water
- blue food colouring
- red food colouring
- hand lens

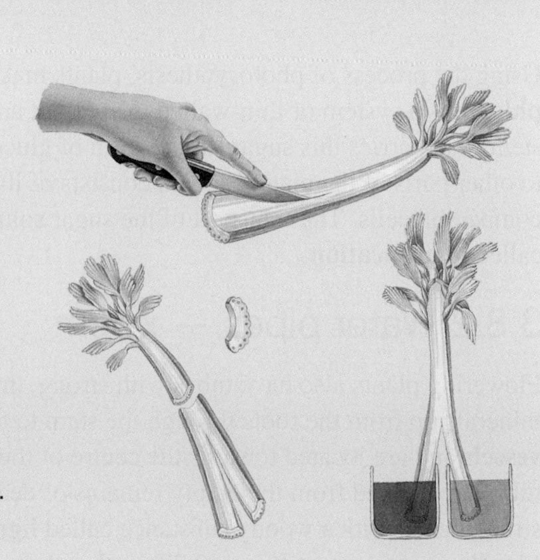

Method

1. Slice the celery along the middle to about halfway up the stem.
2. Fill two beakers with 250 mL of water. Colour one blue and the other red with the food colouring.

3. Place the celery so that each 'side' of the celery is in a separate beaker.
4. Leave for 24 hours and then observe the celery.
5. Cut the celery stick across the stem.
6. Use the hand lens to look at the inside of the stem.

Results

1. Look at where the water has travelled in the celery. Draw a diagram to show your observations.
2. Draw a diagram to show what you can see when you cut across the stem.
3. Where are the different colours found in the stem?
4. Where are the different colours found in the leaves?
5. Draw a diagram of the whole celery stick and trace the path of the water through each side to the leaves.

Discussion

1. Explain your observations of where the water was found in the stems and leaves. Make sure you comment on the relationship between the shape of the structures and your findings.
2. Identify strengths and limitations of this investigation and suggest possible improvements.

Conclusion

Summarise the findings for this investigation.

Extension

How could you turn a white carnation blue? Try it.

Xylem for support

The phloem and the xylem vessels are located together in groups called **vascular bundles**. The strong, thick walls of the xylem vessels are also important in helping to hold up and support the plant. The trunks of trees are made mostly of xylem. Did you know that the stringiness of celery is due to its xylem tissues?

FIGURE 3.35 The vascular bundles can appear different depending on the plant. They can either **a.** be in a ring-like pattern around the stem or **b.** have a random arrangement throughout the stem.

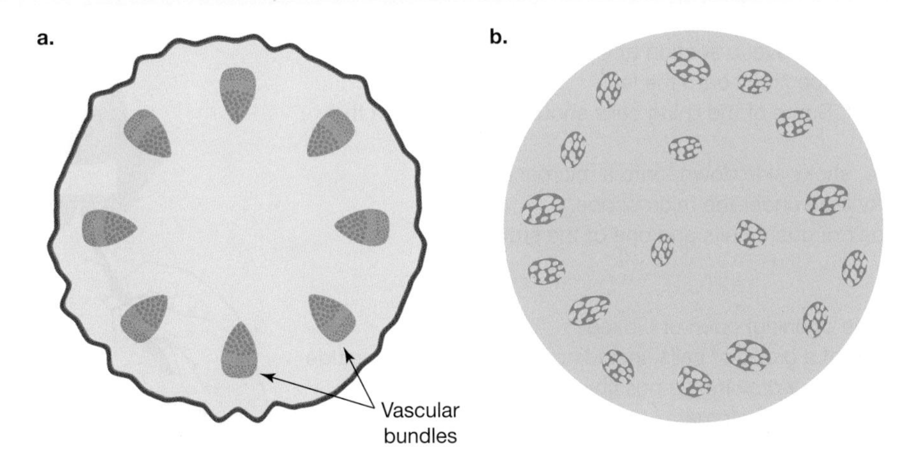

3.8.3 Leaf doorways — stomata

Water transport occurs within the xylem vessels. Some of the water that is transported through the xylem to the leaves is used in photosynthesis. Some water is also lost as water vapour through tiny holes or pores in the leaves. These tiny pores, called stomata (or stomates), are most frequently found on the underside of the leaves. Evaporation of water from the stomata in the leaves helps pull water up the plant. Loss of water vapour through the stomata is called **transpiration**.

vascular bundles groups of xylem and phloem vessels within plant stems

transpiration the loss of water from plant leaves through their stomata

Guard cells in control

Oxygen and carbon dioxide gases also move in and out of the plant through the stomata. Guard cells, which surround each stoma, enable the hole to open and close, depending on the plant's needs. When the plant has plenty of water, the guard cells fill up with water and stretch lengthways. This opens the pore. If water is in short supply, however, the guard cells lose water and they collapse towards each other. The pore is then closed. This is one way in which the plant can control its water loss.

FIGURE 3.36 a. Stomata can open (left) and close (right) to conserve water. **b.** What features can you identify in this image of a leaf under a microscope?

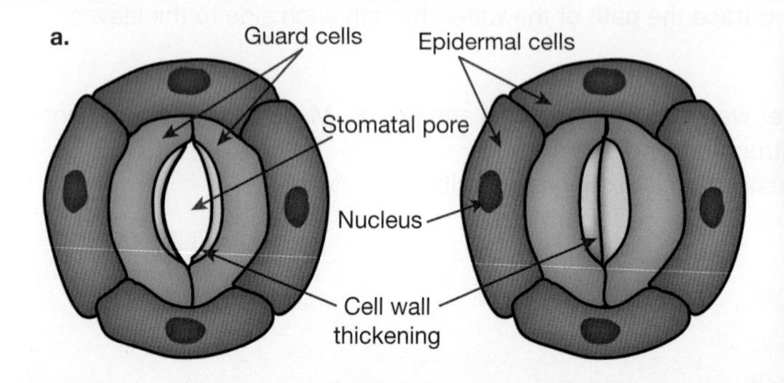

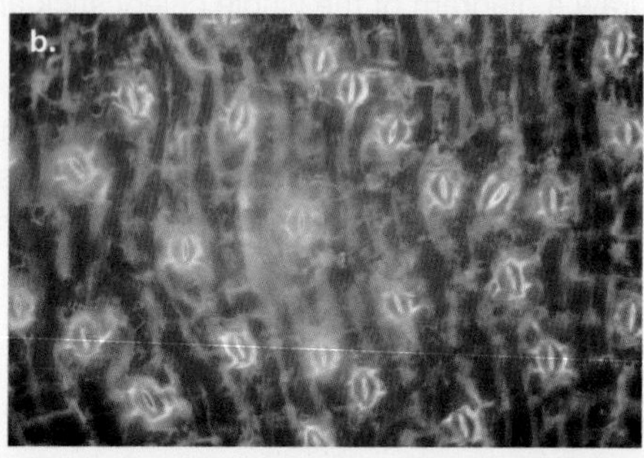

elog-2139

INVESTIGATION 3.8

Observing leaf epidermal cells

Aim

To observe leaf epidermal cells and identify stomata

Materials

- leaf
- clear sticky tape
- microscope slide
- microscope

Method

1. Put some sticky tape over a section of the underside of the leaf.
2. Press the sticky tape firmly onto the leaf.
3. Tear the tape off. Some of the lining cells should come off with the sticky tape.
4. Press the tape, sticky side down, onto a microscope slide.
5. View the sticky tape under the microscope.
6. Try to find a pair of guard cells and one of the stomata.

Results

1. Is the stoma (the opening) open or closed?
2. Make a drawing of a group of cells, including the guard cells. Include as much detail in your drawing as possible.
3. Label the guard cells and stomata.
4. Title and date your drawing. Write down the magnification used.

Discussion

1. Summarise your findings. Include comments on the relationship between the shape of the structures and their function.
2. What causes the changes in guard cells so that stomata open and close?
3. Identify strengths and limitations of this investigation and suggest possible improvements.

Conclusion

Summarise the findings for this investigation.

EXTENSION: Water loss

Although water makes up about 90–95 per cent of the living tissues of plants, water is often being lost to their surroundings. As much as 98 per cent of the water absorbed by a plant can be lost through transpiration.
A variety of factors affect the amount of water that plants lose. Weather is a major factor, as high temperatures, wind and low humidity can increase the evaporation of water from the stomata. It has been recorded that large trees may lose more than 400 litres of water in a day.

elog-2141

INVESTIGATION 3.9

Looking at chloroplasts under a microscope

Aim

To observe chloroplasts under a light microscope

Materials

- tweezers
- water
- moss, spirogyra or elodea
- dilute iodine solution
- light microscope, slides, coverslips

Method

1. Using tweezers, carefully remove a leaf from a moss or elodea plant or take a small piece of spirogyra.
2. Place the plant material in a drop of water on a microscope slide and cover it with a coverslip.
3. Use a light microscope to observe the leaf. Complete a drawing of a cell. Include (a) title, (b) magnification, (c) scale bar and (d) appropriate labels.
4. Put a drop of dilute iodine solution under the coverslip. (Iodine stains starch a blue-black colour.)
5. Using the microscope, examine the leaf again. Complete a drawing of a cell. Include (a) title, (b) magnification, (c) scale bar and (d) appropriate labels.

Results

1. Draw what you see before staining.
2. Label any chloroplasts that are present.

Discussion

1. Describe the colour of the chloroplasts before staining.
2. What gives chloroplasts their colour?
3. Did the iodine stain any part of the leaf a dark colour?, If so, what does this suggest?
4. Identify strengths and limitations of this investigation and suggest possible improvements.

Conclusion

Summarise the findings for this investigation about chloroplasts.

Flaccid or firm?

If too much water is lost or not enough water is available, the plant may **wilt**. When this occurs, water has moved out of the cell vacuoles and the cells have become soft or **flaccid**. The firmness in the petals and leaves is due to their cells being firm or **turgid**.

wilt refers to when plant stems and leaves droop due to insufficient water in their cells

flaccid refers to cells that are not firm due to loss of water

turgid refers to cells that are firm

FIGURE 3.37 If the cells of a plant do not contain enough water, they become flaccid and the plant wilts.

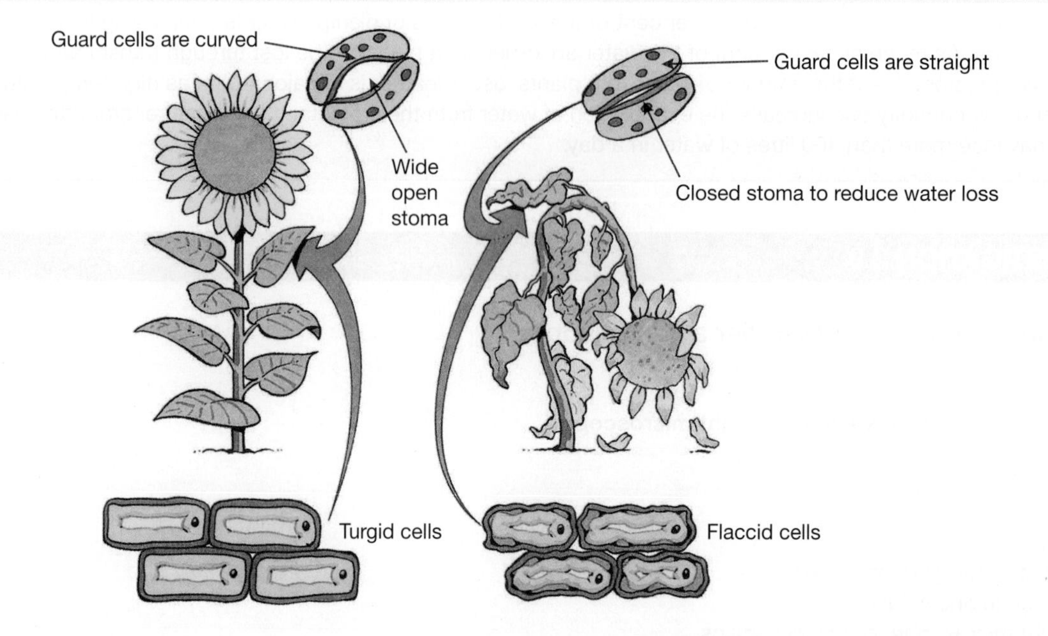

elog-2143

INVESTIGATION 3.10

Moving in or out?

Water moves from areas of high water concentration to areas of low water concentration. In this way, water can move from the xylem cells into photosynthetic cells so that photosynthesis can occur. In this investigation you will see how water moves across a membrane.

Aim

To make a model of a cell membrane to simulate the effect of water moving in and out of a cell

Hypothesis

What do you expect will happen to Bag A?

What do you expect will happen to Bag B?

Materials

- two 20 cm lengths of dialysis tubing
- scales
- starch solution
- iodine solution
- two beakers
- string or food packaging pegs

Method

1. Soak the dialysis tubing in water so it becomes soft.
2. Tie a knot at one end of each piece of dialysis tubing. This will form two small bags.
3. Pour water into Bag A until it is one third full. Pour the same amount of starch solution into Bag B and add ten drops of iodine solution.
4. Tie a knot at the top of each bag to seal them or use the string/food packaging pegs.
5. Weigh each bag and record the weights in the table in the results section of this investigation.

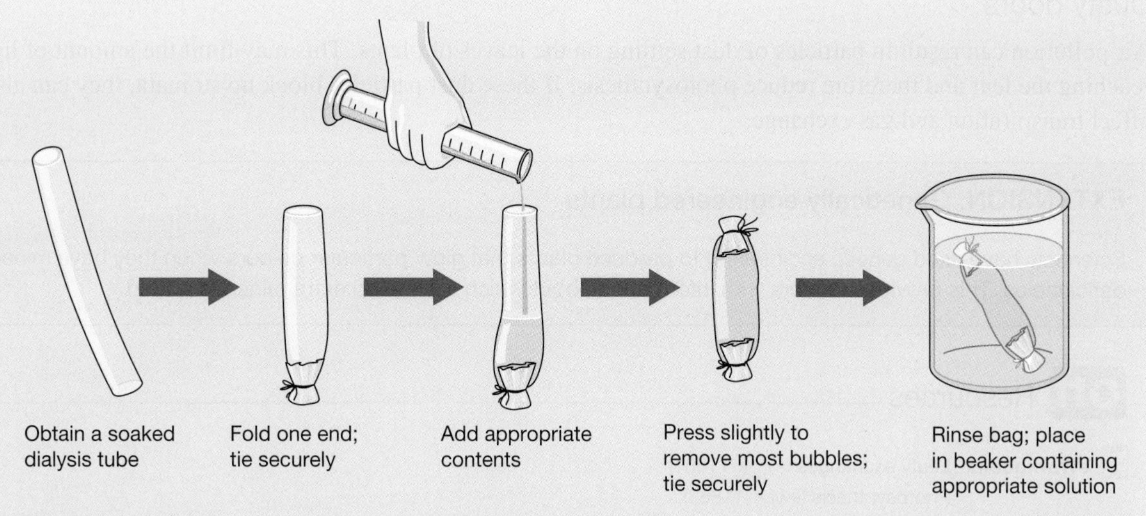

6. Put Bag A in a beaker of starch solution. Add enough iodine to the starch solution to produce a dark blue colour.
7. Put Bag B in a beaker of water.

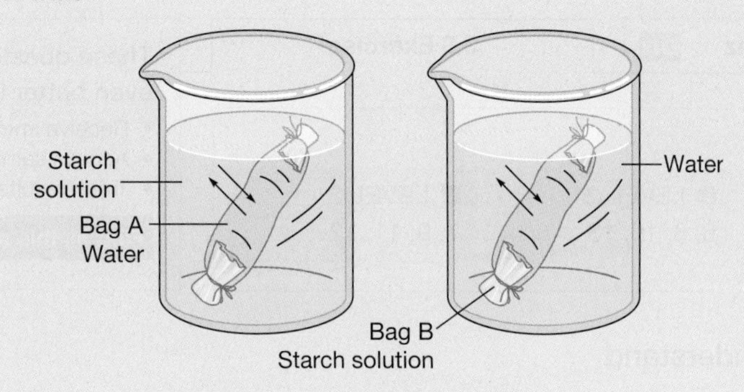

8. Leave the two bags undisturbed for at least two hours (or overnight).
9. Weigh the bags again. Record in the table in the results section.
10. Draw Bags A and B in the beakers they were left in. On your diagram, label where blue and yellow colour can be seen.

Results

Copy and complete a table such as the one shown. Give your table an appropriate title.

Weight	Bag A (bag of water in starch solution)	Bag B (bag of starch in water solution)
Before		
After		

Discussion

1. In this experiment, we made a model of a cell. Which part represented the cell membrane?
2. What happens to iodine when it is added to starch solution?
3. Dialysis tubing allows some substances, but not others, to pass through. Which of the following substances could pass through the dialysis tubing and which could not? What evidence supports this?
   a. Starch b. Water c. Iodine
4. Did the masses of the two bags change? What caused the change or lack of change?
5. When water moves in or out of cells by osmosis, it moves in the direction that balances the concentrations of substances inside and outside the cell. Use this information to explain why the masses of the bags changed.
6. Identify the strengths and limitations of this investigation and suggest possible improvements.

Conclusion

Summarise the findings for this investigation.

Dusty doors

Air pollution can result in particles of dust settling on the leaves of plants. This may limit the amount of light reaching the leaf and therefore reduce photosynthesis. If these dust particles block up stomata, they can also affect transpiration and gas exchange.

EXTENSION: Genetically engineered plants

Scientists have used genetic engineering to produce plants that glow particular colours when they have mineral deficiencies. This provides farmers with information about which soils need extra minerals added.

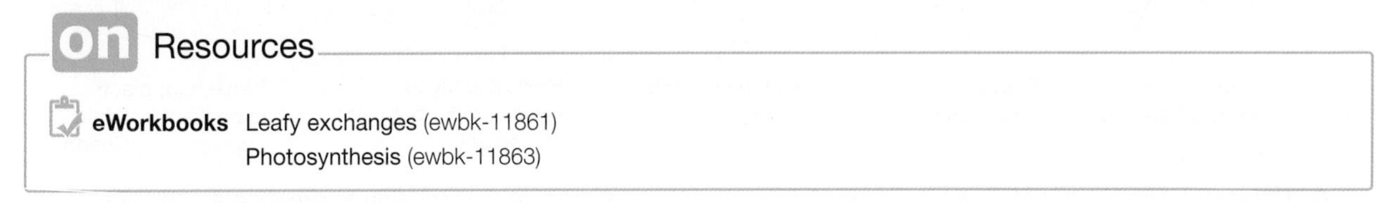

on Resources

eWorkbooks Leafy exchanges (ewbk-11861)
Photosynthesis (ewbk-11863)

3.8 Activities

learnon

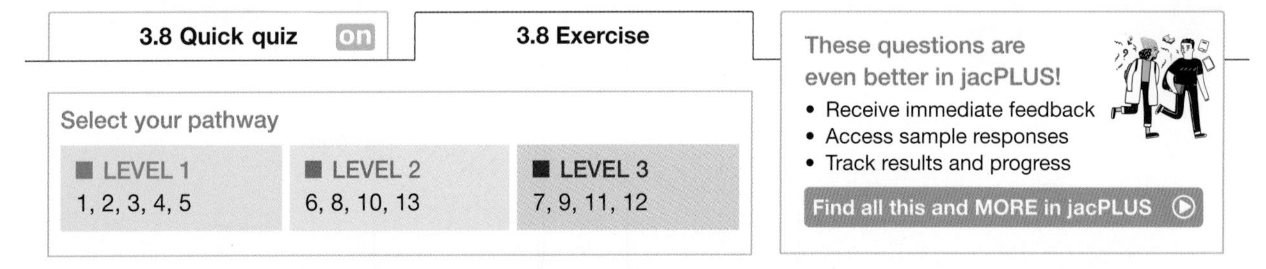

3.8 Quick quiz on | **3.8 Exercise**

Select your pathway

LEVEL 1	LEVEL 2	LEVEL 3
1, 2, 3, 4, 5	6, 8, 10, 13	7, 9, 11, 12

These questions are even better in jacPLUS!
- Receive immediate feedback
- Access sample responses
- Track results and progress

Find all this and MORE in jacPLUS

Remember and understand

1. State the name used for the tubes that carry sugar solution around a plant.
2. Describe the difference between:
   a. phloem and xylem vessels
   b. sugar and water transport in plants.
3. In what ways are the vascular bundles important to plants?
4. State two things that may happen to water in a plant.
5. On which part of the plant are stomata usually found? Can you suggest why?
6. What helps 'pull' water up a plant?

Apply and analyse

7. Describe how the guard cells assist the plant in controlling water loss.
8. Describe the difference between flaccid cells and turgid cells.
9. Copy and complete the table given.

TABLE A comparison of xylem and phloem

Tissue	What it carries	Direction of movement	Name of cells that form tubes	Are cells that form tubes living?
Xylem				
Phloem				

Evaluate and create

10. Carefully examine the reaction for photosynthesis as shown.

$$\text{Carbon dioxide + water} \xrightarrow[\text{chlorophyll}]{\text{light energy}} \text{glucose + oxygen (+ water)}$$

   a. Suggest why water and carbon dioxide are so important to plants.
   b. Suggest why guard cells are important to plants.
   c. Predict consequences for a plant if the guard cells close the stomata for long periods of time.

11. **SIS** Grasses and trees have a very different arrangement of their vascular bundles.

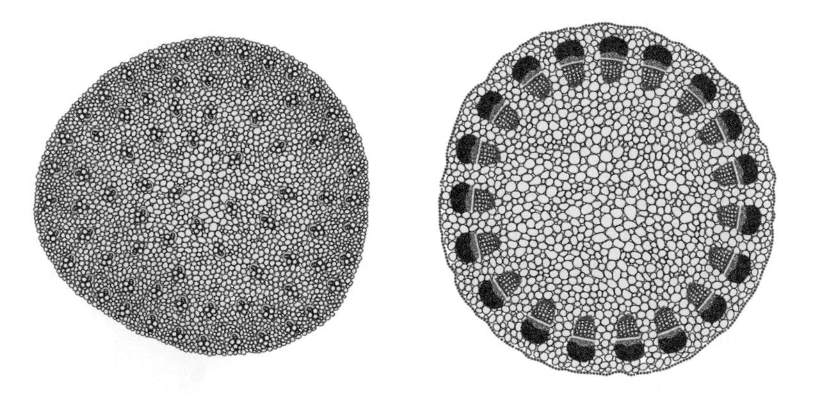

The image on the left is the organisation in a grass stem (random) and the image on the right is that of a stem that eventually becomes a tree (organised in a ring).

MC Choose the feature of trees that you think this specialised arrangement leads to.

A. Roots B. Leaves C. Wood D. Flowers

12. **SIS** This diagram demonstrates transpiration, the loss of water from the leaves of a plant.

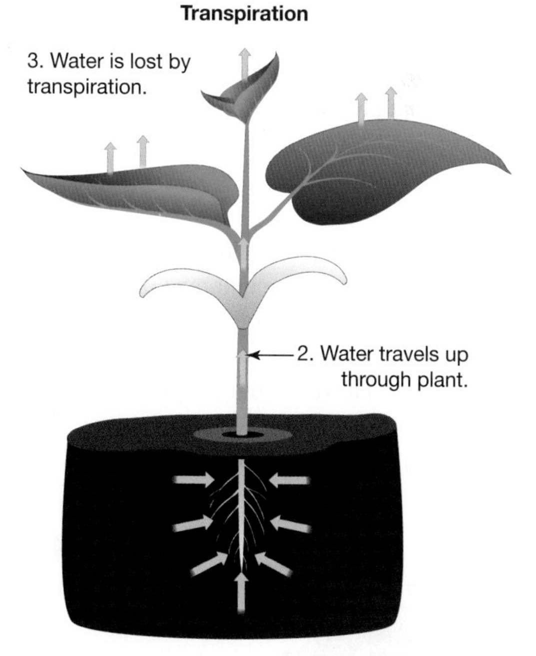

   a. Hypothesise what you would expect the rate of transpiration to do in the following scenarios:
      i. Wind speed increases
      ii. Temperature increases
      iii. Humidity increases.
   b. When a plant experiences water stress because there is not enough water, the guard cells lose water and become flaccid, thereby closing the stomata.
      i. Draw a diagram of this process.
      ii. Hypothesise what you would expect to happen to photosynthetic rate.
      iii. What dependent variable could you measure to test your hypothesis from part **b ii**?
      iv. Design an experiment to measure the amount of water lost through the leaves of a plant. Remember to include the following:
         - What will change?
         - What will stay the same?
         - What will be measured?
         - How will the test be repeatable?

13. Write a story about a group of water molecules that travels from the soil, through a plant and then into the atmosphere as water vapour.

Fully worked solutions and sample responses are available in your digital formats.

LESSON
3.9 Understanding cells to help humans

LEARNING INTENTION

At the end of this lesson you will be able to outline how understanding the way in which cells work can lead to innovations and treatments for diseases.

3.9.1 Cell reproduction

Understanding how cells work and, in particular, how they reproduce has led to a number of innovations that improve the quality of life of other organisms, particularly humans.

Bacteria are very good at reproducing by creating exact replicas of the parent cell very quickly. This can make them particularly nasty **pathogens** as they can affect a number of the cells of the body rapidly, and because they are so small they are easily transferred from one person to another.

FIGURE 3.38 Bacteria growing on an agar plate.

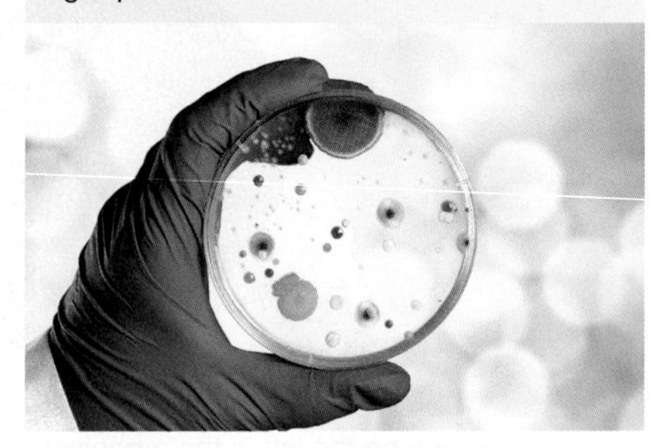

Preventing and treating disease

Scientists use their knowledge of cell division of disease-causing organisms to control or kill them. **Antibiotics** can be used to kill bacteria inside your body. **Disinfectants** can be used to kill bacteria on surfaces of non-living objects. Disinfectants should not be used on your skin as they can damage your cells. **Antiseptics** can be used on your skin. Antiseptics that kill bacteria are referred to as **bactericidal**, and those that stop bacteria from growing or dividing (but do not kill them) are called **bacteriostatic**.

pathogen a microorganism such as bacteria or a virus that can cause disease

antibiotic a substance derived from a microorganism and used to kill bacteria in the body

disinfectant a chemical used to kill bacteria on surfaces and non-living objects

antiseptic a mild disinfectant used on body tissue to kill microbes

bactericidal describes an antiseptic that kills bacteria

bacteriostatic describes an antiseptic that stops bacteria from growing or dividing but doesn't kill them

Soap

Have you ever wondered why soap is so effective against disease? It is because soap can break apart the cell membrane of the bacteria or virus (if there are enough soap molecules) as seen in figure 3.39. It also helps separate the cell from your skin so it can be washed away. Since all cells have a cell membrane, targeting this area makes it very effective. However, our skin cells are not harmed when we use soap.

FIGURE 3.39 Soap is a cheap and effective way to stop the spread of bacteria and viruses.

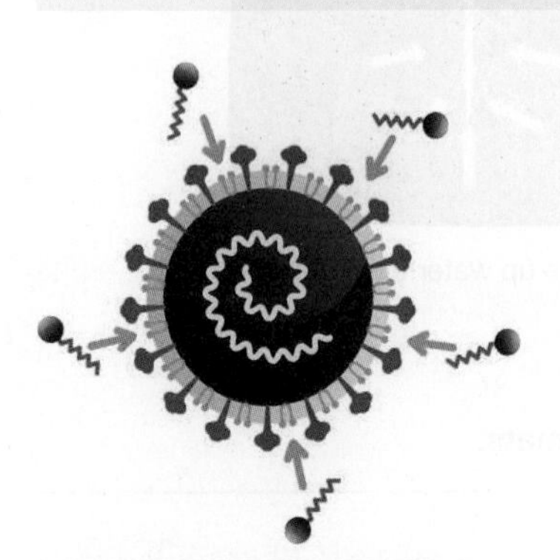

Soap bonds to the membrane of the bacteria or virus

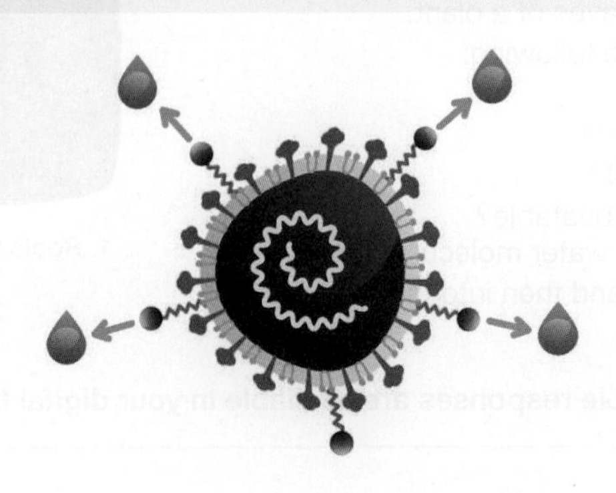

The soap molecules bond to water

The cell membrane breaks apart if there is enough soap molecules

Using bacteria to make human proteins

The rapid reproduction of these simple cells has also led scientists to use them to make human hormones, which are then produced very quickly in large quantities. By inserting a gene into the bacterial DNA we have been able to get bacteria to create human insulin, a hormone used to keep blood sugar levels under control in diabetics.

Treating burns

In order for a person to grow and repair damaged cells, body cells must replicate. They do this in much the same way as bacteria, by producing genetic replicas of each cell type. This information was used to create 'spray-on skin', which was created by Dr Fiona Wood (Australian of the Year 2005) to treat burns. Skin cells harvested from the patient are grown in a culture (replicated in a test tube) and then sprayed onto the patient's skin at the burn site to reduce scarring.

Treating cancer

Cancer is a disease that is caused when replication of cells is out of control — in other words, when too many cells are produced. Understanding the mechanism through which cells replicate has allowed treatments for cancer to be designed that target specific cells.

elog-2145

INVESTIGATION 3.11

Where are those germs?

Aim

To observe a variety of microorganisms from your local environment

Materials

- sterile cotton buds
- sticky tape
- nutrient agar plates in Petri dishes (3 per group)
- sterile Pasteur pipette
- marker pen

CAUTION

Agar plates should not be opened after incubation.

Method

1. Swipe a sterile cotton bud across a surface of your choice (such as a canteen counter, computer keyboard, phone mouthpiece or bin lid).
2. Swipe the cotton bud across the surface of the agar. Be careful not to push down too hard. The cotton bud should not leave a mark on the agar.
3. Use sticky tape to seal the plate around the edge.
4. Use a marker pen to write your group's name and where you collected the sample from.
5. Use a different cotton bud to swipe a part of your body (such as the inside of your nose, your teeth, inside your ear or your scalp).
6. Swipe the cotton bud on the surface of the second agar plate, then seal and label it as before.
7. Use the sterile Pasteur pipette to collect about 1 mL of water from a location of your choice (such as a fish tank, puddle, local creek, school swimming pool or drain pipe).
8. Pour the sample of water over the surface of the agar and swish it around. Seal and label the agar as before.
9. Incubate the three plates upside down at 30 °C for 48 hours. Remove the plates from the incubator and observe the colonies of bacteria through the lid of the Petri dishes. (Do not open the Petri dishes.)

Results

Draw a diagram of each Petri dish showing the location and size of the colonies.

Discussion

1. Colonies of bacteria tend to be smooth, whereas colonies of fungi appear furry and are often larger. Do you have colonies of bacteria or fungi or both on your plates?
2. Look at the other groups' plates.
   a. Which of the surfaces tested by your class had the most microbes? How can you tell?
   b. Which body part tested had the most microbes?
   c. Which of the water samples tested contained the most microbes?
3. Explain why it would be dangerous to unseal the agar plates and lift the lid to look at the colonies of microbes.
4. Design an experiment to test whether antibacterial surface spray really does kill bacteria.

Conclusion

Summarise the findings for this investigation.

ACTIVITY: Investigating bacteria

- Design an experiment to test whether antibacterial surface spray really does kill bacteria. Justify your strategy.
- Research examples of genetic engineering in which bacteria have foreign DNA inserted into them so that they produce human proteins. Communicate your findings as a newspaper article. What beliefs and values do you hold about this type of engineering? How is this different to what others think?

3.9 Activities

learnon

3.9 Quick quiz on | **3.9 Exercise**

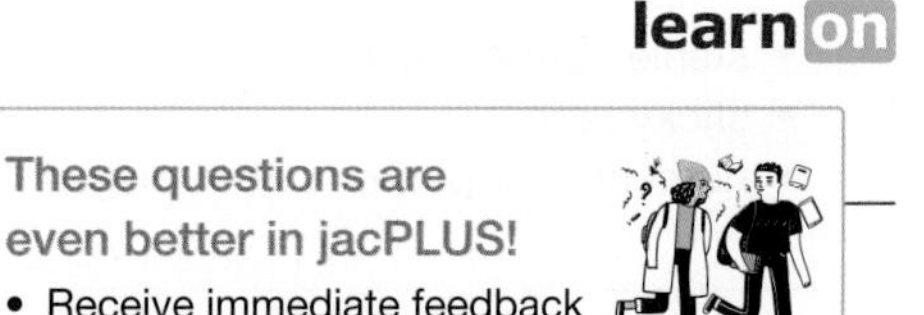

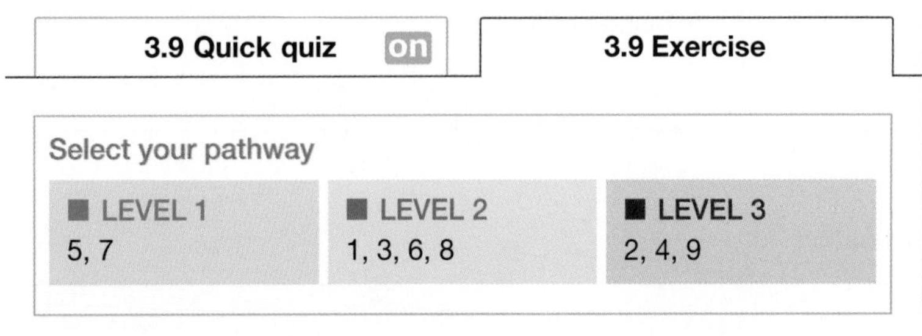

Remember and understand

1. Suggest a way that scientists can apply their knowledge of cell reproduction to benefit humans.
2. Outline the differences between disinfectants, antiseptics and antibiotics.

Apply and analyse

3. SIS Charlotte wanted to find out if antibacterial soap really works. She prepared two agar plates. She swiped her fingers over the surface of plate A. She then washed her hands with antiseptic soap and swiped her fingers over the surface of plate B. She incubated both plates. Her results are shown here.

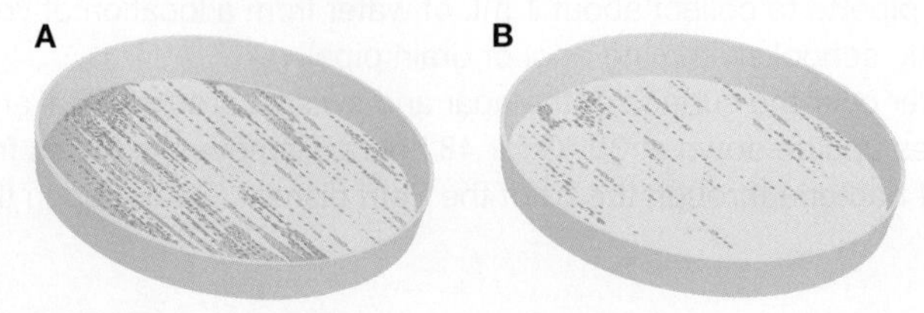

   a. Write a conclusion for Charlotte's experiment.
   b. Which plate was the control?
   c. What were the independent and dependent variables in this experiment?
   d. Which variables need to be controlled in this experiment so that it is a fair test?

4. *Clostridium perfringens* is one of the fastest growing bacteria, having an optimum generation time of about 10 minutes, which means the population of bacteria doubles every 10 minutes under optimal conditions.
   a. If you started with 200 bacteria, plot on a graph how many bacteria there would be over a one-hour period.
   b. Find out more about the structure and reproduction of this bacterium.
   c. Find out why it is sometimes referred to as a 'flesh-eating' bacterium.
   d. Write a story that includes features of *Clostridium perfringens* as a key part of the storyline.
5. **SIS** This is a graph showing the amount of bacteria grown after being heat treated and then left to incubate on a Petri dish for 24 hours.

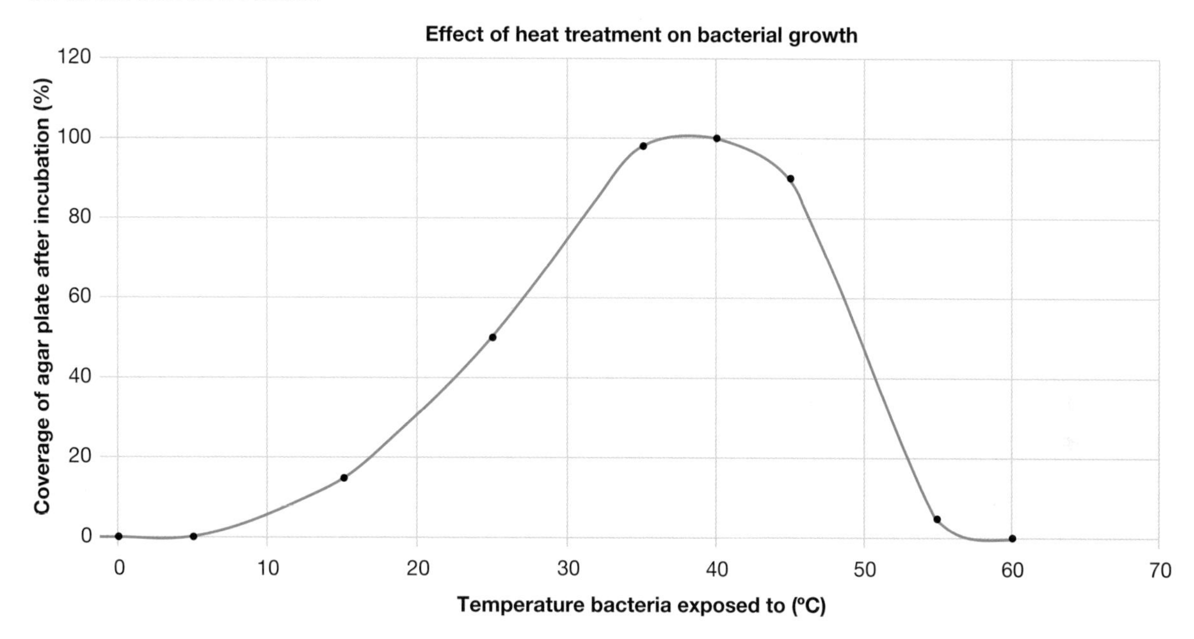

   a. What is the best temperature for bacterial growth?
   b. Suggest a reason food is cooked to reduce the likelihood of being affected by this bacterium.
6. Before mitosis begins, the DNA in the cell is replicated. Suggest why this replication step needs to occur.

Evaluate and create

7. *Entamoeba histolytica* is a unicellular organism that is a cause of diarrhoea among travellers to developing countries.
   a. Find out more about the disease that this organism causes, its life cycle and what you can do to avoid being infected by it.
   b. Prepare a brochure, poster or PowerPoint presentation that could be used to inform travellers.
8. **SIS** Investigate cell division in amoeba, *Euglena* or *Paramecium* and create an animation to show how they reproduce.
9. **a.** Find out why *Escherichia coli* (*E. coli*) counts at beaches are often reported in newspapers.
   b. How is the concentration of *E. coli* measured?
   c. Find out more about the structure and reproduction of *E. coli*.
   d. Create a model of this organism.

Fully worked solutions and sample responses are available in your digital formats.

LESSON
3.10 Thinking tools — Target maps and single bubble maps

3.10.1 Tell me

What is a target map?

A target map is used to help identify (target) what is part of (relevant to) the topic and what is not part of the topic. They are sometimes called circle maps. Target maps are similar to single bubble maps, in that they both identify and describe the range of the content. However, single bubble maps do not identify the non-relevant material.

For example, you might use a target map to show:
- the characteristics of cells
- the relevant features of an animal cell
- the relevant features of a plant cell.

Why use a target map instead of a single bubble map?

Target maps are a good tool to use for distinguishing what is relevant in a topic, and what is not relevant.

FIGURE 3.40 Target map

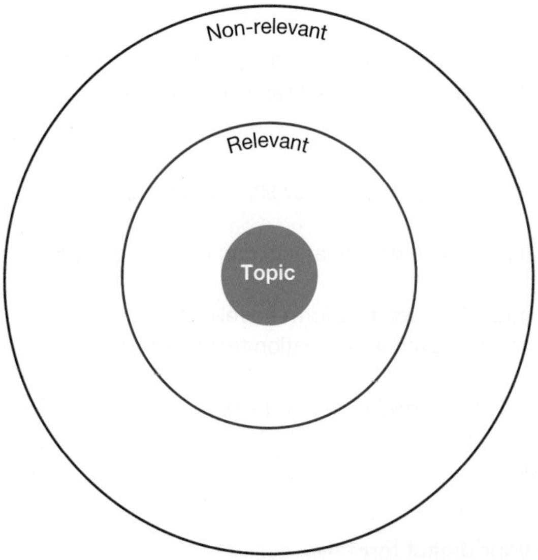

Non-relevant
Relevant
Topic

FIGURE 3.41 Single bubble map

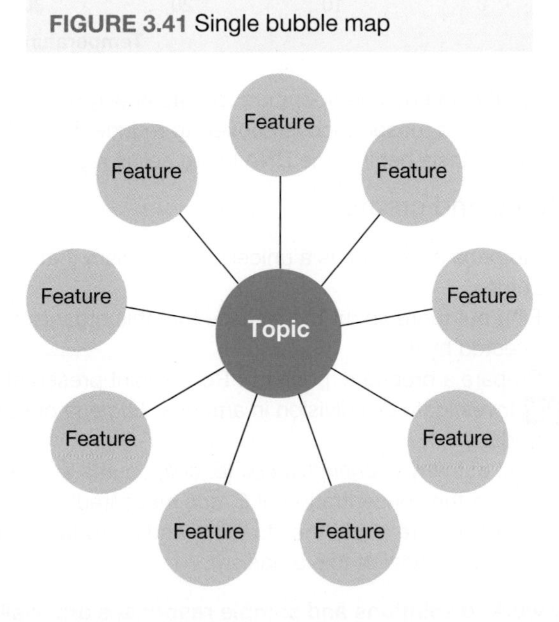

3.10.2 Show me

To create a target map:
1. Draw three concentric circles on a sheet of paper.
2. Write the topic in the centre circle.
3. In the next circle, write words and phrases that are relevant to the topic.
4. In the outer circle, write words and phrases that are not relevant to the topic.

The example in figure 3.40 shows a target map of animal cells.

FIGURE 3.42 Target map of animal and plant cells

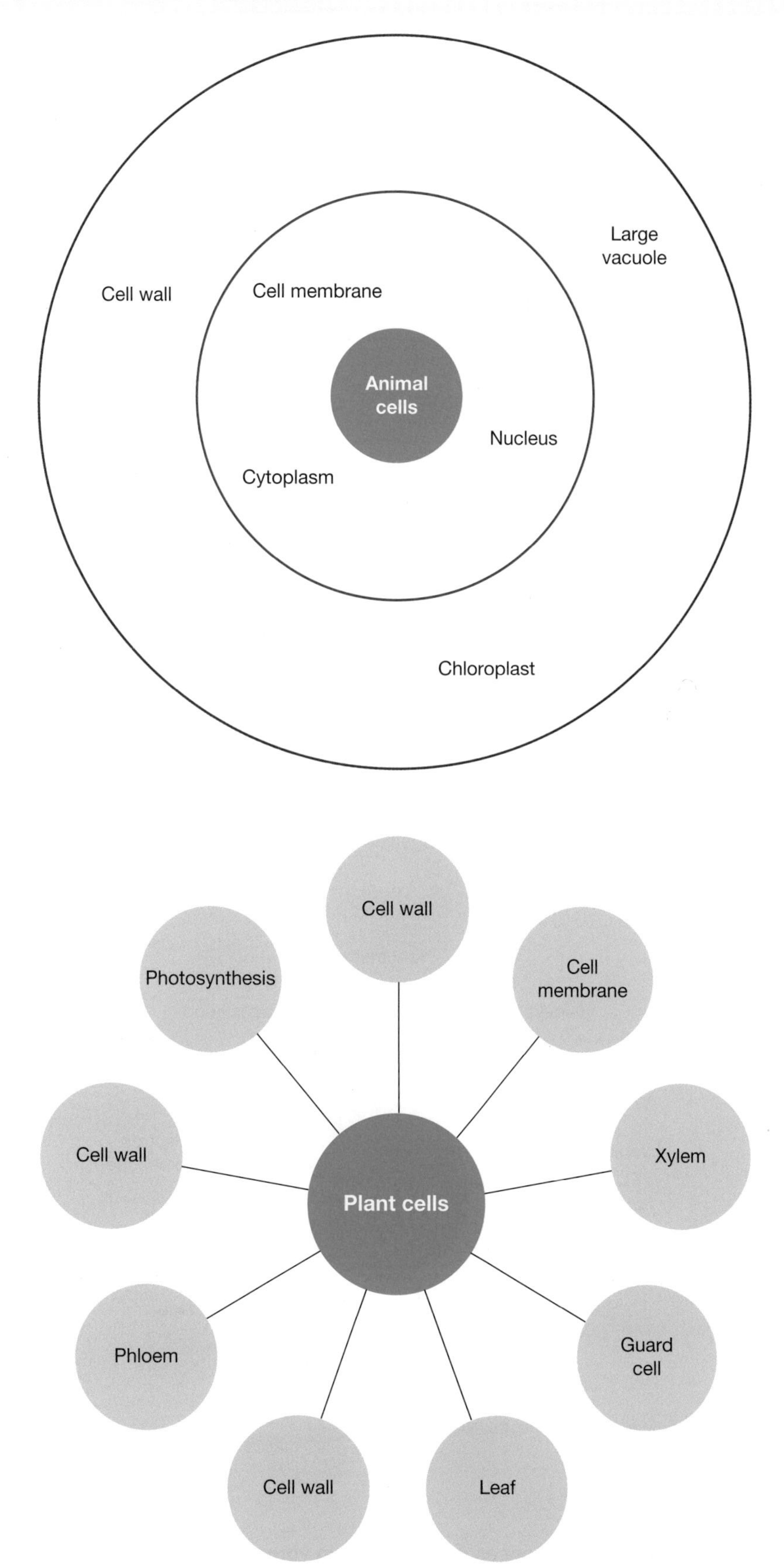

3.10.3 Let me do it

3.10 Activity

These questions are even better in jacPLUS!

- Receive immediate feedback
- Access sample responses
- Track results and progress

Find all this and MORE in jacPLUS 

1. Use the target map to answer the following questions.
   a. List the content that is relevant to animal cells.
   b. List the content that is not relevant to animal cells.
   c. Using the words in the target map, construct a target map that is relevant to plant cells.
   d. Identify which words are relevant to both plant cells and animal cells.
   e. Suggest why plant and animal cells both have these features in common.

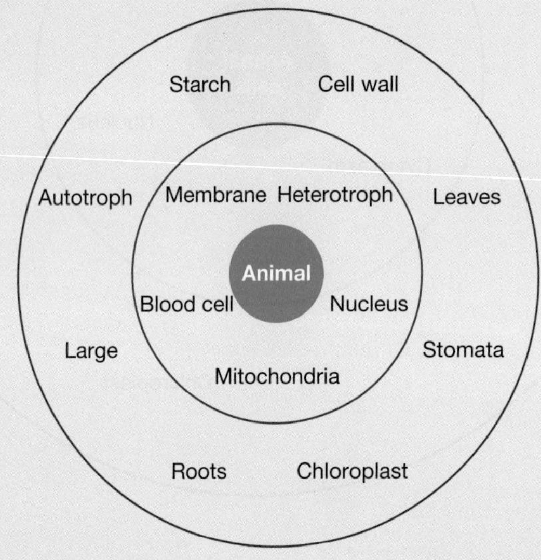

2. Use the information in the single bubble map to construct a target map of the parts of a light microscope.

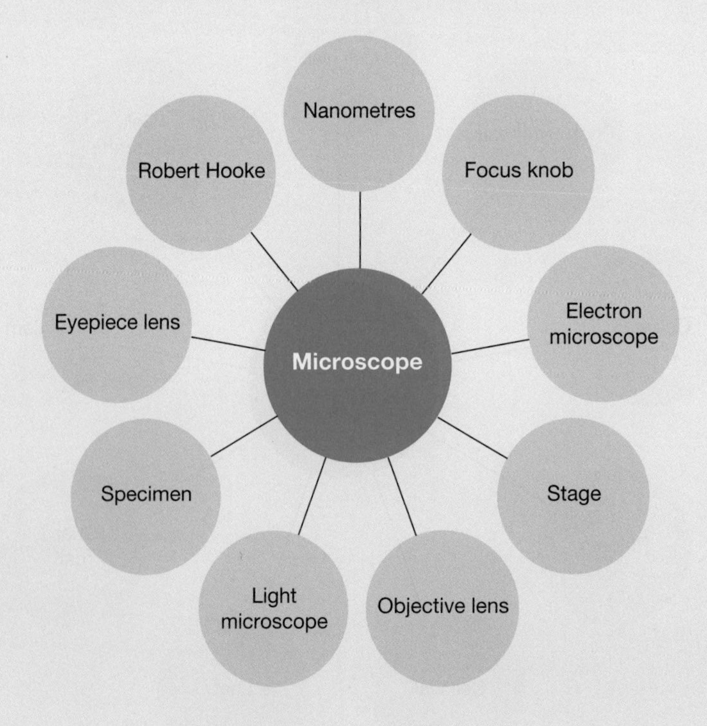

Fully worked solutions and sample responses are available in your digital formats.

LESSON
3.11 Review

Hey students! Now that it's time to revise this topic, go online to:

Review your results

Watch teacher-led videos

Practise questions with immediate feedback

Find all this and MORE in jacPLUS

Access your topic review eWorkbooks

Topic review Level 1	Topic review Level 2	Topic review Level 3
ewbk-11865	ewbk-11867	ewbk-11869

3.11.1 Summary

A whole new world

- The work of many scientists leads to the creation of scientific theories such as the cell theory.
- From the initial microscope developed by Robert Hooke in the seventeenth century, microscopes have improved greatly and allow for the observation of smaller scales such as microns and nanometres.

Focusing on a small world

- Two main types of microscopes (light and electron microscopes) are used for different purposes. They differ in terms of their illumination source, magnification, use of living or dead cells and expense.
- Light microscopes are commonly used in school science laboratories and include a stage to place a slide (with a specimen to be viewed), an eyepiece lens, objective lenses, an in-built light source and focus knobs.
- Magnification is determined by multiplying the magnification of the eyepiece lens with the magnification of the objective lens.

Form and function — cell make-up

- All organisms are made of cells; some are made of only one cell (unicellular) and others are made of many cells (multicellular).
- Prokaryotes, such as bacteria, are organisms that are made up of cells that do not contain membrane-bound organelles such as a nucleus.
- Eukaryotes, such as plants, animals, fungi and protists, are made up of cells that contain membrane-bound organelles such as a nucleus.
- Nearly all cells possess a cytoplasm (where hundreds of chemical reactions occur), cell membrane and ribosomes.
- Living things can be divided into six main kingdoms on the basis of cellular differences — Animalia, Plantae, Fungi, Protista, Eubacteria and Archaebacteria.
- Cells have different structures within them called organelles; their different structures may be due to their different functions.
- Three examples of organelles are:
  - the nucleus, which is the control centre of the cell
  - the mitochondria, which are involved in cellular respiration and the conversion of energy into a form that the cells can use
  - chloroplasts, which are involved in photosynthesis and the conversion of light energy into chemical energy.

Zooming in on life

- When preparing specimens for viewing under a microscope, dyes allow for different structures to be more visible.
- When sketching microscope specimens, it is important to include the magnification, title and clear labels, and use pencil to sketch the image. The image should be one third to one half of a page in size.

Focus on animal cells

- There are various cell types in animals that differ in shape, size and function.
- Some examples of different animal cells include muscle cells, nerve cells, red blood cells and skin cells.

Focus on plant cells

- All plant cells contain a cell wall and a large central vacuole filled with cell sap.
- Some examples of different plant cells include guard cells (on the surface of leaves), root hair cells and epidermal cells.
- Plant cells contain chloroplasts that contain chlorophyll for photosynthesis.

Plant cells — holding, carrying and guarding

- Phloem are tissues that transport sugar through a plant through translocation.
- Xylem are tissues that transport water up through plants (from the roots) through a process called transpiration.
- Stomata are tiny pores on the surface of leaves where gas is moved and water is lost. They are surrounded by guard cells, which can allow the stomata to open and close.

Understanding cells to help humans

- Cell division allows cells to divide and produce new cells.

3.11.2 Key terms

Animalia the kingdom of organisms that have cells with a membrane-bound nucleus, but no cell wall, large vacuole or chloroplasts (e.g. animals)
antibiotic a substance derived from a microorganism and used to kill bacteria in the body
antiseptic a mild disinfectant used on body tissue to kill microbes
bactericidal describes an antiseptic that kills bacteria
bacteriostatic describes an antiseptic that stops bacteria from growing or dividing but doesn't kill them
binocular microscope a microscope with two eyepieces through which the specimen is seen using both eyes
cell the smallest unit of life; cells are the building blocks of living things and can be many different shapes and sizes
cell membrane the structure that encloses the contents of a cell and allows the movement of some materials in and out
cell theory the theory that states that all living things are made up of cells and that all cells come from pre-existing cells
cellular respiration a series of chemical reactions in which the chemical energy in molecules such as glucose is transferred into ATP molecules, which is a form of energy that the cells can use
chlorophyll the green-coloured chemical in plants, located in chloroplasts, that absorbs light energy so that it can be used in the process of photosynthesis
chloroplasts oval-shaped organelles that are involved in the process of photosynthesis, which results in the conversion of light energy into chemical energy
cytoplasm the jelly-like material inside a cell; it contains many organelles, such as the nucleus and vacuoles
cytosol the fluid found inside cells
disinfectant a chemical used to kill bacteria on surfaces and non-living objects
electron microscope an instrument used for viewing very small objects; an electron microscope is much more powerful than a light microscope and can magnify things up to a million times
eukaryote any cell or organism with a membrane-bound nucleus (e.g. plants, animals, fungi and protists)
flaccid refers to cells that are not firm due to loss of water

Fungi the kingdom of organisms made up of cells that possess a membrane-bound nucleus and cell wall, but no chloroplasts (e.g. mushrooms); some fungi can help to decompose dead and decaying matter
guard cells cells on either side of a stoma that work together to control the opening and closing of the stoma
light microscope an instrument used for viewing very small objects; a light microscope can magnify things up to 1500 times
lignin a hard substance in the walls of dead xylem cells that make up the tubes carrying water up plant stems; lignin forms up to 30 per cent of the wood of trees
magnification the number of times the image of an object has been enlarged using a lens or lens system; for example, a magnification of two means the object has been enlarged to twice its actual size
metabolism the chemical reactions occurring within an organism that enable the organism to use energy and grow and repair cells
micrometre one millionth of a metre
microscope an instrument used for viewing small objects
mitochondria small, rod-shaped organelles that are involved in the process of cellular respiration, which results in the conversion of energy into a form that the cells can use
monocular microscope a microscope with a single eyepiece through which the specimen is seen using only one eye
multicellular made up of many cells
nanometre one billionth of a metre
nanotechnology a science and technology that focuses on manipulating the structure of matter at an atomic and molecular level
nucleus a roundish structure inside a cell that acts as its control centre
organelle any specialised structure in a cell that performs a specific function
pathogen a microorganism such as bacteria or a virus that can cause disease
phloem a type of tissue that transports sugars made in the leaves to other parts of a plant
photosynthesis a series of chemical reactions that occur within chloroplasts in which the light energy is converted into chemical energy; the process also requires carbon dioxide and water, and produces oxygen and sugars, which the plant can use as 'food'
Plantae the kingdom of organisms that have cells with a membrane-bound nucleus, cell wall, large vacuole and chloroplasts (e.g. plants)
prokaryote any cell or organism without a membrane-bound nucleus (e.g. bacteria)
Protista the kingdom of organisms made up of cells that possess a membrane-bound nucleus but vary in other features and do not fit into other groups (e.g. protozoans); also called Protoctista
ribosomes small structures within a cell in which proteins such as enzymes are made
stereo microscope a type of binocular microscope through which the detail of larger specimens can be observed
stomata openings mainly on the lower surface of leaves; these pores are opened and closed by guard cells; singular = stoma
tissue a group of cells of similar structure that perform a specific function
translocation the process in which sugars and amino acids are transported within a plant by phloem tissue
transpiration the loss of water from plant leaves through their stomata
transpiration stream the movement of water through a plant as a result of loss of water from the leaves
turgid refers to cells that are firm
unicellular made up of only one cell
vacuoles sacs within a cell used to store food and wastes; plant cells usually have one large vacuole, while animal cells have several small vacuoles or none at all
vascular bundles groups of xylem and phloem vessels within plant stems
wilt refers to when plant stems and leaves droop due to insufficient water in their cells
xylem vessels pipelines for the flow of water up plants, made up of the remains of dead xylem cells fitted end to end with the joining walls broken down; lignin in the cell walls gives them strength

Resources

eWorkbooks	Study checklist (ewbk-11871) Literacy builder (ewbk-11872) Crossword (ewbk-11874) Word search (ewbk-11876) Reflection (ewbk-11923)
Solutions	Topic 3 Solutions (sol-1115)
Practical investigation eLogbook	Topic 3 Practical investigation eLogbook (elog-2123)
Digital document	Key terms glossary (doc-39974)

3.11 Activities

learnon

3.11 Review questions

Select your pathway

LEVEL 1	LEVEL 2	LEVEL 3
1, 3, 4, 6, 7, 10	2, 5, 8, 11, 12	9, 13, 14, 15

These questions are even better in jacPLUS!

- Receive immediate feedback
- Access sample responses
- Track results and progress

Find all this and MORE in jacPLUS

Remember and understand

1. Copy and complete the table provided.

Cell feature	Plant cells	Animal cells	Fungal cells
Cell wall	✓	X	
Cytoplasm			
Cell membrane			
Chloroplast			
Nucleus			
Large vacuole			

2. Draw and label a typical plant cell and a typical animal cell.
3. Which of the following types of microscopes were used to take the photos shown? Give reasons for your answers.
   - Light microscope
   - Scanning electron microscope

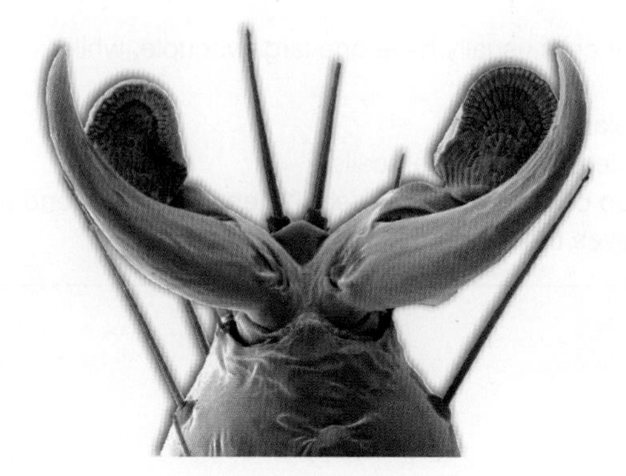

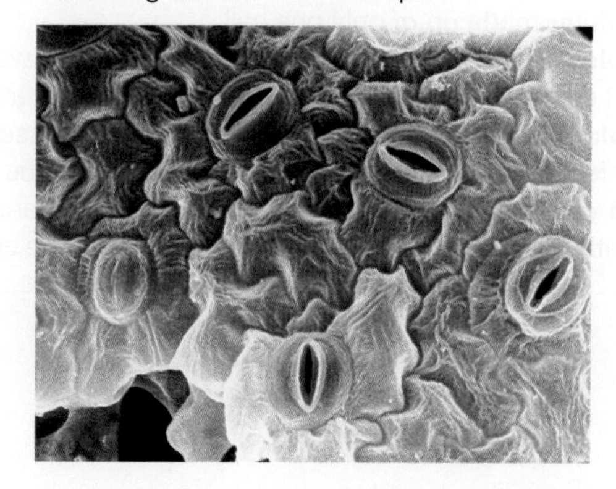

4. Make a sketch of these human cheek cells.

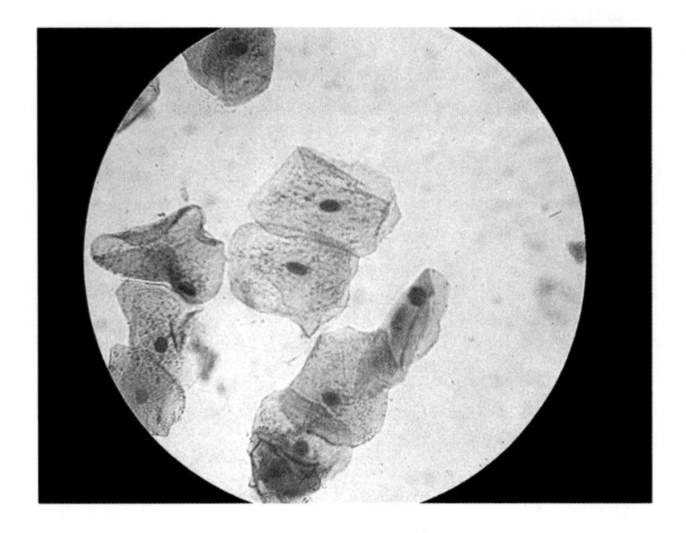

5. **a.** Match the following cell names to the diagrams provided.
   - *Euglena*
   - *Paramecium*
   - Onion epidermal cells
   - Nerve cell
   - Sperm cell
   - Guard cells
   - Root hair cell
   - Bacterium

   b. To which kingdom does each of these cells belong?

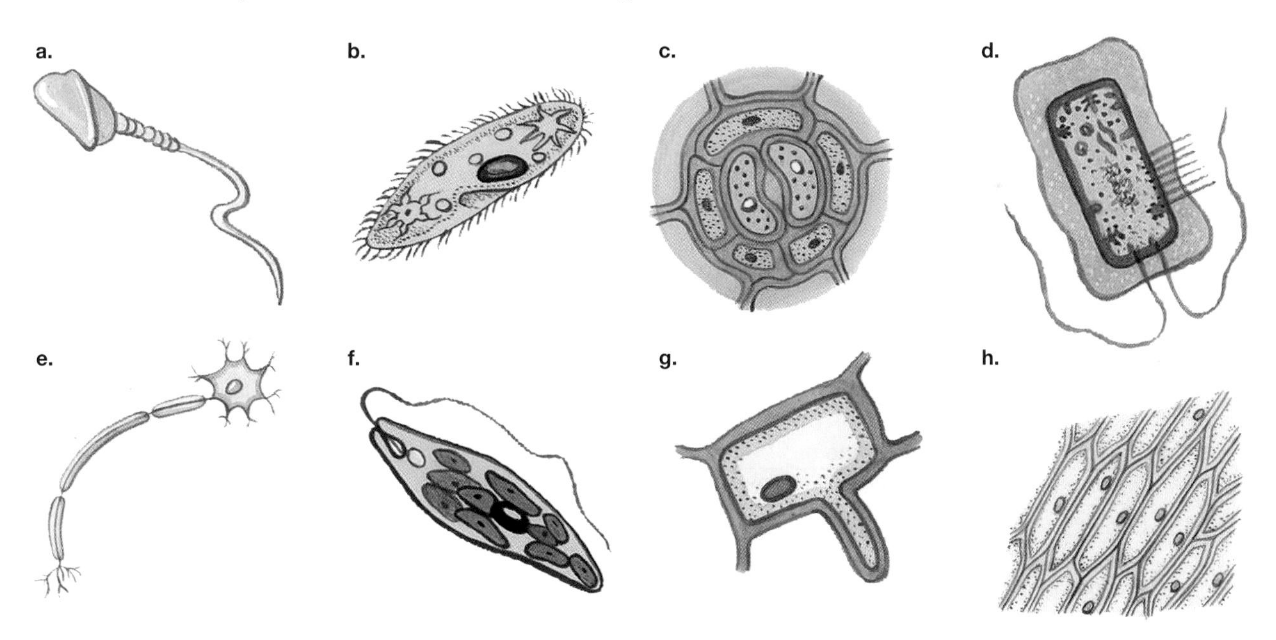

Apply and analyse

6. **a.** Brainstorm as many 'cell'-related words as you can, writing them on a piece of paper.
   b. Pair up with another class member and add any of their words that you missed. Ask your partner what these words mean if you are unsure.
   c. On a new piece of paper, work with your partner to group or link words to make a concept, cluster or mind map.
   d. Compare your map with that of another pair in the class, adding as many more bits and pieces as you can.

7. Construct a Venn diagram to show the similarities and differences between:
   a. light microscopes and electron microscopes
   b. plant, animal and bacteria cells.
8. Explain the significance of the invention of the microscope in terms of how we see the world.
9. Suggest why the invention of microscopes led to the development of new scientific language and classifications.
10. Unscramble the words using the clues provided.
    a. Control centre of the cell SEUNCLU
    b. Surrounds the cell ERAMMBNE
    c. Contains cell sap OCVAUEL
    d. Part of the cell between the cell membrane and the nucleus CATOPLMYS
    e. Building blocks of all living things LELSC
    f. Living things ASMOGNIRS
11. What's green and eats porridge? Identify the parts of the microscope shown and use the code provided to find out the answer to this riddle.
    Code:
    O = revolving nose piece; U = objective lenses;
    S = coarse focus knob; K = fine focus knob;
    D = microscope slide; L = stage slide clip;
    C = base; O = mirror; L = iris adjustment;
    I = stage; M = eyepiece lens.

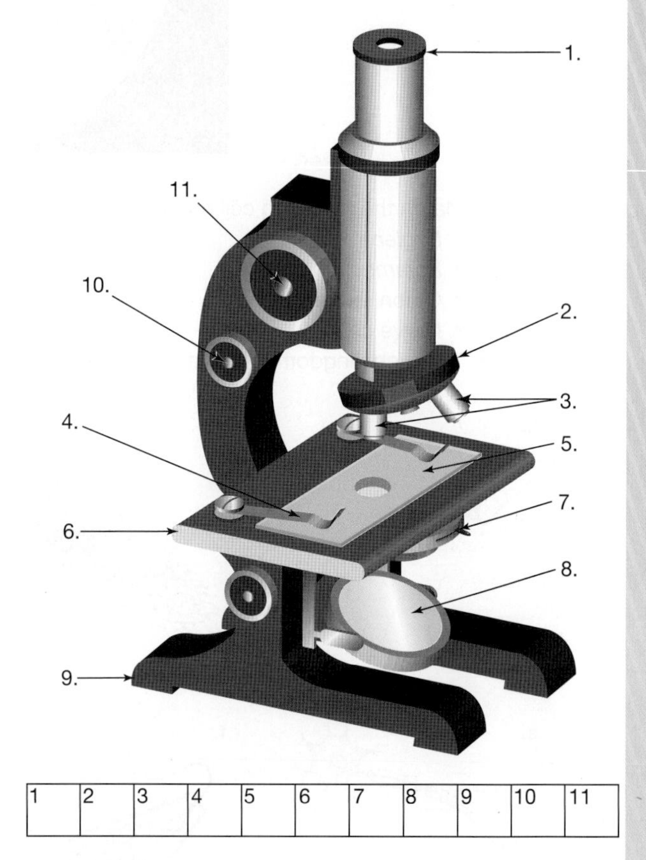

1	2	3	4	5	6	7	8	9	10	11

12. Use the terms in the box provided to construct target maps that are relevant to:
    a. plants
    b. animals
    c. fungi
    d. protists
    e. prokaryotes.

Multicellular	Chloroplast	*Euglena*
Prokaryote	Bacteria	Mushroom
Eukaryote	Fern	Yeast
Nucleus	Alga	Lizard
Cell wall	*Paramecium*	Sponge
Large vacuole	Unicellular	Moss
Xylem	Cell membrane	Blood cells
Possum	Stomata	Phloem

13. Construct a single bubble map to identify:
    a. types of plant cells
    b. types of animal cells
    c. scientists who have contributed to our knowledge of cells
    d. examples of body systems
    e. issues related to stem cells.

Evaluate and create

14. a. Why do you think that cells have been described as 'living factories'?
    b. Think of a typical plant or animal cell. Make a list of all of the different parts and organelles. If the cell was a living factory, what might be the job of each listed part?
    c. Write a play to act out what happens in cells and perform it with others in your class. What sorts of things were easy to show? What sorts of things were hard to show? If you were to rewrite the play, what might you change and why?
    d. Convert the classroom into a giant cell! Take photos and then add information to them on a poster.

15. **SIS** Students are exploring the use of biological stains on cells. They have access to some bacterial cells, some plant cells and animal cells, as well as the three biological stains listed.

TABLE Biological stains and the organelle they stain

Stain	Organelle stained
Crystal violet	Stains cell walls purple
Methylene blue	Stains nuclei blue
Eosin	Stains cell membranes pink

a. What organelle is present in a plant cell but not in an animal cell?
b. What organelle is present in a plant cell but not in a bacterial cell?

Unfortunately, the containers containing the cells are unlabelled and have been mixed up. No one can determine which is which.

c. Outline an experimental design that would allow you to determine which sample is which so that the correct labels can be applied.

The following observations were recorded.

TABLE Results obtained using different stains on three samples

Stain	Sample A	Sample B	Sample C
Crystal violet	Nothing was stained	Purple structure around outer of cell	Purple structure around outer of cell
Methylene blue	Blue organelle present in middle of cell	Nothing was stained	Blue organelle present in middle of cell
Eosin	Pink line outlining cell	Pink line outlining cell	Pink line outlining cell

d. From these results, which do you think is:
i. sample A
ii. sample B
iii. sample C?

This diagram is a drawing of one of the cells taken during the experiment. The student forgot to add the scale (although they remembered to put in a scale bar).

e. If the field of view was 45 μm in diameter, approximately how big is this cell?
f. What type of cell is it most likely to be — plant, animal or bacteria — based on its structure? Give evidence to support your conclusion.

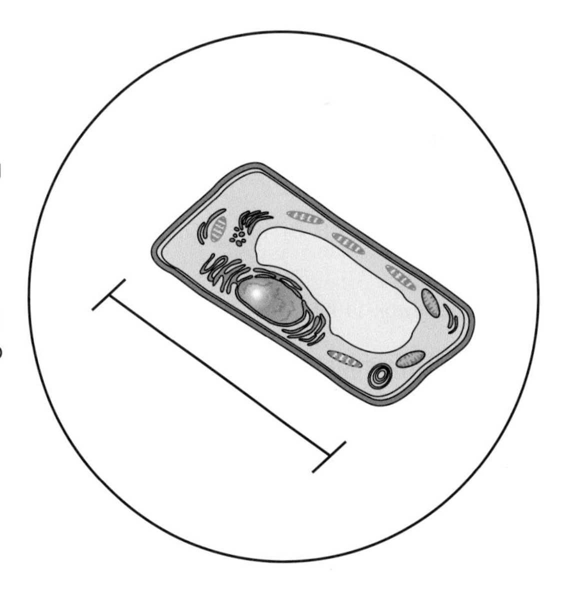

Fully worked solutions and sample responses are available in your digital formats.

Online Resources

Below is a full list of **rich resources** available online for this topic. These resources are designed to bring ideas to life, to promote deep and lasting learning and to support the different learning needs of each individual.

3.1 Overview

eWorkbooks
- Topic 3 eWorkbook (ewbk-11848)
- Starter activity (ewbk-11850)
- Student learning matrix (ewbk-11852)

Solutions
- Topic 3 Solutions (sol-1115)

Practical investigation eLogbook
- Topic 3 Practical investigation eLogbook (elog-2123)

Video eLesson
- Robert Hooke and cells (eles-1780)

Weblink
- The wacky history of cell theory

3.2 A whole new world

eWorkbook
- History of the light microscope (ewbk-11853)

Video eLesson
- Historic bacteriologists Van Leeuwenhoek, Pasteur and Koch (eles-2026)

Interactivity
- Development of microscopes and cell theory (int-3392)

3.3 Focusing on a small world

eWorkbook
- In focus (ewbk-11855)

Practical investigation eLogbooks
- Investigation 3.1: Getting in focus with an 'e' (elog-2125)
- Investigation 3.2: Can you tell the difference? (elog-2127)

Interactivities
- The monocular light microscope (int-3390)
- Field of view (int-5702)

3.4 Form and function — cell make-up

Video eLesson
- Inside cells (eles-0054)

Weblinks
- Zoom on an animal cell
- Virtual plant cell

Interactivity
- Animal and plant cells (int-3393)

3.5 Zooming in on life

eWorkbook
- Preparing a stained wet mount (ewbk-11857)

Practical investigation eLogbooks
- Investigation 3.3: Preparing a wet mount (elog-2129)
- Investigation 3.4: Preparing stained wet mounts (elog-2131)

Teacher-led video
- Investigation 3.4: Preparing stained wet mounts (tlvd-10736)

3.6 Focus on animal cells

Interactivity
- Body cells (int-3395)

Practical investigation eLogbook
- Investigation 3.5: Animal cells — what's the difference? (elog-2133)

3.7 Focus on plant cells

eWorkbook
- Plant transport highways (ewbk-11859)

Interactivity
- Plant cells (int-3396)

Practical investigation eLogbook
- Investigation 3.6: Plant cells in view (elog-2135)

3.8 Plant cells — holding, carrying and guarding

eWorkbooks

- Leafy exchanges (ewbk-11861)
- Photosynthesis (ewbk-11863)

Practical investigation eLogbooks

- Investigation 3.7: Stem transport systems (elog-2137)
- Investigation 3.8: Observing leaf epidermal cells (elog-2139)
- Investigations 3.9: Looking at chloroplasts under a microscope (elog-2141)
- Investigation 3.10: Moving in or out? (elog-2143)

3.9 Understanding cells to help humans

Practical investigation eLogbook

- Investigations 3.11: Where are those germs? (elog-2145)

3.11 Review

eWorkbooks

- Topic review Level 1 (ewbk-11865)
- Topic review Level 2 (ewbk-11867)
- Topic review Level 3 (ewbk-11869)
- Study checklist (ewbk-11871)
- Literacy builder (ewbk-11872)
- Crossword (ewbk-11874)
- Word search (ewbk-11876)
- Reflection (ewbk-11923)

Digital document

- Key terms glossary (doc-39974)

To access these online resources, log on to **www.jacplus.com.au**

4 Systems — living connections

CONTENT DESCRIPTION

Analyse the relationship between structure and function of cells, tissues and organs in a plant and an animal organ system and explain how these systems enable survival of the individual (AC9S8U02)

LESSON SEQUENCE

SCIENCE INQUIRY AND INVESTIGATIONS

Science inquiry is a central component of the Science curriculum. Investigations, supported by a **Practical investigation eLogbook** and **teacher-led videos**, are included in this topic to provide opportunities to build Science inquiry skills through undertaking investigations and communicating findings.

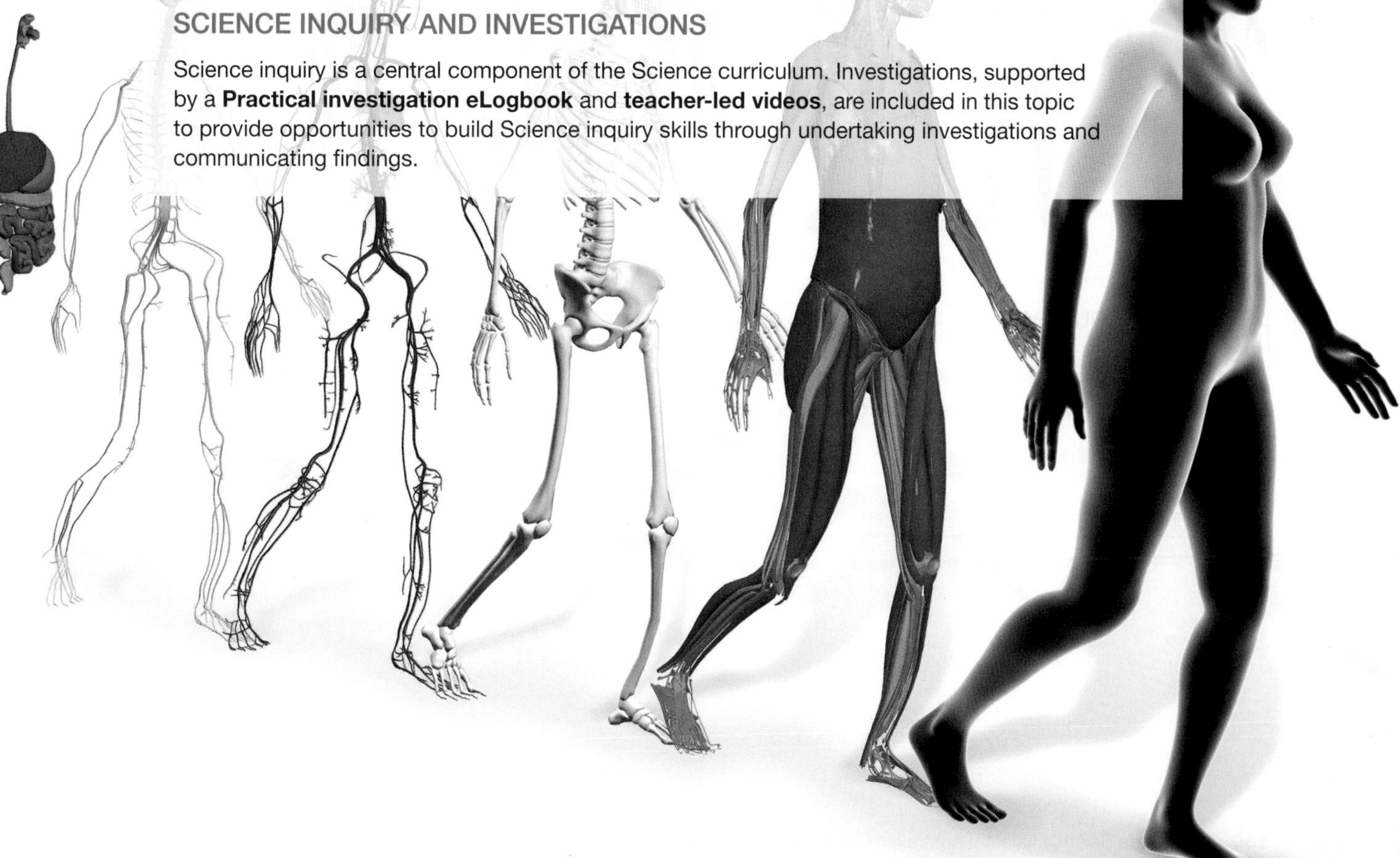

LESSON
4.1 Overview

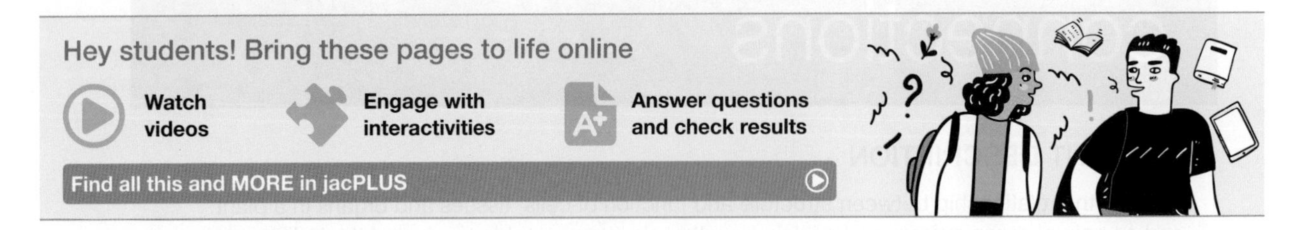

4.1.1 Introduction

Throughout history, humans have asked questions about their bodies and how they work. Our current knowledge has resulted from curiosity, imagination, and persistence. Findings have shown that your body, like that of many other multicellular organisms, is very complex. It consists of body systems that work together to keep you alive. Each of your systems is made up of organs, which are made up of tissues, which are made up of cells — which cannot survive independently of each other. Differences in the structure of the cells, tissues and organs within these body systems well suit them to their specific functions. What questions do you have about your body and how it works?

FIGURE 4.1 In organisms, like humans, many body systems work together to keep them alive.

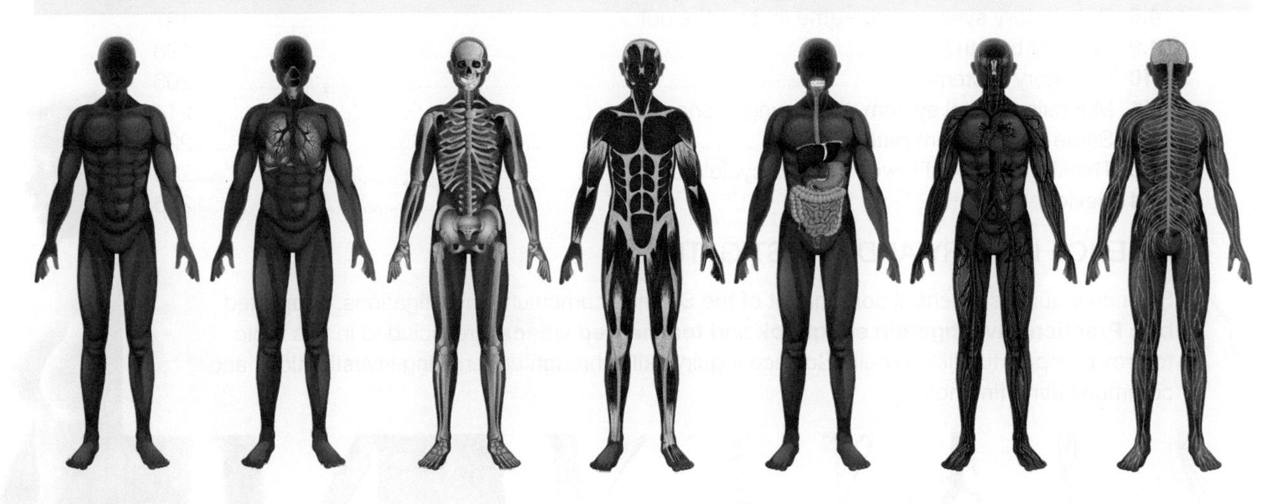

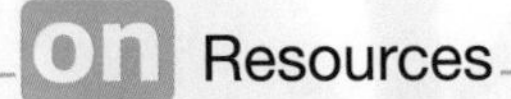

on Resources

Video eLesson The human body and internal organs (eles-2040)

This animation shows the major human body systems. How many can you recognise?

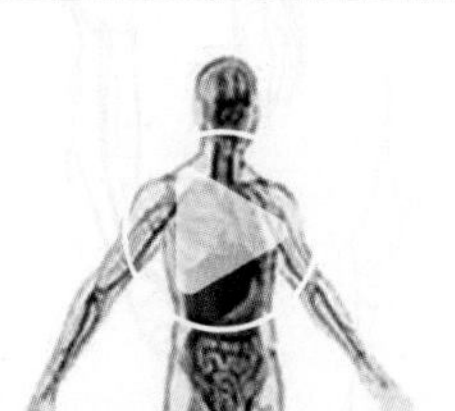

4.1.2 Think about the human body

1. Why is it unusual for herbivores to have canine teeth?
2. Which human blood group is the most common?
3. What do intestinal villi in humans have in common with root hairs in plants?
4. How can burping give you heartburn?
5. What causes asthma?
6. What is the body's largest organ?
7. Why are red blood cells red?
8. What causes the 'lub dub' sound that your heart makes?
9. What is special about cardiac muscle?
10. Why aren't all of your teeth the same shape?

4.1.3 Science inquiry

Getting below the surface

Have a look at the other students in your classroom. How different from one another are you? Which features do you all have in common? Perhaps there are differences on the outside, but inside you are made up of all of the same bits and pieces organised in the same way.

Think and create

Some of the things that you have in common with your classmates are your body systems.

1. Use a mind map to summarise all that you know about human body systems.
2. Compare your mind map with those of at least three team members.
3. Create a new team mind map that combines all your ideas and compare that with the mind map of another team. Add any comments that you think will help you learn more about human body systems.

Think and investigate

4. In your team, make a list of ten questions about human body systems. Select four questions and place these on the class noticeboard with those of other teams to make a 'class question gallery'. Arrange these questions into groups or themes.
5. Browse the class question gallery and select one question that interests you most. Research your selected question and report back to the class on your findings in an interesting and creative way.

 Resources

eWorkbooks	Topic 4 eWorkbook (ewbk-11878) Starter activity (ewbk-11880) Student learning matrix (ewbk-11882)
 Solutions	Topic 4 Solutions (sol-1116)
 Practical investigation eLogbook	Topic 4 Practical investigation eLogbook (elog-2147)

LESSON
4.2 Driven by curiosity?

LEARNING INTENTION

At the end of this lesson you will have learned that scientific knowledge and understanding of the world changes as new evidence becomes available. You will also be able to provide examples of how our knowledge and understanding of the human body has changed over time.

Science as a human endeavour

4.2.1 Intensely curious ...

Resources

 **Video eLesson** Leonardo's sketches and anatomy (eles-1769)

Leonardo da Vinci was one of the best scientific minds of his time. He was intensely curious and painstaking in his observations. He used close observation, repeated testing and precise illustrations with explanatory notes. Using pen, chalk and brush, his scientific illustrations offered visual answers to mysteries that had escaped others for centuries. His volumes of amazing notes of scientific and technical observations in his handwritten scripts led to the birth of a new systematic and descriptive method of scientific study (figure 4.2).

FIGURE 4.2 Leonardo da Vinci spent hours amid rotting corpses to draw amazingly detailed observations of body structures.

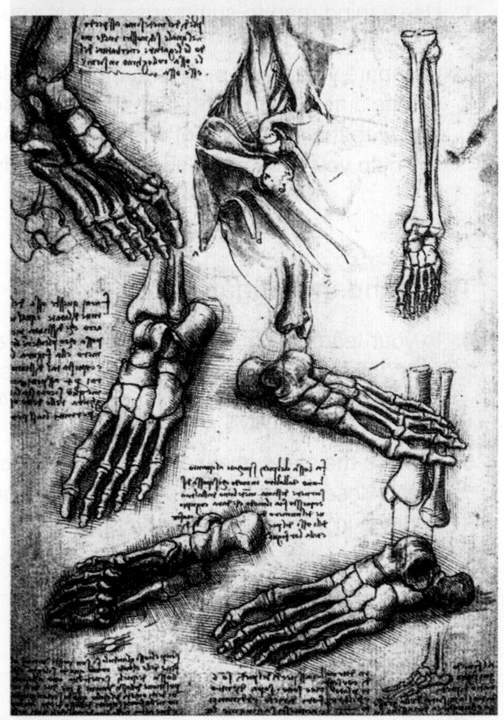

Leonardo da Vinci questioned everything. He may have been the most relentlessly curious man in history. He asked questions such as: Why do birds fly? Why can seashells be found in mountains? What is the origin of the wind and clouds? Why do people die? Where is the human soul found?

4.2.2 Dissecting, details and drawing

Leonardo's anatomical studies of human muscles and bones began around 1490. His exploration of embryology (study of the formation and development of an embryo) and cardiology (study of the heart and cardiovascular system) came later, with his astonishingly detailed image of a foetus within the womb (around 1505) providing details for obstetric surgery hundreds of years later. His observations were not just of bodies — later generations have been in awe of his sketches of inventions that were centuries ahead of their time.

... in the medical faculty he learned to dissect the cadavers of criminals under inhuman, disgusting conditions ... because he wanted [to examine and] to draw the different deflections and reflections of limbs and their dependence upon the nerves and the joints. This is why he paid attention to the forms of even very small organs, capillaries and hidden parts of the skeleton.

Paolo (the first biographer of Leonardo da Vinci), 1520

FIGURE 4.3 Is this a self-portrait of a young Leonardo da Vinci (1452–1519)? 'Hidden' under handwriting on a page of his *Codex on the Flight of Birds* for about 500 years, a combination of scientific techniques were used to 'unveil' it.

4.2.3 Challenging 'knowledge'

Knowledge of the human body was very different in Leonardo's day from what we accept today. The heart was thought to be made up of two chambers and its function to warm the blood, which was thought to be made in the liver. It was also thought that sperm were produced in the marrow of the spinal column and that the human soul may be located in the spine. Leonardo had questions he wanted to answer. He wanted to find out more. His investigations challenged the accepted knowledge of his day.

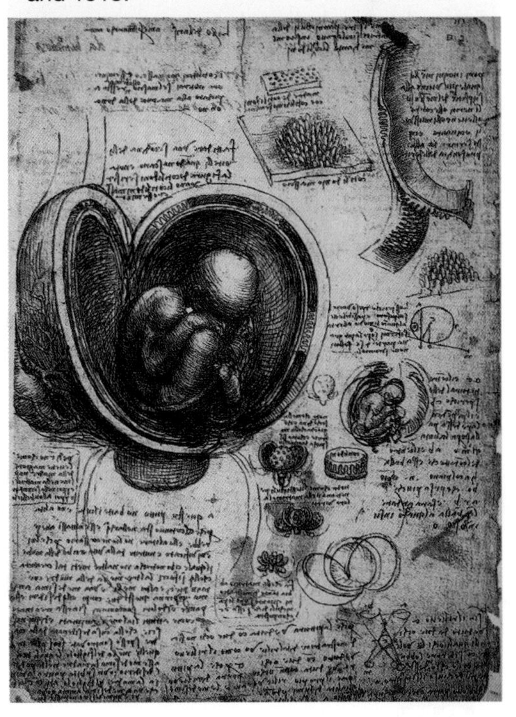

FIGURE 4.4 Leonardo's sketches of a fetus in the womb were completed between 1510 and 1513.

4.2.4 Visions and models

Leonardo also emphasised the significance of visual observations and model making — he believed that reality needed to be reconstructed before it could be represented. His models of hands and legs were used to reveal the structural relationships between different layers of arteries, muscles and bones. Leonardo also made a glass model of the heart and used water with different coloured dyes to trace its flow through the heart. His investigations linked anatomy (structure) and physiology (function).

Analogies are sometimes used to help people to connect new learning to previous knowledge. Leonardo used analogies to compare arteries in human bodies to 'underground rivers in the earth' and described the bursting of blood from a vein like 'water rushing out a burst vein of the earth'.

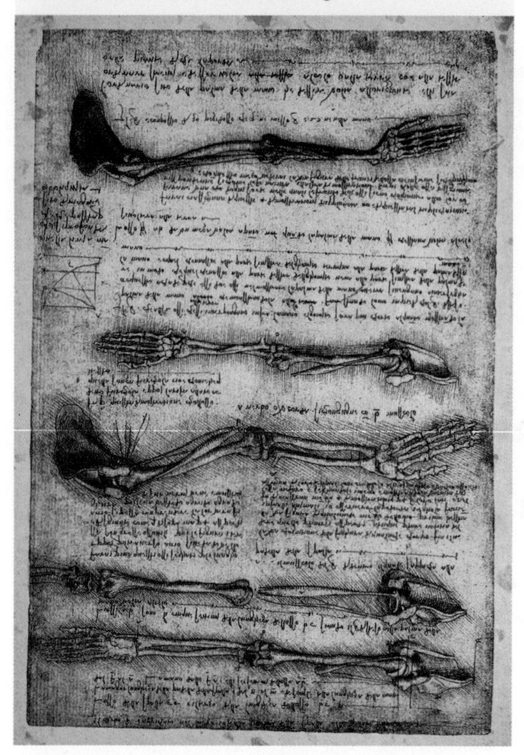

FIGURE 4.5 Leonardo drew this diagram around 1510. Can you see his secretive, reversed form of handwriting?

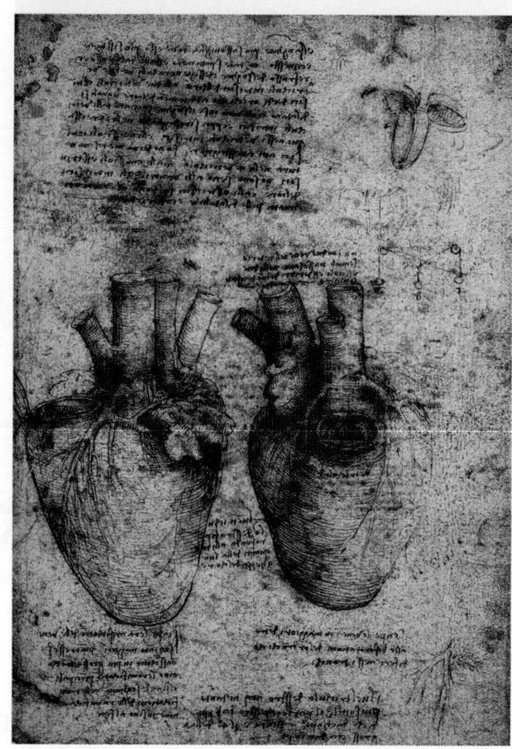

FIGURE 4.6 Leonardo da Vinci was a master of detail with his sketches of body parts.

Leonardo's dissections led to changes in the knowledge and understanding of the structure and function of the heart (figure 4.6), including that:

- the heart was a muscle
- the heart did not warm the blood
- the heart had four chambers
- left ventricle contractions were connected to the pulse in the wrist.

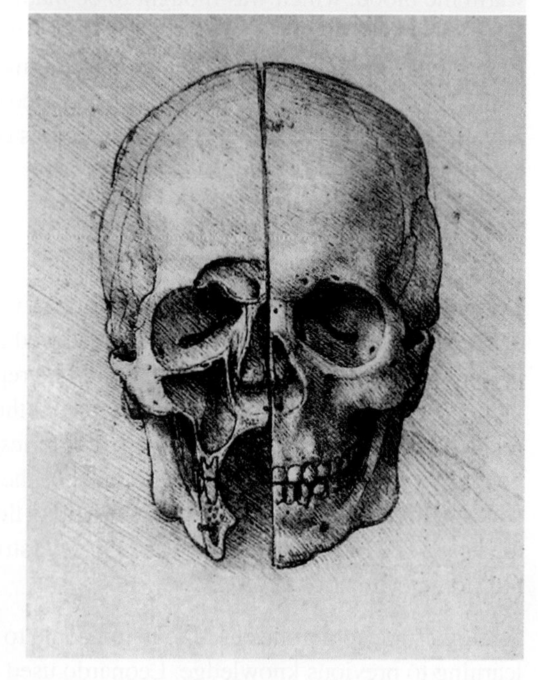

FIGURE 4.7 Leonardo sketched this skull in 1489.

To locate cavities around the brain and cranium (figure 4.7), Leonardo used innovative techniques, such as injecting molten wax into them. Although Leonardo did not find the location of the human soul, his studies led him to the discovery that the brain and spine were connected.

Leonardo's curiosity, determination, creativity and persistence did more than make an amazing contribution to our current scientific knowledge of our bodies. These features also helped mould the way in which scientific frameworks were developed to structure our investigations to explore our questions.

on Resources

Weblink Leonardo da Vinci's inventions turned into models

4.2.5 Curiosity throughout time and space

Curiosity is one of the features of humans that has contributed to our survival. Some of this curiosity has been about the structure and function of our own bodies. Evidence of this curiosity is woven throughout history and is often found in art. While Leonardo da Vinci provides one example of curiosity driving a search to find out more, he is not the only example. Nor is human curiosity limited to the place or time in which you live.

Knowledge of the internal biology and physiological processes in art appears in rock paintings in caves in Australia that are thousands of years old. Examples of First Nations Australian x-ray art (figure 4.8) provide evidence that this type of knowledge dates back more than 6000 years.

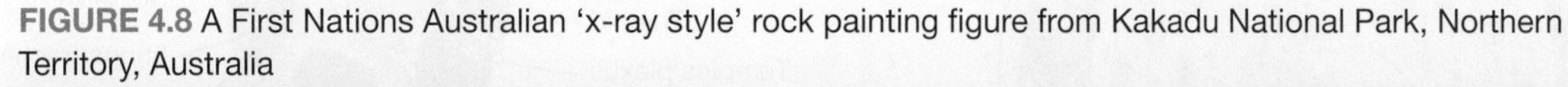
FIGURE 4.8 A First Nations Australian 'x-ray style' rock painting figure from Kakadu National Park, Northern Territory, Australia

The culture and scientific knowledge of the times often determines the types of treatment given for various diseases of the human body. In medieval times, astrology played a key role in medicine and medical prognosis. It was believed that the 'movement of the heavens' could influence human physiology, with each part of the body being associated with a different astrological sign. An image of the 'Zodiac Man' (figure 4.9) in the medical texts of the time was used to assist practitioners in their medical treatments. It was based on astrology and the position of the Moon, and provided practitioners guidance for the correct time to do certain procedures.

Chinese traditional medicine is an ancient medical system that has been practised for over 5000 years and applies understanding of the laws and patterns of nature to the human body. It views health as the changing flow throughout the body of vital energy (*qi*) that, if hindered, can lead to illness. Acupressure is an application of this theory that aims to release blocked energy by stimulating specific points along the body's energy channels.

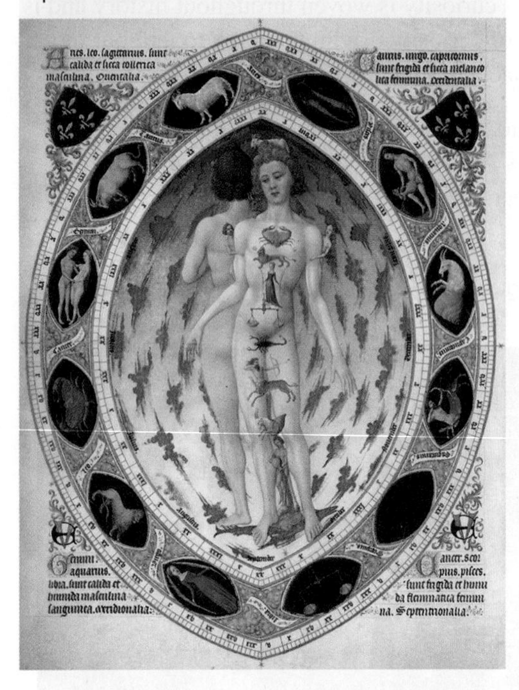

FIGURE 4.9 The 'Zodiac Man' provided advice on when to perform certain medical procedures.

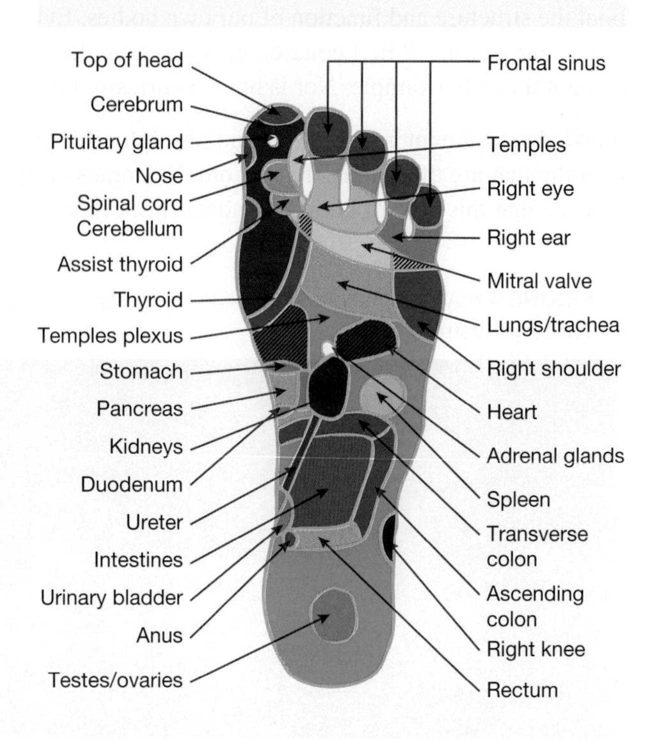

FIGURE 4.10 An example of an acupressure reflexology chart

Scientists are curious

Scientists are also often driven by the thirst to find answers to their questions. With increased advances in technology and knowledge, the answers to these questions often result in even more questions.

Compared with the situation in Leonardo's day, there are now an amazing number of different types of careers that involve investigations, explorations and applications of science to the human body. Australian scientists are involved in medical research and intervention, and development of medical equipment that helps with our understanding of our **body systems**.

body systems groups of organs within organisms that carry out specific functions

4.2.6 Australian scientists: creative inventors and explorers

Australian scientists have made significant contributions to medical discoveries and inventions. Howard Florey and his team discovered how penicillin could be extracted, purified and produced to be used as an antibiotic to help fight bacterial infections. Barry Marshall and Robin Warren showed that a certain type of bacteria caused stomach ulcers that could be treated with antibiotics. Professor Graeme Clark and his team were involved in the invention of an effective 'bionic ear'. Dr Fiona Wood pioneered a new treatment for burns in her development of 'spray-on skin' that uses the patient's own skin cells. Professor Ian Frazer developed the world's first vaccine against cervical cancer.

DISCUSSION

Research and report on the role of Ngangkari healers, their traditional health systems and how they work with people and medical professionals.

SCIENCE AS A HUMAN ENDEAVOUR: The bionic ear

The cochlear implant, also known as the bionic ear, has allowed some people with inner-ear problems to hear sound for the first time. When deafness results from serious inner-ear damage, no sounds are heard at all. Normal hearing aids, which make sound louder, do not help in these cases because the cochlea cannot detect the vibrations. However, the cochlear implant can often help by changing sound energy from outside the ear into electrical signals that can be sent to the brain.

FIGURE 4.11 An enlarged x-ray of the cochlear implant, showing the experimental electrode array inside

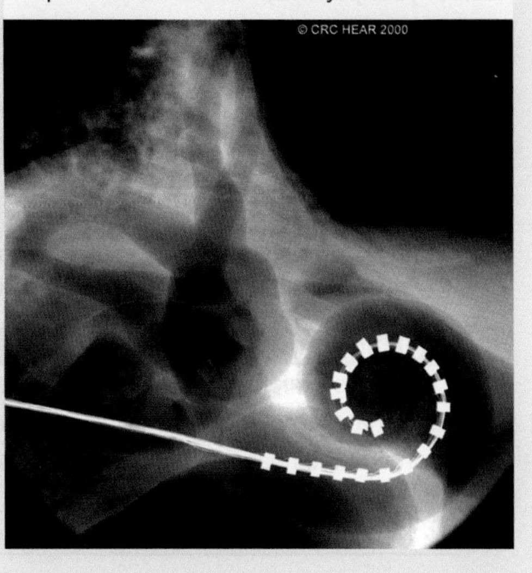

FIGURE 4.12 How a cochlear implant works

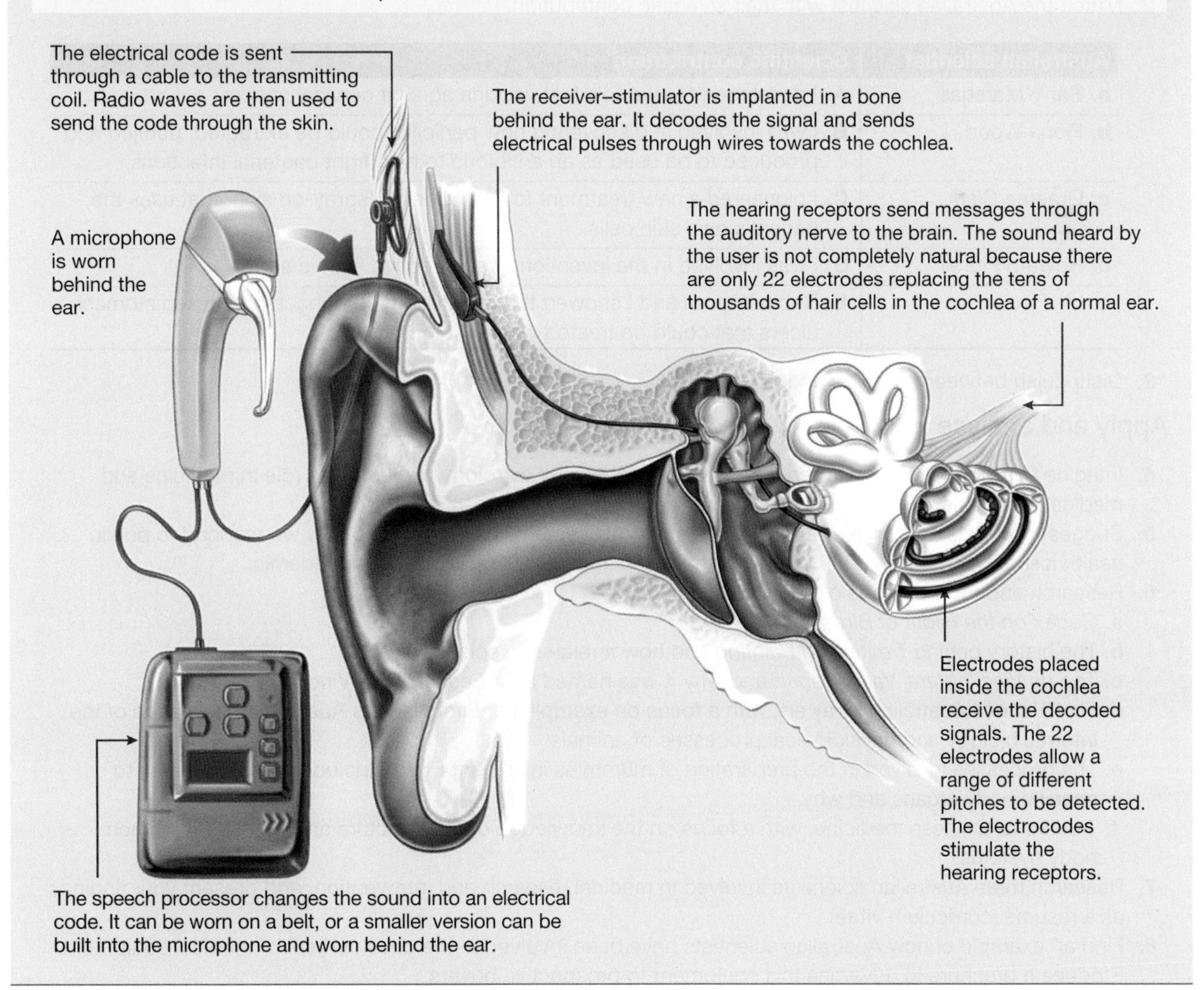

4.2 Activities

learnon

4.2 Quick quiz on	4.2 Exercise

Select your pathway

LEVEL 1	LEVEL 2	LEVEL 3
1, 2, 3, 11	4, 6, 7, 12	5, 8, 9, 10

Remember and understand

1. a. State whether the following statements are true or false.
   Leonardo da Vinci's dissections led to changes in the knowledge and understanding:

Statement	True or false?
i. of the structure and function of the heart	
ii. in that the heart was a muscle with three chambers that warmed the blood	
iii. in that the heart's left ventricle contractions were connected to the pulse in the wrist	
iv. in that the location of the human soul was in the spine.	

   b. If false, justify your response.
2. Match the Australian scientist with their scientific contribution.

Australian scientist	Scientific contribution
a. Barry Marshall	A. I developed the world's first vaccine against cervical cancer.
b. Fiona Wood	B. I was involved in discovering how penicillin could be extracted, purified and produced to be used as an antibiotic to help fight bacterial infections.
c. Graeme Clark	C. I pioneered a new treatment for burns using 'spray-on skin' that uses the patient's own skin cells.
d. Howard Florey	D. I was involved in the invention of an effective 'bionic ear'.
e. Ian Frazer	E. Robin Warren and I showed that a certain type of bacteria caused stomach ulcers that could be treated with antibiotics.

3. Distinguish between anatomy and physiology.

Apply and analyse

4. Imagine you were alive in the medieval times. Suggest why astrology played a key role in medicine and medical prognosis.
5. Suggest how scientific understanding of human body systems can determine how we respond to public health issues such as the 2009 swine flu pandemic and the 2020 coronavirus pandemic.
6. Research and report on one of the following:
   a. *Codex on the Flight of Birds*
   b. The history behind *Treatise on Painting* and how it relates to science
   c. The *da Vinci® Mitral Valve Repair* and why it was named after Leonardo da Vinci
   d. First Nations Australian x-ray art, with a focus on examples of First Nations Australian knowledge of the internal biology and physiological processes of animals
   e. The processes involved in the preparation of mummies in ancient Egypt. Include what happened to specific body organs and why.
   f. Traditional Chinese medicine, with a focus on the knowledge of the structure and function of human body systems.
7. Research three Australian scientists involved in medical research and intervention, and present your findings as a resumé (curriculum vitae).
8. Find an example of how Australian scientists have been involved in the development of medical equipment. Produce a brochure to advertise this equipment to prospective buyers.

Evaluate and create

9. **SIS** **a.** Do you believe that acupressure should be available as a medical treatment? Justify your response.
   b. Does scientific knowledge support your stance? Explain.
10. **SIS** Evaluate the analogies used by Leonardo da Vinci to describe arteries and the bursting of blood from a vein. Incorporate current scientific knowledge into your evaluation.
11. **SIS** Research Leonardo da Vinci's sketches and then select one of the following creative tasks:
   a. Construct a tree diagram to show how one of Leonardo's inventions is related to something that we use today.
   b. Construct a model of one of Leonardo's inventions.
12. **SIS** **a.** Research Leonardo da Vinci's sketches of some of his inventions.
   b. Create your own variation of one of Leonardo's inventions, presenting it as a series of annotated sketches.
   c. Construct your own PMI (Plus, Minus, Interesting) chart to evaluate your invention.
   d. Share and discuss your sketches with two other students and add any relevant comments to your invention PMI.
   e. If you were to create another variation of Leonardo's inventions, what would you do differently to improve your final outcome?

Fully worked solutions and sample responses are available in your digital formats.

LESSON
4.3 Working together?

LEARNING INTENTION

At the end of this lesson you will be able to provide examples of how the different body systems — made up of specialised organs, tissues and cells — work together to keep multicellular organisms alive.

4.3.1 The building blocks of life

Like all matter in the universe, you are made up of **atoms**. Collections of atoms make up **molecules**; molecules make up **organelles**, which make up **cells**, which make up **tissues**, which make up **organs**, which make up systems, which make up you. This progression is shown in figure 4.14.

FIGURE 4.13 We are all made up of atoms.

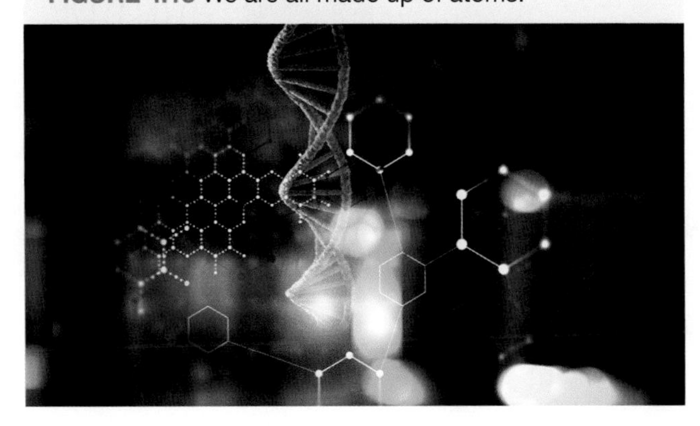

atoms very small particles that make up all things; atoms have the same properties as the objects they make up

molecule two or more atoms joined (bonded) covalently together

organelle any specialised structure in a cell that performs a specific function

cell the smallest unit of life; cells are the building blocks of living things and can be many different shapes and sizes

tissue a group of cells of similar structure that perform a specific function

organs structures, composed of tissue, that perform specific functions

int-6581

FIGURE 4.14 The building blocks of life

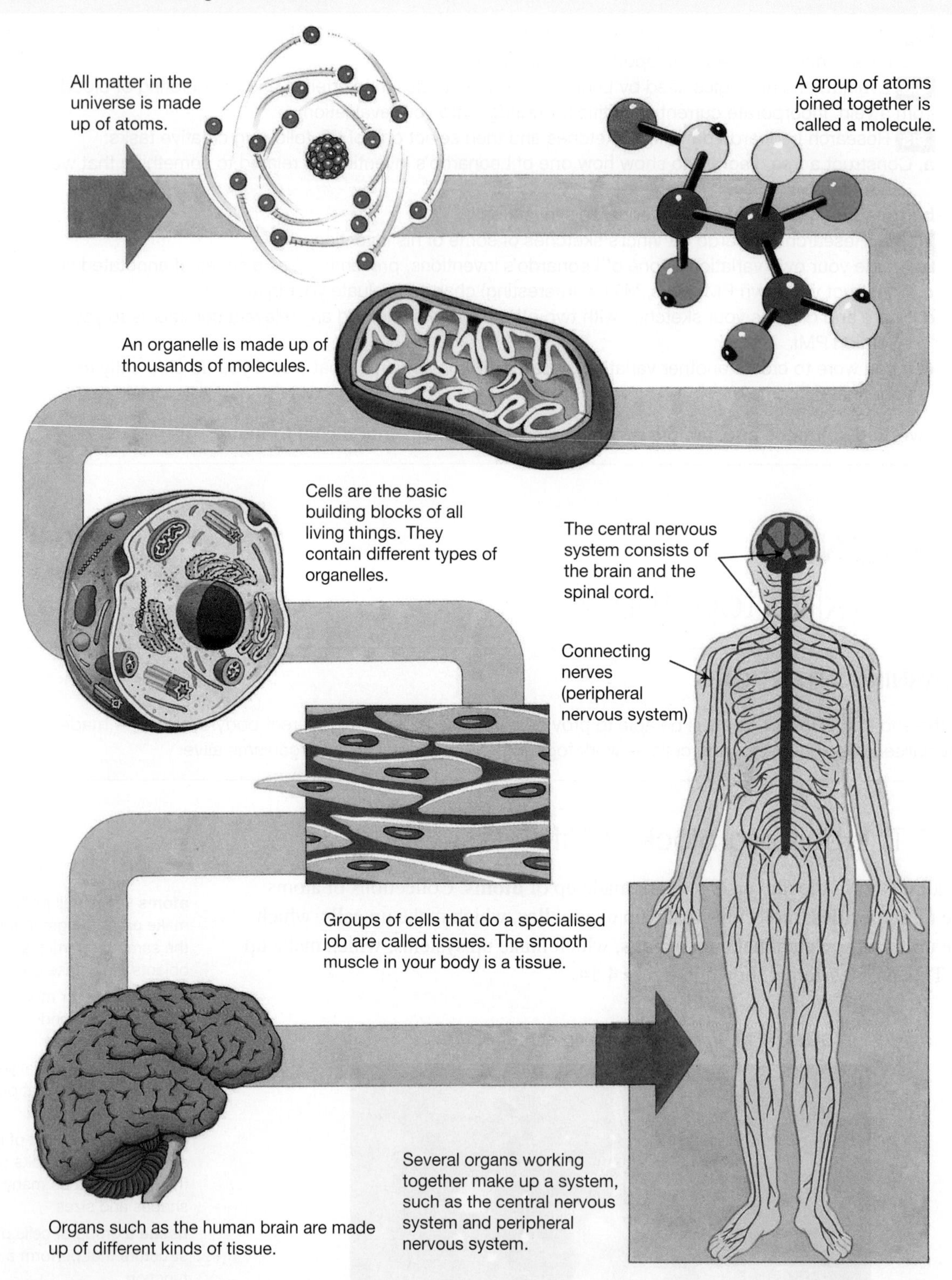

4.3.2 All alone? Independent!

Unicellular organisms are made up of only one cell that must do all of the jobs that are required to keep the organism alive. These single-celled organisms are small enough that essential substances (e.g. oxygen) and wastes (e.g. carbon dioxide) can be exchanged with their environment through simple **diffusion**.

FIGURE 4.15 Useful substances (e.g. oxygen) can move into cells and wastes (e.g. carbon dioxide) can move out through a process called diffusion.

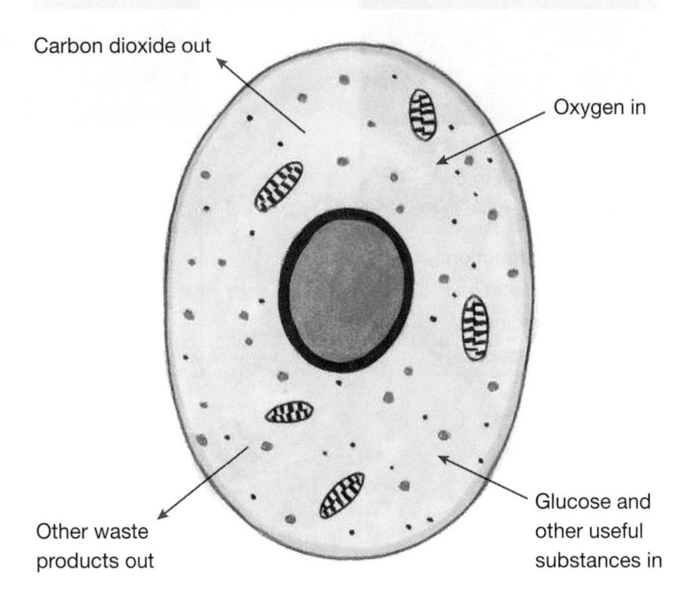

4.3.3 One of many? Better get organised!

Like other multicellular organisms, you are made up of many cells. These cells cannot survive independently of each other. They depend on each other and work together. Working together requires organisation.

Pattern, order and organisation

Multicellular organisms are made up of a number of body systems that work together to keep them alive. Body systems are made up of organs, which are made up of tissues, which are made up of particular types of cells.

FIGURE 4.16 Organisation of systems

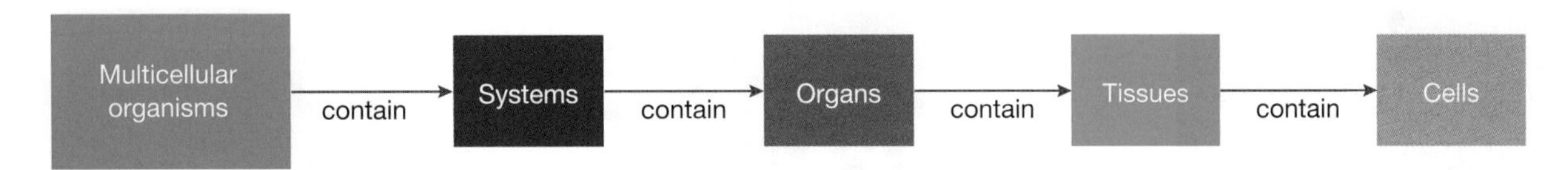

Cells

Within each cell there are structures called organelles. Each organelle has a particular job to do. Mitochondria, for example, are organelles in which the chemical energy in glucose is transformed into energy that our cells can use.

Multicellular organisms are made up of many different types of cells, each with a different job to do. Although these cells may have similar basic structures, they differ in size, shape, and in the number and types of organelles they contain. The different make-up of different types of cells and structures within them makes them well suited to their function.

Tissues

Groups of similar cells that perform a specialised job are called tissues. Muscle tissue contains cells with many mitochondria so that the energy requirements of the tissue can be met. Nerve tissue consists of a network of nerve cells with extensions to help carry messages throughout your body. Figure 4.17 shows some examples of tissues that make up your body, what they look like and what their main functions are.

diffusion movement of molecules through the cell membrane

multicellular organisms living things comprised of specialised cells that perform specific functions

int-6582

FIGURE 4.17 There are different types of tissues, each with structural features that suit them to their function.

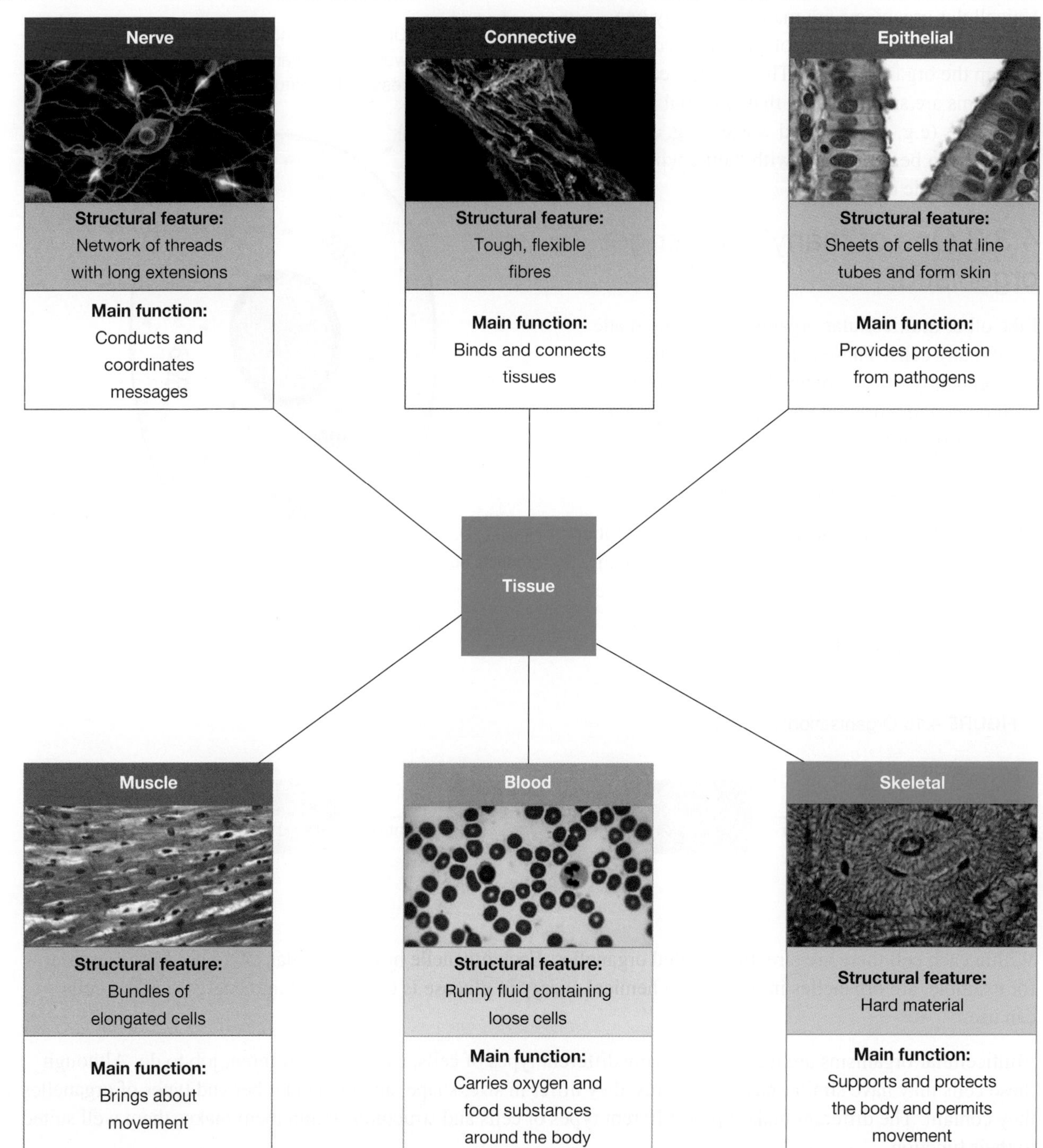

ewbk-11883

int-3397

FIGURE 4.18 You are made up of many different body systems that contain organs that work together to keep you alive.

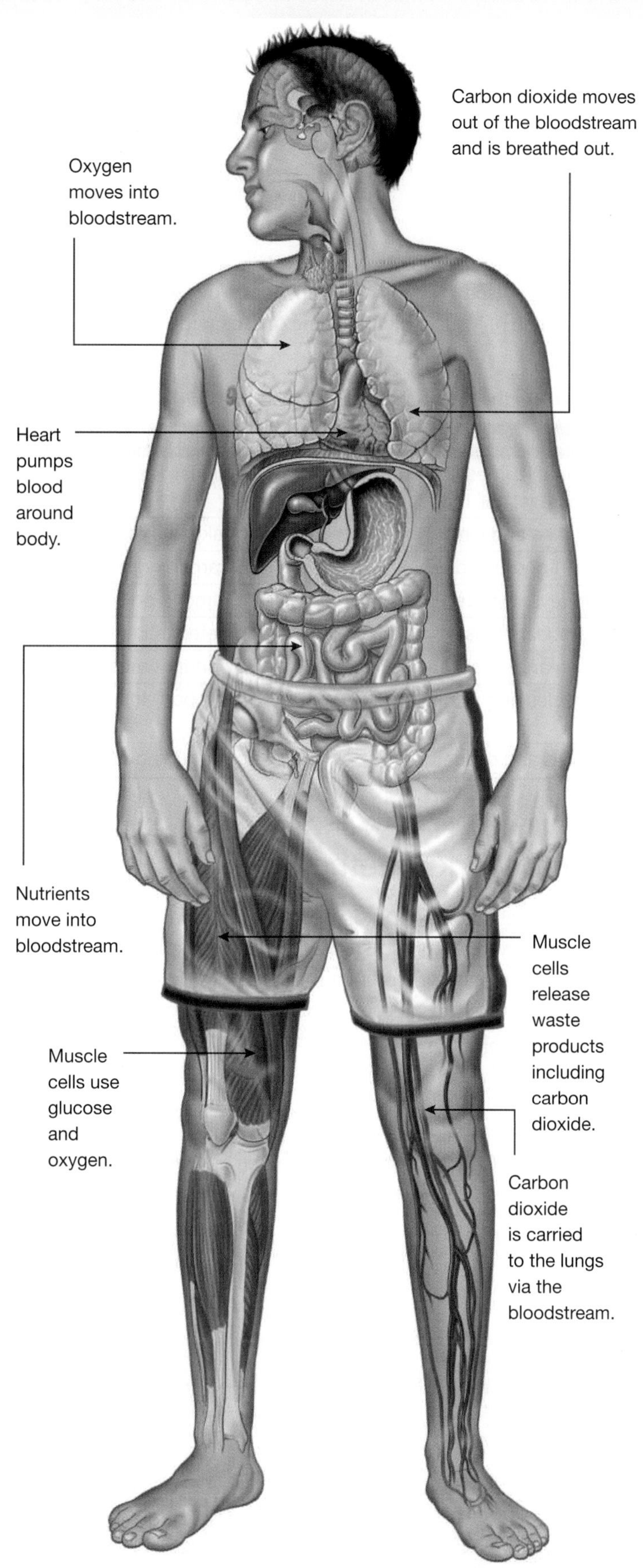

Organs

Organs are made up of one or more different kinds of tissue and perform one (or sometimes more) main function or job. Examples of your organs include:

- brain
- stomach
- lungs
- heart
- skin
- kidneys.

Systems

Multicellular organisms contain organised systems of organs that work together to perform specialised functions. Table 4.1 provides examples of some of your systems, some organs within them and their main functions.

TABLE 4.1 Examples of systems and their main functions

Name of system	Organs in system	Main functions
Digestive system	Stomach, intestine, liver, pancreas, gall bladder	To digest and absorb food
Respiratory system	Trachea and lungs	To take in oxygen and remove carbon dioxide
Circulatory system	Heart and blood vessels	To carry oxygen and food around the body
Excretory system	Kidneys, bladder, liver	To remove poisonous waste substances
Sensory system	Eyes, ears, nose	To detect stimuli
Nervous system	Brain and spinal cord	To conduct messages between body parts
Musculoskeletal system	Muscles and skeleton	To support and move the body
Reproductive system	Testes and ovaries	To produce offspring

4.3.4 Systems need to work together

Body systems within multicellular organisms work together to keep them alive. For example, cells need energy to survive. A process called **cellular respiration** (figure 4.19) breaks down glucose to release energy in a form that your cells can then use. This process also requires oxygen and produces carbon dioxide, a waste product. Your digestive, circulatory, respiratory and excretory systems work together to provide your cells with nutrients and oxygen, and to remove wastes such as carbon dioxide.

cellular respiration a series of chemical reactions in which the chemical energy in molecules such as glucose is transferred into ATP molecules, which is a form of energy that the cells can use

FIGURE 4.19 Your respiratory system is involved in getting oxygen into your body.

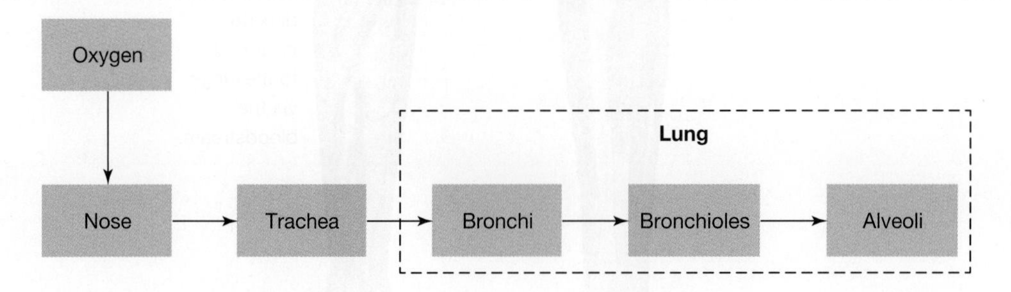

Respiratory system

The **respiratory system** is responsible for getting oxygen into your body and carbon dioxide out (figure 4.20). This occurs when you inhale (breathe in) and exhale (breathe out).

FIGURE 4.20 Your respiratory system is involved in getting carbon dioxide out of your body.

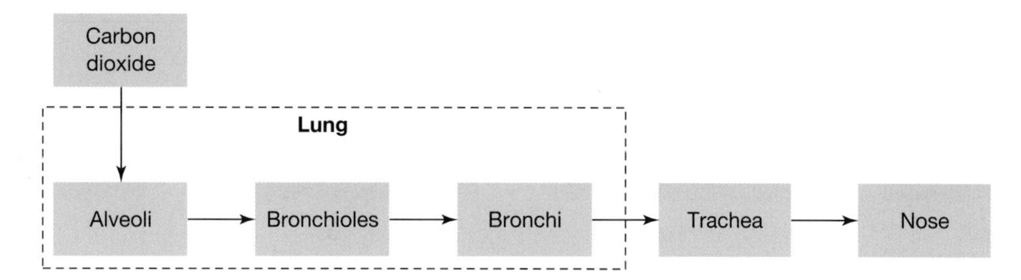

Circulatory system

The **circulatory system** is responsible for transporting oxygen and nutrients to your body's cells, and wastes such as carbon dioxide away from them. This involves **blood cells** that are transported in your **blood vessels** and **heart**. The major types of blood vessels are **arteries**, which transport blood from your heart; **capillaries**, through which materials are exchanged with cells; and **veins**, which transport blood back to the heart. This is seen in figure 4.21.

FIGURE 4.21 Your circulatory system is involved in transporting blood cells in blood vessels to and from your body cells and your heart.

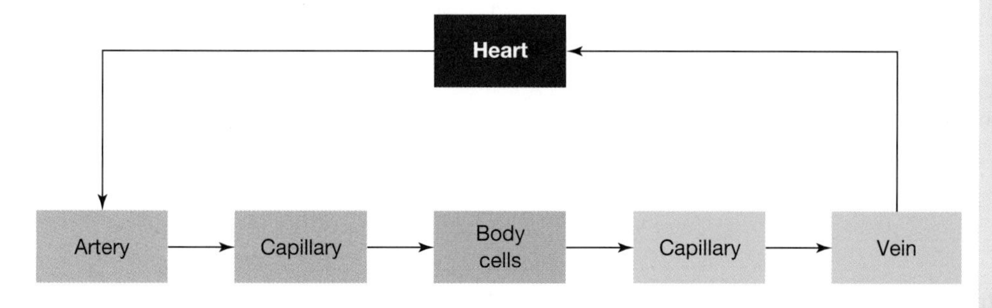

Digestive and excretory systems

The **digestive system** plays a key role in supplying your body with the nutrients it requires to function effectively. You ingest food, digest it, then egest (excrete) it. Your digestive system is involved in breaking food down so nutrients are small enough to be transported to, and used by, your cells.

The **excretory system** removes waste products from a variety of chemical reactions that your body needs to stay alive. The main organs of your excretory system are your skin, lungs, liver and kidneys. Your skin excretes salts and water as sweat, and your lungs excrete carbon dioxide when you breathe out. Your liver is involved in breaking down toxins for excretion. Your kidneys are involved in excreting the used waste products of chemical reactions (e.g. urea) and any other chemicals that may be in excess (including water), so that a balance within your blood is maintained. Some of the organs of the excretory system are shown in the flowchart in figure 4.22.

respiratory system the lungs and associated structures that are responsible for getting oxygen into the organism and carbon dioxide out

circulatory system the heart, blood and blood vessels, which are responsible for circulating oxygen and nutrients to body cells, and carbon dioxide and other wastes away from them

blood cells living cells in the blood

blood vessels the veins, arteries and capillaries through which the blood flows around the body

heart a muscular organ that pumps deoxygenated blood to the lungs to be oxygenated and then pumps the oxygenated blood to the body

arteries hollow tubes (vessels) with thick walls carrying blood pumped from the heart to other body parts

capillaries numerous tiny blood vessels that are only a single cell thick to allow exchange of materials to and from body cells; every cell of the body is supplied with blood through capillaries

veins blood vessels that carry blood back to the heart; they have valves and thinner walls than arteries

digestive system a complex series of organs and glands that processes food to supply the body with the nutrients it needs to function effectively

excretory system the body system that removes waste substances from the body

int-6583

FIGURE 4.22 Different organs in your excretory system are involved in the removal of different types of wastes.

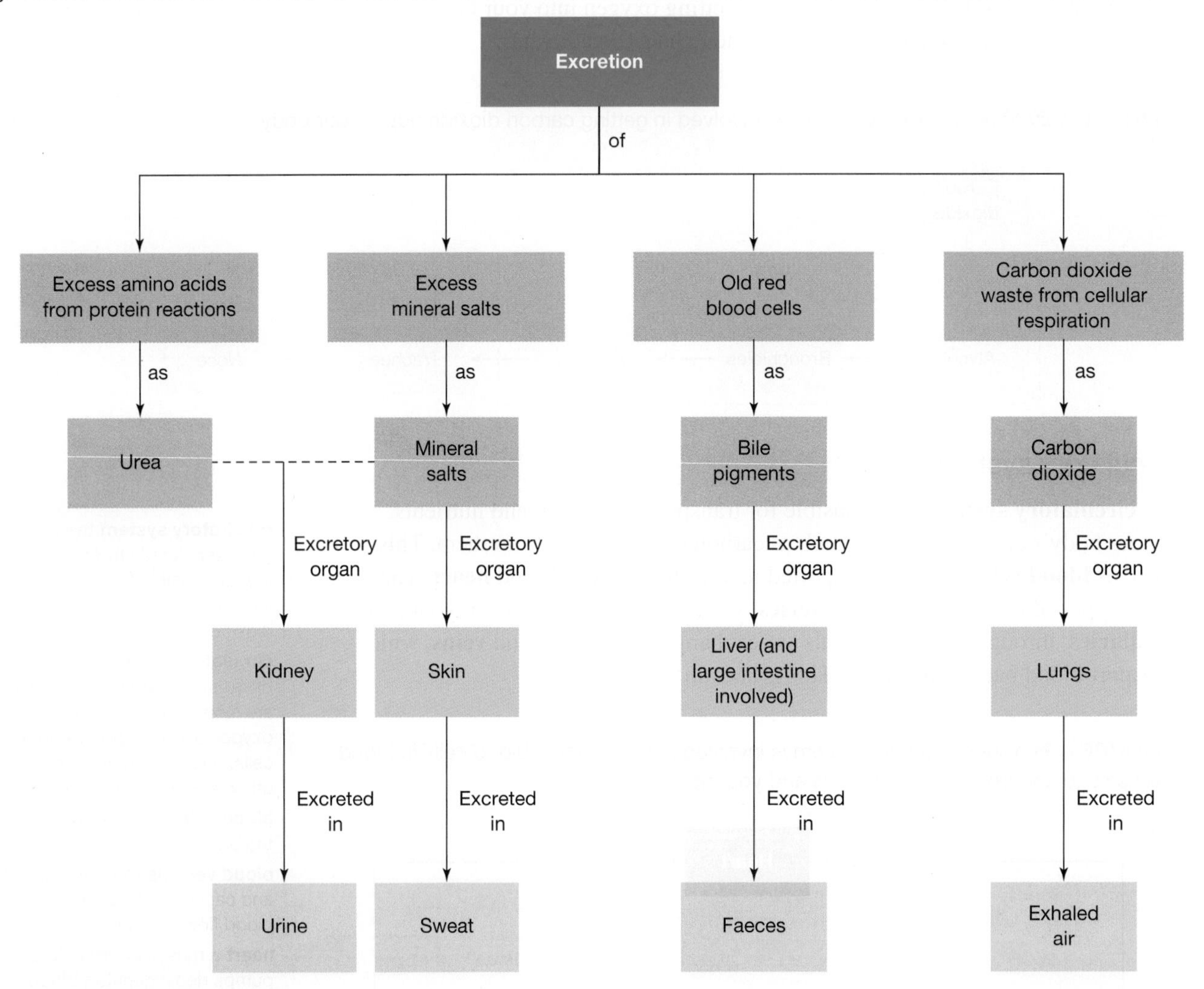

Other body systems

The **musculoskeletal system** consists of both your bones and various types of muscles throughout your body. Bones and muscles provide both support and protection for your organs. While the **reproductive systems** of males and females contain different organs in each sex, they both play a key role in the continuation of our species. Other systems such as the **nervous system** and **endocrine system** are also involved in coordinating and regulating processes in your body. You will find out more about these later in your studies.

musculoskeletal system consists of the skeletal system (bones and joints) and the skeletal muscle system (voluntary or striated muscle); working together, these two systems protect the internal organs, maintain posture, produce blood cells, store minerals and enable the body to move

reproductive system the different reproductive organs required by many organisms to reproduce and create offspring

nervous system consists of neurons, nerves and the brain, which are responsible for detecting and responding to both internal and external stimuli

endocrine system the body system of glands that produce and secrete hormones into the bloodstream to regulate processes in various organs

elog-2149

INVESTIGATION 4.1

Mapping your organs

Aim

To draw a diagram to map out the positions and shapes of some human body organs

Materials

- large sheets of paper (e.g. butcher paper)
- pencils and marker pens
- sticky tape
- scissors
- optional: light coloured material, sewing thread and needle (or stapler or craft glue), 'stuffing', various other bright coloured materials

Method

1. Use the sticky tape to join the paper together so that it is the size of a student's body outline.
2. One member of the team lies down on the paper with their arms away from their body.
3. Another team member carefully draws (about 5 cm away from their body) an outline of their partner's body.
   - Once the outline is drawn, the person on the paper can join the rest of the team for the remainder of the activity.
   - As a team, decide where in the body outline the following organs are located: heart, lungs, small intestine, nose, oesophagus, liver, stomach, ears, kidney, large intestine, pancreas, eyes, bladder, brain, trachea, mouth.
   - Once the location of each organ has been agreed upon, discuss their shape and size.
4. Once consensus is reached within the group, draw each of these organs onto the paper body outline.
   - Compare your diagram to reference materials to judge its accuracy.

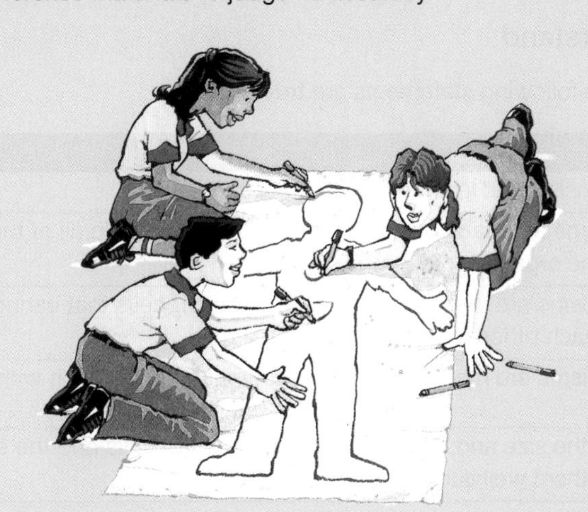

5. Using these references as a guide, use different coloured pens to draw in more accurate organ shapes, sizes or locations onto your paper body outline.
   Optional: Use the final version of your organ body outline as a pattern to make human body organ stuffed toys or a human body organ blanket.

Results

How accurate was your team's first attempt at drawing the body organ outline?

a. Which organs were located correctly and which were not?
b. How closely did your team's estimate of shape and size compare to that referenced for each organ?

Discussion

1. Identify the system to which each of the organs on your outline belong.
2. As an individual learner, identify which organs had a size, shape and location that you expected, and which did not.

Conclusion

Summarise the findings for this investigation.

DISCUSSION

Did you take a particular role in this team? Assess how well you supported other members of your team.

In pairs, apply your understanding of the structure and function of two different body systems by writing ten trivial pursuit questions (with answers) for each of your selected systems.

4.3 Activities

learnon

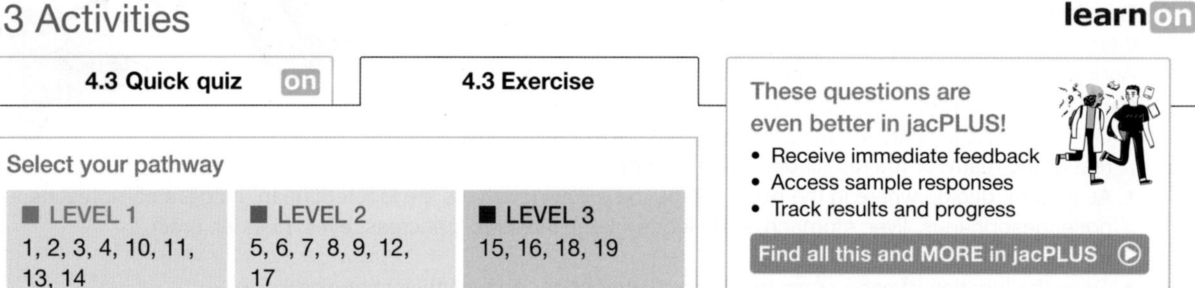

Remember and understand

1. **a.** Identify whether the following statements are true or false.

Statement	True or false?
i. All living things consist of at least one cell.	
ii. Unicellular organisms are made up of only one cell that must do all of the jobs that are required to keep the organism alive.	
iii. Multicellular organisms are made up of different types of cells that cannot survive independently of each other so need to work together.	
iv. Multicellular organisms are made up of different types of cells, each with different jobs to do.	
v. The differences in the size and shape of different types of cells and the structures within them make them well suited to their specific function.	
vi. Cellular respiration involves production of glucose.	
vii. The respiratory system takes oxygen into your body and removes carbon dioxide from your body.	
viii. The circulatory system transports carbon dioxide and nutrients to your body cells, and transports wastes such as oxygen away from them.	
ix. Arteries transport blood to your heart and veins transport blood away from your heart.	
x. Your kidneys, skin, liver and lungs all play a role in removing wastes from your body.	

b. Justify any false responses.

2. Recall a feature that organisms have in common with other matter in the universe.
3. Order the following from most complex to least complex: molecules, organelles, organs, multicellular organisms, systems, tissues, atoms, cells.
4. What am I? Identify the most appropriate term by matching it to the corresponding description in the table shown.

Term	Description
a. Tissue	**A.** A structure within a cell with a specific job to do
b. Organ	**B.** A collection of similar cells that perform a specific function
c. System	**C.** Different types of tissues grouped together to perform a specific function
d. Organelle	**D.** Different organs working together to perform a specialised function to keep an organism alive

5. List six:
   a. types of tissues **b.** examples of organs **c.** systems.
6. Name two organs in the:
   a. respiratory system **b.** circulatory system **c.** digestive system.
7. Name an example of an organelle and state its function.
8. Match the type of tissue with its function in the table shown.

Tissue	Function
a. Blood	**A.** Conducts and coordinates messages
b. Connective tissue	**B.** Brings about movement
c. Muscle tissue	**C.** Binds and connects tissues
d. Nervous tissue	**D.** Lines tubes and spaces, and forms skin
e. Skeletal tissue	**E.** Carries oxygen and food substances around the body

9. Match the system with its organs in the table shown.

System	Organs
a. Circulatory system	**A.** Liver, kidney, skin, lungs
b. Digestive system	**B.** Lungs, trachea
c. Excretory system	**C.** Stomach, liver, gall bladder, intestines, pancreas
d. Nervous system	**D.** Heart, blood vessels
e. Respiratory system	**E.** Brain, spinal cord

10. Describe two ways in which unicellular organisms differ from multicellular organisms.
11. Suggest why different types of cells within a multicellular organism may differ in their size and shape.
12. Explain why cells in muscle tissue contain many mitochondria.
13. Describe the relationship between:
    a. atoms, molecules, organelles and cells
    b. cells, tissues, organs and systems.
14. Outline the overall function of the:
    a. digestive system **b.** respiratory system **c.** circulatory system.

Apply and analyse

15. Suggest how scientific understanding of human body systems can help us to diagnose and treat a variety of illnesses.
16. Construct Venn diagrams to compare the:
    a. digestive system and respiratory system
    b. respiratory system and circulatory system
    c. excretory system and reproductive system
    d. circulatory system and excretory system.

Evaluate and create

17. **SIS** **a.** Select a body system and construct a PMI chart on how its structural features assist it in achieving its function.
    b. Propose ways in which the body system could be improved.
    c. Justify your proposed improvements.
18. **SIS** **a.** Design and construct a model of one of the following human body systems: respiratory, excretory, digestive or circulatory.
    b. Share your model with your team.
    c. With your team, construct a PMI chart on the accuracy of your model in effectively describing the structure and function of your selected body system.
    d. Identify three ways in which it could be improved.
19. **SIS** Select one of the following research questions to investigate and present your findings as a labelled model(s), informative animation, picture story book or interesting class presentation. For each question, select animal (i), (ii) or (iii) to compare it with a human.
    a. In which ways are the respiratory systems of (i) a fish, (ii) an earthworm OR (iii) an insect and a human similar, and how are they different?
    b. In which ways are the digestive systems of (i) a starfish, (ii) a snake OR (iii) a bird and a human similar, and how are they different?
    c. In which ways are the circulatory systems of (i) an insect, (ii) a frog OR (iii) a snake and a human similar, and how are they different?

Fully worked solutions and sample responses are available in your digital formats.

LESSON
4.4 Digestive system — break it down

LEARNING INTENTION

At the end of this lesson you will be able to describe the structure and function of the digestive system.

4.4.1 The gastrointestinal tract

The **gastrointestinal tract** (or digestive tract or **alimentary canal**) may be considered as your main digestive highway (figure 4.23). It consists of a long tube with coils, large caverns and thin passageways. Other organs that provide chemicals to break down food or absorb nutrients are attached to the gastrointestinal tract. The gastrointestinal tract begins at the mouth and ends at the anus, where waste products are removed.

FIGURE 4.23 The human digestive system in 3D

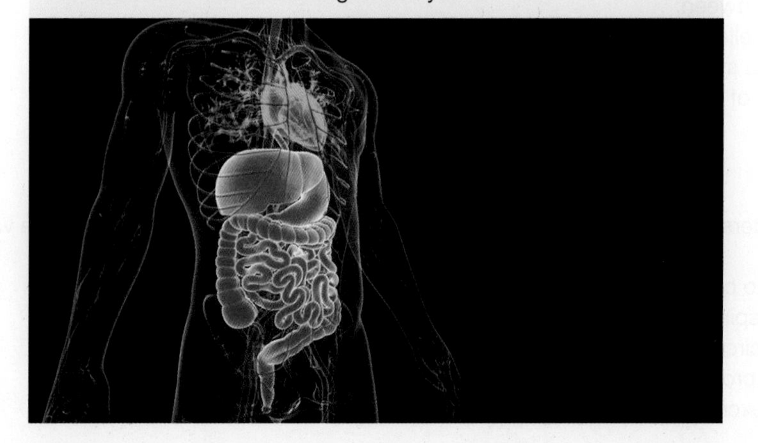

gastrointestinal tract also called the digestive tract or the alimentary canal, it is a tubular passage that starts with the mouth and ends with the anus; it intakes and digests food (absorbing energy and nutrients) and expels waste

alimentary canal see **gastrointestinal tract**

4.4.2 Digestion

Digestion involves the breaking down of food so that the nutrients it contains can be absorbed into your blood and carried to each cell in your body.

digestion the breakdown of food into a form that can be used by an animal; it includes both mechanical digestion and chemical digestion

Five key processes are important in supplying nutrients to your cells. These are:

- ingestion — taking food into your body
- mechanical digestion
- chemical digestion
- absorption of the broken-down food into your cells
- assimilation — converting the broken-down food into chemicals in your cells.

ewbk-11885

int-3398

FIGURE 4.24 The human digestive system 2D

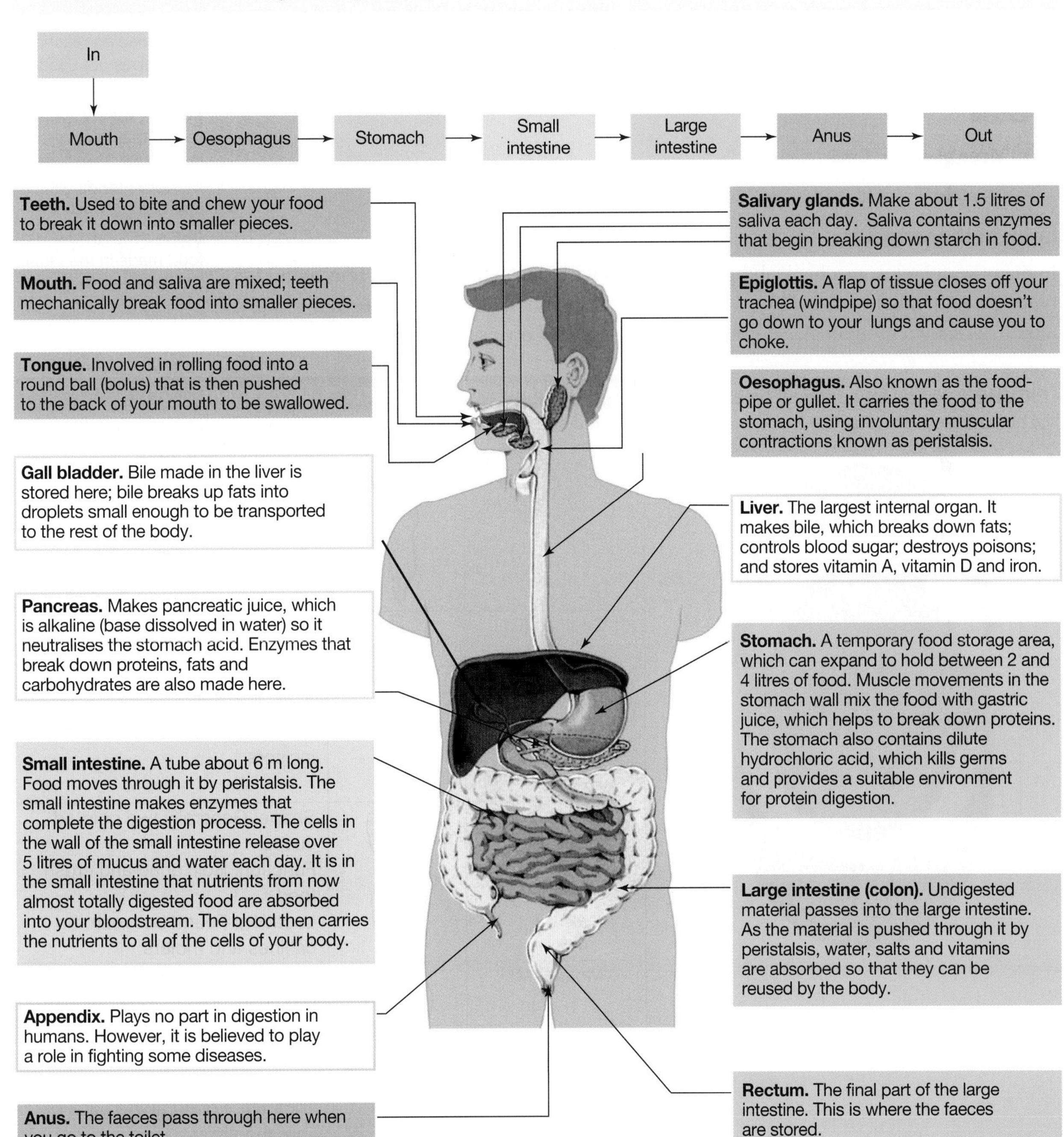

4.4.3 Mechanical and chemical digestion

Mechanical digestion (also known as physical digestion) involves physically breaking down food into smaller pieces. Most of this process takes place in your **mouth** when your **teeth** bite, tear, crush and grind food. **Chemical digestion** involves the use of chemicals called **enzymes** to break down food into small molecules.

Mouth

You ingest food, digest it, then egest it. The whole process of digestion starts with you taking food into your mouth. Enzymes (such as amylases) in your **saliva** are secreted by your **salivary glands** and begin the process of chemical digestion of some carbohydrates. Your teeth physically break down food in a process called mechanical digestion, then your tongue rolls the food into a slimy, slippery ball-shape called a **bolus**.

mechanical digestion digestion that uses physical factors such as chewing with the teeth

mouth the opening of the gastrointestinal tract through which food is taken into the body

teeth hard structures within the mouth that allow chewing

chemical digestion the chemical reactions that change food into simpler substances that are absorbed into the bloodstream for use in other parts of the body

enzymes special chemicals that speed up reactions but are themselves not used up in the reaction

saliva a watery substance in the mouth that contains enzymes involved in the digestion of food

salivary glands glands in the mouth that produce saliva

bolus a round, chewed-up ball of food made in the mouth that makes swallowing easier

Look at those teeth!

In many vertebrates, mechanical digestion begins with the teeth. There are four main types of teeth in humans, each type with a different function and position in your mouth, as shown in figure 4.25. Your teeth are your very own set of cutlery.

FIGURE 4.25 The four different types of human teeth

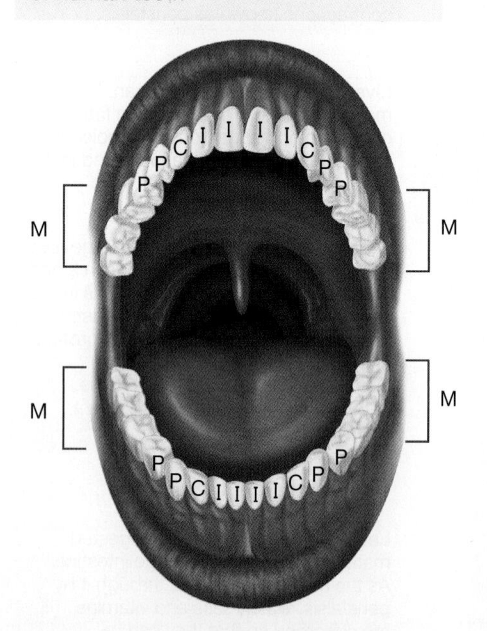

Type of teeth	Structure	Description
Incisors (I)		Shape: Spade-shaped with straight, sharp edge Function: Cutting and biting food Location: Found at the front of your mouth
Canines (C)		Shape: Sharp points and fang-like Function: Shearing and tearing through food Location: Found on each side of incisors
Premolars (P)		Shape: Generally two pointed cusps Function: Roll, grind and crush food Location: Found between the canines and molars
Molars (M)		Shape: Have between three and five cusps that fit together with those in the upper and lower jaws Function: Grind food Location: Found at the back of your mouth

Oesophagus to stomach

The bolus is then pushed through your **oesophagus** by muscular contractions known as **peristalsis**. From here it is transported to your **stomach** for temporary storage and further digestion.

FIGURE 4.26 Partly digested food is forced along the oesophagus by peristalsis — a wave of involuntary muscular contractions.

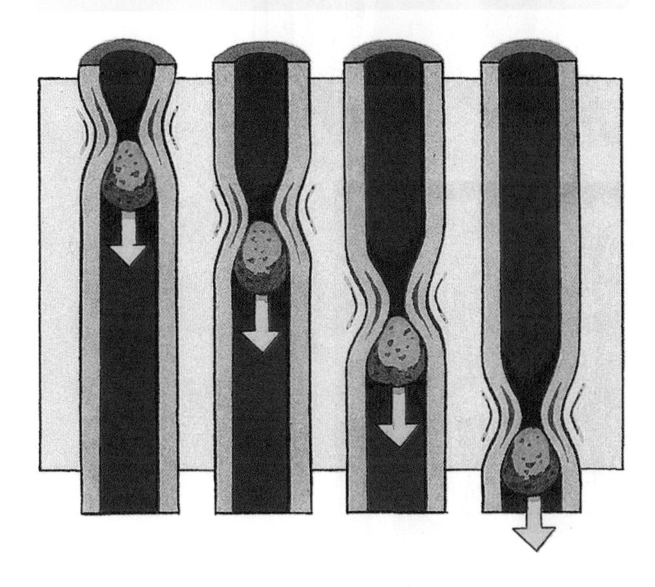

Stomach to small intestine

Once the food gets from your stomach to your **small intestine**, more enzymes (including amylases, proteases and lipases) break it down into molecules that can be absorbed into your body. The **absorption** of these **nutrient** molecules occurs through finger-shaped **villi** in the small intestine (figure 4.27). Villi are shaped like fingers to maximise surface area, which increases the efficiency of nutrient absorption into the surrounding capillaries. Capillaries are tiny blood vessels that transport the nutrients from the villi into your bloodstream. Once absorbed into the capillaries (of your circulatory system), these nutrients are transported to cells in the body that need them.

Large intestine

On its way through the gastrointestinal tract, undigested food moves from the small intestine to the **colon** of the **large intestine**. It is here that water and any other required essential nutrients still remaining in the food mass may be absorbed into your body. **Vitamin D** manufactured by bacteria living within this part of the digestive system is also absorbed. Any undigested food, such as the **cellulose** cell walls of plants (which we refer to as fibre), also accumulates here and adds bulk to the undigested food mass.

The **rectum** is the final part of the large intestine, where the faeces is stored before being excreted through the **anus** as waste.

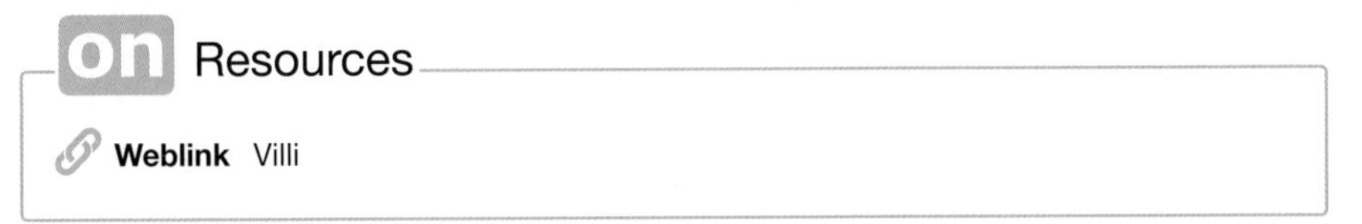

oesophagus part of the digestive system, composed of a tube connecting the mouth with the stomach

peristalsis the process of pushing food along the oesophagus or small intestine by the action of muscles

stomach a large muscular organ that churns and mixes food with gastric juice to start to break down protein

small intestine the part of the digestive system between the stomach and large intestine, where much of the digestion of food and absorption of nutrients takes place

absorption the taking in of a substance; for example, from the intestine to the surrounding capillaries

nutrients substances that provide the energy and chemicals that living things need to stay alive, grow and reproduce

villi tiny, finger-like projections from the wall of the intestine that maximise the surface area of the structure to increase the efficiency of nutrient absorption; singular = villus

colon the part of the large intestine where food mass passes from the small intestine, and where water and other remaining essential nutrients are absorbed into the body

large intestine the penultimate part of the digestive system, where water is absorbed from the waste before it is transported out of the body

vitamin D a nutrient that regulates the concentration of calcium and phosphate in the bloodstream and promotes the healthy growth and remodelling of bone

cellulose a natural substance that keeps the cell walls of plants rigid

rectum the final section of the digestive system, where waste food matter is stored as faeces before being excreted through the anus

anus the end of the digestive system, through which faeces are passed as waste

FIGURE 4.27 The absorption of most nutrients into your body occurs in the ileum, the last section of the small intestine. The finger-like villi on its walls give it a large surface area that speeds up nutrient absorption. Undigested material continues on to the large intestine, where water and vitamins may be removed, and then the remainder is pushed out through the anus as faeces.

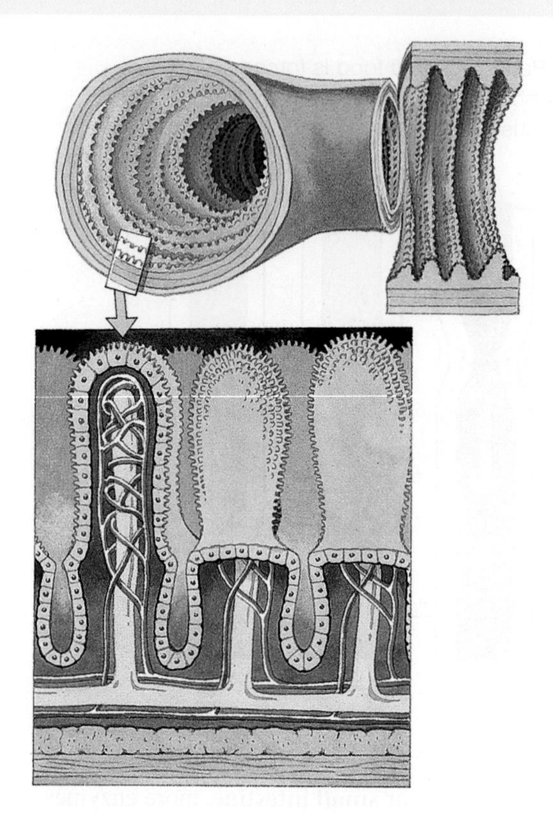

4.4.4 The liver and pancreas

Liver

Your liver is an extremely important organ with many key roles. One of these is the production of **bile**, which is transported to your **gall bladder** via the bile ducts to be stored until it is needed. Bile is transported from the gall bladder to the small intestine, where it breaks down **lipids** such as fats and oils.

Pancreas

Enzymes such as **lipases**, **amylases** and **proteases**, which break down lipids, carbohydrates and proteins respectively, are made by the **pancreas** and secreted into the small intestine to chemically digest these components of food.

4.4.5 Enzymes

Chemical digestion is usually assisted by compounds called enzymes that increase the rate of the chemical reactions. Without enzymes, a single meal could take many years to break down. Mechanical digestion increases the rate of chemical digestion because it increases the surface area of the food particles. This exposes more of the food surface to the digestive chemicals and enzymes.

Chemical digestion begins in your mouth, where enzymes in saliva begin to break down some of the carbohydrates in the food that you eat.

bile a substance produced by the liver that helps digest fats and oils

gall bladder a small organ that stores and concentrates bile within the body

lipids a class of nutrients that include fats and oils

lipases enzymes that break fats and oils down into fatty acids and glycerol

amylases enzymes found in saliva that break starch down into sugar

proteases enzymes that break proteins down into amino acids

pancreas a large gland in the body that produces and secretes the hormone insulin and an important digestive fluid containing enzymes

Fat stuff

Breaking down lipids, such as fats and oils, is hard work! Because lipids are insoluble in water, they tend to clump together into large blobs. Bile assists in solving this problem, as it helps to **emulsify** or separate the lipids so the lipase enzymes can gain access to them and do their job. This is an example of bile and lipase working together.

FIGURE 4.28 Bile emulsifies fat so that lipases can break it down.

FIGURE 4.29 Certain molecules are broken down by specific enzymes.

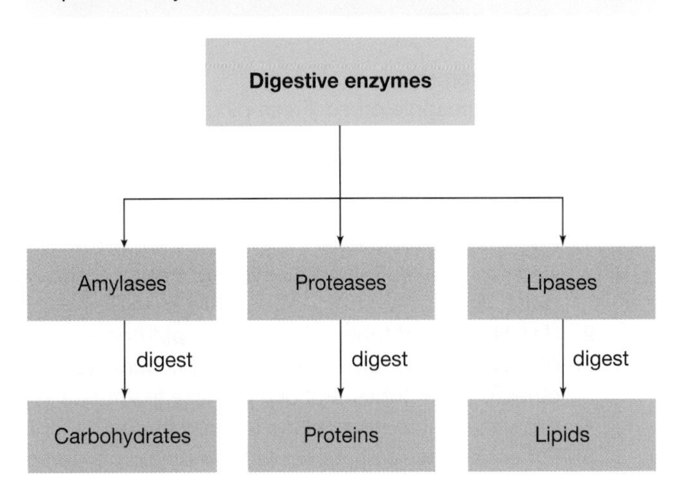

FIGURE 4.30 Amylases in the saliva and stomach break starch down into glucose molecules.

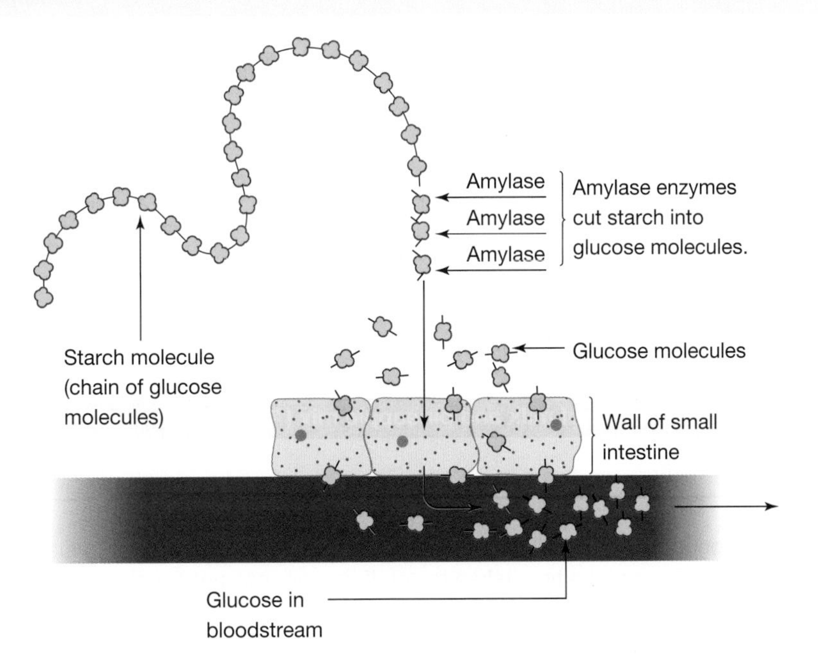

Not too hot!

Enzymes are made of protein. That is why it is important that they are not overheated. If they are too hot, they can become **denatured**. It's the same as cooking an egg — once they are denatured, they can't go back to how they were before, so they can't work as enzymes anymore. Different enzymes operate best within specific temperature ranges.

emulsify combine two liquids that do not normally mix easily

denatured describes the condition of proteins after they have been overheated

INVESTIGATION 4.2

Does temperature affect enzymes?

Aim

To investigate the effect of temperature on enzyme activity

Materials

- 4 beakers
- 8 test tubes
- milk
- 4 thermometers
- fresh pineapple puree (Fresh pineapple can be pureed using a food processor. If fresh pineapple is not available, use junket powder or a junket tablet dissolved in 10 mL water.)

Method

1. Add water to the beakers so that they are two-thirds full. Use cold tap water and ice for beaker 1, cold tap water for beaker 2, hot tap water for beaker 3 and boiling water (from a kettle) for beaker 4. These are the 'water baths'.
2. Half-fill four test tubes with milk and put one test tube in each water bath.
3. Place one teaspoon of fresh pineapple puree (or 1 mL of junket solution dissolved in 10 mL of water) in the other four test tubes. Put one of these test tubes in each water bath.
4. Allow the test tubes to stand in the water baths for at least 5 minutes.
5. For each water bath, pour the fresh pineapple puree into the milk and stir briefly.

Results

1. Copy the table provided and complete it with your results. Remember to include a title for your table.

Water bath	Temp. of milk and pineapple mix (°C)	Time taken to set (minutes)

2. Quickly record the temperature of the milk and pineapple mixture and then allow it to stand undisturbed. The mixture will eventually set. Record the time taken to set. If the milk has not set after 15 minutes, record the time as 15+.

Discussion

1. Pineapple juice and junket contain an enzyme that causes a protein in milk (casein) to undergo a chemical reaction and change texture; that is why the milk sets. At what temperature did the enzyme work best? Explain.
2. Did the enzyme work well at very high temperatures? Explain your answer.
3. Which variables were controlled in this experiment?

4. Do you think that the same results would be obtained if tinned pineapple puree was used instead of fresh pineapple puree? Explain your answer.
5. Identify the strengths and limitations of this investigation, and suggest ways to improve it.
6. Propose a research question about enzymes that could be investigated.

Conclusion

Summarise the findings for this investigation.

4.4.6 Personal explosions

Well, excuse me! Have you burped or passed wind recently? Have you had diarrhoea or vomited? These 'personal explosions' are related to the processing of nutrients by your body.

Burping, or belching, occurs when air is swallowed or sucked in. This may happen when you talk while you eat, eat or drink too quickly, or drink fizzy drinks (such as those with carbon dioxide gas dissolved in them). When you eat too fast and don't chew your food enough, more acid can be produced in your stomach. When you burp, some of this acid can rise up into your oesophagus, resulting in a burning sensation called **heartburn**.

Resources

 **Video eLesson** Stomach acid reflux, heartburn (eles-2531)

Flatulence refers to the release of gases when you 'pass wind' through your anus. These gases are produced by bacteria in your large intestine. The odour and composition of the gases depend on the foods you have eaten and the amount of air you have swallowed.

Diarrhoea is the excessive discharge of watery faeces. It occurs when the muscles of the large intestine contract more quickly than normal, usually in an effort to rid your body of an infection. As a result, the undigested food moves through too rapidly for enough water to be absorbed into your body.

Green vomit? Messages from your stomach wall travel to the 'vomiting centre' of your brain, resulting in the forceful ejection of your stomach contents (and occasionally also contents from your small intestine). **Vomiting** can be caused by eating or drinking too much, anxiety, infections or chemicals that irritate your stomach wall. If the vomit is green, it may be due to the colour of food ingested or the presence of bile.

burping the release of swallowed gas through the mouth

heartburn a burning sensation caused by stomach acid rising into the oesophagus

flatulence the release of gas through the anus; this gas is produced by bacteria in the large intestine

diarrhoea excessive discharge of watery faeces

vomiting the forceful ejection of matter from the stomach through the mouth

elog-2153

INVESTIGATION 4.3

Making a burp model

Aim

To construct a burp model, and to design and construct a model that demonstrates the functioning of a process related to the digestive system

Materials

- vinegar
- baking soda
- medium/large balloon
- funnel

Method

1. Pour a small amount of vinegar into the bottom of the balloon 'stomach'.
2. Add some baking soda to the balloon 'stomach' using a funnel.
3. Using your fingers, pinch the balloon closed at its neck.
4. Watch as your 'stomach' expands with gas.
5. Unpinch the top of the balloon (or 'oesophagus/food tube') to release the gas (or burp).
6. Try to make your model sound like the real thing!

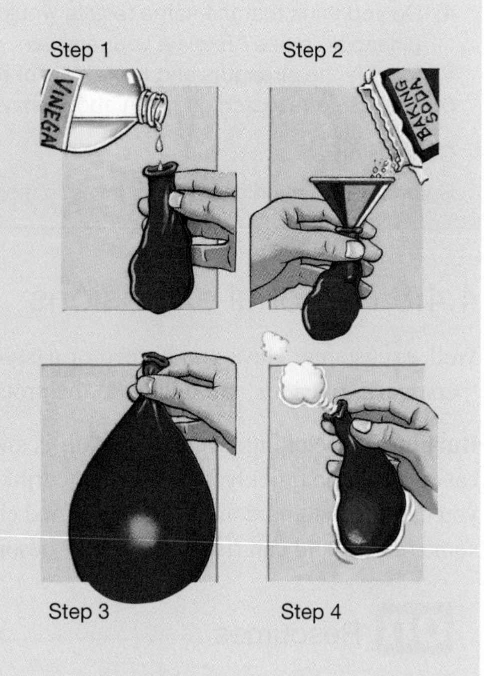

Results

1. Summarise your observations in a flowchart that includes labelled diagrams or digital/photographic images.
2. Select an organ belonging to an animal of your choice. Find out more about the structure and function of your selected organ and how it does its job. Summarise your findings.
   - Design and make a simple model (such as the one used for this experiment) to show how your selected organ achieves its function, or what happens when something goes wrong.
3. Summarise your design plans and labelled diagrams or digital images into an advertising brochure or digital multimedia advertisement.

Discussion

Comment on the challenges you experienced during the design and construction of your model, and suggest ways that you could overcome these if you were to do it again.

Conclusion

Summarise the findings for this investigation.

FIGURE 4.31 Summary: Digestion occurs in a systematic and organised manner.

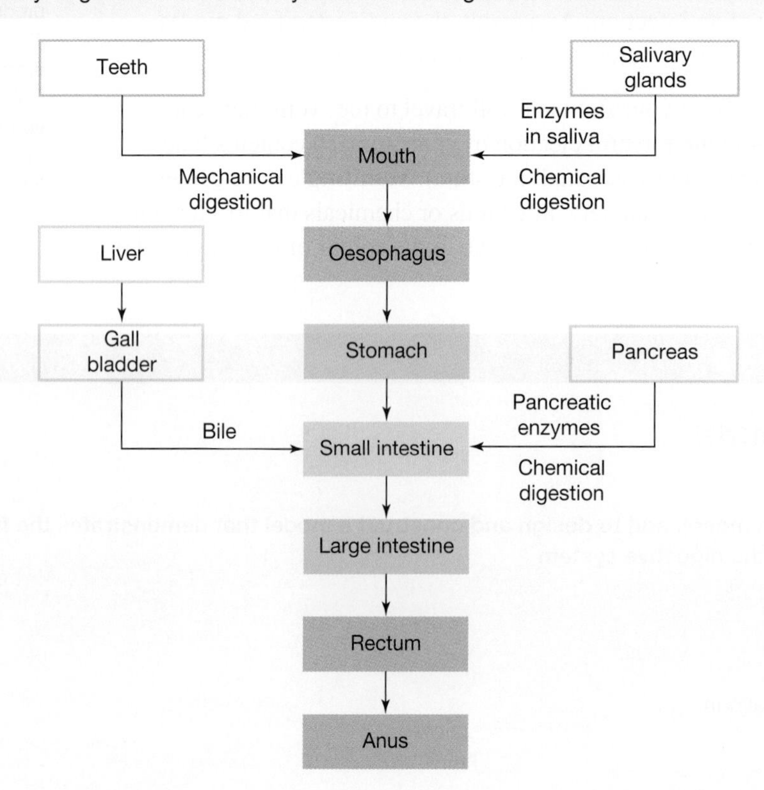

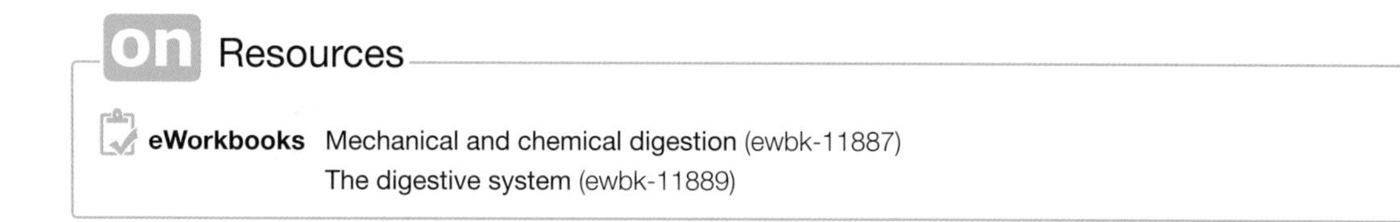

4.4 Activities

learnon

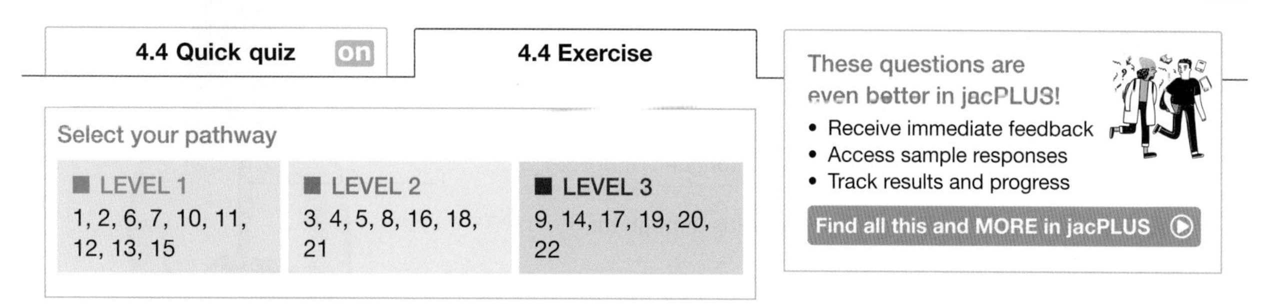

Remember and understand

1. a. State whether the following statements are true or false.

Statement	True or false?
i. Mechanical digestion occurs when chemicals in your body react with food to break it down.	
ii. Ingestion involves taking food into your body, whereas digestion involves breaking food down.	
iii. Many enzymes have names that end with the suffix 'ase'.	
iv. 'Bolus' is the term used to describe the muscular contractions that push food down your oesophagus to your stomach.	
v. Plant cell walls make up much of the fibre that accumulates in our large intestines.	
vi. The process of denaturing enzymes kills them.	
vii. Proteases are enzymes that break down carbohydrates.	
viii. Heartburn can be caused by acid from your stomach rising up your oesophagus.	
ix. Flatulence refers to the release of gases when you 'pass wind' through your anus.	
x. The green colour of vomit may suggest the presence of bile, which has been produced by the gall bladder.	

   b. Justify any false responses.

2. Match the types of teeth with their specific function in the table shown.

Types of teeth	Function
a. Canines	A. Biting and cutting food
b. Incisors	B. Grinding and crushing food
c. Molars and premolars	C. Tearing and grasping food

3. Identify the different types of teeth as being one of the following: incisors, canines, premolars or molars.

a. 

b. 

c. 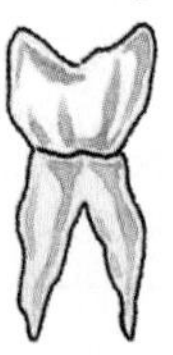

d. 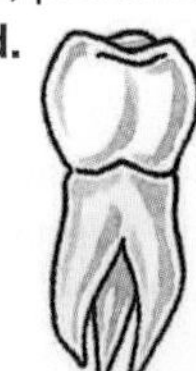

4. Identify the organs (a–m) in the figure given.
5. State the name of the:
   a. organ in which the digestive process begins
   b. type of digestion in which enzymes secreted by your salivary glands are involved in
   c. slimy, slippery ball-shape in which your tongue rolls food
   d. muscular contractions that push the bolus through the oesophagus to the stomach
   e. organ in which most of the absorption of nutrients occurs
   f. finger-shaped structures through which nutrients are absorbed in the small intestine.
6. Order the following organs into the correct sequence: stomach, large intestine, oesophagus, anus, mouth, small intestine.

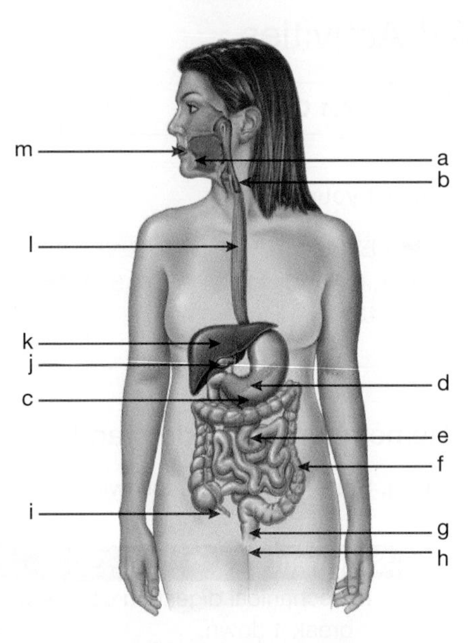

7. Match the organ with its function in the table shown.

Organ	Function
a. Gall bladder	**A.** Where the breakdown of starch and protein is finished and fat breakdown occurs
b. Large intestine	**B.** Temporary storage of food and where protein digestion begins
c. Liver	**C.** Tube that takes food from mouth to stomach
d. Oesophagus	**D.** Stores undigested food and waste while bacteria make some vitamins
e. Pancreas	**E.** Stores faeces
f. Rectum	**F.** Makes enzymes used in the small intestine
g. Small intestine	**G.** Makes bile, stores glycogen and breaks down toxins
h. Stomach	**H.** Stores bile made in the liver until needed in the small intestine

8. State the name of the:
   a. type of digestion that involves enzymes
   b. enzymes that break down fats
   c. enzymes that break down proteins
   d. substances that enzymes act on.
9. List the five key processes that are important in the supply of nutrients to your cells.
10. Explain why it is important to break down food that we eat.
11. Describe what happens to enzymes when they get too hot.
12. Describe the process of peristalsis and suggest why it occurs.
13. Suggest why it is necessary to drink fluids when you suffer from diarrhoea.

14. Describe the relationship between:
    a. teeth and mechanical digestion
    b. the pancreas and the small intestine
    c. the liver, gall bladder and the small intestine
    d. the villi, small intestine and capillaries
    e. bile, lipases and fats.

Apply and analyse

15. Which teeth are used to:
    a. bite into a pear
    b. crush and grind nuts?
16. Suggest how you can still swallow food if you are positioned upside down.
17. Take a small piece of bread into your mouth. Although at first you don't taste much, after a while, it may taste sweet. Suggest why.
18. Create Venn diagrams to compare:
    a. mechanical and chemical digestion
    b. lipases and proteases
    c. the small intestine and large intestine.
19. Use information in this lesson and other resources to relate structural features to the functions of the following parts of the digestive system.

TABLE Parts of the digestive system and their functions

Part of digestive system	Structural features	Function
Oesophagus		
Stomach		
Small intestine		
Villi		
Large intestine		

Evaluate and create

20. **SIS** When cows burp, they release methane gas into the air. This gas is believed to be one of the major causes of global warming. It has been suggested that cows could be responsible for about 20 per cent of the methane in the atmosphere. Research these claims.
    a. On the basis of your findings, do you agree? Justify your response.
    b. Design an experiment that could be used to test the claim that cows contribute to increased methane gas in the atmosphere.
21. **SIS** Design an investigation to test the following hypotheses:
    - Fresh pineapple results in a faster enzyme reaction than canned pineapple.
    - The length of time that pineapple puree is kept in ice affects the rate of enzyme reaction.
    - Different coloured junket tablets result in different rates of enzyme reaction.
22. **SIS** a. Imagine that you are either a cheese and tomato sandwich or a hamburger.
    b. List the ingredients of the food you chose in part **a**.
    c. Research what happens (and where) to each of these ingredients when eaten.
    d. Construct a flowchart to show the process of digestion in the human body, including events and locations.
    e. Use this information to write a story in either a cartoon or picture book format.
    f. Convert your story into a play.
    g. Perform your play to the class using animations, team members or puppets.

Fully worked solutions and sample responses are available in your digital formats.

LESSON
4.5 Digestive endeavours

LEARNING INTENTION

At the end of this lesson you will be able to give examples of ways in which science and technology have contributed to scientific knowledge and understanding of your digestive system, and have led to related improved medical treatments.

Science as a human endeavour

4.5.1 The digestive system as a scientific human endeavour

When your digestive system is healthy, it actively works along, busily doing its job without you even having to think about it. But sometimes, things can go wrong. For example, tooth decay, gum disease, intestinal polyps and a variety of digestive system diseases may result in some form of intervention. Research and developments in science and technology have not only increased our understanding and knowledge about our digestive system, but have also led to improved medical treatments to help us when things go wrong.

4.5.2 Do you look after your teeth?

It is very important to look after your teeth. Damaged or missing teeth can make it difficult for you to chew your food properly and therefore may affect digestion of foods.

Ouch! Does your tooth hurt?

Your teeth can decay when bacteria in your mouth turn sugar from your food into acid. This acid can 'eat' a hole in your tooth enamel and dentine. Once this hole reaches a nerve, you get a toothache. Figure 4.32 shows the structure of a tooth and where decay usually occurs — at the top of large back teeth and at the side, where one tooth touches another.

How many times do you clean your teeth each day?

If you don't brush your teeth and floss between them regularly (at least once a day), they can become covered with a thin film of food, saliva and bacteria. This is called plaque. As this plaque rots, it causes your gums to swell and bleed. This is known as gum disease.

Do you drink tap or bottled water?

Our water supply and toothpaste often contain fluoride, which helps prevent tooth decay. Fluoride protects the enamel and helps repair or rebuild the enamel in your teeth. If you have replaced drinking tap water with bottled water, read the label and find out if it has fluoride in it. In which other ways is bottled water different from tap water? Can drinking bottled water instead of tap water affect the health of your teeth?

Clean your teeth for good health

Poor teeth hygiene has been linked to poor general health and a number of systematic conditions, including heart disease, endocarditis, pregnancy and birth complications, and pneumonia. This is because the mouth is an entry point for bacteria.

eles-2532

FIGURE 4.32 The structure of a tooth, showing where tooth decay occurs

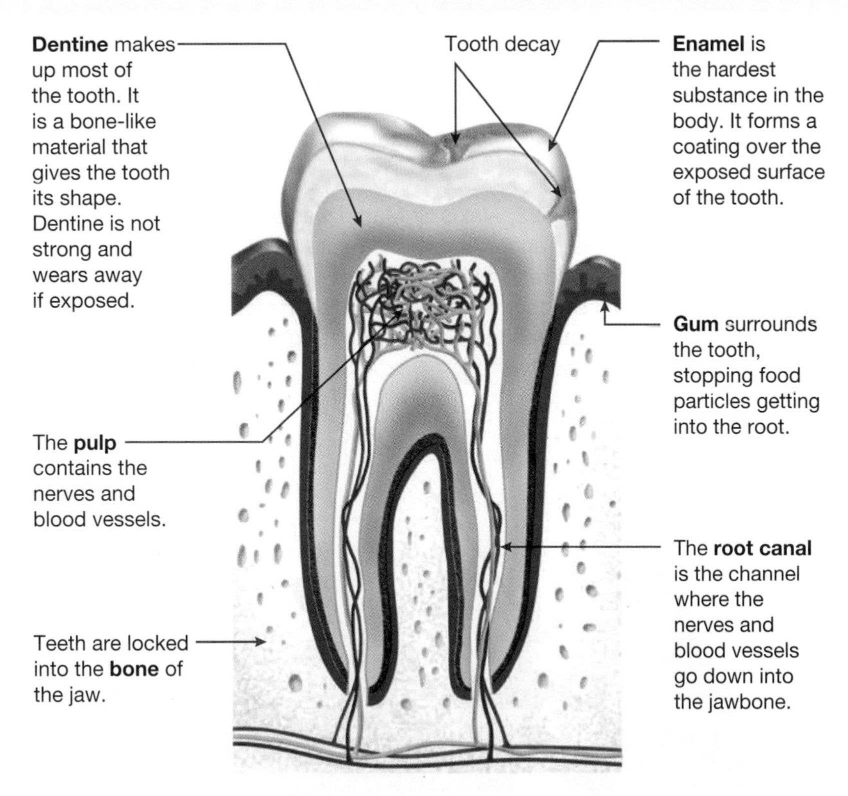

4.5.3 A future in teeth?

Dentistry is only one example of many different 'tooth pathway' careers that you may be aware of. Examples of other dental specialities include oral and maxillofacial surgeons, dental–maxillofacial radiologists, endodontists, oral physicians, oral pathologists, orthodontists, paediatric dentists, periodontists, prosthodontists, public health dentists and special needs dentists.

FIGURE 4.33 Oral surgery is one of the many tooth-related careers you can choose from.

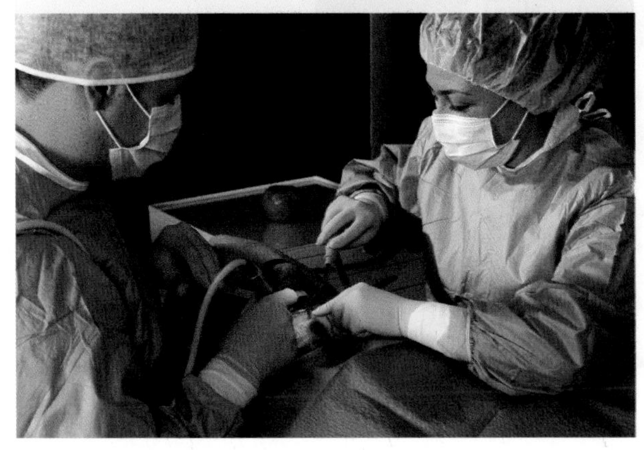

FIGURE 4.34 Missing a tooth? A synthetic replacement for a tooth root is used in a tooth implant.

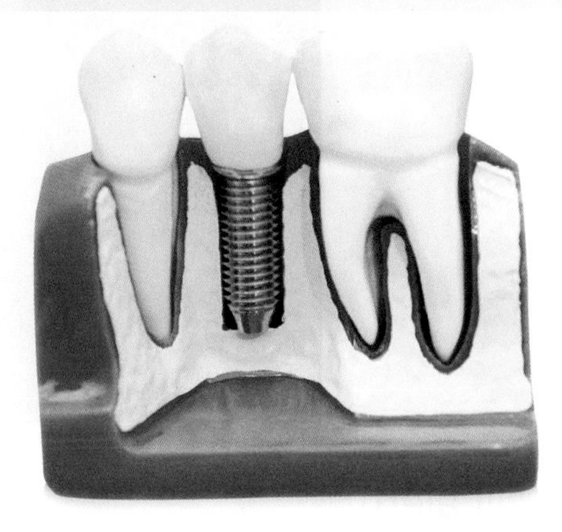

4.5.4 Do you have the stomach for it?

In the early 1800s, Dr William Beaumont made discoveries about how the stomach worked by dangling food on a silk thread into the wounded stomach of Alexis St Martin (figure 4.35). Almost 200 years later, Australian scientists Dr Barry J. Marshall and Dr Robin Warren (figure 4.36) made the discovery that linked *Helicobacter pylori* bacteria to gastroduodenal disease and, as a result, radically improved how peptic ulcer disease is treated.

FIGURE 4.35 In 1822, Alexis St Martin was shot in the stomach at close range. His wound healed, but left an open hole that showed the inside of his stomach. By dangling food suspended on a silk thread, his doctor William Beaumont made some breakthrough discoveries on how our stomachs work.

FIGURE 4.36 Australian scientists Dr Barry J. Marshall and Dr Robin Warren received the 2005 Nobel Prize in Medicine for their discovery that linked *Helicobacter pylori* bacteria to gastroduodenal disease.

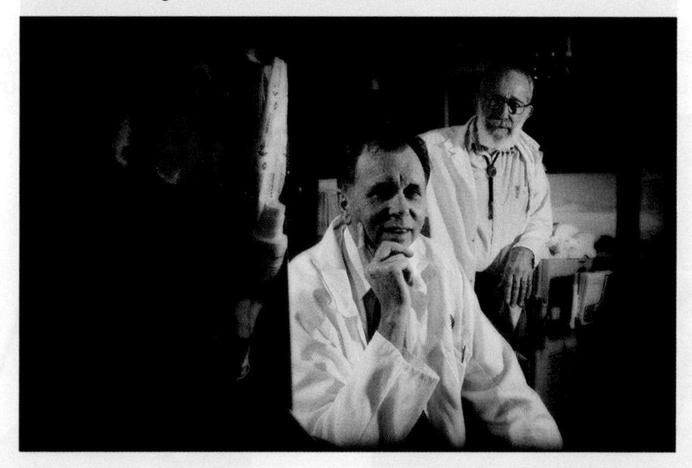

4.5.5 Villi alert!

The last section of your small intestine is lined with finger-like projections called villi. Their shape increases the surface area available for nutrients to diffuse through them into tiny blood vessels called capillaries. Once in the bloodstream, the nutrients are transported to other parts of your body.

Coeliac disease

Coeliac disease is an auto-immune disease — that is, one in which your body produces antibodies to attack your own tissues. In this case, the immune system reacts abnormally to gluten. As a result, the villi of the small intestine become inflamed and flattened (figure 4.37b). This reduces the surface area available for the absorption of nutrients.

FIGURE 4.37 Biopsies of **a.** normal intestine and **b.** coeliac intestine

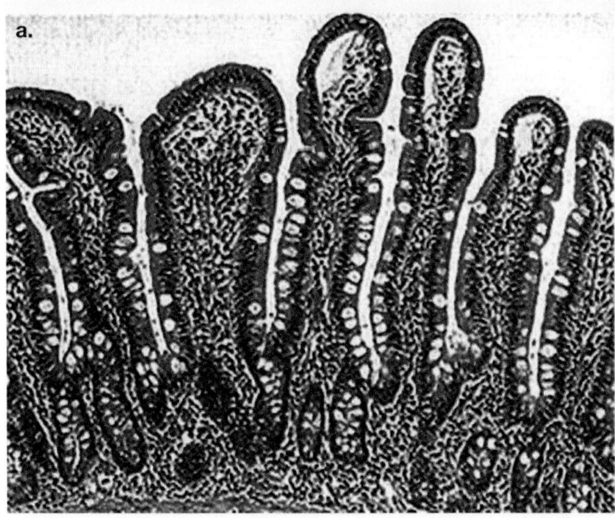

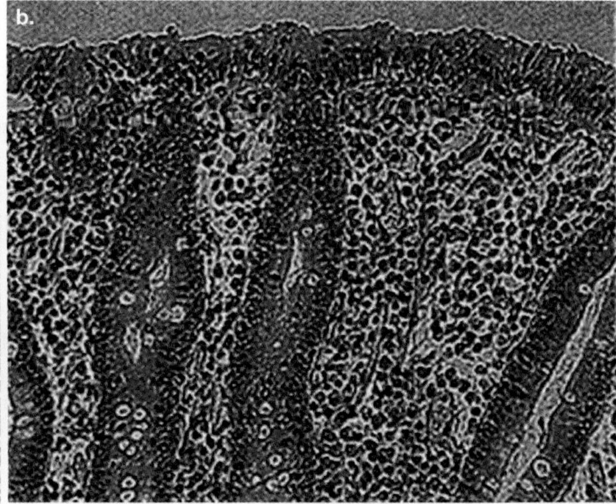

Coeliac Australia refers to coeliac disease as a 'hidden epidemic'. Although about 1 in 70 people in Australia have been diagnosed with the condition, many (about 80 per cent) do not know that they have it. If left undiagnosed, more severe consequences — such as a variety of nutritional deficiencies, bowel cancer and osteoporosis — may result.

SCIENCE AS A HUMAN ENDEAVOUR: Coeliac disease under focus

People with coeliac disease are intolerant to gluten. This protein is found in wheat, rye, barley and oats.

In 2019, within hours of eating gluten, distinct markers were discovered in the blood of people affected by coeliac disease. This discovery involved an international collaboration of the world's leading coeliac disease experts. The team included Associate Professor Jason Tye-Din, Head of coeliac research at Australia's Walter and Eliza Hall Institute of Medical Research. Although there are already blood tests available for diagnosis, this discovery has triggered research with a focus of developing a simple blood test to diagnose coeliac disease.

In 2009, the world's first trials of a coeliac disease vaccine developed by Australian researchers began. Ten years later, the phase 2 trials for Nexvax2® were discontinued. Although the quest to discover and develop an effective medical treatment for coeliac disease continues, a strict, gluten-free diet is still the only way to manage it.

FIGURE 4.38 Associate Professor Jason Tye-Din, Head of coeliac research at Australia's Walter and Eliza Hall Institute of Medical Research and a gastroenterologist at The Royal Melbourne Hospital

elog-2155

INVESTIGATION 4.4

Observing villi

Aim

To investigate the internal structure of the lining of the small intestine

Materials

- prepared slides of the walls of the small intestine
- monocular light microscope

Method

Use a light microscope to observe the prepared slide of the walls of the small intestine.

Results

Draw a diagram of your observations. Record the magnification used, label a villus and use descriptive labels to record your detailed observations.

Discussion

1. Describe the function of a villus. (Read through the information previously given in this lesson if you are unsure.)
2. With reference to your observations, suggest how the shape of a villus suits its function.

Conclusion

Summarise the findings for this investigation.

4.5 Activities

learnon

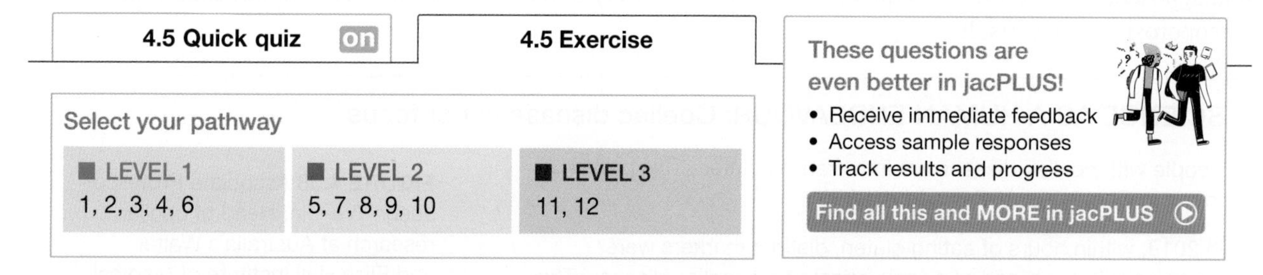

Remember and understand

1. **a.** State whether the following statements are true or false.

Statement	True or false?
i. Acid produced from sugar in food may 'eat' a hole in your tooth enamel and dentine.	
ii. Plaque refers to a thin film of food, saliva and bacteria that may cover your teeth.	
iii. Rotting plaque can cause your gums to swell and bleed.	
iv. Fluoride may be added to the water supply or to toothpaste to increase tooth decay.	
v. It is in the last section of your large intestine where most of the nutrients are absorbed into your bloodstream.	
vi. The shape of the villi in the small intestine increases the surface area available for absorption of nutrients into capillaries.	
vii. People with coeliac disease can eat foods containing gluten.	

   b. Justify any false responses.

2. Label the structures (a–e) in the diagram of the tooth using the following terms: enamel, dentine, gum, pulp, root canal.

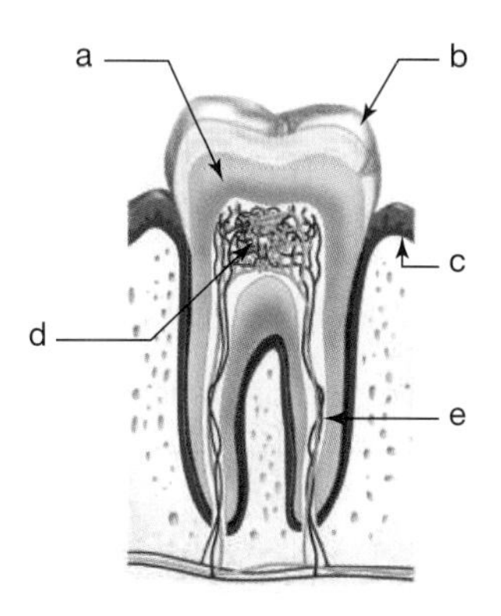

3. Match the tooth part to its description in the table shown.

Tooth part	Description
a. Dentine	**A.** The name of the channel where the nerves and blood vessels go down into the jawbone
b. Enamel	**B.** The bone-like material that gives the tooth its shape and makes up most of the tooth but, if exposed, wears away
c. Pulp	**C.** This forms a coating over the exposed surface of the tooth and is the hardest substance in your body.
d. Root canal	**D.** This part of the tooth contains most of the nerves and blood vessels.

4. Describe the discovery that led to two Australian scientists winning the 2005 Nobel Prize in Medicine.
5. Outline the relationship between diet, coeliac disease and the digestive system.
6. Approximately how many people in Australia are affected by coeliac disease?

Apply and analyse

7. Imagine that you have invited two friends over for a sleepover. One of them has coeliac disease and the other is lactose intolerant.
   a. Find out the cause and symptoms associated with each of these conditions.
   b. Find out what sorts of foods you could offer your friends.
   c. Design a dinner and breakfast menu that includes foods that each of your friends would be able to eat.
   d. Share and discuss your menus with those of other class members.
8. Select one of the 'tooth pathway' careers in section 4.5.3 'A future in teeth'. Find out details of the training required and what a career in this pathway would entail. Present your findings in a brochure and include a section that describes what 'a day in the life of a …' would be like.
9. Research one of the following digestion-related diseases and report on the cause, symptoms, treatment or cure, possible consequences and current research.
   - Heartburn
   - Inflammatory bowel disease
   - Irritable bowel syndrome
   - Appendicitis
   - Constipation
   - Crohn's disease
   - Diverticulosis
   - Gallstones
   - Haemorrhoids
   - Pancreatitis
   - Peptic ulcer
   - Colonic polyps

Evaluate and create

10. **SIS** Recently, scientists have suggested a link between the presence of the bacteria *Helicobacter pylori* and cancer protection. Find out more about this research and suggest possible implications that it may have.
11. **SIS** **a.** Identify claims made about the coeliac vaccines such as Nexvax2®.
    b. Do you think that the Nexvax2® clinical trials should have been discontinued? Justify your response.
12. **SIS** Design an investigation to test the following hypotheses:
    - Drinking fluoridated water reduces tooth decay.
    - Mouthwash prevents the growth of bacteria that cause tooth decay.
    - Drinking bottled water rather than tap water increases tooth decay.

Fully worked solutions and sample responses are available in your digital formats.

LESSON
4.6 Circulatory system — blood highways

LEARNING INTENTION

At the end of this lesson you will be able to describe the structure and function of the circulatory system.

4.6.1 What's in blood?

An average-sized human has about five litres of blood; that's about a bucketful. Blood is made up of red blood cells (**erythrocytes**), white blood cells (**leucocytes**), blood platelets and the straw-coloured fluid they all float in, called **plasma** (figure 4.39).

int-6584

FIGURE 4.39 You have all of this in your blood.

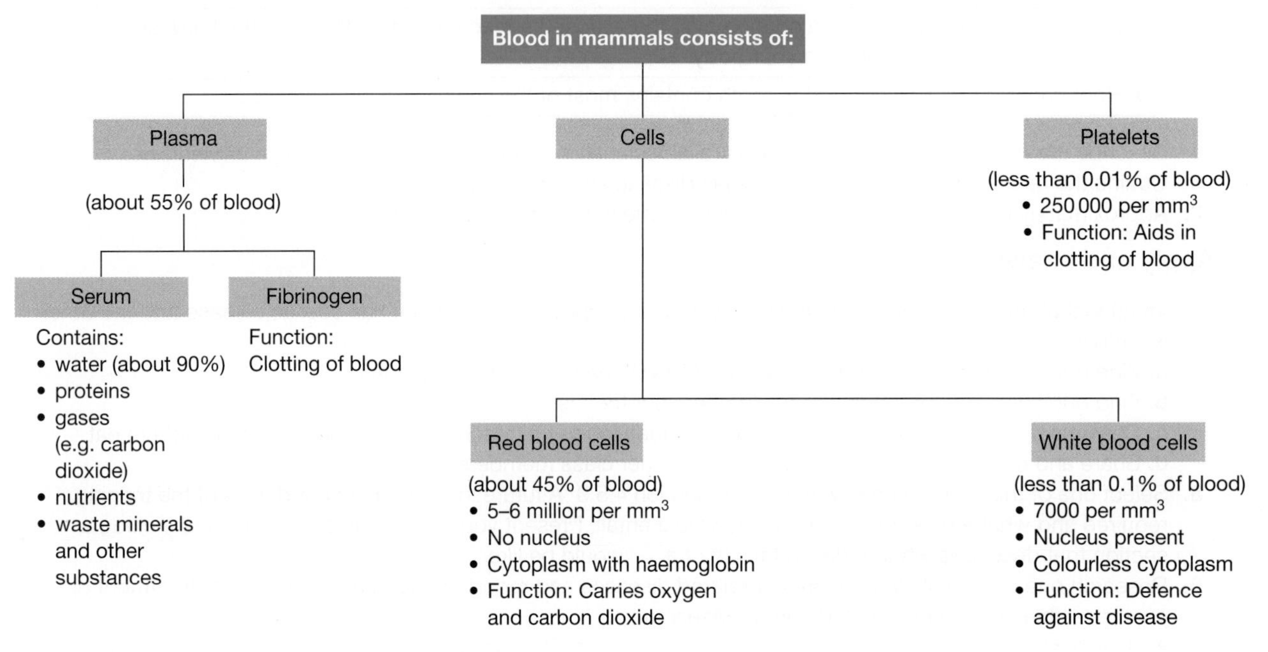

Ready to carry!

Each drop of blood contains about 300 million **red blood cells** with the important job of carrying oxygen around your body. Red blood cells are red because they contain an iron-containing pigment called **haemoglobin**. Oxygen reacts with haemoglobin in red blood cells to form oxyhaemoglobin, which makes the blood an even brighter red. This change in colour intensity can indicate the amount of oxygen being transported in blood at a particular time.

The shape and size of red blood cells make them well suited to their function. Their small size allows them to fit inside tiny capillaries. When mature, red blood cells lack a nucleus, increasing space available to carry haemoglobin and hence oxygen. Their biconcave shape means that they have a large surface area for their size, which also assists in their important oxygen-transporting role.

erythrocytes red blood cells

leucocytes white blood cells

plasma the yellowish liquid part of blood that contains water, minerals, food and wastes from cells

red blood cells living cells in the blood that transport oxygen to all other living cells in the body

haemoglobin the red pigment in red blood cells that carries oxygen

FIGURE 4.40 These red blood cells (erythrocytes) travel around the body up to 300 000 times (or for about 120 days). After this they literally wear out and die. Fortunately, each second, you are manufacturing about 1.7 million replacement red blood cells in your bone marrow.

EXTENSION: The effect of altitude on oxygen levels in blood

The amount of oxygen carried by haemoglobin varies with altitude. At sea level, about 100 per cent of haemoglobin combines with oxygen. At an altitude of 13 000 metres above sea level, however, only about 50–60 per cent of haemoglobin combines with oxygen.

Fit to fight!

White blood cells contain a nucleus, and are larger and fewer in number than red blood cells. They are often referred to as the 'soldiers' in the blood as they are involved in fighting disease. Some white blood cells produce chemicals called antibodies; others engulf and 'eat' bacteria and other foreign matter. When you are ill or fighting an infection, the number of white blood cells in your blood increases for this reason.

white blood cells living cells that fight bacteria and viruses as part of the human body's immune system

platelets small bodies involved in blood clotting; they are responsible for healing by clumping together around a wound

Clot and cover ...

If you cut yourself, you bleed. This is because a blood vessel has been cut. **Platelets** in the blood help it to clot and plug the damaged blood vessel. This seal prevents germs getting in.

Resources

Video eLesson Bleeding (eles-2535)

EXTENSION: Blood pigments

Insect blood looks a little like raw egg white, because it contains no pigment. The blood of crabs and crayfish, however, contains the pigment haemocyanin. This pigment has copper in it and is blue when combined with oxygen. This differs from haemoglobin in humans, which is red when combined with oxygen.

4.6.2 Mix and match?

How much do you know about the red stuff that flows throughout your body? Did you know that your blood might not mix too well with that of your friends? Blood can be grouped into eight types using the ABO system and the Rhesus (Rh) system. Your blood type is inherited from your parents.

FIGURE 4.41 How common is your blood?

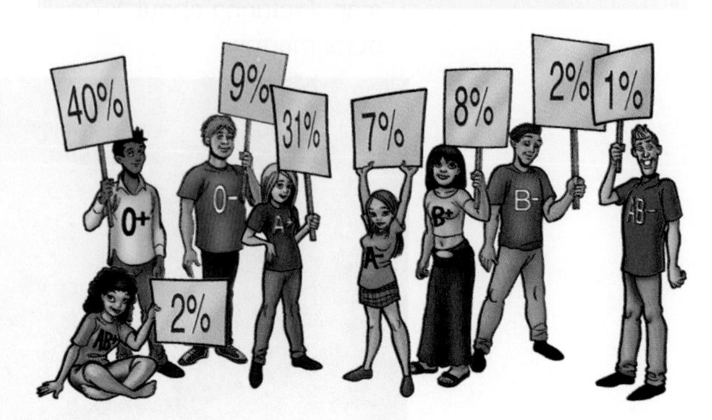

The blood-type classification systems are based on whether particular chemicals are present or absent on your red blood cells. If you are Rh-negative, you do not have the Rhesus factor on your red blood cells; if you do, you are Rh-positive.

The ABO system divides blood into groups A, B, AB and O. If you need a blood transfusion, it is very important to know your blood type and that of the donor because some blood types cannot be mixed. If the wrong types are mixed, the blood cells may clump together and cause fatal blockages in blood vessels.

4.6.3 Connected pathways

Your circulatory system is responsible for transporting oxygen and nutrients to your body's cells, and wastes such as carbon dioxide away from them. This involves interactions between blood cells, blood vessels and your heart.

Blood vessels called arteries transport blood from your heart, whereas others, called veins, transport blood back to the heart, as seen in figure 4.42. Materials are exchanged between blood and cells through tiny blood vessels called capillaries that are located between arteries and veins.

FIGURE 4.42 Your circulatory system consists of your heart, blood vessels and blood. Arteries, capillaries and veins are the major types of blood vessels through which your blood travels.

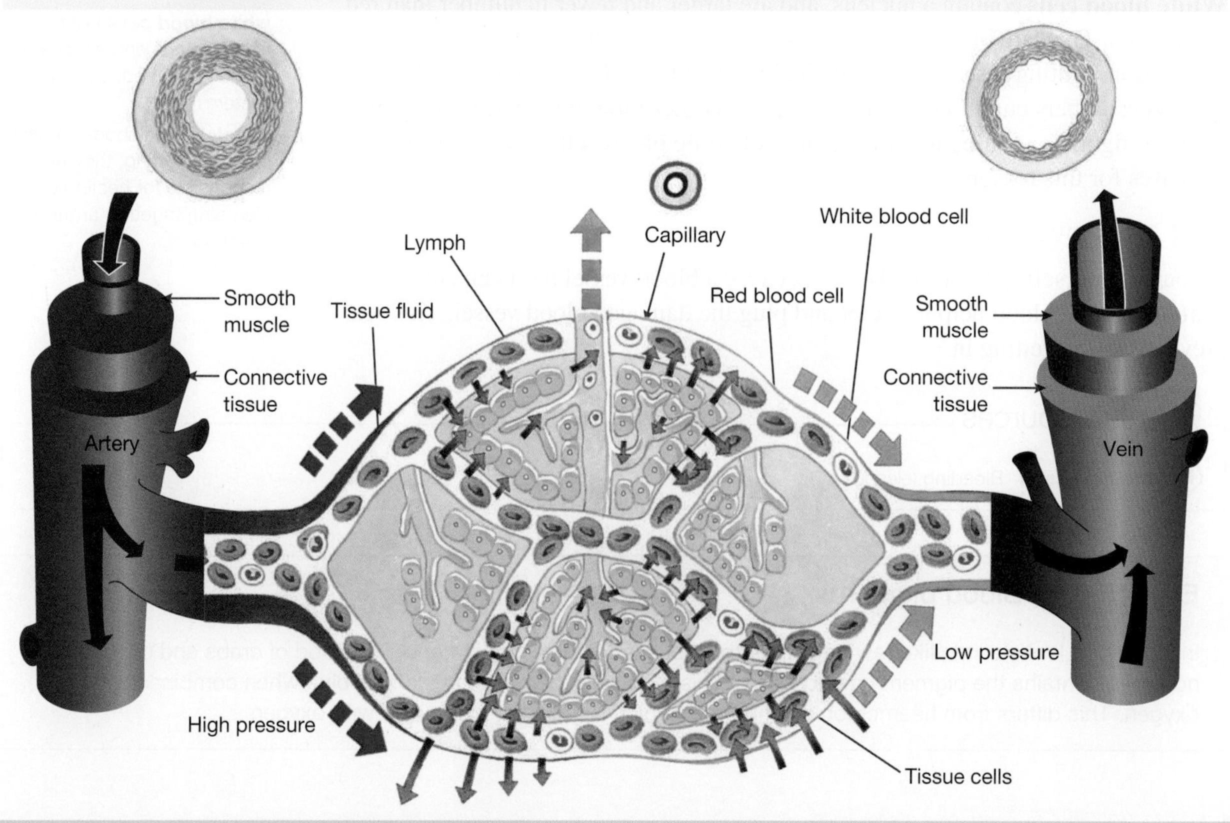

Arteries, veins and capillaries

Arteries have thick, elastic, muscular walls and carry blood under high pressure away from your heart. Veins have thinner walls and possess valves that prevent the blood from flowing backwards as they take blood to your heart.

FIGURE 4.43 The circulatory system

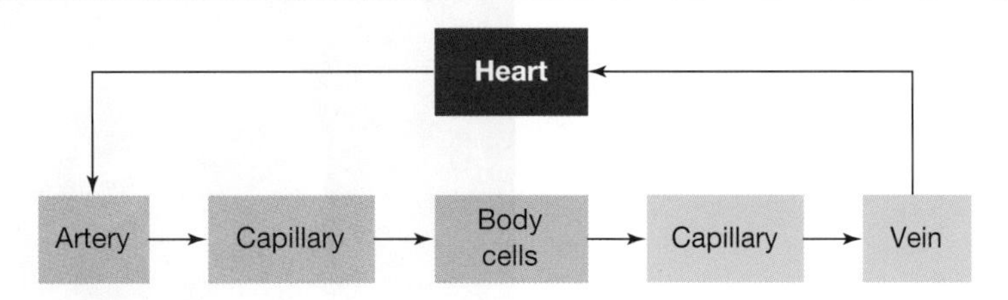

eles-2047

FIGURE 4.44 Veins have valves to ensure that blood flows in only one direction.

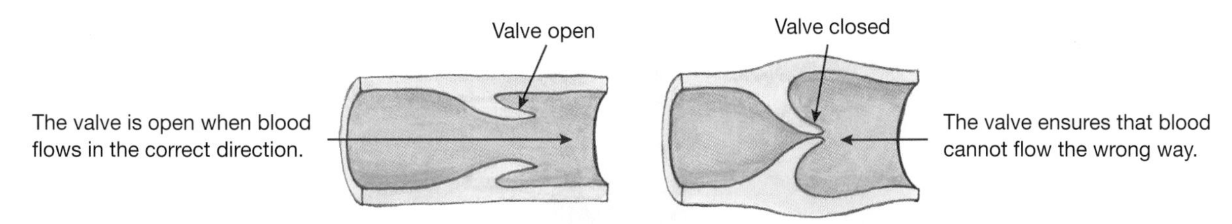

Capillaries are the most numerous and smallest blood vessels. Your body contains about 100 000 km of capillaries, which penetrate almost every tissue, so no cell is very far away from one. Capillaries are very important because they transport substances such as oxygen and nutrients to cells and remove wastes such as carbon dioxide.

CASE STUDY: Bruising

If you bump yourself but haven't cut your skin, a bruise may form. Bruises are caused by burst blood capillaries under your skin. The bruise changes from black to purple to yellow as the blood clears away.

FIGURE 4.45 As the blood clears away from a bruise the colour changes.

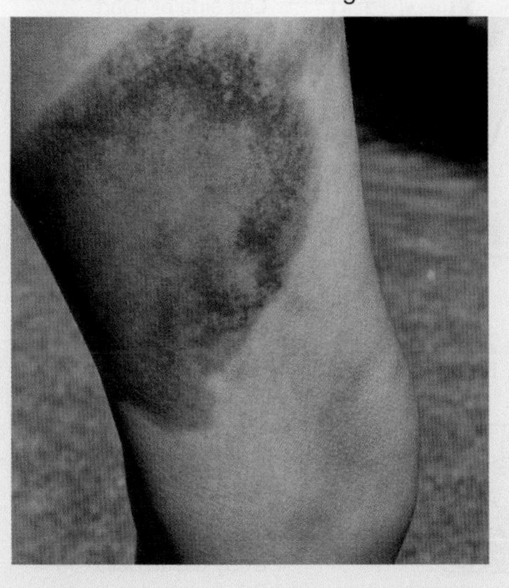

FIGURE 4.46 In the capillaries, oxygen diffuses out of the blood and waste produced by cells diffuses into the bloodstream.

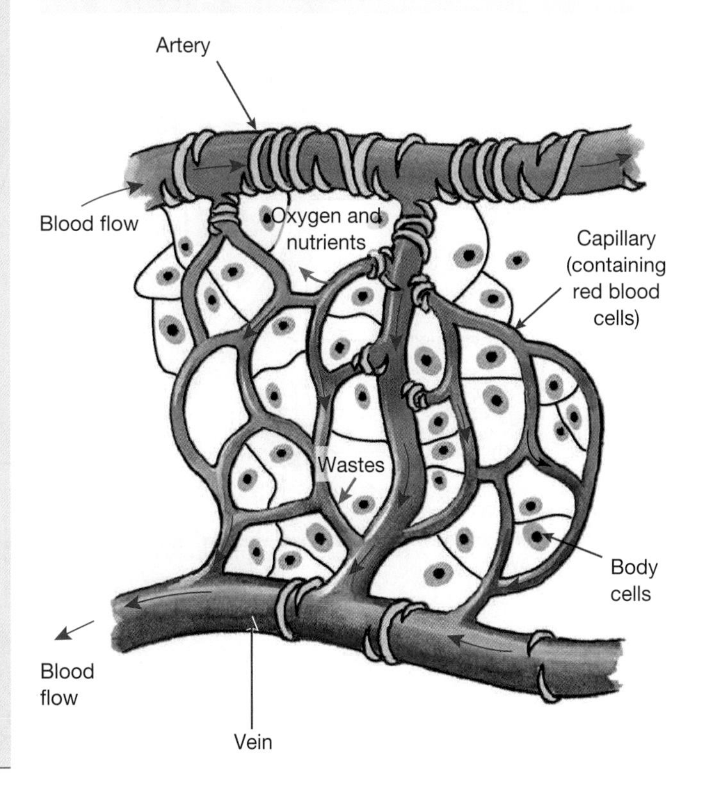

4.6.4 Have a heart

Often linked with emotions, love and courage, the heart has a special meaning for most of us. In a clinical sense, however, it is merely a pump about the size of your clenched fist. It is located between the lungs, slightly towards the left-hand side, which explains why the left lung is slightly smaller.

The heart is responsible for the movement of blood throughout the body, circulating oxygen, carbon dioxide and other molecules in a continuous cycle as seen in figure 4.48.

FIGURE 4.47 A 3D image of the location of the heart between the lungs

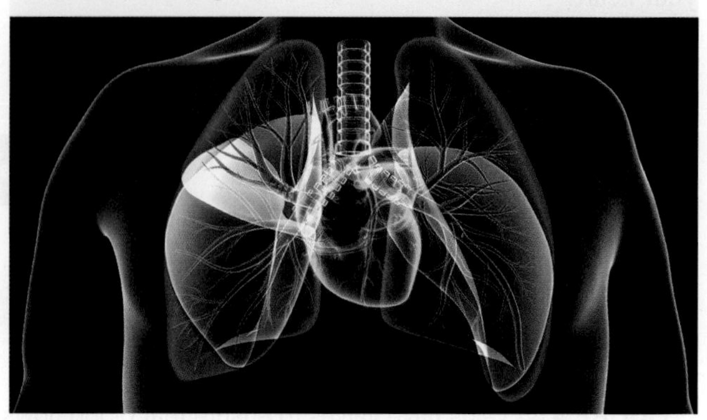

ewbk-11891

eles-2049

int-0210

FIGURE 4.48 The movement of blood through the heart

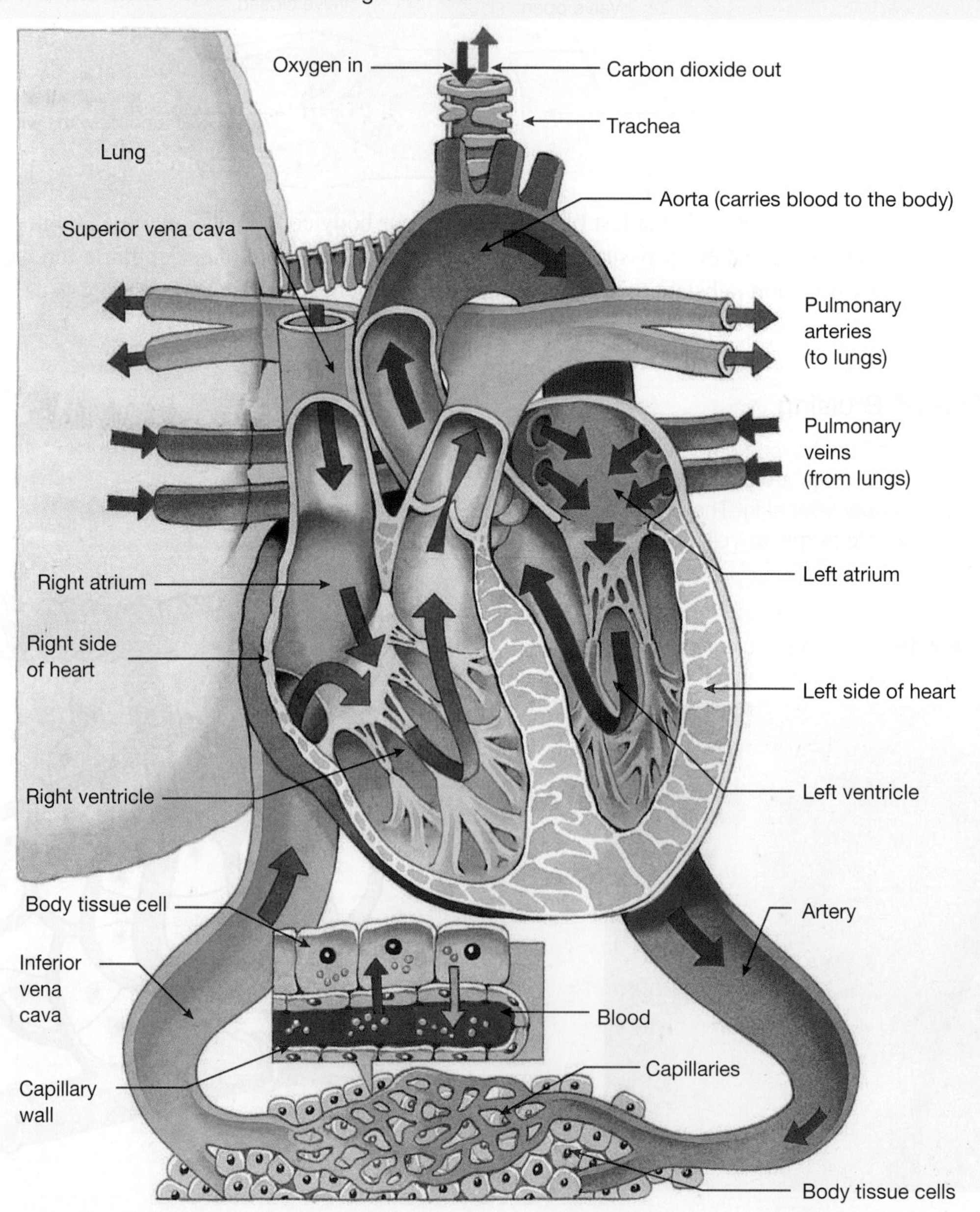

Two pumps in one

To be more precise, the human heart is actually *two* pumps. One side contains **oxygenated blood** and the other **deoxygenated blood**. Veins bring 'used' deoxygenated blood (stripped of oxygen and bluish in colour) from cells in your body back to your heart. All of these veins join up into a larger vein called the **vena cava**. Entering the top-right chamber of your heart, blood is pumped into the bottom-right chamber. It is then pumped out to your lungs, where it picks up oxygen and becomes oxygenated and more reddish in colour. It also loses some of its carbon dioxide. The oxygenated blood then returns via a vein from your lungs to the left-hand side of your heart to be pumped out through arteries to your body tissues, where it delivers oxygen and nutrients. The deoxygenated blood then returns to the right-hand side of the heart for the cycle to be repeated. This can be seen in figure 4.49 which is a simplified flow chart of how blood is pumped around the body.

Four chambers

The human heart has four chambers. The upper two chambers are called the **left atrium** and **right atrium** (plural = atria), and the lower two chambers are called the **left ventricle** and **right ventricle**. The two sides of the heart are different. The walls of the left side are thicker and more muscular because they need to have the power to force the blood from the heart to the rest of the body.

Flap-like structures attached to the heart walls, called **valves**, prevent the blood from flowing backwards and keep it going in one direction. If you listen to your heart beating, you will hear a **'lub dub'** sound. The 'lub' sound is due to the valves between the ventricles and atria shutting. The 'dub' sound is due to the closing of the valves that separate the heart from the big blood vessels that lead to the lungs and the rest of the body.

oxygenated blood the bright red blood that has been supplied with oxygen in the lungs

deoxygenated blood blood from which some oxygen has been removed

vena cava the large vein leading into the top-right chamber of the heart

left atrium the upper-left section of the heart where oxygenated blood from the lungs enters the heart

right atrium the upper-right section of the heart where deoxygenated blood from the body enters

left ventricle the lower-left section of the heart, which pumps oxygenated blood to all parts of the body

right ventricle the lower-right section of the heart, which pumps deoxygenated blood to the lungs

valves flap-like folds in the lining of a blood vessel or other hollow organ that allow a liquid, such as blood, to flow in one direction only

'lub dub' the sound made by the heart valves as they close

FIGURE 4.49 The heart is actually two pumps. One side pumps oxygenated blood and the other deoxygenated blood.

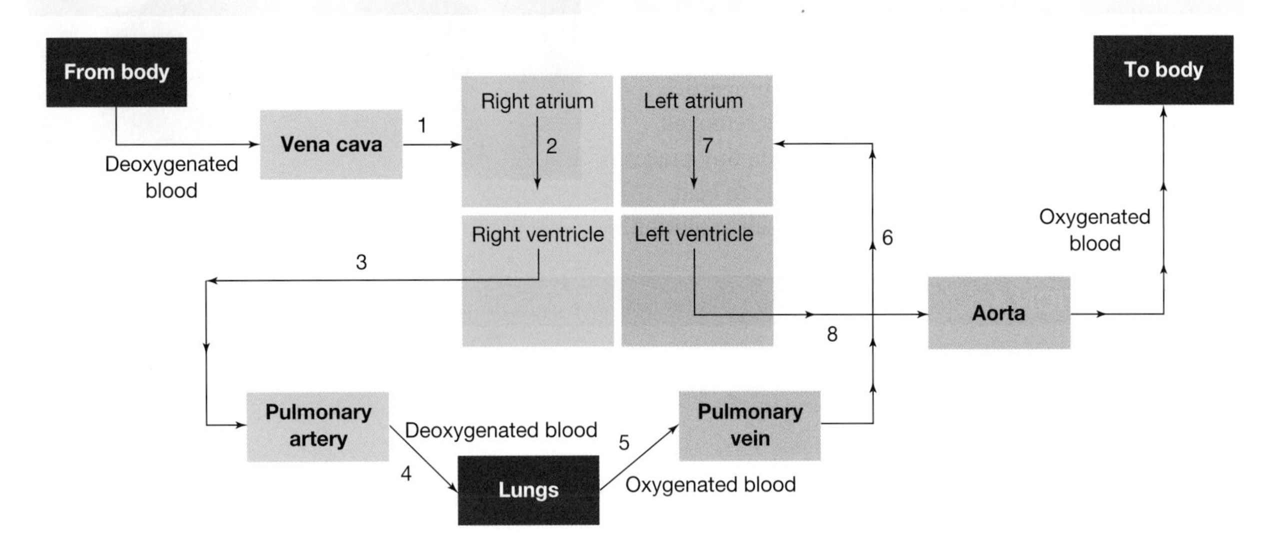

4.6.5 Blood pressure

The heart's pumping action and the narrow size of the blood vessels result in a build-up of considerable pressure in the arteries. The force with which blood flows through the arteries is called **blood pressure**. It is affected by different activities and moods. It also goes up and down as the heart beats, being highest when the heart contracts (**systolic pressure**) and lowest when the heart relaxes (**diastolic pressure**). A person's blood pressure is expressed as a fraction. This fraction is the systolic pressure over the diastolic pressure, such as 120/70.

blood pressure measures how strongly the blood is pumped through the body's main arteries

systolic pressure the higher blood pressure reading during contraction of the heart muscles

diastolic pressure the lower blood pressure reading during relaxation of the heart muscles

heartbeat a contraction of the heart muscle occurring about 60–100 times per minute

pulse the alternating contraction and expansion of arteries due to the pumping of blood by the heart

pacemaker an electronic device inserted in the chest to keep the heart beating regularly at the correct rate; it works by stimulating the heart with tiny electrical impulses

cardiac muscle a special kind of muscle in the heart that never tires; it is involved in pumping blood through the heart

4.6.6 Keeping the pace

During each minute that you are sitting and reading this, about 5–7 litres of blood completes the entire circuit around your body and lungs. In a single day, your heart may beat about 100 000 times and pump about 7000 litres of blood around your body.

A normal human heart beats about 60–100 times a minute, with this rate increasing during exercise or stress. With each **heartbeat**, a wave of pressure travels along the main arteries. If you put your finger on your skin just above the artery in your wrist, you can feel this **pulse** wave as a slight throb. Your pulse rate immediately after exercise can be used as a guide to your physical fitness. The fitter you are, the less elevated your heart rate will be after vigorous exercise.

The regular rhythmic beating of the heart is maintained by electrical impulses from the heart's **pacemaker**, which is located in the wall of the right atrium. Some people with irregular heartbeats are fitted with artificial, electronic pacemakers to regulate the heart's actions and correct abnormal patterns.

FIGURE 4.50 A person fitted with an artificial pacemaker

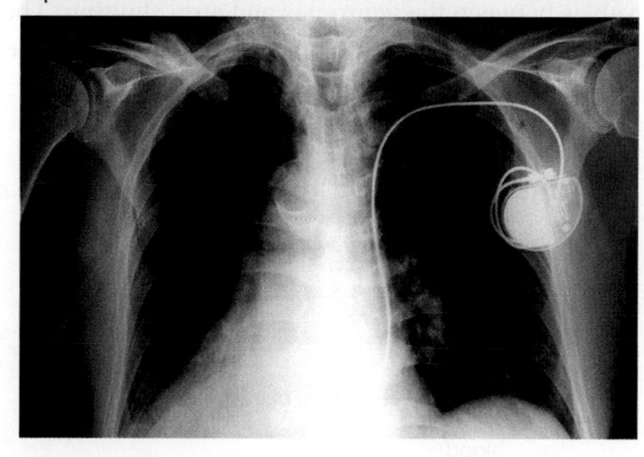

Try clenching your fist every second for five minutes. Getting a little tired? The heart is made up of special muscle called **cardiac muscle**, which never tires. Imagine having a 'cramp' or 'stitch' in your heart after running to catch the bus! Owing to its unique electrical properties, heart muscle will continue to beat even if it is removed from the body. Scientists have shown that even tiny pieces of this muscle cut from the heart will continue to beat when they are placed in a test tube of warm salty solution.

elog-2157

INVESTIGATION 4.5

Viewing blood cells

Aim

To observe blood cells under a light microscope

Materials

- prepared slide of blood smear
- microscope

Method

1. Place the prepared slide onto the microscope stage.
2. Use low power to focus, then carefully adjust to high power.
3. Find examples of red blood cells and white blood cells on the slide.

Results

1. Draw diagrams of representative red blood cells and white blood cells. On your diagram, include descriptive labels and the magnification used.
2. Estimate (a) how many red blood cells could fit inside a white blood cell and (b) how many of each cell type could fit across the field of view.

Discussion

1. Summarise the similarities and differences between the structures of red blood cells and white blood cells.
2. Suggest reasons for the differences.
3. Find out more about the structural differences between red blood cells and white blood cells.
4. Describe how the structure of each type of blood cell well suits it to its function.

Conclusion

Summarise the findings for this investigation.

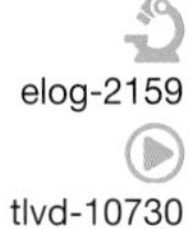

elog-2159

tlvd-10730

INVESTIGATION 4.6

Heart dissection

Aim

To observe the structure of a mammalian heart

Materials

- sheep's heart, preferably with the blood vessels still attached
- dissecting instruments
- newspaper or paper to cover dissection board

Method

1. Place newspaper on the dissection board, then place the heart on top of the paper.
2. Use the diagram shown to identify the parts of the heart.
3. Try to locate where blood enters and leaves the heart:
   a. to and from the lungs
   b. to and from the rest of the body.
4. Cut the heart in two so that both halves show the two sides of the heart (similar to the illustration in figure 4.48).

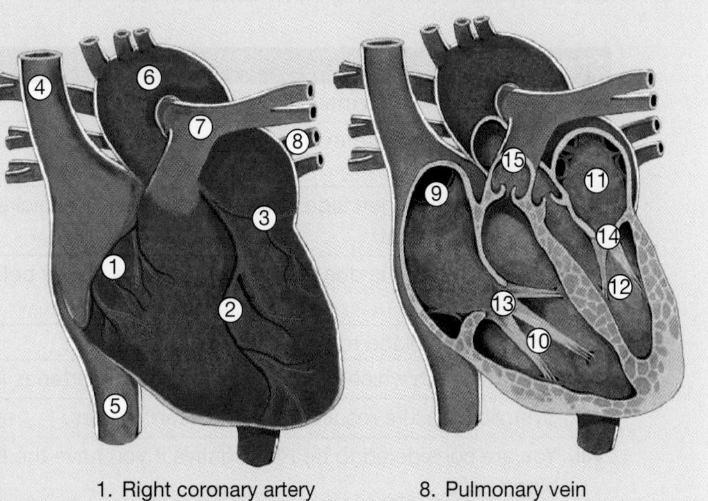

1. Right coronary artery
2. Left anterior descending coronary artery
3. Circumflex coronary artery
4. Superior vena cava
5. Inferior vena cava
6. Aorta
7. Pulmonary artery
8. Pulmonary vein
9. Right atrium
10. Right ventricle
11. Left atrium
12. Left ventricle
13. Tricuspid valve
14. Mitral valve
15. Pulmonary valve

Results

1. Sketch and label the heart and use arrows to show the direction of blood flow.
2. In a diagram, record your observations of the thickness of the walls on the left side of the heart compared with the right side.
3. Suggest reasons for the differences observed.
4. Try to locate the valves in the heart.

Discussion

1. Describe the valves and suggest their function.
2. Write a summary paragraph about the structure and function of the heart.

Conclusion

Summarise the findings for this investigation.

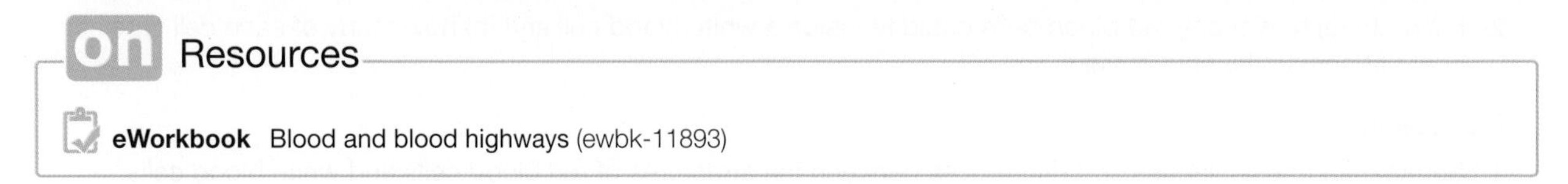

on Resources

eWorkbook Blood and blood highways (ewbk-11893)

4.6 Activities

learnon

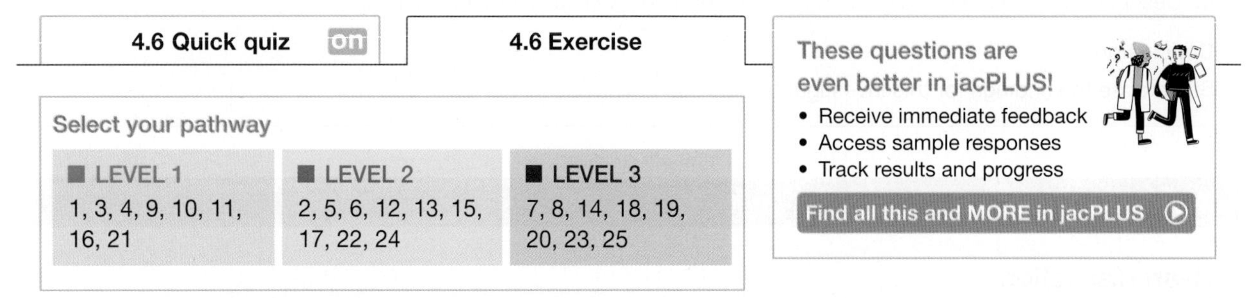

4.6 Quick quiz on | 4.6 Exercise

Select your pathway

■ LEVEL 1
1, 3, 4, 9, 10, 11, 16, 21

■ LEVEL 2
2, 5, 6, 12, 13, 15, 17, 22, 24

■ LEVEL 3
7, 8, 14, 18, 19, 20, 23, 25

These questions are even better in jacPLUS!
- Receive immediate feedback
- Access sample responses
- Track results and progress

Find all this and MORE in jacPLUS

Remember and understand

1. a. State whether the following statements are true or false.

Statement	True or false?
i. The human heart is made up of three chambers.	
ii. Valves in the heart prevent blood from flowing backwards and keep it going in one direction.	
iii. The walls on the left side of the human heart are thicker and more muscular than those on the right.	
iv. The 'dub' sound is due to the closing of the valves between the ventricles and atria shutting.	
v. Arteries take blood to and from the heart.	
vi. The force with which blood flows through the arteries is called blood pressure.	
vii. Systolic pressure results when the heart relaxes.	
viii. You are considered to be Rh-negative if you have the Rh antigen in your red blood cells.	
ix. The right side of the human heart pumps deoxygenated blood to the lungs.	

   b. Justify any false responses.

2. What am I? State another name for each of the following.
   a. Red blood cell
   b. Leucocyte
   c. The straw-coloured fluid in which blood cells float
   d. A cell fragment involved in clotting of the blood
   e. The iron-containing pigment that gives red blood cells their colour

3. Match the circulatory system term with its description in the table shown.

Term	Description
a. Artery	A. The bottom two chambers of the heart
b. Atria	B. Cell involved in transporting oxygen around the body
c. Capillary	C. Blood vessel that takes blood to the heart
d. Heart	D. Cell involved in protection against infection
e. Red blood cell	E. Blood vessel that takes blood away from the heart
f. Vein	F. The top two chambers of the heart
g. Ventricles	G. Organ that pumps blood around the body
h. White blood cell	H. Blood vessel that exchanges substances with cells

4. List the following in the order that a red blood cell would travel after leaving the aorta: pulmonary artery, left ventricle, right atrium, intestine, lung, pulmonary vein, left atrium, liver, right ventricle.
5. Outline what blood is and what blood does.
6. Name and describe the types of blood vessels in which blood travels around your body.
7. Describe the relationship between arteries, capillaries and veins.
8. a. Describe what is unusual about cardiac muscle.
   b. Describe what blood pressure is caused by.
   c. Explain why there are valves in the heart.
9. a. How many times does a normal human heart beat each minute?
   b. Suggest what may cause the heart rate to increase.
   c. Explain how the rhythmic beating of the heart is maintained.

Apply and analyse

10. Carefully examine figure 4.41. Which blood type is the most common? Which is the least common?
11. **SIS** Construct an appropriate graph to show the different proportions of blood components.
12. The higher the altitude, the less oxygen there is in the air. Propose a reason people living at high altitudes usually have more red blood cells than people living at low altitudes.
13. Think of other ways that information about the components of blood could be organised visually. Organise the material in one of these ways.
14. a. Copy figure 4.49 into your workbook.
    b. Use red and blue coloured pencils to show the path taken for a red blood cell to travel from the vena cava to the aorta. Use a red coloured pencil to indicate oxygenated blood and a blue coloured pencil to show deoxygenated blood.
    c. In your diagram, indicate which two blood vessels transport:
       i. deoxygenated blood
       ii. oxygenated blood.
    d. Use an 'X' to indicate which blood vessel you would expect to have the highest blood pressure.
15. Find out more about how blood circulates in insects and lobsters.
16. Find out what happens when people donate their blood at a blood bank. How often can you donate blood, how long does it take and how much blood do they take? Prepare a brochure, storyboard, PowerPoint presentation or cartoon to share your findings.
17. Observe the image of human blood cells shown.
    a. Identify which are white blood cells and which are red blood cells.
    b. Describe how you distinguished between the two types of blood cells.
    c. Which are in the greatest abundance? Suggest a reason for this.

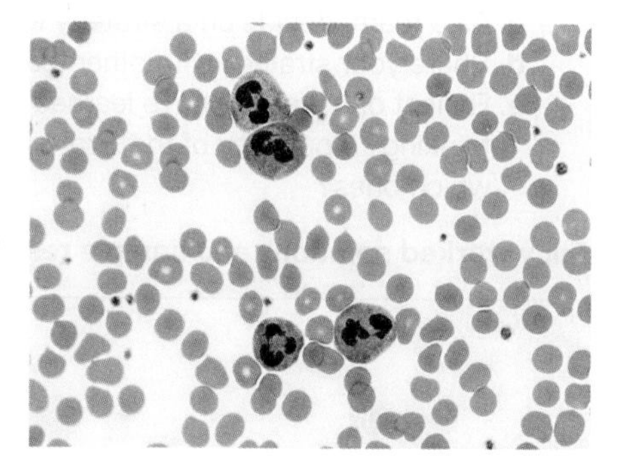

18. What is the difference between:
    a. the blood in the two sides of the heart
    b. the structure of the two sides of the heart
    c. systolic and diastolic pressure?

Evaluate and create

19. Compare red blood cells, white blood cells and blood platelets.

20. Blood transfusions involve an injection of a volume of blood, previously taken from a healthy person, into a patient. This has to be done carefully, as if the wrong type of blood is injected, the blood will clump and the patient could die. Carefully examine the figure, which shows which blood group combinations may be compatible for a blood transfusion.

a. Which blood group(s), A, B, AB or O, can be accepted by:

i. all blood groups

ii. blood group AB

iii. blood group A?

b. Which blood group, A, B, AB or O, can receive transfusions from all blood types?

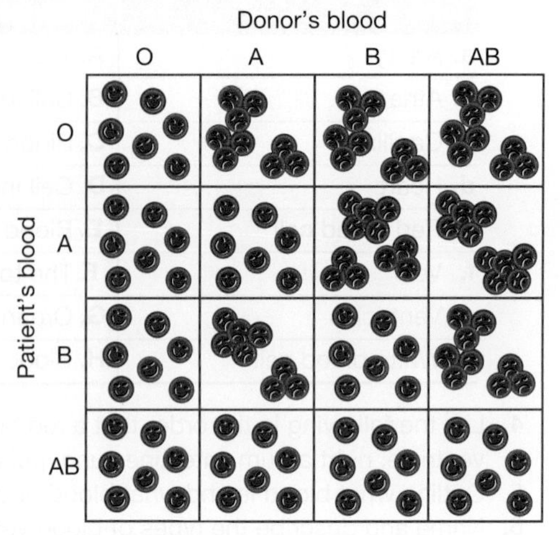

21. a. Use information in this lesson and other resources to relate structural features to the functions of the following parts of the circulatory system.

TABLE Parts of the circulatory system and their functions

Part of circulatory system	Structural features	Function
Arteries		
Veins		
Capillaries		
Red blood cells		
White blood cells		

b. Convert the information in the table into a Venn diagram, target map or another visual thinking tool.

22. a. Some people have religious grounds for disagreeing with the use of blood transfusions. Imagine a four-year-old child with a life-threatening condition. Her parents will not allow her to have the blood transfusion that she needs. What should the doctors do? Discuss this with your team and report your decision to the class. If there are any differences of opinion, organise a class debate on the issue.

b. Would your response be different if the child was 18 years old and wanted the blood transfusion, but her parents would not allow it?

23. A day after donating blood, a person finds that they have an infectious disease that can be transmitted by blood. What should they do? Discuss this with your team, giving reasons for your opinions.

24. With a partner, construct a PMI chart for a law that makes it compulsory for everyone over 16 to donate blood at least once a year.

25. Imagine that you have a friend who is anaemic. They are constantly tired and very pale.

a. Using the internet and other resources, find out what you could do to help your friend improve their health.

b. Report back to your team, sharing your ideas and any other relevant information. Have your team summarise your ideas in a cluster map or mind map.

c. As a team, decide on a strategy for helping your anaemic friend.

d. Share your strategy with other teams as a mind map, flowchart, concept map or another visual tool.

e. Reflect on what you have learned during this activity. How might it influence your future behaviour or thinking? Could any of the strategies designed by the teams be used to solve any other problems? If so, which ones?

Fully worked solutions and sample responses are available in your digital formats.

LESSON
4.7 Transport technology

LEARNING INTENTION

At the end of this lesson you will be able to provide examples of how, due to discoveries made using new and improved technologies, our understanding of the circulatory system has changed over time.

Science as a human endeavour

4.7.1 Scientific theories can change over time

Our understanding of the circulatory system has been built by scientists and physicians throughout human history. With new observations and evidence, some theories have been discarded and others developed or modified. New technologies have enabled new observations to be made, which have resulted in new ways of thinking about the structure and functioning of the human body.

Claudius Galen (*c*.129–*c*.199 AD)

For over a thousand years, the key training books used for doctors were based on the ideas of the Greek physician Claudius Galen. Galen's ideas were based on his observations of dissections of animals (other than humans). Galen described the human heart as being made up of two chambers and also being the source of the body's heat. He believed that blood was made by the liver and travelled to the right chamber of the heart, and that the left chamber made 'vital spirits' which were then transported by arteries to body organs. He was the first to use the pulse as a diagnostic aid.

FIGURE 4.51 Claudius Galen (*c*.129–*c*.199 AD)

Andreas Vesalius (1514–1564)

Hundreds of years later another physician, Andreas Vesalius (figure 4.52), began to transform medical knowledge — by questioning all previous theories. He believed that it was necessary to dissect bodies to find out how they worked. As the Church did not allow this, he took bones from graves and even stole a body from the gallows. His drawings showed the positions and workings of the muscles and organs in the body (figure 4.53). Vesalius's observations proved that some of Galen's theories were wrong and he discovered anatomical structures previously unknown. His findings helped establish surgery as a separate medical profession.

FIGURE 4.52 Andreas Vesalius (1514–1564), Belgian anatomist, dissecting a cadaver. Vesalius was made professor of anatomy and surgery at Padua University, Italy, in 1537.

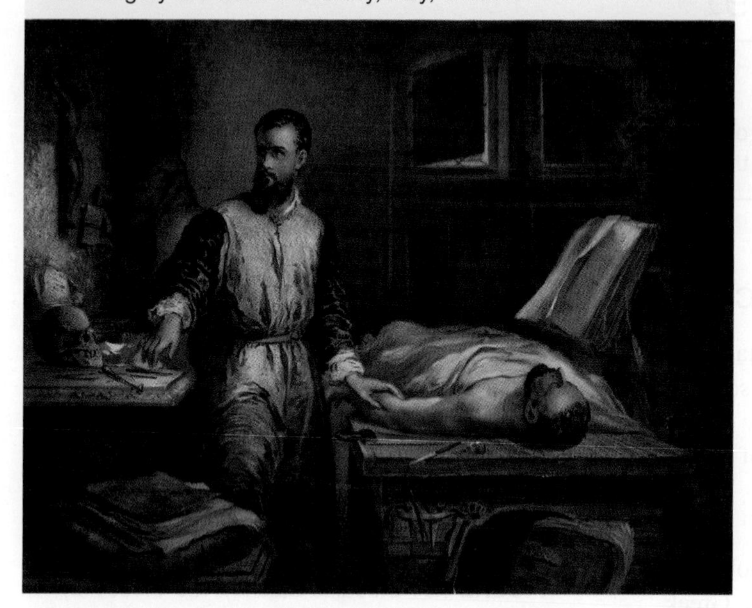

FIGURE 4.53 Artwork by Andreas Vesalius in 1543, showing the circulatory system

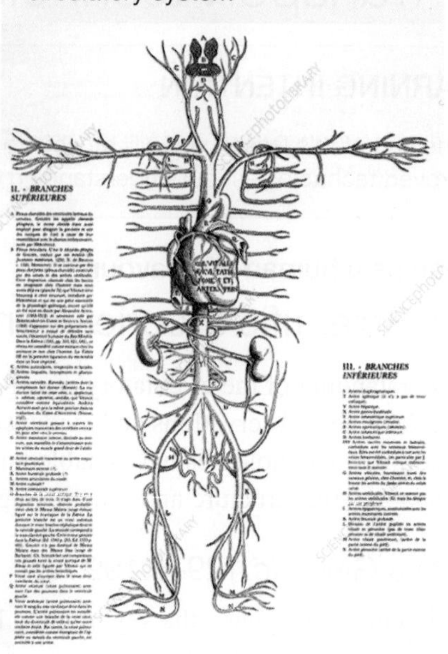

William Harvey (1578–1657)

Although Vesalius had assisted in revising the structure of the human heart, there was still confusion about its function. About 100 years later, William Harvey (figure 4.54), an English physician, conducted a series of circulation experiments (figure 4.55) that showed the way valves in the veins control the flow of blood back to the heart. The publication of this work in *On the Motion of the Heart and Blood in Animals* (1628) led to another change in how we think about the heart and our circulatory system.

FIGURE 4.54 English physician, William Harvey (1578–1657)

FIGURE 4.55 William Harvey's artwork of an arm with a tourniquet shows the way the valves in the veins control the flow of blood back to the heart.

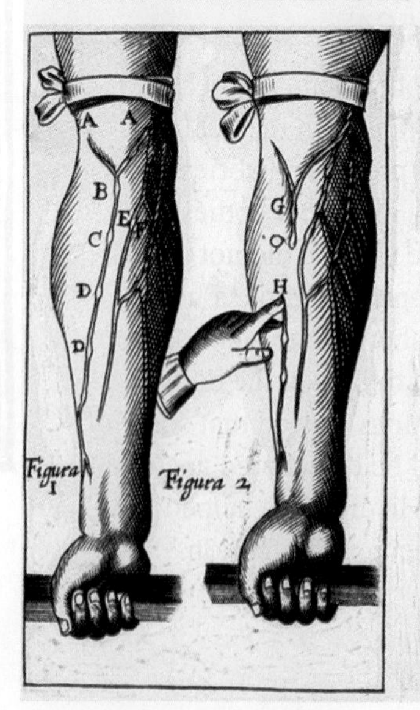

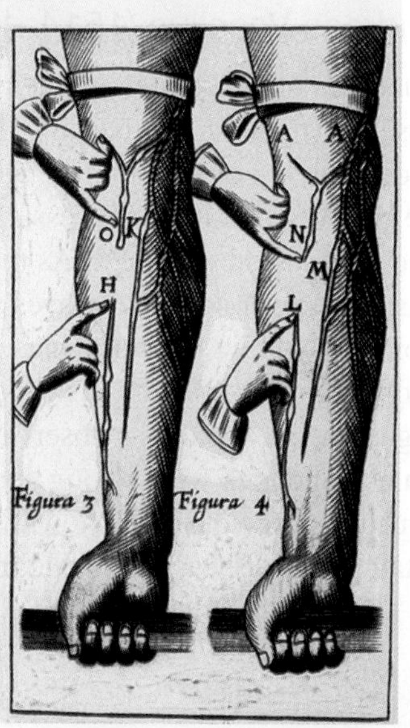

4.7.2 Heart technology

Heart and blood vessel diseases are a key cause of death for many Australians. Medical research and new technologies strive to minimise the effects of diseases and disorders of the circulatory system.

Faulty heart and vein valves

The heart, like many other pumps, depends on a series of valves to work properly. These valves open and close to receive and discharge blood to and from the chambers of the heart. They also stop the blood from flowing backwards. If any of the four heart valves becomes faulty, the function of the heart may be impaired.

varicose veins expanded or knotted blood vessels close to the skin, usually in the legs; they are caused by weak valves that do not prevent blood from flowing backwards

Veins throughout the body may also contain valves that keep the blood flowing in one direction. Defective valves in leg veins can cause blood to drain backwards, and to pool in the veins closest to the skin surface. These veins can become swollen, twisted and painful, and are called **varicose veins**.

FIGURE 4.56 A faulty heart valve may be replaced by an artificial valve. Why are the heart valves so important to the functioning of the heart?

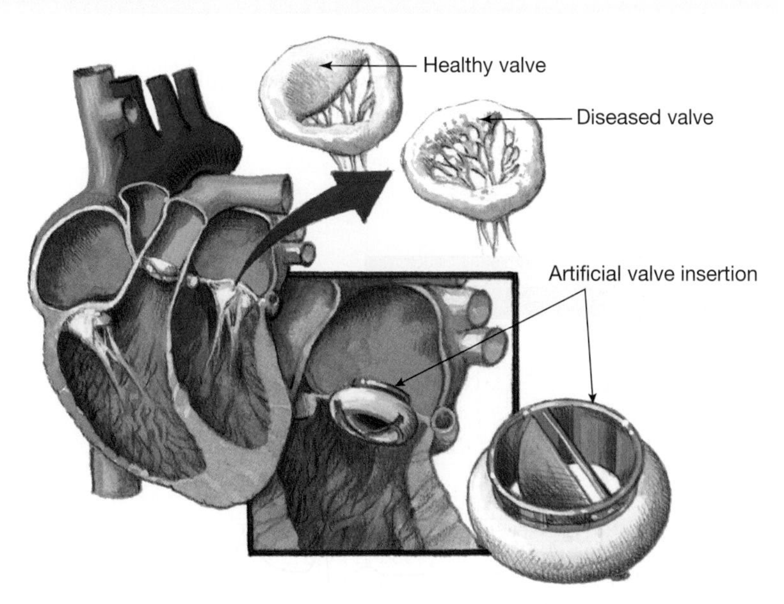

on Resources

Video eLesson Heart valve (eles-0858)

4.7.3 'If I only had a heart ...'

For those with a heart that no longer works, a transplant is the best option, but this requires a compatible donor. In Australia there are more than 100 people waiting for a heart transplant, which means there are waiting times of up to two years.

Artificial hearts

The tin man from *The Wizard of Oz* would have been very happy with the development of an artificial heart (figure 4.57). This mechanical device is made of titanium and plastic. Surgeons also implant a small electronic device in the abdominal wall to monitor and control the pumping speed of the heart. An external battery is strapped around the waist and can supply about 4–5 hours of power. An internal rechargeable battery is also implanted inside the wearer's abdomen. This is so they can be disconnected from the main battery for about 30–40 minutes for activities such as showering.

FIGURE 4.57 An artificial heart can be made from metals, ceramics and polymers.

Genetically modified animal hearts

In January 2022 the first pig heart was inserted into a man in the United States. The pig embryo had been genetically modified so that it would not cause the human body to reject the heart as foreign. While the man only lived for an additional six weeks, there are hopes that this technology can be refined to overcome the present organ shortage.

4.7.4 A heart — but no pulse?

If only the left ventricle is damaged, and the rest of the heart is in good working order, a back-up pump may be implanted alongside the heart. One model of these devices results in its wearers having a gentle whirr rather than a pulse. This is the sound of the propeller spun by a magnetic field to force a continuous stream of blood into the aorta.

Getting the beat!

An **electrocardiogram (ECG)**, as seen in figure 4.58, shows the electrical activity of a person's heart. ECG patterns are valuable in diagnosing heart disease and abnormalities.

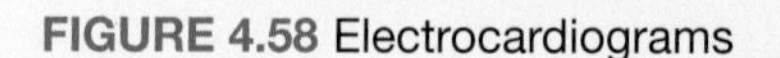

FIGURE 4.58 Electrocardiograms

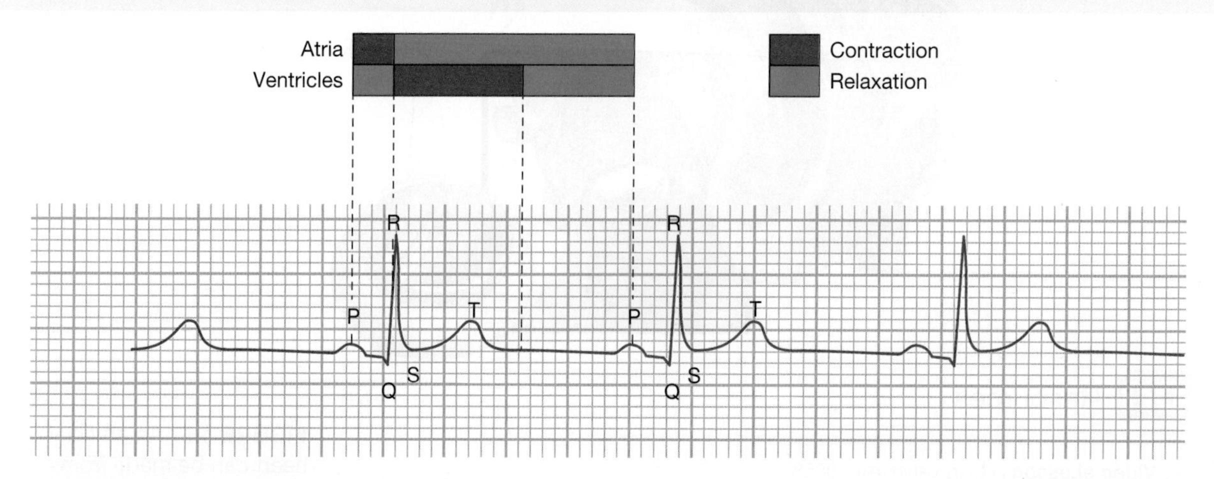

a. Normal electrocardiogram

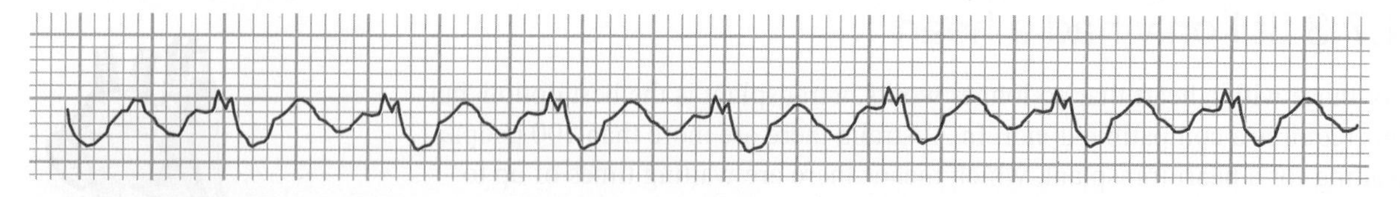

b. Abnormal electrocardiogram

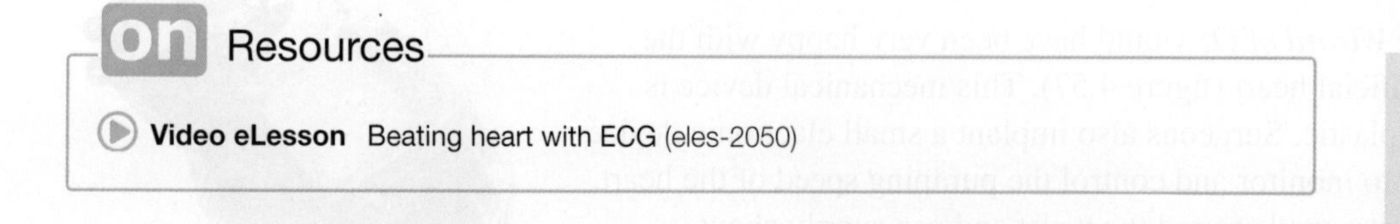

on Resources

Video eLesson Beating heart with ECG (eles-2050)

electrocardiogram (ECG) a graph made using the tiny electrical impulses generated in the heart muscle, giving information about the health of the heart

4.7.5 Artificial blood

A current wave of interest in vampire movies and books has brought with it discussion about the merits of artificial blood sources. The interest in artificial blood, however, is not new; people have thought about its use in blood transfusions for hundreds of years. William Harvey's description in 1628 of how blood circulated through the body prompted a variety of unsuccessful investigations into the use of alternative fluid substitutes. A shortage of blood supplies during war and disease epidemics has fired up the quest for an artificial blood substitute. Currently, the two most promising red blood cell substitutes are haemoglobin-based oxygen carriers (HBOCs) and perfluorocarbon-based oxygen carriers (PFCs).

PFCs are usually white, whereas HBOCs are a very dark red. Although PFCs are entirely synthetic, HBOCs are made from sterilised haemoglobin. The haemoglobin may be from human or cow blood, human placentas or bacteria that have been genetically engineered to produce haemoglobin. As the haemoglobin doesn't have a cell membrane to protect it, various techniques are used to make it less fragile. Some scientists are even investigating the idea of wrapping it in an artificial membrane.

FIGURE 4.59 Human red blood cells are much larger than HBOCs and PFCs.

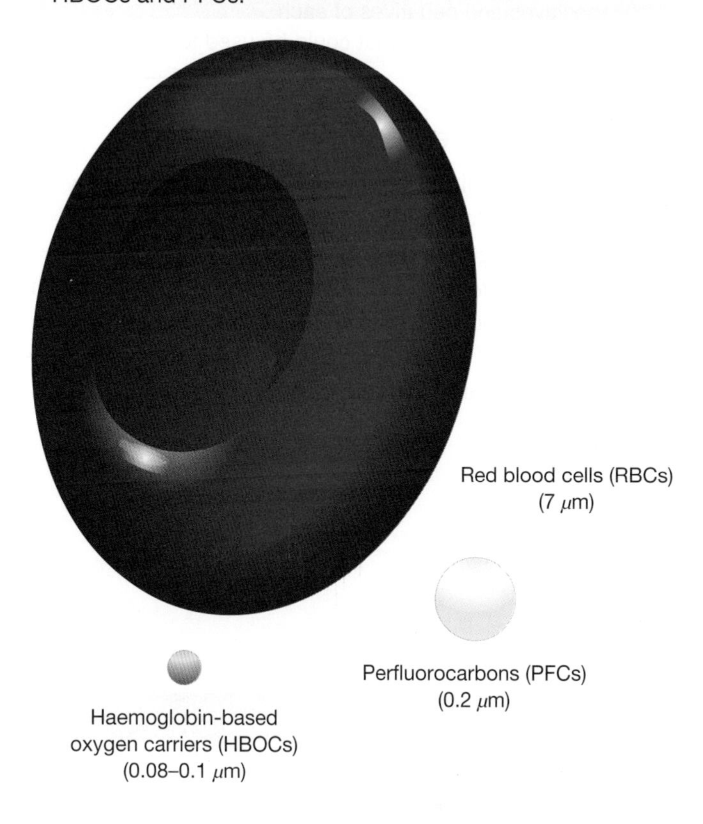

CASE STUDY: Transplant pioneer

If your heart or lungs were not working properly and you had needed a heart or lung transplant in the 1980s, the doctor to see was Victor Chang.

Victor Chang was an Australian doctor who was awarded a Companion of the Order of Australia for his contribution to medicine. Dr Chang played an important role in establishing the heart transplant unit at St Vincent's Hospital in Sydney. He set up a team of 40 health professionals who were the finest in their field, and developed new procedures and techniques that led to an improved rate of success. Of his patients, 92 per cent were still alive one year after their heart or lung transplant operation and 85 per cent were still alive five years later.

FIGURE 4.60 The late Dr Victor Chang, pioneering heart transplant surgeon

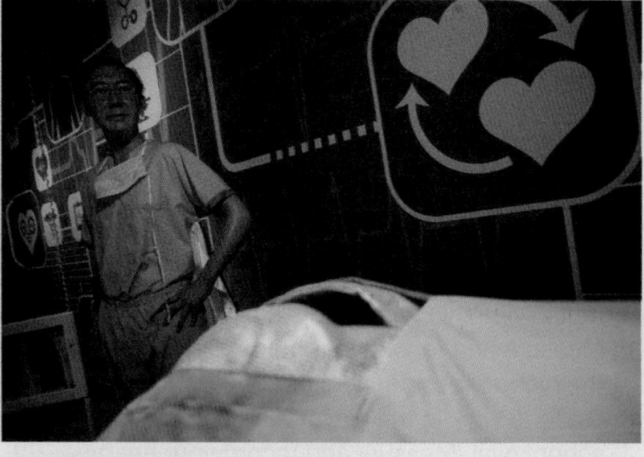

The first heart transplant operation that Victor Chang carried out at St Vincent's Hospital was in 1984 on a young girl called Fiona Coote. Fiona is now an adult and, although she has since needed a second heart transplant, she owes her life to Dr Chang.

Dr Victor Chang also developed an artificial heart valve, called the St Vincent heart valve, and was working on developing an artificial heart. Unfortunately, his life was tragically cut short in 1991 when he was shot.

ACTIVITY: Organ donation

- Research campaigns used to increase the rates of organ donation.
- For each campaign, run a class poll to determine if you would donate an organ based on it. Discuss the positives and negatives of each.
- Design a campaign that could be used to increase the rates of organ donation.

Artificial blood vessels?

Will the artificial blood vessels of the future be made by bacteria? Molecular biologist Helen Fink, working in Sweden, has suggested this may be the case. The cellulose produced by *Acetobacter xylinum* bacteria is strong enough to cope with blood pressure and function within our bodies, and could be used for artificial blood vessels in heart bypass operations in the future (figure 4.61).

FIGURE 4.61 In the future, will artificial blood vessels like this one be made by bacteria?

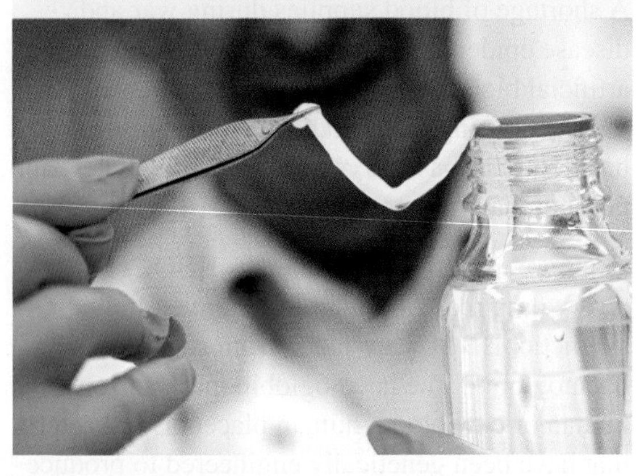

elog-2161

INVESTIGATION 4.7

Check your heart

Aim

To investigate the short-term effects of exercise on heart rate and blood pressure

Materials

- stopwatch
- blood pressure monitor
- optional: data logging or digital measuring devices

Method

1. Find your pulse, either on the inside of your wrist or in your neck (see the illustrations). Make sure you use two fingers, not your thumb, to find your pulse.
2. Measure and record your heart rate in beats per minute (bpm) by counting the number of times your heart beats in 15 seconds and then multiplying this number by 4.
3. Measure and record your blood pressure using the blood pressure monitor.
4. Go for a walk in the playground or around the school oval. Measure and record your heart rate and blood pressure again.
5. Run up and down a flight of stairs. Measure and record your heart rate and blood pressure again.

Two places where your pulse should be easy to find:
a. radial location (wrist) **b.** carotid location (neck)

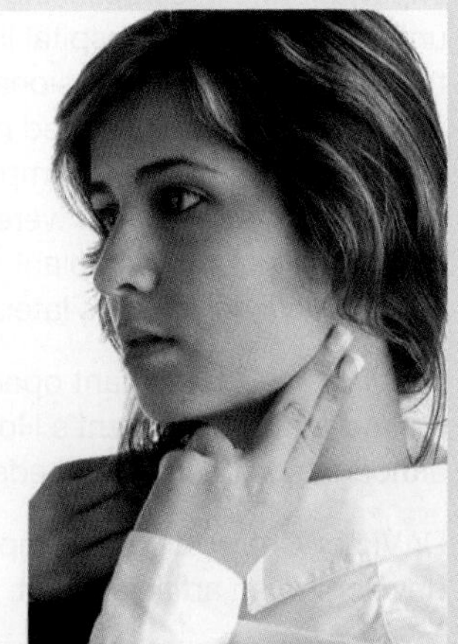

Results

Record your answers in the table provided.

TABLE Heart rate and blood pressure at rest and during exercise

Test	Heart rate (bpm)	Blood pressure (mmHg)
Before exercise		
After walking		
After running up stairs		

Discussion

1. What effect does exercise have on heart rate and blood pressure?
2. Identify strengths and limitations of this investigation and suggest improvements.
3. Design and carry out an experiment to test the following hypothesis: 'There is a link between a person's resting heart rate and the number of hours the person spends exercising each week'.

Conclusion

Summarise the findings for this investigation.

4.7 Activities

learnon

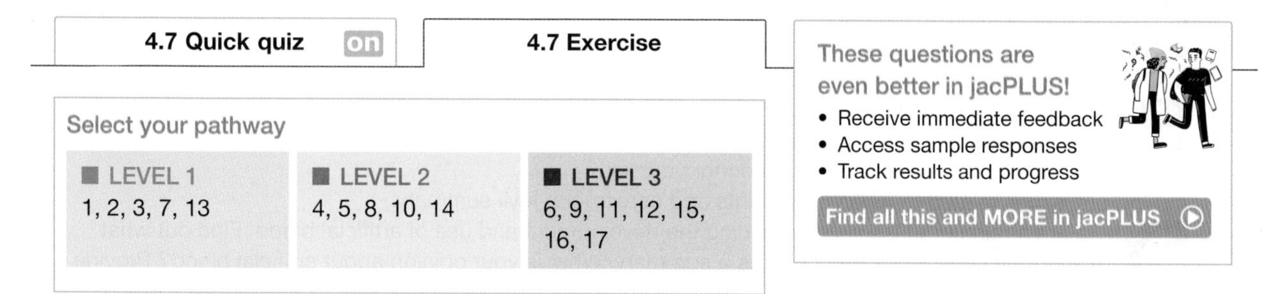

Remember and understand

1. **a.** State whether the following statements are true or false.

Statement	True or false?
i. Over 1700 years ago, Andreas Vesalius described the human heart as being made up of two chambers and the source of the body's heat.	
ii. Over 350 years ago, William Harvey published his findings on blood circulation, which contributed to our present-day understanding of the heart and the circulatory system.	
iii. Over 30 years ago, Victor Chang developed an artificial heart.	
iv. An electrocardiogram (ECG) shows the electrical activity of a person's heart.	
v. Defective valves in leg arteries can cause varicose veins.	

   b. Justify any false responses.
2. What are varicose veins and what causes them?
3. What is an electrocardiogram and when is it useful?
4. **a.** Explain why valves are important to the functioning of the heart.
   b. Outline how heart valves are similar to the valves in veins.
5. Describe how an electrocardiogram is used to detect heart abnormalities.

Apply and analyse

6. **SIS** Look at the electrocardiograms in figure 4.58.
   a. At 'P', are the muscle cells of the atria contracted or relaxed?
   b. After the 'QRS' wave, is the ventricle relaxed or contracted?
   c. How does the normal electrocardiogram differ from the abnormal electrocardiogram?
   d. Suggest what might be wrong with the heart activity shown on the abnormal electrocardiogram.
7. **SIS** Construct a matrix table to show the differences between red blood cells, HBOCs and PFCs.
8. **a.** Which organs are most successfully transplanted into humans?
   b. List sources of the organs for transplant and identify associated issues.
   c. Describe how donors and organ recipients are matched.
   d. Organ recipients can require specific treatment after the operation. Outline what this involves and why it is needed.
9. Dr Mary Kavurma and Dr Seana Gall are Tall Poppy Science Award winners. This award recognises young scientists who excel at research, leadership and communication. Dr Kavurma is a scientist at the University of New South Wales involved in research into atherosclerosis and cardiovascular disease. Dr Gall is based at the Menzies Research Institute, University of Tasmania, and her research field is cardiovascular epidemiology.
   a. Find out more about their research and that of other scientists in this field of science.
   b. Find out more about Australia's Tall Poppy Science Awards and other winners.

Evaluate and create

10. **SIS** **a.** Identify issues associated with organ transplantation.
    b. As a team, select one of these issues and find out why it is an issue.
    c. What is your opinion on the issue?
    d. Share your opinion and reasons for it with other members of your team.
    e. Construct a team PMI chart on the organ transplant issue.
11. **SIS** **a.** If you required a new heart, would you prefer an artificial one or one from a human or other natural source? Provide reasons for your response.
    b. Outline your opinion on being an organ donor yourself.
    c. Find out issues related to organ transplants and construct a PMI summary.
12. **SIS** There are a number of issues surrounding the development and use of artificial blood. Find out what these are and then construct a PMI chart as a summary. What is your opinion about artificial blood? Provide reasons for your opinion.
13. **SIS** Find articles in the media that advertise foods or drinks that can reduce heart disease. As a team, research the claims and summarise your findings in a SWOT diagram. As a class, be involved in a debate that includes members from different interest groups or with different perspectives or biases.
14. **SIS** Doctors use a stethoscope to listen to heartbeats. Make and test your own stethoscope using rubber tubing and a plastic funnel.
15. **SIS** **a.** Use the internet to identify problems relating to the circulatory system.
    b. Select one of these problems and construct a model or animation to demonstrate its effect on normal body function.
16. **SIS** In a team, design and perform an experiment to investigate the effect of different types of activities on your heart rate.
17. Find out more about Galen, Vesalius and Harvey and their work and discoveries. Suggest how they were influenced by the times in which they lived. Why didn't they just accept the ideas of their times? Why did they ask questions? Propose a question or hypothesis that you may have asked if you lived in each of their times.

Fully worked solutions and sample responses are available in your digital formats.

LESSON
4.8 Respiratory system — breathe in, breathe out

LEARNING INTENTION

At the end of this lesson you will be able to describe the structure and function of the respiratory system.

4.8.1 Cells need energy!

Breathe in deeply … now breathe out. You have exchanged gases with your environment. You have supplied your body with some essential oxygen and removed some unwanted carbon dioxide. You do this about 15–20 times a minute without even having to think about it. Where does this oxygen go and where did the carbon dioxide come from?

Your cells need **oxygen** as it is essential for cellular respiration. This process involves breaking down **glucose** so that energy is released in a form that your cells can use. This reaction produces **carbon dioxide** as a waste product that needs to be removed.

oxygen an atom that forms molecules (O_2) of tasteless and colourless gas; it is essential for cellular respiration for most organisms and is a product of photosynthesis

glucose a six-carbon sugar (monosaccharide) that acts as a primary energy supply for many organisms

carbon dioxide a colourless gas (CO_2) made up of one carbon and two oxygen atoms; it is essential for photosynthesis and is a waste product of cellular respiration; the burning of fossil fuels also releases carbon dioxide

trachea the narrow tube from the mouth to the lungs through which air moves

lungs the organ for breathing air; gas exchange occurs in the lungs

epiglottis a leaf-like flap of cartilage behind the tongue that closes the air passage during swallowing

Cellular respiration

Glucose + oxygen → carbon dioxide + water + energy

Respiratory system

The main role of your respiratory system is to get oxygen into your body and carbon dioxide out. This occurs when you inhale (breathe in) and exhale (breathe out). The respiratory system is made up of your **trachea** (or windpipe) and your **lungs** (figure 4.62).

EXTENSION: Epiglottis

Wrong way, turn back! There is a flap of tissue at the top of the trachea called the **epiglottis**. This tissue's job is to stop food 'going down the wrong way'. If food does go the wrong way, a cough moves the food back up, to either be removed or travel its correct pathway — down your oesophagus to your stomach.

FIGURE 4.62 The epiglottis stops food from going down your trachea.

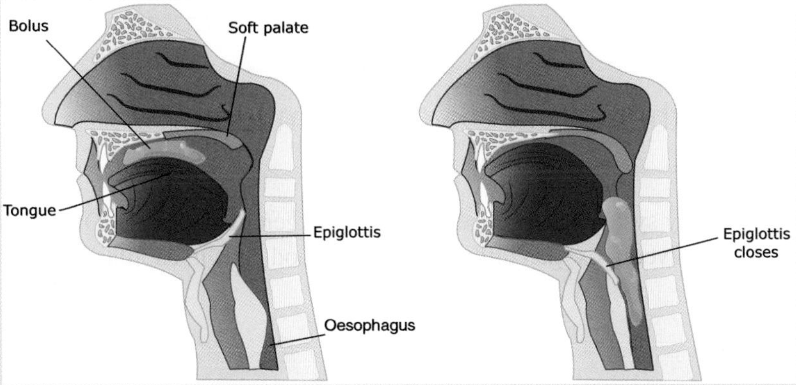

ewbk-11895
int-8233

FIGURE 4.63 **a.** Organs of the respiratory system with **b.** a portion of the lung expanded to show details

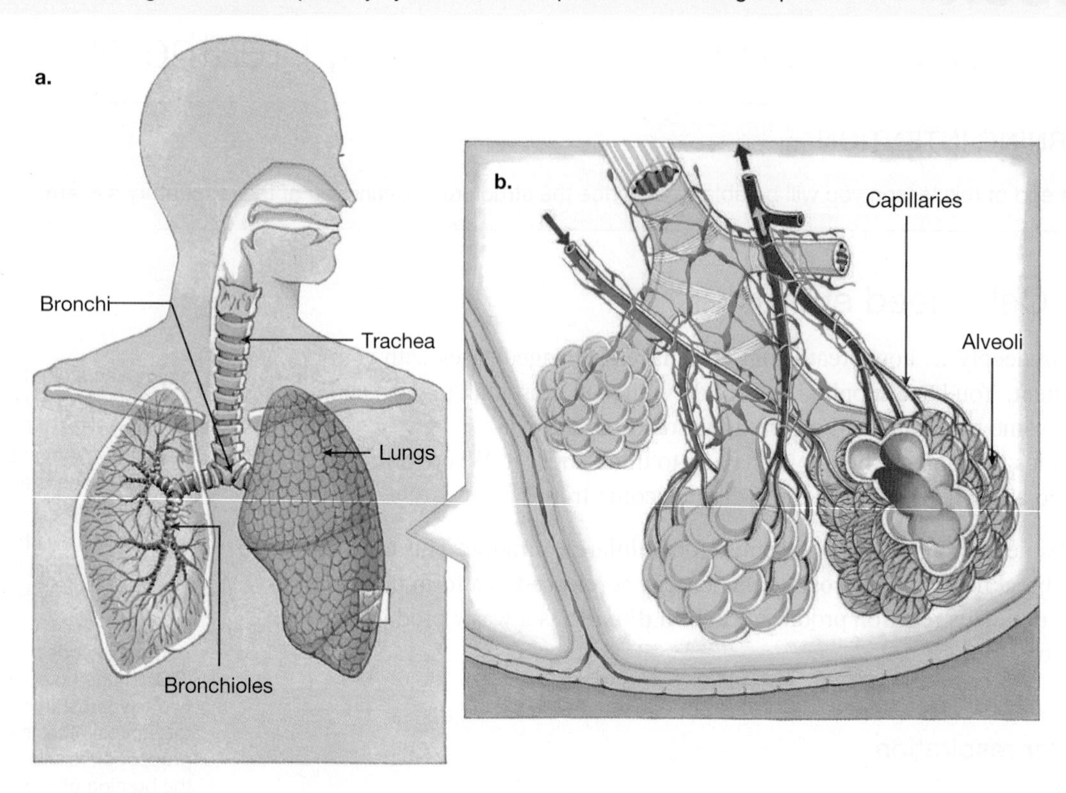

Getting oxygen to your lungs

When you breathe in, air moves down your trachea, then down into one of two narrower tubes called **bronchi** (bronchus). After that, the air moves into smaller branching tubes called **bronchioles**, which end in tiny air sacs called **alveoli** (alveolus) (figures 4.63 and 4.64). It is at the alveoli that gases (such as oxygen and carbon dioxide) are exchanged between the respiratory system and the circulatory system.

FIGURE 4.64 The pathway oxygen travels to your lungs when you inhale

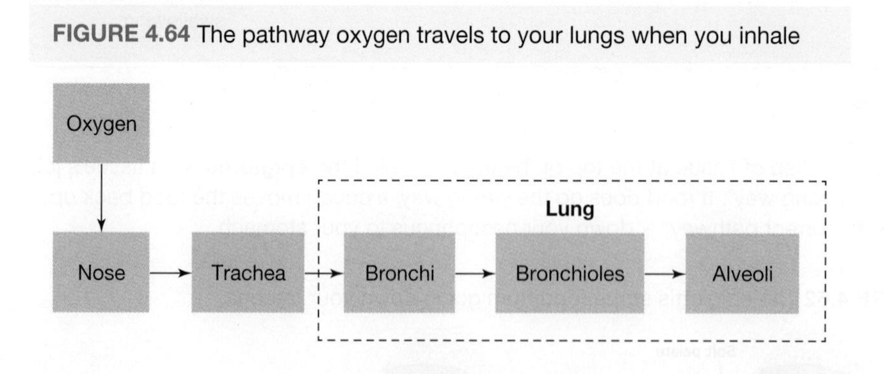

bronchi the narrow tubes through which air passes from the trachea to the smaller bronchioles and alveoli in the respiratory system; singular = bronchus

bronchioles small, branching tubes in the lungs leading from the two larger bronchi to the alveoli

alveoli tiny air sacs in the lungs at the ends of the narrowest tubes; oxygen moves from alveoli into the surrounding blood vessels, in exchange for carbon dioxide; singular = alveolus

4.8.2 Working together to get oxygen from lungs to cells

Your circulatory and respiratory systems work together to get oxygen to your cells. Once you have breathed in and oxygen has reached your alveoli, oxygen diffuses into red blood cells in capillaries that surround the alveoli (figure 4.65).

FIGURE 4.65 In an alveolus, oxygen diffuses into the blood and carbon dioxide diffuses out of the blood.

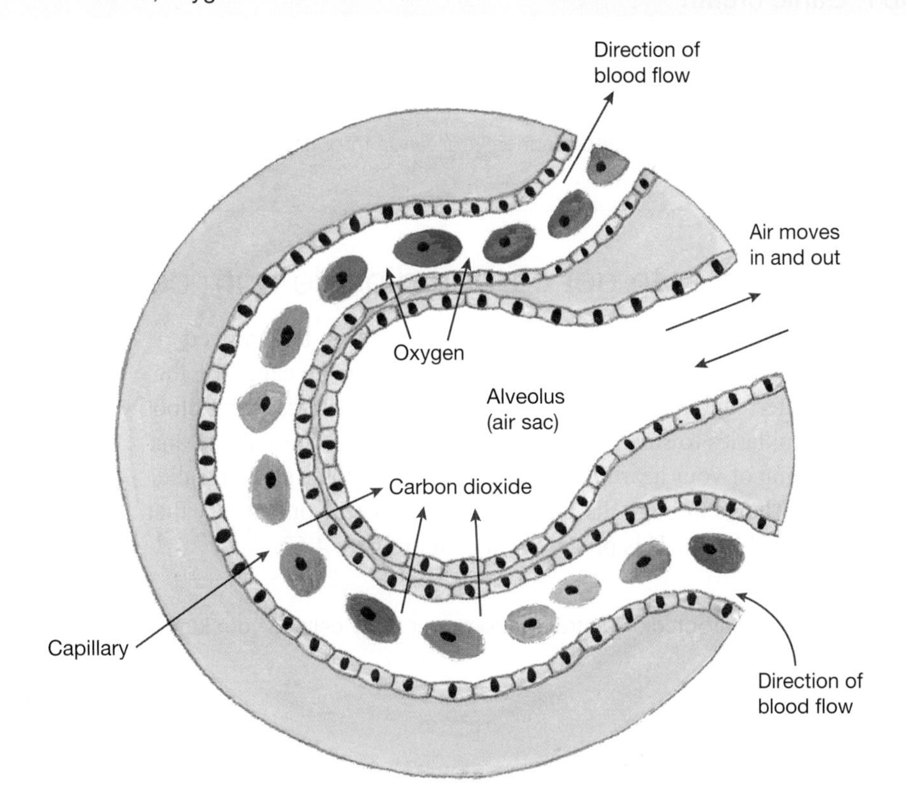

The oxygen diffuses into the red blood cells because there is a higher concentration of oxygen inside the alveoli than inside the blood cells. Once inside the red blood cells, oxygen binds to haemoglobin to form oxyhaemoglobin. It is in this form that oxygen travels throughout your body. The blood that it travels in is referred to as oxygenated blood.

In figure 4.66, the oxygenated blood travels from your lungs via the **pulmonary vein** to the left atrium of your heart. From here, it travels to the left ventricle where it is pumped under high pressure to your body through a large artery called the **aorta**. The oxygenated blood is then transported to smaller vessels (**arterioles**) and finally to capillaries, through which it diffuses into body cells for use in cellular respiration.

pulmonary vein the vessel through which oxygenated blood travels from your lungs to the heart
aorta a large artery through which oxygenated blood is pumped at high pressure from the left ventricle of the heart to the body
arterioles vessels that transport oxygenated blood from the arteries to the capillaries

FIGURE 4.66 The pathway oxygen travels from your lungs to your body cells

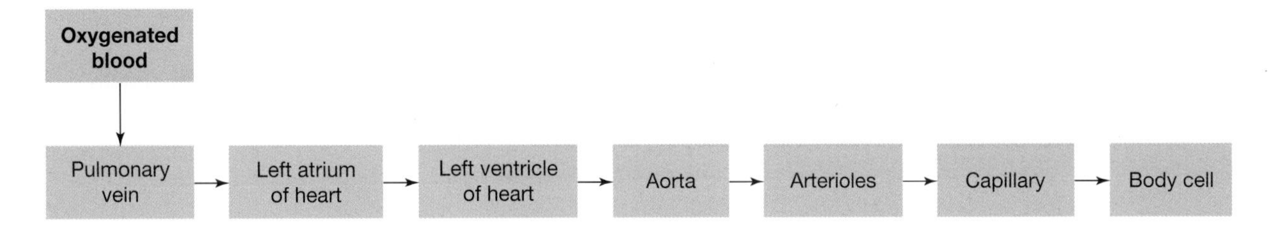

CASE STUDY: Garlic breath

Phew ... garlic breath! Have you ever heard someone say this? Garlic or onion breath comes from further down than your mouth! It has travelled through a number of your body systems. After you have eaten food containing either of these, and it has been digested, it is absorbed through the walls of your intestines and then into your blood. When the smelly onion or garlic blood reaches your lungs through your circulatory system, you breathe out the smelly gas.

4.8.3 Working together to get carbon dioxide from cells

Carbon dioxide is a waste product of cellular respiration and needs to be removed from the cell. When carbon dioxide has diffused out of the cell into the capillary, the blood in the capillary is referred to as deoxygenated blood. This waste-carrying blood is transported from the capillaries to small veins (**venules**), to large veins called vena cava, then to the right atrium of your heart. From here it travels to the right ventricle, where it is pumped to your lungs through the **pulmonary artery** (the only artery that does not contain oxygenated blood). This process is shown in figure 4.67.

venules small veins

pulmonary artery the vessel through which deoxygenated blood, carrying wastes from respiration, travels from the heart to the lungs

FIGURE 4.67 The pathway carbon dioxide travels from your body cells to your lungs

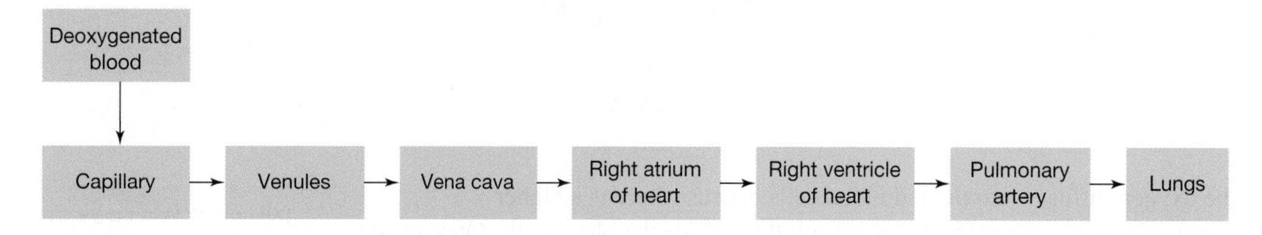

Exhaling carbon dioxide from lungs

Once the deoxygenated blood reaches the alveoli of the lungs, carbon dioxide diffuses out of the capillaries. This occurs because there is a higher concentration of carbon dioxide inside the capillaries than in the alveoli. Carbon dioxide is then transported into the bronchiole, then bronchi and trachea, until it finally exits through your nose (or mouth) when you exhale, as seen in figure 4.68.

FIGURE 4.68 The pathway carbon dioxide travels from your lungs to be exhaled

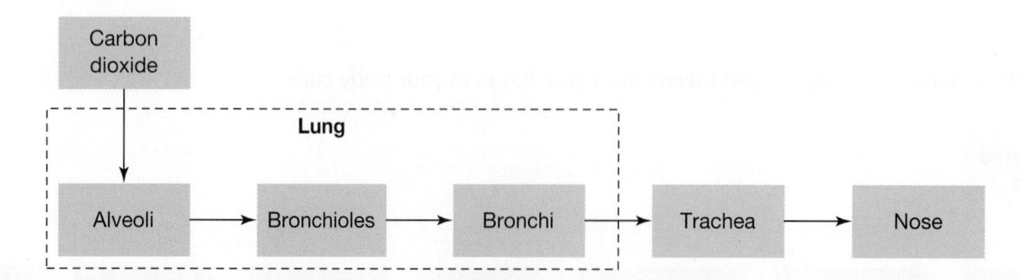

FIGURE 4.69 Your respiratory, circulatory and digestive systems form connected highways that provide your cells with what they need and remove what they don't.

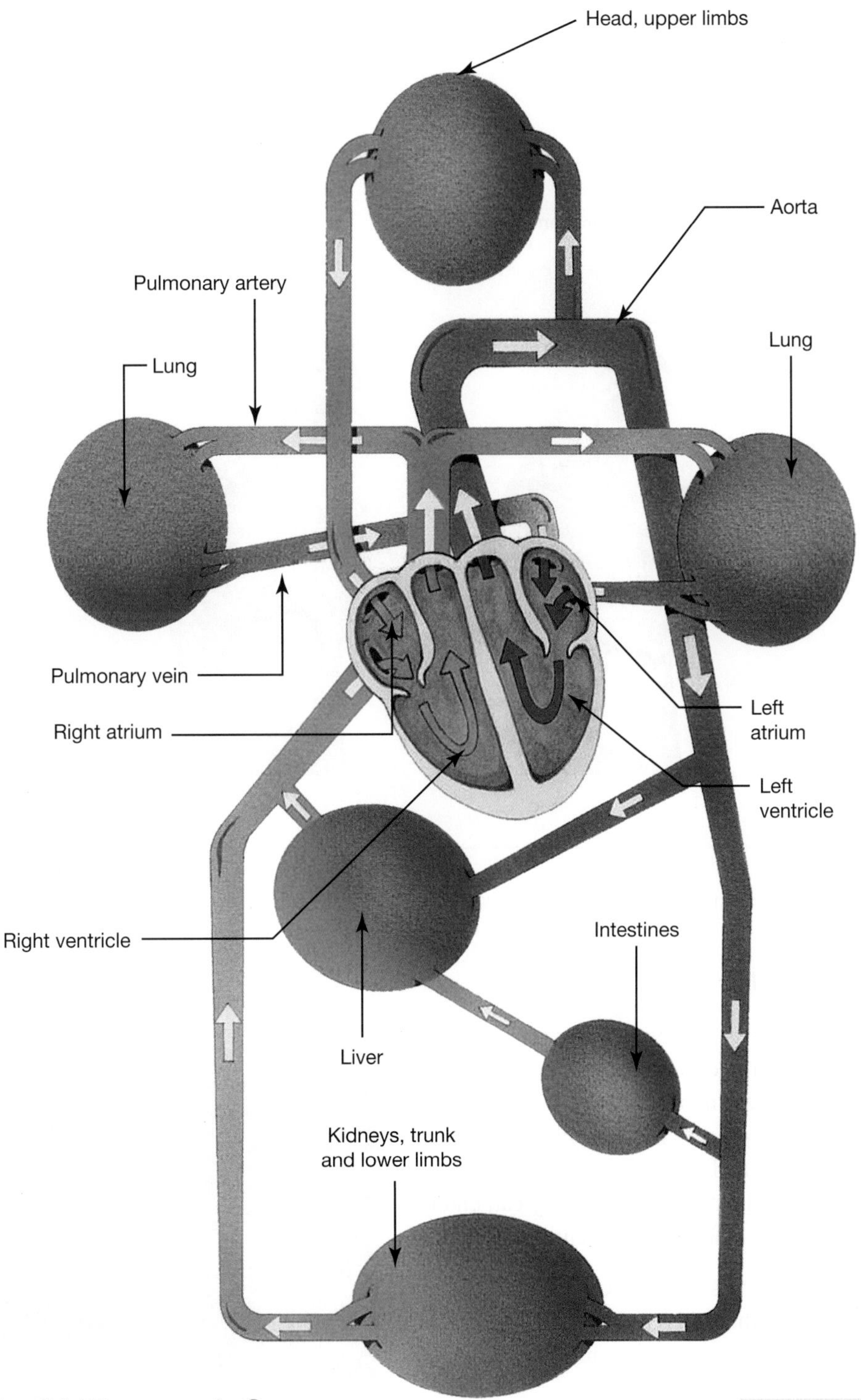

4.8.4 Brain AND muscle?

The respiratory system also relies on organs from other systems. Figure 4.70 shows that when you breathe in, a muscle beneath your rib cage called the **diaphragm** tightens. This allows the lungs to expand and air to be pulled into them. When you breathe out, the diaphragm relaxes, which reduces the size of the lungs and pushes air out. The largest volume of air that you can breathe in or out at one time is called your **vital capacity**.

diaphragm a flexible, dome-shaped, muscular layer separating the chest and the abdomen; it is involved in breathing

vital capacity the largest volume of air that can be breathed in or out at one time

Breathing involves muscle movements that are automatic and controlled by the brainstem. The brainstem is the part of the brain that controls subconscious body functions such as breathing and heart rate.

breathing the movement of muscles in the chest causing air to enter the lungs and the altered air in the lungs to leave; the air entering the lungs contains more oxygen and less carbon dioxide than the air leaving the lungs

FIGURE 4.70 When you breathe in, your diaphragm tightens; when you breathe out, it relaxes.

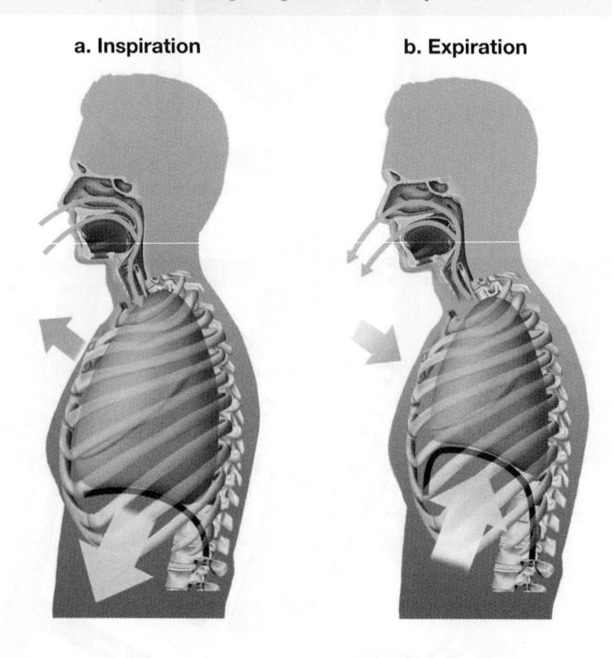

on Resources

Video eLesson The action of the diaphragm and lungs (eles-2045)

4.8.5 Cleaning the air

The surfaces of the upper respiratory tract are covered in cilia. These hair-like projections work to move the mucus that also covers these membranes up and out of the airway. This is a vital process in protecting the body against pathogens, pollen and particulates such as smoke. The items get trapped in the mucus before they can enter the cells of the respiratory tract, thus preventing disease.

Mucus from the lungs is transported back up the trachea and is swallowed, where the acids of the stomach can kill anything harmful. This also occurs with mucus produced in the nose, unless you blow your nose!

FIGURE 4.71 Cilia of the nose with inhaled pollen

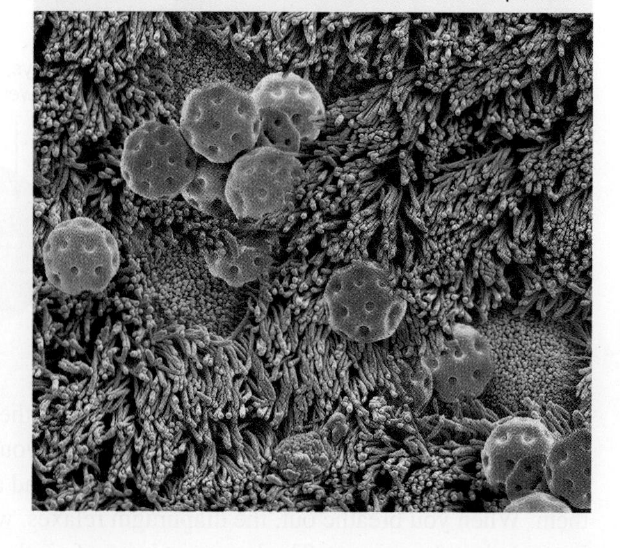

elog-2163

INVESTIGATION 4.8

Hands on pluck

Aim

To investigate the trachea, lungs, heart and liver of a mammal

Materials

- sheep's pluck (heart and lungs) with part of the liver and trachea attached
- newspaper and tray to place the pluck on
- plastic disposable gloves
- balloon pump and rubber tubing

Method

1. Carefully observe and record the shape, size, colour and texture of the sheep's trachea, lungs, heart and liver. Include notes on how they are connected. Can you see any blood vessels?
2. Push a piece of rubber tubing down the trachea to the lungs and use a balloon pump to blow some air into the trachea.

CAUTION

For hygiene reasons, do not use your mouth to blow into the tube inserted in the trachea.

3. Cut through the lung, heart and liver tissue. Make a record of your observations describing how they are similar and how they are different. Discuss possible reasons for the differences with your team members.
4. Using a scalpel or scissors, cut off a small piece of heart, lung and liver. Place each piece into a beaker of water and observe what happens. Discuss possible reasons for your observations with your team members.

Results

Record observations in the table provided.

TABLE Observations of dissection of a mammal's pluck

Organ	Shape (sketch)	Approx. size (cm)	Colour	Texture	Other comments	System to which the organ belongs
Trachea						
Lung						
Heart						
Liver						

Discussion

1. Could you see any blood vessels? Try to find out their names and what sort of blood they carry.
2. Suggest why there are rings of cartilage around the trachea.
3. Suggest reasons for the differences in texture between the heart and lungs.
4. Suggest reasons for the differences in the shapes of the organs that you observed.
5. Comment on something that you learned or found particularly interesting from this investigation. Share your comment with others.
6. Research and report on the following points for each of the organs in this investigation:
   - Its function and how it carries this out
   - The system to which it belongs
   - A disease relevant to it.

Conclusion

Summarise the findings for this investigation.

elog-2165

INVESTIGATION 4.9

Measuring your vital capacity

Aim

To investigate the vital capacity of lungs

Materials

- balloon
- ruler

Method

1. Blow up a balloon to about 20 cm in diameter two or three times to stretch it. Release the air each time.
2. Take the biggest breath you can, then blow out all the air into the balloon. Tie up the end of the balloon to hold in your 'blown out' air.
3. Use a ruler to measure and record the diameter of the balloon as shown in the following figure.

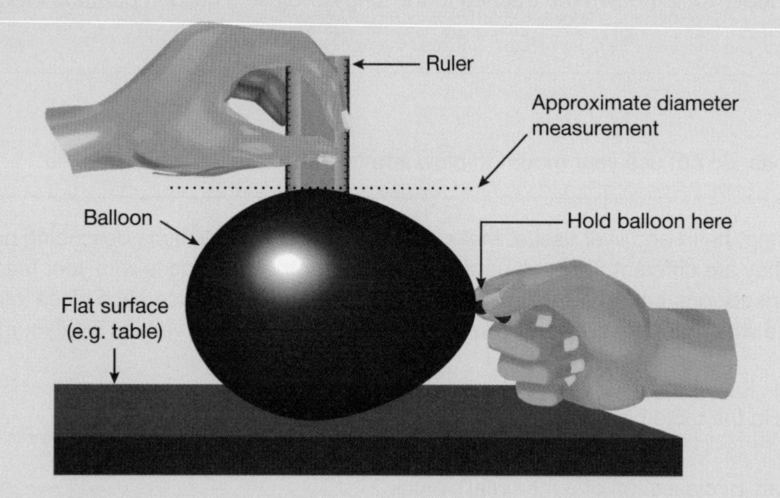

4. Use the following table to determine your approximate vital capacity in litres.
5. Release the air from the balloon and repeat your measurement of vital capacity three more times. Average your results to get your best estimate of the maximum 'blow-out' of your lungs.

TABLE How to measure the diameter of the balloon

Balloon diameter (cm)	8	9	10	11	12	13	14	15	16	17	18	19	20	21
Approx. vital capacity (L)	0.3	0.4	0.5	0.7	0.9	1.2	1.4	1.8	2.1	2.6	3.0	3.6	4.2	4.8

Results

1. Draw a table with the following headings.

TABLE Vital capacity results

Name	Male or female?	Does this student play a wind instrument?	Lung capacity (L)

2. Collect results from all the students in your class and complete the table.
3. Calculate the average lung capacity for all the females and for all the males in your class.
4. Calculate the average lung capacity for all students in your class who play a wind instrument.

Discussion

1. Suggest why you were asked to stretch the balloon first.
2. Suggest why you measured your vital capacity four times.
3. With reference to your results, do females have a bigger or smaller lung capacity than males in your class?
4. Compare the average lung capacity for students who play a wind instrument with the average value for students who do not. Do your results suggest that playing a wind instrument has an effect on lung capacity? Explain.
5. Reflect on the method used to investigate vital capacity and identify improvements that could be made if you were to do the investigation again.

Conclusion

Summarise the findings for this investigation.

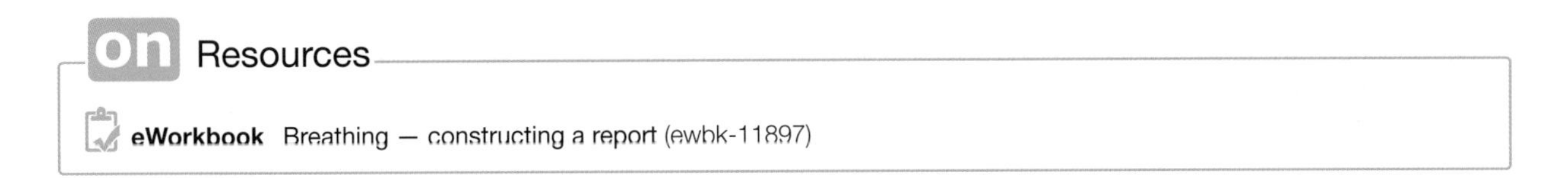

4.8 Activities

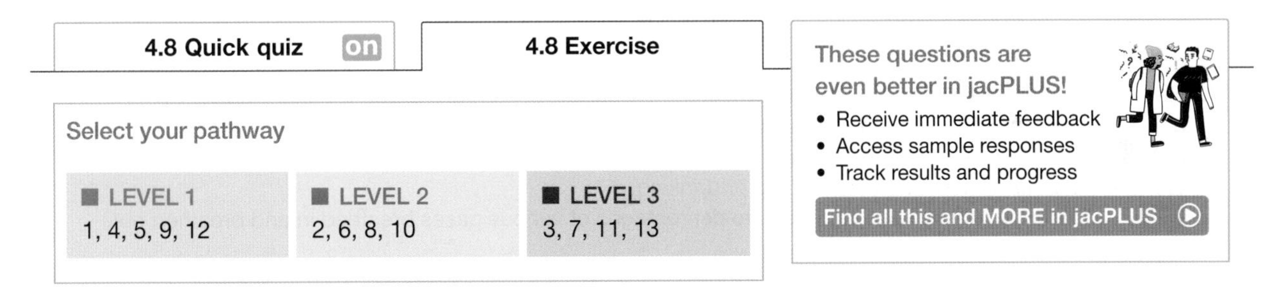

These questions are even better in jacPLUS!
- Receive immediate feedback
- Access sample responses
- Track results and progress

Find all this and MORE in jacPLUS

Remember and understand

1. a. State whether the following statements are true or false.

Statement	True or false?
i. Body cells need oxygen for the process of cellular respiration.	
ii. The process of cellular respiration produces energy in a form that the cell can use and oxygen is a waste product.	
iii. The role of the respiratory system is to get carbon dioxide into your body and oxygen out.	
iv. The respiratory system is made up of your oesophagus and your lungs.	
v. Oxygen diffuses from the alveoli into red blood cells in capillaries that surround them.	

b. Justify any false responses.

2. Match the terms associated with the respiratory system with their description in the table shown.

Term	Description
a. Alveoli	**A.** Blood vessel that carries deoxygenated blood from the heart to the lungs
b. Bronchiole	**B.** One of two narrower tubes that leads off the trachea
c. Bronchus	**C.** A muscle that allows the lungs to expand so that air can be pulled in
d. Diaphragm	**D.** A red pigment that binds to oxygen
e. Haemoglobin	**E.** Blood vessel that carries oxygenated blood from the lungs to the heart
f. Pulmonary artery	**F.** Tube through which air moves from your mouth to your lungs
g. Pulmonary vein	**G.** Tiny air sac through which oxygen diffuses into capillaries
h. Trachea	**H.** Small branching tube with alveoli at its end

3. Use flowcharts to identify the pathway that:
   a. oxygen travels to get from the air outside your body to the alveoli of your lungs
   b. oxygen travels to get from your lungs to your body cells
   c. carbon dioxide travels to get from your body cells to the alveoli of your lungs
   d. carbon dioxide travels to get from your lungs to the air outside your body.
4. Describe how oxygen gets from the alveoli of your lungs into blood cells in your capillaries.
5. Differentiate between the terms 'cellular respiration', 'respiratory system' and 'breathing'.

Apply and analyse

6. Some people describe the structure of the lungs as an upside-down hollowed-out tree. To which parts of the lungs might the following be referring?
   a. Trunk
   b. Branches
   c. Twigs
   d. Leaves
7. Give reasons for the following pieces of advice.
   a. It is better to breathe through your nose than your mouth.
   b. You should blow your nose when you have a cold rather than sniff it back.
   c. You should not talk while you are eating or drinking.
8. **SIS** The following table shows approximate percentages of various gases breathed in and breathed out.

TABLE Composition of air inhaled and exhaled

	Oxygen (%)	Carbon dioxide (%)	Water vapour (%)	Nitrogen (%)
Air breathed in	21	0.04	1	78
Air breathed out	15	4	5	76

   a. i. Compare the percentage of oxygen breathed in to that breathed out.
   ii. Suggest a reason for this pattern.
   b. i. Compare the percentage of carbon dioxide breathed in to that breathed out.
   ii. Suggest a reason for this pattern.
   c. The percentages in the table can vary in different weather conditions and at different heights above sea level. Research these variations and the possible implications this may have on humans.
9. Find out what a spirometer is.
10. Some singers can hold a musical note for a very long time. Investigate what muscles and techniques they use to be able to do this.
11. Did you know that mountain climbers often find it difficult to breathe? Some wear oxygen tanks to allow them to climb very high mountains. Research the effects of high altitude on breathing and report your findings.

Evaluate and create

12. Use information in this lesson and other resources to relate structural features to the functions of the following parts of the respiratory system.

TABLE Parts of the respiratory system and their functions

Part of respiratory system	Structural features	Function
Trachea		
Alveoli		
Lungs		
Capillaries		

13. **SIS** Construct a model lung as shown in the diagram. You can use the following items:
- Two clear 1-litre plastic bottles with tops
- Four balloons
- Two plastic drinking straws
- Rubber bands or very sticky tape
- Plasticine or Blu-Tack
- Scissors.

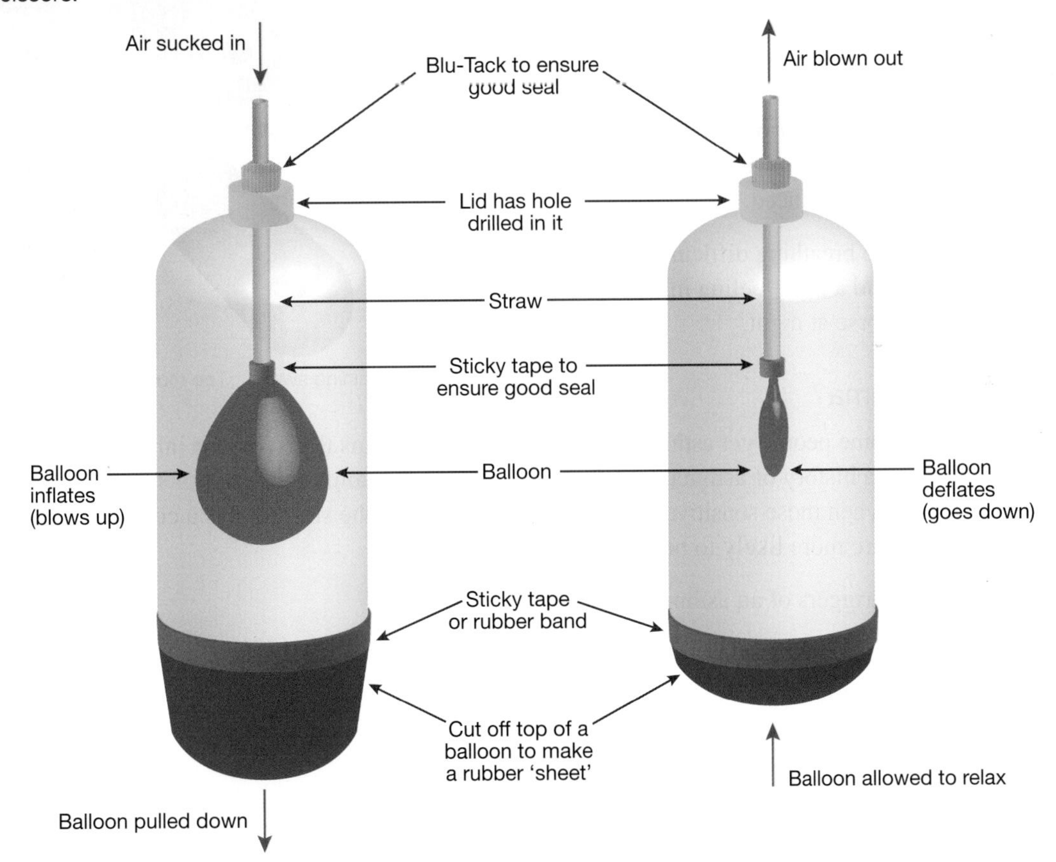

a. Identify which body parts are represented in the model lung by the:
   i. straw
   ii. rubber sheet at the bottom of the bottle
   iii. balloon connected to the straw
   iv. plastic bottle.

b. Pull the rubber sheet at the bottom of your model downwards. Record your observations. Release the rubber sheet. Record your observations. Discuss how your observations relate to how a lung works. Suggest how the model could be improved.

Fully worked solutions and sample responses are available in your digital formats.

LESSON
4.9 Short of breath?

LEARNING INTENTION

At the end of this lesson you will be able to describe examples of illnesses and problems associated with the respiratory system.

4.9.1 Asthma

int-5351

If you do not suffer from **asthma**, it is very likely that you know someone who does. Asthma is a very common condition and the number of people who suffer from it has increased over the years.

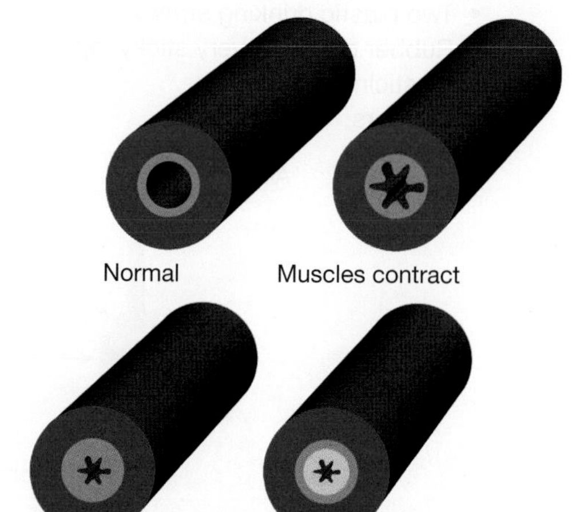

FIGURE 4.72 Asthma is a narrowing of the air pipes.

What is asthma?

Asthma is a narrowing of the air pipes that join the mouth and nose to the lungs. The pipes most affected are the bronchi. They become narrower as:

- the muscle walls of the air pipes contracts
- the lining of the air pipes swells
- too much mucus is produced.

The narrow pipes make breathing difficult and can result in wheezing, coughing and a tight feeling in the chest. The coughing is usually worse at night.

What causes asthma?

It is not known why some people get asthma and others do not. It seems that it can be inherited, but many people from families without a history of asthma are affected. Asthma is certainly the result of sensitive airways. An asthma attack occurs when those sensitive airways are 'triggered'. If the sufferer has a cold, the airways are already inflamed and are more likely to be triggered.

Some of the common triggers of an asthma attack are:

- vigorous exercise
- cold weather
- cigarette smoke
- dust and dust mites
- moulds
- pollen
- air pollution
- some foods and food additives
- some animals.

Not all asthma sufferers are affected by the same triggers. Some people suffer attacks only as a result of exercise. Others might be affected by any one or more of the triggers. It is important that those who get asthma try to find out what triggers the attacks. Many of the triggers can be avoided.

asthma narrowing of the air pipes that join the mouth and nose to the lungs

4.9.2 Controlling the triggers

The best way to control asthma is to avoid the triggers. While this is not always possible, it is worthwhile to recognise the triggers so that you can minimise them.

Pollen and moulds

Pollen from some grasses and trees is very light and becomes airborne on even slightly windy days. The inhaling of pollen can be reduced by avoiding outdoor activities and keeping windows and doors closed on breezy spring days. Moulds live in warm, humid conditions and thrive in bathrooms, kitchens and bedrooms. Their spores are easily breathed in, triggering attacks in some asthma sufferers. Moulds can be reduced by airing the house regularly.

Air pollution

Those asthma sufferers whose attacks are triggered by air pollution are warned to remain indoors as much as possible and avoid vigorous activity on smoggy days. If tobacco smoke is a trigger, the cigarette smoke of others needs to be avoided.

Dust mites

Dust mites are a common trigger of asthma attacks. Dust mites are microscopic animals that live in their thousands in warm, moist and dark places like doonas, sheets, pillows, carpets and curtains. Dust mite droppings float in the air and are easily inhaled.

Since you share so much of yourself and where you live with this fellow Australian, you should probably know its name. It is the most common dust mite (a relative of spiders and ticks) in Australia, *Dermatophagoides pteronyssinus.* The good news is that it is half a millimetre long and doesn't bite. The bad news is that there may be thousands of them living in your pillow, each defecating about 20 faecal pellets a day, reproducing (each female laying about 30 eggs in her lifetime), dying and decomposing. The fact that dust mites mate for 24 hours at a time (perhaps because their penis is only about as wide as their sperm) may make this particularly disturbing!

FIGURE 4.73 A house dust mite

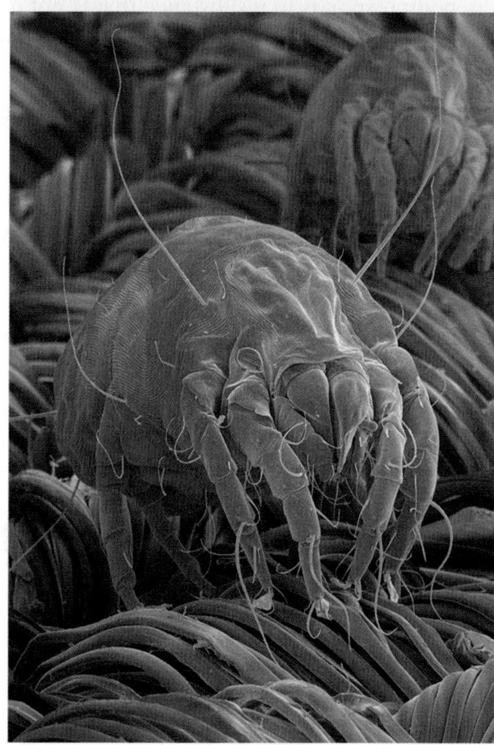

Our skin scales are the main food source for these dust mites, so wherever we are, they are. Dr Janet Rimmer (a respiratory physician and Director of the National Asthma Council Australia) also suggests that, of the 45 per cent of Australians who are affected by allergies, about 80 per cent are allergic to dust mites. But not all researchers have bad news about dust mites. Dr Matthew Colloff, a CSIRO researcher, has found them so interesting that he wrote a book (called *Dust mites*) about them.

Even the cleanest house has dust mites, but their numbers can be reduced by:
- exposing your mattress to the sun, because dust mites are susceptible to drying out
- washing bedding materials and bedclothes with tea-tree or eucalyptus oil (however, note that essential oils may also trigger asthma), or in hot water (above 55 °C)
- removing soft toys that collect dust, or hot washing them weekly
- regularly vacuuming curtains and carpets
- airing bedrooms by keeping doors and windows open
- replacing carpets with hard flooring.

EXTENSION: Shedding our skin

Dust mites thrive best in bedding and carpets because these contain plenty of human dead skin cells. Humans shed a complete layer of dead skin cells every month. That amounts to about 1 kilogram of skin cells each year. In fact, most of the dust in your house consists of dead skin cells.

4.9.3 Asthma medication

Asthma medications can be divided into two main groups: preventers and relievers. Preventers make the lining of the airways less sensitive and therefore less likely to be triggered. Relievers open up the airways once an attack has commenced. Most asthma medications are applied with inhalers or 'puffers', as seen in figure 4.74, which direct the medication straight into the lungs for fast action. Severe attacks of asthma require other drugs and sometimes extra oxygen needs to be supplied.

FIGURE 4.74 Asthma medication is usually delivered through an inhaler.

4.9.4 Allergies

Allergies can also trigger a response in the airways. Allergies occur when the body responds to a substance that is not harmful as if it is. Common substances that cause reactions are called **allergens** and include pollen (causing hayfever), animal dander, dust mites, insect bites or stings, as well as some foods — particularly nuts, shellfish, eggs and milk.

Sometimes allergies can result in hives, itchy and watering eyes, or a runny or blocked nose. If the reaction is very severe, parts of the respiratory system may swell. A severe reaction is called **anaphylaxis** and may cause death.

FIGURE 4.75 The symptoms of anaphylaxis

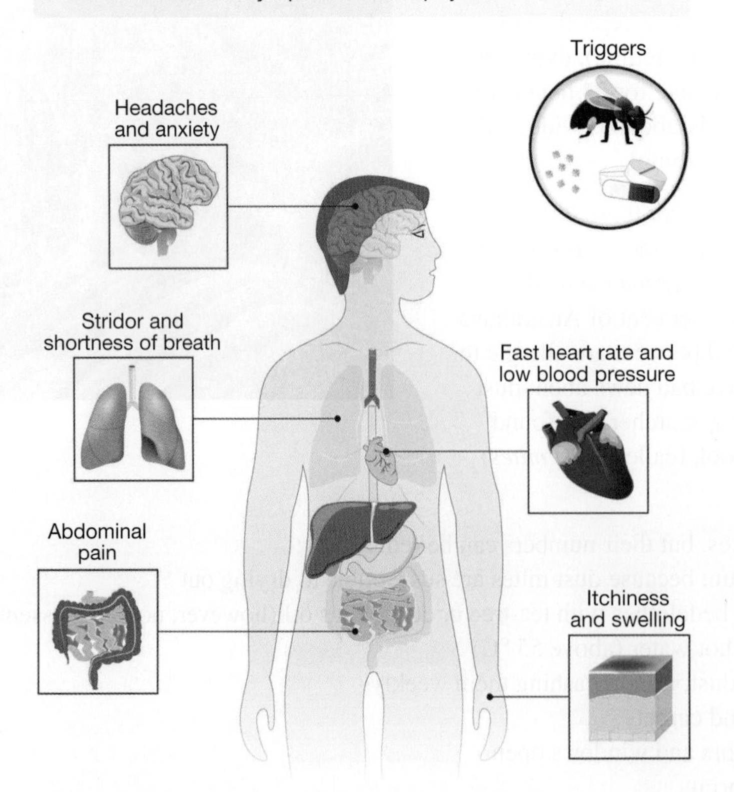

allergy an abnormal immune response to a substance that is harmless for most people

allergen an antigen that elicits an allergic response

anaphylaxis an acute and potentially lethal allergic reaction to an allergen to which a person has become hypersensitive

As the parts of the respiratory system swell, it becomes harder for the sufferer to get oxygen into their body. The treatment for anaphylaxis is an injection of adrenaline via an EpiPen (figure 4.76).

4.9.5 Up in smoke

Asthma is not the only condition that can interfere with your lungs functioning as they should. Some human activities can damage not only your lungs, but also those of others around you. Smoking is one example of such an activity. About 15 000 Australians die each year as a result of diseases caused by smoking. Smoking is actually the largest preventable cause of death and disease in Australia.

4.9.6 Just one cigarette

There are clearly many long-term effects of smoking. However, figure 4.78 shows what happens to you after smoking just one cigarette.

There are some more obvious effects such as bad breath, body odour and watery eyes. After several cigarettes, your teeth and fingers become stained. Your sense of taste is reduced. Even your stomach is affected as acid levels increase.

Smoking and your lungs

Lung cancer is the most well-known disease caused by smoking. Chemicals that cause cancer are called **carcinogens**. Cigarette tobacco contains a number of carcinogens. The chemicals in cigarettes also clog up the fine hairs in your air tubes with a mixture of mucus and foreign chemicals.

Cough it up

Coughing is the body's way of trying to clear the air tubes. However, not all of the clogging can be cleared by coughing. A dirty mixture remains in the air tubes, causing swelling, making them sensitive and slowing down the passage of air. Eventually, the sticky mixture sinks down into the lungs where it blocks some of the pathways to the alveoli, where freshly breathed air should deliver oxygen to the blood.

FIGURE 4.76 An EpiPen

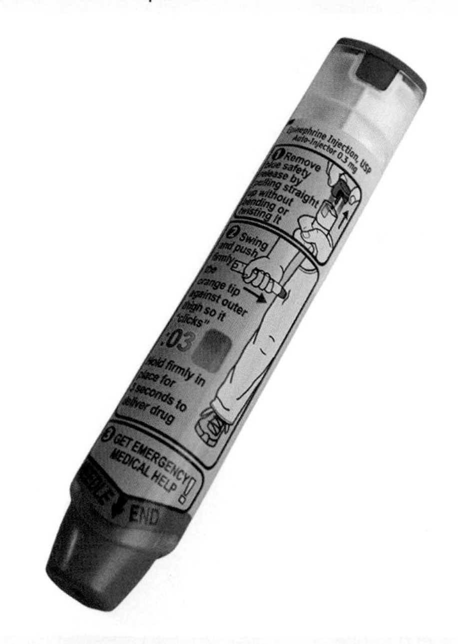

FIGURE 4.77 With over 4000 chemicals in each cigarette, smoking can lead to any of these conditions and effects.

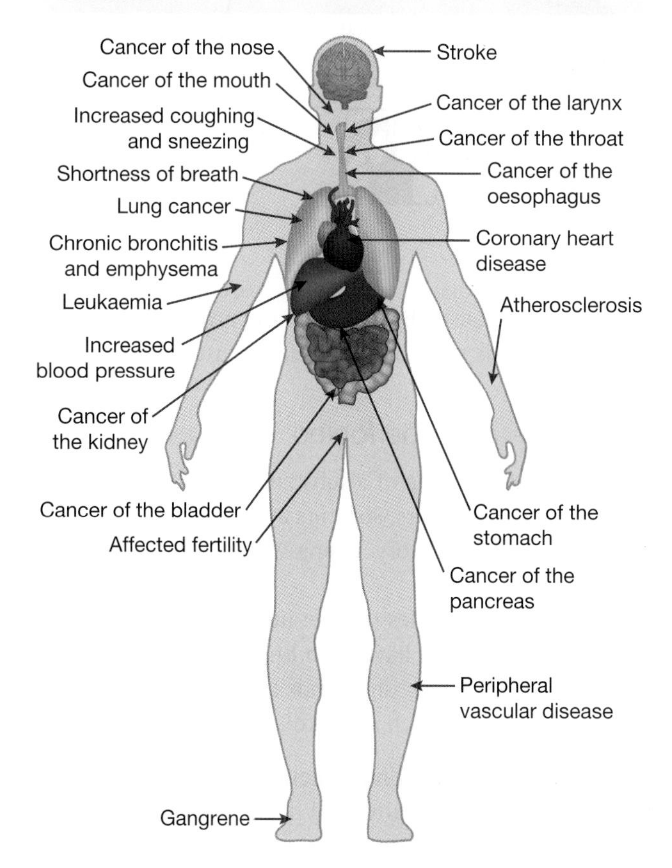

carcinogens chemicals that cause cancer

FIGURE 4.78 The results of just one cigarette

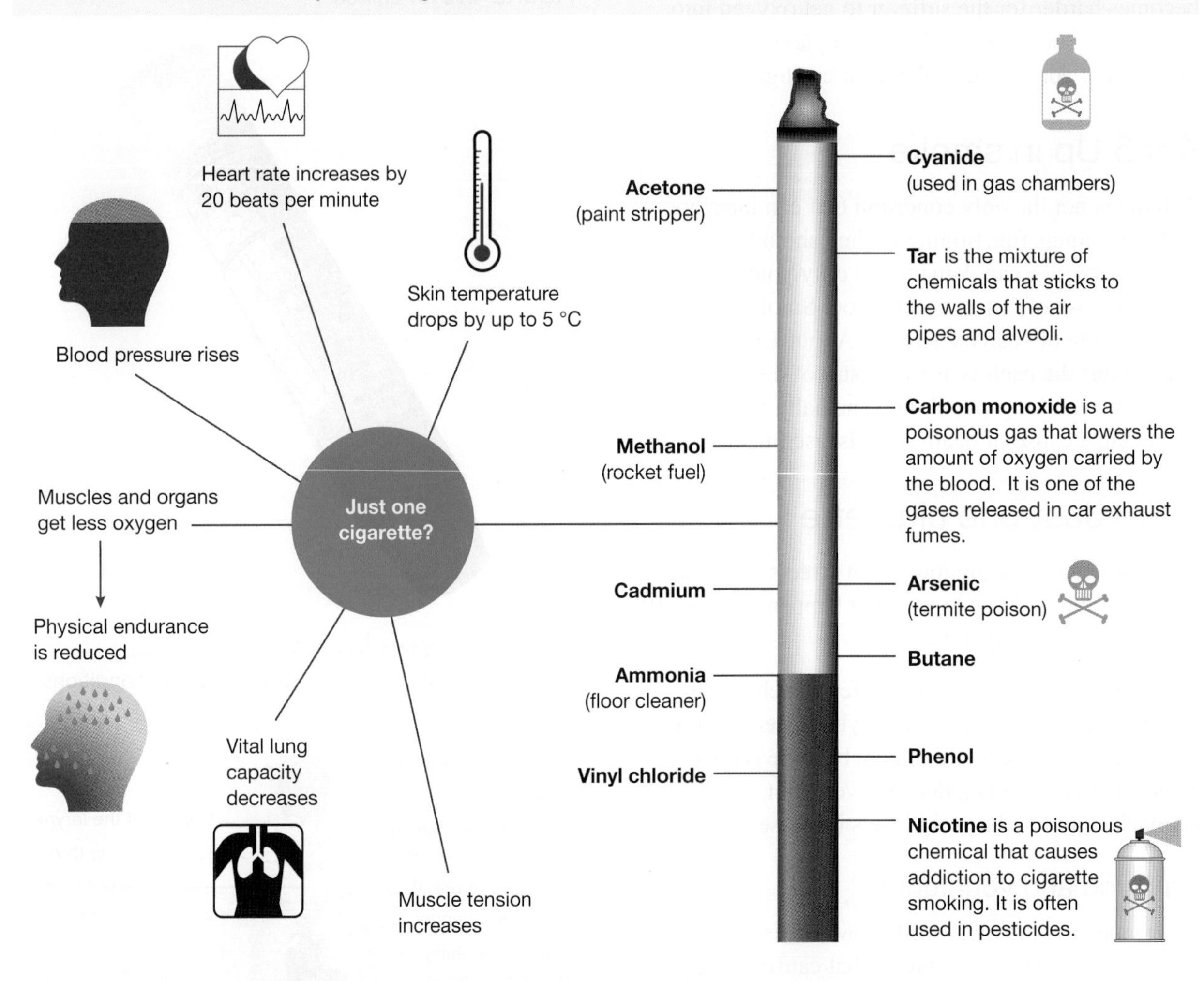

The diseases caused by this blocking process are called chronic obstructive pulmonary diseases, or COPD. **Emphysema** is the worst of these diseases and results in the eventual destruction of the alveoli.

Vaping — don't be fooled

Vaping is the inhaling of a substance that has been heated and turned into a vapour. These substances generally contain water, flavours, solvents and nicotine. The obviously harmful chemical in this list is the nicotine — but this isn't the whole story. Along with nicotine (which has been found in vapours that are supposedly nicotine-free) there may be:

- ultra-fine particles that get inhaled deep into the lungs and stay there
- flavourants that have been linked to lung disease
- volatile organic compounds (phenol and toluene are two examples also found in cigarettes)
- heavy metals such as nickel, tin and lead.

This makes vaping not much better for you than smoking a regular cigarette. Vaping has been linked to a number of deaths from acute lung disease. This was linked to a compound in e-cigarettes that is safe if ingested, but toxic if inhaled.

emphysema a condition in which the air sacs in the lungs break open and join together, reducing the amount of oxygen taken in and carbon dioxide removed

Short-term and medium-term effects of vaping have been seen in the lungs and hearts of those who vape, including a higher prevalence of asthma, emphysema and chronic bronchitis. Longer-term effects, such as a relationship to cancer, are still being studied, as this is a relatively new technology.

SCIENCE AS A HUMAN ENDEAVOUR: Professor Robyn O'Hehir BSc, MBBS (Hons I), FRACP, PhD, FRCP, FRCPath

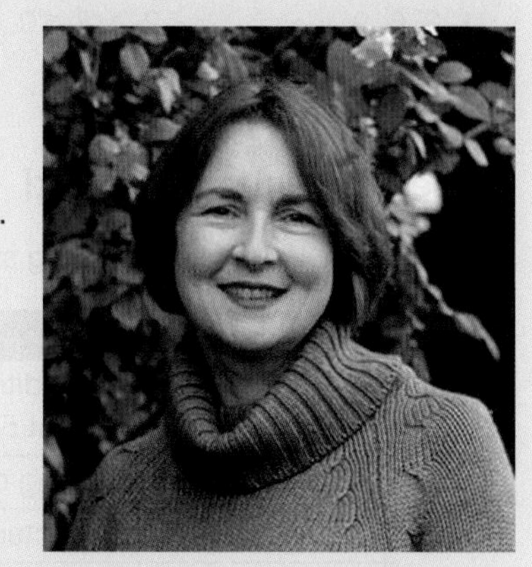

1. **What is your current science-related title?**
   I am a Professor of Medicine, with particular responsibilities for allergy, clinical immunology and respiratory medicine, at Monash University, Melbourne. I am also the Director of the Department of Allergy, Immunology and Respiratory Medicine at the Alfred Hospital in Melbourne.
2. **What field of science are you in?**
   Allergy, cellular immunology and respiratory medicine. I was appointed to the first Chair in Allergy and Clinical Immunology in Australia.
3. **Describe some science that you are involved in at the moment.**
   Millions of people around the world suffer from allergies. I am sure you know several friends who have asthma or hay fever, or you may even have them yourself. Asthma and hay fever are usually triggered by proteins called allergens, from house dust mites or grass pollens. Allergies to peanuts and shellfish are less common but often more serious, because they can trigger life-threatening allergic reactions called anaphylaxis. Allergies are caused by reactions between white blood cells ('T cells') and environmental proteins that are usually harmless. My research group is trying to find ways to damp down the allergic T-cell responses.

   Allergen immunotherapy (allergy shots) is the only treatment that can prevent allergic diseases, but currently it can't be used for peanut allergies, even though this is one of the most serious allergens. To develop a safe and effective vaccine against peanut allergies, we are identifying parts of critical peanut proteins that can build up tolerance in allergic patients without risking anaphylaxis.
4. **What do you enjoy about being a scientist?**
   I enjoy the fact that my research not only is laboratory-based, exploring novel methods for switching off allergic responses, but also lets me see patients and train other doctors in how to do research from bench to bedside to the community. I head an active clinical department, still carry out clinics with patients, and am actively engaged in national and international tests of new preventions and treatments for allergies. My combined research and clinical duties allow translation of our research findings into better clinical practice.
5. **What triggered your interest in science?**
   I decided to specialise in allergy and respiratory medicine, focusing on asthma, following my experiences as a young trainee physician at the Alfred Hospital in Melbourne. Asthma was a huge problem in Australia at that time, and many times I resuscitated young adults in the hospital emergency room — and I watched them return, with appropriate medication and careful education, to confident, full lives. Some remain my patients today. The ability to dissect underlying mechanisms of disease and then work towards new therapeutics and practices to benefit patients is a great excitement and honour. The diversity of patients and their needs ensures that every day is quite different.
6. **Do you have any other comments that may be of interest to Year 8 Science students?**
   A career in science combined with medicine may take a bit longer in terms of training, but it gives you a fantastic ability to do interesting work that is intellectually demanding and also involves working with lots of people who need your help. I am very glad that I chose a career in science and medicine.

Resources

eWorkbook Smoking and diseases (ewbk-11899)

4.9 Activities

learnon

4.9 Quick quiz on	4.9 Exercise

Select your pathway

■ LEVEL 1	■ LEVEL 2	■ LEVEL 3
1, 3, 6, 9	2, 4, 7, 10, 11	5, 8, 12, 13, 14

These questions are even better in jacPLUS!
- Receive immediate feedback
- Access sample responses
- Track results and progress

Find all this and MORE in jacPLUS

Remember and understand

1. a. State whether the following statements are true or false.

Statement	True or false?
i. Asthma is a rare condition and the number of people suffering from it has decreased in the past 50 years.	
ii. Asthma is a narrowing of the air pipes that join the mouth and nose to the lungs.	
iii. The tubes most affected by asthma are the bronchi.	
iv. Breathing can be restricted during an asthma attack due to swelling of the bronchi.	
v. All asthma sufferers are affected by the same triggers.	
vi. Fungal spores from moulds can be a trigger for asthma attacks.	
vii. Fewer asthma attacks are likely to occur on a windy day than on a day without wind.	
viii. Dust mites are a common trigger for asthma attacks.	
ix. Asthma medications may be classified as being preventers if they make the lining of the airways more sensitive.	
x. Asthma medications may be classified as being relievers if they open up the airways once an attack has commenced.	
xi. Smoking just one cigarette may increase your blood pressure.	
xii. There is no link between smoking and lung cancer.	
xiii. Chemicals in cigarettes can block some of the pathways to the alveoli in your lungs, which can reduce the amount of oxygen delivered to your cells.	

   b. Justify any false responses.
2. a. What happens to the air pipes during an asthma attack to make breathing difficult?
   b. Why is an asthma attack more likely to be triggered in a person with a cold?
   c. What is an asthma trigger?
   d. What are the two major types of asthma medication and how are they different from each other?

Apply and analyse

3. Create a poster that sends one single important message about smoking or vaping.
4. a. Find out more about Allergy & Anaphylaxis Australia.
   b. Outline the topics covered in a first aid course for management of anaphylaxis.
   c. What is an EpiPen and how is it used?
5. SIS In a team, brainstorm ideas about the common triggers of asthma that can be controlled. Summarise your discussion in a bubble map.
6. SIS a. If you suffer from asthma, prepare a talk for the rest of your class explaining:
   i. what it is
   ii. how it affects you
   iii. how you control it or try to prevent attacks.

   b. If you do not suffer from asthma, write a set of at least five questions that you could ask an asthma sufferer in an interview. If possible, conduct the interview and record the answers in writing, or as audio or video.

7. **SIS** Propose a series of questions to find out more about each of the areas listed. Investigate them, and then share your findings with others in your class.
   a. Allergies
   b. Asthma
   c. Anaphylaxis
   d. Allergen immunotherapy
   e. Clinical immunology and respiratory medicine
8. a. Describe the structure and function of an alveolus.
   b. Suggest how the structure of an alveolus is related to its function.
   c. Suggest how smoking affects the ability of an alveolus to perform its function.

Evaluate and create

9. **SIS** Draw up a two-column table. The first column should be headed 'Reasons for smoking/vaping'; the second column should be headed 'Reasons for not smoking/vaping'. With at least one other person, complete the table. Then compare your table with others. You might be able to construct a large table for the whole class.
10. **SIS** Smoking-related diseases cost taxpayers many millions of dollars because hospitals are mostly paid for by governments. Write down your opinion of each of the proposals given. Give reasons for your opinion.
    a. The cost of hospital treatment for diseases caused by smoking should be paid for by the patient because it was their fault that they got sick.
    b. Cigarettes should cost more. The extra money made from them could then be given to hospitals to help pay for treating smoking-related diseases.
    c. Cigarette companies who make profits from smoking should be made to pay for hospital treatment of patients with diseases caused by smoking.
11. **SIS** The following table shows how the popularity of smoking has changed over the past 70 years or so.

TABLE Percentage of adult Australians who smoke

Year	1945	1964	1969	1974	1976	1980	1983	1986
Males (%)	72	58	45	41	40	40	37	33
Females (%)	26	28	28	29	31	31	30	28

Year	1989	1992	1998	2001	2004	2011	2014	2017
Males (%)	30	28	29	28	26	18.2	16.9	16.5
Females (%)	27	24	24	21	20	14.4	12.1	11.1

   a. Draw a line graph of the data in the table. Use 'Year' on the *x*-axis and '% of adult Australians who smoke' on the *y*-axis. Draw lines for males and females in different colours.
   b. Why do you think that the percentage of females who smoke has changed little while the percentage of males who smoke has declined greatly?
   c. Use dotted lines to show your prediction of the trends up to the year 2024. What percentage of males and females do you predict will be smoking in 2024?
12. **SIS** Study the graph provided, which shows that the risk of getting lung cancer increases with the number of cigarettes smoked daily.
    a. Copy and complete the following statements.
       i. People who smoke 10 cigarettes a day are ____________ times more likely to develop lung cancer than non-smokers.
       ii. People who smoke 30 cigarettes a day are ____________ times more likely to develop lung cancer than people who smoke 10 cigarettes a day.
    b. If a packet of 20 cigarettes costs $30, calculate how much a person smoking 40 cigarettes a day spends on smoking:
       i. each day
       ii. each week
       iii. each year.

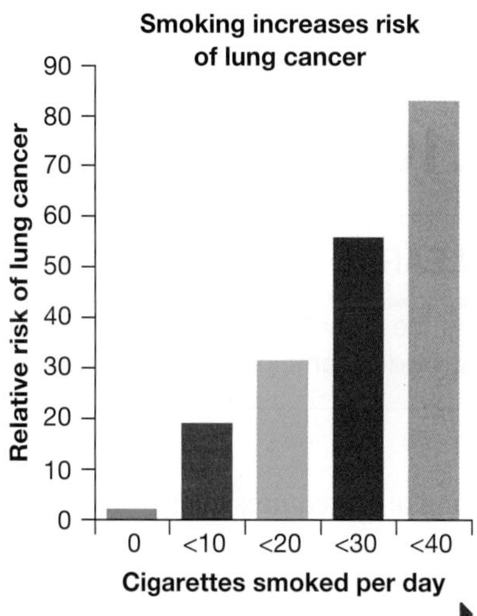

13. **SIS** Study the graph provided, which shows that the number of deaths from lung cancer has risen as cigarette consumption has increased, but there is a 20-year lag time because lung cancer takes years to develop.

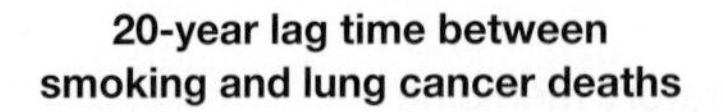

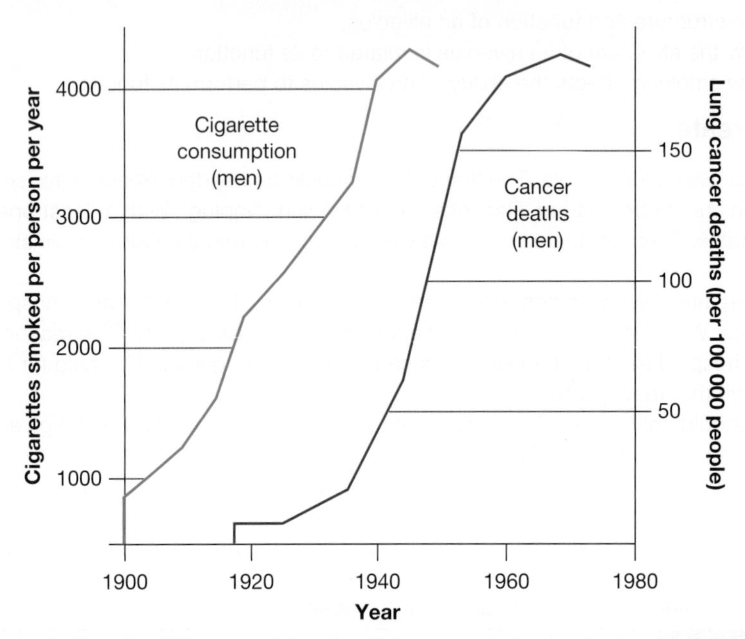

a. Describe how the incidence of lung cancer deaths changed between 1900 and 1980.
b. Identify when the number of male smokers peaked.
c. Identify when the number of deaths from lung cancer peaked.
d. Explain why there is a 20-year gap between the two numbers.
e. The graph shows data for male smokers only. Predict when the number of cases of lung cancer deaths in women peaked (use the graph you drew for question **11** to answer this).

14. **SIS** There are claims that vaping is safer that smoking cigarettes.
a. Research the claim that vaping is safer than smoking. Create a table that shows the source of your information, whether the source claims that vaping is safer and the reasons given. Look to find five reliable sources.
b. What are the risks of e-cigarettes to teenagers?

Fully worked solutions and sample responses are available in your digital formats.

LESSON
4.10 Excretory system

LEARNING INTENTION

At the end of this lesson you will be able to describe the structure and function of the excretory system and provide examples of problems associated with it.

Being alive requires energy and nutrients. It also results in the production of wastes that need to be removed.

4.10.1 Excretion

Excretion is any process that gets rid of unwanted products or waste from the body. The main organs involved in human excretion are your **skin**, lungs, **liver** and **kidneys**. Your skin excretes salts and water as sweat, and your lungs excrete carbon dioxide (produced by cellular respiration) when you breathe out. Your liver is involved in breaking down toxins for excretion, and your kidneys are involved in excreting the unused waste products of chemical reactions (e.g. urea) and any other chemicals that may be in excess (including water) so that a balance within our blood is maintained.

FIGURE 4.79 The four main organs involved in human excretion are the kidneys, skin, liver and lungs.

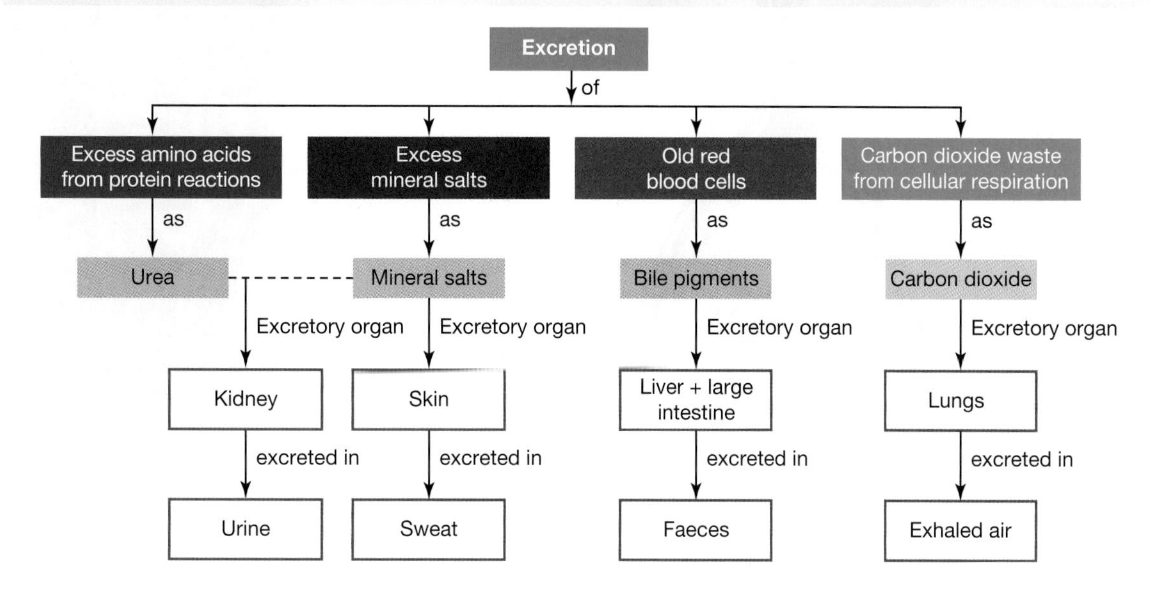

4.10.2 Kidneys

If you put your hands on your hips, your kidneys are close to where your thumbs are (figure 4.80). You have two of these reddish-brown, bean-shaped organs. Without them you would survive only a few days.

Organs, tubes and urine

Your kidneys play a key role in filtering your blood and keeping the concentration of various chemicals and water within appropriate levels. Each of your kidneys is made up of about one million **nephrons**. These tiny structures filter your blood, removing waste products and chemicals that may be in excess. Chemicals that your body needs are reabsorbed into capillaries. The fluid remaining in your nephrons travels through tubes called **ureters** to your **bladder** for temporary storage. As it fills, your bladder expands like a balloon. It can hold about 400 mL of this watery fluid that contains unwanted substances, called **urine**. **Urination** occurs when urine moves from your bladder through a tube called the **urethra** and out of your body.

excretion the removal of wastes from the body

skin the external covering of a vertebrate's body

liver the largest gland in the body; it secretes bile for digestion of fats, builds proteins from amino acids, breaks down many substances harmful to the body and has many other essential functions

kidneys body organs that filter the blood, removing urea and other wastes

nephrons the filtration and excretory units of the kidney

ureters tubes from each kidney that carry urine to the bladder

bladder a sac that stores urine

urine a yellowish liquid, produced in the kidneys; it is mostly water and contains waste products from the blood such as urea, ammonia and uric acid

urination the passing of urine from the bladder to the outside of the body

urethra the tube through which urine is emptied from the bladder to the outside of the body

ewbk-11901
eles-2051
int-8234

FIGURE 4.80 Your kidneys have an important role in the excretion of wastes from your body. **a.** A 2D image of the kidneys and bladder **b.** A rear 3D image of the location of the kidneys and bladder relative to organs in the digestive system

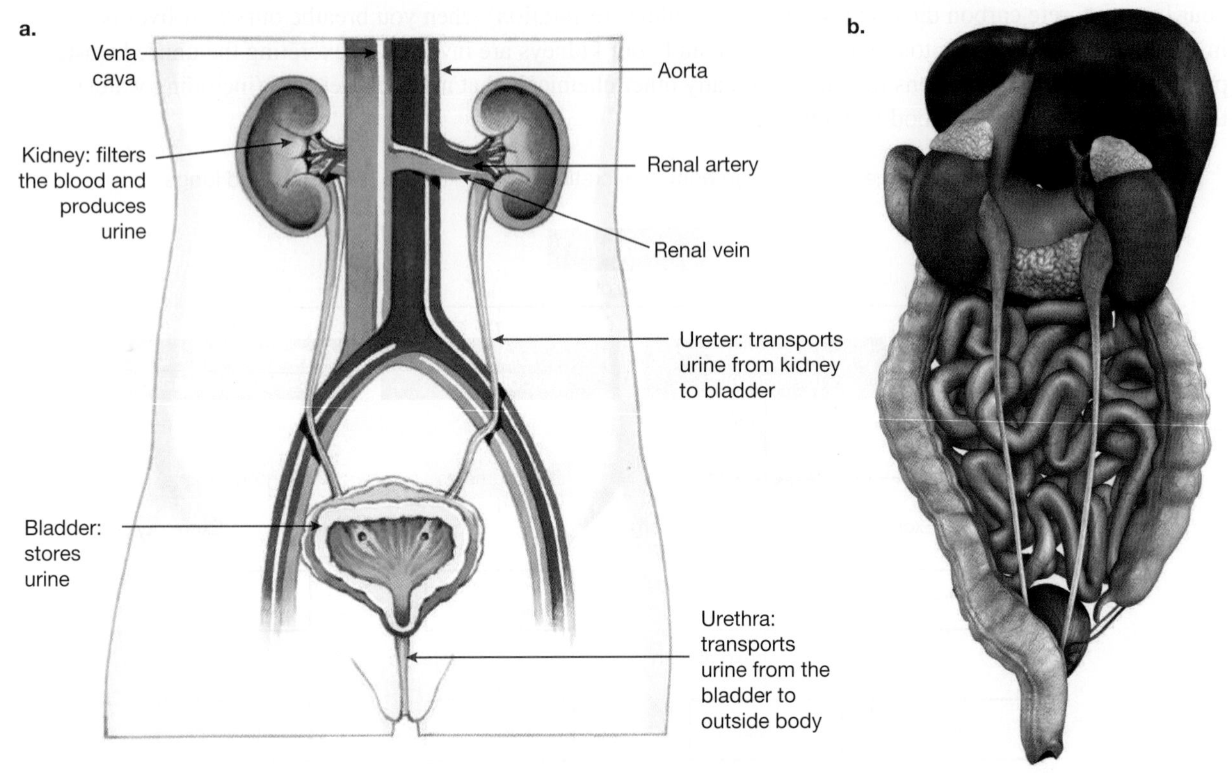

Nephrons — how their structure suits their function

Each nephron is made up of a long tubule (very fine tube) that forms a cup-like structure at one end called the **Bowman's capsule**. This structure surrounds a cluster of capillaries called the **glomerulus** (from an ancient Greek word meaning 'filter'), as see in figure 4.81.

Blood containing wastes travels to the glomerulus within each nephron in your kidneys, where the blood is filtered. Wastes and excess water move into the surrounding Bowman's capsule. As this 'waste' fluid moves along the tubules, any useful substances are reabsorbed back into capillaries that are 'twisted' around the tubules, and hence back into circulation. The remaining fluid becomes urine, which eventually travels in your ureters to your bladder prior to urination.

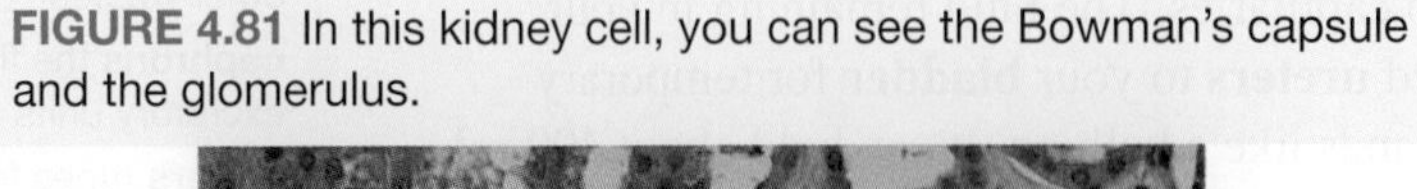
FIGURE 4.81 In this kidney cell, you can see the Bowman's capsule and the glomerulus.

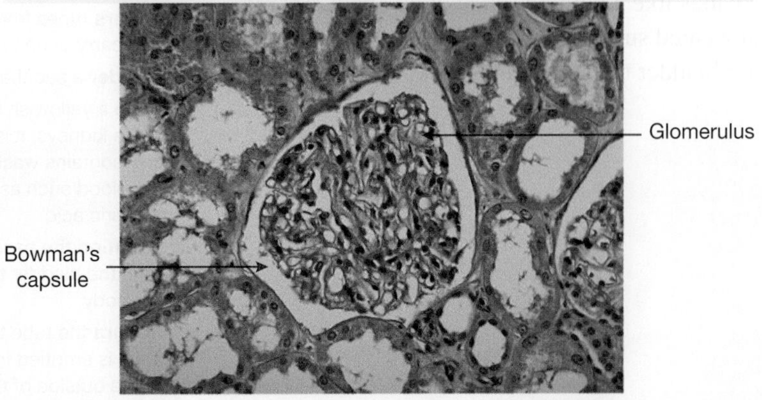

Bowman's capsule a cup-like structure at one end of a nephron within the kidney, surrounding the glomerulus; it serves as a filter to remove wastes and excess water

glomerulus a cluster of capillaries in the kidney that acts as a filter to remove wastes and excess water

4.10.3 Have a drink!

Both blood and urine are mostly made up of water. Water is very important because it assists in the transportation of nutrients within and between the cells of the body. The concentration of substances in blood is also influenced by the amount of water in it.

Water helps the kidneys do their job because it dilutes toxic substances and absorbs waste products so that they can be transported out of the body. If you drink a lot of water, more will be absorbed from your large intestines, and your kidneys will produce a greater volume of dilute urine. If you do not consume enough liquid, you will urinate less and produce more concentrated urine.

FIGURE 4.82 How your kidneys work to remove waste from your body.

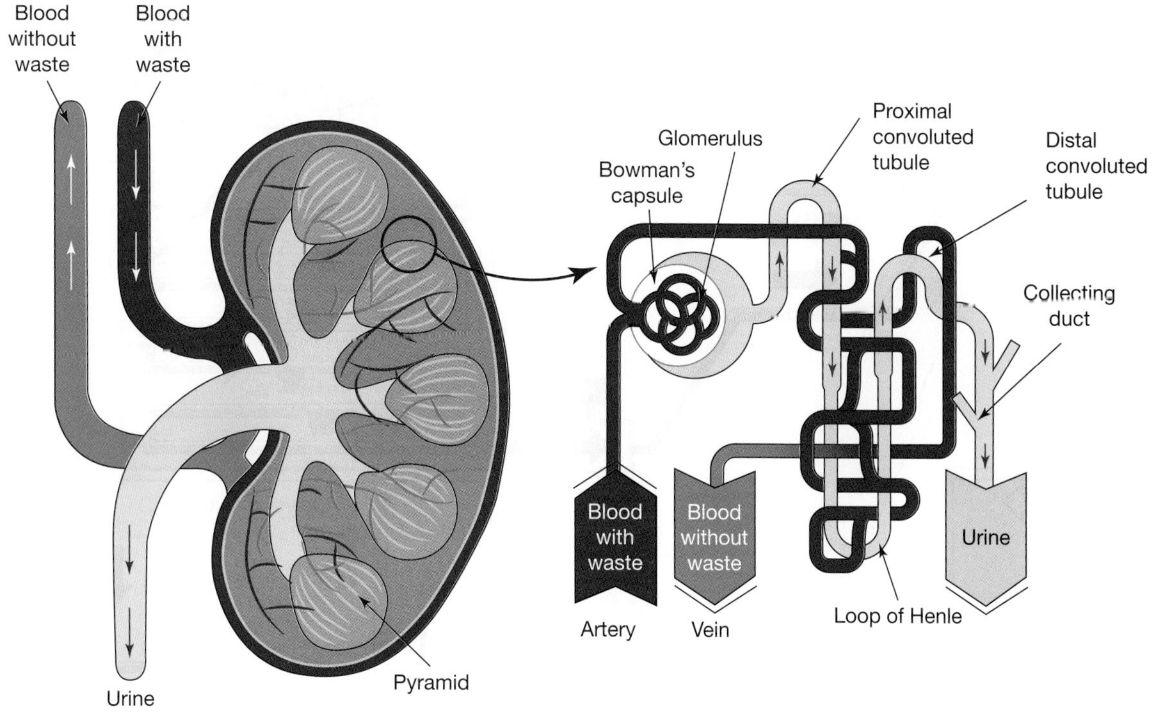

FIGURE 4.83 An organised approach is required for your excretory system to work effectively. Although each structure has its own specific job to do, it is also dependent on other structures doing their jobs as well.

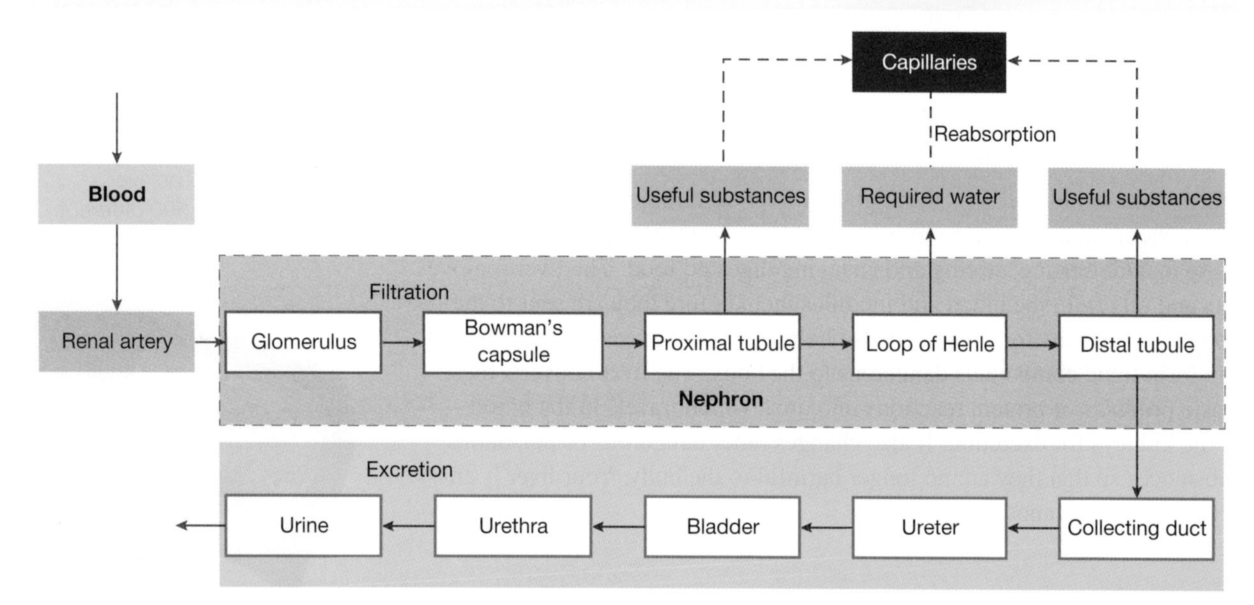

4.10.4 Haemodialysis

People with kidney disease may not be able to remove the waste materials from their blood effectively. They may need to be linked up to a machine that does this job for them. Their blood is passed along a tube that lets wastes, such as urea, pass out of it. However, useful substances, such as glucose, proteins and red blood cells, stay in the tube and are kept in the blood. This process is called **haemodialysis** and is seen in figure 4.84.

haemodialysis the process of passing blood through a machine to remove wastes

FIGURE 4.84 Haemodialysis

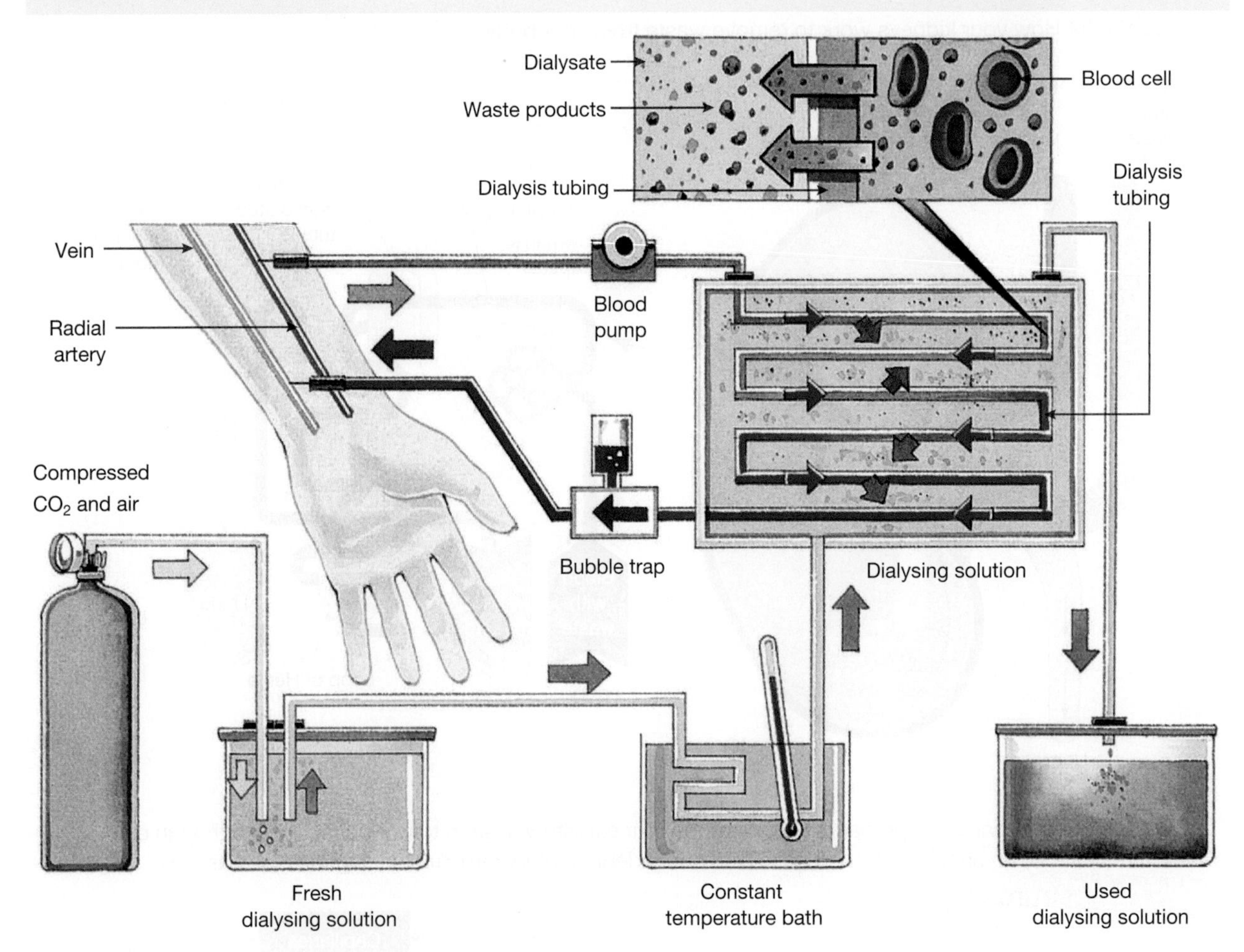

4.10.5 Liver

Livers are busy places!

Over a litre of blood passes through your liver each minute. Your liver is like a chemical factory, with more than 500 different functions. Some of these include sorting, storing and changing digested food. The liver removes fats and oils from the blood and modifies them before they are sent to the body's fat deposits for storage. It also helps get rid of excess protein, which can form toxic compounds dangerous to the body. The liver converts these waste products of protein reactions into urea, which travels in the blood to the kidneys for excretion. It also changes other dangerous or poisonous substances so that they are no longer harmful to the body. Your liver is an organ that you cannot live without.

FIGURE 4.85 The liver performs over 500 different functions.

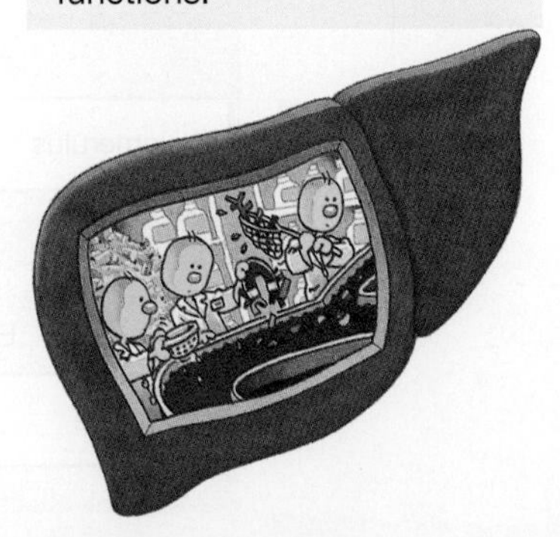

Too much alcohol?

The liver is also involved in breaking down alcohol. Alcohol is converted into a substance called acetaldehyde, which is then converted to acetate and finally into carbon dioxide and water. The carbon dioxide is transported from the liver to the lungs and then exhaled out of the body. The water may be removed as vapour in breath, sweat on skin or as urine.

Alcohol can also affect the amount of urine produced by the kidneys. Reabsorption of water may be reduced in the kidneys, resulting in the production of more urine. Increased urination can result in dehydration and consequently impair other body functions.

 Resources

 **Video eLesson** The liver (eles-2536)

EXTENSION: Removing excess salt

The human kidneys remove excess salt from the blood to help keep levels constant. Different types of animals have other ways of removing excess salt from their bodies. Turtles, for example, have salt-secreting glands behind their eyes. Hence, you may see a turtle 'shedding tears'. On the other hand, penguins and some other seabirds, such as the southern giant petrel, may appear to have runny noses because that is where their salt-secreting glands are located.

 Resources

 **eWorkbook** Removing waste from blood (ewbk-11903)

SCIENCE AS A HUMAN ENDEAVOUR: Organoids

Scientists have developed organoids to be used in the biomedical research field. This includes using the organoids to test drugs, organ development, model diseases such as cancer and potentially use for transplants. These organoids are made from stem cells (cells that can develop into different cells), and form a smaller and simpler version of an organ, designed to mimic its structure and function.

There are many benefits of using organoids, such as:

- replacing the use of live animals for medical testing
- being a more effective testing method, given animals do not have the same genetic make-up as humans
- being cheaper
- being faster to grow and use
- reducing the environmental impact of live animals testing, given resources would not be needed for maintaining and storing live animals.

4.10 Activities

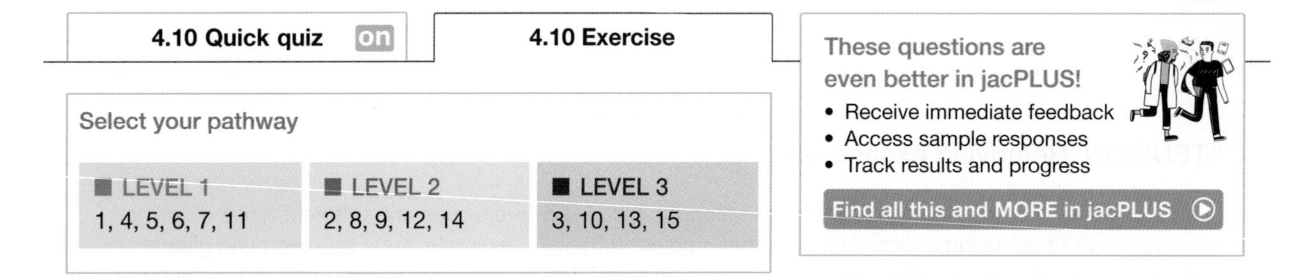

Remember and understand

1. a. State whether the following statements are true or false.

Statement	True or false?
i. Excretion involves getting rid of unwanted products or wastes from the body.	
ii. Skin, lungs, liver and kidneys are the main organs involved in human excretion.	
iii. Human skin is involved in the excretion of carbon dioxide.	
iv. Human lungs excrete urea.	
v. Human kidneys remove excess salt from the blood to help keep levels constant.	
vi. The liver converts waste products of protein into urea, which is then excreted via the kidney.	
vii. Blood and urine are mostly made up of water.	
viii. If you drink a lot of water, less will be absorbed from your large intestines and your kidneys, which will result in a greater volume of dilute urine.	
ix. The filtering of blood in each nephron of your kidney occurs in the glomerulus.	
x. Blood in your renal artery contains less 'waste' than blood in your renal vein.	

   b. Justify any false responses.

2. Match the terms associated with the excretory system with their description in the table shown.

Term	Description
a. Bladder	A. Watery fluid produced by kidneys that contains unwanted substances
b. Kidney	B. Transports urine from the bladder to outside the body
c. Ureter	C. Stores urine
d. Urethra	D. When urine moves from the bladder, through the urethra and out of the body
e. Urination	E. Transports urine from kidneys to bladder
f. Urine	F. Filters the blood and produces urine

3. Define the term *excretion*.
4. Draw and label a diagram of the kidneys, showing the following attachments: renal arteries, renal veins, ureters, bladder.

5. Outline what happens when you drink a lot of water.
6. Describe one way in which excess salt is removed from your body.
7. Explain how haemodialysis assists people with kidney disease.
8. Describe the relationship between:
   a. a kidney and a nephron
   b. kidneys and urine
   c. alcohol, lungs and kidneys.

Apply and analyse

9. **SIS** **a.** Carefully observe the haemodialysis diagram in figure 4.84. Suggest reasons the following are included in the process.
      i. Blood pump
      ii. Bubble trap
      iii. Constant temperature bath
   b. Suggest what you would expect to find in used dialysis solution.
   c. Suggest why red blood cells don't pass through the dialysis tubing.
   d. Use a Venn diagram to compare haemodialysis with real kidneys.
10. Distinguish between:
    a. the ureter and the urethra
    b. a Bowman's capsule and a glomerulus
    c. the bladder and the kidney.
11. Research and report on one of the following conditions: urinary incontinence, kidney stones, kidney transplants, cystitis, blood in urine, proteinuria, nephritis.
12. Find out and report on:
    a. the differences between the urethra in human males and human females
    b. why pregnant women often need to urinate more frequently
    c. how the prostate gland in males may affect urination in later life
    d. which foods can change the colour or volume of urine
    e. which tests use urine in the medical diagnosis of diseases.

Evaluate and create

13. Research the nephrons of animals that live in different environments (e.g. deserts, oceans or rivers). Comment on similarities and differences in their structure. Suggest reasons for the differences.
14. **SIS** Use the table provided and the other information in this lesson to answer the following questions.

TABLE Substances in blood and urine

	Quantity (%)	
Substance	**In blood**	**In urine**
Water	92	95
Proteins	7	0
Glucose	0.1	0
Chloride (salt)	0.37	0.6
Urea	0.03	2

    a. Draw two bar graphs to show the quantity of water, proteins, glucose, salt and urea in blood and in urine.
    b. Which substance is in the greatest quantity? Suggest a reason for this.
    c. Which substances are found only in blood?
    d. Which substances are found in urine in a greater quantity than in blood? Suggest a reason for this.
    e. When would the amount of these substances in the urine become greater or less than in the blood?
15. **SIS** Find out more about nephrons and how they work. Construct a model of a nephron that shows how it is linked to blood vessels and how urine gets to your bladder from it.

Fully worked solutions and sample responses are available in your digital formats.

LESSON

4.11 Musculoskeletal system — keeping in shape

LEARNING INTENTION

At the end of this lesson you will be able to describe the structure and function of the musculoskeletal system.

4.11.1 Bones

Your musculoskeletal system consists of your **skeletal system** (bones and joints) and your **skeletal muscle system**. Working together, these two systems protect your internal organs, maintain posture, produce blood cells, store minerals and enable your body to move.

skeletal system consists of the bones and joints

skeletal muscle system voluntary or striated muscle

skeleton the bones or shell of an animal that support and protect it as well as allowing movement

bones the pieces of hard tissue that make up the skeleton of a vertebrate

Did you know that an adult human **skeleton** contains over 200 separate **bones**? Without a skeleton we may resemble jelly-like blobs! Not only does it provide a structure for our muscles to attach to, allowing us to move, but it also provides support, protects vital organs (e.g. brain and heart) and forms a frame that gives our body shape.

ewbk-11905

eles-2537

int-8256

FIGURE 4.86 An adult human skeleton contains over 200 separate bones.

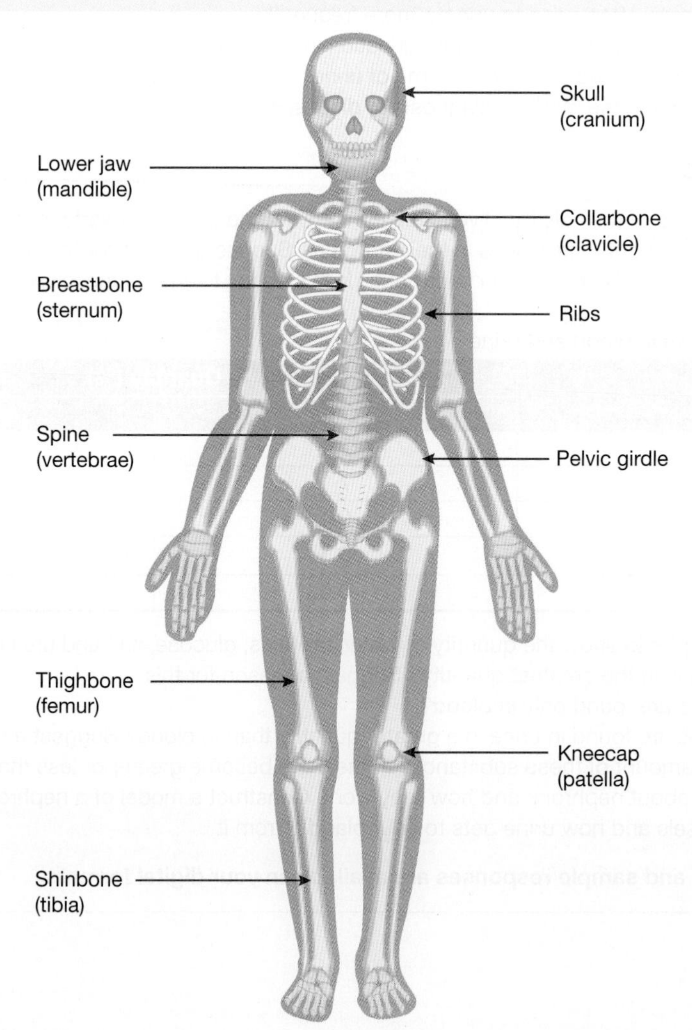

Your bones are many different shapes and sizes, depending on the job that they have to do. They can be short, thick, round or flat. The longest bone in your body is the femur, or thigh bone. A feature that all bones have in common, however, is that they are all light and strong. Why do you think they share this feature?

It's hard being a bone

Bones are alive. If bones were not alive, how would you grow taller? How would a broken arm or leg mend? Bones contain living cells and need a blood supply to provide them with oxygen and other nutrients.

Not only are bones busy providing you with support and movement, they are also busy within. Bones contain soft tissue called **bone marrow**. This is very important because it is where blood cells are made.

Throughout the first 20 years of your life, most of the soft and rubbery **cartilage** that made up your skeleton is gradually replaced with bone. Your trachea, nose and ears, however, are made mostly of cartilage, and the ends of your bones remain covered in cartilage.

Compact bones, such as the long bone of your femur, have a strong and hard outer layer that contains **calcium** and **phosphorus**. This is why you need an adequate supply of these minerals. Investigation 4.10 shows what could happen to your bones without a supply of these important **minerals**.

The hardening of your bones as you get older is called **ossification**. After ossification, the bone is made up of about 70 per cent non-living matter and 30 per cent living matter. As you get even older, your bones may get dry and **brittle**, which is why older people break their bones more easily.

bone marrow a substance inside bones in which blood cells are made

cartilage a waxy, whitish, flexible substance that lines or connects bone joints or, in some animals such as sharks, replaces bone as the supporting skeletal tissue; the ears and tips of noses of humans are shaped by cartilage

calcium an element occurring in limestone and chalk, and also present in vertebrates and other animals as a component of bone and shell; it is necessary for nerve conduction, heartbeat, muscle contraction and many other physiological functions

phosphorus a substance that plays an important role in almost every chemical reaction in the body; together with calcium, it is required by the body to maintain healthy bones and teeth

mineral any of the inorganic elements that are essential to the functioning of the human body and are obtained from foods

ossification hardening of bones

brittle can easily break if hit; the opposite of malleable

elog-2167

INVESTIGATION 4.10

Rubbery bones

Aim

To investigate the effect of calcium and phosphorus deficiency on bone

Background information

Vinegar is an acid that dissolves minerals such as calcium and phosphorus, removing them from bones.

Materials

- 2 chicken or turkey bones
- water
- 2 jars (or beakers)
- vinegar

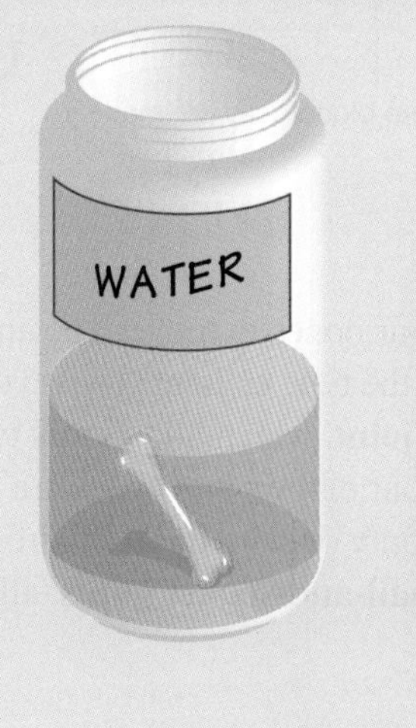

Method

1. Clean the two chicken or turkey bones and leave them to dry overnight.
2. Observe the bones and then place one bone in a jar of vinegar and the other in a jar of water.
3. Allow the bones to soak for at least three days. Then remove the bones and observe any changes.
4. Return the bones to their previous jars for another week, then remove and observe any further changes in the bones. Can you tie either bone into a knot?

Results

1. Construct a table to record your observations.
2. Record your observations of the bones:
   a. before placing them in the solutions
   b. after soaking for three days
   c. a week after your observation in part **b**.

Discussion

1. Suggest a reason for the inclusion of the jar of water in the investigation.
2. Describe changes that you observed for each bone.
3. Provide reasons for the changes that you observed.
4. Relate this investigation to your bones and your diet.

Conclusion

Summarise the findings for this investigation.

4.11.2 Joints

A **joint** is the region where two bones meet. Your knees and elbows are examples of joints. Bones at a joint are held together by bundles of strong fibres called **ligaments**. The cartilage that covers the end of each bone is itself covered with a liquid called **synovial fluid**. The cartilage and synovial fluid work together to stop bones from scraping against each other.

FIGURE 4.87 The region where bones meet is called a joint. Cartilage and synovial fluid stop bones scraping against each other.

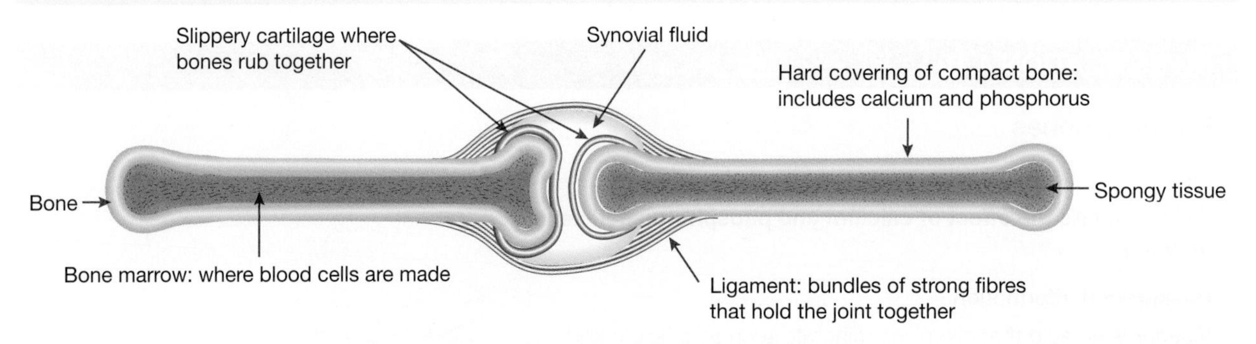

Most joints allow your bones to move. The amount and direction of movement allowed depends on the type of joint. Twist your neck. The joint between your skull and spine is a **pivot joint**, which allows this twisting type of movement (figure 4.88a). Bend your elbow. Your elbows and knees are **hinge joints**, like those of a door. They allow movement in only one direction (figure 4.88b). Roll your shoulder. Your hip and shoulder joints are **ball-and-socket joints**, allowing movement in many directions (figure 4.88c).

joint a region where two bones meet

ligament a band of tough tissue that connects the ends of bones or keeps an organ in place

synovial fluid the liquid inside the cavity surrounding a joint that helps bones to slide freely over each other

pivot joints joints that allows a twisting movement

hinge joints joints in which two bones are connected so that movement occurs in one plane only

ball-and-socket joints joints in which the rounded end of one bone fits into the hollow end of another

eles-2052

FIGURE 4.88 Different types of joints: a. pivot joint b. hinge joint c. ball-and-socket joint

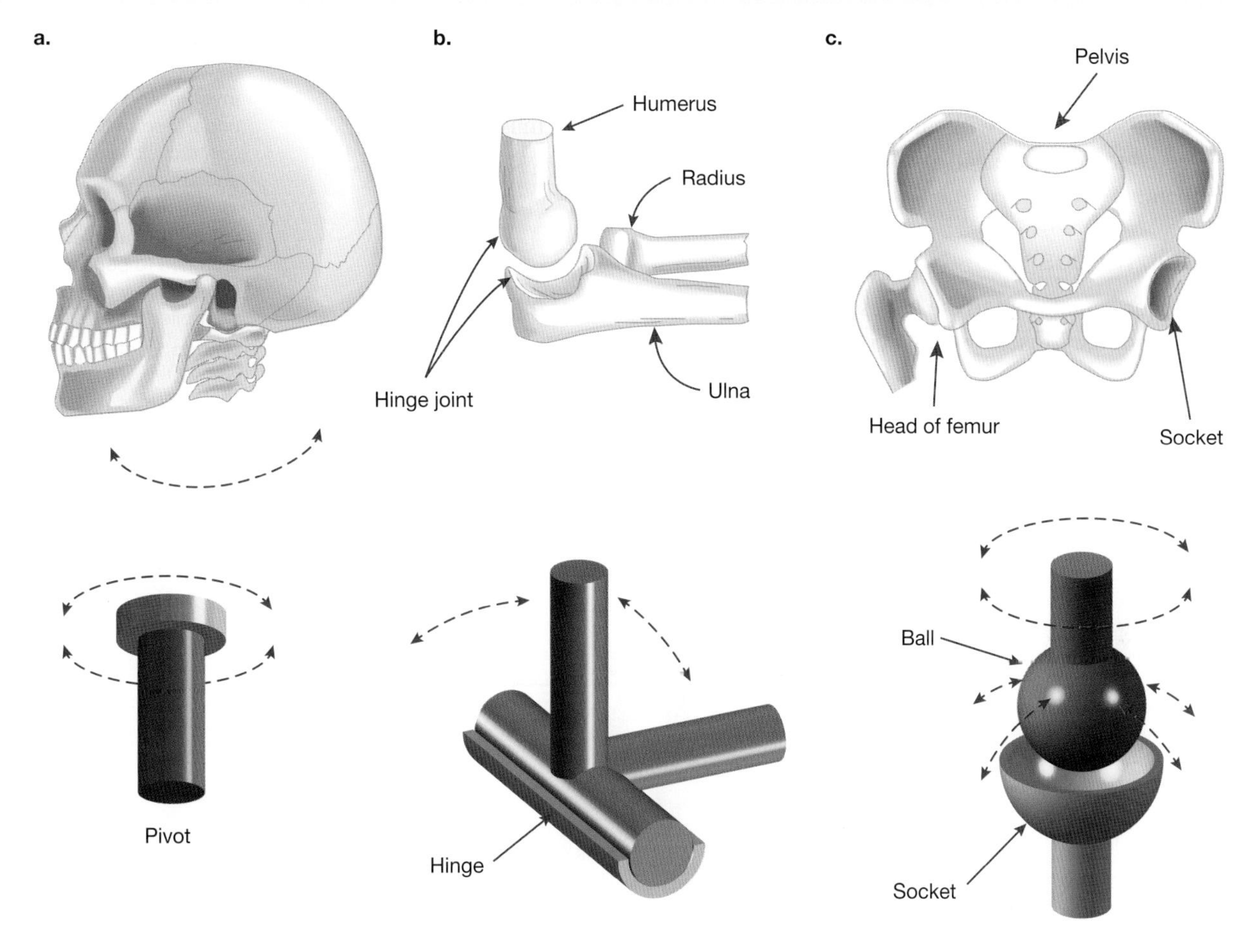

Some joints, such as those that join the plates in your skull, do not move. These are called **immovable joints**. While not allowing movement, these joints provide a thin layer of soft tissue between bones. Their job is to absorb enough energy from a severe knock to prevent the bone from breaking.

immovable joints joints that allow no movement except when absorbing a hard blow

FIGURE 4.89 The plates of the skull are immovable joints.

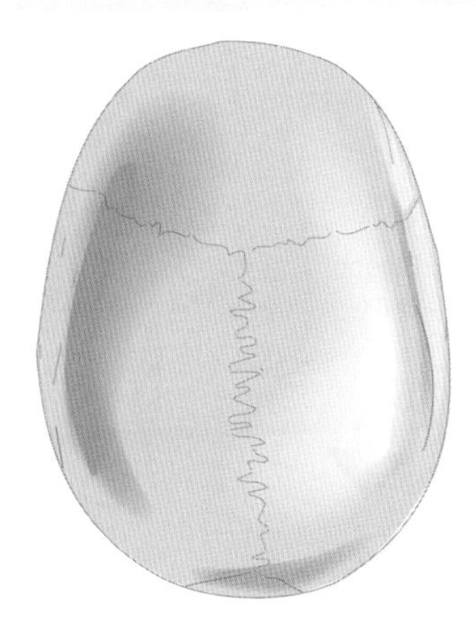

4.11.3 Muscles

Muscles are tough and elastic fibres. The movement of muscles is controlled by the brain, which sends signals through your nerves. Muscles such as those that make your heart pump and those that control your breathing rate are called **involuntary muscles** — they work without you having to think about it. The muscles that are connected to bones are called **voluntary muscles** because you have to choose to use them.

muscles tissue consisting of cells that can shorten

involuntary muscles muscles not under the control of the will; they contract slowly and rhythmically, and are at work in the heart, intestines and lungs

voluntary muscles muscles attached to bones; they move the bones by contracting and are controlled by an animal's thoughts

tendon tough, rope-like tissue connecting a muscle to a bone

All pull, no push!

Muscles are connected to the bones of your skeleton by bundles of tough fibres called **tendons**. Muscles pull on bones by contracting or shortening. Muscles never push.

FIGURE 4.90 When your biceps contract, your arm bends upwards. When your triceps contract, your arm straightens.

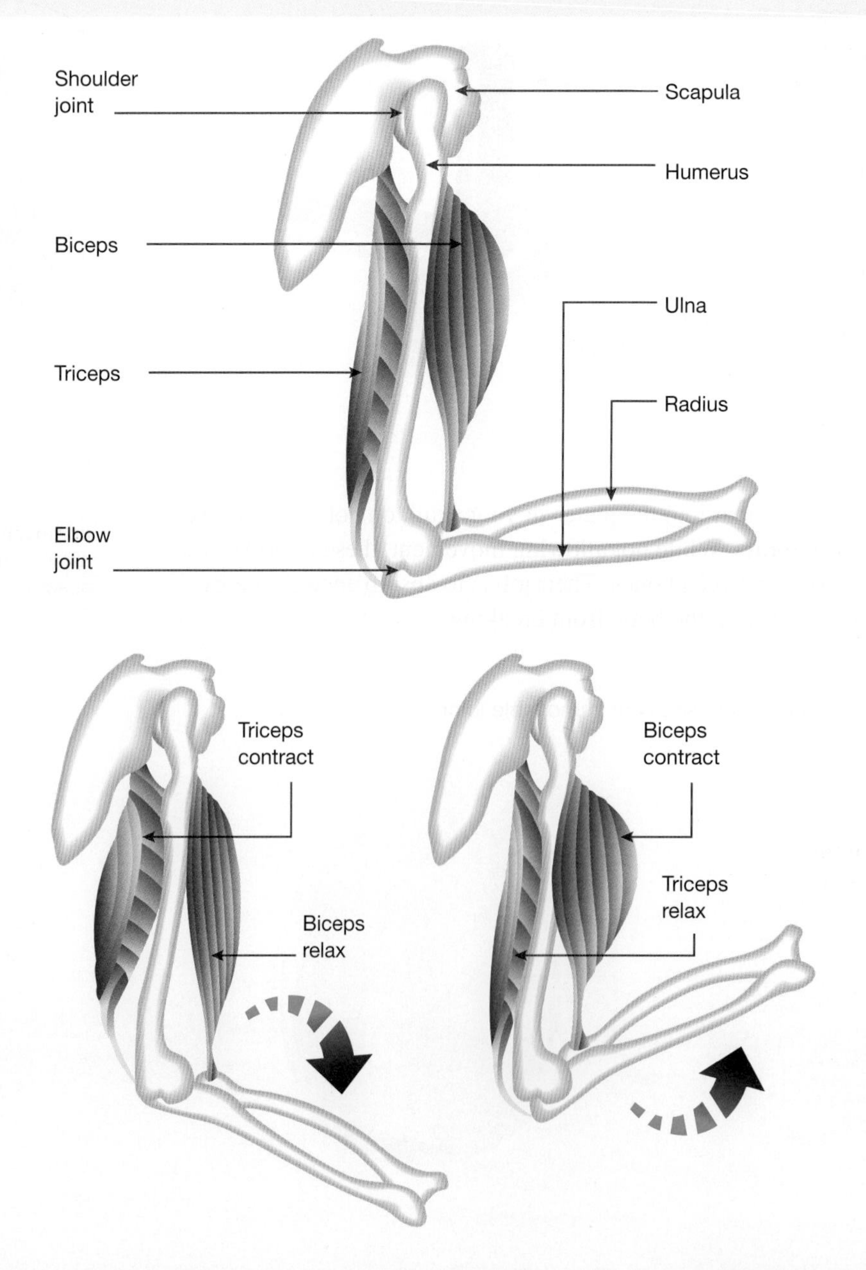

4.11.4 Broken bones

Breaks and fractures

When a bone breaks, the ends of the bone may need to be put back into place (set) so that they can grow together. If a bone is shattered into several pieces, it is sometimes possible to use pins or wire to hold the pieces in place while the bone heals. A **greenstick fracture** occurs when the bone cracks but does not break. This type of **fracture** is common in children because their bones are more flexible.

On the mend?

New technologies are being researched and developed to help fix broken bones. Some of these involve special cells called **stem cells**, while others involve the use of special 'glues' that hold bones together and aid the healing process. Scientists at the CSIRO are currently working on a liquid gel called NovoSorb® that glues the fractured bone together so that it is supported while it heals. As this gel degrades naturally, it does not require follow-up surgery to remove pins as is needed with older technologies.

Osteoporosis

Osteoporosis is a loss of bone mass that causes bones to become lighter, more fragile and more easily broken. In Australia, over one million people are estimated to have osteoporosis and a further six million to have low bone density. Osteoporosis is more common in older age groups. It is also more common in females than males.

In your teenage years, you can help protect yourself from getting osteoporosis later in life by having a healthy diet and exercising. Your diet should include dairy products such as milk, cheese and yoghurt, as well as other foods high in calcium. Such a diet will help ensure that your bone mass is adequate as an adult.

Resources

Video eLesson Osteoarthritis (eles-2053)

4.11.5 Ouch! Torn or swollen?

Sprains

Sprains occur when ligaments joining bones at a joint are torn or stretched. Sprains usually happen when you fall onto a joint, such as an elbow or an ankle, and twist it.

Arthritis

Arthritis is a swelling of the joints that makes movement difficult. Osteoarthritis occurs mainly in elderly people and is caused by wear and tear of the joints. The cartilage gradually breaks down, thus allowing bare bones to grate against each other instead of sliding or turning smoothly. Rheumatoid arthritis is a swelling of the tissue between the joints. The swelling causes the joints to slip out of place, which then causes great pain and deformities.

Tennis elbow

Tennis elbow can result from repeated grasping and bending back of your wrist. This repeated action can lead to the inflammation of the tendon that connects the muscles of your forearm to the bone in your upper arm. As these muscles are used to bend your wrist backwards, any activity (not just playing tennis or other racquet sports) with repetitive actions (e.g. painting, texting, or using your computer keyboard or mouse) may result in tennis elbow.

greenstick fracture a break that is not completely through the bone, often seen in children

fracture a break in a bone

stem cells undeveloped cells found in blood and bone marrow that can reproduce themselves indefinitely

osteoporosis loss of bone mass that causes bones to become lighter, more fragile and more easily broken

sprain an injury caused by tearing a ligament

arthritis a condition in which inflammation of the joints causes them to swell and become painful

tennis elbow occurs when repeated grasping and bending back of your wrist leads to the inflammation of the tendon that connects the muscles of your forearm to the bone in your upper arm, causing pain

Torn hamstrings

Torn hamstrings are a common sporting injury. The hamstring muscle joins the pelvis to the bottom of the knee joint, running along the back of the thigh. It controls the bending of the knee and straightening of the hips. A sudden start or turn in sport often stretches the hamstring muscle too far. It tears, causing great pain. Cold and unprepared muscles are more likely to tear. Proper warming up before strenuous sporting activity is one way to reduce the chances of tearing a muscle.

FIGURE 4.91 Torn hamstrings are a common but painful sporting injury.

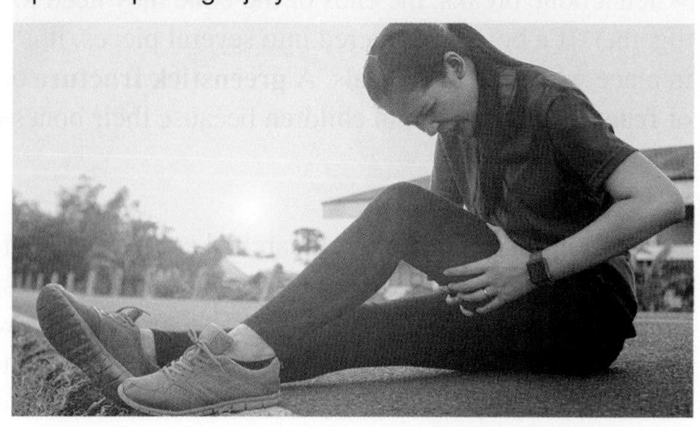

torn hamstring a common sporting injury caused by overstretching the hamstring muscle, which joins the pelvis to the knee joint

elog-2169

INVESTIGATION 4.11

Chicken wing dissection

Aim

To investigate the structure of a chicken wing

CAUTION

Take special care when using scissors and scalpels.

Materials

- chicken wing
- dissection tray or board
- newspaper
- disposable gloves
- scalpel
- scissors

Method

1. Use the scalpel carefully to make a small incision in the middle of the upper part of the wing.
2. Use the scissors to cut through the chicken skin in order to remove it but keep the bones, muscles (meat) and tendons in place. Do not cut the tendons.
3. Observe the wing without its skin.

Results

1. Sketch one of the joints in the chicken wing. Label the bones, the tendons and the muscles. Show clearly where the muscle inserts (attaches to the bones). Use arrows to show how the bones move when the muscle is shortened.
2. Is cartilage harder or softer than bone?
3. Feel the cartilage with a gloved hand. Does the cartilage feel rough or slippery? Why does it need to be slippery?

Discussion

1. Describe differences that you observed between cartilage, tendons, muscles and bones.
2. Relate this investigation to your own joints, muscles, tendons and bones.

Conclusion

Summarise the findings for this investigation.

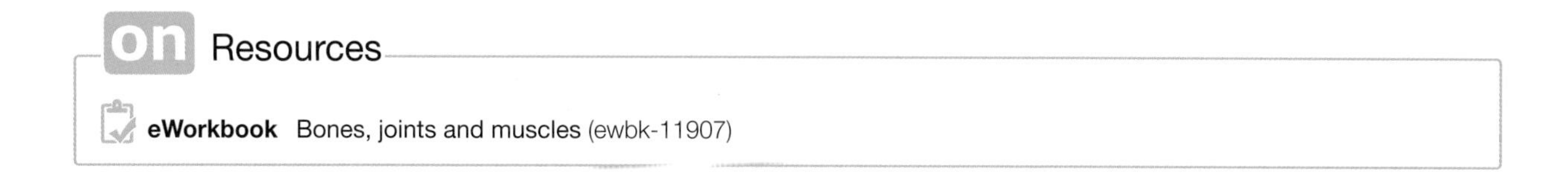

4.11 Activities

learnon

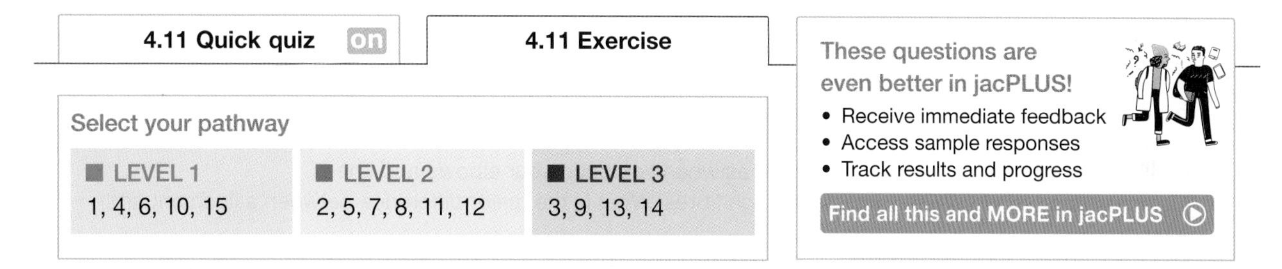

Remember and understand

1. a. State whether the following statements are true or false.

Statement	True or false?
i. The musculoskeletal system consists of your skeletal system (bone and joints) and your skeletal muscle system.	
ii. Working together, the skeletal system and the skeletal muscle system protect internal organs, maintain posture, produce blood, store minerals and enable your body to move.	
iii. Bones contain hard tissue called bone marrow in which blood cells are made.	
iv. Your trachea, nose and ears are mostly made up of bone.	
v. The job of immovable joints, such as those in your skull, is to absorb enough energy from a severe knock to prevent the bone from breaking.	
vi. Bones at a joint are held together by tendons, whereas muscles are connected to bones by ligaments.	
vii. New technologies are being researched and developed to help fix broken bones that include using stem cells and special 'glues'.	
viii. Voluntary muscles are connected to bones and you can choose when and how to use them.	
ix. Involuntary muscles only work when you consciously decide to use them.	

b. Justify any false responses.

2. Match the common name with its scientific term in the table shown.

Common name	Scientific term
a. Breastbone	**A.** Mandible
b. Kneecap	**B.** Cranium
c. Lower jaw	**C.** Vertebrae
d. Shinbone	**D.** Tibia
e. Skull	**E.** Sternum
f. Spine	**F.** Femur
g. Thighbone	**G.** Patella

3. Some joints are referred to as immovable joints. What is the use of having joints that don't move?
4. Write down an example of each of the following types of joint.
   a. Hinge
   b. Ball-and-socket
   c. Pivot
   d. Immovable
5. Identify the type of joint (pivot, hinge, ball-and-socket or immovable) that:
   a. allows movement in many directions, such as when you roll your shoulder
   b. does not move, such as those that join the plates in your skull
   c. allows a twisting type of movement, like when you twist your neck
   d. allows movement in one direction, such as when you bend your elbows and knees.
6. Ligaments and tendons are bundles of tough fibres. What is the major difference between a ligament and a tendon?
7. Describe the action of the biceps and triceps muscles as you bend your elbow to raise your forearm.
8. Describe the job done by each of the following parts of a joint.
   a. Ligament
   b. Cartilage
   c. Synovial fluid

Apply and analyse

9. Look carefully at each of the skeletons shown. Three of them are incomplete. Identify the incomplete skeletons and name the missing parts.

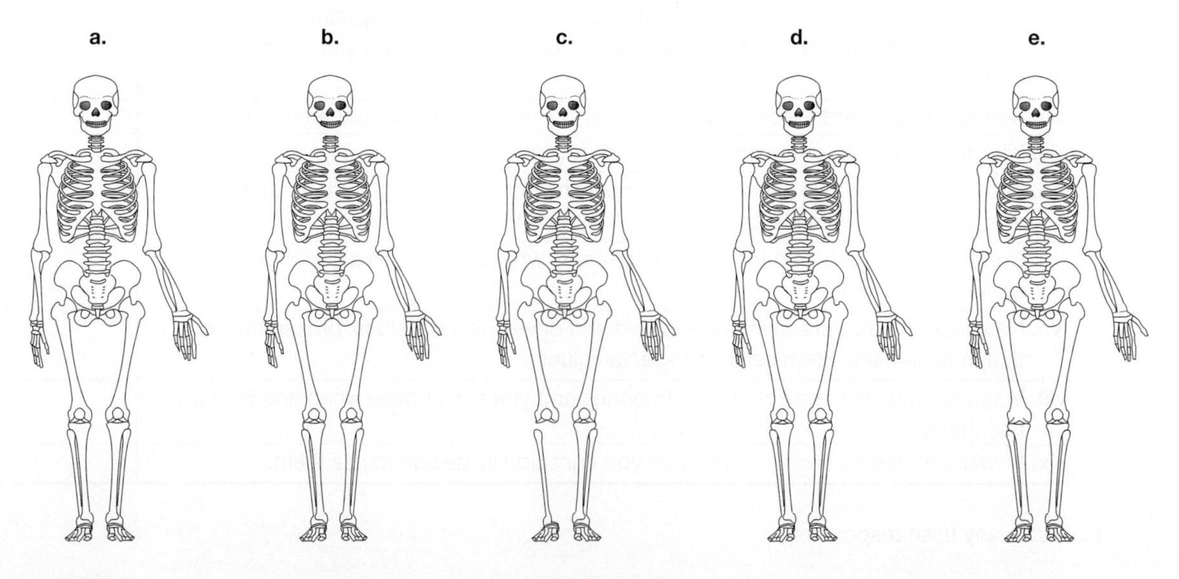

10. Apart from warming up just before a game, how do the best basketball and netball players reduce the likelihood of torn muscles and tendons?
11. What would happen if the cartilage in your knee joint wore out?

12. Research and report on one of the following science careers: orthopaedic surgeon, physiologist, physiotherapist, occupational therapist, rheumatologist, fitness trainer.
13. Find out more about the structure and function of skeletal, smooth and cardiac muscle tissue.
14. What is dietary rickets and how is it caused?
15. Your musculoskeletal system consists of your skeletal system (bones and joints) and your skeletal muscle system (voluntary or striated muscle). Working together, these two systems protect your internal organs, maintain posture, produce blood cells, store minerals and enable your body to move.
Use information in this lesson and other resources to relate structural features to the functions of the following parts of the musculoskeletal system.

TABLE Parts of the musculoskeletal system and their functions

Part of musculoskeletal system	Structural features	Function
Bones		
Cartilage		
Joints		
Skeletal muscles		

Fully worked solutions and sample responses are available in your digital formats.

LESSON
4.12 Same job, different path

LEARNING INTENTION

At the end of this lesson you will be able to provide examples of how variations in the structures of the organs in the respiratory and digestive systems make them well suited to perform their specific tasks.

4.12.1 Patterns, order and organisation

Similar, but different? While organisms can have different solutions to life's challenges, these differences share similar patterns, order and organisation.

Organisms possess a variety of structures that help them to obtain the resources that they need to survive. While there are similarities and patterns in some of these structures, there are also differences. These differences provide examples of wonderful and creative solutions to the continual challenge of staying alive.

FIGURE 4.92 Similar structures? What might their function be?

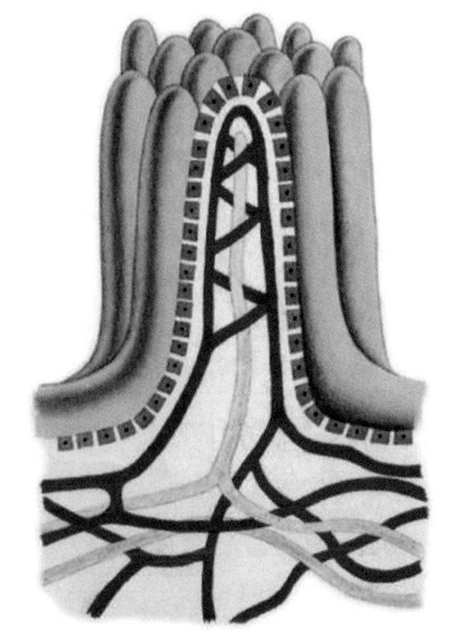

Intestinal villi

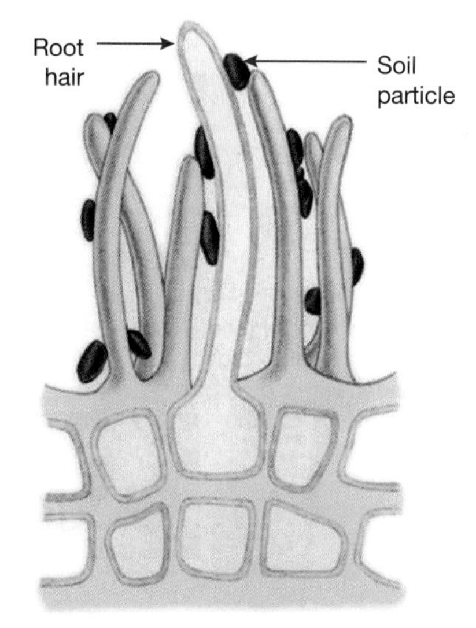

Plant root hairs

Shaping clues

The structures of cells and tissues often provide clues to their function. For example, structures that are involved in absorption often have shapes that increase their surface-area-to-volume ratio. Intestinal villi in humans and plant root hairs are examples of this. Can you see the similarities in figure 4.92? Can you think of other cells or tissues that also share this pattern?

4.12.2 Respiratory routes

Cellular respiration

Glucose + oxygen → carbon dioxide + water + energy

Cellular respiration is essential for life. Organisms require a supply of oxygen and a way to remove the carbon dioxide that is produced as waste. Although this gaseous exchange is essential, different types of organisms achieve it in different ways.

FIGURE 4.93 A flatworm

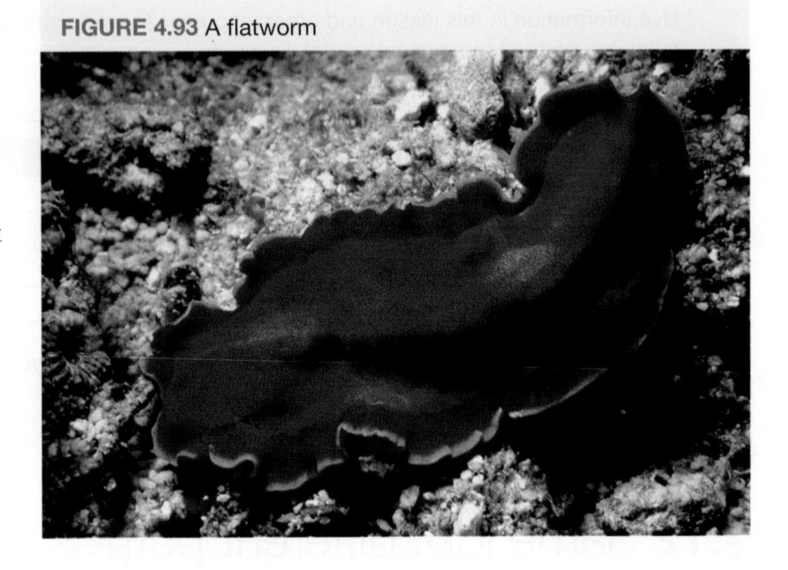

Unicellular organisms are small enough that gases such as oxygen and carbon dioxide can simply diffuse in and out of their cell. Likewise, some very thin multicellular organisms have many of their cells in direct contact with their environment. These organisms rely simply on diffusion for their exchange of gases. Flatworms (figure 4.93), for example, do not need a respiratory system, as they use their whole body surface to obtain the oxygen they require from the water in which they live. Some other small animals, such as worms living on land, can exchange gases through their mucus-covered skin. Oxygen from the air dissolves in the mucus, while carbon dioxide seeps out. Tiny blood vessels in their skin transport the gases to and from the rest of the worm's body.

Other animals may have specialised gas exchange organs. Three main kinds of these organs are lungs in mammals and amphibians, gills in fish, and tracheae in insects. Examine figure 4.94 to compare the structure of these organs. How are they similar? How are they different?

FIGURE 4.94 Notice any similarities or differences in these gas exchange surfaces?

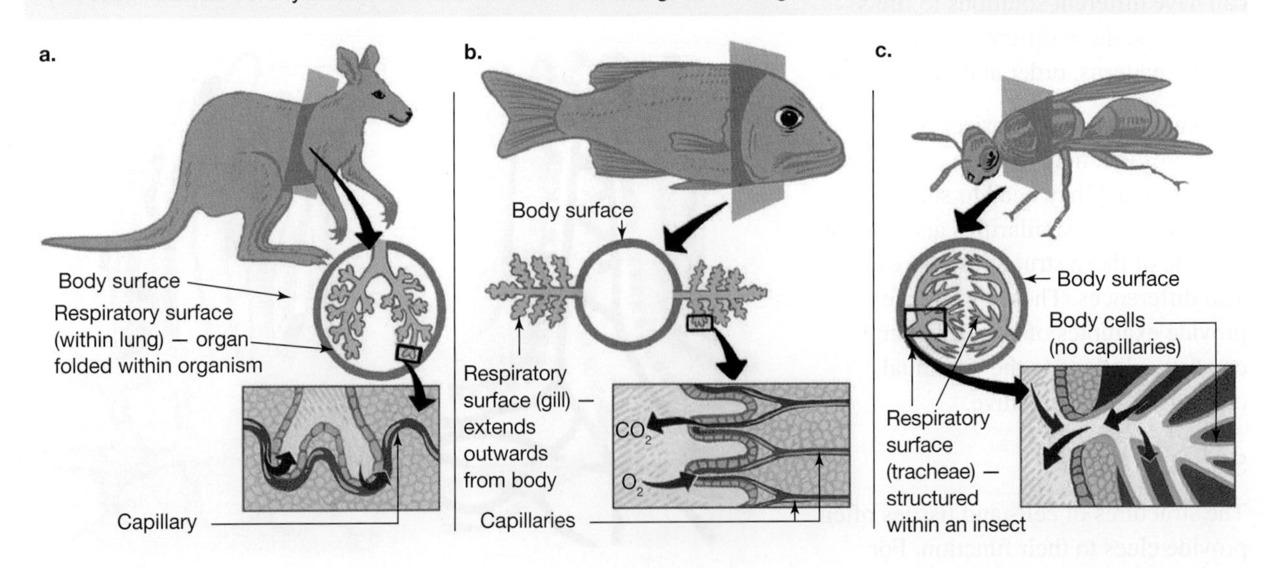

In topic 3 (section 3.7.10), it was seen that plants undergo a process called photosynthesis, where carbon dioxide from the air and water are converted to oxygen, glucose and energy in chloroplasts.

$$\text{Carbon dioxide} + \text{water} \rightarrow \text{oxygen} + \text{glucose} + \text{energy}$$

This is how plants 'breathe'; however, production of oxygen can only occur during the day when there is sunlight. This is very different to animals who undergo cellular respiration through gas exchange, as shown previously, and this occurs at all times of the day.

4.12.3 What do they eat?

Animals can be classified on the basis of their diet. **Herbivores** eat plants; **carnivores** eat other animals; **omnivores** eat both plants and animals. An animal's teeth can provide hints to the types of food that it eats. Observe the teeth in the skulls of the vertebrates shown in figure 4.95. Based on your observations, predict the types of food that they may eat.

FIGURE 4.95 Skulls of vertebrates

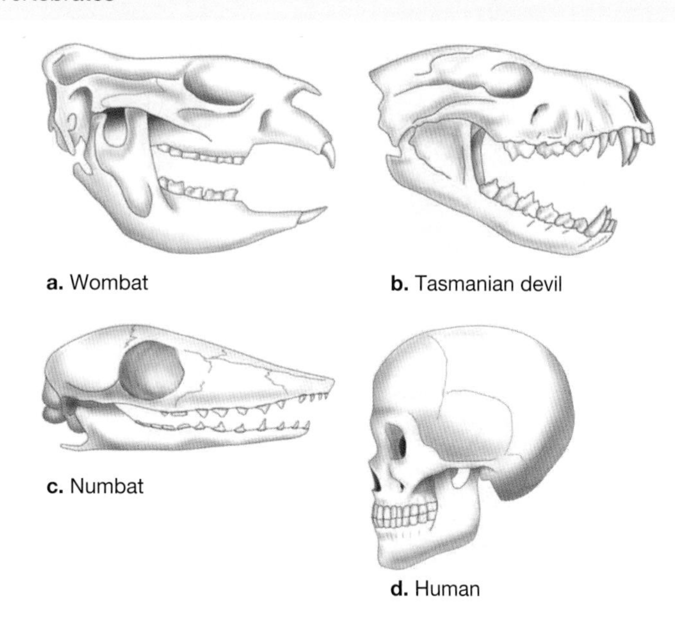

Wombats are herbivores.

- They have large incisors for biting and cutting, but no canines.
- They also have large premolars and molars because the fibrous plant materials they eat need a lot of grinding.

Tasmanian devils are carnivores.

- They possess large canines for stabbing and holding on to prey that is alive and moving.
- Their incisors are used for tearing meat.
- The molars and premolars in carnivores have cutting edges.

Insectivores are carnivores that eat only insects.

- Their teeth are small and pointed so that they can crush the exoskeleton of the insect, which they then swallow whole.

Humans are omnivores.

- We possess all of the different types of teeth needed to break down both meat (from animal tissue) and plants.

herbivore an animal that eats only plants

carnivore an animal that eats other animals

omnivore an animal that eats plants and other animals

insectivore a carnivore that eats only insects

4.12.4 Digestive differences

Although most vertebrates possess a digestive system that has a similar pattern, order and organisation, there may be differences that are related to nutritional needs and diet. Consider, for example, differences in the digestive systems of herbivores with diets that are high in plant material with lots of cellulose, compared with those of carnivores with lots of animal flesh, high in protein. How would these compare with the digestive system of an organism that ate only nectar and pollen?

DISCUSSION

Some animals that live in water, such as sea anemones, have a digestive sac that acts as both a mouth and an anus. Find out more about the digestive system of sea anemones.

a. Describe how this digestive system is similar to that of humans and how it is different.
b. Suggest reasons for the differences.

FIGURE 4.96 Notice any similarities or differences in these digestive systems?

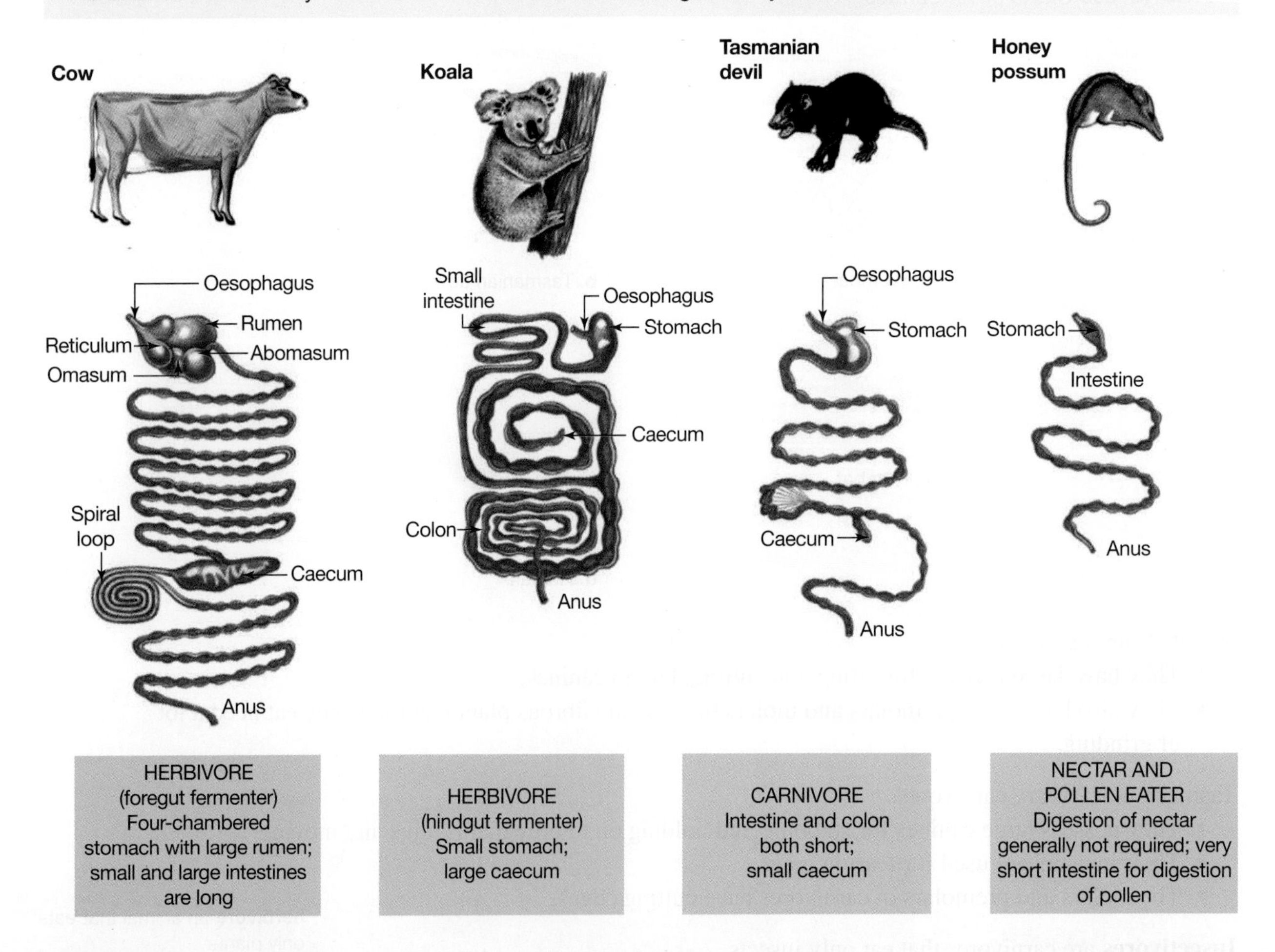

As the process of photosynthesis also creates the food that most plants need, they do not have a specific digestive system. However, there are some plants, like *Dionaea muscipula* (Venus flytrap, figure 4.97), that also catch prey for essential nutrients, including nitrogen and phosphorus. Once the trap closes, the digestive glands release enzymes to break down the prey. In this instance, the mouth is also the stomach. Once the enzymes have done their job, the nutrients are absorbed into the leaf to be used. After a few days, the trap will re-open, leaving behind the exoskeleton of the prey.

FIGURE 4.97 An unlucky fly being eaten by a Venus flytrap

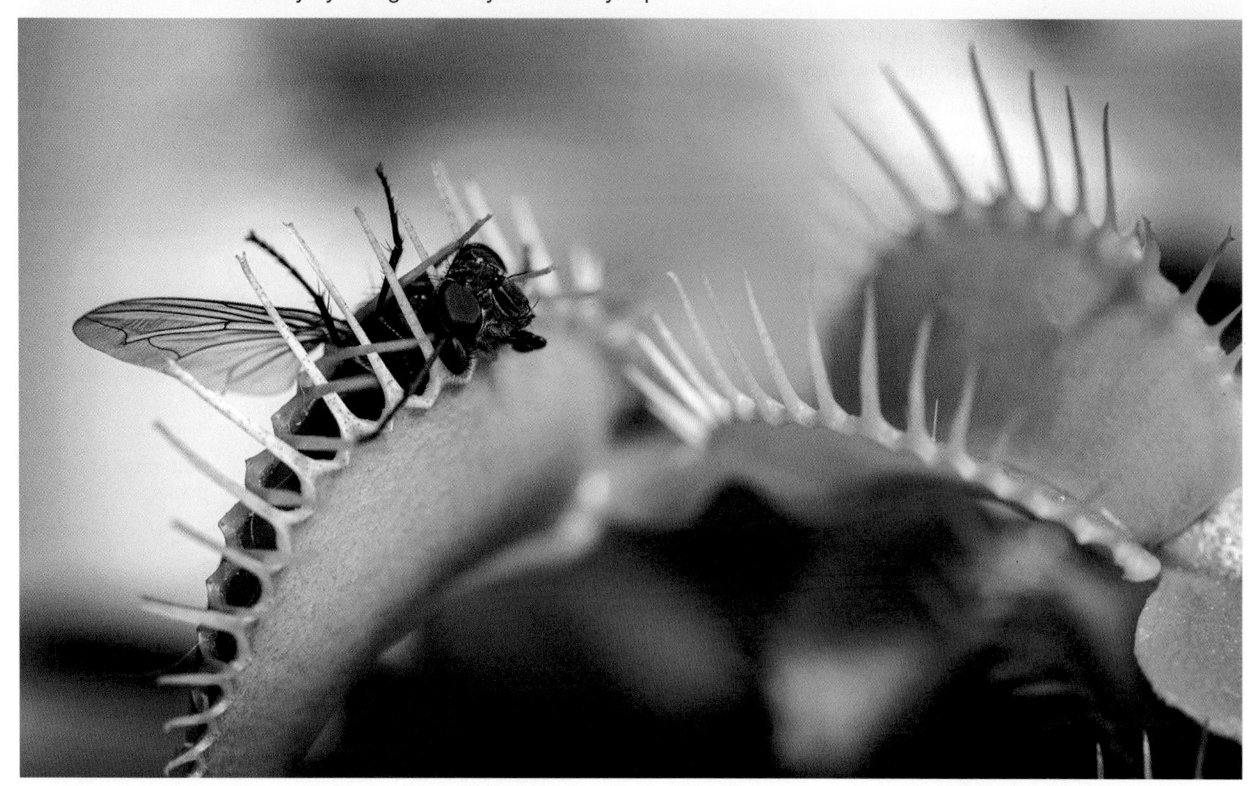

4.12.5 Heart count?

Two, three or four? Not all animals have a four-chambered heart like you. Fish have a heart with two chambers and blood passes through the heart only once each time around the body. The hearts of amphibians and most reptiles are three-chambered and allow oxygenated and deoxygenated blood to mix. Birds and mammals are similar to amphibians and most reptiles in that blood flows through the heart twice in each circulatory trip, but they possess a heart with four chambers that does not allow the mixing of blood. What do they share? How are they different?

FIGURE 4.98 Around we go … but which route do we take?

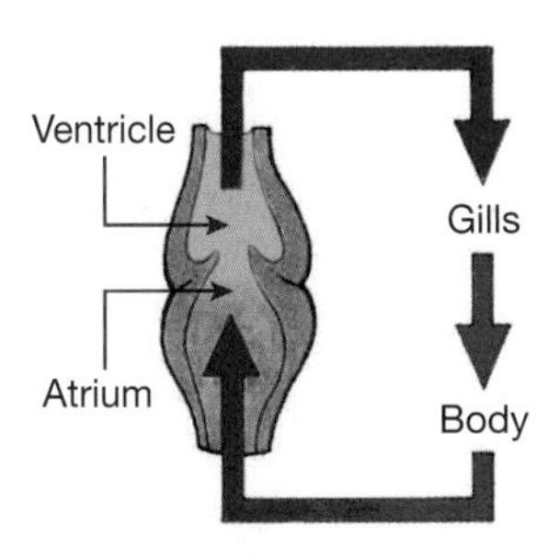

a. A fish heart has two chambers. Note that blood passes through the heart only once for every circulation within the body.

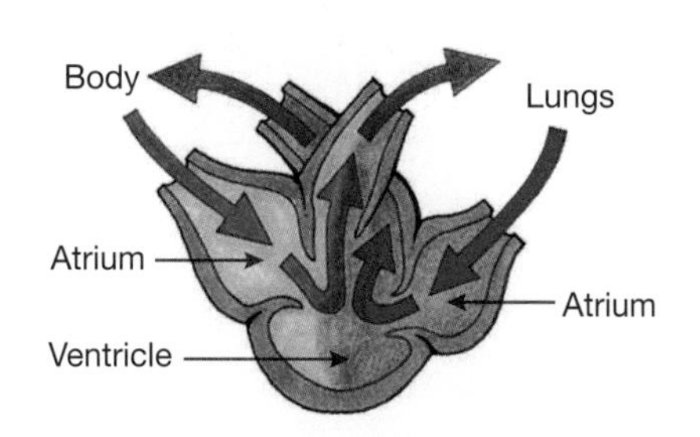

b. Amphibians and most reptiles have a three-chambered heart. Oxygenated and non-oxygenated blood mixes in the single ventricle as blood flows through the heart twice for every circulation within the body.

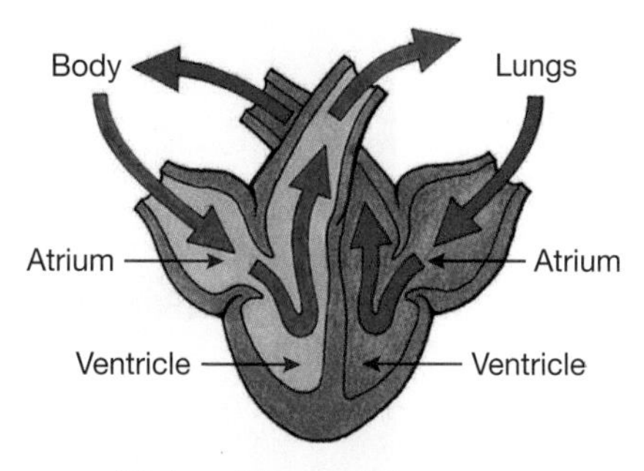

c. Birds and mammals have a four-chambered heart and blood flows through the heart twice for every circulation within the body.

4.12.6 Throwing out the trash!

Different types of fish, living in different environments, can also differ in how they maintain their salt balance.

Proteins are involved in a variety of different chemical reactions that keep animals alive. **Ammonia** is formed when proteins break down. Ammonia is toxic to cells and requires either lots of water to release it into, or conversion into a less toxic form (such as urea or uric acid). Conversion into other forms costs the animal energy. Whichever form these **nitrogenous wastes** are in, they need to be removed from the animal's body.

FIGURE 4.99 Saltwater fish, such as snapper, drink seawater constantly and produce a small volume of urine. However, freshwater fish, such as Murray cod, rarely drink, but make lots of urine.

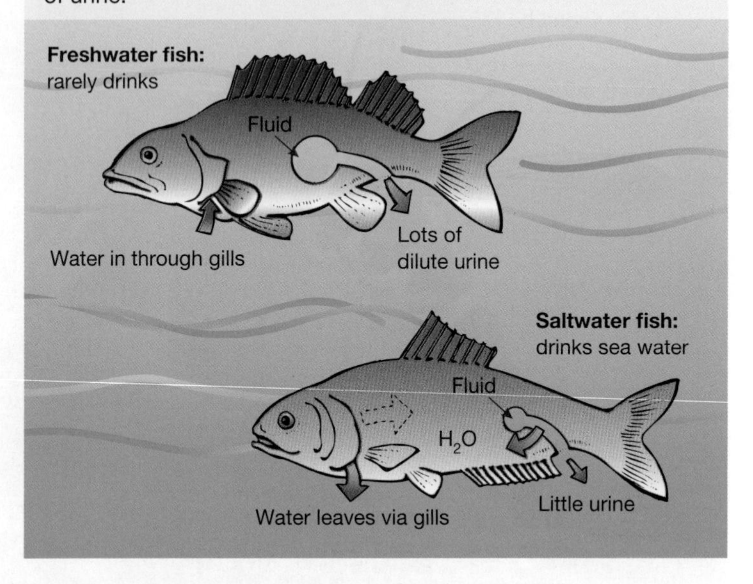

Different types of animals use different strategies to remove nitrogenous wastes. This is linked to the amount of water available in the environments in which they live. Fish, for example, have a ready supply of water, so most fish release their nitrogenous waste as ammonia. The main nitrogenous waste excreted by humans is **urea**.

Uric acid requires the least water for excretion. Insects, spiders and birds excrete their wastes as uric acid. The uric acid produced by birds is solid; it is stored in their bodies without diluting it with water and is excreted with their faeces. Animals living in dry environments, such as insects and snakes, also excrete their wastes in this form to conserve water.

FIGURE 4.100 Birds produce nitrogenous waste in a form of solid uric acid. What advantage to they gain from this?

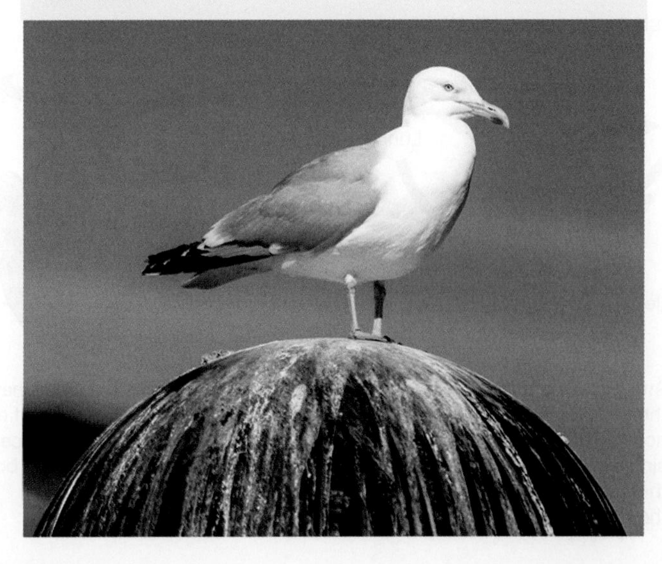

protein a chemical made up of amino acids needed for growth and repair of cells in living things

ammonia a nitrogenous waste product of protein break down

nitrogenous wastes waste products from protein breakdown, including ammonia, urea and uric acid

urea a nitrogen-containing substance produced by the breakdown of proteins and removed from the blood by the kidneys

uric acid a nitrogenous waste product of protein breakdown

Have you noticed the pattern? The environment in which an organism lives, in this case the amount of water available, can play a role in how different species have evolved different strategies to the same problem of removal of their wastes.

In plants, the waste products are removed in a variety of ways. Water vapour and gaseous wastes, including carbon dioxide, are released through the stomata (topic 3, section 3.8.3). Some of the waste is also stored in the vacuole of the leaf cell, which then eventually falls off the plant.

FIGURE 4.101 a. The stomata are important for allowing waste products to be removed from the plant. **b.** Stomata can be found on both the top and underside of a leaf.

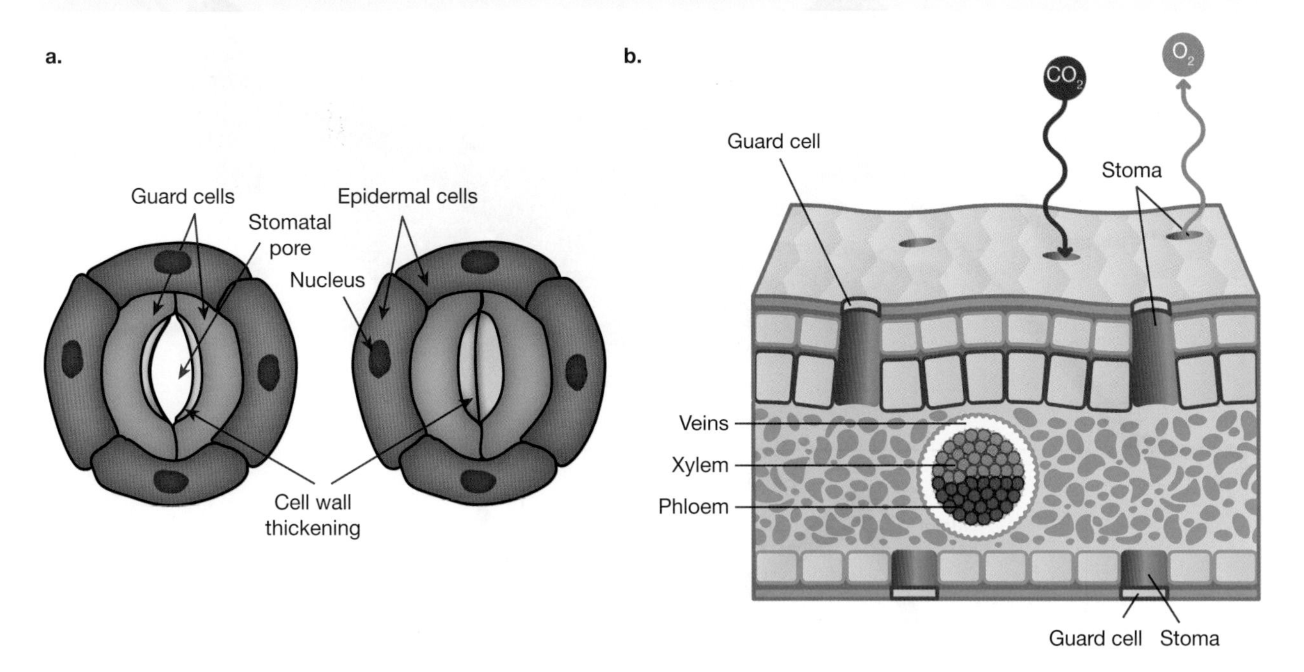

4.12.7 Inside or out?

Are you wearing your skeleton on the inside or outside? Vertebrates (such as humans) have an internal backbone. They also have an internal skeleton called an **endoskeleton**. You can read more about our endoskeleton in the previous lesson. Invertebrates, however, do not possess a backbone. Some invertebrates, such as grasshoppers and ants, have their skeleton on the outside of their bodies. Their external skeleton is called an **exoskeleton**. Other invertebrates, such as worms and jellyfish, do not have any skeleton at all.

Due to their different body structures, invertebrates can use their muscles in different ways to achieve movement. Worms and slugs, for example, can stretch and shorten muscles in certain parts of their body to bring about movement. Even though the muscles of jellyfish have no bones or other hard parts to attach to, they propel themselves through water by pumping water into their body cavities and releasing it suddenly. The muscles in insects, such as grasshoppers, are attached to their exoskeleton. They can extend their legs by contracting the extensor muscles and relaxing their flexor muscles.

endoskeleton a skeleton that lies inside the body

exoskeleton a skeleton or shell that lies outside the body

FIGURE 4.102 The muscles in insects are attached to the exoskeleton, the outer covering of the body. This grasshopper can extend its leg by contracting the extensor muscle and relaxing the flexor muscle.

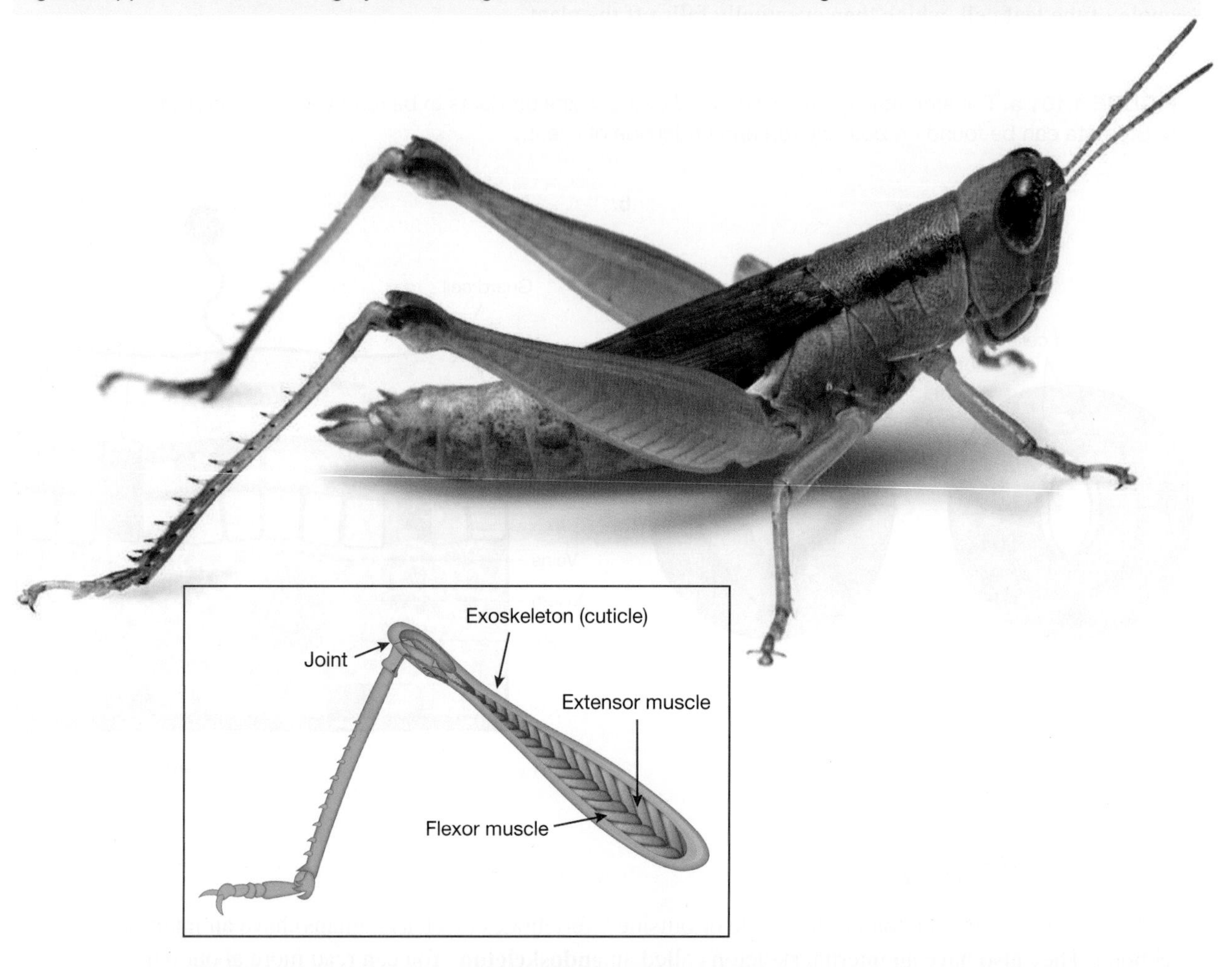

elog-2171

INVESTIGATION 4.12

Inside or out?

Aim

To use models to investigate the differences between how muscles join to bones in animals with endoskeletons and exoskeletons

Materials

- 2 cardboard tubes, each at least 30 cm long
- sticky tape
- rubber bands
- large nail or other pointed object

Method

1. Cut each cardboard tube into two pieces about 15 cm long.
2. Using the nail, make two holes on opposite sides of each tube. These should be about 5 cm from one end of each piece.
3. Label two pieces 'Endo A' and 'Endo B' and the other two pieces 'Exo A' and 'Exo B'.
4. Tape Endo A and Endo B together on one side, so that they form a hinge at the ends with the small holes.
5. Cut two rubber bands and thread the cut ends through the holes from the outside.
6. Tie knots so that the rubber bands can't pull back through the holes.

7. Tape Exo A and Exo B together in the same way as Endo A and Endo B.
8. Cut another two rubber bands and thread the cut ends through the holes so that they run *inside* the tube.
9. Make sure that they are stretched very tightly, and then tie knots on the outside of the tubes.

The rubber bands are like the muscles in your arm. They are attached to the bones on either side of your elbow. The arm bends at the joint when the muscle contracts.

The rubber bands are like the muscles in an insect's limb. When a muscle contracts, the joint on which it operates straightens.

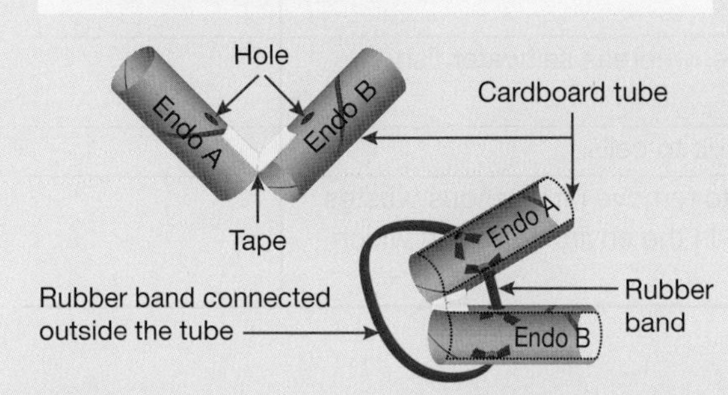

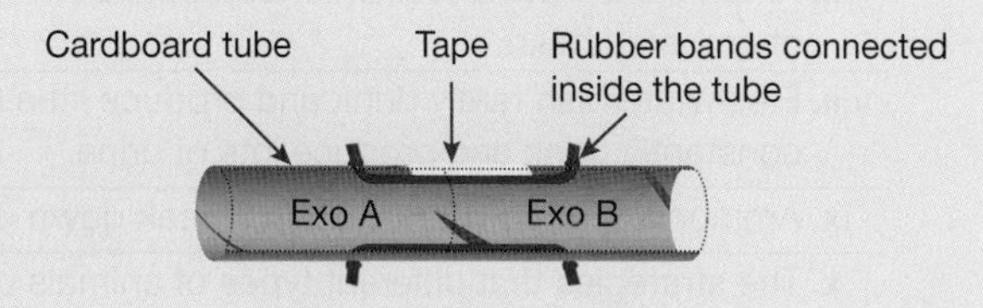

Results

1. When one rubber band contracts, what happens to the one on the opposite side?
2. Draw sketches of each tube and record your observations when the joint is moved.

Discussion

1. Describe how the two skeletons are different.
2. Identify the strengths and limitations of the method and how you could improve it.
3. Research and construct models that demonstrate how two different types of organisms have developed different strategies to solve the same 'problem' that is related to their survival.

Conclusion

Summarise the findings for this investigation.

4.12 Activities

learnon

4.12 Quick quiz on	4.12 Exercise

Select your pathway

■ LEVEL 1	■ LEVEL 2	■ LEVEL 3
1, 2, 4, 7, 8, 10, 19	3, 6, 9, 11, 14, 16, 17, 20	5, 12, 13, 15, 18, 21, 22

These questions are even better in jacPLUS!

- Receive immediate feedback
- Access sample responses
- Track results and progress

Find all this and MORE in jacPLUS

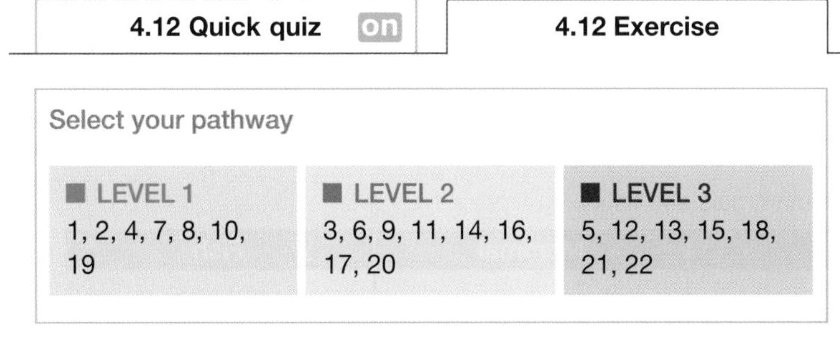

Remember and understand

1. **a.** State whether the following statements are true or false.

Statement	True or false?
i. The structures of cells and tissues often provide clues to their function.	
ii. Structures involved in absorption often have shapes that increase their surface-area-to-volume ratio.	
iii. Unicellular organisms are small enough that gases such as oxygen and carbon dioxide can simply diffuse in and out of their cell.	

(continued)

(continued)

iv. Respiratory surfaces in different animals can have shapes that make them efficient at exchanging gases such as oxygen and carbon dioxide.	
v. Intestinal villi in humans and root hairs in plants have shapes that reduce the surface-area-to-volume ratio available for exchange of materials.	
vi. Humans excrete their nitrogenous wastes as uric acid, whereas birds excrete it as urea.	
vii. Amphibians have a four-chambered heart, whereas birds have a two-chambered heart.	
viii. Freshwater fish rarely drink and produce little urine, whereas saltwater fish constantly drink and produce lots of urine.	
ix. Ammonia formed when proteins break down is toxic to cells.	
x. The strategies that different types of animals use to remove nitrogenous wastes is not influenced by the amount of water available in the environments in which they live.	

b. Justify any false responses.

2. Place the following terms in order of simplest to most complex: cell, organ, system, multicellular organism, tissue.

3. Provide an example of how structure can give clues about function.

4. Write the word equation for cellular respiration.

5. Describe two key functions of gaseous exchange.

6. Suggest why there are differences between herbivores and carnivores in the structures of their digestive systems.

7. Name two organs belonging to each of the following systems.
- **a.** Respiratory system
- **b.** Circulatory system
- **c.** Excretory system

Apply and analyse

8. Why don't herbivores have canine teeth?

9. How do we know what different types of dinosaurs ate, even though they haven't existed for about 65 million years?

10. Select an organ (e.g. heart, lungs or stomach) and find out the answers to the following questions.
- **a.** What is the function of the organ?
- **b.** Which system does it belong to?
- **c.** What other organs are in the same system?

11. Complete the following table.

TABLE Features of the body systems of mammals and fish

Feature	Mammal	Fish
Number of chambers in the heart		
Times blood travels through the heart in each circulatory trip		
Name of gaseous-exchange respiratory organ		

12. **a.** Outline the key differences between the structures of the digestive systems of a cow, a koala, a Tasmanian devil and a honey possum.
b. Suggest reasons for the differences.

13. Some animals that live in water, such as sea anemones (shown in the figure), have a digestive sac that acts as both a mouth and an anus.

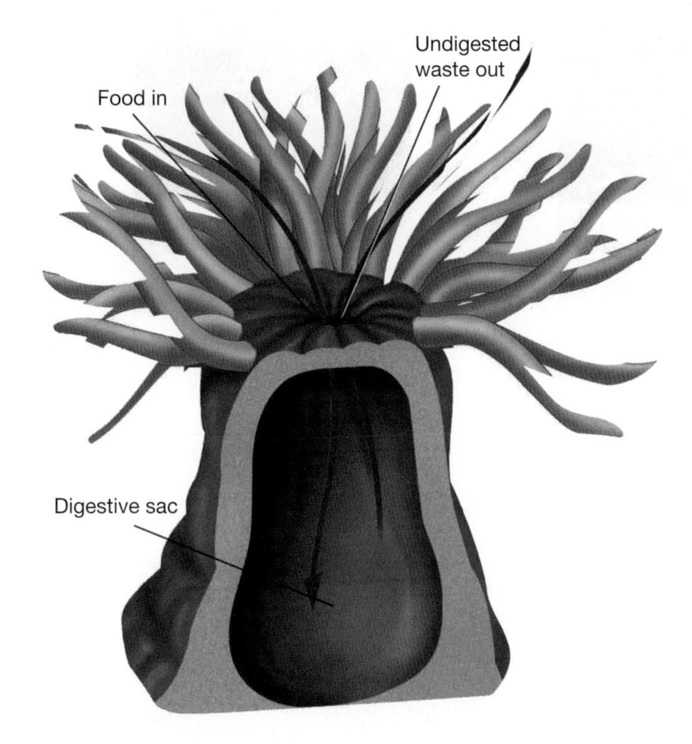

   a. Find out more about the digestive system of sea anemones and construct a model to demonstrate your understanding of how it works.
   b. Describe how this digestive system is similar to that of humans and how it is different.
   c. Suggest reasons for the differences.
14. **SIS** Construct a matrix table to compare the diets and teeth of herbivores, carnivores, omnivores and insectivores.
15. **SIS** Construct a table to summarise similarities and differences between the hearts and circulation of fish, amphibians and mammals.

Evaluate and create

16. Research and prepare a poster on the hearts of different types of animals.
17. Find out about the different tissues and systems that exist in plants. Present your information using diagrams and lots of colour, on a poster or in a PowerPoint presentation. Be as creative as you can.
18. **SIS** In a small team, formulate scientific questions about how the structure of the heart, kidney or lungs is related to the function of that organ. Research these questions and present your findings to the class.
19. **SIS** Design and construct a model of one of the following systems: respiratory, excretory, reproductive, digestive.

20. **a.** Select one of the tissues in the mind map shown and find out more about what it does and how it works.

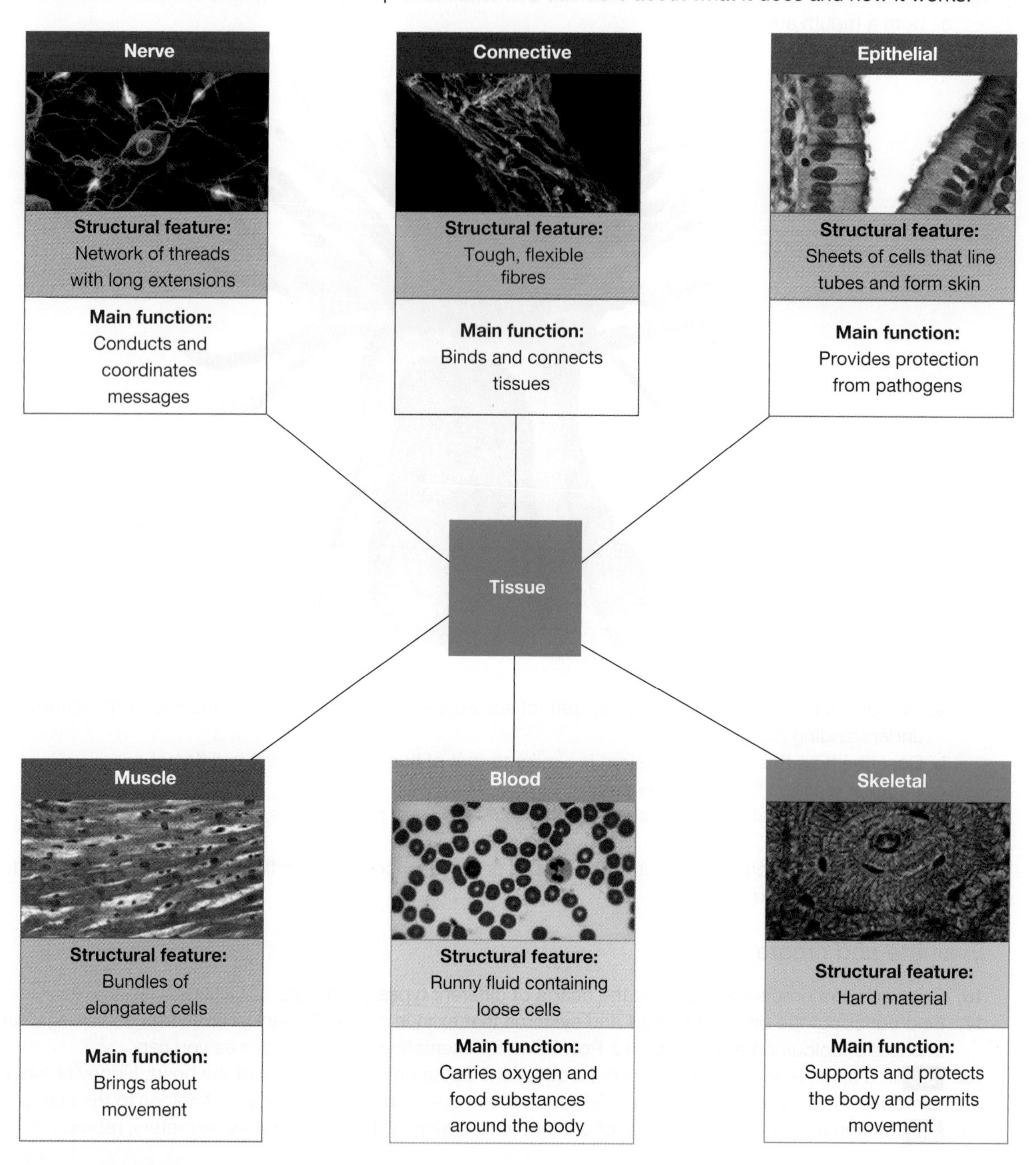

b. Use your findings to write a brief play that other students in your class can act out.

21. Select an animal of your choice.
    a. Find out how it:
    - **i.** detects stimuli
    - **ii.** supports itself and moves
    - **iii.** takes in oxygen and removes carbon dioxide
    - **iv.** conducts messages from one part of its body to another.

    b. Construct a model or animation to demonstrate one of the functions listed in part **a**.
22. There are differences in the form in which groups of animals excrete nitrogenous wastes. Find out the differences between humans, freshwater fish, saltwater fish and insects. Communicate your findings using models or in a puppet play, documentary or animation.

Fully worked solutions and sample responses are available in your digital formats.

LESSON
4.13 Thinking tools — Flowcharts and cycle maps

4.13.1 Tell me

Flowcharts are a useful thinking tool that can help to show the order in which events happen. They help you to diagrammatically display the order that stages occur. Sometimes they are called a flow map, a sequence chart or a chain of events.

FIGURE 4.103 A flowchart

Flowcharts are similar to cycle maps in that the both show a sequence of events. However, cycle maps show the sequence of events repeated in the same order; flowcharts show only a linear sequence.

For example, you might use a flowchart to show:
- the respiratory system
- the excretory system.

But you may choose a cycle map to show:
- the movement of blood through the body.

FIGURE 4.104 A cycle map

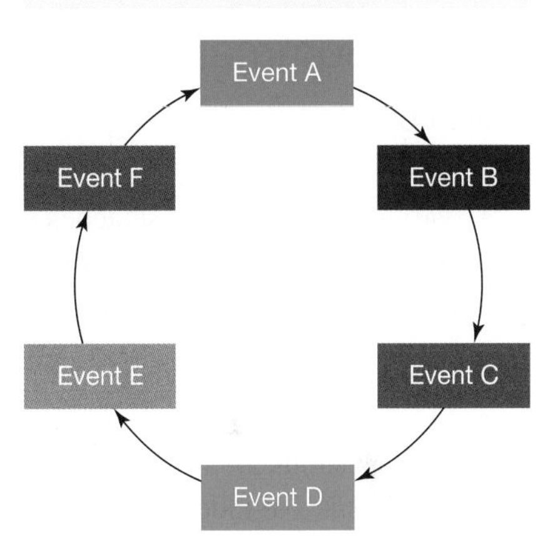

4.13.2 Show me

To create a flowchart:
1. Decide in which direction your flowchart will be read — from left to right, from the bottom up or from the top down.
2. Write the first action of the process you are describing inside a box.
3. Write the next event in another box and join this by an arrow to the first box.
4. Repeat until you have reached the final event.

4.13.3 Let me do it

4.13 Activity

1. a. Read through the information in lesson 4.6 to refresh your memory on the structure and function of your heart.
   b. Use a flowchart to show the movement of blood through your body using the following labels: left atrium, right atrium, right ventricle, left ventricle, pulmonary artery, pulmonary vein, lungs, aorta, vena cava, from body, to body.
2. a. Construct a cycle map to outline the overall movement of blood through the heart.
   b. Use this information to design a working model of a human heart.
   c. Use a flowchart to plan the construction of your heart model.
3. a. Find out more about Leonardo da Vinci's models of the human heart.
   b. Use a flowchart to map changes in scientific ideas about the heart throughout history.

Fully worked solutions and sample responses are available in your digital formats.

LESSON
4.14 Review

Hey students! Now that it's time to revise this topic, go online to:

Review your results

Watch teacher-led videos

Practise questions with immediate feedback

Find all this and MORE in jacPLUS

Access your topic review eWorkbooks

Topic review Level 1	Topic review Level 2	Topic review Level 3
ewbk-11909	ewbk-11911	ewbk-11913

4.14.1 Summary

Driven by curiosity?

- Advances in our understanding of anatomy has been driven by the curiosity of people.
- Various observations, sketches and models were created to enhance and further their understanding of how organs and body systems work.
- Different cultures have applied their scientific knowledge at the time to medical treatments.

Working together?

- Multicellular organisms are made up of body systems that work together to keep them alive.
- Body systems are made up of organs, which are made up of tissues, which are made up of cells, which cannot survive independently of each other.
- Although the cells within a multicellular organism may have similar basic structures they differ in size, shape and in the number and types of organelles they contain.
- The variations in different types of cells and structures within them make them well suited to their function.
- The structures of cells and tissues often provide clues to their function. For example:

TABLE Structural features of body parts and their function

Structure	Structural details	Well suited to function
Muscle tissue	Contains cells with many mitochondria	Ensures that the energy requirements of the tissue can be met
Human intestinal villi and plant root hairs	Have shapes that increase their surface-area-to-volume ratio	Both involved in absorption

- Body systems within multicellular organisms work together to keep them alive.
- Your digestive, circulatory, respiratory and excretory systems work together to provide your cells with nutrients and oxygen, and to remove wastes such as carbon dioxide.

Digestive system — break it down

- Your digestive system plays a key role in supplying your body with the nutrients it requires to function effectively. It is involved in breaking food down, so nutrients are small enough to be transported to, and used by, your cells.
- The main organs in your digestive system are your mouth (including teeth and tongue), oesophagus, stomach, small intestine, large intestine, pancreas, gall bladder and liver.

- Mechanical digestion involves physically breaking down the food into smaller pieces. Most of this process takes place in your mouth using your different types of teeth. For example:

TABLE Structural features of the teeth and their function

Type of teeth	Structure	Well suited to function
Incisors	Chisel edge and thin crown	Biting
Canines	Sharp point and cone-shaped crown	Tearing
Premolars and molars	Points or cusps that fit together in upper and lower jaws	Crushing and grinding

- Chemical digestion involves the use of chemicals called enzymes to break down food into small molecules.
- Most nutrients are absorbed into your bloodstream in the last of your small intestine, which is lined with finger-like projections called villi.
- The shape of villi increases the surface area through which nutrients can diffuse into tiny blood vessels called capillaries, which transport them to other parts of your body.

Digestive endeavours

- It's important to look after your teeth as poor dental hygiene is linked to a number of diseases.
- Damaged villi can reduce the amount of nutrients absorbed into your body and this can occur if you have coeliac disease.

Circulatory system — blood highways

- Your circulatory system is responsible for transporting oxygen and nutrients to your body's cells, and wastes such as carbon dioxide away from them.
- The main organs in your circulatory system are your blood, blood vessels and heart.
- Blood is made up of red blood cells (erythrocytes), white blood cells (leucocytes), blood platelets and the plasma that they all float in.
- Red blood cells are red because they contain an iron-containing pigment called haemoglobin, which reacts with oxygen to form oxyhaemoglobin.
- The structure of red blood cells makes them well suited to perform their oxygen-transporting role. For example:

TABLE Structural features of red blood cells and their function

Structural details	Well suited to function
Biconcave shape	Large surface-area-to-volume ratio
Lack of a nucleus	Increases the available space for haemoglobin and hence oxygen
Small size	Enables them to fit inside tiny capillaries

- Involved in fighting disease, white blood cells contain a nucleus and are larger and fewer in number than red blood cells.
- Platelets in the blood help it to clot and plug the damaged blood vessel.
- Blood cells are transported in blood vessels to and from your heart.
- The major types of blood vessels are arteries, veins and capillaries.

TABLE Structural features of the circulatory system and their function

Structure	Structural details	Well suited to function
Arteries	Thick, elastic, muscular walls	Carry blood under high pressure away from the heart
Veins	Valves	Help prevent the backflow of blood
Capillaries	Only one cell thick	The site of exchange of materials to and from cells

- The structure of your heart is also well suited to its function.

TABLE Structural features of the heart and their function

Structural details	Well suited to function
Two pumps	Right side for deoxygenated blood and left side for oxygenated blood
Left side more muscular than right side	The right side of the heart pumps deoxygenated blood a short distance to the lungs, whereas the left side pumps oxygenated blood out to the whole body.
Valves	Help prevent the back flow of blood

Transport technology

- Understanding of the human body has changed over time with the development of new technologies.
- Artificial blood and hearts can help with the limited supply of organs available.

Respiratory system — breathe in, breathe out

- Your respiratory system is responsible for getting oxygen into your body and carbon dioxide out. This occurs when you inhale and exhale.
- The main organs in your respiratory system are your trachea and lungs (bronchi, bronchioles and alveoli).
- The large surface-area-to-volume ratio of alveoli and their closeness to capillaries make them well suited to their role in gas exchange.

Short of breath?

- Your ability to breathe can be hampered by asthma and allergies, which can close up your airways.
- Smoking and vaping not only damage your lungs, but can have short- and long-term effects on your health.

Excretory system

- The excretory system removes waste products from a variety of chemical reactions that your body needs to stay alive.
- The main organs of your excretory system are your skin, lungs, liver and kidneys.
- Skin is involved in the excretion of salts and water as sweat; lungs are involved in the excretion of carbon dioxide; the liver is involved in breaking down toxins for excretion; and the kidneys are involved in the excretion of the unused waste products of chemical reactions (e.g. urea) and any other chemicals that may be in excess (including water).

Musculoskeletal system — keeping in shape

- Your musculoskeletal system consists of both your bones and various types of muscles throughout your body, which provide both support and protection for your organs.

Same job, different path

- Organisms possess a variety of structures that help them to obtain the resources they need to survive. While there are similarities and patterns in some of those structures, there are also differences.
- While unicellular organisms are small enough to use diffusion to obtain their oxygen and remove carbon dioxide waste, multicellular organisms have evolved a variety of specialised structures to achieve this.
- Some animals have developed specialised gas exchange organs that all have a gas exchange surface with high surface-area-to-volume ratios. For example:

TABLE Gas exchange organs of different animals

Animal	Specialised gas exchange organ
Mammals and amphibians	Lungs
Fish	Gills
Insects	Tracheae

- Likewise, variations in the structures within digestive systems make them well suited to perform their task for the diets of different types of animals. For example:
  - the size and types of teeth
  - size of the stomach
  - length of the small intestine
  - presence of a large caecum.

4.14.2 Key terms

absorption the taking in of a substance; for example, from the intestine to the surrounding capillaries
alimentary canal see **gastrointestinal tract**
allergen an antigen that elicits an allergic response
allergy an abnormal immune response to a substance that is harmless for most people
alveoli tiny air sacs in the lungs at the ends of the narrowest tubes; oxygen moves from alveoli into the surrounding blood vessels, in exchange for carbon dioxide; singular = alveolus
ammonia a nitrogenous waste product of protein break down
amylases enzymes found in saliva that break starch down into sugar
anaphylaxis an acute and potentially lethal allergic reaction to an allergen to which a person has become hypersensitive
anus the end of the digestive system, through which faeces are passed as waste
aorta a large artery through which oxygenated blood is pumped at high pressure from the left ventricle of the heart to the body
arteries hollow tubes (vessels) with thick walls carrying blood pumped from the heart to other body parts
arterioles vessels that transport oxygenated blood from the arteries to the capillaries
arthritis a condition in which inflammation of the joints causes them to swell and become painful
asthma narrowing of the air pipes that join the mouth and nose to the lungs
atoms very small particles that make up all things; atoms have the same properties as the objects they make up
ball-and-socket joints joints in which the rounded end of one bone fits into the hollow end of another
bile a substance produced by the liver that helps digest fats and oils
bladder a sac that stores urine
blood cells living cells in the blood
blood pressure measures how strongly the blood is pumped through the body's main arteries
blood vessels the veins, arteries and capillaries through which the blood flows around the body
body systems groups of organs within organisms that carry out specific functions
bolus a round, chewed-up ball of food made in the mouth that makes swallowing easier
bone marrow a substance inside bones in which blood cells are made
bones the pieces of hard tissue that make up the skeleton of a vertebrate
Bowman's capsule a cup-like structure at one end of a nephron within the kidney, surrounding the glomerulus; it serves as a filter to remove wastes and excess water
breathing the movement of muscles in the chest causing air to enter the lungs and the altered air in the lungs to leave; the air entering the lungs contains more oxygen and less carbon dioxide than the air leaving the lungs
brittle can easily break if hit; the opposite of malleable
bronchi the narrow tubes through which air passes from the trachea to the smaller bronchioles and alveoli in the respiratory system; singular = bronchus
bronchioles small, branching tubes in the lungs leading from the two larger bronchi to the alveoli
burping the release of swallowed gas through the mouth
calcium an element occurring in limestone and chalk, and also present in vertebrates and other animals as a component of bone and shell; it is necessary for nerve conduction, heartbeat, muscle contraction and many other physiological functions
capillaries numerous tiny blood vessels that are only a single cell thick to allow exchange of materials to and from body cells; every cell of the body is supplied with blood through capillaries
carbon dioxide a colourless gas (CO_2) made up of one carbon and two oxygen atoms; it is essential for photosynthesis and is a waste product of cellular respiration; the burning of fossil fuels also releases carbon dioxide
carcinogens chemicals that cause cancer
cardiac muscle a special kind of muscle in the heart that never tires; it is involved in pumping blood through the heart
carnivore an animal that eats other animals

cartilage a waxy, whitish, flexible substance that lines or connects bone joints or, in some animals such as sharks, replaces bone as the supporting skeletal tissue; the ears and tips of noses of humans are shaped by cartilage
cell the smallest unit of life; cells are the building blocks of living things and can be many different shapes and sizes
cellular respiration a series of chemical reactions in which the chemical energy in molecules such as glucose is transferred into ATP molecules, which is a form of energy that the cells can use
cellulose a natural substance that keeps the cell walls of plants rigid
chemical digestion the chemical reactions that change food into simpler substances that are absorbed into the bloodstream for use in other parts of the body
circulatory system the heart, blood and blood vessels, which are responsible for circulating oxygen and nutrients to body cells, and carbon dioxide and other wastes away from them
colon the part of the large intestine where food mass passes from the small intestine, and where water and other remaining essential nutrients are absorbed into the body
denatured describes the condition of proteins after they have been overheated
deoxygenated blood blood from which some oxygen has been removed
diaphragm a flexible, dome-shaped, muscular layer separating the chest and the abdomen; it is involved in breathing
diarrhoea excessive discharge of watery faeces
diastolic pressure the lower blood pressure reading during relaxation of the heart muscles
diffusion movement of molecules through the cell membrane
digestion the breakdown of food into a form that can be used by an animal; it includes both mechanical digestion and chemical digestion
digestive system a complex series of organs and glands that processes food to supply the body with the nutrients it needs to function effectively
electrocardiogram (ECG) a graph made using the tiny electrical impulses generated in the heart muscle, giving information about the health of the heart
emphysema a condition in which the air sacs in the lungs break open and join together, reducing the amount of oxygen taken in and carbon dioxide removed
emulsify combine two liquids that do not normally mix easily
endocrine system the body system of glands that produce and secrete hormones into the bloodstream to regulate processes in various organs
endoskeleton a skeleton that lies inside the body
enzymes special chemicals that speed up reactions but are themselves not used up in the reaction
epiglottis a leaf-like flap of cartilage behind the tongue that closes the air passage during swallowing
erythrocytes red blood cells
excretion the removal of wastes from the body
excretory system the body system that removes waste substances from the body
exoskeleton a skeleton or shell that lies outside the body
flatulence the release of gas through the anus; this gas is produced by bacteria in the large intestine
fracture a break in a bone
gall bladder a small organ that stores and concentrates bile within the body
gastrointestinal tract also called the digestive tract or the alimentary canal, it is a tubular passage that starts with the mouth and ends with the anus; it intakes and digests food (absorbing energy and nutrients) and expels waste
glomerulus a cluster of capillaries in the kidney that acts as a filter to remove wastes and excess water
glucose a six-carbon sugar (monosaccharide) that acts as a primary energy supply for many organisms
greenstick fracture a break that is not completely through the bone, often seen in children
haemodialysis the process of passing blood through a machine to remove wastes
haemoglobin the red pigment in red blood cells that carries oxygen
heart a muscular organ that pumps deoxygenated blood to the lungs to be oxygenated and then pumps the oxygenated blood to the body
heartbeat a contraction of the heart muscle occurring about 60–100 times per minute
heartburn a burning sensation caused by stomach acid rising into the oesophagus
herbivore an animal that eats only plants
hinge joints joints in which two bones are connected so that movement occurs in one plane only
immovable joints joints that allow no movement except when absorbing a hard blow

insectivore a carnivore that eats only insects
involuntary muscles muscles not under the control of the will; they contract slowly and rhythmically, and are at work in the heart, intestines and lungs
joint a region where two bones meet
kidneys body organs that filter the blood, removing urea and other wastes
large intestine the penultimate part of the digestive system, where water is absorbed from the waste before it is transported out of the body
left atrium the upper-left section of the heart where oxygenated blood from the lungs enters the heart
left ventricle the lower-left section of the heart, which pumps oxygenated blood to all parts of the body
leucocytes white blood cells
ligament a band of tough tissue that connects the ends of bones or keeps an organ in place
lipases enzymes that break fats and oils down into fatty acids and glycerol
lipids a class of nutrients that include fats and oils
liver the largest gland in the body; it secretes bile for digestion of fats, builds proteins from amino acids, breaks down many substances harmful to the body and has many other essential functions
'lub dub' the sound made by the heart valves as they close
lungs the organ for breathing air; gas exchange occurs in the lungs
mechanical digestion digestion that uses physical factors such as chewing with the teeth
mineral any of the inorganic elements that are essential to the functioning of the human body and are obtained from foods
molecule two or more atoms joined (bonded) covalently together
mouth the opening of the gastrointestinal tract through which food is taken into the body
multicellular organisms living things comprised of specialised cells that perform specific functions
muscles tissue consisting of cells that can shorten
musculoskeletal system consists of the skeletal system (bones and joints) and the skeletal muscle system (voluntary or striated muscle); working together, these two systems protect the internal organs, maintain posture, produce blood cells, store minerals and enable the body to move
nephrons the filtration and excretory units of the kidney
nervous system consists of neurons, nerves and the brain, which are responsible for detecting and responding to both internal and external stimuli
nitrogenous wastes waste products from protein breakdown, including ammonia, urea and uric acid
nutrients substances that provide the energy and chemicals that living things need to stay alive, grow and reproduce
oesophagus part of the digestive system, composed of a tube connecting the mouth with the stomach
omnivore an animal that eats plants and other animals
organelle any specialised structure in a cell that performs a specific function
organs structures, composed of tissue, that perform specific functions
ossification hardening of bones
osteoporosis loss of bone mass that causes bones to become lighter, more fragile and more easily broken
oxygen an atom that forms molecules (O_2) of tasteless and colourless gas; it is essential for cellular respiration for most organisms and is a product of photosynthesis
oxygenated blood the bright red blood that has been supplied with oxygen in the lungs
pacemaker an electronic device inserted in the chest to keep the heart beating regularly at the correct rate; it works by stimulating the heart with tiny electrical impulses
pancreas a large gland in the body that produces and secretes the hormone insulin and an important digestive fluid containing enzymes
peristalsis the process of pushing food along the oesophagus or small intestine by the action of muscles
phosphorus a substance that plays an important role in almost every chemical reaction in the body; together with calcium, it is required by the body to maintain healthy bones and teeth
pivot joints joints that allows a twisting movement
plasma the yellowish liquid part of blood that contains water, minerals, food and wastes from cells
platelets small bodies involved in blood clotting; they are responsible for healing by clumping together around a wound
proteases enzymes that break proteins down into amino acids
protein a chemical made up of amino acids needed for growth and repair of cells in living things
pulmonary artery the vessel through which deoxygenated blood, carrying wastes from respiration, travels from the heart to the lungs

pulmonary vein the vessel through which oxygenated blood travels from your lungs to the heart
pulse the alternating contraction and expansion of arteries due to the pumping of blood by the heart
rectum the final section of the digestive system, where waste food matter is stored as faeces before being excreted through the anus
red blood cells living cells in the blood that transport oxygen to all other living cells in the body
reproductive system the different reproductive organs required by many organisms to reproduce and create offspring
respiratory system the lungs and associated structures that are responsible for getting oxygen into the organism and carbon dioxide out
right atrium the upper-right section of the heart where deoxygenated blood from the body enters
right ventricle the lower-right section of the heart, which pumps deoxygenated blood to the lungs
saliva a watery substance in the mouth that contains enzymes involved in the digestion of food
salivary glands glands in the mouth that produce saliva
skeletal muscle system voluntary or striated muscle
skeletal system consists of the bones and joints
skeleton the bones or shell of an animal that support and protect it as well as allowing movement
skin the external covering of a vertebrate's body
small intestine the part of the digestive system between the stomach and large intestine, where much of the digestion of food and absorption of nutrients takes place
sprain an injury caused by tearing a ligament
stem cells undeveloped cells found in blood and bone marrow that can reproduce themselves indefinitely
stomach a large muscular organ that churns and mixes food with gastric juice to start to break down protein
synovial fluid the liquid inside the cavity surrounding a joint that helps bones to slide freely over each other
systolic pressure the higher blood pressure reading during contraction of the heart muscles
teeth hard structures within the mouth that allow chewing
tendon tough, rope-like tissue connecting a muscle to a bone
tennis elbow occurs when repeated grasping and bending back of your wrist leads to the inflammation of the tendon that connects the muscles of your forearm to the bone in your upper arm, causing pain
tissue a group of cells of similar structure that perform a specific function
torn hamstring a common sporting injury caused by overstretching the hamstring muscle, which joins the pelvis to the knee joint
trachea the narrow tube from the mouth to the lungs through which air moves
urea a nitrogen-containing substance produced by the breakdown of proteins and removed from the blood by the kidneys
ureters tubes from each kidney that carry urine to the bladder
urethra the tube through which urine is emptied from the bladder to the outside of the body
uric acid a nitrogenous waste product of protein breakdown
urination the passing of urine from the bladder to the outside of the body
urine a yellowish liquid, produced in the kidneys; it is mostly water and contains waste products from the blood such as urea, ammonia and uric acid
valves flap-like folds in the lining of a blood vessel or other hollow organ that allow a liquid, such as blood, to flow in one direction only
varicose veins expanded or knotted blood vessels close to the skin, usually in the legs; they are caused by weak valves that do not prevent blood from flowing backwards
veins blood vessels that carry blood back to the heart; they have valves and thinner walls than arteries
vena cava the large vein leading into the top-right chamber of the heart
venules small veins
villi tiny, finger-like projections from the wall of the intestine that maximise the surface area of the structure to increase the efficiency of nutrient absorption; singular = villus
vital capacity the largest volume of air that can be breathed in or out at one time
vitamin D a nutrient that regulates the concentration of calcium and phosphate in the bloodstream and promotes the healthy growth and remodelling of bone
voluntary muscles muscles attached to bones; they move the bones by contracting and are controlled by an animal's thoughts
vomiting the forceful ejection of matter from the stomach through the mouth
white blood cells living cells that fight bacteria and viruses as part of the human body's immune system

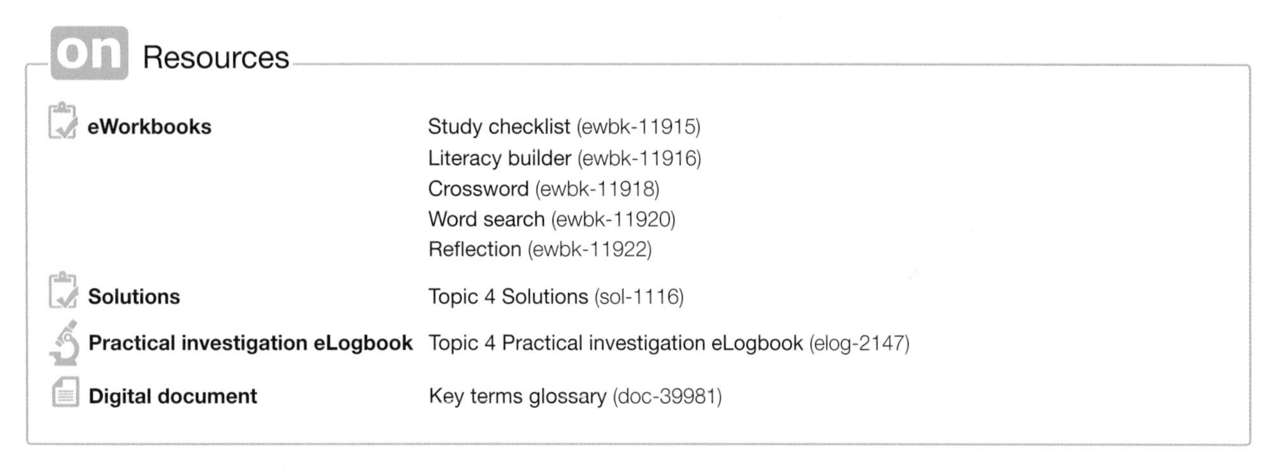

on Resources

eWorkbooks	Study checklist (ewbk-11915) Literacy builder (ewbk-11916) Crossword (ewbk-11918) Word search (ewbk-11920) Reflection (ewbk-11922)
Solutions	Topic 4 Solutions (sol-1116)
Practical investigation eLogbook	Topic 4 Practical investigation eLogbook (elog-2147)
Digital document	Key terms glossary (doc-39981)

4.14 Activities

learn**on**

4.14 Review questions

Select your pathway

■ LEVEL 1	■ LEVEL 2	■ LEVEL 3
1, 2, 3, 4, 5, 6, 7, 23, 24	8, 10, 11, 12, 14, 15, 18, 20, 21, 25	9, 13, 16, 17, 19, 22, 26

These questions are even better in jacPLUS!

- Receive immediate feedback
- Access sample responses
- Track results and progress

Find all this and MORE in jacPLUS

Remember and understand

1. **a.** Identify whether the following statements are true or false.

Statement	True or false?
i. Multicellular organisms are made up of different types of cells that cannot survive independently of each other so they need to work together.	
ii. The cells that make up multicellular organisms are all the same.	
iii. The shape and size of different types of cells within a multicellular organism, and differences in structures within them, make them well suited to their specific function.	
iv. Tissues are structures within cells that have a particular job to do.	
v. Organelles are a collection of similar cells that perform a particular function.	
vi. Organs are made up of different types of tissues grouped together to perform a particular function.	
vii. Systems are made up of different organs working together to perform a specialised function to keep an organism alive.	
viii. The excretory system supplies your body with the nutrients it requires to function effectively.	
ix. The stomach, liver, gall bladder, intestines and pancreas are all organs of the respiratory system.	
x. When you burp, some of your stomach acid can rise into your oesophagus and cause heartburn.	

   b. Justify any false responses.

2. Use a flowchart to show the relationship between the following, from most complex to least complex: cells, systems, multicellular tissues, organisms, organs.

3. Identify which description matches the term in the table shown.

Term	Description
a. Tissue	**A.** A collection of similar cells that perform a specific function
b. Organ	**B.** Different organs working together to perform a specialised function to keep an organism alive
c. System	**C.** Different types of tissues grouped together to perform a specific function
d. Organelle	**D.** Structures within cells that have a specific function

4. Match the type of tissue with its function in the table shown.

Tissue	Function
a. Blood tissue	**A.** To bind and connect tissues together
b. Connective tissue	**B.** To bring about movement
c. Epithelial tissue	**C.** To carry oxygen and food substances around the body
d. Muscle tissue	**D.** To conduct and coordinate messages
e. Nerve tissue	**E.** To line tubes and spaces, and form the skin
f. Skeletal tissue	**F.** To support and protect the body, and permit movement

5. Match the organs with their system in the table shown.

Tissue	System
a. Brain, spinal cord	**A.** Circulatory system
b. Heart, blood vessels	**B.** Digestive system
c. Liver, kidneys, skin, lungs	**C.** Excretory system
d. Lungs, trachea	**D.** Nervous system
e. Stomach, liver, gall bladder, intestines, pancreas	**E.** Reproductive system
f. Testes, ovaries	**F.** Respiratory system

6. Match the system with its function in the table shown.

System	Function
a. Circulatory system	**A.** Removes waste products from your body
b. Digestive system	**B.** Supplies your body with nutrients it requires to function effectively
c. Excretory system	**C.** Takes oxygen in and removes carbon dioxide from your body
d. Respiratory system	**D.** Transports oxygen and nutrients to your body cells and transports wastes such as carbon dioxide from them

7. Use a flowchart to show the correct sequence for the following parts of the digestive system: anus, mouth, oesophagus, stomach, small intestine, large intestine.

8. Label (a–m) this diagram of the human digestive tract.

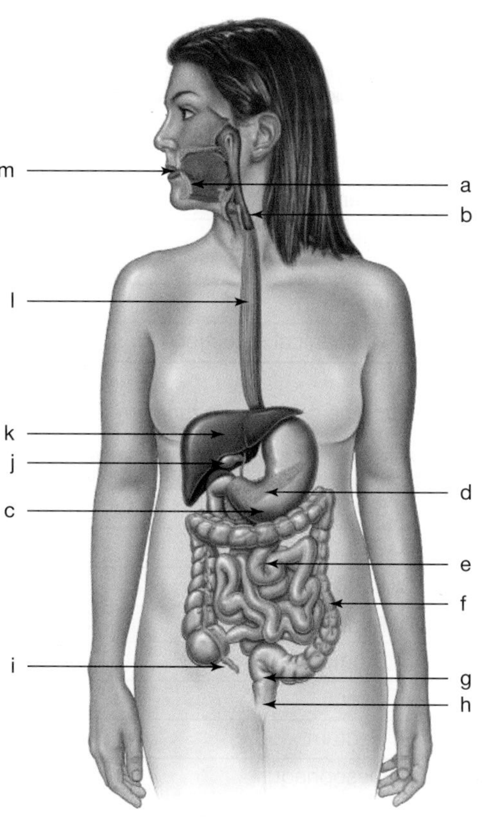

9. Match the labels to the digestive system flowchart figure provided: anus, gall bladder, liver, mouth, pancreas, salivary glands, small intestine, teeth.

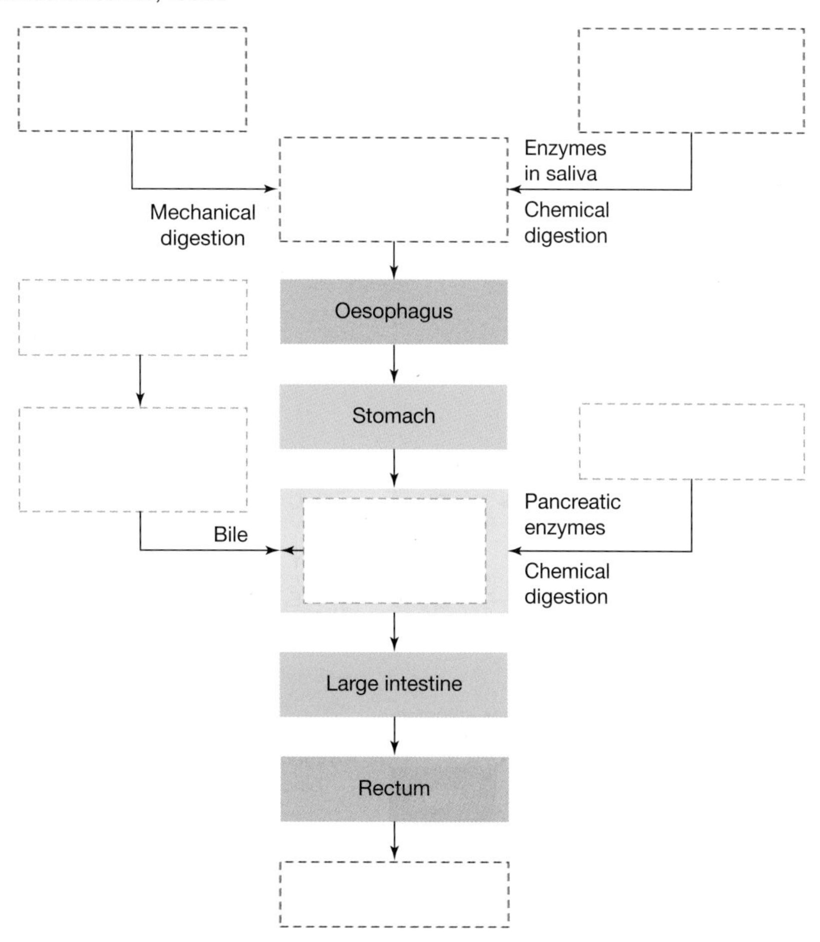

10. Match the process to the description in the table shown.

Process	Description
a. Absorption of nutrients	**A.** Physically breaking down food into smaller pieces
b. Assimilation	**B.** Occurs through villi in the small intestine into capillaries
c. Chemical digestion	**C.** Taking food into your body
d. Egestion	**D.** Use of chemicals called enzymes to break down food into small molecules
e. Ingestion	**E.** Undigested materials and wastes are removed from the body
f. Mechanical digestion	**F.** Conversion of broken-down food into chemicals in your cells

11. Match the organ to its function in the table shown.

Organ	Function
A. Small intestine	**a.** Makes bile, stores glycogen and breaks down toxins
B. Gall bladder	**b.** Makes enzymes used in the small intestine
C. Large intestine	**c.** Stores bile made in the liver until needed by the small intestine
D. Pancreas	**d.** Stores faeces
E. Stomach	**e.** Stores undigested food and waste while bacteria make some vitamins
F. Liver	**f.** Temporary storage of food and where protein digestion begins
G. Rectum	**g.** Tube that takes food from mouth to stomach
H. Oesophagus	**h.** Where the breakdown of starch and protein is finished and fat breakdown occurs

12. Match the labels to their position in the flowchart provided: artery, body cells, artery capillary, vein capillary, vein.

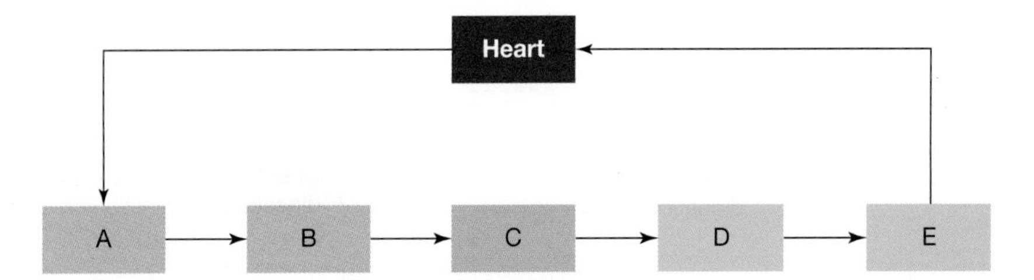

13. Label the figure provided.

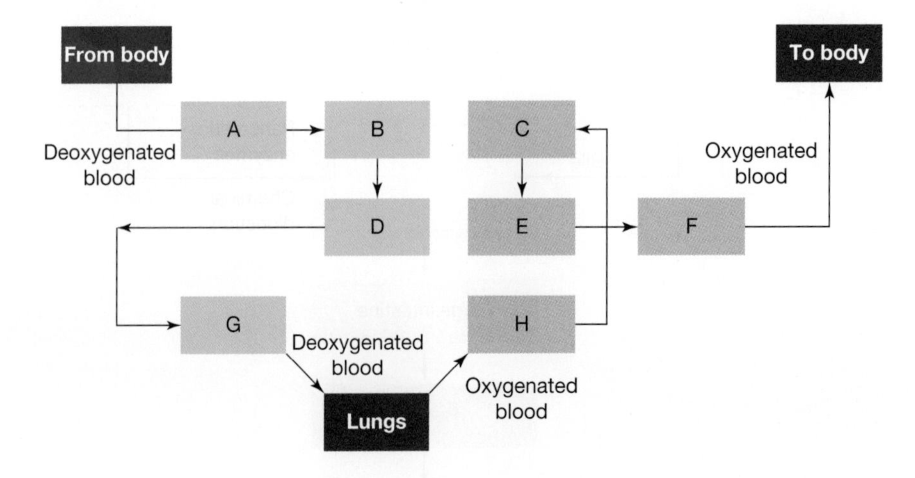

14. Match the term with its most appropriate description in the table shown.

Term	Description
a. Artery	**A.** Cells involved in transporting oxygen throughout the body
b. Atria	**B.** The bottom two chambers of the heart
c. Capillary	**C.** The top two chambers of the heart
d. Heart	**D.** Blood vessel that exchanges substances with cells
e. Red blood cells	**E.** Cells involved in production against infection
f. Vein	**F.** Organ that pumps blood around the body
g. Ventricles	**G.** Blood vessel that takes blood to the heart
h. White blood cells	**H.** Blood vessel that takes blood away from the heart

15. Match each term with the description of its function in the table shown.

Term	Description
a. Kidney	**A.** Transports urine from kidneys to bladder
b. Bladder	**B.** When urine moves from the bladder, through the urethra and out of the body
c. Urethra	**C.** Filters the blood and produces urine
d. Ureter	**D.** Stores urine
e. Urine	**E.** Transports urine from bladder to outside body
f. Urination	**F.** Watery fluid produced by kidneys that contains unwanted substances

16. Suggest labels for the diagrams shown.

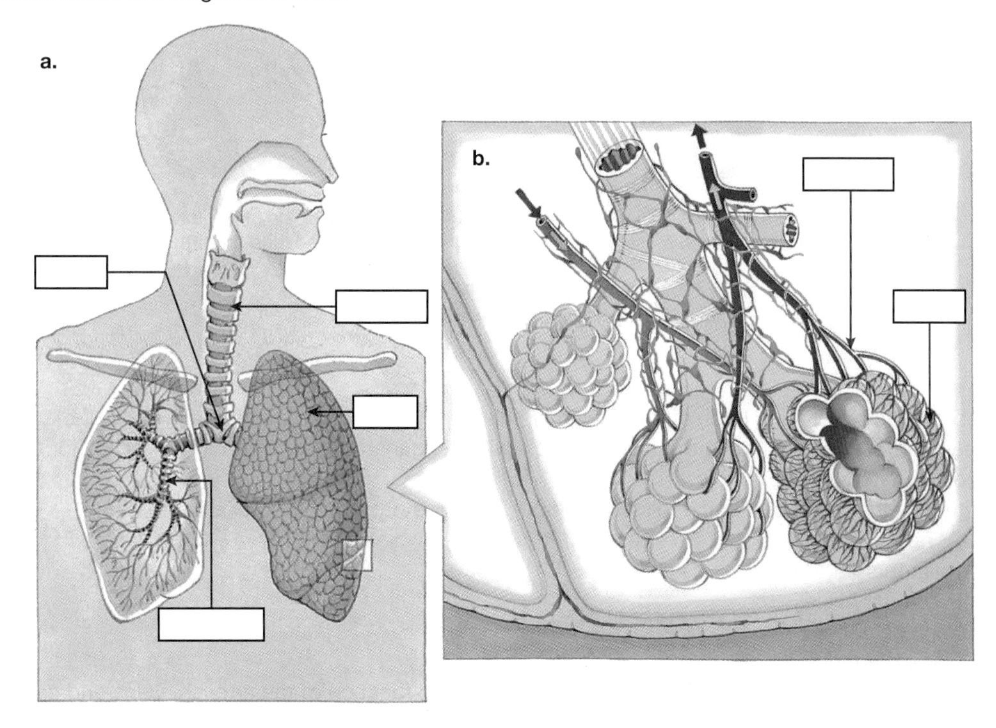

17. Explain how the structure of each of the following makes them well suited to their function.

a. Red blood cells
b. Human intestinal villi
c. Arteries
d. Capillaries
e. Alveoli
f. Molar teeth

Apply and analyse

18. Make a copy of the diagram provided in this question in your workbook.

The human circulatory system

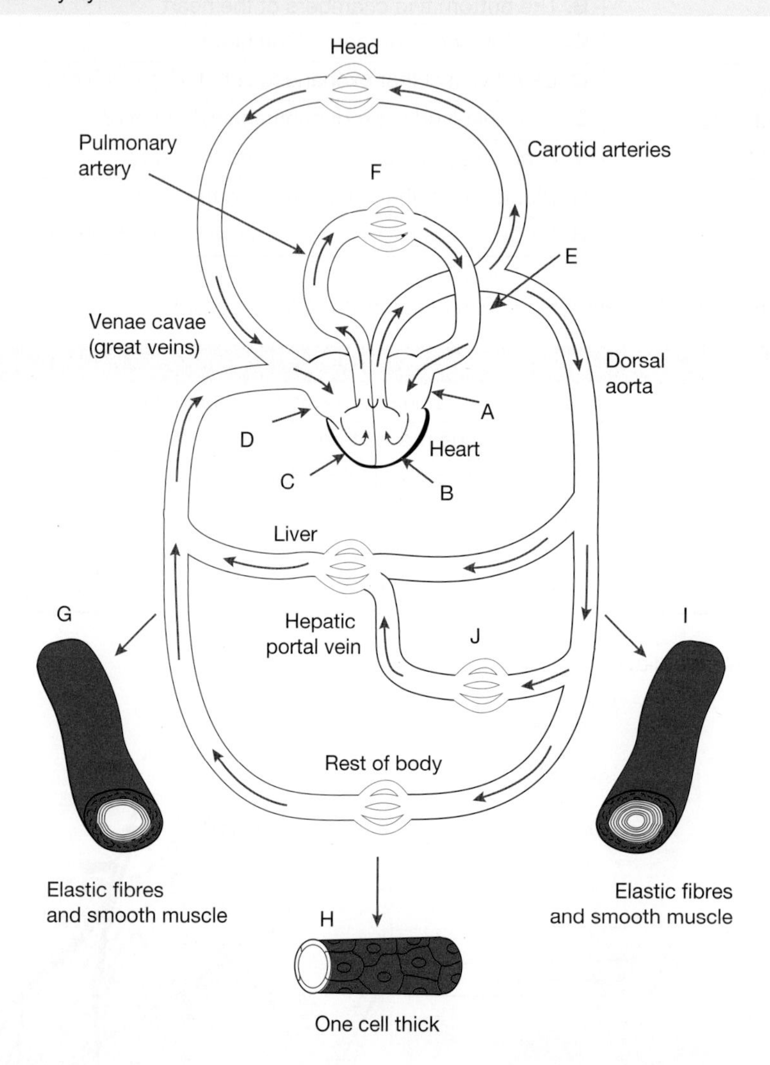

a. Label the lettered parts (A–J) of the human circulatory system and blood vessels on your diagram.

b. Use a red pencil to colour in the blood vessels with oxygenated blood, and a blue pencil for those with deoxygenated blood.

c. State whether the blood in the following blood vessels is deoxygenated or oxygenated.

- **i.** Aorta
- **ii.** Pulmonary artery
- **iii.** Pulmonary vein
- **iv.** Vena cava
- **v.** Carotid arteries

d. Draw a table that shows the differences in structure and function of the arteries, veins and capillaries.

Evaluate and create

19. The process of replacing oxygen with carbon dioxide is called gas exchange. Some animals exchange gases through lungs or gills. Some other animals exchange gases through their skin, and yet others through rows of air holes along both sides of their bodies.
Investigate how animals and insects exchange gases to create a mind map and answer the following questions.
a. Construct a mind map, poster or PowerPoint presentation that summarises your findings on how at least five different animals achieve their exchange of gases.
b. Why do frogs need lungs when they can exchange gases through their skin?
c. Why can't fish survive in the air?
d. Why can't humans breathe under water without air tanks?
e. How is gas exchange in insects similar to that in humans?

20. Carefully examine the diagram shown. What other points about enzymes could you add to this map? Can you suggest any more links between points already on the map?

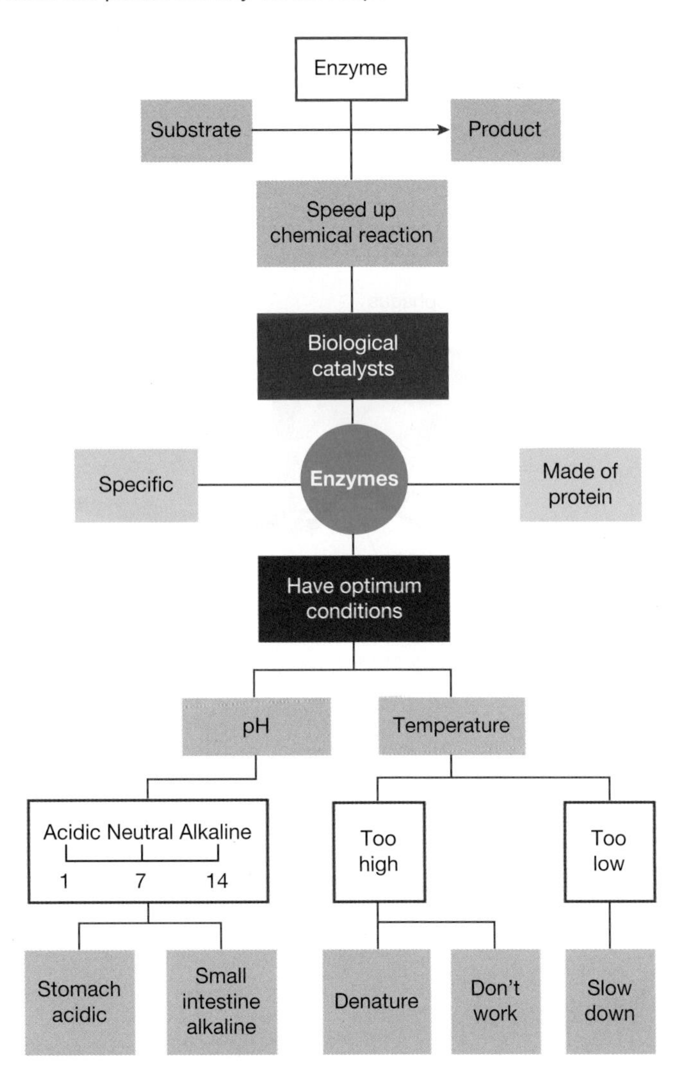

21. Construct Venn diagrams to compare the following.
a. Incisor teeth and canine teeth
b. Intestinal villi and alveoli
c. Red blood cells and white blood cells
d. Trachea and oesophagus
e. Mechanical digestion and chemical digestion

22. Carefully observe the diagrams of the digestive systems of the animals shown. Construct a matrix table that shows the similarities and differences between birds, earthworms, insects and humans.

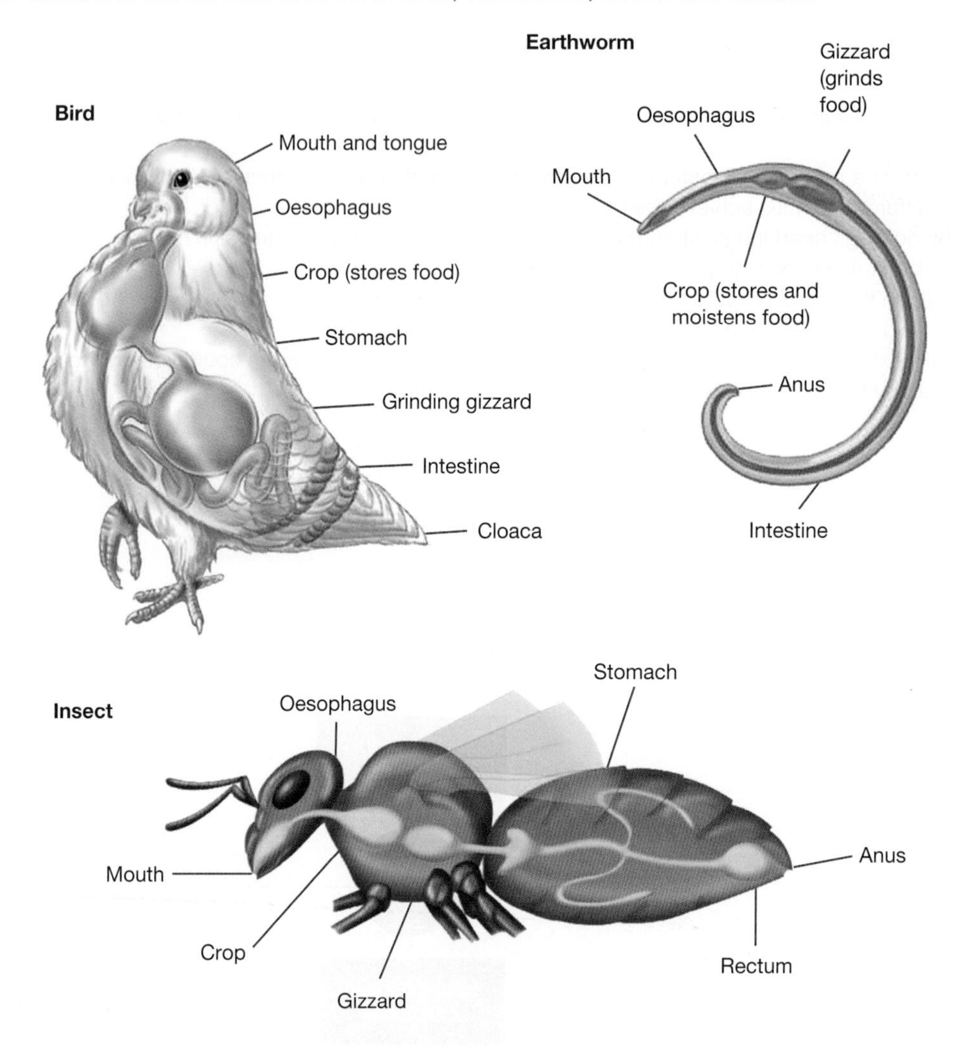

23. Use your six thinking hats (see lesson 2.2) for three of the issues or statements provided.

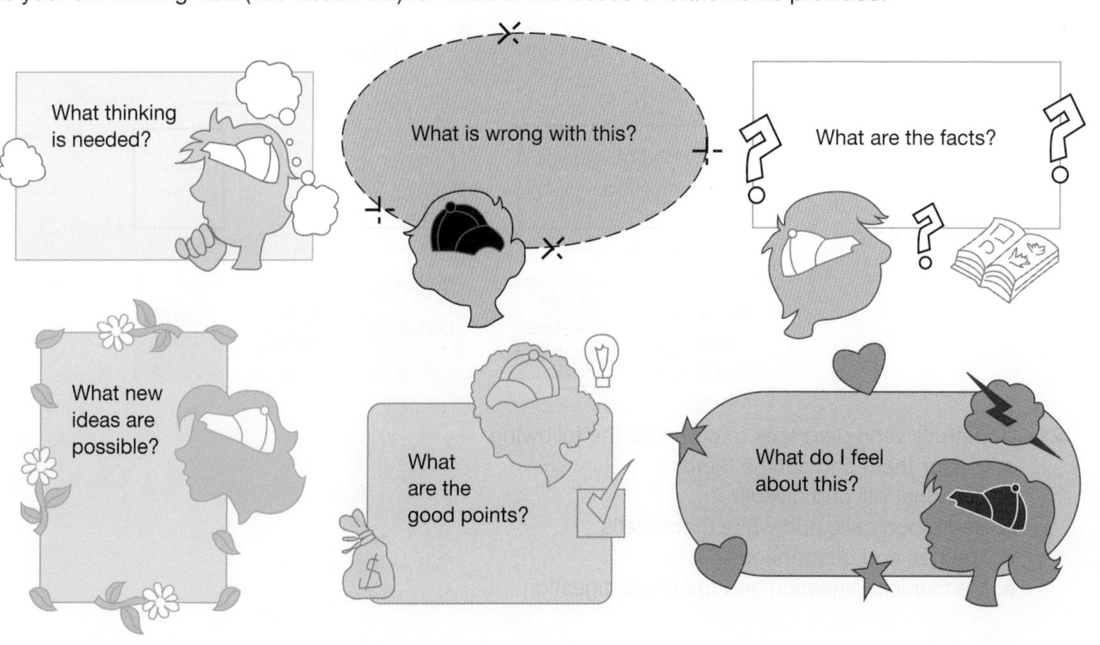

a. Drinking of any alcohol in Australia should be illegal.
b. Smoking in public should be punishable by a 10-year prison sentence.
c. Donating blood at least four times a year should be compulsory for all over the age of 16.
d. Only people under the age of 40 should be allowed to have a heart transplant.
e. Smokers should not be allowed to have surgery.
f. Blood transfusions should be illegal.
g. Everyone should have the right to a blood transfusion.
h. Organ donation should be compulsory.
i. Overweight people should not be allowed to have surgery on their circulatory system.

24. Write a story that tells of the life of a red blood cell.
25. As a team, create a song, poem, cartoon or play about something that you have learned.
26. a. Create a game called 'Nutridigest' that summarises your learning about the digestive system.
b. In a team of four, brainstorm as many questions as you can that could be placed on each of the squares. Creatively write these onto your Nutridigest cards.
c. As a team, discuss the rules for your Nutridigest game. Write down those you agree on.
d. Construct a gameboard for your game.
e. Make a brochure that explains how to play your game.
f. Trial playing the game with your team.
g. Make any alterations to your game that you think would improve it.
h. Play the game that has been created by another team.
i. In a team of eight, discuss the good things (strengths) and not-so-good things (limitations) of each game. Also, suggest ways that they could be improved in the next 'edition'.

Fully worked solutions and sample responses are available in your digital formats.

Online Resources

Below is a full list of **rich resources** available online for this topic. These resources are designed to bring ideas to life, to promote deep and lasting learning and to support the different learning needs of each individual.

4.1 Overview

eWorkbooks
- Topic 4 eWorkbook (ewbk-11878)
- Starter activity (ewbk-11880)
- Student learning matrix (ewbk-11882)

Solutions
- Topic 4 Solutions (sol-1116)

Practical investigation eLogbook
- Topic 4 Practical investigation eLogbook (elog-2147)

Video eLesson
- The human body and internal organs (eles-2040)

4.2 Driven by curiosity?

Video eLesson
- Leonardo's sketches and anatomy (eles-1769)

Weblink
- Leonardo da Vinci's inventions turned into models

4.3 Working together?

eWorkbook
- Labelling the different body systems (ewbk-11883)

Practical investigation eLogbook
- Investigation 4.1: Mapping your organs (elog-2149)

Interactivities
- Systems (int-6581)
- Types of tissues (int-6582)
- Body systems (int-3397)
- Excretion (int-6583)

4.4 Digestive system — break it down

eWorkbooks
- Labelling the human digestive system (ewbk-11885)
- Mechanical and chemical digestion (ewbk-11887)
- The digestive system (ewbk-11889)

Practical investigation eLogbooks
- Investigation 4.2: Does temperature affect enzymes (elog-2151)
- Investigation 4.3: Making a burp model (elog-2153)

Video eLessons
- Swallowing (eles-2042)
- Stomach acid reflux, heartburn (eles-2531)

Interactivities
- The digestive system (int-3398)
- Types of teeth (int-5335)

Weblink
- Villi

4.5 Digestive endeavours

Practical investigation eLogbook
- Investigation 4.4: Observing villi (elog-2155)

Video eLesson
- Tooth anatomy and process of decay (eles-2532)

4.6 Circulatory system — blood highways

eWorkbooks
- Labelling the movement of blood through the heart (ewbk-11891)
- Blood and blood highways (ewbk-11893)

Practical investigation eLogbooks
- Investigation 4.5: Viewing blood cells (elog-2157)
- Investigation 4.6: Heart dissection (elog-2159)

Teacher-led video
- Investigation 4.6: Heart dissection (tlvd-10730)

Video eLessons
- Bleeding (eles-2535)
- Veins (eles-2047)
- Blood flow through the heart (eles-2049)

Interactivities
- Components of blood (int-6584)
- Beat it! (int-0210)

4.7 Transport technology

Practical investigation eLogbook
- Investigation 4.7: Check your heart (elog-2161)

Video eLessons
- Heart valve (eles-0858)
- Beating heart with ECG (eles-2050)

4.8 Respiratory system — breathe in, breathe out

eWorkbooks
- Labelling the human respiratory system (ewbk-11895)
- Breathing — constructing a report (ewbk-11897)

Practical investigation eLogbooks
- Investigation 4.8: Hands on pluck (elog-2163)
- Investigation 4.9: Measuring your vital capacity (elog-2165)

Video eLessons
- The action of the diaphragm and lungs (eles-2045)

Interactivity
- Labelling the human respiratory system (int-8233)

4.9 Short of breath?

eWorkbook
- Smoking and diseases (ewbk-11899)

Interactivity
- Asthma (int-5351)

4.10 Excretory system

eWorkbooks
- Labelling the kidneys (ewbk-11901)
- Removing waste from blood (ewbk-11903)

Video eLessons
- Urinary system (eles-2051)
- The liver (eles-2536)

Interactivity
- Labelling the kidneys (int-8234)

4.11 Musculoskeletal system — keeping in shape

eWorkbooks
- Labelling the major bones of the human skeleton (ewbk-11905)
- Bones, joints and muscles (ewbk-11907)

Practical investigation eLogbooks
- Investigation 4.10: Rubbery bones (elog-2167)
- Investigation 4.11: Chicken wing dissection (elog-2169)

Video eLessons
- Skeleton with organs and nervous system (eles-2537)
- Joints in the human body (eles-2052)
- Osteoarthritis (eles-2053)

Interactivity
- Labelling the major bones of the human skeleton (int-8256)

4.12 Same job, different path

Practical investigation eLogbook
- Investigation 4.12: Inside or out? (elog-2171)

4.14 Review

eWorkbooks
- Topic review Level 1 (ewbk-11909)
- Topic review Level 2 (ewbk-11911)
- Topic review Level 3 (ewbk-11913)
- Study checklist (ewbk-11915)
- Literacy builder (ewbk-11916)
- Crossword (ewbk-11918)
- Word search (ewbk-11920)
- Reflection (ewbk-11922)

Digital document
- Key terms glossary (doc-39981)

To access these online resources, log on to **www.jacplus.com.au**

LESSON
5.2 It's elementary

LEARNING INTENTION

At the end of this lesson you will be able to recognise how the work of scientists today compares with the work of the early alchemists. You will also be able to describe the properties of some elements.

5.2.1 Discovering elements

The alchemists

SCIENCE AS A HUMAN ENDEAVOUR: The alchemists

About 1000 years ago, when kings and queens lived in castles and were defended by knights in shining armour, there lived the **alchemists**.

They chanted secret spells while they mixed magic potions in their flasks and melted metals in their furnaces. They tried to change ordinary metals into gold. They also tried to find a potion that would make humans live forever. They studied the movements of the stars and claimed to be able to see into the future. Kings and queens took the advice of the alchemists very seriously.

FIGURE 5.2 Alchemists believed that the four basic elements in nature were air, fire, water and earth.

The alchemists never found the secrets they were looking for, but they did discover many things about substances around us. During the same period, people who worked with materials also helped us to understand many everyday substances. Blacksmiths worked with metals to make stronger and lighter swords and armour, fabric dyers learned how to colour cloth, and potters decorated their work with glazes from the earth. Without the knowledge passed down by these people, the world as we know it would be very different!

Twelve important substances were discovered during these ancient times: **gold, iron, silver, sulfur, carbon, lead, mercury, tin, arsenic, bismuth, antimony** and **copper**. Alchemists discovered five of these.

EXTENSION ACTIVITY: Medieval swords

Explore how swords were made in the middle ages.

Real science

In about the seventeenth century, people stopped thinking about magic and instead carried out **investigations** based on careful **observations**. These new seekers of knowledge were called **scientists**. They found that the 12 substances discovered during ancient times could not be broken down into other substances. Scientists investigated many common everyday substances as well, including salt, air, rocks, water and even urine! They discovered that nearly everything around us could be broken down into other substances. They gave the name '**element**' to the substances that could not be broken down into other substances. Between 1557 and 1925, another 76 elements were discovered. We now know that 92 elements exist naturally. In recent years, scientists working in laboratories have been able to make at least another 26 artificial elements. We'll shortly look at how all these natural and artificial elements are arranged based on their properties.

alchemist an olden-day 'chemist' who mixed chemicals and tried to change ordinary metals into gold; alchemists also tried to predict the future

investigations activities aimed at finding information

observations information obtained by the use of our senses or measuring instruments

scientists people skilled in or working in the fields of science; scientists use experiments to find out about the material world around them

elements pure substances made up of only one type of atom

FIGURE 5.3 Seventeenth-century scientists carried out investigations based on careful observations.

Scientists use investigations based on observations to make discoveries.

 Resources

Video eLesson Lavoisier and hydrogen (eles-1772)

DISCUSSION

Discuss the similarities and differences between the work of the alchemists and the real scientists of the seventeenth century.

EXTENSION: Mercury poisoning

In days gone by, substances containing the element mercury were used to make hats. In those days it was not known that mercury is a very poisonous element. Poisoning by mercury can affect your nervous system and your mind. This sometimes happened to hat makers who were exposed to mercury for a long time; hence the expression 'mad as a hatter'!

FIGURE 5.4 Hat makers at work

5.2.2 Examining elements

Most of the substances around you are made up of two or more elements. Most of the elements are not stable in their pure elemental form and so you will not be able to find many of the 92 naturally occurring elements in their pure, uncombined form. It is possible, however, to examine many of the elements in the school laboratory.

Many elements are safe to handle; however, there are also many that are not. The elements sodium, potassium and mercury, for example, need special care and handling. Sodium and potassium are soft metals that can be cut with a knife. They both get very hot and can cause an explosion when they come into contact with water. Therefore, they are stored under oil so that water in the atmosphere cannot reach them.

elog-2255

INVESTIGATION 5.2

Checking out appearances of elements

Aim

To examine and describe the properties of a selection of elements

Materials

- samples of chemical elements (e.g. carbon, sulfur, copper, iron, aluminium, silicon)
- magnifying glass or stereo microscope
- iron nail

Method

1. Copy the table in the results section into your workbook or obtain a copy from your teacher.
2. Using a periodic table, write the names and symbols of each of the elements.
3. Carefully examine the appearance of each of the elements in the set (look for colour, texture).
4. Test the hardness by scratching with a nail where possible.
5. Find out where the substance might be found.
6. Complete the table by filling in the description. Research or discuss with other students the substances that might include the element. One example is completed for you.

Results

TABLE Observations and predictions of elements observed

Element	Symbol	State	Observations	In which substances might the element be present?
Hydrogen		Gas	Clear, colourless	Acids, water

Discussion

1. Describe any similarities between the elements.
2. Divide the elements into groups according to one of the properties that you observed. Give the groups names.
3. Which element was the hardest to classify?
4. Discuss the accuracy of the test for hardness.
5. Suggest another test that could be performed to find out more about the elements.

Conclusion

Summarise the properties that you observed.

Resources

 **eWorkbook** How big is an atom? (ewbk-12289)

 **Weblink** Alkali metals reacting in water

5.2 Activities

learnon

5.2 Quick quiz on	5.2 Exercise

Select your pathway

■ LEVEL 1	■ LEVEL 2	■ LEVEL 3
1, 2, 5, 6, 7	3, 8, 9	4, 10, 11

Remember and understand

1. **MC** An element is:
   A. a substance that cannot be combined with other substances.
   B. a substance that cannot be broken down.
   C. a substance that is only stable in its pure form.
   D. a substance that is only stable in a combined form.
2. Complete the following table by identifying which types of substances blacksmiths worked with and helped us to understand.

Substance	Worked with	Didn't work with
Alloys		
Non-metals		
Metalloids		
Metals		
Compounds		

3. Fill in the blanks to complete the sentence.
   Despite __________ never finding the answers they were looking for, they did discover many things about the substances around them. While mixing potions and chanting spells, they tried to change ordinary __________ into __________.
4. Explain why sodium and potassium need to be stored under oil.
5. Identify whether the following statements are true or false. Justify any false responses.

Statement	True or false?
a. Mercury is a poison that affects your nervous system and can cause brain damage.	
b. Substances containing mercury were once used in the process of hat-making.	
c. Mercury is a solid at room temperature.	

6. Classify the following as either correct or incorrect observations of the seventeenth-century scientist shown in figure 5.3.
   a. The scientist has long hair and a beard.
   b. The scientist has his right hand on the table.
   c. The scientist is wearing green robes.
   d. The scientist is wearing a baseball cap.
   e. The scientist is sitting in front of a mortar and pestle.
   f. There are three jugs on the table.

Apply and analyse

7. **MC** Select one reason for displaying chemical safety symbols at the entrances of many buildings.
   A. Chemical safety signs let manufacturers know which chemicals are used in the building so they can be more easily ordered when they run out.
   B. Chemical safety signs let cleaners know which chemicals are stored and used in the building.
   C. Chemical safety signs warn people of the dangers of chemicals stored and used in the building.
   D. Chemical safety signs let delivery people know which chemicals to deliver to the building.
8. Is water an element? Give a reason for your answer.
9. How do the scientists of today differ from the alchemists in the seventeenth century?

Evaluate and create

10. **SIS** Many years ago, balloons were filled with hydrogen so that they could float high in the sky. However, hydrogen is no longer used in balloons because it explodes too easily. At fairs, carnivals and florist shops, you can often buy colourful gas-filled balloons that fly high into the sky if you let them go. These balloons are filled with an element called helium. Find out who discovered helium, where it was discovered and when.
11. **SIS** The element mercury was known to ancient people and was very important to the alchemists. Find out all you can about this liquid metal. What does its name mean? Where is it found? What has it been used for in the past? What is it used for now? What is the safety procedure if mercury is spilt?

Fully worked solutions and sample responses are available in your digital formats.

LESSON
5.3 Elements — the inside story

LEARNING INTENTION

At the end of this lesson you will be able to describe how the model of the atom developed. You will be able to describe sub-atomic particles and apply chemical symbols.

5.3.1 Atoms and elements

SCIENCE AS A HUMAN ENDEAVOUR: Developing models of the atom

Democritus

About 2500 years ago a teacher named Democritus lived in ancient Greece. He walked around the gardens with his students, talking about all sorts of ideas.

Democritus suggested that everything in the world was made up of tiny particles so small that they couldn't be seen. He called these particles *atomos*, which means 'unable to be divided'. Other thinkers at the time disagreed with Democritus. It took about 2400 years for evidence of the existence of these atoms (as we now call them) to be found.

FIGURE 5.5 Democritus (*c*.460 – *c*.370 BCE), an ancient Greek philosopher

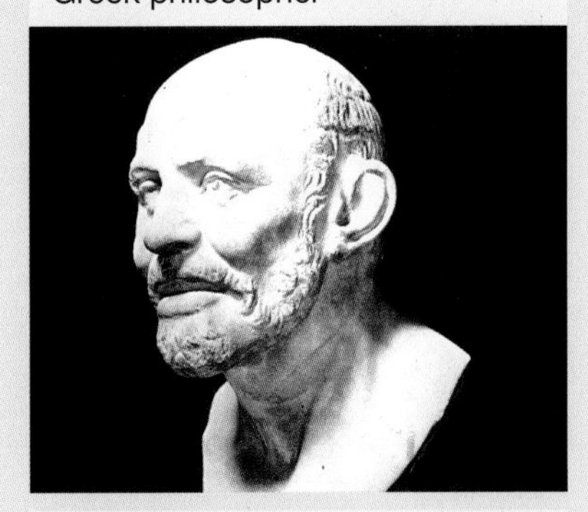

 Resources

 **Interactivity** Democritus and the atom (int-5744)

John Dalton

Even though the atom couldn't be seen, scientists did experiments over many years and they thought carefully about the information they gathered.

Finding evidence for the existence of atoms was not possible until Galileo wrote about the need for controlled experiments, and the importance of accurate observations and mathematical analysis, in the sixteenth century. Galileo's 'scientific method', along with the development of more accurate weighing machines, was used by John Dalton in 1803 to show that matter was made up of atoms. He proposed that atoms could not be divided into smaller particles and that atoms of different elements had different masses.

FIGURE 5.6 John Dalton (1766–1844), an English chemist, physicist and meteorologist

Ernest Rutherford

For the next 100 years, scientists thought the atom was a solid sphere, but discoveries including radioactivity and electric current, and new technology such as the vacuum tube and Geiger counters, allowed scientists to 'peek' inside. In 1897 Joseph Thomson, a British physicist, discovered the electron and a few years later in 1911, New Zealander Sir Ernest Rutherford used some of the new discoveries and inventions to prove that atoms were not solid particles.

Rutherford fired extremely tiny particles at a very thin sheet of gold. Most of the particles went straight through. Only sometimes did they bounce off as if they had hit something solid. He concluded that the tiny particles could be getting through only if each atom consisted of mostly empty space with a positive nucleus at the centre.

FIGURE 5.7 Rutherford's model of the atom

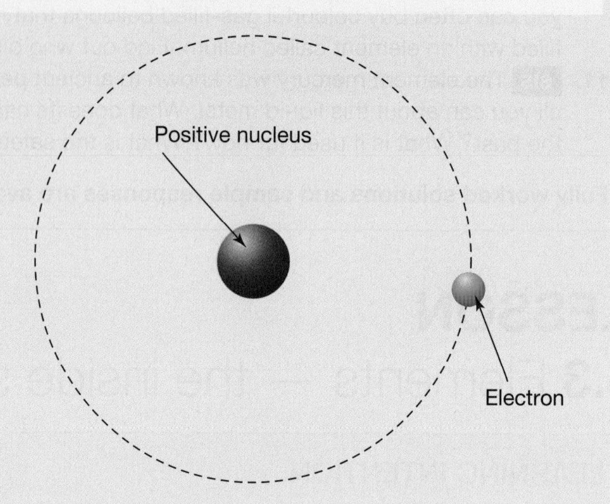

Niels Bohr

Niels Bohr proposed the next model of the atom. He suggested that the electrons circled the nucleus in shells, and that in the first shell there were two electrons and in the second shell up to eight electrons.

In 1932 James Chadwick found another type of particle in the nucleus of the atom — the neutron.

FIGURE 5.8 Bohr's model of the atom

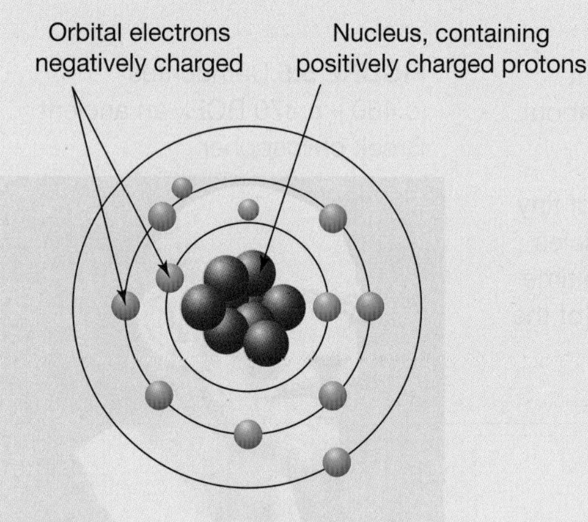

FIGURE 5.9 The Chadwick–Bohr model of the atom

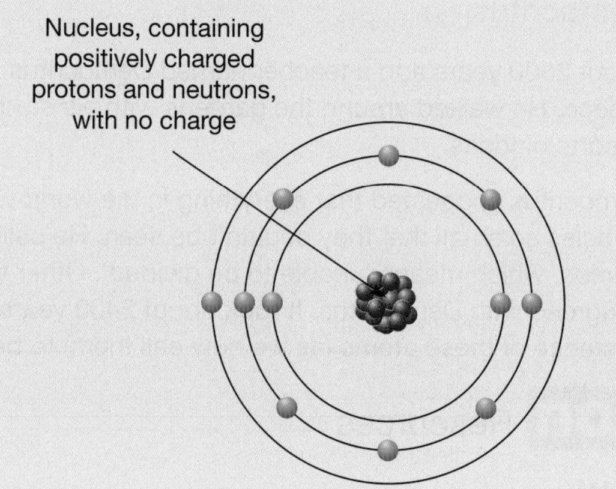

5.3.2 Inside the atom — protons, neutrons and electrons

We now know that each element is made of its own particular kind of **atom**. Gold contains only gold atoms, oxygen contains only oxygen atoms, carbon contains only carbon atoms and so on. But what is it that makes atoms different from one another? To answer this question, we need to know a little bit more about the atom. It is now understood that all atoms are made up of even smaller particles (table 5.1):

- **Protons**
- **Neutrons**
- **Electrons**.

atoms very small particles that make up all things; atoms have the same properties as the objects they make up

protons tiny, but heavy, positively charged particles found in the nucleus of an atom

neutrons tiny, but heavy, particles found in the nucleus of an atom; they have no electrical charge

electrons very light, negatively charged particles inside an atom; electrons orbit around the atom's nucleus

FIGURE 5.10 If helium atoms are lighter than all others except for hydrogen atoms, why do you think that helium is used in blimps?

TABLE 5.1 Sub-atomic particles and their properties

Sub-atomic particle	Charge		Relative mass	Location
Proton	Positive	+1	1	Nucleus
Neutron	Neutral	0	1	Nucleus
Electron	Negative	–1	0	Energy shells (orbits; electron cloud)

The amount of negative charge carried by each electron is equal but opposite to the amount of positive charge carried by each proton. Therefore, in an atom, the number of protons is equal to the number of electrons, so there is no overall electric charge.

int-3387

FIGURE 5.11 Diagram of the atom

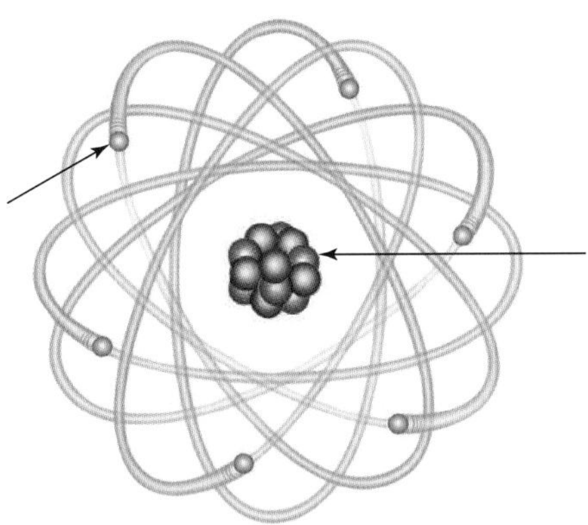

Matter is made of atoms that contain protons, neutrons and electrons. Hydrogen is an exception because most hydrogen atoms do not have a neutron.

5.3.3 Atomic numbers

The number of protons in an atom is called its **atomic number**. Each element has a different atomic number. The blimp in figure 5.10 is filled with helium, which has an atomic number of 2. Helium atoms are lighter than all others except for hydrogen atoms (which have an atomic number of 1). Carbon atoms have six protons inside the **nucleus**, so the atomic number of carbon is 6. The number

atomic number the number of protons in the nucleus of an atom, which identifies the element to which the atom belongs

nucleus the central part of an atom, made up of protons and neutrons; plural = nuclei

of neutrons in an atom can vary; most carbon atoms have six neutrons, but some have seven or eight in their nuclei. For each proton in the carbon atom there is also one electron, meaning a carbon atom has six electrons. The lightest element is hydrogen, which has one proton in each atom and an atomic number of 1. The heaviest natural element is uranium, with 92 protons in each atom.

TABLE 5.2 Protons, neutrons and electrons in the first 12 elements

Name	Symbol	Protons (atomic number)	Electrons	Neutrons*
Hydrogen	H	1	1	0
Helium	He	2	2	2
Lithium	Li	3	3	4
Beryllium	Be	4	4	5
Boron	B	5	5	6
Carbon	C	6	6	6
Nitrogen	N	7	7	7
Oxygen	O	8	8	8
Fluorine	F	9	9	10
Neon	Ne	10	10	10
Sodium	Na	11	11	12
Magnesium	Mg	12	12	12

*The number of neutrons can vary but this is the most common number of neutrons for these elements.

Resources

Video eLessons The hydrogen atom (eles-2269)
An atom of carbon (eles-2031)

Protons identify the element

Each of the elements differ in their number of protons. It is the number of protons in an atom that identifies the element to which it belongs.

EXTENSION ACTIVITY: What is mass number of an element?

Research to find out what 'mass number' refers to.
Calculate the mass number of the first 12 elements in the periodic table.

5.3.4 Chemical names and symbols

As the early scientists discovered more and more elements, it became increasingly important that they all agreed on what to call them. Each element was given a name and a **chemical symbol**.

chemical symbol the standard way that scientists write the names of the elements, using either a capital letter or a capital followed by a lower-case letter; for example, carbon is C and copper is Cu

The chemical symbols of most elements are very easy to understand. The symbol sometimes starts with the capital letter that is the first letter of the element's name. For some elements that is the complete symbol.

For example:

O = oxygen C = carbon N = nitrogen H = hydrogen

When there is more than one element starting with the same capital letter, a lower-case letter is also used.

For example:

Cl = chlorine	Ca = calcium	Cr = chromium	Cu = copper

FIGURE 5.12 Element 21 is scandium and has the chemical symbol Sc.

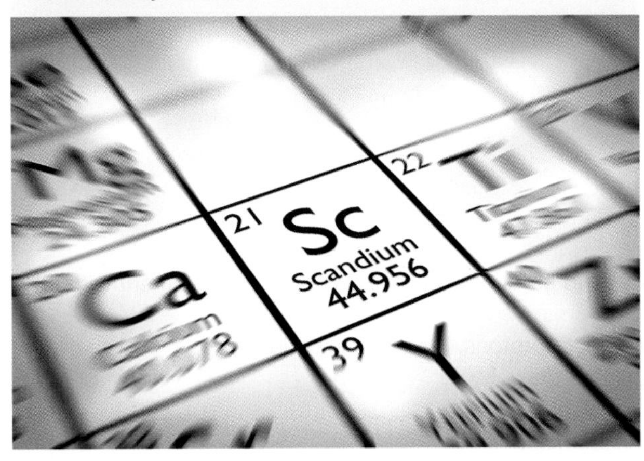

If an element has a symbol that doesn't match its modern name, that's because the symbol is taken from the original Greek or Latin name.

For example:

Na = sodium (*natrium*)	Pb = lead (*plumbum*)	Ag = silver (*argentum*)
Hg = mercury (*hydro argyros*)	Fe = iron (*ferrum*)	K = potassium (*kalium*)

SCIENCE AS A HUMAN ENDEAVOUR: Naming the elements

The names and symbols of some of the elements have some interesting origins.

- Einsteinium (Es) is named after the famous scientist Albert Einstein.
- Polonium (Po) was discovered by another famous scientist, Marie Curie. She named polonium after Poland, the country of her birth.
- Helium (He) was first discovered in the Sun. It is named after Helios, the Greek god of the Sun.
- Sodium (Na) was first called by the Latin name *natrium*.
- Lead (Pb) also used to have a Latin name, *plumbum*. That's where the word 'plumber' comes from. The ancient Romans, who spoke Latin, used lead metal to make their water pipes.

FIGURE 5.13 A statue of Polish scientist Marie Curie holding a model of a polonium atom in Warsaw, Poland

elog-2257

INVESTIGATION 5.3

Getting to know atoms

Aim

To investigate and prepare models of atoms

Materials

A selection of craft materials; for example:

- thin card
- coloured paper
- wool
- buttons
- counters
- pipe cleaners
- string
- small beans

Method

1. In a group, discuss, design and produce a 2D or 3D model of an element with an atomic number between 3 and 12.
2. Include the correct placement of protons, neutrons and electrons.
3. Think about the size of the particles.
4. Include labels or a key.
5. Include the symbol and full name.
6. Write 5–7 sentences about the discovery, source and use of your element.

Results

Draw a diagram or take a photo of your model. Your teacher may ask you to present your element to the class.

Discussion

1. Outline what you learned about working in a group.
2. Describe how the atoms that the students in the class prepared are similar and how they are different.
3. In what ways is your model different from a real atom?

Conclusion

Summarise what you learned about atoms.

5.3 Activities

learnon

5.3 Quick quiz on	5.3 Exercise

Select your pathway

■ LEVEL 1	■ LEVEL 2	■ LEVEL 3
1, 2, 3, 4, 5, 12	6, 7, 10, 11, 13	8, 9, 14, 15, 16, 17

These questions are even better in jacPLUS!
- Receive immediate feedback
- Access sample responses
- Track results and progress

Find all this and MORE in jacPLUS

Remember and understand

1. Recall the ideas that Democritus had around 2500 years ago about what substances were made up of, and fill in the blanks to complete the sentence.
   Democritus thought that all __________ was made of tiny __________ that could not be __________ and that these particles could not be divided into anything __________. He called these particles __________.
2. How many types of naturally occurring atoms are there?

3. Draw and label a diagram of an atom. Ensure you state the location of each of the three parts of an atom.
4. What does the atomic number of an element tell you?

Apply and analyse

5. State the symbols of each of the following elements: hydrogen, carbon, oxygen, nitrogen, iron, calcium, copper, lead, mercury.
6. Explain why carbon atoms have six electrons.
7. **MC** Why don't electrons fly off their atoms?
   A. They are repelled from the external environment.
   B. They are attracted to the neutrons in the nucleus.
   C. They are directly bonded to the protons.
   D. They are attracted to the protons in the nucleus.
8. Attribute each of the following discoveries/theories, which were used to learn more about the atom, to the appropriate scientist — Democritus, Dalton, Rutherford or Bohr.
   a. Experimented with shooting tiny particles at a thin sheet of gold and observed that most particles went straight through
   b. Used Galileo's scientific method with an accurate weighing machine to show that matter was made up of atoms
   c. Said 'everything in the world is made up of tiny particles so small they can't be seen'
   d. Determined that electrons circled the nucleus in shells
   e. Proved that atoms were not solid particles
   f. Said 'atoms of different elements had different masses'
   g. Proposed that atoms could not be divided into small particles
   h. Determined that the nucleus had a positive charge
   i. Called the tiny particles 'atomos'
   j. Determined that atoms were made up of mostly space
9. Describe what makes up most of every atom.
10. State the atomic number of uranium.

Evaluate and create

11. In what ways is an atom of carbon different from an atom of uranium?
12. a. Describe the nucleus of an atom.
    b. What type of electric charge does the nucleus of every atom have?
    c. Explain why atoms have no electric charge.
13. To which element does the atom illustrated in figure 5.11 belong?
14. Explain why it is important for scientists around the world to agree on the names and chemical symbols of the elements.
15. State the names and atomic numbers of the elements with the following symbols.

Symbol	Name	Atomic number
Sn		
Au		
Cu		
N		
Ne		
Sr		
Ca		

16. Draw and label a diagram of an atom that has:
    - three protons
    - one neutron
    - an appropriate number of electrons.
17. **SIS** Research and report on what nanotechnology is and what connection it has with atoms.

Fully worked solutions and sample responses are available in your digital formats.

LESSON
5.4 Types of elements and the periodic table

LEARNING INTENTION

At the end of this lesson you will recognise the properties of the types of elements in the periodic table, and be able to explain how these metal and non-metal elements are organised in the periodic table into groups and periods.

5.4.1 Grouping elements in common

It is often convenient to group objects that have features in common. Shops provide a good example of this. In a department store, the goods are grouped so that you know where to buy them. You go to the clothing section for a new pair of jeans, to the jewellery section for a new watch and to the food section for a packet of potato chips.

Scientists also organise objects into groups. Biologists organise living things into groups. Animals with backbones are divided into mammals, birds, reptiles, amphibians and fish. Geologists organise rocks into groups. The elements that make up all substances can also be organised into two main groups of elements with common characteristics: **metals** and **non-metals**.

FIGURE 5.14 Mercury is the only metal that is liquid at room temperature.

The two main types of elements in the periodic table are metals and non-metals.

Metals

The metals have several features in common (figure 5.15). They:
- are solid at room temperature, except for mercury, which is a liquid
- can be polished to produce a high shine or **lustre**
- are good conductors of electricity and heat
- can all be beaten or bent into a variety of shapes; we say they are **malleable**
- can be made into a wire; we say they are **ductile**
- usually melt at high temperatures. Mercury, which melts at –40 °C, is one exception.

metals elements that conduct heat and electricity; shiny solids that can be made into thin wires and sheets that bend easily; mercury is the only liquid metal at room temperature

non-metals elements that do not conduct electricity or heat; they melt and turn into gases easily, and are brittle and often coloured

lustre the high shine and sheen of a substance caused by the way it reflects light

malleable able to be beaten, bent or flattened into shape

ductile capable of being drawn into wires or threads; a property of most metals

Resources

Video eLesson Malleability (eles-2033)

FIGURE 5.15 Metals have many characteristic properties.

Non-metals

Only 22 of the elements are non-metals. At room temperature, 11 of them are gases, 10 are solid and 1 is liquid. The solid non-metals have most of the following features in common. They:

- cannot be polished to give a shine like metals; they are usually dull or glassy
- are **brittle**, which means they shatter when they are hit
- cannot be bent into shape
- are usually poor conductors of electricity and heat
- usually melt at low temperatures.

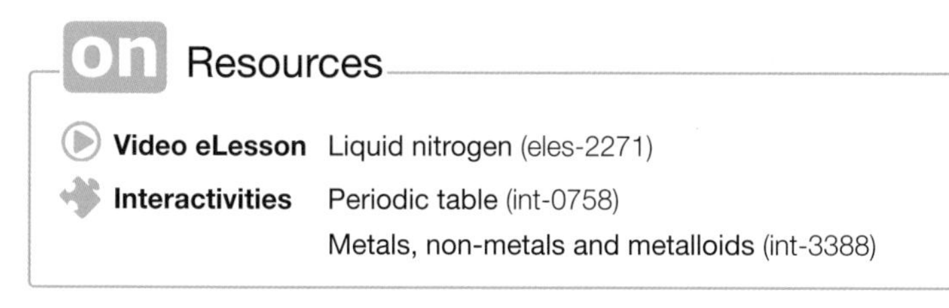

on Resources

Video eLesson Liquid nitrogen (eles-2271)

Interactivities Periodic table (int-0758)

Metals, non-metals and metalloids (int-3388)

FIGURE 5.16 Common examples of non-metals are sulfur, carbon and oxygen.

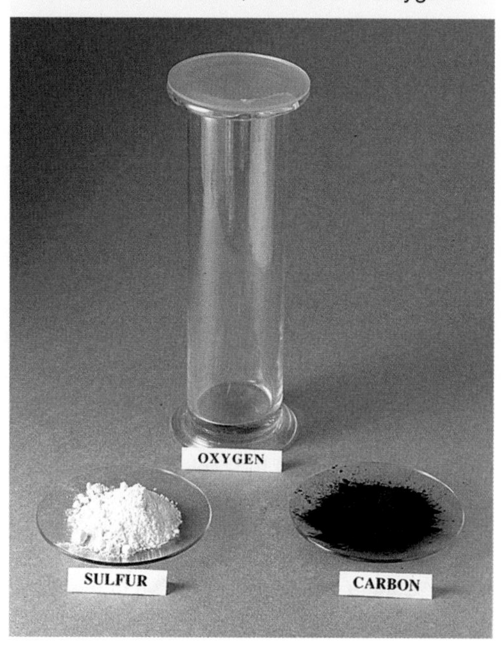

Metalloids

Some of the elements in the non-metal group look like metals. One example is silicon. While it can be polished like a metal, silicon is a poor conductor of heat and electricity, and cannot be bent or made into wire. Those elements that have features of both metals and non-metals are called **metalloids**. There are eight metalloids altogether: boron, silicon, arsenic, germanium, antimony, polonium, astatine and tellurium.

FIGURE 5.17 Metalloids are important materials often used in electronic components of computer circuits.

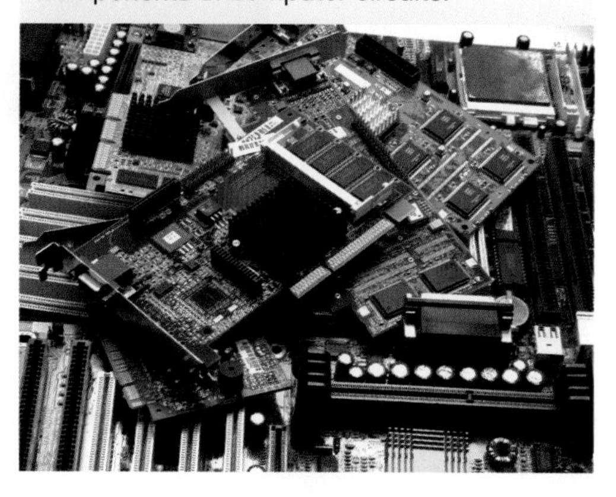

DISCUSSION

Make a list of five items in the classroom that are made of metal. Explain what the property is that lead to its use.

brittle can easily break if hit; the opposite of malleable

metalloids elements that have the appearance of metals but not all the other properties of metals

EXTENSION ACTIVITY: Exploring metalloids

Prepare a fact sheet on the source, properties and use of one of the eight metalloids previously listed.

elog-2259

INVESTIGATION 5.4

Looking for similarities

Aim

To describe the characteristics of a variety of elements

Materials

- safety glasses
- samples of sulfur, zinc, tin, carbon, silicon, copper
- steel wool or very fine sandpaper
- battery or power pack
- wires with alligator clips
- light globe

CAUTION

Power pack safety: Ensure all wires are connected whilst the power pack is turned off. Set the voltage, and only then turn the power pack on.

Method

1. Make a copy of the table in the results section of this investigation and use it to record your observations for each of the substances.
2. Rub each of the elements with the fine sandpaper and observe whether they are shiny or dull.
3. Try to bend the elements.
4. Connect the circuit as shown in the diagram to determine whether electricity passes through each of the elements.
5. Connect each element sample into this circuit.

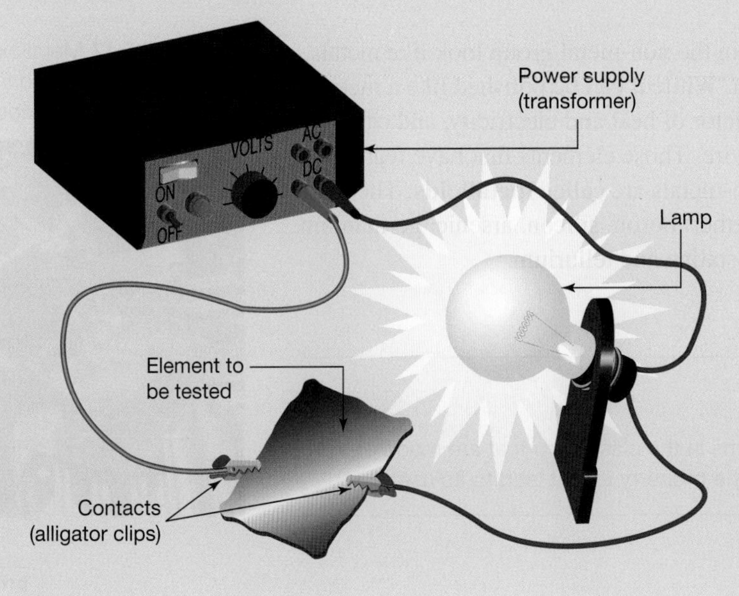

Results

TABLE Characteristics of some elements

Element	Shiny or dull?	Does it bend?	Does it conduct electricity?
Sulfur			
Zinc			
Tin			
Carbon			
Silicon			
Copper			

Discussion

1. Identify the elements that have a shiny surface when polished.
2. Identify the elements that do not have a shiny surface when polished.
3. Identify the elements that can be bent.
4. Identify the elements that cannot be bent.
5. Identify the elements that allow electricity to pass through.
6. Identify the elements that do not conduct electricity.
7. Attempt to divide the elements into two groups based on your observations. Suggest names for these groups.
8. Which of the six elements tested does not seem to fit into either of these two groups?
9. Refer to the aim of this investigation and suggest how the design of this experiment might be improved.

Conclusion

Write sentences that state which substances:
- just had properties of metals
- just had properties of non-metals
- had properties of both metals and non-metals.

Resources

eWorkbook Metals and non-metals (ewbk-12292)

5.4.2 Developing the periodic table

As more and more elements were being discovered, the early scientists began to find that some of them had things in common.

SCIENCE AS A HUMAN ENDEAVOUR: Developing the periodic table

The search began for an organised system to classify the chemical elements as they were being discovered. By comparing to the timeline for the discovery of sub-atomic particles, we must remember that the development of the periodic table occurred before the discovery of protons, neutrons and electrons.

1862 — Alexandre-Émile Béguyer de Chancourtois

Alexandre-Émile Béguyer was a French geology professor who had a small (often not remembered) influence in the field of chemistry in 1862 with his *vis tellurique*, the first periodic representation of known elements (figure 5.18). He identified a repeating, or periodic, pattern of properties with every atomic mass increase of 16.

FIGURE 5.18 a. Alexandre-Émile Béguyer de Chancourtois **b.** His *vis tellurique* from his original publication

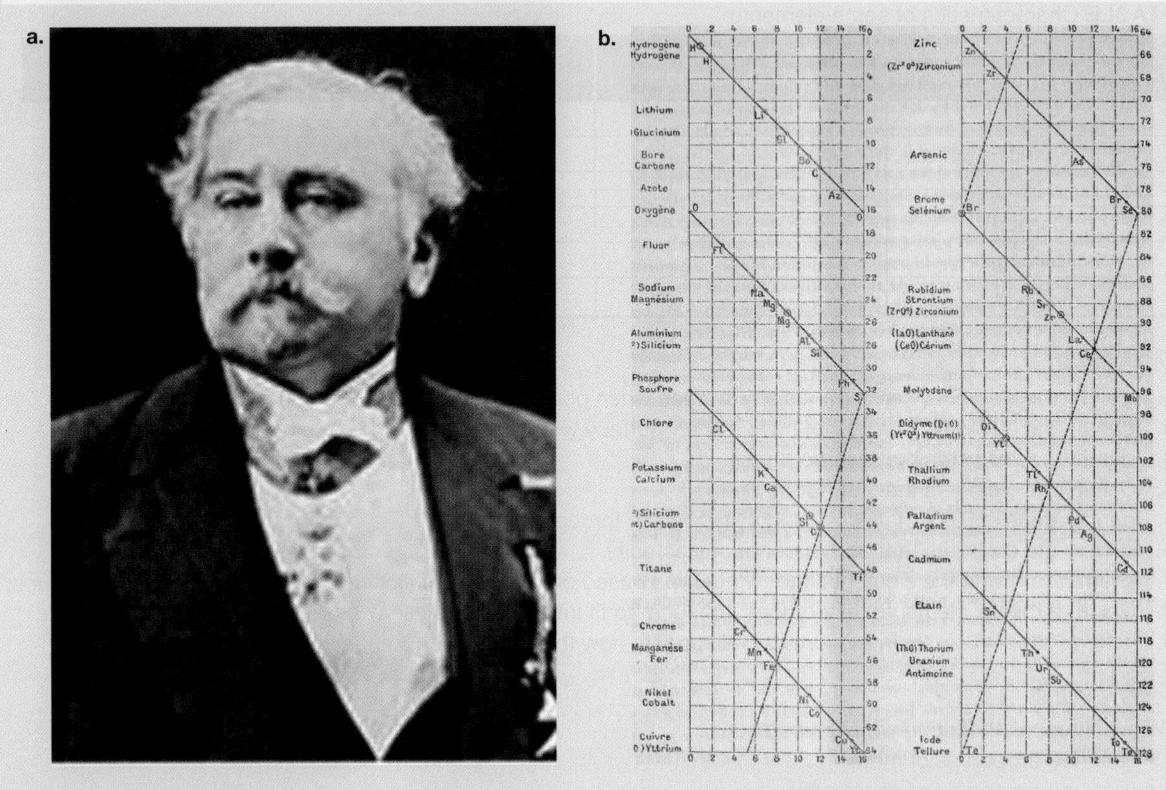

1864 — John Newlands

John Newlands, a British chemist known for his Law of Octaves, noticed similarities in the characteristics of elements that differed by an atomic mass of 7. Despite an octave signifying a difference of 8, it was the similarity of periodicity to music that provided the name. Coincidentally, the noble gases were later discovered and enabled the Law of Octaves to represent a difference in atomic mass by 8.

Unfortunately, there were limitations to Newlands's representation as he did not provide any gaps for elements that were not yet discovered. These limitations led the Chemical Society to not publish his paper, resulting in Mendeleev taking more credit for the discovery. It was not until 1998, over 100 years later, that the Royal Society of Chemistry in England finally recognised Newlands' discovery.

FIGURE 5.19 a. John Newlands **b.** His proposal for the Law of Octaves in the 18th August 1865 edition of *Chemical News* **c.** His arrangement of elements

a.

b.

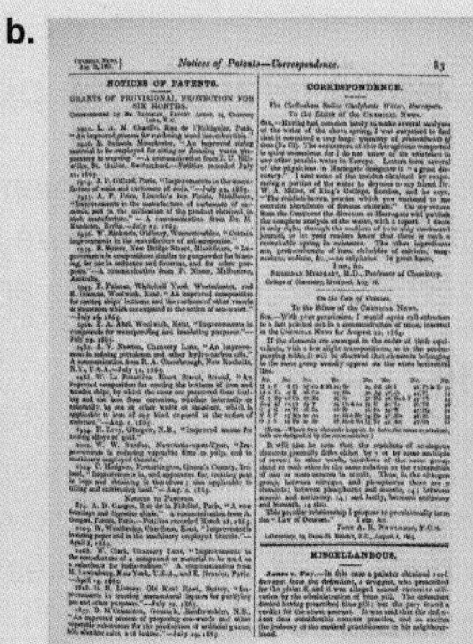

c.

H	Li	Be	B	C	N	O
F	Na	Mg	Al	Si	P	S
Cl	K	Ca	Cr	Ti	Mn	Fe

1864 — Julius Lothar von Meyer

Julius Lothar von Meyer, a German scientist, created several representations of the organisation of the chemical elements, with his first table produced in 1864. He identified the relationship between the atomic mass and the atomic volume for the first 28 elements and this was later extended to 53 elements in 1870 when his paper was published — unfortunately a year after Mendeleev (figure 5.21).

FIGURE 5.20 Julius Lothar von Meyer

FIGURE 5.21 Julius Lothar von Meyer periodic table graph

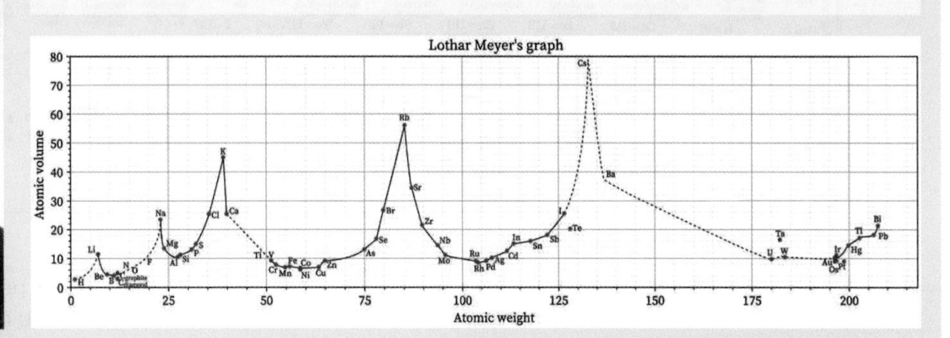

1869 — Dmitri Mendeleev

Dmitri Mendeleev, a Russian chemist (figure 5.22), is well known for his representation of the periodic table that formed the basis for the modern periodic table we know today. As opposed to Newlands's octave model, Mendeleev worked out a system for grouping the elements that did take into account gaps where yet undiscovered (predicted) elements should be placed. One predicted element was named 'Eka-aluminium' as it was predicted to be placed after aluminium. Mendeleev had predicted its chemical properties, including the atomic mass, density and melting point, which matched extremely closely to gallium when it was discovered in 1875 by Paul-Émile Locoq de Boisbaudran.

FIGURE 5.22 a. Dmitri Mendeleev **b.** His rough periodic table notes in 1869 **c.** His table as revised in 1871

a.

b.

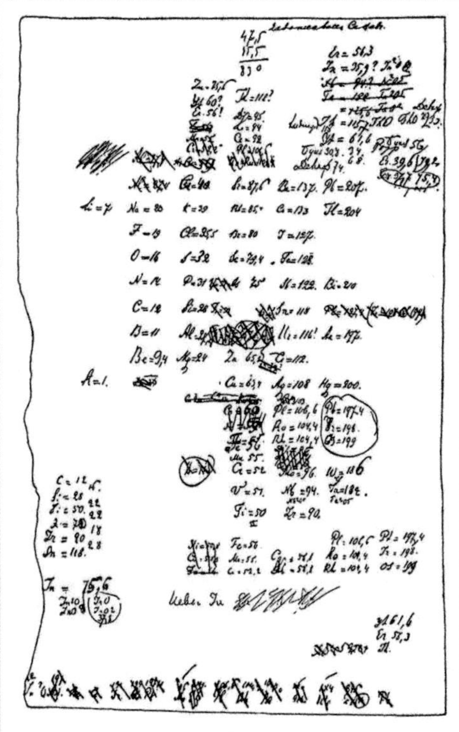

c.

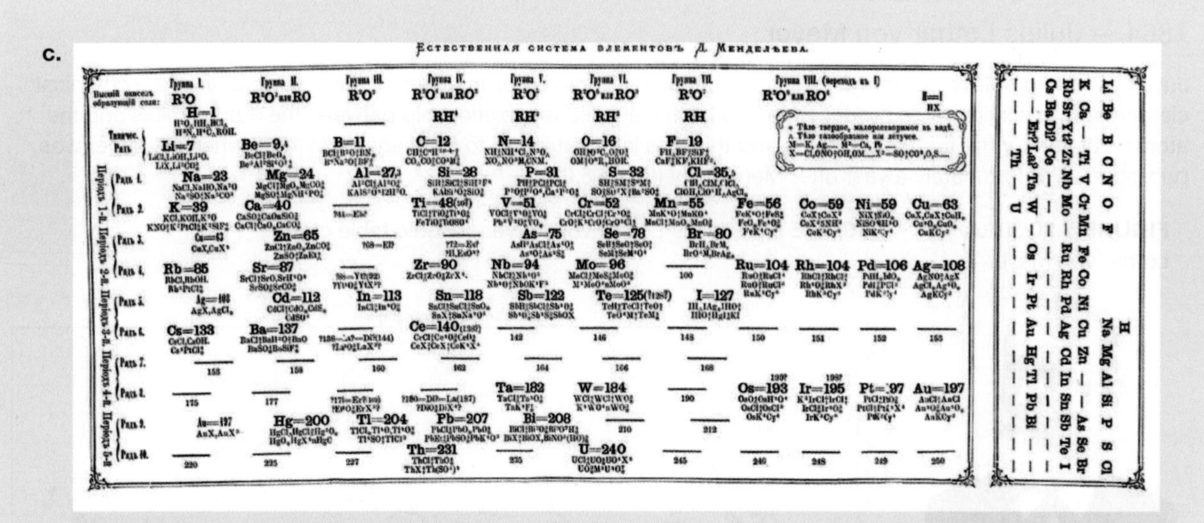
Естественная система элементовъ Д. Менделѣева.

Since Mendeleev's periodic table, there have been numerous attempts to recreate the periodic table as an organised system of chemical elements that takes into consideration both physical and chemical properties; however, it is still Mendeleev's periodic table that forms the basis of the modern periodic table.

Alternative versions of the periodic table

FIGURE 5.23 Left-step periodic table (1928)

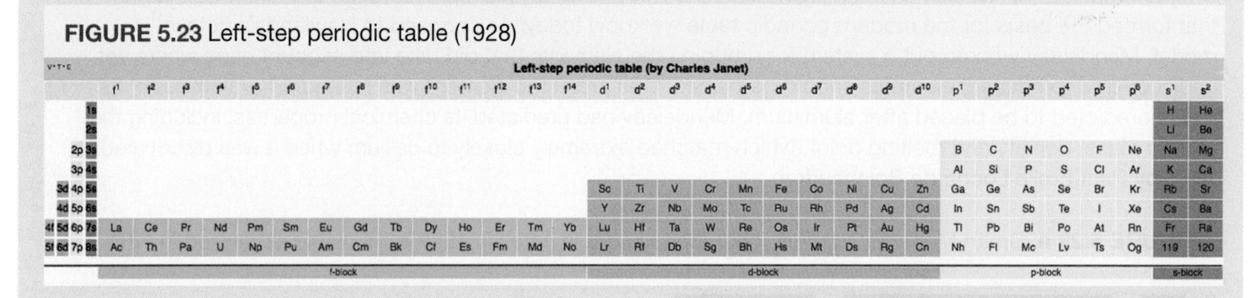
Left-step periodic table (by Charles Janet)

	f^1	f^2	f^3	f^4	f^5	f^6	f^7	f^8	f^9	f^{10}	f^{11}	f^{12}	f^{13}	f^{14}	d^1	d^2	d^3	d^4	d^5	d^6	d^7	d^8	d^9	d^{10}	p^1	p^2	p^3	p^4	p^5	p^6	s^1	s^2
1s																															H	He
2s																															Li	Be
2p 3s																									B	C	N	O	F	Ne	Na	Mg
3p 4s																									Al	Si	P	S	Cl	Ar	K	Ca
3d 4p 5s															Sc	Ti	V	Cr	Mn	Fe	Co	Ni	Cu	Zn	Ga	Ge	As	Se	Br	Kr	Rb	Sr
4d 5p 6s															Y	Zr	Nb	Mo	Tc	Ru	Rh	Pd	Ag	Cd	In	Sn	Sb	Te	I	Xe	Cs	Ba
4f 5d 6p 7s	La	Ce	Pr	Nd	Pm	Sm	Eu	Gd	Tb	Dy	Ho	Er	Tm	Yb	Lu	Hf	Ta	W	Re	Os	Ir	Pt	Au	Hg	Tl	Pb	Bi	Po	At	Rn	Fr	Ra
5f 6d 7p 8s	Ac	Th	Pa	U	Np	Pu	Am	Cm	Bk	Cf	Es	Fm	Md	No	Lr	Rf	Db	Sg	Bh	Hs	Mt	Ds	Rg	Cn	Nh	Fl	Mc	Lv	Ts	Og	119	120
	f-block														d-block										p-block						s-block	

FIGURE 5.24 Periodic snail (1964)

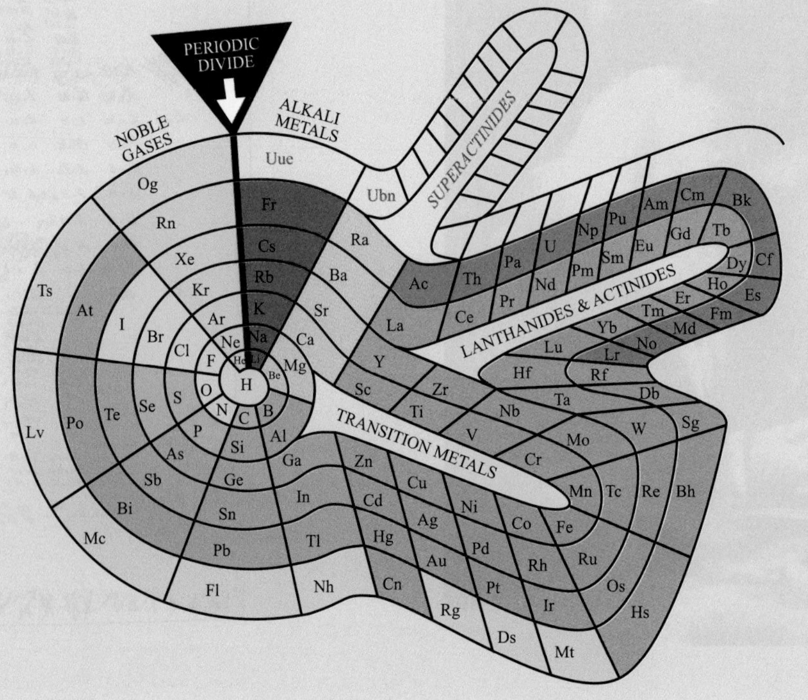

FIGURE 5.25 Curled ribbon (1975)

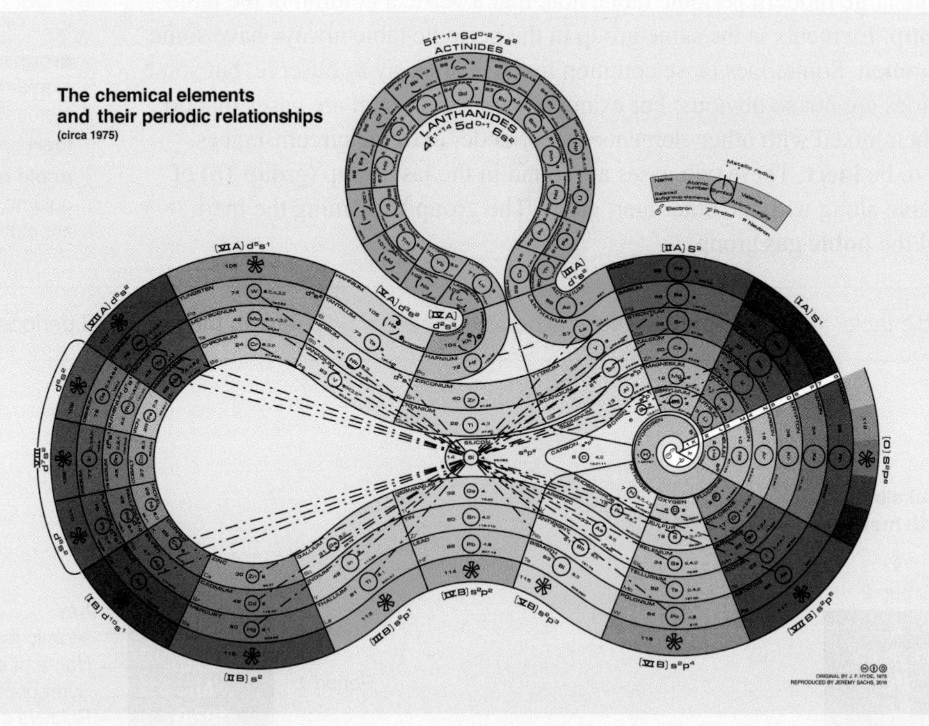

FIGURE 5.26 Circular representations

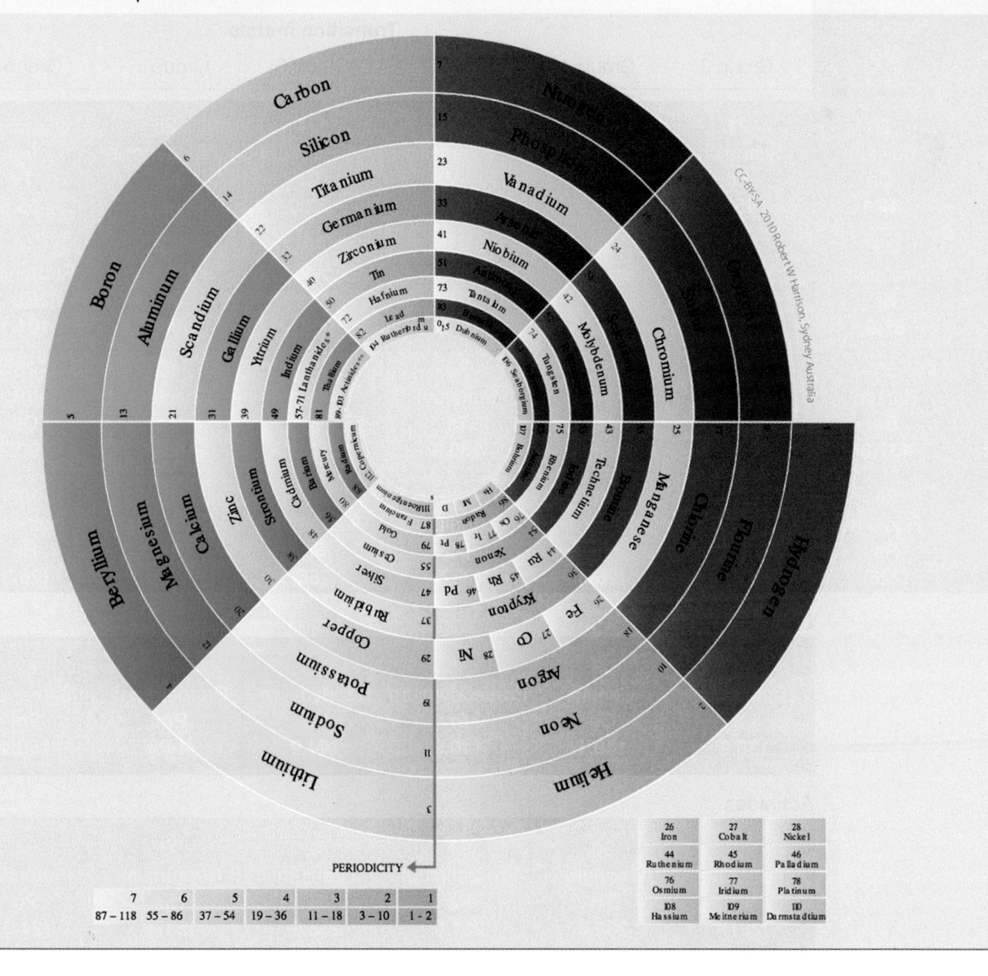

5.4.3 The modern periodic table

Referring to the large modern periodic table, note that a vertical column in the table is called a **group**. Elements in the same group in the periodic table always have some features in common. Sometimes these common features are easy to observe, but some of the similarities are not so obvious. For example, neon and argon are gases that do not change when mixed with other elements except under extreme circumstances. They are said to be **inert**. These two gases are found in the last group (group 18) of the periodic table along with the other inert gases. The group containing the inert gases is called the **noble gas** group.

group in the periodic table of elements, a single vertical column of elements with a similar nature

inert not reactive

noble gases elements in the last column of the periodic table; they are extremely inert gases

In the periodic table, 'groups' refer to the vertical columns and 'periods' refer to the horizontal periods.

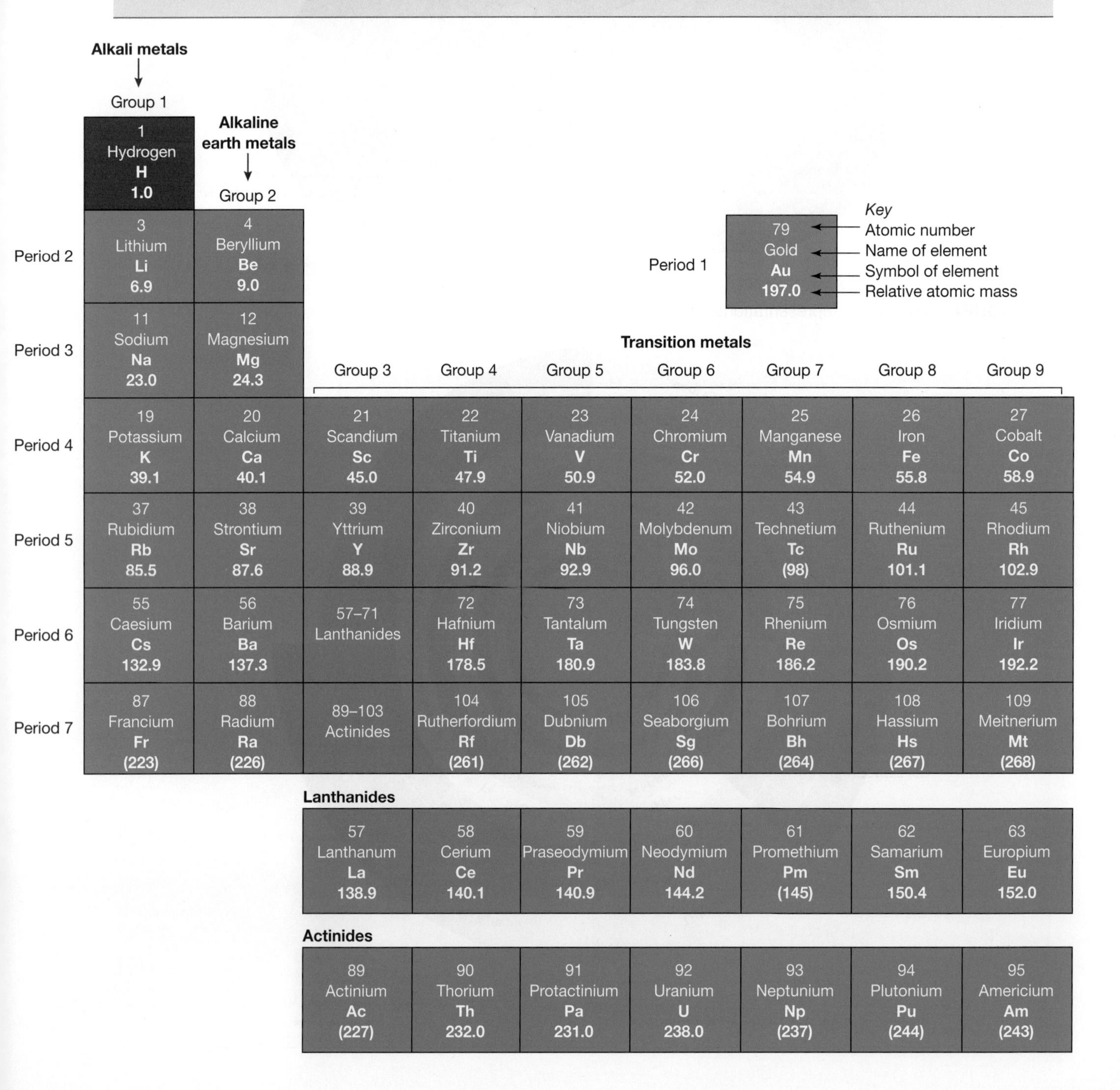

Alkali metals → Group 1; Alkaline earth metals → Group 2; Transition metals: Groups 3–9

Key (Period 1): 79 Gold Au 197.0 — Atomic number; Name of element; Symbol of element; Relative atomic mass

	Group 1	Group 2	Group 3	Group 4	Group 5	Group 6	Group 7	Group 8	Group 9
	1 Hydrogen **H** 1.0								
Period 2	3 Lithium **Li** 6.9	4 Beryllium **Be** 9.0							
Period 3	11 Sodium **Na** 23.0	12 Magnesium **Mg** 24.3							
Period 4	19 Potassium **K** 39.1	20 Calcium **Ca** 40.1	21 Scandium **Sc** 45.0	22 Titanium **Ti** 47.9	23 Vanadium **V** 50.9	24 Chromium **Cr** 52.0	25 Manganese **Mn** 54.9	26 Iron **Fe** 55.8	27 Cobalt **Co** 58.9
Period 5	37 Rubidium **Rb** 85.5	38 Strontium **Sr** 87.6	39 Yttrium **Y** 88.9	40 Zirconium **Zr** 91.2	41 Niobium **Nb** 92.9	42 Molybdenum **Mo** 96.0	43 Technetium **Tc** (98)	44 Ruthenium **Ru** 101.1	45 Rhodium **Rh** 102.9
Period 6	55 Caesium **Cs** 132.9	56 Barium **Ba** 137.3	57–71 Lanthanides	72 Hafnium **Hf** 178.5	73 Tantalum **Ta** 180.9	74 Tungsten **W** 183.8	75 Rhenium **Re** 186.2	76 Osmium **Os** 190.2	77 Iridium **Ir** 192.2
Period 7	87 Francium **Fr** (223)	88 Radium **Ra** (226)	89–103 Actinides	104 Rutherfordium **Rf** (261)	105 Dubnium **Db** (262)	106 Seaborgium **Sg** (266)	107 Bohrium **Bh** (264)	108 Hassium **Hs** (267)	109 Meitnerium **Mt** (268)

Lanthanides

57 Lanthanum **La** 138.9	58 Cerium **Ce** 140.1	59 Praseodymium **Pr** 140.9	60 Neodymium **Nd** 144.2	61 Promethium **Pm** (145)	62 Samarium **Sm** 150.4	63 Europium **Eu** 152.0

Actinides

89 Actinium **Ac** (227)	90 Thorium **Th** 232.0	91 Protactinium **Pa** 231.0	92 Uranium **U** 238.0	93 Neptunium **Np** (237)	94 Plutonium **Pu** (244)	95 Americium **Am** (243)

The elements in the **periodic table** are arranged in order of increasing atomic number, beginning with hydrogen. An atom of hydrogen has just one proton in its nucleus. Metals are found on the left and in the centre of the periodic table, and non-metals can be found in the top right-hand side.

periodic table a table listing all known elements; the elements are grouped according to their properties and in order of the number of protons in their nucleus

DISCUSSION

1. Why is the periodic table called the periodic table?
2. Explain why women have been under-represented in the history of science.

EXTENSION ACTIVITY: Alternative periodic tables

Research to see whether you can find any other representations of the periodic table. Can you determine how they grouped the elements?

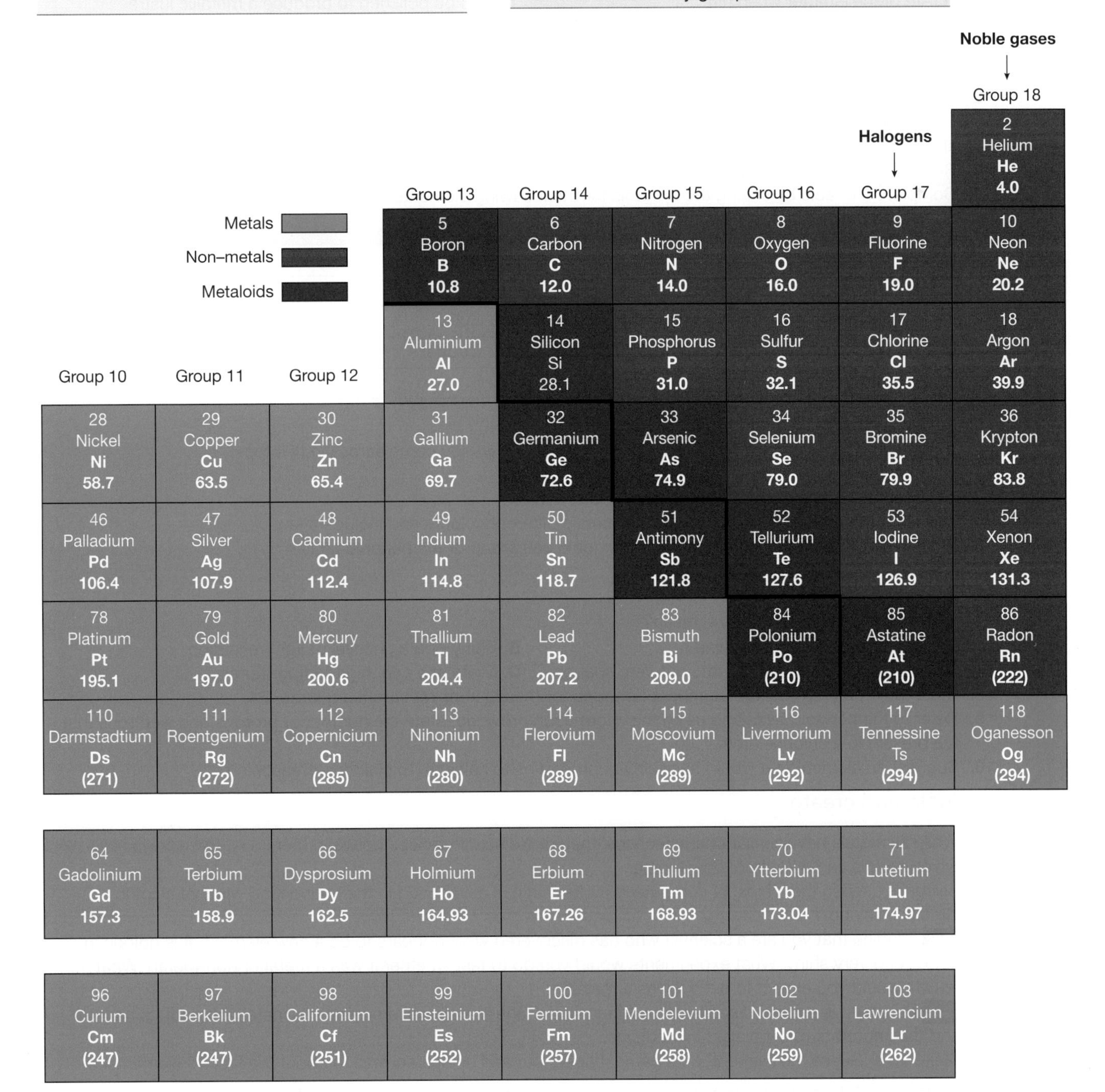

5.4 Activities

learnon

5.4 Quick quiz on | 5.4 Exercise

These questions are even better in jacPLUS!
- Receive immediate feedback
- Access sample responses
- Track results and progress

Find all this and MORE in jacPLUS

Select your pathway

LEVEL 1	LEVEL 2	LEVEL 3
1, 2, 3, 4, 8	5, 7, 9, 11, 15	6, 10, 12, 13, 14

Remember and understand

1. Identify whether each of the following properties are consistent with metals or non-metals.
   a. Are brittle (shatter when hit)
   b. Can be polished to produce a metallic lustre
   c. Can be made into a wire (ductile)
   d. Melt at low temperatures
   e. Are poor conductors of electricity and heat
   f. Can be beaten into a shape (malleable)
   g. Do not have a metallic lustre
   h. Are good conductors of electricity and heat
2. Fill in the blanks to complete the sentence about metalloids.
   Metalloids are ___________ that have features of both ___________ and ___________. Examples include silicon, ___________, arsenic, germanium, antimony, astatine and tellurium.
3. Recall which metal is liquid at room temperature.
4. Complete the table by stating the symbols for the following elements.

Element	Symbol	Element	Symbol
a. Hydrogen		**f.** Tin	
b. Carbon		**g.** Calcium	
c. Oxygen		**h.** Sulfur	
d. Nitrogen		**i.** Copper	
e. Iron		**j.** Krypton	

5. Review a copy of the periodic table.
   a. Write in it the symbols of the elements that you have already come across in this topic.
   b. Label the groups and periods.
   c. Label the names of groups 1, 2, 17 and 18.
   d. Colour the metals, metalloids and non-metals different colours and include a key.
6. What is similar about all of the gases in the noble gas group of the periodic table?

Apply and analyse

7. MC What does 'metallic lustre' mean?
   A. Able to be drawn out into wires
   B. Shiny or a high level of light reflection
   C. Able to be beaten or bent into different shapes
   D. Melts at a very high temperature
8. State the name, symbol and number of protons for the element with atomic number 74.
9. Refer to the periodic table and using the information provided, state the number of protons and electrons that are present in a chlorine atom.
10. Suggest an element that would have similar properties to calcium. Explain your reasoning.

Evaluate and create

11. While all metals have similar characteristics, there are also differences between them. List three ways in which metals can differ from one another.
12. Silicon is used in the 'chips' of computer circuits, but it is never used in the connecting wires of electric circuits. Why not?
13. SIS Imagine that you are a scientist who has discovered what appears to be a new element. It is golden in colour and very shiny. What experiments would you do to test whether it was a metal or non-metal? What results would you expect to get if it was a metal?
14. SIS Polonium is a metal discovered by Marie Curie. She also discovered another metal. Find out its name and the important role it played in medicine.
15. Make up a 'Guess the element' card game, finding out and using information about the first 18 elements.

Fully worked solutions and sample responses are available in your digital formats.

LESSON
5.5 Compounding the situation

LEARNING INTENTION

At the end of this lesson you will be able to describe the difference between elements, compounds and mixtures and that, unlike compounds, there cannot be a fixed chemical formula for mixtures.

5.5.1 Elements, compounds and mixtures

There are millions and millions of different substances in the world. They include the paper of a book, the ink in the print, the air in the room, the glass in the windows, the wool of your jumper, the cotton and polyester in your shirt or dress, the wood of your desk, the paint on the walls, the plastic of your pen, the hair on your head, the water in the taps and the metal of your chair legs. The list could go on and on.

All substances can be placed into one of three groups:

- Elements
- **Compounds**
- **Mixtures**.

It is not always possible to tell whether a substance is an element, compound or mixture by just looking at it; you need to have more information. Oxygen and carbon dioxide, for example, are both colourless gases, but oxygen is an element and carbon dioxide is a compound. Water and seawater look the same, but water is a compound and seawater is a mixture.

compound a substance made up of two or more different types of atoms that are chemically bonded (covalent or ionic) together

mixture a combination of substances in which each keeps its own properties (i.e. not chemically bonded)

bonded joined by a force that holds particles of matter, such as atoms, together

- Elements are made up of one type of atom.
- Compounds are made up of two or more different types of atoms that are chemically combined.

Elements

As you have just seen in lesson 5.4, elements are substances that contain only one type of atom, as it is the elements that make up the periodic table.

- Very few substances exist as pure elements (as most elements are not stable on their own — they bond with other atoms to form compounds that make them stable).
- Most substances around us are either compounds or mixtures.
- Examples of elements are hydrogen, oxygen, carbon and iron.

Compounds

Compounds are usually very different from the elements of which they are made.

- In compounds, the atoms of one element are **bonded** very tightly to the atoms of another element or elements.
- The elements that make up a compound are completely different substances from the compound itself. For example, pure salt (sodium chloride) that you might put on your French fries is a compound made up of the elements sodium (a silvery metal) and chlorine (a green, poisonous gas).

FIGURE 5.27 A compound is completely different from the elements of which it is made. Pure table salt is a compound that consists of the elements sodium and chlorine.

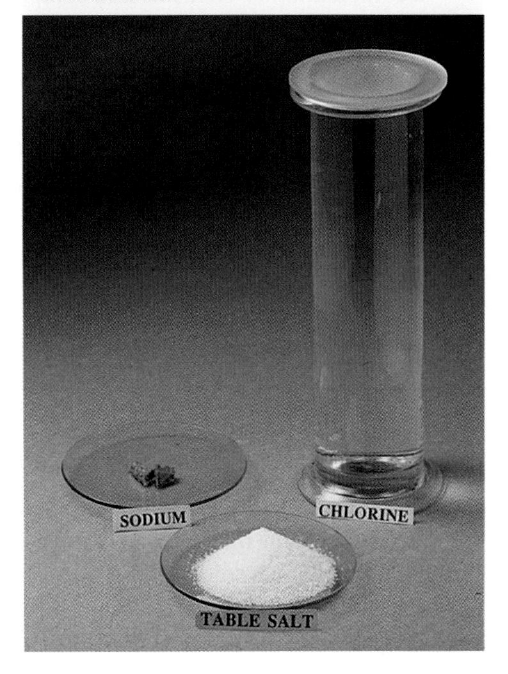

- A compound always contains the same relative amounts of each element. For example, the compound carbon dioxide is always made up of two atoms of oxygen for each atom of carbon. Its **chemical formula** is therefore CO_2. The compound sodium chloride always has one sodium atom for each chlorine atom and its formula is simple: NaCl.
- Every compound has a formula comprising the symbols of the elements that make it up. Unlike mixtures, the elements within a compound cannot easily be separated from each other. When the atoms of different elements bond together during a chemical reaction, a compound is formed. When the elements iron and sulfur are heated together, they form a new compound called iron sulfide. Iron sulfide has the formula FeS.
- Elements can be separated from compounds in several ways. These include:
  - passing electricity through a compound (see section 5.5.3)
  - burning the compound
  - mixing the compound with other chemicals.

  Each of these methods involves a chemical reaction in which completely different substances are formed.

chemical formula shows the ratio of the atoms of each element present in a molecule or compound

solute a dissolved substance in a solution

dissolved refers to when a solid substance integrates into the liquid solvent

solvent the substance in which the solute is dissolved

homogeneous has a uniform composition throughout

aqueous a solution with water as the solvent

> **WHAT DOES IT MEAN?**
>
> The word 'compound' comes from the Latin word *componere*, meaning 'to put together'.

Mixtures

A mixture is a combination of substances in which each keeps its own properties; that is, they are not chemically bonded.

Mixtures can be made up of:
- two or more elements
- two or more compounds
- a combination of elements and compounds in any states of matter (solid, liquid or gases).

FIGURE 5.28 A mixture of lollies

Consider the image of lollies shown in figure 5.28. There are more than two types of lollies pictured, all in the solid state. We can therefore say that it is a mixture of lollies. Can you determine how many different types of lollies are in the mixture?

Different types of mixtures include:
- solutions
- suspensions
- colloids.

Solutions

A solution is a liquid mixture in which the **solute** is the **dissolved** substance in the solution, while the **solvent** is the substance that the solute dissolves into. Together the solute and the solvent form a **homogeneous** solution. This means that it has uniform composition and properties throughout the whole solution. If the solvent is water, the solution is known as an **aqueous** solution. When sugar is dissolved in a cup of tea, for example, an aqueous solution is formed. Our survival is dependent on reactions that occur in aqueous solutions inside and outside of the cells in our body.

FIGURE 5.29 Dissolving sugar (solute) in a cup of tea (solvent), forming an aqueous solution

Suspensions

A **suspension** is a mixture in which solid substances do not dissolve and are dispersed throughout the volume of the liquid. This is an example of a **heterogeneous** mixture. For example, looking at the glass of muddy water in figure 5.30, you can see the solid sediment at the bottom of the glass and some of the dirt still suspended in the solution as it is slowly settling to the bottom. You can see the suspended particles with your naked eye.

FIGURE 5.30 Muddy water forms a suspension.

Colloids

A **colloid** is a mixture in which a microscopically insoluble substance is dispersed and suspended throughout another substance. The particles cannot be seen by the naked eye. A common example of a liquid colloid is milk, and in 1857 Michael Faraday made a gold colloid sample (figure 5.31).

FIGURE 5.31 A sample of gold colloid made by Michael Faraday in 1857.

Properties of mixtures

- Unlike compounds, the parts of a mixture are not always in the same proportion (because in a mixture the components are not bonded together, so you can add as much or as little of each substance). For example:
  - seawater is the most common mixture on Earth's surface, but the percentage of salt is not always the same. Seawater can also include a variety of other elements and compounds in different quantities.
  - a coffee drink is a mixture that can contain different relative amounts of water, milk, coffee and sugar
  - brass is a mixture of metals that can have different relative amounts of copper and zinc.
- As the components are not bonded to each other, there can be no unique chemical formula for mixtures; however, you can determine the percentage of each component in the mixture.
- The substances that make up mixtures can usually be easily separated from each other (except for in colloids).
- When the parts of a mixture are separated, no new substances have been formed as no chemical reaction has taken place. For example, fizzy soft drink contains water, gas, sugar and flavourings. If you shake the soft drink, the gas bubbles separate from the water and go into the air. You still have the water in the bottle and the gas in the air; they are just not mixed together anymore.The parts of the mixture can be separated relatively easily. The gas escapes when the lid of the container is opened, and the water can be separated by evaporation, leaving behind sugar and some other substances.

There can be no unique chemical formula for mixtures because, unlike compounds, the parts of a mixture are not always in the same proportion.

DISCUSSION

Substance X is a blue colour.

When it is heated a gas is given off, leaving a black solid. Is substance X an element or a compound?

suspension a mixture in which solid substances do not dissolve and are dispersed throughout the volume of the liquid

heterogeneous has a non-uniform composition throughout

colloid a mixture in which a microscopically insoluble substance is dispersed and suspended throughout another substance

EXTENSION ACTIVITY: What is air a mixture of?

Air is a mixture. Find out what substances are present, whether they are elements or compounds, and how much of each substance is present.

Table 5.3 lists some common substances and identifies whether they are an element, compound or mixture.

TABLE 5.3 Some common substances

Substance	Type	Composed of	Scientific name
Gold	Element	Gold	Gold
Diamond	Element	Carbon	Carbon
Water	Compound	Hydrogen and oxygen	Dihydrogen monoxide
Pure salt	Compound	Sodium and chlorine	Sodium chloride
Brass	Mixture	Copper and zinc	Brass
Soft drink	Mixture	Water, sugar, carbon dioxide and other compounds	
Seawater	Mixture	Water, sodium chloride and other compounds	

INVESTIGATION 5.5

elog-2261

tlvd-10738

Making a compound from its elements

Aim

To use a chemical reaction to make a compound from its elements

Materials

- 4–5 cm strip of clean, shiny magnesium ribbon. It can be coiled to fit in the crucible.
- crucible with lid
- pipeclay triangle, tongs and safety glasses
- Bunsen burner, heatproof mat and matches
- emery paper
- electronic scales

CAUTION

Do not stare at the magnesium ribbon while it is burning as it can cause temporary blindness.

A lab coat and safety glasses should be worn while conducting this experiment.

Method

1. Examine the piece of magnesium and note its appearance before placing it in the crucible and covering it with the lid.
2. Put the crucible on the pipeclay triangle as shown in the diagram.
3. Heat the crucible with a strong blue flame, monitoring the reaction by occasionally lifting the lid a little with tongs.
4. When all the magnesium ribbon has been changed, turn off the flame and leave the crucible on the tripod to cool.

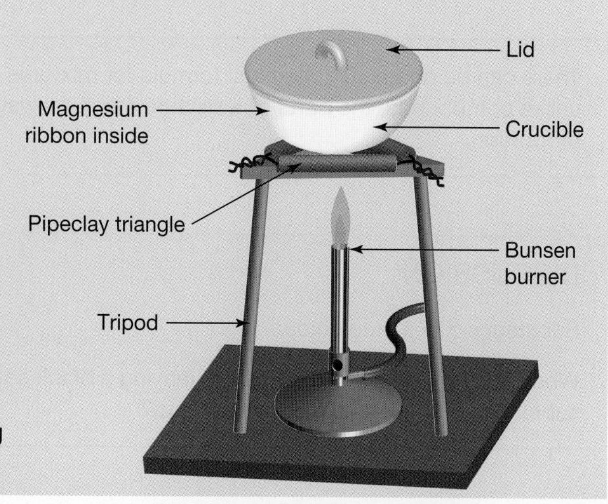

Results

Record your observations of the magnesium before heating and the substance in the crucible after heating.

Discussion

1. Is magnesium an element or a compound? Give a reason for your decision.
2. Magnesium is one of the reactants in this experiment. What is the other reactant?
3. Is the substance remaining in the crucible an element or a compound? What is its name?
4. What is the evidence that a new substance has been made?
5. Apart from observing whether the reaction is complete, give another reason for lifting the lid of the crucible a little with tongs during the burning.
6. Do you think that the mass of the contents of the crucible will be the same, higher or lower after heating? Explain your answer.

Conclusion

Describe and name the compound formed.

5.5.2 Bonding

The naturally occurring elements (of metals, non-metals and metalloids) are the building blocks of everything in our world. The atoms of various elements join together (bond) in a wide variety of ways to produce many compounds. Elements and compounds can be combined in many ways to make countless mixtures.

Atoms can join, or bond, in many different ways. In some substances, atoms are joined in groups called **molecules**. For example, in oxygen gas (O_2), oxygen atoms are joined in groups of two. In the compound carbon dioxide (CO_2), one carbon and two oxygen atoms are joined in every molecule. Atoms can join to form small or large molecules of many different shapes.

There are three main types of bonds between atoms within a compound:

- Metallic bonds
- Ionic bonds
- Covalent bonds.

Note: In Year 8, you don't need to know the difference between these bonds.

molecule two or more atoms joined (bonded) covalently together

TABLE 5.4 Models representing the molecules of some compounds

Molecule	Chemical formula	Visual representation
Carbon dioxide	CO_2	
Water	H_2O	
Methane	CH_4	

KEY: Red = oxygen, white = hydrogen, black = carbon

Some compounds are not made up of molecules. Instead, the atoms bond by lining up one after the other. This is known as a lattice. In the lattice in figure 5.32, sodium bonds to chlorine, which bonds to sodium and so on. Common table salt is an example of a substance that is bonded in this way.

FIGURE 5.32 The atoms of common table salt bond by lining up one after the other.

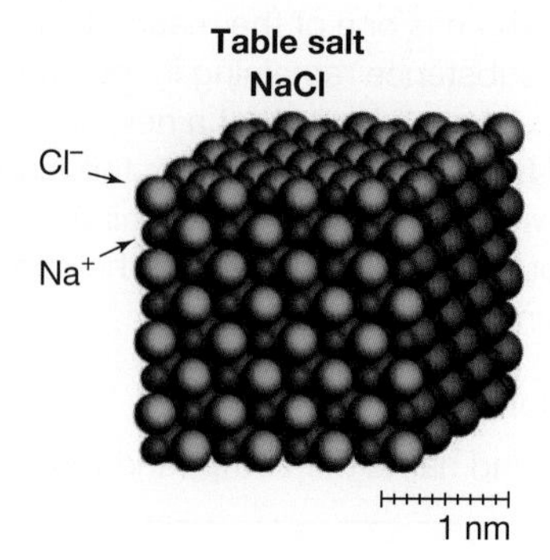

A molecule consists of two or more atoms chemically joined (bonded) together.

 Resources

 **Video eLesson** Methane (eles-2272)

elog-2263

INVESTIGATION 5.6

Modelling elements, compounds and mixtures

Aim

To make models of elements, compounds and mixtures

Materials

- LEGO® blocks that are all the same size, or a molecular kit
- Each block (or ball) represents an atom of the following elements (possible colour key but others are acceptable):
  - White = hydrogen (H)
  - Red = oxygen (O)
  - Black = carbon (C)
  - Blue = nitrogen (N)
  - Green = chlorine (Cl)
  - Yellow = helium (He)

Method

Remember:

- Elements contain only one type of atom.
- When two or more atoms are chemically joined (bonded) together they are called molecules. They can be the same or different atoms.
- Mixtures contain different elements and/or compounds but they are not chemically joined.

1. Draw a key showing the colours used for each element.
2. Draw the table shown in the results section to record your results.
3. Keep two of each model that you make.

Elements

4. A single block represents an atom; a LEGO® helium atom is drawn and coloured in the table.
5. Pick up a handful of yellow blocks; this represents the element helium (He), a noble gas.
6. Join two white blocks together (hydrogen atoms are found in pairs); this is a molecule of hydrogen.
7. Make a few more hydrogen molecules; these represent the element hydrogen (H_2).
8. Oxygen gas also contains atoms that are found in pairs. Make and draw an oxygen molecule (O_2).

Compounds

9. Take a black and a red block and join them together; this is a molecule of carbon monoxide. Draw this molecule in the table.
10. Make a few more carbon monoxide molecules. These represent the compound carbon monoxide.
11. Make a water molecule, H_2O. Draw it in the table. You can place the elements in any order until you learn more about molecules in later years.
12. Make a hydrogen chloride molecule, HCl. Draw it in the table.
13. Now make the following molecules and draw them in the table: CO_2, NO_2, NH_3.

Extension

Make the following molecules:

- Dinitrogen monoxide (laughing gas, N_2O)
- Methane (natural gas, CH_4)
- Benzene (in petrol, C_6H_6).

Mixtures

14. Collect the models of elements that you made but DO NOT join them together. This is a mixture of elements. A mixture of elements is shown in the table.
15. Collect the models of compounds that you made but DO NOT join them together. This is a mixture of compounds. Draw it in the table.
16. Collect a few of the models of any of the atoms or molecules that you have made but DO NOT join them together. This is a mixture. Draw it in the table.

Results

Elements

He (helium atom)		H_2 (hydrogen molecule)		O_2 (oxygen molecule)	

Compounds

CO (carbon monoxide)		H_2O (water)		HCl (hydrogen chloride)	
CO_2 (carbon dioxide)		NO_2		NH_3 (ammonia)	

Mixtures

Mixture of elements		Mixture of compounds		Mixture	

Discussion

1. Nitrogen gas is also made up of atoms in pairs.
   a. What would the formula for nitrogen gas be?
   b. What would the LEGO® model of a nitrogen molecule look like?
2. a. Name the molecule with the formula NO_2.
   b. How many atoms are present in this molecule?
   c. How many elements are present in this molecule?
3. Dihydrogen monoxide is the chemical name of a familiar substance. What is it?
4. Hydrogen peroxide (found in hair bleach) contains two oxygen atoms and two hydrogen atoms. What would a LEGO® molecule of hydrogen peroxide look like?
5. How would you explain elements, compounds and mixtures to a friend using fruit as examples?

Conclusion

1. What do these models show about molecules?
2. How do these models of molecules compare with those in section 5.5.2?

- Elements are made up of one type of atom and you will only use one chemical symbol to identify the element. A chemical formula will therefore include at least two different chemical symbols to show they are bonded together as a compound.
- Mixtures contain two or more different elements, compounds or ions; however, they are not chemically bonded and can exist in different amounts, altering their percentage composition.

5.5.3 Splitting water

Water is essential for life and we are surrounded by water. It is in our taps, in our bodies, in the rivers, in the sea, in the air and it comes down as rain. We wash in it, cook with it and drink it. We cannot live without water. Water is not an element; it is a molecule made up of two hydrogen atoms and one oxygen atom (H_2O), and it can be broken down into simpler substances. Figure 5.33 shows an apparatus called a Hofmann voltameter. Water is placed in the voltameter, which is connected to a battery. The electricity splits the water into the elements of which it is made: the colourless and odourless gases **hydrogen** and **oxygen**. Hydrogen and oxygen are both elements and have quite different properties from water.

FIGURE 5.33 Water is split in a Hofmann voltameter. The clear gas in the left tube is hydrogen (H_2). The gas in the right tube is oxygen (O_2). What do you notice about the amounts of hydrogen and oxygen that are produced?

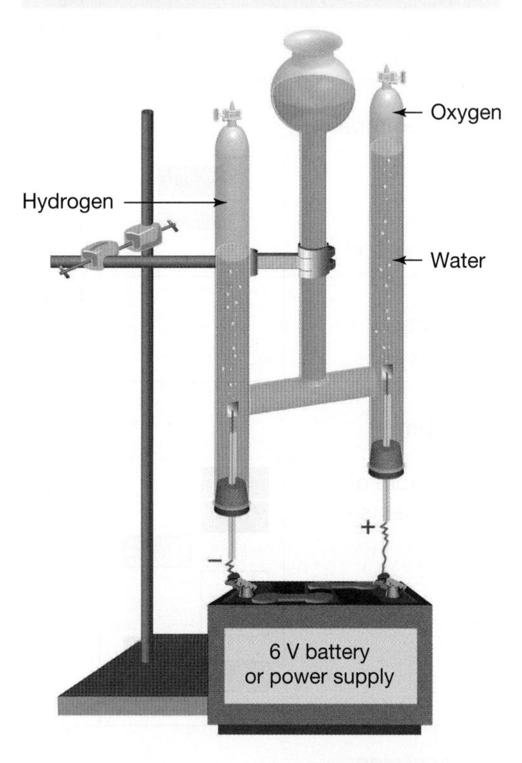

Hydrogen

- It is a much less-dense gas than oxygen. This means that a balloon filled with hydrogen will float up very high, but one filled with oxygen will not.
- It is present in almost all acids. By placing a piece of certain metals in an acid, the hydrogen is forced out. The hydrogen can be collected and tested with a flame. This is called the **pop-test**.
- When burned, it combines with oxygen in the air to form water. This releases a lot of energy. If large amounts of hydrogen and oxygen are used, enough energy can be released to lift a space rocket.
- It is a possible fuel for the future as it produces water when it burns; however, it needs a cheap form of energy to separate it from water.

Oxygen

- It is the gas that all living things need to stay alive.
- It is necessary for substances to burn — even hydrogen does not burn without it.
- It is present in water, air, rocks and even hair bleach.

Resources

Interactivity Hofmann voltameter (int-3389)

hydrogen the element with the smallest atom and the most common element in living things; by itself, it is a colourless gas and combines with other elements to form a large number of substances, including water

oxygen an atom that forms molecules (O_2) of tasteless and colourless gas; it is essential for cellular respiration for most organisms and is a product of photosynthesis

pop test a test that uses a flame to test for the presence of hydrogen; a 'pop' sound will be heard on ignition if the gas has been produced

elog-2265

INVESTIGATION 5.7

Let's collect an element

Aim

To observe a chemical reaction between a metal and an acid

Materials

- safety glasses
- 2 test tubes and a test-tube rack
- matches
- dilute hydrochloric acid
- measuring cylinder
- magnesium metal

Method

1. Measure 10 mL of hydrochloric acid and pour it into the test tube.
2. Add a piece of magnesium and place the second test tube on top of the first as shown in the diagram. Carefully observe what happens.
3. After one minute, take the second test tube off the first. While it is still inverted, immediately light the gas in the second test tube with a match.

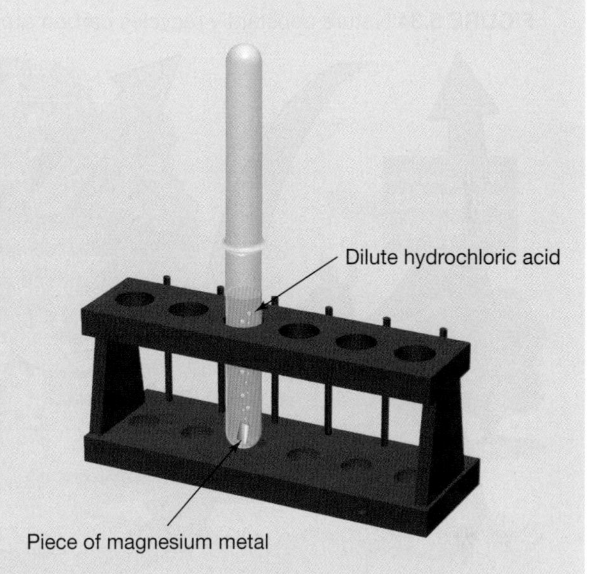

Results

1. Record your observations of what happened when you put the magnesium strip in the test tube with the acid.
2. Record your observations of what happened when you put the match below the upside-down test tube.

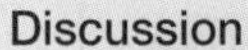

Discussion

1. Describe what happened in the test tube containing the metal and the acid.
2. What does hydrogen gas look like?
3. What happened when you lit the gas?
4. Look closely at the second test tube. Describe what you see inside it.
5. Compare the properties of the elements that reacted and the compound that was formed.
6. Do you think that all metals and acids would react the same way? Describe some steps that would help you answer this question.

Conclusion

Write a sentence describing what happens when a metal, like magnesium, is added to an acid.

Resources

eWorkbook Pure substances and mixtures (ewbk-12294)

EXTENSION: Carbon — it's everywhere

Carbon is a most amazing element. It is found naturally in three different forms. One form is diamond, another is graphite (the 'lead' in lead pencils) and the third is called amorphous carbon (coal, charcoal and soot).

Carbon is also found combined with other elements in a huge range of compounds. No other element forms as many different types of compounds as carbon. Carbon is found in everything from the skin of an elephant to paint on the walls!

The chemistry of life

All living things are made up of compounds including proteins, fats and carbohydrates. The main element in these compounds is carbon. The carbon atoms in living things will eventually become carbon atoms in the air or carbon atoms in limestone under the sea. Figure 5.34 shows how nature constantly recycles carbon atoms.

FIGURE 5.34 Nature constantly recycles carbon atoms.

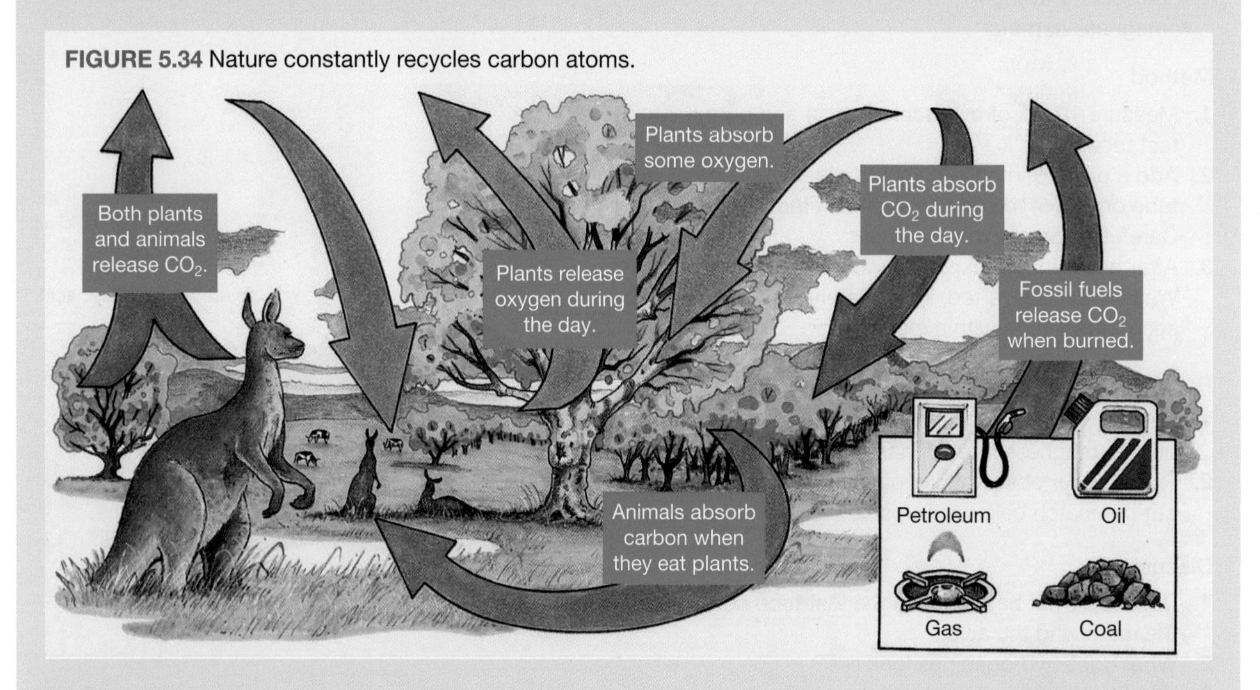

Plants take in carbon dioxide through their leaves and, in a process known as **photosynthesis**, use the carbon dioxide and water to make sugar. Sugar is a compound made up of carbon, hydrogen and oxygen atoms. Plants use the sugar to make other substances and for energy to grow. Animals eat plants and the carbon atoms then become part of the animals' bodies.

Carbon atoms in the bodies of living things return to the air in several ways: **cellular respiration**, **decomposition** and **burning**.

- Cellular respiration is a process that occurs in the cells of every living thing. It uses glucose and oxygen, and releases energy and produces carbon dioxide. The carbon dioxide that you breathe out contains carbon atoms that were once part of your body.
- Decomposition is what occurs when plant or animal material breaks down, such as in a compost heap or after something is buried. Microscopic living creatures called decomposers absorb some of the substances in the dead material and release carbon dioxide to the air by respiration.

photosynthesis a series of chemical reactions that occur within chloroplasts in which the light energy is converted into chemical energy; the process also requires carbon dioxide and water, and produces oxygen and sugars, which the plant can use as 'food'

cellular respiration a series of chemical reactions in which the chemical energy in molecules such as glucose is transferred into ATP molecules, which is a form of energy that the cells can use

decomposition the breaking up of a substance into smaller parts

burning combining a substance with oxygen in a flame

- When substances containing carbon are burned, carbon dioxide is released. Coal, natural gas and oil are all **fossil fuels** formed from living things, and contain carbon atoms. Fossil fuels undergo **combustion**, a reaction with oxygen that releases energy, which you know as burning. The burning of these fuels, as well as bushfires, releases carbon dioxide back to the air.

EXTENSION ACTIVITY: Carbon dioxide levels

Find out why the amount of carbon dioxide in the atmosphere is increasing.

Make fives slides summarising your findings.

5.5.4 Compounds of today and tomorrow

Polymer is the name given to a compound made of molecules that are long chains of atoms. Most polymers are made up of chains containing carbon atoms. **Plastics** are synthetic polymers, whereas cotton and rubber are examples of natural polymers. Although scientists first developed polymers in laboratories in the 1800s, it was not until after World War II that most modern polymers were invented. Modern polymers are used in food wrapping, paint, plastic 'glass', polystyrene foam for packaging and cups, banknotes, cases for electronic appliances such as computers and televisions, clothing, glues, shopping bags, sports equipment and even tea bags!

fossil fuel a substance, such as coal, oil or natural gas, that has formed from the remains of ancient organisms; coal, oil and natural gas are often used as fuels — that is, they are burnt in order to produce heat

combustion the process of combining with oxygen, most commonly burning with a flame

polymer a substance made by joining smaller identical units; all plastics are polymers

plastic a synthetic substance capable of being moulded

WHAT DOES IT MEAN?

The word 'polymer' comes from the Greek word *polymeres*, meaning 'of many parts'.

CASE STUDY: The elements nitrogen and gold

- Nitrogen is an element. It is a clear, colourless gas made up of molecules. Each molecule is made up of a pair of atoms. Nitrogen makes up 80 per cent of the atmosphere, which means that four-fifths of each breath you take is nitrogen. Our bodies cannot use this nitrogen so we breathe it straight out again! The gases oxygen, hydrogen and chlorine also exist as molecules made up of pairs of atoms.
- Gold is the only metal element found in large amounts in its pure form, rather than bonded in compounds with other elements.

FIGURE 5.35 An almost two-kilogram nugget of gold found in Australia

DISCUSSION

What are the advantages and disadvantages of the use of plastics in current society?

EXTENSION ACTIVITY: Nanomaterials

Why are very tiny particles called nanomaterials? Find out what they are and where they might be used.

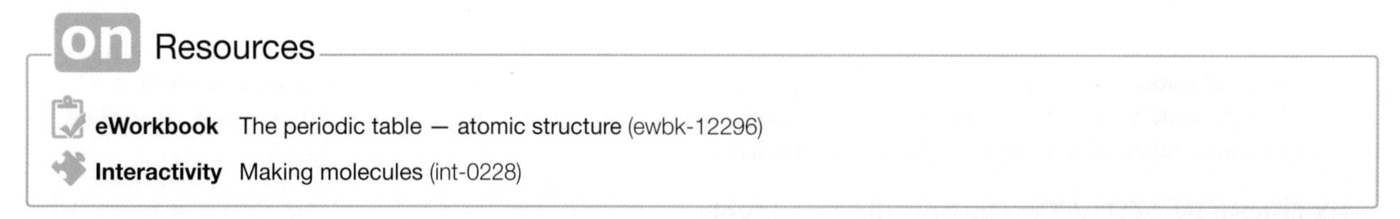

Resources

eWorkbook The periodic table — atomic structure (ewbk-12296)

Interactivity Making molecules (int-0228)

5.5 Activities

learnon

5.5 Quick quiz on | **5.5 Exercise**

Select your pathway

■ LEVEL 1	■ LEVEL 2	■ LEVEL 3
1, 2, 3, 5, 6	4, 7, 11, 12	8, 9, 10, 13, 14, 15, 16

These questions are even better in jacPLUS!

- Receive immediate feedback
- Access sample responses
- Track results and progress

Find all this and MORE in jacPLUS

Remember and understand

1. Fill in the blanks to complete the sentence.
   The elements that are ionically bonded together to form table salt are __________ and __________.
2. MC How do compounds differ from elements?
   A. Compounds contain atoms of more than one type of element.
   B. Compounds contain more than one atom.
   C. Compounds are always solid.
   D. Compounds are not reactive.
3. What are the important differences between a mixture and a compound?
4. Which of the diagrams shown represent:
   a. elements
   b. compounds
   c. mixtures?

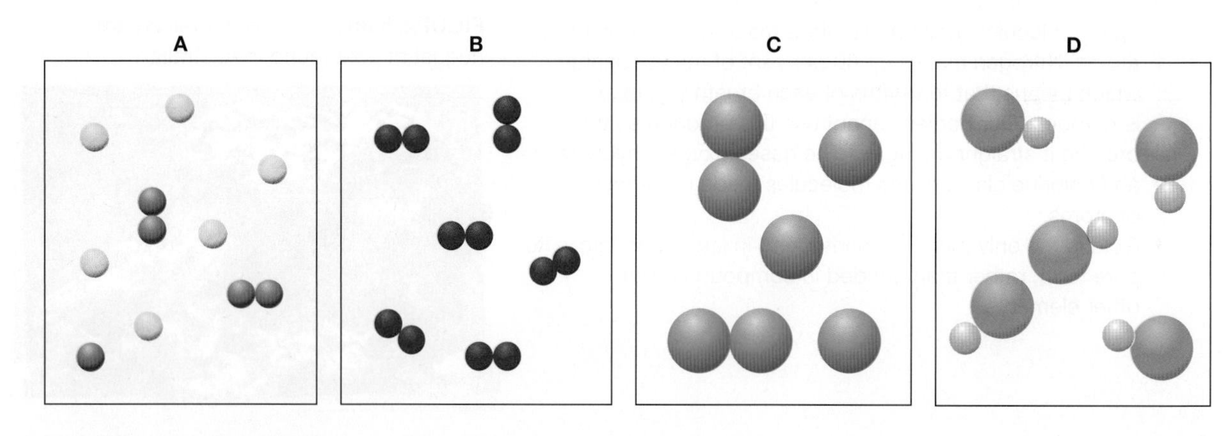

5. **a.** Define the term *molecule*.
   b. Name two elements made of molecules.
   c. Name two compounds that are made up of molecules.
6. **a.** What are polymers?
   b. What is the difference between a natural polymer and a synthetic polymer? State two examples of natural polymers and two examples of synthetic polymers.
7. Are all compounds made up of molecules? Explain.

Apply and analyse

8. SIS Fizzy soft drink is a mixture of several compounds. List three of the compounds and suggest how each of them could be separated from the mixture.
9. Fill in the blanks to complete the sentence, using the following terms:

   chemical, bonded, physical

   If atoms are __________ together, they cannot be separated by __________ methods such as sieving or filtering. They can be separated only by __________ processes.
10. List three ways in which elements can be separated from their compounds.
11. Complete the table provided. Use the formula of each compound to work out how many elements are present and which ones they are. (The formula of a compound not only tells you which elements are present, but also indicates the ratio of atoms of the different elements. For example, in the compound NH_3 there are three hydrogen atoms for each nitrogen atom.)

TABLE Composition of different compounds

Compound	Formula	Number of elements	Names of elements
Copper sulfate	$CuSO_4$	3	Copper, sulfur, oxygen
Zinc sulfide	ZnS		
Ammonia	NH_3		
Sulfuric acid	H_2SO_4		
Hydrochloric acid	HCl		
Table salt	NaCl		

Evaluate and create

12. How do you know that water is a compound and not simply a mixture of hydrogen and oxygen?
13. Magnesium oxide is a compound of magnesium and oxygen. How do you know that it is a completely different substance from each of the two elements it is made up of?
14. SIS Joseph Priestley was one of the first scientists to discover the element oxygen. He also discovered many compounds that are gases. Research and report on the life of Joseph Priestley.
15. What is the difference between an atom and a molecule?
16. SIS Australia has led the way in the production of polymer banknotes. Find out what you can about how these notes are made.

Fully worked solutions and sample responses are available in your digital formats.

LESSON
5.6 Thinking tools — Affinity diagrams

5.6.1 Tell me

What is an affinity diagram?

An affinity diagram is a useful thinking tool that allows you to become aware of both your and others' feelings and thoughts about issues. They can assist you to show what you understand about a particular topic. They graphically demonstrate the hierarchical structure of concepts related to a topic, and they also explain the links or relationships between the concepts and lessons.

They are sometimes called the 'JK method', named after its developer Jiro Kawakita.

FIGURE 5.36 Affinity diagram

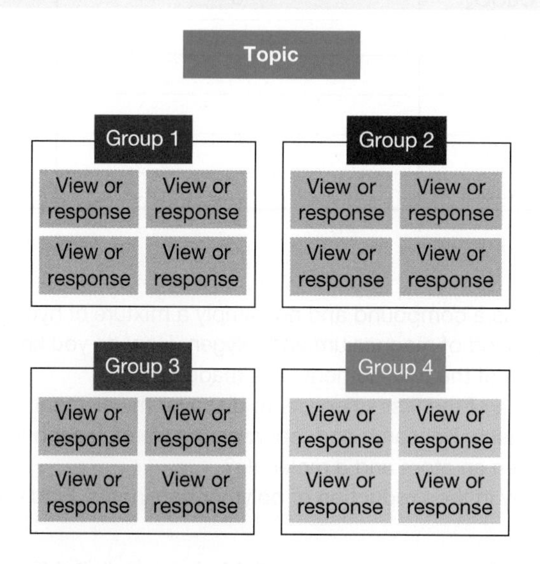

Comparing affinity diagrams to cluster maps

Both infinity diagrams and cluster maps organise ideas or features into groups. Related features radiate out of a cluster map; however, they are organised into boxes in affinity diagrams.

FIGURE 5.37 Cluster map

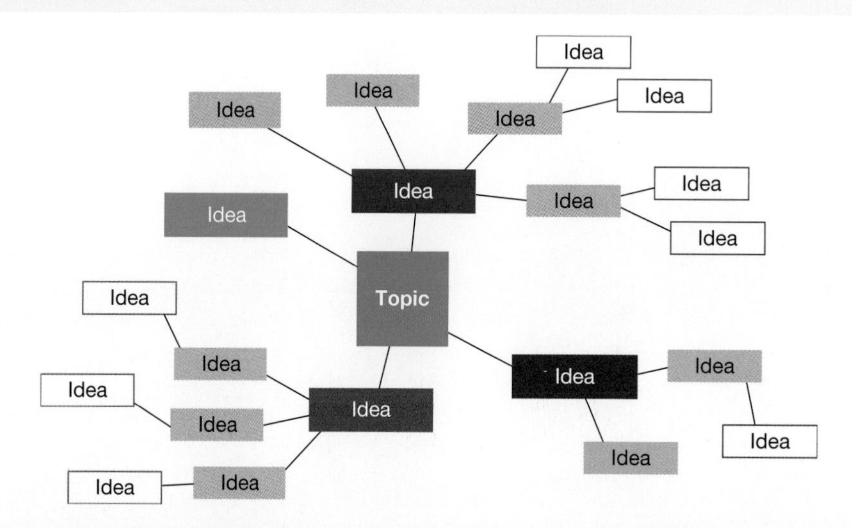

5.6.2 Show me

To create an affinity diagram:

1. Think about a topic and write any ideas you have onto small pieces of paper. You could also use software or apps to create your affinity diagram.
2. Examine your pieces of paper and put similar ideas into groups. Feel free to rearrange your groups until you are happy with them.
3. Think of names for your groups.
4. Now you are ready to draw an affinity diagram like the one shown.

The affinity diagram in figure 5.38 represents some of our knowledge about the properties and uses of metals, fuels and fabrics.

FIGURE 5.38 Affinity diagram comparing properties and uses of metals, fuels and fabrics

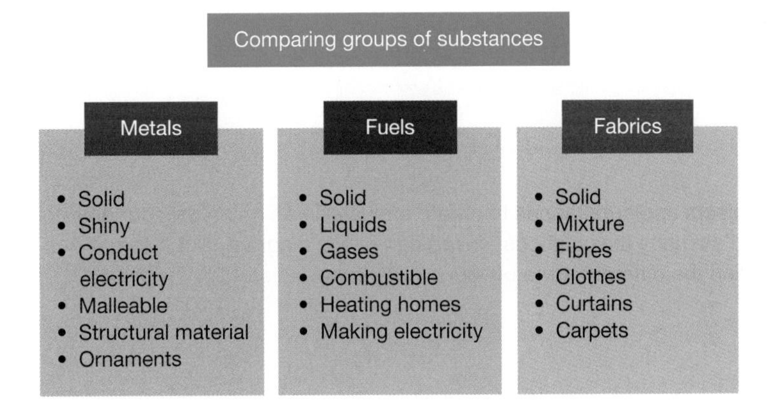

5.6.3 Let me do it

5.6 Activity

These questions are even better in jacPLUS!

- Receive immediate feedback
- Access sample responses
- Track results and progress

Find all this and MORE in jacPLUS

1. Consider each of the ideas, objects or substances listed.

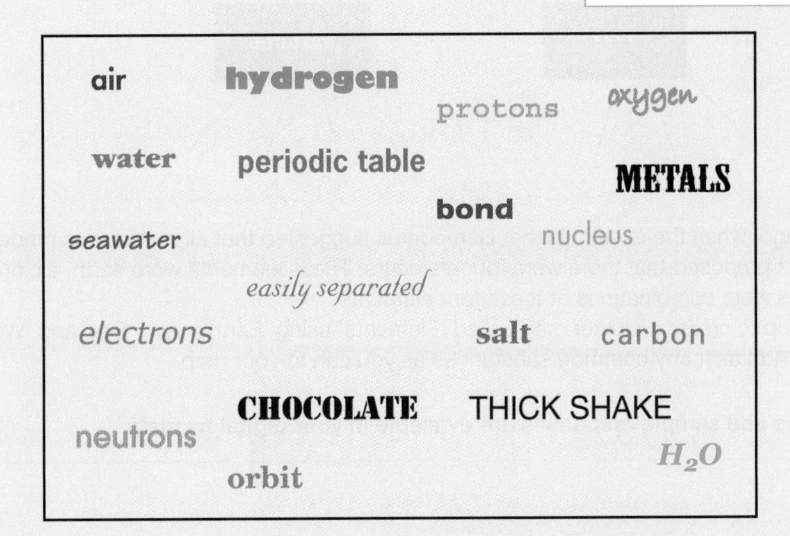

a. Arrange the ideas, objects or substances into four categories in an affinity diagram like the one provided. You will need to work out the name of the missing category.

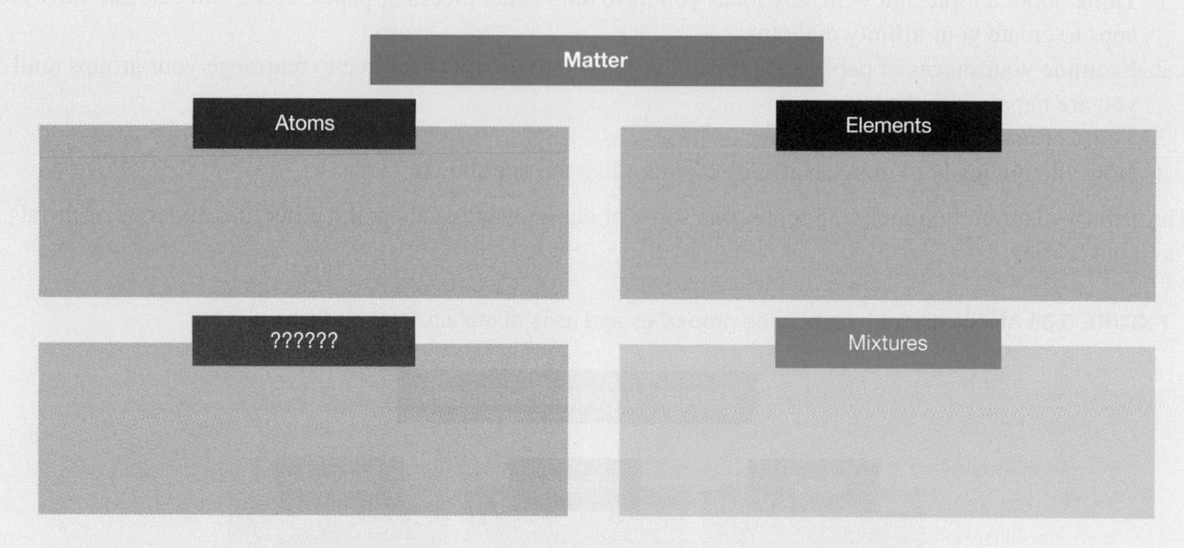

b. Use the ideas, objects and substances to create a cluster map using the four categories as the main associations. Add as many associations as you can to the diagram. Don't forget that you can sometimes make links between the different arms of your cluster map.

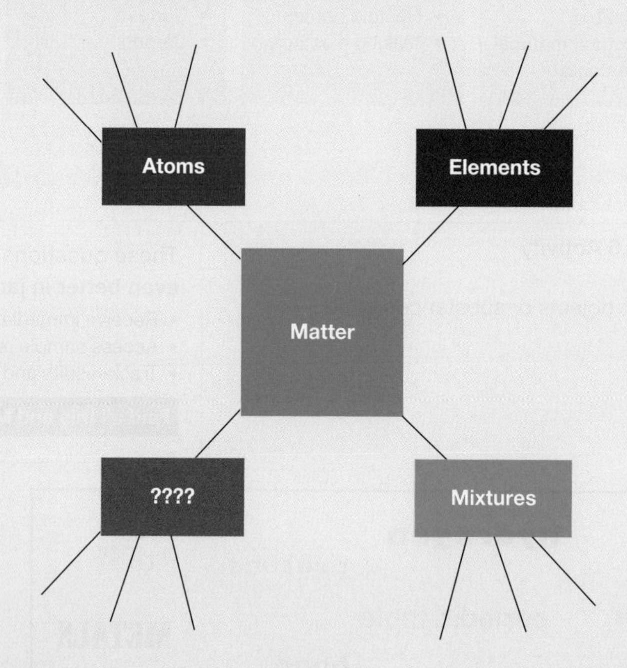

2. About 2500 years ago, when the Greek teacher Democritus suggested that all matter was made of atoms, other Greek thinkers proposed that there were four elements. These elements were earth, air, fire and water. All other substances were combinations of these four elements.
Work in a small group to create a cluster map called 'Elements' using 'Earth', 'Air', 'Fire' and 'Water' as the main associations. Add as many common substances as you can to your map.

Fully worked solutions and sample responses are available in your digital formats.

LESSON
5.7 Project — *Science TV*

Scenario

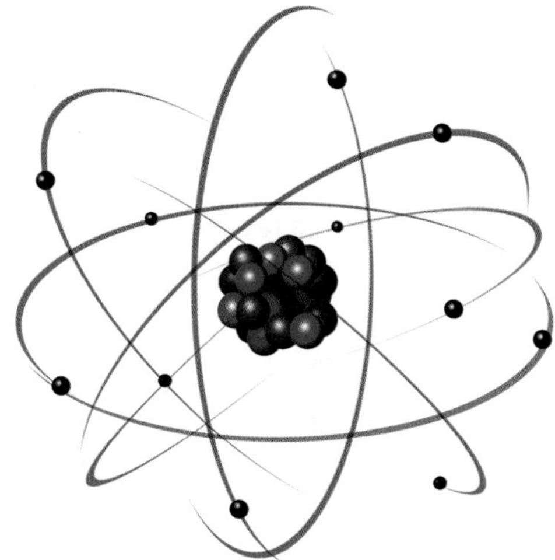

In the media world, programs that combine entertainment and education are known as 'edutainment'. With the success of edutainment programs such as *Mythbusters* (SBS/7Mate), *Scope* (Network 10) and *The ExperiMentals* (ABC), it seems that science is attracting a bigger share of the television market than many network executives would have expected. Now, your local TV network — Channel 55 — has decided to jump on the 'science as edutainment' bandwagon and has announced that next year it will develop a program called *Science TV*.

To make *Science TV* more appealing to a younger audience, the developing executives of the program want it to be presented by a team of school students, who will do all of the introductions, explanations and experiments for each of the segments. It is important that the right team of students is found or the program will be canned after only a few episodes, so Channel 55 has announced that it is accepting online audition files from groups of students who think they have what it takes to be the *Science TV* stars.

Your task

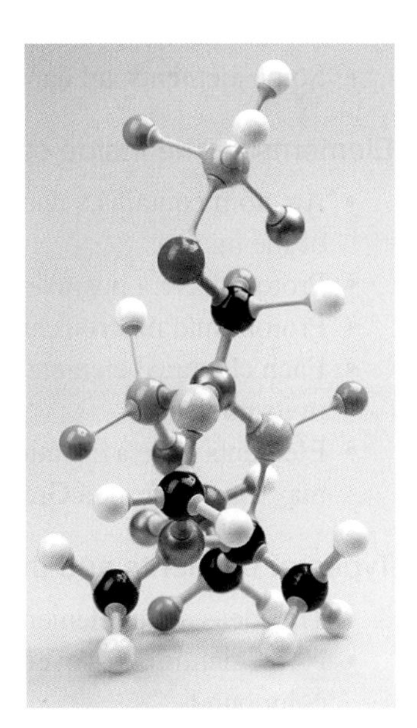

Your group is going to put together a video submission that you could send to the Channel 55 developers to showcase how suitable you would be as the stars of *Science TV*.

The guidelines for the video submission from the Channel 55 website are as follows:

- The video must be between four and five minutes in length.
- The target audience of *Science TV* is between 8 and 14 years old.
- At least two people must be shown on camera.
- The video must be in the form of a chemistry segment that explains ONE of the following:
  - **A.** The big mix up: What are elements, compounds and mixtures?
  - **B.** Setting the periodic table: Why is the periodic table important?
  - **C.** Famous molecules: Small and/or large
  - **D.** Concerning carbon: How is carbon important to our lives?
- At least one experiment must be performed in the segment — the experiment must be relevant to the segment and safe to perform (i.e. no explosions and no dangerous fumes produced).

The segment should be engaging and informative. It should have an introduction (either a scenario played out or a discussion between the presenters), an experiment to either test or demonstrate an idea, an explanation of the main concepts involved and a resolution that ties back into the original scenario or discussion. Remember, the main idea is to show that science is FUN!

Resources

ProjectsPLUS *Science TV* (pro-0090)

LESSON
5.8 Review

Hey students! Now that it's time to revise this topic, go online to:

Review your results

Watch teacher-led videos

Practise questions with immediate feedback

Find all this and MORE in jacPLUS

Access your topic review eWorkbooks

Resources

Topic review Level 1	Topic review Level 2	Topic review Level 3
ewbk-12298	ewbk-12300	ewbk-12302

5.8.1 Summary

It's elementary

- Alchemists were early scientists who tried to manipulate matter, but it wasn't until about the seventeenth century that the development of the scientific method resulted in many important discoveries.
- There are 92 naturally occurring elements.
- An element is a substance made of one type of atom.
- Properties describing elements include colour, texture, state at approximately 25 °C (room temperature), crystalline, lustre and surface.
- Some elements are dangerous (e.g. sodium is very reactive and mercury is poisonous).

Elements — the inside story

- An atom contains a nucleus, which contains protons and neutrons, and electrons move around the outside of the nucleus.
- Protons have a positive charge, neutrons have no charge and electrons are negatively charged.
- Protons and neutrons are similar in size; electrons are much smaller.
- Each chemical element is identified with a unique atomic number, which is equal to the number of protons in its nucleus.
- Elements have a chemical symbol, which may be the first letter or letters of their name (e.g. C, S, N), but may also be from a Greek or Latin name (e.g. Na, K, Pb).

Types of elements and the periodic table

- The ideas about elements and the atom have changed over time.
- New scientific discoveries and technology have impacted on our understanding of the atom, elements and compounds
- About 2500 years ago Democritus proposed the existence of atoms, but these ideas were not developed until Dalton expanded on them in 1803. Rutherford proposed the idea of the nuclear atom and Joseph Thomson experimented with electrons, which he suggested existed in shells moving around the nucleus.
- Early scientists began to find that different elements had things in common and proceeded to design various formats to organise this information. The periodic table went through many iterations before Dmitri Mendeleev's modern periodic table.

- Elements are arranged in the periodic table in order of increasing atomic number. Elements with similar properties are arranged in vertical groups and elements are arranged in horizontal periods according to the number of electron shells in their atoms.
- Most of the elements are metals and are found on the left and in the centre of the periodic table; they are separated from the non-metals by the metalloids.

Compounding the situation

- A compound is a substance made up of two or more different types of atoms that have been joined (bonded) together.
- Elements are usually found combined with other elements.
- Compounds have a huge variety of uses.
- The atoms in compounds are bonded very tightly together and can be separated from compounds only through a chemical reaction.
- The properties of compounds are different from the properties of the elements that make them up.
- In compounds there is a definite amount of each element present. In water, for example, there are twice as many hydrogen atoms as oxygen atoms. In mixtures, however, the amounts of each element are variable, so it is not possible to write a chemical formula for mixtures.
- Elements can be separated from a compound by different means, including passing electricity through the compound, burning the compound or reacting the compound with other substances.
- A mixture is a combination of substances that keep their own properties. The substances are relatively easy to separate from a mixture because of their different properties.
- Atoms can be joined (bonded) in groups called molecules; for example: water, H_2O, and carbon dioxide, CO_2. The formula shows the ratio of the atoms of each element present in the molecule.

5.8.2 Key terms

alchemist an olden-day 'chemist' who mixed chemicals and tried to change ordinary metals into gold; alchemists also tried to predict the future
aqueous a solution with water as the solvent
atomic number the number of protons in the nucleus of an atom, which identifies the element to which the atom belongs
atoms very small particles that make up all things; atoms have the same properties as the objects they make up
bonded joined by a force that holds particles of matter, such as atoms, together
brittle can easily break if hit; the opposite of malleable
burning combining a substance with oxygen in a flame
cellular respiration a series of chemical reactions in which the chemical energy in molecules such as glucose is transferred into ATP molecules, which is a form of energy that the cells can use
chemical formula shows the ratio of the atoms of each element present in a molecule or compound
chemical symbol the standard way that scientists write the names of the elements, using either a capital letter or a capital followed by a lower-case letter; for example, carbon is C and copper is Cu
colloid a mixture in which a microscopically insoluble substance is dispersed and suspended throughout another substance
combustion the process of combining with oxygen, most commonly burning with a flame
compound a substance made up of two or more different types of atoms that are chemically bonded (covalent or ionic) together
decomposition the breaking up of a substance into smaller parts
dissolved refers to when a solid substance integrates into the liquid solvent
ductile capable of being drawn into wires or threads; a property of most metals
electrons very light, negatively charged particles inside an atom; electrons orbit around the atom's nucleus
elements pure substances made up of only one type of atom
fossil fuel a substance, such as coal, oil or natural gas, that has formed from the remains of ancient organisms; coal, oil and natural gas are often used as fuels — that is, they are burnt in order to produce heat
group in the periodic table of elements, a single vertical column of elements with a similar nature
heterogeneous has a non-uniform composition throughout

homogeneous has a uniform composition throughout
hydrogen the element with the smallest atom and the most common element in living things; by itself, it is a colourless gas and combines with other elements to form a large number of substances, including water
inert not reactive
investigations activities aimed at finding information
lustre the high shine and sheen of a substance caused by the way it reflects light
malleable able to be beaten, bent or flattened into shape
metalloids elements that have the appearance of metals but not all the other properties of metals
metals elements that conduct heat and electricity; shiny solids that can be made into thin wires and sheets that bend easily; mercury is the only liquid metal at room temperature
mixture a combination of substances in which each keeps its own properties (i.e. not chemically bonded)
molecule two or more atoms joined (bonded) covalently together
neutrons tiny, but heavy, particles found in the nucleus of an atom; they have no electrical charge
noble gases elements in the last column of the periodic table; they are extremely inert gases
non-metals elements that do not conduct electricity or heat; they melt and turn into gases easily, and are brittle and often coloured
nucleus the central part of an atom, made up of protons and neutrons; plural = nuclei
observations information obtained by the use of our senses or measuring instruments
oxygen an atom that forms molecules (O_2) of tasteless and colourless gas; it is essential for cellular respiration for most organisms and is a product of photosynthesis
periodic table a table listing all known elements; the elements are grouped according to their properties and in order of the number of protons in their nucleus
photosynthesis a series of chemical reactions that occur within chloroplasts in which the light energy is converted into chemical energy; the process also requires carbon dioxide and water, and produces oxygen and sugars, which the plant can use as 'food'
plastic a synthetic substance capable of being moulded
polymer a substance made by joining smaller identical units; all plastics are polymers
pop test a test that uses a flame to test for the presence of hydrogen; a 'pop' sound will be heard on ignition if the gas has been produced
protons tiny, but heavy, positively charged particles found in the nucleus of an atom
scientists people skilled in or working in the fields of science; scientists use experiments to find out about the material world around them
solute a dissolved substance in a solution
solvent the substance in which the solute is dissolved
suspension a mixture in which solid substances do not dissolve and are dispersed throughout the volume of the liquid

Resources

eWorkbooks
Study checklist (ewbk-12304)
Literacy builder (ewbk-12305)
Crossword (ewbk-12307)
Word search (ewbk-12309)
Reflection (ewbk-12311)

Solutions Topic 5 Solutions (sol-1117)

Practical investigation eLogbook Topic 5 Practical investigation eLogbook (elog-2253)

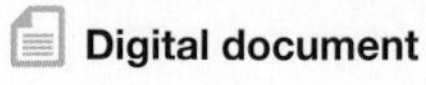

Digital document Key terms glossary (doc-40123)

5.8 Activities

5.8 Review questions

Select your pathway

■ LEVEL 1	■ LEVEL 2	■ LEVEL 3
1, 2, 3, 6, 8, 13, 18	4, 5, 9, 10, 12, 15, 19	7, 11, 14, 16, 17, 20, 21

These questions are even better in jacPLUS!

- Receive immediate feedback
- Access sample responses
- Track results and progress

Find all this and MORE in jacPLUS

Remember and understand

1. Complete the table provided, describing the structure of atoms.

TABLE Features of sub-atomic particles

Part of atom	Location	Size and mass (relative)	Electric charge
		Large	Positive
Neutron			
	Outside the nucleus		

2. Which element is shown in the diagram?

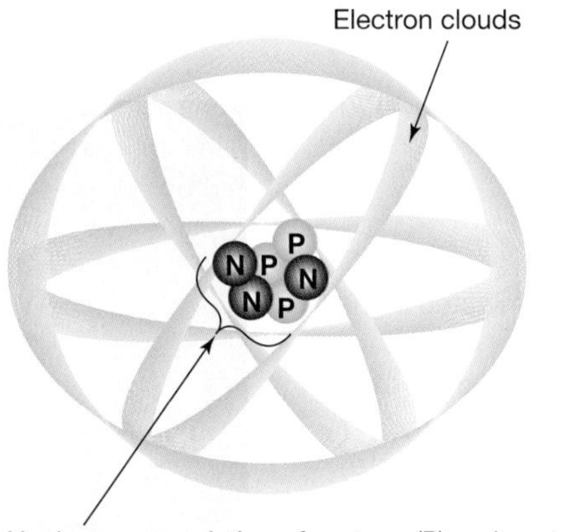

3. If a neutral atom has 12 protons, how many electrons does it have?
4. What takes up most of the space in an atom?
5. **MC** Identify the one feature that every single atom of the element sodium has in common.
   A. Eleven electrons in the electron orbits
   B. Eleven neutrons in the nucleus
   C. Eleven protons in the nucleus
   D. Eleven protons and eleven electrons
6. State the atomic number and the number of protons and electrons in each of following elements listed in the table.

	Atomic number	Protons	Electrons
Hydrogen			
Oxygen			
Carbon			
Uranium			

7. Consider the diagram given.
   a. Label an electron and the nucleus on the diagram.
   b. i. How many protons does this atom have?
      ii. How many neutrons does this atom have?
      iii. How many electrons does this atom have?
      iv. What is the atomic number of this atom?
   c. Describe one use of the element that is made up of these atoms.

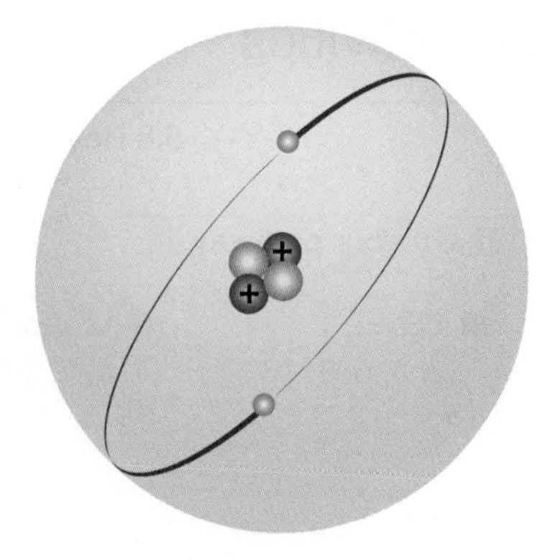

8. Construct a table like the one shown to summarise what you know about metals and non-metals.

TABLE Properties of metals and non-metals

Property	Metals	Non-metals
Conduct electricity well		
Conduct heat well		
Surface features		
State at room temperature		
Malleable		
Ductile		
Brittle		

9. Which of the elements iron, lead, hydrogen, oxygen, silicon, uranium, sodium and zinc are:
   a. metals
   b. metalloids
   c. non-metals?
10. a. Which element is used inside illuminated signs like the one shown?
    b. To which group in the periodic table does this element belong?
11. What event must take place in order to separate a compound into separate elements?

Apply and analyse

12. **MC** How are the molecules in polymers different from the molecules of other compounds?
    A. They are very small and consist of repeating sub-units or monomers.
    B. They are very large and consist of repeating sub-units or monomers.
    C. They consist of single elements.
    D. They consist of repeating elements.

13. **a.** State whether the substances listed in the table are elements, compounds or mixtures.
    b. Indicate why you made each decision in part **a**.

TABLE Classifying substances

Substance	Element, compound or mixture	Why do you think so?
Gold		
Diamond		
Carbon dioxide		
Air		
Seawater		
Pure water		
Iron		
Ammonia		
Table salt (NaCl)		

14. **MC** Why doesn't water appear in the periodic table?
    A. Water is a different kind of element.
    B. Water cannot be classified into the groups.
    C. Water is a compound.
    D. Water is naturally occurring.
15. Consider each of the comments below and identify whether the person is thinking of an atom, element, compound or molecule.

16. What do diamonds, the 'lead' in pencils and coal have in common?
17. Explain why, unlike compounds, mixtures cannot be represented by chemical formulas.
18. About 2500 years ago, Democritus suggested what all substances were made up of. In what way was Democritus' idea about substances the same as the model that scientists currently use to describe substances?
    Suggest why most thinkers of the time disagreed with Democritus.

Evaluate and create

19. Each of the diagrams shown represents one type of substance.

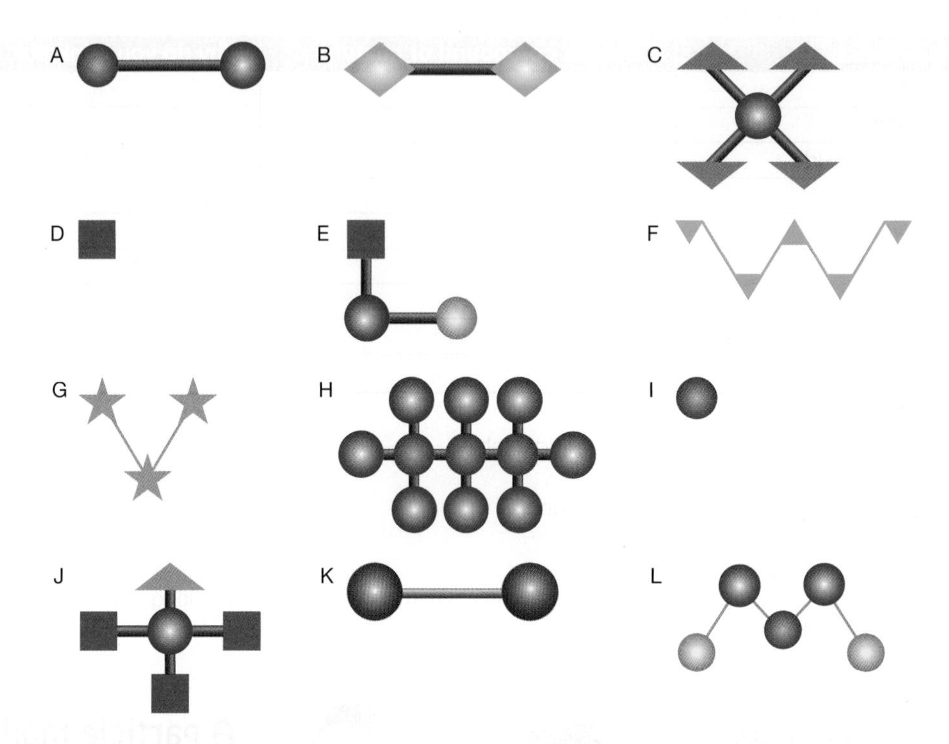

Which of the diagrams represents:

a. an atom of an element
b. a molecule of an element
c. a molecule of a compound?

20. Most of the substances around you are compounds and mixtures.

a. What differences could be observed between a mixture of hydrogen and oxygen, and a compound of hydrogen and oxygen?
b. Explain the difference between a compound and a mixture in your own words.

21. Respiration is a chemical reaction in which carbon dioxide is produced.

a. Where in your body does respiration take place?
b. What is released during respiration apart from carbon dioxide?
c. Suggest how the carbon atoms in carbon dioxide enter your body.

Fully worked solutions and sample responses are available in your digital formats.

Online Resources

Below is a full list of **rich resources** available online for this topic. These resources are designed to bring ideas to life, to promote deep and lasting learning and to support the different learning needs of each individual.

5.1 Overview

eWorkbooks
- Topic 5 eWorkbook (ewbk-12285)
- Starter activity (ewbk-12287)
- Student learning matrix (ewbk-12187)

Solutions
- Topic 5 Solutions (sol-1117)

Practical investigation eLogbooks
- Topic 5 Practical investigation eLogbook (elog-2253)
- Investigation 5.1: How big is an atom? (elog-2251)

Video eLesson
- A 3D view of the structure of DNA (eles-2030)

5.2 It's elementary

eWorkbook
- How big is an atom? (ewbk-12289)

Practical investigation eLogbook
- Investigation 5.2: Checking our appearances of elements (elog-2255)

Video eLesson
- Lavoisier and hydrogen (eles-1772)

Weblink
- Alkali metals reacting in water

5.3 Elements — the inside story

Practical investigation eLogbook
- Investigation 5.3: Getting to know atoms (elog-2257)

Video eLessons
- The hydrogen atom (eles-2269)
- An atom of carbon (eles-2031)

Interactivities
- Democritus and the atom (int-5744)
- Diagram of the atom (int-3387)

5.4 Types of elements and the periodic table

eWorkbook
- Metals and non-metals (ewbk-12292)

Practical investigation eLogbook
- Investigation 5.4: Looking for similarities (elog-2259)

Video eLessons
- Malleability (eles-2033)
- Liquid nitrogen (eles-2271)

Interactivities
- Periodic table (int-0758)
- Metals, non-metals and metalloids (int-3388)

5.5 Compounding the situation

eWorkbooks
- Pure substances and mixtures (ewbk-12294)
- The periodic table — atomic structure (ewbk-12296)

Practical investigation eLogbooks
- Investigation 5.5: Making a compound from its elements (elog-2261)
- Investigation 5.6: Modelling elements, compounds and mixtures (elog-2263)
- Investigation 5.7: Let's collect an element (elog-2265)

Teacher-led video
- Investigation 5.5: Making a compound from its elements (tlvd-10738)

Video eLesson
- Methane (eles-2272)

Interactivities
- Hofmann voltameter (int-3389)
- Making molecules (int-0228)

5.7 Project — *Science TV*

ProjectsPLUS
- *Science TV* (pro-0090)

5.8 Review

eWorkbooks
- Topic review Level 1 (ewbk-12298)
- Topic review Level 2 (ewbk-12300)
- Topic review Level 3 (ewbk-12302)
- Study checklist (ewbk-12304)
- Literacy builder (ewbk-12305)
- Crossword (ewbk-12307)
- Word search (ewbk-12309)
- Reflection (ewbk-12311)

Digital document
- Key terms glossary (doc-40123)

To access these online resources, log on to **www.jacplus.com.au**

6 Chemical change

CONTENT DESCRIPTION

Compare physical and chemical changes and identify indicators of energy change in chemical reactions (AC9S8U07)

LESSON SEQUENCE

SCIENCE INQUIRY AND INVESTIGATIONS

Science inquiry is a central component of the Science curriculum. Investigations, supported by a **Practical investigation eLogbook** and **teacher-led videos**, are included in this topic to provide opportunities to build Science inquiry skills through undertaking investigations and communicating findings.

LESSON
6.1 Overview

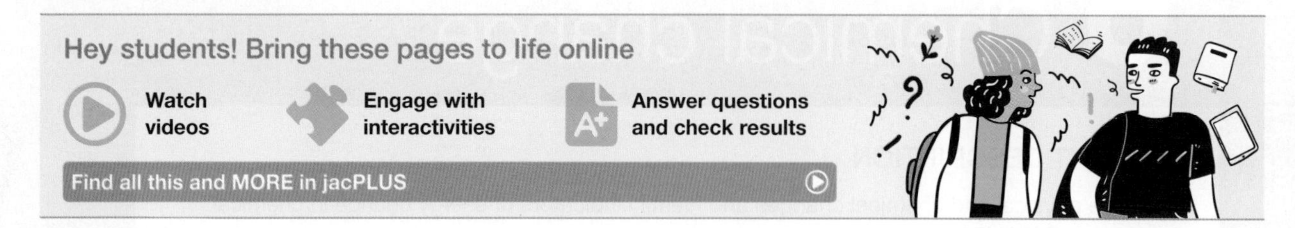

6.1.1 Introduction

Chemical reactions are happening all the time, everywhere — even in your body. Within your body, chemical reactions digest food, decay your teeth, use oxygen to convert sugar to energy, convert excess energy to fats, make enzymes and much more. Chemical reactions occur in batteries to provide electricity; in the oven when you bake a cake; in your hair when it is bleached or coloured; and in your car when it burns fuel.

In this topic, we will investigate the difference between physical and chemical changes. Consider the image on the opening page of this topic. Sucrose is being heated, which produces caramel and steam. Do you think this is a chemical change or a physical change?

FIGURE 6.1 Explosions are very fast chemical reactions. What signs of chemical change are noticeable in an explosion?

on Resources

Video eLesson A strong acid is poured into a solution containing glucose (eles-2584)

When acid is added to a glucose solution, a series of colour changes can be observed, which shows that a chemical reaction has taken place.

6.1.2 Think about chemical reactions

1. Why does a half-eaten apple go brown?
2. How is an explosion different from other chemical reactions?
3. What makes a nail rust?
4. Why is the Sydney Harbour Bridge continually being painted?
5. What is a backdraught and what causes it?
6. What makes Lycra® so special?
7. Why is recycling so important?

6.1.3 Science inquiry

What is a chemical reaction?

Chemical reactions occur all around us every day. For example, there are thousands of chemical reactions occurring in our body every second, allowing us to breathe, digest and survive.

There are many different types of chemical reactions. Some are easy to see, such as colour changes, but others are not easy to detect, such as the production of a gas.

There are many signs of chemical reactions that allow us to know that a change in chemical structure of the elements, compounds and mixtures has occurred. It is important to know the changes in a chemical reaction and how vital chemical reactions are to our lives.

What is a chemical reaction and how do you know whether a chemical reaction has taken place?

Look at figures 6.2 and 6.3. Consider if chemical reactions are taking place in each image.

1. Write down your opinion on whether or not a chemical reaction is taking place.
2. Explain how you know if a chemical reaction has taken place.

FIGURE 6.2 The boiling liquid began as a mixture of reds, yellows and blues. After stirring, it is changing into a dangerous-looking green soup.

FIGURE 6.3 Runners in long-distance races sweat heavily. Is the loss of water from skin through sweating a chemical reaction?

elog-2267

INVESTIGATION 6.1

Investigating chemical reactions

Aim

To investigate changes that occur during a chemical reaction

Materials

- Space Rocks (popping candy) or Fruit Tingles
- glow stick
- aspirin tablet
- instant ice pack
- plastic cup
- well plate
- bread
- iodine solution
- yeast
- sugar
- conical flask
- water
- balloon

Method

Complete the following tasks and note down your observations.

1. Eat some Space Rocks or Fruit Tingles.
2. Crack a glow stick.
3. Carefully drop an aspirin tablet into a clear cup of water.
4. Hit an instant ice pack.
5. Place a small piece of bread in a well plate. Put two drops of iodine solution onto the bread.
6. Combine a packet of yeast, 100 mL of warm water and 2 tablespoons of sugar in a 50 mL conical flask. Place the neck of a balloon over the top of the flask as quickly as possible.

Results

TABLE Observations of investigation 6.1

Reaction	Observations
Space Rocks in your mouth	
Cracking a glow stick	
Aspirin in water	
Hitting an instant ice pack	
Placing iodine on bread	
Combining yeast, water and sugar	

Discussion

1. Explain why each of these reactions is classed as a chemical reaction.
2. Name three signs that a chemical reaction has occurred.
3. Melting an ice cube is classed as a physical change instead of a chemical change. Explain why you think this might be the case.
4. Name three other examples of chemical reactions.

Conclusion

Write a conclusion for this investigation. Your conclusion should state what you discovered about chemical reactions from this investigation.

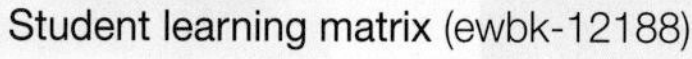

Resources

eWorkbooks	Topic 6 eWorkbook (ewbk-12312) Starter activity (ewbk-12314) Student learning matrix (ewbk-12188)
 **Solutions**	Topic 6 Solutions (sol-1118)
 **Practical investigation eLogbook**	Topic 6 Practical investigation eLogbook (elog-2269)

LESSON
6.2 Physical and chemical properties

LEARNING INTENTION

At the end of this lesson you will be able to identify the difference between the chemical and physical properties of various substances.

6.2.1 Physical and chemical properties

Thousands and thousands of different substances are used in the objects that surround you. All objects are made up of matter. Matter cannot be created nor destroyed. Matter has both physical and chemical properties. Each substance shown in figure 6.4 has physical properties that make it useful for a particular purpose.

ductile capable of being drawn into wires or threads; a property of most metals

malleable able to be beaten, bent or flattened into shape

elasticity the property that allows a material to return to its original size after being stretched

FIGURE 6.4 The objects around you are made of different substances that make those objects useful for a particular purpose. All of the properties shown in the labels for this photo are physical properties.

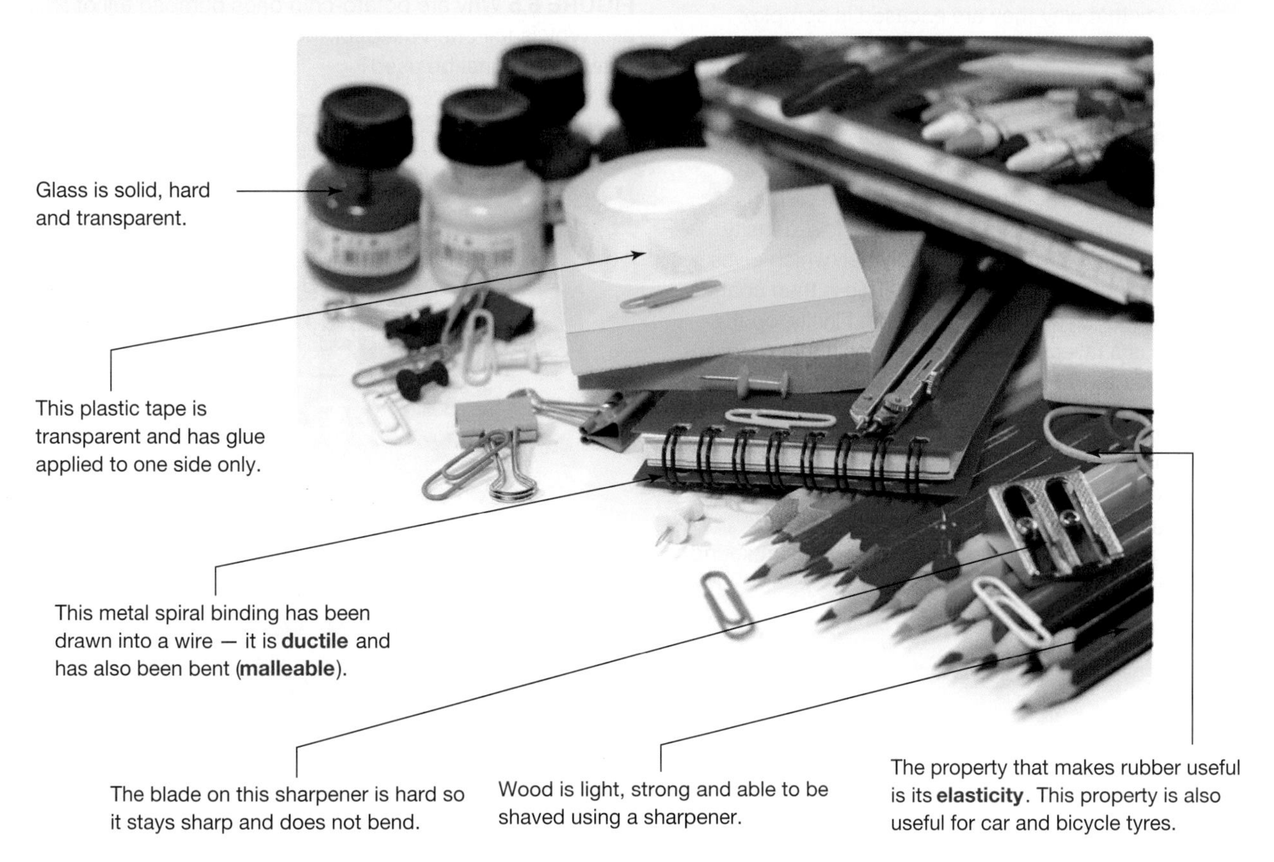

Physical properties are those that you can observe using one or more of your five senses — seeing, hearing, touching, smelling and tasting — or by measuring directly. Examples include colour, size, shape, texture, temperature, malleability, flammability and ductility, but there are many, many more.

physical properties properties that you can either observe using your five senses – seeing, hearing, touching, smelling and tasting – or measure directly

chemical properties properties that describe how a substance combines with other substances to form new chemicals, or how a substance breaks up into two or more different substances

flammability an indicator of how easily a substance catches fire

reactivity a measure of how likely a particular substance reacts to make new substances

toxicity the danger to your health caused when poisonous substances combine with chemicals in your body to produce new substances with damaging effects

Chemical properties are those that describe how a substance combines with other substances to form new chemicals, or how a substance breaks up into two or more different substances. Examples of chemical properties include **flammability**, **reactivity** and **toxicity**.

- Flammability refers to how easily a substance catches fire. When a substance burns, it creates new substances.
- Reactivity refers to how easily a substance combines with other substances to produce new substances.
- Toxicity refers to the damage caused to an organism when environmental or ingested substances combine with chemicals in its body and produce new substances that their immune system identifies as toxic. Toxic or foreign substances are always metabolised first by your body. These chemicals can have damaging, short- or long-term effects on your body.

CASE STUDY: What are potato-chip bags made from?

Potato chips are delicious, but to make sure they stay that way they are encased in complex packaging made of several layers. The packaging must be strong because it will need to be handled, but it also needs to be easy enough to open. The inner layer next to the potato chips is usually a polymer called polypropylene, which locks oil in and also keeps moisture and gases, which can spoil the chips, out. Next is a layer of low-density polyethylene (LDPE), another polymer, which gives some strength to the packaging. This is then coated with another layer of polypropylene. Finally, on the outside is a layer of thermoplastic resin that can be printed onto and coloured.

Some manufacturers are experimenting with compostable packaging that has a layer of polyactic acid polymer; but there is one drawback – these packages are much nosier to open than LDPE packaging. You would notice the difference in sound in a quiet movie theatre!

FIGURE 6.5 Why are potato-chip bags pumped full of air? Why is foil often used for the packaging? Could a different material be used?

elog-2271

INVESTIGATION 6.2

Describing properties

Aim

To describe the physical properties of a variety of substances

Materials

A range of small items that might include a tennis ball, a table-tennis ball, a table-tennis paddle, a dishwashing sponge, assorted fabrics (e.g. wool from a jumper, nylon socks and stockings, polyester, cotton), a magnifying glass or lens, a roll of sticky tape, a candle, paper clips, small springs, polystyrene cups, foam rubber, aluminium foil, a clear plastic bottle of dishwashing detergent or a bottle of perfume

Method

Work in groups of three or four so that you can discuss the properties of the objects provided. Work on one item at a time.

Results

1. For each item, list all of the physical properties that you can think of. Some items will consist of more than one substance. In those instances, list the physical properties of each substance.
2. For each physical property of each substance examined, explain how that property makes the substance useful for its purpose.

Discussion

List various tests that you could perform to discover some of the chemical properties of the substances provided.

Conclusion

Write a conclusion for this investigation. Your conclusion should state what you discovered about the physical properties of substances.

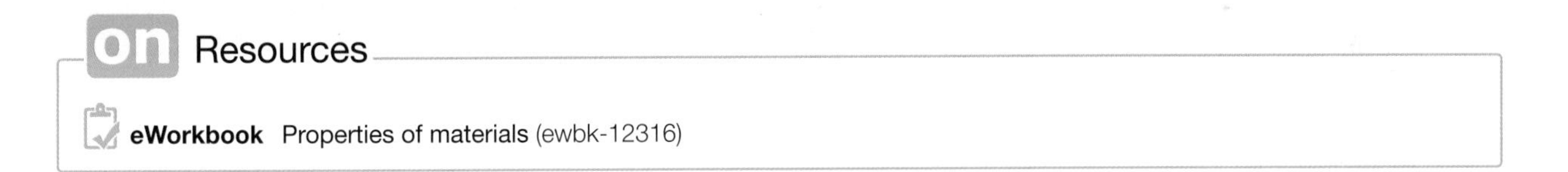

on Resources

eWorkbook Properties of materials (ewbk-12316)

6.2 Activities

6.2 Quick quiz on | **6.2 Exercise**

Select your pathway

■ LEVEL 1	■ LEVEL 2	■ LEVEL 3
1, 2, 3, 7	4, 6	5, 8, 9

These questions are even better in jacPLUS!
- Receive immediate feedback
- Access sample responses
- Track results and progress

Find all this and MORE in jacPLUS

Remember and understand

1. Recall information about physical properties to complete the following sentences.
   a. Most metals can be described as ductile and malleable. Being ductile means that a metal is able to be ______________.
   b. Being malleable means that a metal is able to be ______________.

2. **MC** Some substances have the chemical property of toxicity. This means that
   A. there is a danger to your health when the substance combines with chemicals in your body to produce new substances that have damaging effects.
   B. there is a danger to your health when the substance combines with chemicals in your body to cause fire.
   C. the substance is very reactive and can cause fire.
   D. the substance is explosive.
3. Consider potato-chip packaging.
   a. List the properties of potato-chip packaging.
   b. Compare these properties to the plastic packaging used for bags of lollies.
   c. Explain why plastic is used for lollies.
4. **MC** Identify from the following options two physical properties that you can describe using your sense of touch.
   A. Texture
   B. Density
   C. Colour
   D. Solubility
5. Flammability, reactivity and toxicity are three examples of chemical properties. Can you think of another one? Explain that chemical property using an example.

Apply and analyse

6. **SIS** Identify the properties of leather that makes a soccer ball easy to grip. Design an experiment to test which has better grip: a leather ball or a synthetic leather (PVC) ball.

7. **SIS** Complete the following table, listing the properties of the materials that make them suitable for their purpose.

Material	Malleable	Ductile	Able to be coated	Flexible	Adhesive	Able to be cut
Paperclip						
Sticky tape						
Aluminium foil						

8. Imagine that you are designing a spacecraft that will take astronauts to the Moon and back. List the properties that the outer surface of the spacecraft would need to have. Include at least two chemical properties.

Evaluate and create

9. **SIS** Road-bike frames for serious and competitive cyclists are made from aluminium or carbon fibre, or a combination of both. Find out:
   a. what properties both aluminium and carbon fibre have that make them suitable for the frames of road-racing bikes. Create a table to summarise your findings.
   b. which properties make aluminium bikes more suitable than carbon fibre bikes for some purposes. (Note that cost is not a property!)

Fully worked solutions and sample responses are available in your digital formats.

LESSON
6.3 Chemical and physical changes

LEARNING INTENTION

At the end of this lesson you will be able to identify the difference between chemical and physical changes, and how these can be described using word equations.

6.3.1 Chemical changes

When you hard-boil an egg, a **chemical change** takes place. At about 60 °C the egg white and yolk undergo chemical changes that alter their chemical make-up. Bonds between components of the egg are broken and new bonds between particles are formed. Unlike cooling melted chocolate, which brings about a physical change, cooling the egg will not return it back to its raw state. Most chemical changes are irreversible.

FIGURE 6.6 Cooling a boiled egg will not change it back to the raw state.

When paper is burnt, it combines with oxygen to form ash (carbon) and smoke (water vapour and potentially other gases, such as carbon dioxide and carbon monoxide). This is a chemical reaction, because new substances are formed. Burning gas in a Bunsen burner is also a chemical change. The methane gas reacts with oxygen in the air to form two new substances: carbon dioxide and water vapour. During this **chemical reaction**, heat is also produced.

How does a candle burn?

When you light a piece of solid wax, it melts, but does not burn. If solid wax doesn't burn, how does a candle burn? It is the wick in the middle of the candle that burns.

When you light the wick of a candle, the wax at the top of the candle melts. The molten (melted) wax is drawn up the wick just as water soaks into a paper towel. As the liquid wax flows up the wick and gets closer to the heat of the flame it **evaporates** (becomes a vapour). The wax vapour then mixes with oxygen in the air and burns.

FIGURE 6.7 The physical and chemical changes that occur when a candle burns

Wax vapour reacts with oxygen to produce carbon dioxide and water vapour

Liquid wax turns to vapour

Solid wax becomes liquid

FIGURE 6.8 a. When a glass is placed over a burning candle it initially continues to burn, using up the oxygen in the glass and producing carbon dioxide. b. A few seconds later all the oxygen is used and the candle goes out.

a.

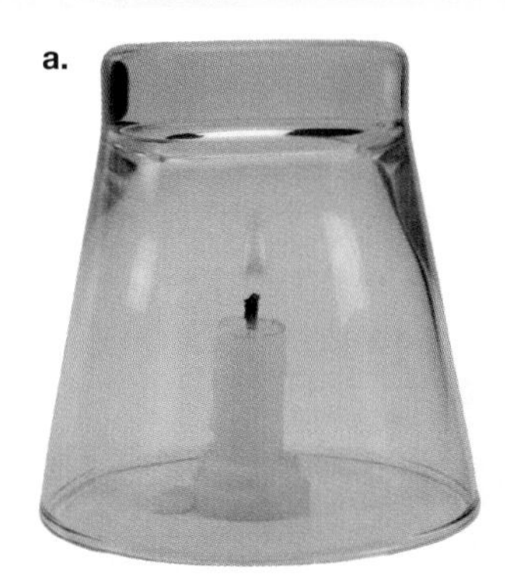

b.

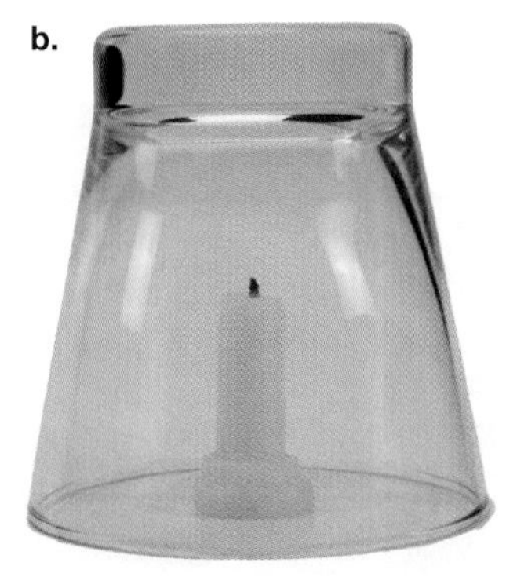

chemical change a change that results in at least one new substance being formed due to the breaking and forming of chemical bonds and rearrangement of atoms in a reaction

chemical reaction a chemical change between two or more substances in which one or more new chemical substances are produced

evaporates changes state from a liquid to a gas

elog-2273

INVESTIGATION 6.3

A burning candle

Aim

To observe and describe the changes that take place when a candle burns

Materials

- safety glasses
- candle
- jar lid
- matches
- heatproof mat

Method

1. Place a jar lid on a heatproof mat.
2. Light a candle and allow a drop of wax to drip onto the lid. Place the candle on the drop of wax and fix it to the lid.

Results

1. Observe the candle and write down as many observations of the burning candle as you can.
2. Discuss your observations with others in your group.
3. Blow out your candle and you will see a white vapour rising from the top of the wick.

CAUTION

Do not smell the vapour directly. Fan the odour to your nose with your hand.

To confirm that the white vapour is not smoke, carry out the following test:

Relight the candle. Once it is burning properly, blow it out. Quickly light the top of the vapour trail. The flame should run down the vapour to the wick and relight the candle.

Discussion

1. How far is the flame from the solid wax?
2. The solid wax forms a little pool of liquid wax around the wick. Why does this happen?
3. Describe the odour of the vapour that is present after the candle is blown out.
4. Draw a diagram of a candle and its flame. Label this diagram to explain how a candle burns.
5. Explain why lighting the wax vapour causes the candle to relight.
6. Which of the observations you have made show evidence of chemical changes?

Conclusion

Summarise the findings of the experiment in three or four sentences.

6.3.2 Physical changes

If you enjoy eating chocolate you will know that it melts on a hot summer's day. This occurs because heat energy from the hot environment is transferred to the chocolate, causing it to melt. The chocolate changes from a solid **state** to a molten state. This means that it is no longer a solid, but it is not (by scientific definition) a true liquid. This change in state is reversible. If the temperature is cooled down, the chocolate will solidify again.

FIGURE 6.9 Chocolate in **a.** solid and **b.** molten state

a.

b.

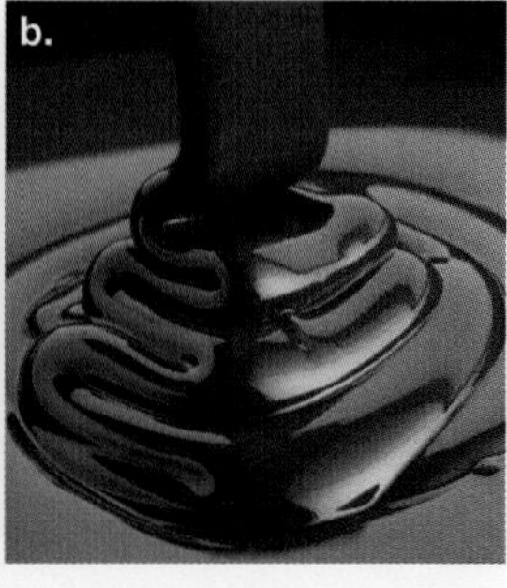

state the condition or phase of a substance; the three main states of matter are solid, liquid and gas

These changes to the chocolate are **physical changes**. Melting, evaporation, condensation and freezing are all physical changes. Changes of state are reversible, physical changes.

FIGURE 6.10 Changes of state are reversible, physical changes.

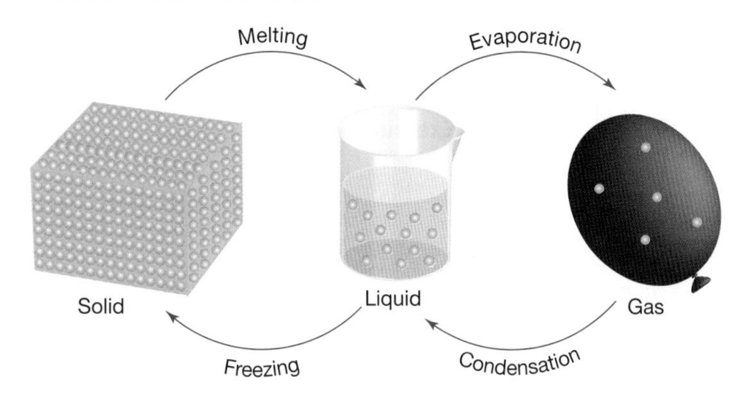

Changes in the shape or size of a substance can be physical or chemical changes. These changes are not always reversible. For example, if you drop an egg and it breaks, its shape is changed forever. But when you stretch an elastic band, it can quickly return to its original shape when you let it go.

A physical change does not break any bonds between the atoms of a substance, nor does it create any new bonds. No new substances are formed. However, in a chemical change, bonds are broken and new bonds are formed. New substances are formed and, in general, the reaction is irreversible.

FIGURE 6.11 When water is added to dried copper sulfate, it turns blue. But has a chemical reaction taken place? No reaction has occurred since the solid turns white again when dried.

6.3.3 Word equations

When a candle is burned, there are both physical and chemical changes. The melting of the solid wax forms liquid wax and evaporation of liquid wax forms wax vapour. These are physical changes. The burning of the wax vapour is a chemical change. The wax vapour reacts with oxygen in the air to form new substances, including carbon dioxide and water.

Using word equations to describe changes

Physical and chemical changes can be described using word equations.

Physical change: Melting chocolate can be described by the equation:

Solid chocolate → liquid chocolate

This equation is reversible by decreasing the temperature:

Liquid chocolate → solid chocolate

Chemical change: The burning of paper can be described by the equation:

Paper + oxygen → carbon dioxide and water vapour

This equation cannot be reversed.

physical changes changes in which no new chemical substances are formed; a physical change may be a change in shape, size or state, and many of these changes are easy to reverse

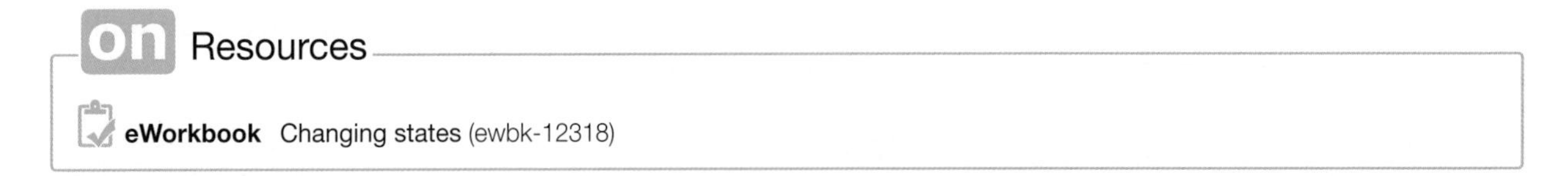

6.3 Activities

learnon

6.3 Quick quiz on | **6.3 Exercise**

Select your pathway

■ LEVEL 1	■ LEVEL 2	■ LEVEL 3
1, 2, 5	3, 4, 6, 9	7, 8, 10

Remember and understand

1. **MC** Identify the statement that describes the difference between a physical and a chemical change.
   A. During a chemical change, no bonds are broken.
   B. During a physical change, bonds are broken.
   C. During a chemical change, bonds are broken.
   D. During a physical change, the reactants do not change.
2. Describe two examples of a physical change.
3. Describe two examples of a chemical change.
4. Match the change of state with the physical change.

Change of state	Physical change
a. Change from solid to liquid	**A.** Freezing
b. Change from gas to liquid	**B.** Melting
c. Change from liquid to solid	**C.** Condensation
d. Change from liquid to gas	**D.** Evaporation

5. **MC** Identify which type of physical change can always be reversed by heating or cooling.
   A. Change of the size of particles
   B. Change of state
   C. Change of colour
   D. Change of shape
6. Complete the following word equations to describe the changes of state that take place when a candle burns.
   a. Solid wax ⟶ ____________________ **b.** Liquid wax ⟶ ____________________
7. **MC** When you hard-boil an egg, the inside of the egg gets hard. Identify the most appropriate statement that describes the change that has occurred.
   A. It is a physical change as heat has been applied to the egg.
   B. It is a physical change as the egg has changed from mostly liquid to a solid.
   C. It is a chemical change as the reaction cannot be reversed.
   D. It is a chemical change as the egg has changed colour.

Apply and analyse

8. **SIS** Consider the observations in the following table. Complete the table by classifying whether the observation is a physical or a chemical change.

Observation	Physical change	Chemical change
Water freezing to form snow		
A cake baking		
Lighting the gas on the stove		
Petrol evaporating at the petrol pump		
Lighting a match		
Steam condensing on the bathroom mirror		
Melting gold to cast gold bars		
Dynamite exploding		
Bleaching a stain		
Dissolving eggshell in ethanoic acid		

9. **SIS** Sort the following list so that the first step is at the top to describe the chemical change that takes place when methane burns in a Bunsen burner.
   - Carbon dioxide and water vapour produced
   - Methane enters the burner
   - Methane mixes with air
   - Methane burns in oxygen

Evaluate and create

10. **SIS** Create a labelled scientific diagram of a burning piece of wood, showing both the chemical and physical changes occurring.

Fully worked solutions and sample responses are available in your digital formats.

LESSON
6.4 Chemical reactions

LEARNING INTENTION

At the end of this lesson you will be able to recognise that chemical reactions begin with reactants, which are changed into products. You will be able to write word equations for chemical reactions.

6.4.1 Chemical reaction

A chemical reaction is a chemical change in which a completely new substance or substances are produced.

Almost all the products you use or wear each day are made by chemical reactions. Examples include cosmetics, concrete, plastics, paper, glass, graphite, stainless steel, shampoo, fibres, food additives, margarine, medicines and many, many more.

6.4.2 Food and chemical reactions

A cheese-and-lettuce sandwich is an incredible mixture of chemicals. Every part of it has been produced by chemical reactions. The most important chemical reaction in growing the lettuce is photosynthesis, in which the reactants are carbon dioxide and water. The products are glucose (a type of sugar) and oxygen. This chemical reaction cannot take place without sunlight and a chemical called **chlorophyll**, which gives plants their green colour. In fact, none of the other components of the sandwich could be grown or produced without photosynthesis.

FIGURE 6.12 Photosynthesis is the process by which plants convert light energy into chemical energy.

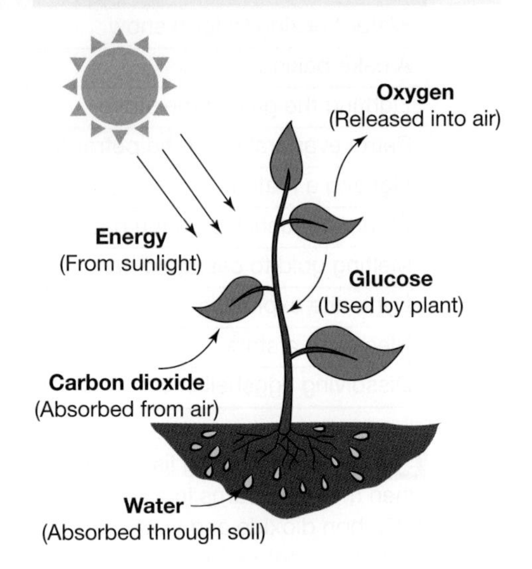

The substance used to make cheese is the product of a chemical reaction in which a protein in cow's milk called casein reacts with ethanoic acid when heated. Ethanoic acid is found in orange and lemon juice and is more commonly known as acetic acid or vinegar.

6.4.3 Reactants and products

The substances that you begin with in a chemical reaction are called the **reactants**; the substances that are produced are called the **products**. When you wash the dishes, a chemical reaction occurs between the detergent and the mess on the dishes. When you shampoo your hair, some of the chemicals in the shampoo react with the greasy substances on your scalp that contain dust, dirt and tiny organisms such as bacteria that can make your hair unhealthy.

FIGURE 6.13 In a chemical reaction, reactants are substances that change into products.

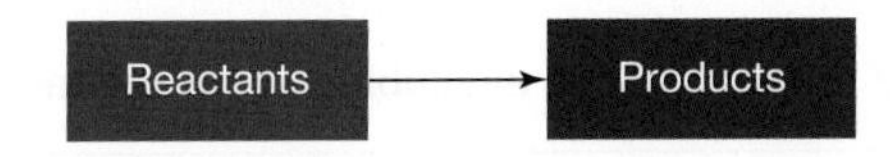

WHAT DOES IT MEAN?

The word 'product' comes from the Latin word *productum*, meaning 'thing produced'.

Where's the evidence?

You can usually tell whether a chemical reaction has taken place by identifying one or more of these clues:

- A **precipitate** is produced. A precipitate is a solid substance, which may or may not be of the same colour as the original mixture. Sometimes this is seen as sediment at the bottom of the reaction flask, or the reaction flask may look cloudy, which is indicative of tiny solid particles having formed.
- An odour is detected.
- Bubbles appear.
- There is an increase or decrease in temperature.
- Light is emitted or a flame appears.
- There is a change in colour.
- A reactant 'disappears'.
- You cannot reverse the reaction and get back to the reactants.

chlorophyll the green-coloured chemical in plants, located in chloroplasts, that absorbs light energy so that it can be used in the process of photosynthesis

reactants chemical substances used in a chemical reaction; chemical bonds of the reactants are broken during a chemical reaction

products new chemical substances that result from a chemical reaction; new chemical bonds are formed to make the products during a chemical reaction

precipitate the new, solid product produced when reactants are mixed together; a precipitate is insoluble in water

FIGURE 6.14 Ways to identify that a chemical reaction has occurred

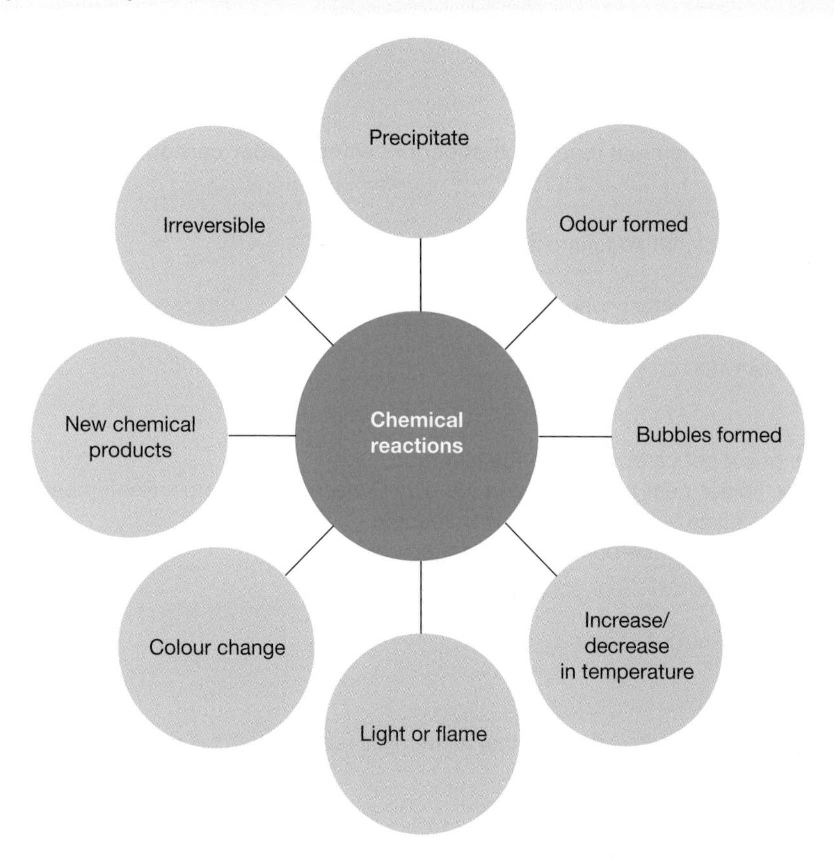

However, the only way to be certain that a chemical reaction has taken place is to identify if any new substances (products) have been formed.

6.4.4 Chemical reaction experiments

Before you start each of the following three investigations, design a suitable table for recording your observations.

As you perform the experiments:

1. make a note of the appearance of each of the reactants you start with
2. carry out the experiment and observe carefully to detect any changes that occur
3. describe the changes that take place and the products of the reaction.

FIGURE 6.15 Safety glasses should always be worn during experiments involving chemical reactions.

Resources

Video eLessons Precipitation (eles-2058)
Magnesium metal burning (eles-2303)
Baking and carbon dioxide (eles-2059)

elog-2275

INVESTIGATION 6.4

Heating copper carbonate

Aim

To observe and record the chemical reaction that occurs when copper carbonate is heated

Materials

- Bunsen burner, heatproof mat and matches
- safety glasses
- test tube, test-tube rack and test-tube holder
- spatula
- copper carbonate powder

Method

1. Add two spatulas of copper carbonate into the test tube.
2. Using the test-tube holder, heat the test tube in the blue Bunsen burner flame. Remember to move the test tube in and out of the flame and point it away from people.
3. Stop heating when the copper carbonate has changed colour.

Results

Record your observations.

Discussion

Describe which observation provides evidence that a chemical reaction has taken place. Explain your reasoning.

Conclusion

Summarise the findings of the experiment in three or four sentences.

Extension

Light a match over the mouth of the test tube once you have stopped heating the copper carbonate. Observe and record what happens to the flame. A gas has been produced in this reaction; suggest what this gas could be and explain your reasoning. Write a chemical equation for the reaction.

elog-2277

INVESTIGATION 6.5

Magnesium metal in hydrochloric acid

Aim

To observe and describe the chemical reaction between magnesium metal and hydrochloric acid

Materials

- heatproof mat
- safety glasses
- test tube and test-tube rack
- 1 cm piece of magnesium ribbon
- dropping bottle of 0.5 M hydrochloric acid

Method

1. Put the magnesium into the test tube. Place the test tube in the test-tube rack.
2. Add 20 drops of hydrochloric acid to the test tube.

CAUTION

The test tube may become quite hot.

Results

Record your observations.

Discussion

Describe which observation provides evidence that a chemical reaction has taken place. Explain your reasoning.

Conclusion

Summarise the findings of the experiment in three or four sentences.

Extension

Light a match over the mouth of the test tube after the magnesium and hydrochloric acid have reacted. Observe and record what happens to the flame. A gas has been produced in this reaction; suggest what this gas could be and explain your reasoning. Write a chemical equation for the reaction.

INVESTIGATION 6.6

Steel wool in copper sulfate solution

Aim

To observe and record the chemical reaction between steel wool and copper sulfate

Materials

- heatproof mat
- safety glasses
- test tube and test-tube rack
- glass stirring rod
- 1 cm ball of steel wool
- dropping bottle of 0.5 M copper sulfate solution

Method

1. Put the steel wool in the test tube. Using the glass stirring rod, push it gently to the bottom of the test tube.
2. Add copper sulfate solution to the test tube to a depth of 2 cm.

Results

Record your observations.

Discussion

Describe which observation provides evidence that a chemical reaction has taken place. Explain your reasoning.

Conclusion

Summarise the findings of the experiment in three or four sentences.

6.4.5 Writing word equations

Each of the chemical reactions in investigations 6.4–6.6 can be described by a chemical word equation. In each case the reactants are on the left-hand side of the equation and the products are on the right-hand side.

1. When magnesium metal reacts with hydrochloric acid, hydrogen gas and magnesium chloride are formed:

$$\text{Magnesium} + \text{hydrochloric acid} \longrightarrow \text{hydrogen} + \text{magnesium chloride}$$

eles-2294

FIGURE 6.16 The chemical reaction when magnesium is placed into hydrochloric acid

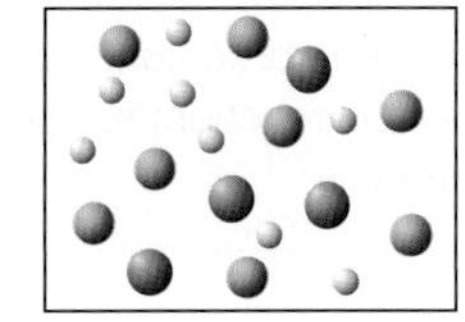

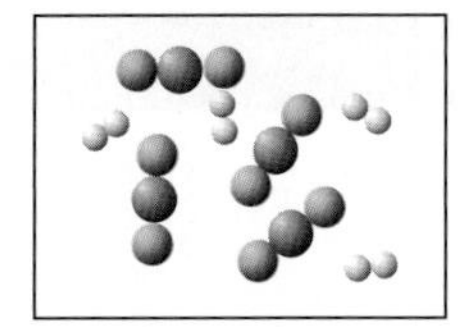

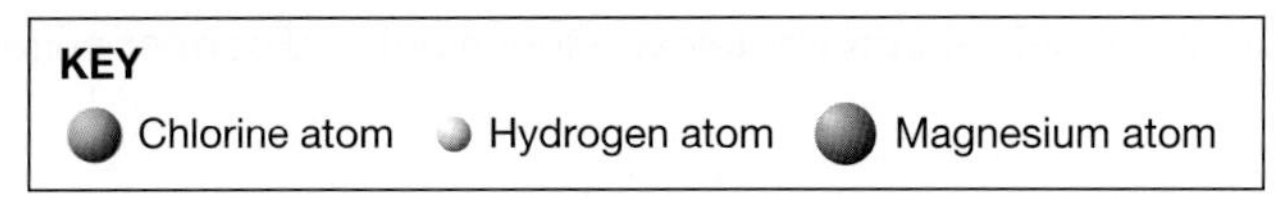

2. Heating copper carbonate forms copper oxide and carbon dioxide:

$$\text{Copper carbonate} \xrightarrow{\text{heat}} \text{copper oxide} + \text{carbon dioxide}$$

   Although heat is required for this chemical reaction to take place, it is not a substance and therefore is not a reactant. It is written above the arrow for this reason.

3. Sodium sulfate and barium chloride in solution react to form solid barium sulfate and sodium chloride, which remains dissolved in the solution:

$$\text{Sodium sulfate solution} + \text{barium chloride} \longrightarrow \text{solid barium sulfate} + \text{sodium chloride solution}$$

4. Steel wool (which is made of iron) dissolves in copper sulfate solution to form iron sulfate solution and copper metal:

$$\text{Iron} + \text{copper sulfate solution} \longrightarrow \text{iron sulfate solution} + \text{copper}$$

   When writing word equations, it does not matter in which order the reactants are written and the same is true for writing the products. The word equation in the steel wool example above could also be written as:

$$\text{Copper sulfate solution} + \text{iron} \longrightarrow \text{copper} + \text{iron sulfate solution}$$

Resources

eWorkbooks Physical and chemical changes (ewbk-12320)
Describing chemical changes (ewbk-12322)

6.4 Activities

learnon

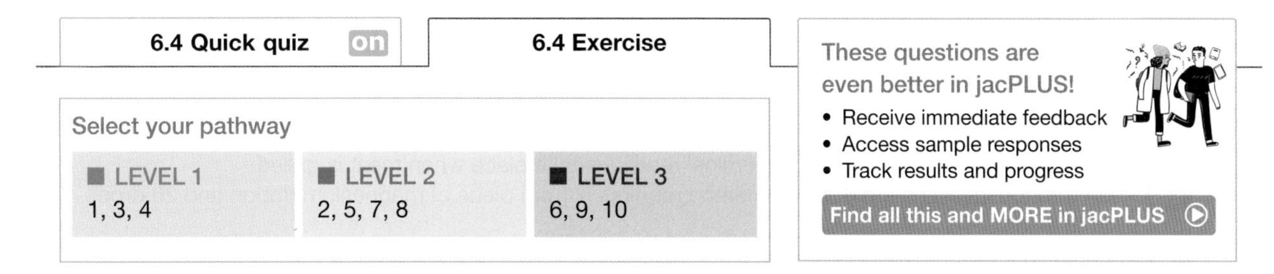

Remember and understand

1. Identify which of the following observations provide evidence that a chemical reaction has taken place.

Observation	Is there evidence of a chemical reaction?
a. Change in colour	
b. Heat or light produced	
c. Melting of a substance	
d. Gas produced when boiling a kettle	
e. Stretching a substance	
f. Formation of a precipitate	
g. Gas given off	

2. When magnesium metal reacts with hydrochloric acid, hydrogen gas and magnesium chloride are formed.
   a. Identify the reactants in this reaction.
   b. Identify the products of this reaction.
3. **MC** Identify which of the following is the provided 'real' proof that a chemical reaction has taken place.
   A. Production of a precipitate
   B. Formation of new products
   C. Gas formation
   D. Change in colour
4. Word equations can be very useful to represent chemical reactions.
   a. Fill in the blanks to represent the following reaction as a word equation:
   Octane gas is burnt with oxygen in a car engine to produce carbon dioxide and water.

   ____________ + oxygen gas ⟶ ____________ + water vapour

   b. **MC** Identify which word equation correctly describes the following reaction.
   Sodium metal reacts with chlorine gas to form sodium chloride.
   A. Salt + chlorine gas ⟶ sodium chloride
   B. Sodium gas + chlorine metal ⟶ sodium chloride
   C. Sodium metal + chlorine gas ⟶ sodium chloride
   D. Sodium chloride ⟶ chlorine gas + sodium metal
   c. True or false? The reaction in which hydrogen gas and oxygen gas combine to form water can be described by the following word equation:

   Hydrogen gas + oxygen gas ⟶ water

   d. True or false? The reaction in which zinc metal dissolves in hydrochloric acid to form hydrogen gas and zinc chloride can be represented by the following word equation:

   Zinc metal + hydrogen gas ⟶ hydrochloric acid solution + zinc chloride solution

5. True or false? The reaction that takes place when copper carbonate is heated is called a precipitation reaction.
6. Explain why the tomato, cheese, bread and meat in a hamburger cannot be grown or produced without photosynthesis.

Apply and analyse

7. Describe the evidence that one or more chemical reactions take place when meat is grilled.
8. **SIS** Read through investigation 6.5. The method requires a 1 cm piece of magnesium ribbon and 20 drops of hydrochloric acid.
   a. How could you tell if the reaction had been completed?
   b. Explain how you would know if all the hydrochloric acid had been used up in the reaction.
9. **SIS** Draw a diagram (or create a model using an atom modelling kit) to show sodium metal reacting with hydrochloric acid. Remember to add a key to your diagram.

Evaluate and create

10. **SIS** Performing some chemical reactions can be dangerous. Design a safety poster for one of the experiments you have done. Be sure to list all the safety precautions. Your teacher can provide you with the relevant risk assessments for the experiment you have chosen.

Fully worked solutions and sample responses are available in your digital formats.

LESSON
6.5 Chemical reactions and energy

LEARNING INTENTION

At the end of this lesson you will understand the difference between endothermic and exothermic reactions.

6.5.1 Starting a chemical reaction

Simply placing two chemicals together does not always mean they will react. For example, hydrogen and oxygen react violently, yet a mixture of these two gases can be stored indefinitely if kept cool in a secure container. Energy must be supplied to start the reaction. Sometimes only a small amount of energy is needed to start (or initiate) the reaction. Heat transferred from the surroundings may be enough.

Energy may also be supplied by an electric current, a beam of light, a match or a Bunsen burner flame. This energy is needed to begin the process of breaking the bonds in the reactants, which allows the atoms to rearrange and form new bonds in the products.

Energy must be supplied to start the reaction between hydrogen and oxygen. This is shown by the word 'heat' over the reaction arrow.

$$\text{Hydrogen} + \text{oxygen} \xrightarrow{\text{heat}} \text{water}$$

$$2H_2 + O_2 \xrightarrow{\text{heat}} 2H_2O$$

6.5.2 Energy and chemical reactions

When fuels such as petrol are burned in motor vehicles, energy is released and used to keep the vehicle in motion. Burning is a chemical reaction in which fuel reacts with oxygen, producing carbon dioxide, water and several other products.

The energy released comes from the rearrangement of atoms. There is less energy stored in the chemical bonds in the products than there was in the reactants. Chemical reactions that release energy are called **exothermic** reactions (figure 6.18a).

Chemical reactions in which energy is absorbed from the surroundings are called **endothermic** reactions. There is more energy 'stored' in the chemical bonds of the products than there was in the reactants (figure 6.18b).

FIGURE 6.17 The energy that powers this space shuttle comes from an exothermic chemical reaction.

FIGURE 6.18 a. Exothermic reactions release energy. **b.** Endothermic reactions absorb energy.

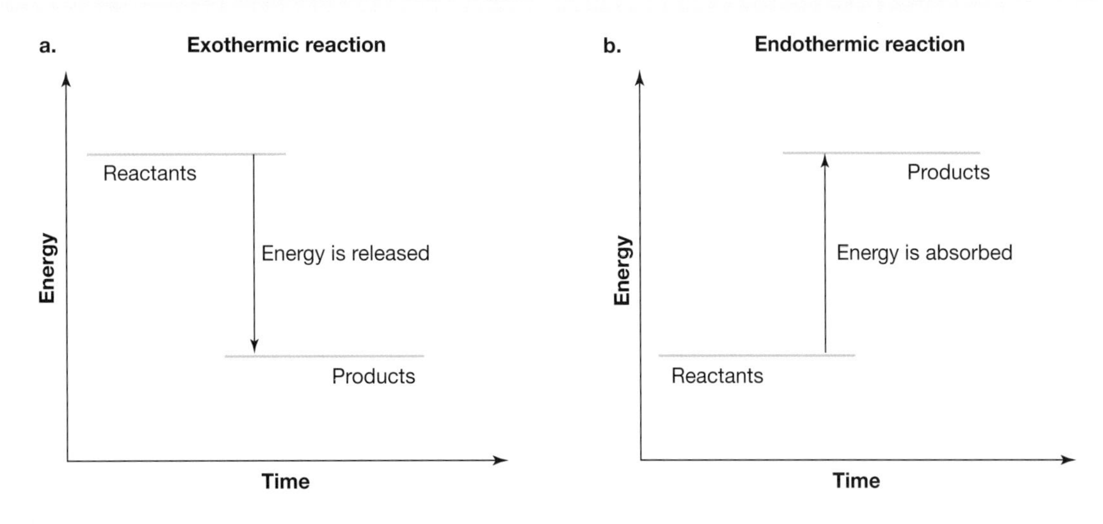

Whether energy is absorbed or released during a chemical reaction can be observed by comparing the temperature of the substance before the reaction with the temperature after the reaction.

exothermic refers to chemical reactions that give out heat energy to the surroundings

endothermic refers to chemical reactions that absorb heat energy from the surroundings

FIGURE 6.19 Endothermic and exothermic reactions

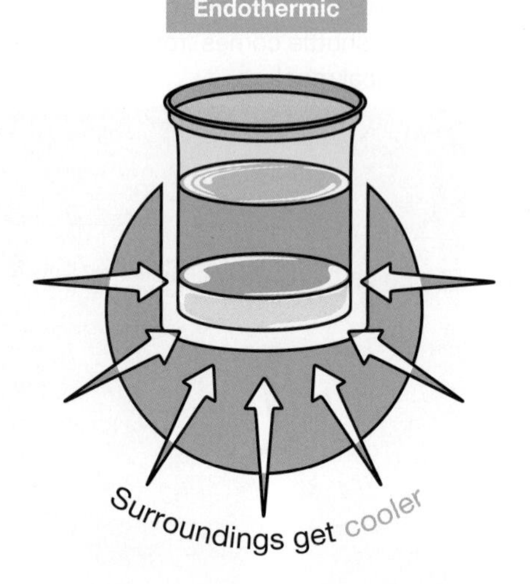

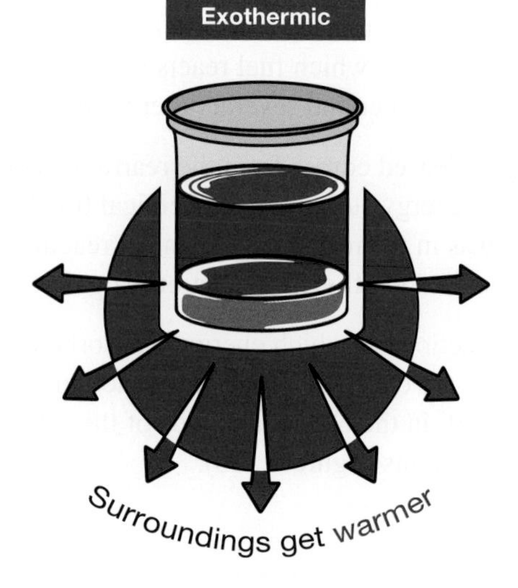

Melting/evaporation

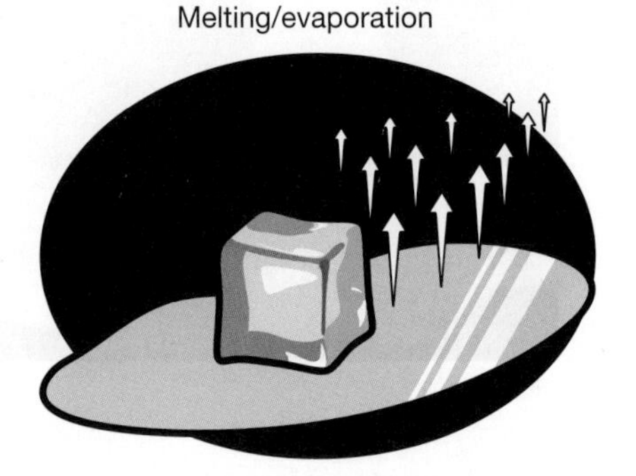

Burning/combustion

WHAT DOES IT MEAN?

The words 'exothermic' and 'endothermic' come from the Greek words *exo*, meaning 'out', *endo*, meaning 'in', and *therme*, meaning 'heat'.

6.5.3 Exothermic reactions

Portable hand warmers, commonly used by skiers and campers, become hot when shaken due to an exothermic chemical reaction in which energy is released to the surroundings. One type of hand warmer contains iron, water, salt and sawdust. When the contents of the packet are shaken quickly, the powdered iron reacts with oxygen to form iron oxide. During this chemical reaction, some of the **chemical energy** of the substances is transformed into heat energy that is transferred to the hands, increasing their temperature. We can show this chemical reaction with a word equation.

chemical energy energy stored in chemical bonds that is released during chemical reactions

Exothermic reactions produce heat.
For example:

Iron + oxygen ⟶ iron oxide + heat

6.5.4 Endothermic reactions and processes

Endothermic reactions and processes absorb energy. Athletes use instant ice packs to treat injuries. The ice pack may consist of a plastic bag containing ammonium nitrate or ammonium chloride powder and an inner bag of water. Squeezing the bag breaks the weaker inner bag and immediately causes the powder to dissolve in the water. The reaction that takes place absorbs energy from the injured area, thus lowering its temperature.

FIGURE 6.20 Endothermic reactions can be used to treat injuries.

Photosynthesis is the most widely known endothermic reaction. Plants take in energy from the Sun, as well as carbon dioxide from the air and water from the soil, and produce glucose (a sugar) and oxygen.

We can describe this chemical process with a word equation.

Endothermic reactions absorb energy.
For example:

Carbon dioxide + water + heat ⟶ glucose + oxygen

elog-2281
tlvd-10789

INVESTIGATION 6.7

Exothermic and endothermic processes

Note: Part A is a teacher demonstration only.

Aim

To investigate some exothermic and endothermic reactions

Materials

- safety glasses
- bench mat
- 4 large test tubes and a test-tube rack
- 10 mL measuring cylinder
- balance
- thermometer (–10 °C to 110 °C)
- stirring rod
- magnesium ribbon
- sandpaper
- 0.5 M hydrochloric acid
- lithium chloride
- sodium thiosulfate
- potassium chloride
- spatula

Method

Part A: Magnesium in hydrochloric acid

1. Pour 10 mL of 0.5 M hydrochloric acid into a test tube in a test-tube rack. Place a thermometer in the test tube and allow it to come to a constant temperature. Record the temperature of the solution.
2. Clean a 10 cm piece of magnesium ribbon using the sandpaper until it is shiny on both sides. Coil the magnesium ribbon and place it into the test tube of hydrochloric acid.
3. Observe the temperature of the solution as the magnesium reacts with the hydrochloric acid and record the final temperature of this solution.

Part B: Lithium chloride in water

4. Pour 10 mL of water into a test tube in a test-tube rack. Place a thermometer in the water in the test tube and allow it to come to a constant temperature. Record the temperature of the water.
5. Use a balance to weigh 2 g of lithium chloride, add it to the water in the test tube and stir gently.
6. Observe the temperature of the solution as the lithium chloride dissolves in the water and record the final temperature of this solution.

Part C: Sodium thiosulfate in water

7. Using a new test tube, repeat part B using 2 g of sodium thiosulfate instead of lithium chloride.

A data logger can be used to record the temperature changes in this investigation.

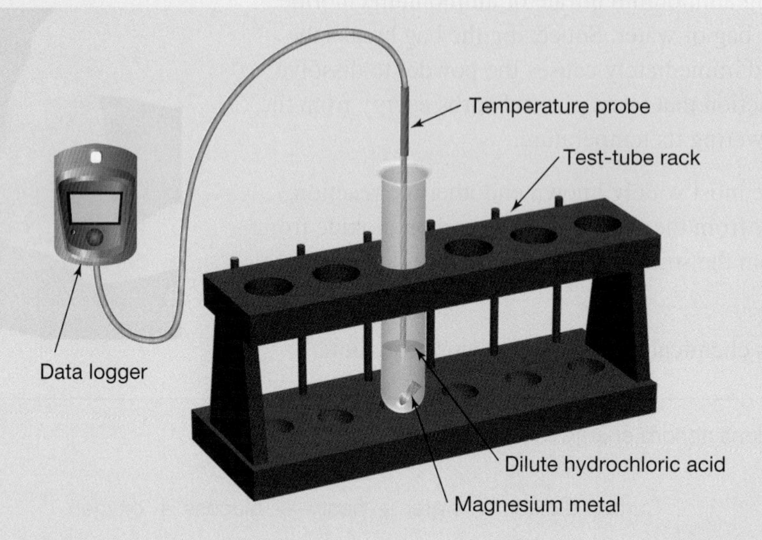

Part D: Potassium chloride in water

8. Using a new test tube, repeat part B using 2 g of potassium chloride instead of lithium chloride.

Results

Construct a table like the one provided to record the temperature changes as the four chemical processes described take place. Complete the table by calculating the change in temperature resulting from each process. Use + or – signs to indicate whether the temperature decreased or increased.

Chemical process	Initial temperature (°C)	Final temperature (°C)	Change in temperature (°C)
Part A			
Part B			
Part C			
Part D			

Discussion

1. Which reactions were exothermic and which were endothermic?
2. Which one or more of the chemical processes above was a chemical reaction? How do you know?

Conclusion

Summarise the findings of the experiment in three or four sentences.

elog-2283

INVESTIGATION 6.8

Instant ice pack

Aim

To investigate energy changes in a chemical reaction within a real-life context

Materials

- safety goggles, lab coat and gloves
- 100 g of ammonium nitrate (NH_4NO_3)
- scales
- 200 mL of tap water
- ziplock sandwich bag
- ziplock snack bag
- measuring cylinder

Method

1. Weigh out 100 g of ammonium nitrate in the ziplock snack bag.
2. Using the measuring cylinder, measure out 200 mL of water.
3. Pour the water into the ziplock sandwich bag.
4. Quickly and carefully, pour the ammonium nitrate into the water and seal the bag. (Try to remove the excess air before sealing the bag.)
5. Gently squeeze the bag to mix the contents together.

Results

Record your observations.

Discussion

1. Describe which observation provides evidence that a chemical reaction has taken place. Explain your reasoning.
2. Identify what type of reaction has occurred.
3. Give some examples of real-life situations where an instant ice pack is useful.

Conclusion

Summarise the findings of the experiment in three or four sentences.

SCIENCE AS A HUMAN ENDEAVOUR: Airbags

Airbags have saved many people from death or serious injury in car accidents. When an airbag inflates, it creates a cushion between the occupant's body and the windscreen, dashboard and other parts of the inside of the car. Airbags, which are made from nylon, may be concealed in the steering wheel, dashboard, doors or seats.

The rapid inflation of an airbag is the result of an explosive exothermic chemical reaction. The reaction is triggered by an electronic device in the car that detects any sudden change in speed or direction of the car. The bag fills with a harmless gas. When the occupants move forwards or sideways into the bag, they push the gas out of the airbag through tiny holes in the nylon. The airbag is usually totally deflated by the time the car comes to rest.

FIGURE 6.21 Airbags inflate as a result of an explosive chemical reaction.

One of the chemical reactions commonly used in airbags produces a massive burst of nitrogen gas. In some airbags, the nitrogen is released when the toxic chemical sodium azide (NaN_3) decomposes:

$$\text{Sodium azide} \longrightarrow \text{sodium} + \text{nitrogen gas}$$

$$2NaN_3 \longrightarrow 2Na + 3N_2$$

Other chemicals, including potassium nitrate, were present to react with the potentially dangerous sodium metal that was produced. In newer airbags, sodium azide has been replaced with less toxic (and less expensive) chemicals.

SCIENCE AS A HUMAN ENDEAVOUR: Alfred Nobel — an explosive career

Alfred Nobel is probably most famous for bequeathing his fortune to establish the Nobel Prizes in Physics, Chemistry, Medicine, Literature and Peace. However, Nobel made his fortune inventing **dynamite** and developing the use of explosives in the 1860s.

FIGURE 6.22 Alfred Nobel bequeathed his fortune to establish the Nobel Prizes.

Alfred Nobel was born in Sweden in 1833. He was educated in Russia. Nobel was fluent in several languages and interested in literature, poetry, chemistry and physics. In Paris he met a young Italian chemist, Ascanio Sobrero, who had earlier invented nitroglycerine, a highly explosive liquid. Alfred Nobel became very interested in nitroglycerine and saw its potential in the construction industry. When he returned to Stockholm in Sweden, he tried to develop nitroglycerine as an explosive. Several explosions, including one in 1864 in which Nobel's younger brother was killed, made the authorities realise that nitroglycerine was extremely dangerous.

Nobel had to move his laboratory out of Stockholm's city limits and onto a barge anchored on a nearby lake. He was determined to make nitroglycerine safe to work with. He discovered that mixing nitroglycerine with silica would turn the liquid into a paste that could be shaped into rods suitable for inserting into drilling holes. In 1866 he patented this material under the name dynamite.

FIGURE 6.23 Dynamite is used in the mining industry.

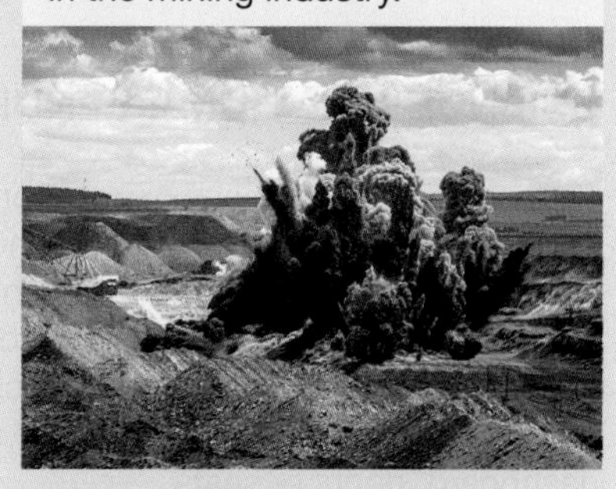

Dynamite is mainly used in the mining and construction industries. Huge areas of rock can be broken apart because the chemical reaction involved in dynamite's explosion releases large amounts of energy and gas, which can exert great pressure. Explosives can release enough energy to cause a small earthquake. One of the largest non-nuclear explosions ever was the Texas City disaster that occurred on 16 April 1947, and it was the deadliest industrial accident in U.S. history (figure 6.24). A fire started on board a ship docked in port, the *SS Grandcamp*, which detonated approximately 2100 metric tonnes of ammonium nitrate. The explosion triggered explosions on other ships and in an oil refinery nearby.

FIGURE 6.24 The Texas City disaster in 1947

The invention of dynamite could not have come at a better time than the middle of the nineteenth century. New mines were being opened to supply coal for heating and steam engines, iron and other building materials. Railways were being laid all over the world and passes had to be blasted through the mountains. Over the years, Nobel set up factories and laboratories in more than 20 countries.

Nobel died in 1896, and when his will was opened it came as a surprise that the interest earned by his $9 million fortune was to be used for the establishment of the Nobel Prizes. The prizes were to be awarded 'for the good of humanity' in the fields of chemistry, physics, physiology or medicine, literature and peace.

 Resources

 **eWorkbook** Exothermic and endothermic reactions (ewbk-12324)

 **Video eLesson** An explosion in a quarry (eles-2587)

dynamite a relatively stable explosive invented by Alfred Nobel in 1866; it is created by mixing nitroglycerine with an absorbent substance such as silica, forming a paste that can be shaped into rods

6.5 Activities

learnon

6.5 Quick quiz on	6.5 Exercise

Select your pathway

■ LEVEL 1	■ LEVEL 2	■ LEVEL 3
1, 3, 7, 8, 11	2, 4, 6, 10, 13	5, 9, 12, 14

Remember and understand

1. MC Energy can be required to start a reaction. Select three possible sources of this energy.
   A. Electricity B. Ice C. A flame
   D. Wind E. Heat from a beam of light
2. Fill in the blanks to complete the sentences.
   How are exothermic reactions different from endothermic reactions? An exothermic reaction is one in which ____________. Examples include the ____________ or the reaction of hydrochloric acid and magnesium. An endothermic reaction is one in which ____________. An example is ____________.
3. MC In a chemical reaction in which energy is absorbed from the surroundings, where does the extra energy go?
   A. It disappears
   B. Into the chemical bonds of the products
   C. To the surrounding air
   D. To the space between the bonds of the products
4. MC An endothermic reaction
   A. releases energy into the surroundings but can be reversed.
   B. absorbs energy from the surroundings but can be reversed.
   C. is a chemical reaction that cannot be reversed.
   D. is not a chemical reaction and cannot be reversed.
5. Explain why the chemical reaction that takes place in an ice pack containing ammonium chloride is not a chemical reaction.
6. Write a word equation to describe one chemical reaction that occurs to inflate an airbag.
7. MC What was Alfred Nobel's most famous invention?
   A. Nitroglycerine B. Airbags
   C. Dynamite D. Sodium azide

Apply and analyse

8. Create a flow chart of four steps to describe how an explosive is able to split large volumes of rock using some of the following options.
   a. Release small amounts of energy and gas
   b. Release large amounts of energy and gas
   c. Breaks apart huge areas of rock
   d. A chemical reaction involved in dynamite's explosion
   e. The pressure exerted
   f. The energy escalates to create more gas

9. Determine whether each of the chemical reactions described are exothermic or endothermic.
   a. Dilute hydrochloric acid is added to dilute sodium hydroxide in a test tube. They react to produce sodium chloride and water. After the reaction, the test tube feels very warm.
   b. As garden compost decomposes, the compost heap gets warmer.
   c. Barium hydroxide and ammonium thiosulfate solutions are mixed and the temperature drops enough to freeze water.

10. **SIS** Instant hot compresses are used by athletes to warm torn muscles. They relieve pain and speed up the healing process. The hot compresses contain calcium chloride powder and an inner bag of water. When the inner bag bursts, the calcium chloride dissolves in the water and releases energy.
    a. Is the chemical reaction that takes place in the compress endothermic or exothermic?
    b. How does the energy stored in the chemical bonds of the product compare with the energy stored in the chemical bonds of the calcium chloride and water?
    c. Write a word equation to describe this chemical reaction.
    __________ powder ⟶ ____________ solution
11. **a.** True or false? Explosions are exothermic reactions.
    b. Explain your answer to part **a**.
12. **SIS** Read through investigation 6.7 (you may have already completed some or all of this investigation). Predict the change in temperature of one of the reactions if twice the amount of reactants were added. Give a reason for your answer.

Evaluate and create

13. Suggest why Alfred Nobel donated his entire fortune to reward those who worked for the 'good of humanity'.
14. **SIS** The winners of Nobel Prizes are referred to as laureates. The Nobel Prizes are announced in October of each year.
    Choose one Australian scientist who has won the Nobel Prize and write a short biography about them. Include in your biography information on when they were awarded the Nobel Prize and the work that they did to receive such a prestigious award.

Fully worked solutions and sample responses are available in your digital formats.

LESSON
6.6 Corrosion

LEARNING INTENTION

At the end of this lesson you will be able to recognise that corrosion is a reaction in which the chemical and physical properties of metals are changed as they break down, which affects their use. You will be able to describe the processes of surface protection and galvanising to prevent corrosion.

6.6.1 Rusting and corrosion

Rusting is an example of **corrosion**. Corrosion is a chemical reaction that occurs when substances in the air or water around a metal 'eat away' the metal and cause it to deteriorate.

There are many examples of corrosion: silver tarnish; the green film that forms on copper or brass objects; and the most common one, the rusting of iron. Corrosion causes enormous damage to buildings, bridges, ships, railway tracks and cars.

rusting the corrosion of iron

corrosion a chemical reaction between air, water or chemicals in the air or water with a metal, which causes the metal to wear away

rust a red-brown substance formed when iron reacts with oxygen and water

6.6.2 Rust

Rust is the flaky substance that forms when iron corrodes. Iron reacts with water and oxygen in the air to form iron oxide and other iron compounds that make up the familiar red-brown substance known as rust.

Rusting is a slow chemical reaction that can be represented by the following word equation:

Iron + water + oxygen ⟶ rust

Even strong buildings and bridges that are made from steel, an alloy of iron, are weakened by rusting. The Sydney Harbour Bridge, for example, is continually painted to protect it from moisture and the air, which would cause its steel girders to rust. Ships and cars are also constructed largely of steel. Despite the strength of steel, it needs to be protected from the corrosive effects of the environment.

FIGURE 6.25 The Sydney Harbour Bridge is continually painted to protect it from moisture and the air, which would cause its steel girders to rust.

elog-2285

INVESTIGATION 6.9

Observing rusting

Steel wool is made from iron. You can observe rusting of the iron in steel wool by performing the following experiment.

Aim

To observe and describe the rusting of steel wool

Materials

- Petri dish
- water
- steel wool (without any soap)
- small glass
- permanent marker

Method

1. Pour some water into the Petri dish.
2. Place the steel wool in the middle of the Petri dish.
3. Cover the steel wool by placing the glass over it upside-down.
4. Mark the level of the water on the outside of the glass with a permanent marker.
5. Leave for several days, adding water as required to keep the level at the mark on the glass.

Observing the rusting of iron

Results

Construct a table to record your observations over several days.

Discussion

1. What did you observe about the level of water inside the glass? Can you explain why this happened?
2. Write down a word equation for the chemical reaction that occurred inside the glass.

Conclusion

What can you conclude about the rusting of steel wool?

6.6.3 Speeding up rusting

Some substances in the environment make rusting happen much more quickly. One of the most effective of these is salt. Steel dinghies that are used in the ocean rust much faster than those that are used only in fresh water. This is because the salt in the seawater speeds up the reaction between oxygen in the air and the iron in the steel.

Some chemicals released from factories also increase the rate of rusting. A CSIRO study conducted in Melbourne found that rusting rates were high near airports and sewage treatment plants.

Rusting is much slower in dry environments such as deserts, where the rainfall is nearly zero and there is very little water vapour in the air. This can be seen in figure 6.27 where some aircrafts are still structurally sound after 20 years in the desert.

FIGURE 6.26 This wrecked car has rusted quickly because of its proximity to the sea.

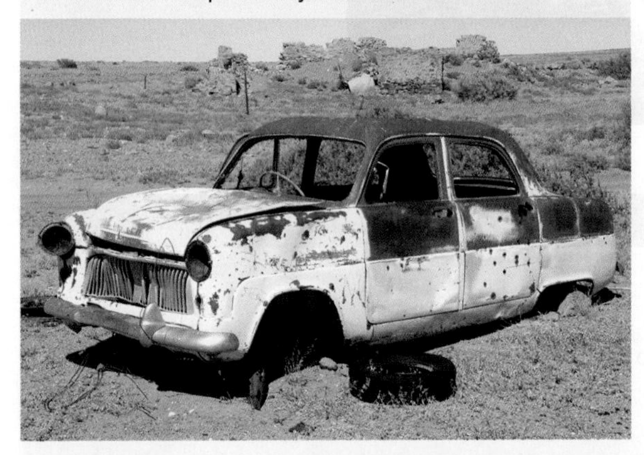

FIGURE 6.27 In the Mojave Desert unused aircrafts are stored out in the open air.

INVESTIGATION 6.10

elog-2287

Investigating the corrosion of different metals

Aim

To investigate the corrosion of a variety of metals

Materials

- small strips of a range of metals such as copper, aluminium, zinc and magnesium
- sandpaper
- other equipment approved by your teacher

Method

Design and carry out an investigation to study the resistance of a selection of different metals to corrosion. Ensure that appropriate variables are controlled. Before commencing, clean the metal strips with sandpaper to ensure that any coatings already caused by corrosion are removed.

Results

Write a report on your investigation that includes your aim, method, results (including a table), discussion and a clear conclusion listing the metals in order of resistance to corrosion, from most resistant to least resistant. Include the answers to the questions below in your discussion.

Discussion

1. Identify the independent and dependent variables in your investigation.
2. Name the variables that you controlled.
3. Suggest how you might be able to improve or speed up the investigation.

Conclusion

What can you conclude about corrosion of different metals?

6.6.4 Rust protection

The layer of rust that forms on an iron object flakes off the metal, allowing air and moisture to get through to the iron below. This causes more rusting to occur and eventually the iron becomes a heap of rust. It is important to protect iron and steel from corrosion, especially if they are part of a bridge or the hull of a ship.

There are several ways to protect iron and steel from rusting. One way is to prevent oxygen or moisture from contacting the metal. This is called **surface protection**. The metal can be protected by coating it with paint, plastic or oil. If the surface protection becomes scratched or worn off, the metal below can be attacked by moisture and oxygen, and rusting will occur. Examine the painted surface of an old car. Wherever the paint has chipped off you will find that corrosion has occurred and rust can be seen.

FIGURE 6.28 Rust is a serious problem in the use of iron. The rust formed is brittle and flakes off, which allows the rusting to continue deeper and deeper into the metal.

Another way to protect iron from rusting is to coat it with a layer of zinc. This is called **galvanising**. Zinc is a more reactive metal than iron, and in the presence of moisture and oxygen the zinc layer corrodes, leaving the iron unaffected. Many roofing materials and garden sheds are made from galvanised iron. You can also buy galvanised nails.

surface protection refers to when a protective coating is applied over a metal surface to prevent corrosion

galvanising protecting a metal by covering it with a more reactive metal that will corrode first

elog-2289

INVESTIGATION 6.11

Rusting and salt water

Aim

To investigate the effect of salt water on the rate of rusting

Materials

- test tubes and test-tube rack
- measuring cylinder
- iron nails
- water
- salt (sodium chloride)

Method

1. Design an experiment to test the effect of the saltiness of water on the time taken for an iron nail to rust.
2. Propose a hypothesis.
3. Discuss your experiment design with a partner. You will need to consider which conditions must be kept the same and which condition will be varied.
4. Consider the purpose of a control and set up a control test tube.
5. Write down your method. It should be clear enough for someone else to follow without any help.

Results

Construct a table in which to record your observations over the next few days.

Discussion

1. Describe the effect of salt on the time taken for the nail to rust.
2. Explain the purpose of a control.
3. Was your hypothesis supported?
4. Outline how your results compare with those of others in your class.
5. Write a report of your findings. Include in your report the aim, materials, method, results and conclusion for your investigation.

Conclusion

Does salt water affect the rate of rusting?

6.6.5 Rusting can be useful

Not all rusting is bad. You can buy hand warmers, which are commonly used by skiers and campers, from pharmacies. These packages will produce heat when you shake them. The contents of the packet include powdered iron, water, salt and sawdust. When the packet is shaken vigorously, the iron rusts quickly, which produces heat.

HOW ABOUT THAT!

City councils face problems caused by the action of dogs on metal lampposts. The corrosive properties of the dogs' urine rusts the steel of the lampposts a few centimetres above the ground.

EXTENSION: Corten steel

Sometimes rust can protect the surface of steel. The steel alloy sculpture shown in figure 6.29 is made from Corten steel. It is a specially developed alloy that is designed to rust over a period of weeks or months to build up a dense protective layer of rust. The rust layer is densely packed and does not let air or water through it, so further rusting cannot occur once the protective layer has rusted. As well as in sculptures, this type of steel can be seen used in modern buildings, garden fences and edging.

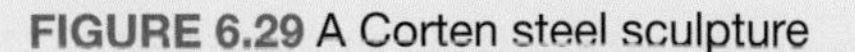

FIGURE 6.29 A Corten steel sculpture

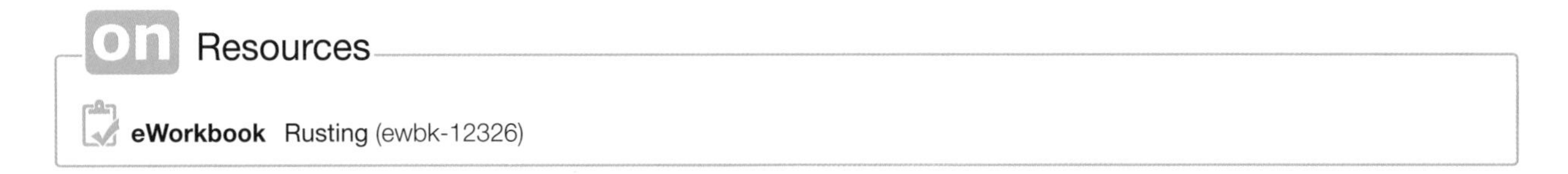

6.6 Activities

learnon

6.6 Quick quiz	6.6 Exercise

Select your pathway

LEVEL 1	LEVEL 2	LEVEL 3
1, 2, 4, 5, 7	3, 8	6, 9, 10

Remember and understand

1. **MC** Identify the statement that describes corrosion.
   A. A physical change in the composition that causes the solid to break down
   B. A chemical reaction with a metal and other substances that causes the metal to deteriorate
   C. A chemical reaction between a fuel and oxygen that produces heat
   D. A chemical reaction with only one reactant that causes the breakdown of a metal
2. **MC** Identify the statement that describes rusting.
   A. A physical change from solid iron to liquid iron
   B. A chemical reaction between iron and carbon dioxide
   C. A physical change from solid iron to gaseous iron
   D. A chemical reaction between iron, water and oxygen that produces rust
3. State which of the following substances can be used for surface protection of metals.

Substance	Provides surface protection of metals?
a. Paint	
b. Water	
c. Oil	
d. Oxygen gas	
e. Plastic	

4. **a.** Complete the sentence: Galvanised iron is iron covered with a layer of _______________.
   b. What advantage does galvanised iron have over iron that is not galvanised?
5. Explain how galvanising protects iron from rusting when the zinc coating corrodes more quickly than the iron.

Apply and analyse

6. Explain why the powdered iron inside hand warmers used by skiers and campers rusts much more quickly than an iron nail.
7. Explain why rusting occurs faster in coastal regions than in areas further away from the sea.
8. **SIS** If you have access to an old car, survey it carefully and record all its rust spots. If you do not, then carefully observe the rusty car in figure 6.26. Why are some parts of the car more likely to rust?
9. **SIS** Corrosion is found in many places. Survey your school for rust spots. Write a report about your findings.
10. **SIS** Aluminium corrodes quite quickly, yet it is used to make soft-drink cans. Write a hypothesis about why aluminium cans are not corroded by the drinks they store. Make sure that your hypothesis can be scientifically tested.

Fully worked solutions and sample responses are available in your digital formats.

LESSON
6.7 Combustion

LEARNING INTENTION

At the end of this lesson you will be able to identify that one familiar reaction that causes a temperature change is combustion, which is the reaction of fuels with oxygen to produce heat, carbon dioxide and water vapour.

6.7.1 Oxidation reactions

Burning is a chemical reaction; it is also known as combustion. It involves the combination of oxygen with a fuel and always produces heat and gases. Reactions that involve combination with oxygen are examples of **oxidation** reactions. One such example is the rusting of iron in the presence of water and oxygen to form iron oxide. There are many other oxidation reactions.

burning combining a substance with oxygen in a flame

oxidation a chemical reaction involving the loss of electrons by a substance

fossil fuel a substance, such as coal, oil or natural gas, that has formed from the remains of ancient organisms; coal, oil and natural gas are often used as fuels — that is, they are burnt in order to produce heat

6.7.2 Burning fossil fuels

Burning a **fossil fuel** is a combustion reaction, in which the fuel reacts with oxygen to produce heat, carbon dioxide and water vapour. Fossil fuels are fuels formed from the remains of living things. Petrol, natural gas, coal, wood and even paper are fossil fuels.

The production of carbon dioxide from the burning of fossil fuels is a major source of greenhouse gas emissions, which are a major contributor to climate change.

If combustion of fuels occurs when there is not enough oxygen, incomplete combustion occurs. The products are a mix of carbon monoxide, carbon dioxide, solid carbon (soot) and water.

The oxyacetylene torch

To obtain temperatures as high as 3000 °C (hot enough to melt iron and weld metals), acetylene fuel is mixed with pure oxygen in an oxyacetylene torch (figure 6.30).

Acetylene + oxygen ⟶ carbon dioxide + water

FIGURE 6.30 An oxyacetylene torch is used in construction work.

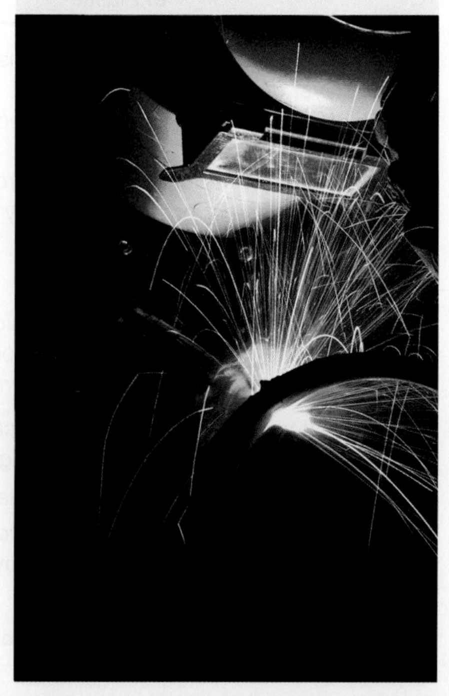

The car engine

Petrol or gas car engines work by the combustion of petrol or gas in the cylinders. A mixture of air and fuel is drawn into each cylinder and ignited by a spark from the spark plug. The fuel reacts rapidly with oxygen in the air. The resulting explosion pushes the piston, which turns the drive shaft. The products of the reaction, carbon dioxide and water vapour, leave the car engine through the exhaust pipe.

EXTENSION: The danger of backdraughts

A backdraught occurs when a fire in a closed room dies down because it has been starved of oxygen, but flammable gases continue to stream out of the hot materials in the room. When a door to the room is opened, air is quickly drawn inside, restoring the supply of oxygen and allowing the fire to reignite. The resulting fire consumes all the flammable gases in a few seconds and produces sufficient heat to ignite any remaining materials in the room. This is very dangerous to firefighters.

Rocket fuels

Liquid and solid fuels are used in the NASA rocket program. When these fuels are burnt, they provide sufficient thrust to place a rocket in orbit hundreds of kilometres from Earth. Liquid hydrogen and liquid oxygen react to power the rocket's main engines. The only product of this reaction is water; it does not produce pollution, unlike the burning of fossil fuels. For this reason, hydrogen is being investigated as an alternative fuel source on Earth.

FIGURE 6.31 Oxidation reactions provide the thrust to launch a rocket.

$$\text{Hydrogen} + \text{oxygen} \longrightarrow \text{water}$$

Most of the thrust required to place the rocket in its desired orbit comes from chemical reactions in the solid fuel, which is located in the solid rocket boosters. In space, a liquid fuel such as hydrazine is oxidised to produce an enormous volume of gas. As the gas is released, the rocket is thrust forward. By controlling the direction of the thrust, it is possible to steer the rocket.

elog-2291

INVESTIGATION 6.12

Burning paper

Aim

To observe and record the combustion of paper

Materials

- safety glasses
- Bunsen burner, heatproof mat and matches
- tongs
- gas jar
- limewater
- paper
- deflagrating spoon

Method

1. Place 10 mL of limewater in the bottom of the gas jar.
2. Put a ball of scrunched-up paper into the deflagrating spoon.
3. Light the paper and lower it into the gas jar.
4. When burning has stopped, remove the deflagrating spoon and cover the jar.
5. Shake the gas jar and observe the colour of the limewater.

Results

What happened to the limewater?

Discussion

1. What gas was given off by the burning paper?
2. Which other substance or substances were produced by the reaction?

Conclusion

Summarise the findings for this investigation.

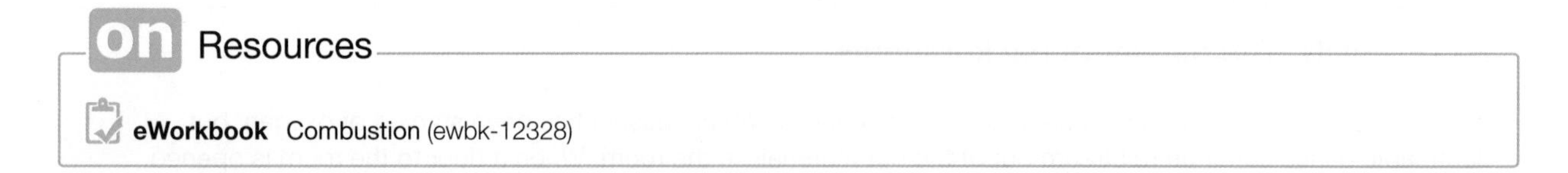

6.7 Activities

learnon

6.7 Quick quiz on | 6.7 Exercise

Select your pathway

LEVEL 1	LEVEL 2	LEVEL 3
1, 3, 5	2, 4, 6, 7	8, 9, 10, 11, 12

These questions are even better in jacPLUS!
- Receive immediate feedback
- Access sample responses
- Track results and progress

Find all this and MORE in jacPLUS

Remember and understand

1. **MC** Identify the statement that describes what burning means.
   A. A physical change that produces a gas by evaporation
   B. A chemical reaction between a fuel and carbon dioxide that produces heat and gases
   C. A chemical reaction between a fuel and oxygen that produces heat and gases
   D. A chemical reaction between a fuel and heat that produces gases
   E. A physical change in which heat is applied to produce gases
2. Identify which of the following are evidence that burning is a chemical reaction.

Evidence	Evidence of chemical reaction?
a. Change of state	
b. Release of heat	
b. There are only two products	
d. Production of gases	
e. Bubbles are produced	

3. What is a fossil fuel? List three examples of fossil fuels.
4. **a.** **MC** What type of reaction is burning?
   A. Reduction
   B. Corrosion
   C. Oxidation
   D. Neutralisation
   b. Write the word equation for each of the following parts **i** to **iii**.
   i. The reaction of acetylene with oxygen to make a flame hot enough to weld metals together

   ____________ + oxygen ⟶ ____________

   ii. The reaction of oxygen and petrol or gas in a car engine that produces a hot gas, which causes the movement of the piston, the drive shaft and the wheels of the car

   ____________ + fuel vapour ⟶ carbon dioxide + water

   iii. Energy is released by the reaction of hydrogen and oxygen that produces the thrust necessary to get rockets into orbit.

   ____________ + ____________ ⟶ water
5. Is rusting an example of burning? Explain your response.
6. Select the correct substances to complete the word equation.

   Fuel + ____________ ⟶ ____________ + water vapour

7. **MC** Which of the following are names given to chemical reactions in which fuels react with oxygen? Select all possible answers from the options given.
   A. Burning
   B. Corrosion
   C. Combustion
   D. Oxidation

Apply and analyse

8. Space agency scientists are constantly searching for better rocket fuels. List the properties for these fuels that are
   a. desirable
   b. undesirable.

9. **SIS** The following table shows the heat of combustion per kilogram of certain fossil fuels.

Fuel	Heat value (MJ per kg)
Methane	50
Petrol	44
Natural gas	42
LPG	46
Diesel fuel	42

   a. Use the data in the table to make a bar chart of the fossil fuels.
   b. Identify which of the fossil fuels in your bar chart produces the most amount of heat per kilogram.
   c. Explain why this fossil fuel is not used to power cars in Australia.

Evaluate and create

10. **SIS** Suggest why in most buildings we have thick fire doors that have to remain closed.
11. A student conducts an experiment in which they burn magnesium in the presence of oxygen. At the end of this experiment, the compound magnesium oxide is produced.
    a. How do you know that a chemical reaction has taken place?
    b. Write a word equation for the chemical reaction.
    c. Would this reaction be classed as a combustion reaction? Justify your response.
12. **SIS** Read through investigation 6.12.
    a. How could you establish that the limewater did not react with another substance in order to turn milky?
    b. Design an investigation to show that limewater reacts with carbon dioxide to turn a milky colour.

Fully worked solutions and sample responses are available in your digital formats.

LESSON
6.8 Indicators of chemical change in our everyday world

LEARNING INTENTION

At the end of this lesson you will be able to discuss where indicators of chemical change are used in order to identify the presence of particular substances, such as in soil, water and medical testing kits.

Indicators of chemical change are used for identifying the presence of particular substances, such as in soil, water and medical testing kits.

6.8.1 Test kits

As our knowledge of our environment increases, so too does our need to test the land, waterways and air for the safety of ourselves and all living things in these habitats. The invention of simple, affordable and easy-to-use test kits has enabled environmental testing to occur quickly and easily. In addition, the development of medical testing kits in response to public demand has resulted in people being more aware of their medical needs, and able to feel more in control of their health. These kits test for a specific chemical or chemicals by the addition of a very small amount of the test substance. A chemical reaction occurs, which shows the presence or absence of the substance being tested for. The kits may show a colour change, or have a line appear that indicates a positive result for the substance being tested.

6.8.2 Indicators of chemical change in soil

Crop production is a vital industry worldwide to feed our growing population. Farmers are now able to monitor the health of the soil by using simple test kits. Soils that are nutrient poor will result in little to no crop production. Testing the pH of soil is an easy chemical reaction to perform that provides important information about the health of the soil and what treatment it may need. If soil is too acidic or too basic, crops may not grow. pH is used as an indicator of the availability of nutrients in the soil. If the pH is less than 6, the soil will be deficient in calcium, magnesium, molybdenum, phosphorus and potassium. If the pH is less than 4, it is most likely that the soil has toxic amounts of aluminium and manganese present. If the pH is greater than 7, the soil may be deficient in boron, copper, iron, manganese and zinc. These test kits generally show a colour change, with a key that explains the colour produced.

FIGURE 6.32 A soil testing kit showing a pH test result

INVESTIGATION 6.13

elog-2293

Examining flower colour under different pH conditions

Aim

To determine the effect of acidic and basic conditions on the colour of the flowers of hydrangea plants

Materials

Two hydrangea plants that have been grown in acidic and basic conditions and are currently flowering

Method

1. Measure and record the pH of the soil of each plant.
2. Record the colour of each plant.

Results

1. Create an appropriate table to tabulate your results and observations.
2. Write a report about your experiment.

Discussion

1. Describe the conditions that led to the various colours of the flowers.
2. What colour would the flowers be if the pH was neutral?

Conclusion

What can you conclude about how pH affects the plant growth and flower colour of hydrangeas?

6.8.3 Indicators of chemical change in water

Measuring the chemical properties of water is a common requirement in many industries. The run-off of water from industrial sites can be tested for the presence of lead, iron, copper, nitrates, chlorine, fluoride, toxins, fertiliser and even microorganisms using water testing kits. These test kits can also be used to analyse drinking water, ponds, lakes and swimming pools. The result is generally indicated by a colour change on a test strip, with a key to explain the result.

FIGURE 6.33 Pool-water test kits can be made up of test strips or a water test with an indicator showing different pH colours.

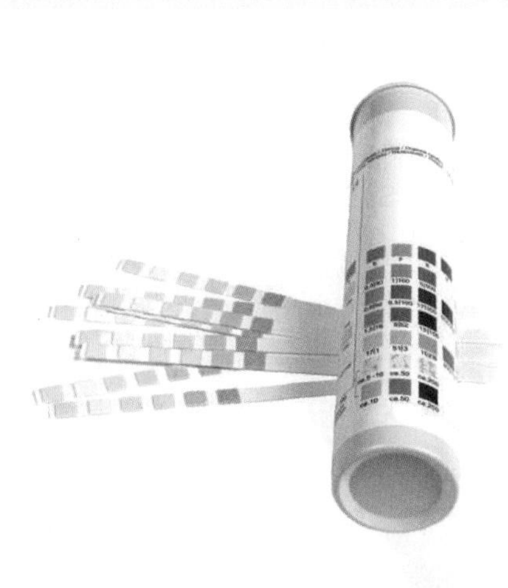

elog-2295

INVESTIGATION 6.14

Water testing

Aim

To test what chemicals are present in the water of our school environment and its pH

Materials

- water samples from various locations in the school (e.g. a pond, taps in different locations, puddles, rain water, tank water)
- water test kit — either test strips or a liquid test kit
- pH test kit

Method

1. Create a table for the results listing the sites where the water samples were obtained from, the chemicals being tested for and a column for the pH.
2. Following the instructions on the test kits, test each water sample for each chemical component and record your results in your table.

Results

1. Record your results and observations in your table of results.
2. Graph your data on the one graph, using different colours for each water sample.

Discussion

1. Were there any trends in the data that you recorded?
2. Were there any specific chemicals that were present or absent for the majority of the water samples? What does this mean in terms of water quality?
3. Describe how the pH varied across the various water samples. What does this mean in terms of water quality?
4. Can you think of any reasons to explain the results that you obtained for the water samples from the various sites?

Conclusion

What can you conclude about the water quality of the various sources in your school?

6.8.4 Indicators of chemical change in medical testing kits

There have been incredible developments in the production of simple and quick medical testing kits over the last decade. All of these kits test for specific chemicals, proteins or hormones. The test substance is added to the kit and a response is observed within a specified time frame. The most significant and broadly used test kits are the rapid antigen tests (RATs) for the presence of the COVID-19 virus.

Testing for COVID-19

COVID-19 is a disease caused by the SARS-CoV-2 coronavirus. There are two kinds of tests available for the testing of COVID-19. The tests detect either of the following:

- The presence of the SARS-CoV-2 virus in the body, by testing if the virus is present in the throat, nose, nasal secretions or sputum
- Whether the body is producing antibodies to the SARS-CoV-2 virus, by testing a blood sample.

FIGURE 6.34 A negative SARS-CoV-2 rapid antigen test

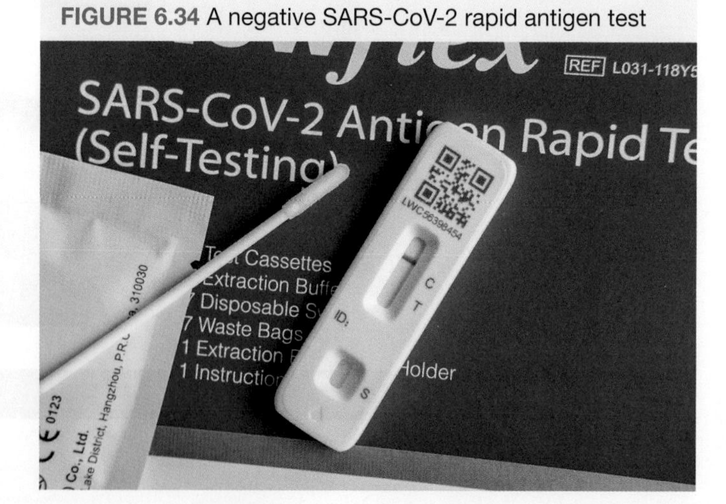

The two types of tests that detect the presence of specific components of the virus are rapid antigen tests and nucleic acid tests. Rapid antigen self-testing kits have allowed for home testing, where a person can collect the sample, perform the test and interpret results, indicated by lines that appear in the panel.

Nucleic acid tests detect the presence of genetic material of the SARS-CoV-2 virus. The tests available include the polymerase chain reaction (PCR) test and a kit called the loop-mediated isothermal amplification (LAMP) test. These tests are more complicated to complete and are performed by trained scientists in pathology and medical laboratories.

Other medical testing kits

There are numerous medical testing kits available, from the blood alcohol and drug testing kits utilised by police, to home testing kits for ovulation, pregnancy, glucose levels in the blood and urine, and genetic data. Pregnancy tests identify the level of a hormone called human chorionic gonadotropin (HCG) in urine. This hormone is only produced during pregnancy and usually can only be detected after approximately two weeks of pregnancy due to the sensitivity of the test.

Glucose test kits can use urine or blood samples to monitor blood sugar levels. A urine test strip is a very basic test for the presence of glucose. To determine whether a person has diabetes, more testing is required. Once a person has been diagnosed with either type 1 or type 2 diabetes, they must monitor their blood sugar levels regularly. If blood glucose levels get too high or too low, the condition may become life-threatening to the individual. There are many different types of devices available for testing blood glucose levels, with technological developments improving their accuracy and ease of use, progressively.

FIGURE 6.35 Two testing methods for body glucose levels: **a.** a simple urine testing kit **b.** a blood testing device

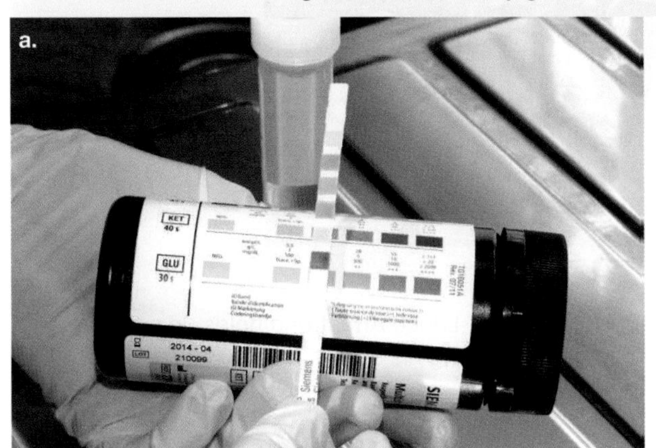

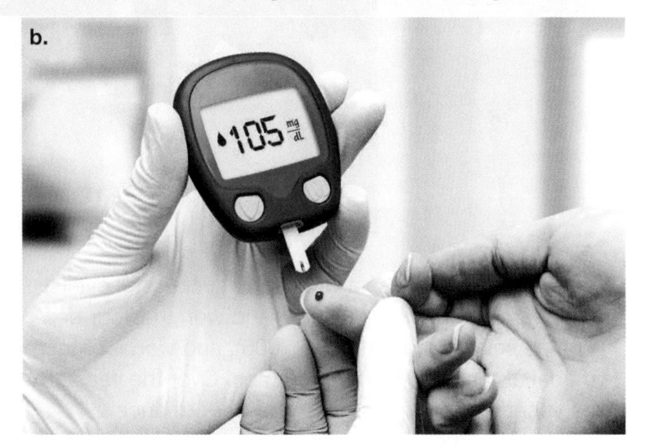

SCIENCE AS A HUMAN ENDEAVOUR: Testing for performance-enhancing drugs in sport

Athletes may accidentally or intentionally ingest banned substances or performance-enhancing drugs that increase their athletic performance. Banned substances include some over-the-counter medicines as well as drugs of abuse. They also include some dietary supplements or foods that are considered to provide an enhancement to athletic performance. Performance-enhancing drugs are illegal substances that provide an unfair advantage to the athlete.

The ingestion of either type of substance may lead to a failed drug test, possibly resulting in the end of the career for the athlete. As a result, performance-enhancing drugs are constantly being developed, produced and modified in order to avoid detection.

Drugs are metabolised by the human body. Scientists can search for the drug or its products of metabolism when analysing biofluids such as the saliva, blood or urine samples of an athlete. The products of metabolism of drugs can stay in the body's system and bloodstream for many weeks. Biofluids are kept by the relevant sporting authorities for many, many years so that scientists are able to test the samples years later for performance-enhancing drugs. The biggest challenge to the World Anti-Doping Agency in the detection of illicit drugs is the fact that they need to know the composition of the drug that they need to screen for. Scientists can only develop a test for a substance once they know what the substance is. Therefore, the sporting authorities are always investigating substances found in biofluids and researching whether they are performance-enhancing drugs or not.

One of the most famous athletes to have competed in track and field events is Tyson Gay. In 2012, he won a silver medal in the Olympic Games, and at the Shanghai Golden Grand Prix he recorded the second-fastest time in the 100-metre race of 9.69 seconds. In 2013, his stored biofluid samples tested positive for a banned substance. He was banned from athletics for a year and stripped of his silver Olympic medal, with his reputation forever tarnished.

FIGURE 6.36 Tyson Gay (centre) running to win at the Shanghai Golden Grand Prix in 2009, equalling the second-fastest 100 m sprint time ever

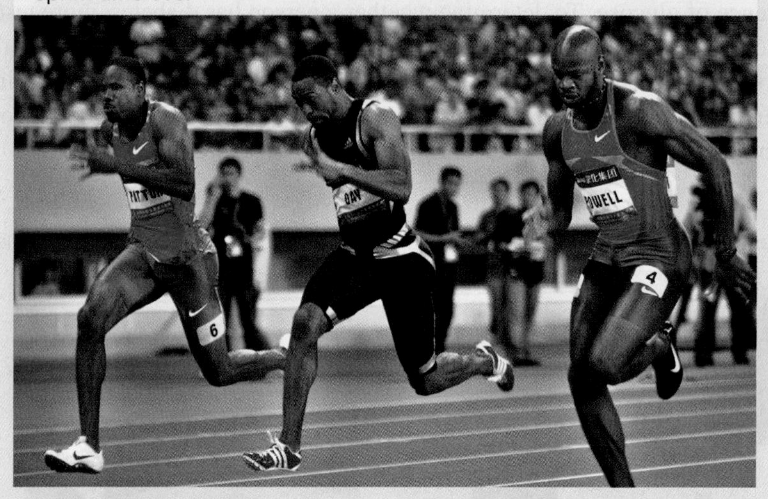

Lance Armstrong was a worldwide successful cyclist who survived testicular cancer and won the Tour de France seven times (1999 to 2005), and won a bronze medal at the 2000 Olympic Games. Over the years, he vehemently denied the use of performance-enhancing drugs. However, in an interview with Oprah Winfrey in 2013, he revealed that he had, in fact, used performance-enhancing drugs. This revelation resulted in worldwide anger and frustration at his repeated lies and deceptions over the years. Armstrong lost millions of dollars in advertising and endorsements, and destroyed his reputation forever. He was stripped of his seven Tour de France titles and his Olympic medal. The United States Anti-Doping Agency described Armstrong as the mastermind of 'the most sophisticated, professionalised and successful doping program that sport has ever seen'.

FIGURE 6.37 Lance Armstrong holds up seven fingers signifying his seven Tour de France titles

6.8 Activities

learnon

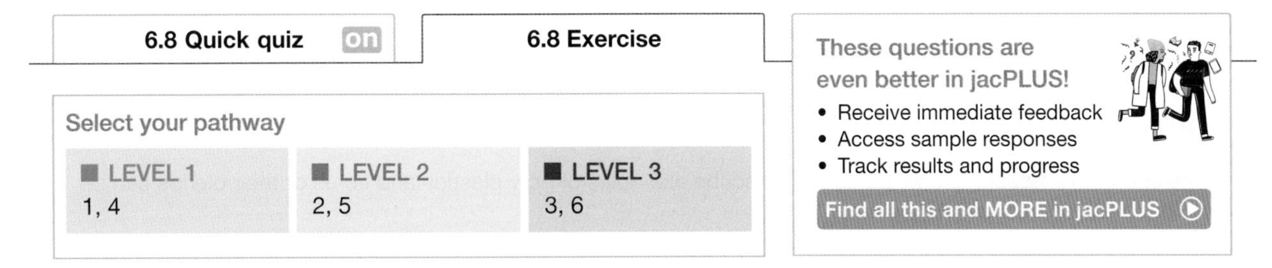

Remember and understand

1. Why do we use chemical test kits in our everyday world?
2. MC Which of the following statements is *incorrect*?
   Chemical test kits can be used to indicate
   A. the presence of sugar or protein in urine.
   B. very small amounts of chemicals in the soil.
   C. how much chlorine needs to be added to a pool.
   D. the amount of ozone high up in the atmosphere.
3. How do chemical test kits work?

Apply and analyse

4. The pH of soil can have a significant effect on plant growth. Describe how soil pH affects the flower colour of hydrangea plants.
5. In the following figure, the pregnancy test kit results show one line for a 'not pregnant' result and two lines for a 'pregnant' result. Why does the 'not pregnant' result show only one line?

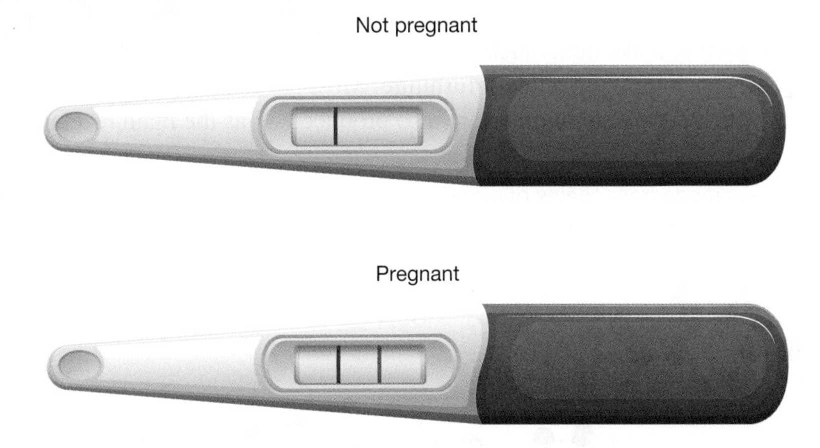

Evaluate and create

6. A new chemical test kit called IntelliGender® has recently been marketed around the world. The product is advertised as a gender verification kit and says that it will turn urine green for a boy or yellow/orange for a girl. The company claims a 90% accuracy rate. However, its website advises against making financial, emotional or family-planning decisions based on the test results, including NOT painting the nursery or buying gender-specific clothing.
   Evaluate and analyse the information provided. Do you agree with the 90% accuracy rate? Justify your response.

Fully worked solutions and sample responses are available in your digital formats.

LESSON
6.9 Plastics and fibres

LEARNING INTENTION

At the end of this lesson you will be able to describe examples of how plastics and fibres or their blends can be developed for a particular purpose.

6.9.1 Plastics

The scientists and engineers who develop new plastics for spacesuits that allow astronauts to walk in space need a knowledge of chemistry to create materials that are strong, light and heat resistant. Developing new materials for a particular purpose requires an assessment of the required properties and an understanding of chemical reactions.

FIGURE 6.38 The spacesuits worn by astronauts when they are walking in space contain many layers of materials developed by scientists and engineers.

Metals, paper and ceramics have been used for thousands of years. But plastics have been around for less than 100 years. Plastics are synthetic (manufactured) materials that can be easily moulded into shape. Some plastics are flexible and soften when they are heated. They can be easily moulded into products such as milk and fruit-juice containers, rubbish bins, spectacle lenses, electrical insulation and laundry baskets. Others are quite hard and rigid. These plastics are used to make items such as toilet seats, electrical switches, bench tops and outdoor furniture. Most plastics are the products of chemical reactions with crude oil — from which petrol and bitumen are also produced — as the main reactant.

FIGURE 6.39 Plastics are made from chains of molecules and are also called polymers (poly means 'many'). This chain of molecules has two repeated units.

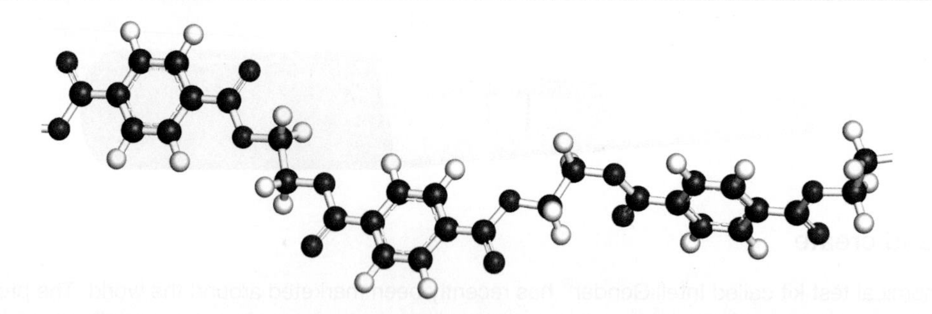

WHAT DOES IT MEAN?

The word 'plastic' comes from the Greek word *plastikos*, meaning 'able to be moulded'.

CASE STUDY: Plastic currency

Australia was the first country in the world to use only plastic notes for currency. The notes are more difficult to forge and last much longer than the old paper notes.

6.9.2 Fibres

Until the development of nylon in 1938, just in time to make parachutes for World War II, the world relied almost completely on fabrics made from **natural fibres**, such as wool, cotton, linen and silk.

Animal-based fibres include wool from sheep and silk from silkworms. Cotton is derived from cotton bushes and linen comes from flax plants. Today, it would be impossible to provide clothing and bedding for the world's population with purely natural fibres because of the amount of land and water that would be needed for crops and sheep.

Synthetic fibres such as those used in compression sports gear have many desirable qualities that natural fibres lack, including easy care, colour-fastness and being light weight.

Of the many synthetic fibres, the most widely used are **nylon** and **polyester**. Synthetic fibres are made by pushing softened plastic materials through tiny holes in a nozzle called a **spinneret**, which looks a little like a shower head.

FIGURE 6.40 Synthetic fibres form when soft plastic is forced through the holes of a spinneret.

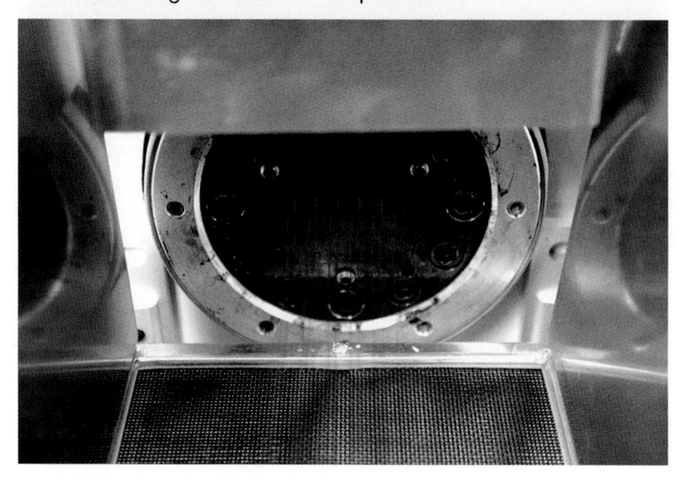

natural fibres fibres that form naturally — that is, they have not been made by humans; they include wool and silk from animals, and cotton from plants

nylon a synthetic fibre; the monomers are joined together by the elimination of water molecules at the joins

polyester a synthetic fibre; the monomers are joined together by the elimination of water molecules at the joins

spinneret a nozzle with small holes through which a plastic material passes, forming threads; also the organ used by spiders to create their webs

EXTENSION: Biomimicry

Biomimicry is the imitation of designs found in nature to solve human problems using technology. The spinneret used to make synthetic fibres is an example of biomimicry. The spinneret gets its name from the organ used by spiders to spin their webs. Liquid silk flows through the spider's spinneret. It hardens into a fibre as it passes through. Most spiders have six spinnerets.

FIGURE 6.41 The spinneret a. on a spider b. manufacturing device

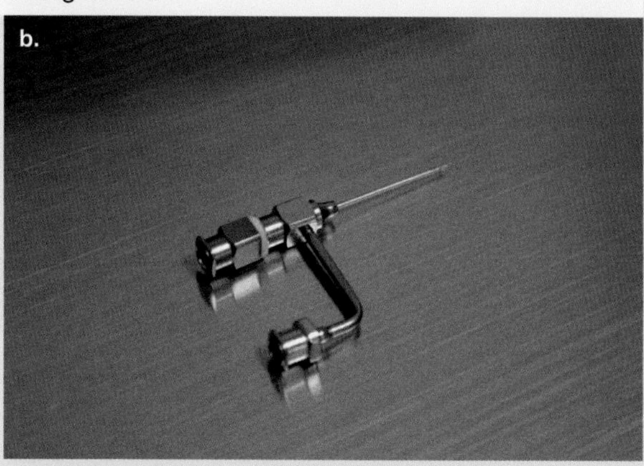

on Resources

Video eLesson The future of clothing (eles-0859)

SCIENCE INQUIRY SKILLS: Variables

In a fair test such as the one you will develop in investigation 6.15, it is important to control and measure variables. For a test to be fair the variables must be controlled. For example, if the elasticity of different fibres was being tested, the fibres should all be the same length and thickness. The fibres should all be tested in the same way; for example, they should be joined to a testing frame in the same way and the load or force used to test them be applied for the same time and at the same rate.

elog-2297

INVESTIGATION 6.15

Testing fibres

Aim

To observe and describe the properties of a range of fibres

Materials

- a range of threads of different fibres (e.g. cotton, polyester, wool, nylon, rayon)
- uniformly sized fabric samples made from different fibres or blends of fibres
- equipment decided upon by the group

Method

Work in groups of three or four to complete this investigation.

1. Start by listing the properties of either the fabric samples or the fibres that can be tested by experiments. Some examples to help get you started include flammability, elasticity and the ability to absorb water.

Devise an experiment that will allow you to compare one property of threads of either different fibres or different fabric samples.

2. Make a list of the equipment you will need.
3. Have your experiment plan and equipment list checked by your teacher.

CAUTION

Obtain your teacher's approval before carrying out any tests. Synthetic fibres or blends should be burned only in a fume cupboard.

Results

1. Carry out your experiment and keep a record of your measurements and observations.
2. Write a report about your experiment.

Discussion

1. List the dependent and independent variables.
2. List the variables that you were able to control in your experiment.
3. Identify any variables that were uncontrolled.
4. Identify the most useful properties of each of the fibres or fabrics that you tested.
5. Suggest at least one improvement that you could make to your experiment.

Conclusion

Summarise the findings of the experiment in three or four sentences.

6.9.3 Blends

Many of today's fabrics are made from blends of natural and synthetic fibres to make the best use of the properties of each fabric in the blend. A blend of polyester and cotton is commonly used for shirts and dresses. The cotton helps keep the wearer cool, while the polyester reduces creasing.

Rayon is shiny, easy to dry and cool in summer. On the 'down' side, it has low **durability** and is not **elastic**. Elasticity describes the ability of a material to return to its original size and shape after being stretched. Rayon is neither a natural nor a synthetic fibre. To make it, cellulose fibres from spruce and eucalypt trees are mixed with chemicals that soften them. The mixture is then passed through a spinneret.

durability the quality of lasting; not easily being worn out

elastic describes a material that is able to return to its original size after being stretched

6.9.4 Lycra®

When you watch the feats of Olympic athletes, cyclists, skiers and skaters, it is almost certain that they are wearing Lycra®. Lycra is not a fabric; it is the registered trademark of a synthetic fibre called spandex. Spandex was invented in 1958. Spandex is lightweight, durable, retains its shape and fits snugly. It even pulls moisture away from the wearer's skin. Spandex is very elastic. It can be stretched to up to seven times its normal length and spring back to its initial length when released. Spandex is always blended with other fibres. As little as two per cent of this material in a blend makes a difference to the properties of the fabric. Lycra suits usually consist of between three per cent and ten per cent spandex.

Each fibre, whether natural or synthetic, has advantages and disadvantages. Some of these are outlined in table 6.1.

TABLE 6.1 Advantages and disadvantages of fibres

Fibre	Advantages	Disadvantages
Wool	• Warm in cold weather • Crease resistant • Burns slowly • Retains its shape well	• Shrinks when washed • Turns yellow in sunlight
Cotton	• Absorbs moisture • Soft • Cool in hot weather	• Creases easily • Burns quickly
Nylon	• Dries quickly • Light • Strong • Elastic	• Builds up static electricity • Melts rather than burns
Polyester	• Dries quickly • Crease resistant • Resistant to many chemicals	• Builds up static electricity • Melts rather than burns

DISCUSSION

Working collaboratively, or independently, make a list of at least ten items in your houses that are made from plastic. Imagine that these could no longer be made from plastic! For each item, think about the most important properties the item must have and discuss which other material could be used to make it.

6.9 Activities

learnon

6.9 Quick quiz on | 6.9 Exercise

Select your pathway

■ LEVEL 1	■ LEVEL 2	■ LEVEL 3
1, 2, 3, 4, 5, 7	6, 8, 10, 12	9, 11, 13, 14

These questions are even better in jacPLUS!
- Receive immediate feedback
- Access sample responses
- Track results and progress

Find all this and MORE in jacPLUS

Remember and understand

1. **MC** Which single property do all plastics have?
   A. Flexible
   B. Can be moulded
   C. Rigid
   D. Cannot be moulded
2. From which substance found beneath the ground are most plastics made?
3. **MC** Identify the source that all natural fibres come from.
   A. Plants or animals
   B. Plants only
   C. Animals only
   D. Soil
4. **MC** Identify which of the following options explains why woollen clothing is popular in winter, whereas cotton clothing is popular in summer.
   A. Wool is cool on warm days, but cotton allows heat and moisture to be trapped against the body in cold weather.
   B. Wool is warm on cold days, but cotton allows heat and moisture to escape from the body in hot weather.
   C. Wool is cool on cold days, but cotton allows heat and moisture to be trapped against the body in cold weather.
   D. Wool is warm on warm days, but cotton allows heat and moisture to escape from the body in cold weather.
5. State the reason cotton and polyester blends are so commonly used for shirts and dresses.
6. Describe what a spinneret is used for.
7. **MC** Which fibre does Lycra® clothing always contain?
   A. Spandex
   B. Nylon
   C. Elastane
   D. Polyester

Apply and analyse

8. State which properties make plastic more suitable for use in outdoor furniture than:
   a. wood
   b. metal.
9. Identify which properties of plastic banknotes make them more suitable than the old paper ones.
10. State whether each of the following are properties of nylon; if yes, explain how the properties contributed to making nylon suitable for use as parachute material in World War II.
    a. Light but strong
    b. Strong and heavy
    c. Shrinks when washed
    d. Elastic
    e. Burns quickly
    f. Waterproof and dries quickly
    g. Builds up static electricity.

11. Explain why rayon is neither a natural fibre nor a synthetic fibre.
12. Describe how the properties of a pure cotton fabric would change by blending it with spandex.

Evaluate and create

13. **SIS** Examine the table provided, which lists some advantages and disadvantages of natural and synthetic fibres.

Fibre	Advantages	Disadvantages
Wool	• Warm in cold weather • Crease resistant • Burns slowly • Retains its shape well	• Shrinks when washed • Turns yellow in sunlight
Cotton	• Absorbs moisture • Soft • Cool in hot weather	• Creases easily • Burns quickly
Nylon	• Dries quickly • Light • Strong • Elastic	• Builds up static electricity • Melts rather than burns • Turns yellow in sunlight

  a. Which of the materials would burn slowly if it were exposed to flames?
  b. Deduce one advantage and one disadvantage of synthetic materials over natural materials.
  c. Deduce one disadvantage that one natural material and one synthetic material both have in common.
14. **SIS** Choose one of the properties in investigation 6.15 (one that you have not investigated) and decide on the equipment you would use to investigate it.

Fully worked solutions and sample responses are available in your digital formats.

LESSON
6.10 Recycling

LEARNING INTENTION

At the end of this lesson you will be able to identify the types of plastics that are easily recyclable and those that are more difficult. You will also understand how glass, metal and paper products can be sorted and recycled in a recycling facility.

6.10.1 Recycling and landfill

The material that you throw out as household rubbish is buried in landfill tips.

The food scraps that make up almost half of your household rubbish are biodegradable. They will be broken down by microbes and other decomposers in the soil, such as worms. Chemical changes take place when these organisms digest the scraps, returning nutrients to the soil. You can use compost bins or compost heaps to allow this to happen in your own backyard. Even paper and cardboard break down fairly quickly in the soil.

However, materials such as plastics, glass and metals take hundreds or even thousands of years to break down. The properties of these materials allow them to be **recycled**.

FIGURE 6.42 Look closely at this photo. This is not garbage. All these things can be recycled, including cardboard, paper, egg cartons, steel cans, plastics and glass.

6.10.2 Packaging

Just about everything you buy at the supermarket comes in a package. Even if it does not, you usually put it in a bag to take it home. The type of packaging needed depends on the properties of the product inside. For example, you cannot package tomato sauce in a paper bag. The most commonly used materials in packaging are paper (or cardboard), plastic, metal and glass. For a consumer, it is not just the properties of the packaging that are important. At least two questions should be asked when you make a choice about buying a product:

- Is the packaging recyclable?
- Is the packaging biodegradable?

If the packaging is glass, aluminium or steel, it is probably recyclable, which can save energy and water. If it is a plastic bottle, it is also likely to be recyclable. If the packaging is not recyclable, think about whether it is **biodegradable**; that is, can it be broken down by natural chemical reactions in the bodies of worms or other small **organisms** that live in the soil? Plastics, metals and glass are non-biodegradable. If they are thrown out with other household rubbish such as food waste, they end up in rubbish tips and will not break down. This creates the need for more rubbish tips. Of course, there is a limit to how much land can be used for rubbish tips in or near major towns and cities.

Paper is mostly biodegradable. Paper packaging that has been contaminated by food or oils, however, cannot be recycled. But at least when it gets to the rubbish tip it can be broken down in the soil. If you have a choice, choose items with packaging that is either recyclable or biodegradable.

recycle to reuse an unwanted substance or object for another purpose

biodegradable describes a substance that breaks down or decomposes easily in the environment

organisms living things

6.10.3 Recycling plastics

There are two very good reasons for recycling plastics:

- Plastics are non-biodegradable; that is, they are not broken down naturally by microorganisms. Plastics add thousands of tonnes of new rubbish to the environment every year.
- Plastics are made from oil — a resource that is expensive and dwindling. This is a chemical reaction. The continued production of new plastics is not **sustainable**. Recycling plastics is usually a physical change involving heat.

FIGURE 6.43 The symbols shown make the sorting of plastics before recycling easier.

Household waste contains many different types of plastic, which need to be separated. The plastics industry has introduced a code system to help consumers identify recyclable plastics. The symbols shown in table 6.2 make the sorting of plastics before recycling easier and cheaper. Some plastics are more easily recycled than others because of differences in the structure of the chains of molecules of which they are made.

sustainable describes the concept of using Earth's resources so that the needs of the world's present population can be met, without damaging the ability of future populations to meet their needs

TABLE 6.2 Symbols for types of plastics and recycling

Symbol and name	Details	Examples of uses	Recycling information
1 PET/PETE Polyethylene terephthalate	• Most used consumer plastic • Mainly for single-use products • Not heat resistant • Difficult to clean properly; should not be reused	Textiles such as fleece garments, carpets, stuffing for pillows and life jackets; soft drink and water bottles	• Widely recycled • Empty PET bottles completely and remove the lids before placing them in a home recycling bin.
2 HDPE High-density polyethylene	• Hard wearing; does not break down • Stronger than PET • Reusable • Suitable for freezing	Compost bins; irrigation pipes and plumbing fittings; household bags; milk, juice, water and detergent bottles	• Easily recycled • Empty HDPE bottles completely and remove the lids before placing them in a home recycling bin.

(continued)

TABLE 6.2 Symbols for types of plastics and recycling *(continued)*

Symbol and name	Details	Examples of uses	Recycling information
3 V PVC	• Can leach toxins • Not suitable for food and drinks • Useful for outdoor products because it doesn't break down • Produces toxic chemicals when heated and this limits the ability to be recycled	Juice and detergent bottles, PVC piping, credit cards	• Not easily recycled but can be made into more PVC products such as flooring • PVC bottles can be placed in a home recycling bin as long as the lids are removed. They are separated from the more easily recyclable plastics and sent to a separate plant for processing.
4 LDPE Low-density polyethylene	• Thin and flexible • Easy and inexpensive to produce • Safe to use with food	Frozen food bags, bin liners, squeezable bottles, flexible container lids, cling film, carry bags and packaging film, bubble wrap	• Bottles and other containers are recyclable and can be placed in a home recycling bin. • Soft, scrunchable plastics can often be returned to the supermarket for recycling (but it is best to avoid using them by using reusable bags for shopping).
5 PP Polypropylene	• Hard and light weight • Withstands heat • Resistant to grease and chemicals • Safe to reuse	Reusable microwave containers, kitchenware, nappies, yoghurt containers, straws, disposable cups and plates, landscaping border stripping, battery cases, margarine tubs	Can be recycled but depends on the product and local council
6 PS Polystyrene	• Lightweight and soft • Flammable • Inexpensive to make • Easily moulded • Can release harmful chemicals, particularly if heated	Packing 'peanuts'; disposable cups, plates and trays; insulation, disposable takeaway containers	Not easily recycled and must be disposed of in general waste; avoid use if possible

Symbol and name	Details	Examples of uses	Recycling information
7 OTHER Other	• Other plastics, including nylon, fibreglass and polycarbonate • Strong and tough • Possible release of hazardous BPA	Beverage bottles, baby milk bottles, electronic casing, lenses for sunglasses and safety goggles	Generally not recyclable and should not be placed in your home recycling bin

6.10.4 Recycling glass

cullet used glass

About 45 per cent of the glass packaging used in Australia is recycled. Used glass bottles, known as **cullet**, are collected and melted down in a furnace to produce new products. This is a physical change. The overall energy saving is only 8 per cent of that used in making new glass. This is because of the high cost of collecting and melting down the bottles. In some countries, milk is sold in bottles that can be sterilised and reused up to 50 times before they need melting down, which saves a large amount of energy.

6.10.5 Recycling paper and cardboard

Over a million tonnes of paper — about a third of our annual consumption — is recycled in Australia. Paper is made out of fibres of the chemical cellulose and is relatively easy to recycle. Waste paper is first mixed with water to separate the fibres. Additives such as ink and adhesives are then removed, producing low-quality fibres that can be used to make cardboard and other products. Steam rollers are used to improve the quality of the finished paper. Recycling paper reduces the amount of new paper needed, saving millions of trees.

6.10.6 Recycling metals

Metals such as steel and aluminium are easily recycled as long as they can be cheaply separated from other rubbish. Steel cans, aerosol containers, jar lids and bottletops can be recycled. The recycling of aluminium cans saves huge amounts of energy. Twenty aluminium cans can be recycled with the same amount of energy needed to produce just one new can.

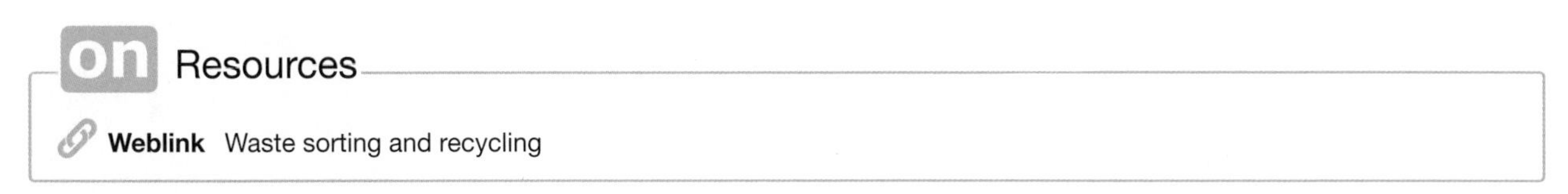

ACTIVITY: Design your own waste disposal system

Your group is responsible for preparing a report on ways to improve the household waste management and disposal for the shire of Green Valley.

The shire currently collects rubbish from its 134 500 ratepayers using large green bins that are emptied by compactor trucks. The rubbish is taken to the local tip and used as landfill, at a cost to the council of $60 per tonne. The tip is nearing capacity and will be closed within 12 months. Waste paper is collected separately by a private recycling company.

Your report could be produced in written form or as audio or video. It should address the following issues:
1. How will the shire encourage each household to produce less waste?
2. Is recycling too costly?
3. If recycling occurs, will recyclable wastes such as plastics, glass and metals be separated at a disposal station after collection, or collected in separate containers from households?
4. What measures will be used to encourage households to use compost bins?
5. How will the shire dispose of rubbish when the landfill site closes?

6.10.7 Sorting out recyclables

The separation of the items in your recycling bin relies on differences in their physical properties, including size, weight, magnetic properties and even colour. For example, items of different weights can be separated using blasts of air or a centrifuge that works like the spin dryer of a washing machine. Steel can be separated from other metals by a large magnet.

Special recycling programs

There are separate recycling programs for some products that cannot be placed in home recycling bins. These recycling programs are generally used to collect products containing substances that would endanger the environment or the community if they were dumped in landfill tips. For example, printer cartridges can be placed in recycling boxes at many Australia Post outlets and retail stores that sell computers and printers. Mobile phones can be left at most mobile phone outlets for recycling. Use the **Recycling weblink** in your Resources tab to find out where computers and other electronic equipment, white goods such as fridges and washing machines, corks, light globes and many other items are collected for recycling. This website also provides information about how to dispose of chemical wastes from home, school or industry. Oil, paints and unused medicines should not be placed in rubbish bins or flushed down the sink.

FIGURE 6.44 This compost bin is made from recycled polypropylene (PP). The compost decreases in volume as it breaks down. Almost 50 per cent of domestic waste in Australia is suitable for composting.

6.10.8 You can make a difference

The three-bin collection system used by many city and shire councils throughout Australia makes it very easy for you to make a difference to the environment by recycling.

For example, recycling paper:
- reduces the amount of energy needed to produce new paper
- prevents tonnes of greenhouse gases from entering the atmosphere from manufacturing new paper
- saves the many litres of water used in paper production; less water is used in recycling it.

on Resources

Weblink Recycling

6.10 Activities

learn on

6.10 Quick quiz on	6.10 Exercise

Select your pathway

■ LEVEL 1	■ LEVEL 2	■ LEVEL 3
1, 2, 3, 4, 5, 6, 8	7, 9, 11, 12	10, 13, 14, 15, 16

Remember and understand

1. MC How are biodegradable substances different from those that are non-biodegradable?
   A. Non-biodegradable substances are referred to as such because they are not made from biological material.
   B. Biodegradable waste cannot be broken down by decomposers.
   C. Biodegradable substances are referred to as such because they are made from biological material.
   D. Biodegradable waste can be broken down by decomposers.
2. Recall and state where the chemical changes that break down biodegradable waste take place.
3. Identify two of the main benefits of recycling plastics.
4. MC Why are plastics such as PET now identified by a code?
   A. To identify how long plastics can be used for
   B. To identify which plastics can be heated
   C. To identify which plastics can be recycled
   D. To identify which plastics are safe for food items
5. **a.** True or false? Cullet is a type of plastic.
   b. Explain your response to part **a**.
6. MC The first part of the process of recycling paper involves mixing it with water. What is the major purpose of mixing the paper with water?
   A. To begin the chemical reaction that breaks down paper
   B. To cause a change of state from solid to liquid
   C. To increase the effect of temperature
   D. To weaken the forces between the fibres
7. List three problems associated with the disposal of waste in landfill sites.

8. **a.** True or false? Only a small amount of energy is saved when glass is recycled.
   b. Explain your response to part **a**.
9. List two benefits of recycling aluminium.
10. Ink is removed from paper when it is recycled. Is the combination of ink and paper a compound or a mixture? Explain your answer.

Apply and analyse

11. Most plastics are non-biodegradable. Metals and glass are also non-biodegradable.
    a. What does biodegradable mean?
    b. If non-biodegradable rubbish cannot be recycled, what happens to it?

12. **SIS** The following graph shows the amount of waste that went to landfill compared to the amount of waste that was recycled between 2006 and 2021.

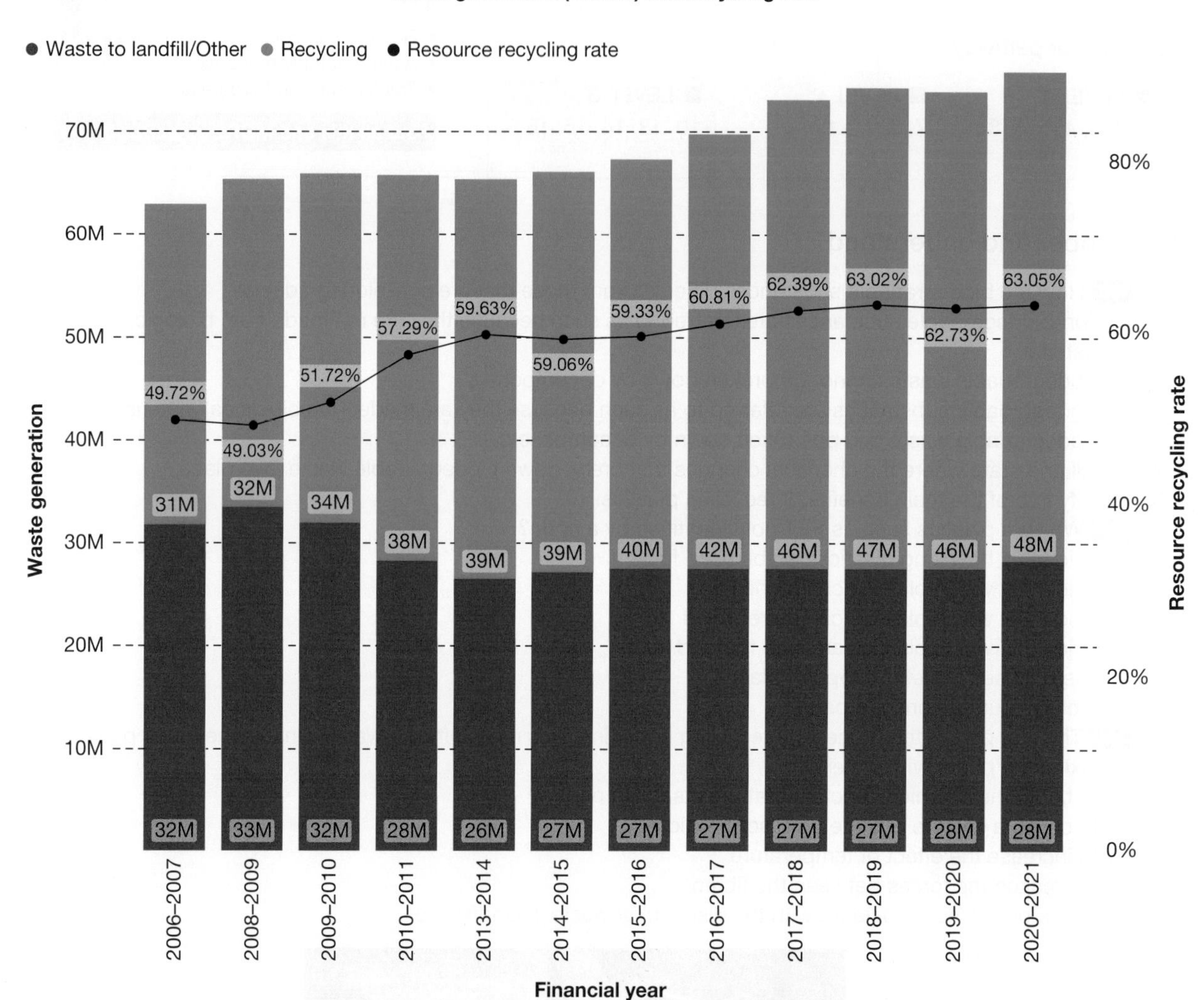

Source: Based on data from Department of Climate Change, Energy, the Environment and Water. 2022. Waste and Resource Recovery Data Hub - National waste data viewer, https://www.dcceew.gov.au/environment/protection/waste/how-we-manage-waste/data-hub/data-viewer.

From this graph:

a. identify which method of dealing with waste was the least used in the years from 2006 until 2021

b. describe the general trend in waste going to landfill.

13. **SIS** If a plastic-bag manufacturer claimed that the bags it produced were biodegradable, what evidence would you need to be satisfied that the claim was correct?

Evaluate and create

14. **a.** What factors influence the decision as to whether it is worth the trouble of recycling a resource?

    b. **SIS** Which of these factors can be investigated scientifically?

15. **SIS** Research and report on what different alternatives to plastics are currently being developed.

16. **SIS** Current projects include making roofing tiles from milk-bottle tops. Investigate how plastics are being recycled in imaginative ways and share your findings.

Fully worked solutions and sample responses are available in your digital formats.

LESSON
6.11 Thinking tools — Target maps

6.11.1 Tell me

What is a target map?

A target map is a very useful thinking tool that can assist you to identify (target) which concepts are relevant to the topic and which concepts are not relevant to the topic. They are sometimes called circle maps.

Why use a target map over a single bubble map?

Similar to single bubble maps, target maps identify and describe the subtopics and concepts that relate to the topic of content. However, single bubble maps do not separate the relevant material from the non-relevant material.

FIGURE 6.43 Target map

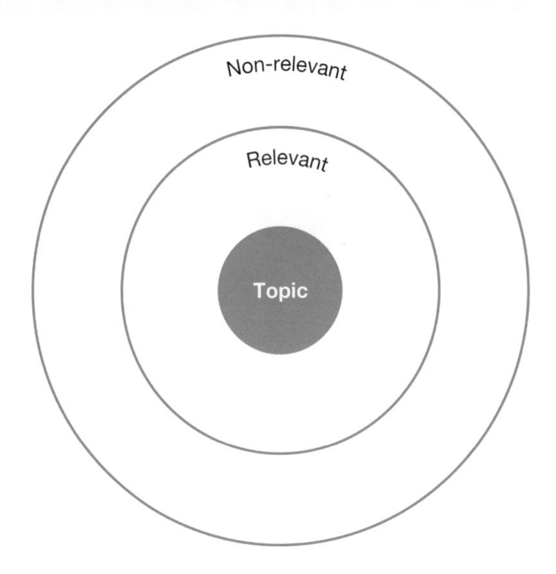

FIGURE 6.44 Single bubble map

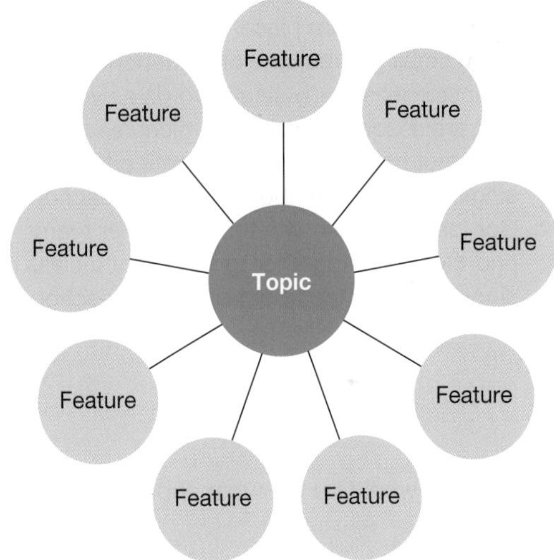

6.11.2 Show me

To create a target map:

1. Draw three concentric circles on a sheet of paper.
2. Write the topic in the centre circle.
3. In the next circle, write words and phrases that are relevant to the topic.
4. In the outer circle, write words and phrases that are not relevant to the topic.

The example given shows a single bubble map of the methods of rusting.

FIGURE 6.45 Target map of rusting

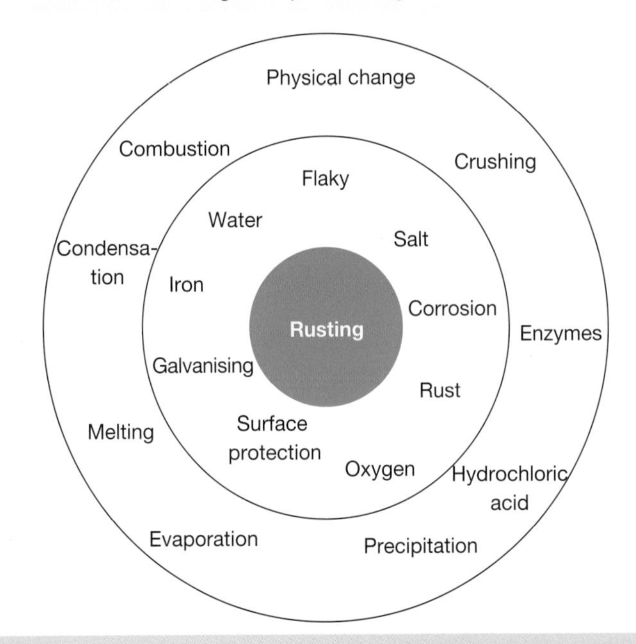

6.11.3 Let me do it

6.11 Activity

These questions are even better in jacPLUS!

- Receive immediate feedback
- Access sample responses
- Track results and progress

Find all this and MORE in jacPLUS

1. Construct a target map about physical change. Use each of the words shown in your target map.

Melting	Burning	Precipitate
Evaporation	Condensation	Explosion
Reactant	Stretching	
Rusting	Freezing	

2. The single bubble map shown identifies some of the ideas associated with a burning candle.
   a. Draw your own single bubble map about the topic 'A burning candle', adding as many additional bubbles as you can.
   b. Construct a single bubble map that identifies clues that provide evidence that a chemical reaction has taken place.
3. a. Form small groups, or work independently, to brainstorm as many single words as you can that are associated with chemical reactions.
   b. Use your brainstorm list to create a team single bubble map about chemical reactions.
   c. Compare your list with those of other teams and then work together to construct a class single bubble map about chemical reactions.

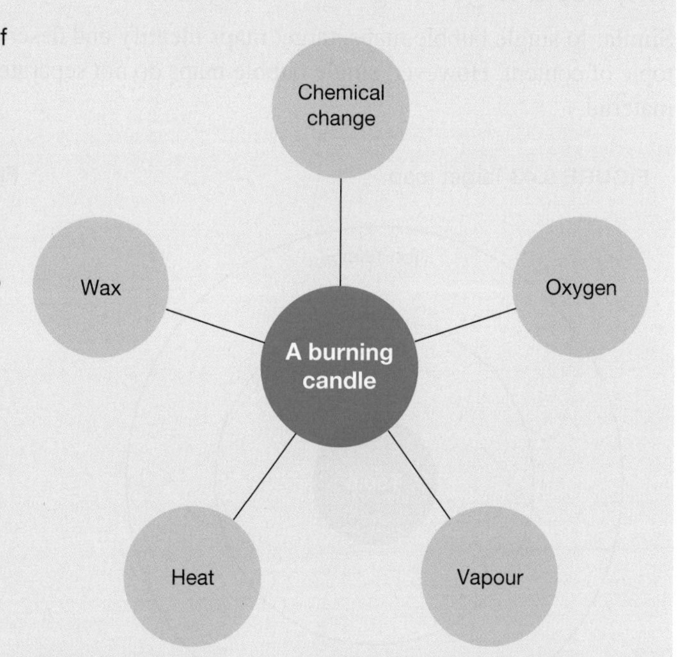

4. What topic would be the most appropriate for the target map shown?

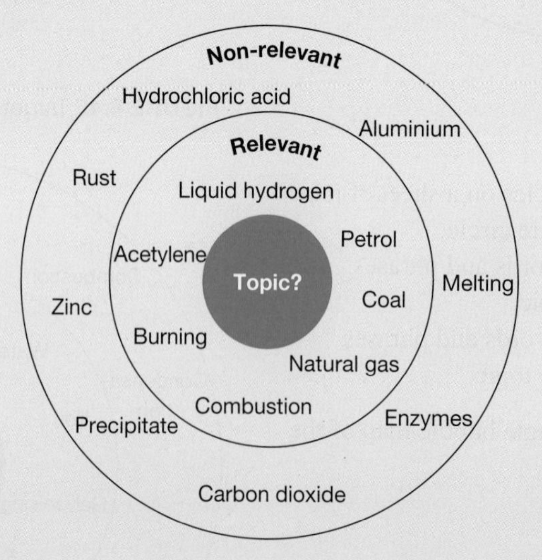

5. Create a target map about recycling.

Fully worked solutions and sample responses are available in your digital formats.

LESSON
6.12 Review

Hey students! Now that it's time to revise this topic, go online to:

Review your results

Watch teacher-led videos

Practise questions with immediate feedback

Find all this and MORE in jacPLUS

Access your topic review eWorkbooks

Resources

Topic review Level 1	Topic review Level 2	Topic review Level 3
ewbk-12331	ewbk-12333	ewbk-12335

6.12.1 Summary

Physical and chemical properties

- Physical properties can either be observed or measured directly.
- Chemical properties are those that describe how a substance combines with other substances to form new chemicals, or how a substance breaks up into two or more different substances.
- There are many chemical properties, including flammability, reactivity and toxicity.

Chemical and physical changes

- Chemical change results in at least one new substance being formed due to the breaking and forming of chemical bonds and rearrangement of atoms in a reaction.
- Evaporation is a change of state from a liquid to a gas.
- During physical changes no new substances are formed.
- Word equations for both physical changes and chemical changes have the starting substances on the left-hand side and the final substances on the right-hand side, separated by an arrow.

Chemical reactions

- A chemical reaction is a chemical change in which a new substance or substances are produced.
- Photosynthesis is a chemical reaction that life depends on. Carbon dioxide and water react with energy from the Sun to form glucose and oxygen.
- Reactants are the chemical substances used up in a chemical reaction.
- The substances that are produced in a chemical reaction are called products.
- Evidence of a chemical reaction occurring includes bubbles, odour, change in temperature, a precipitate, light or a flame, and colour change.
- When writing word equations, it does not matter which of the reactants or which of the products are written first.

Chemical reactions and energy

- Many chemical reactions must be initiated by an input of energy.
- Endothermic reactions absorb heat and exothermic reactions release heat.
- Airbags are powered by a reaction that produces a large amount of gas in a very short time.
- Alfred Nobel invented a stable version of dynamite.

Corrosion

- Rusting is an example of corrosion, which is a chemical reaction that occurs when substances in the air or water around a metal cause it to deteriorate.
- Rusting can be avoided using galvanising or surface protection.

Combustion

- Reactions that involve combination with oxygen, such as burning, are examples of oxidation reactions.
- When fossil fuels are combusted with oxygen, the products are heat, carbon dioxide and water, according to the reaction:

$$\text{Fossil fuel} + \text{oxygen} \longrightarrow \text{carbon dioxide} + \text{water}$$

- The production of carbon dioxide from burning fossil fuels is a major contributor to greenhouse gas emissions and climate change.
- Rocket fuels rely on the combustion of hydrogen and oxygen to produce energy and water according to the reaction:

$$\text{Hydrogen} + \text{oxygen} \longrightarrow \text{water}$$

Indicators of chemical change in our everyday world

- Indicators of chemical change are used for identifying the presence of particular substances, such as in soil, water and medical testing kits.
- Many tests use colour change as evidence of particular substances.
- Tests that measure pH utilise a colour change that gives an indication of the presence and amount of different metals in soil.
- Water testing kits utilise a change in colour to indicate not only pH but the presence of substances such as lead, iron, copper, nitrates, chlorine, fluoride, toxins, fertiliser and even microorganisms.
- Urine test strips are used in the medical industry to identify and measure the amounts of various substances in the body.

Plastics and fibres

- Plastics are polymers produced from a series of monomers.
- Natural fibres include wool and silk from animals, and cotton from plants.

Recycling

- Recycling involves reusing an unwanted substance or object for another purpose.
- Biodegradable substances break down or decompose easily in the environment.

6.12.2 Key terms

biodegradable describes a substance that breaks down or decomposes easily in the environment
burning combining a substance with oxygen in a flame
chemical change a change that results in at least one new substance being formed due to the breaking and forming of chemical bonds and rearrangement of atoms in a reaction
chemical energy energy stored in chemical bonds that is released during chemical reactions
chemical properties properties that describe how a substance combines with other substances to form new chemicals, or how a substance breaks up into two or more different substances
chemical reaction a chemical change between two or more substances in which one or more new chemical substances are produced
chlorophyll the green-coloured chemical in plants, located in chloroplasts, that absorbs light energy so that it can be used in the process of photosynthesis
corrosion a chemical reaction between air, water or chemicals in the air or water with a metal, which causes the metal to wear away
cullet used glass

ductile capable of being drawn into wires or threads; a property of most metals
durability the quality of lasting; not easily being worn out
dynamite a relatively stable explosive invented by Alfred Nobel in 1866; it is created by mixing nitroglycerine with an absorbent substance such as silica, forming a paste that can be shaped into rods
elastic describes a material that is able to return to its original size after being stretched
elasticity the property that allows a material to return to its original size after being stretched
endothermic refers to chemical reactions that absorb heat energy from the surroundings
evaporates changes state from a liquid to a gas
exothermic refers to chemical reactions that give out heat energy to the surroundings
flammability an indicator of how easily a substance catches fire
fossil fuel a substance, such as coal, oil or natural gas, that has formed from the remains of ancient organisms; coal, oil and natural gas are often used as fuels — that is, they are burnt in order to produce heat
galvanising protecting a metal by covering it with a more reactive metal that will corrode first
malleable able to be beaten, bent or flattened into shape
natural fibres fibres that form naturally — that is, they have not been made by humans; they include wool and silk from animals, and cotton from plants
nylon a synthetic fibre; the monomers are joined together by the elimination of water molecules at the joins
organisms living things
oxidation a chemical reaction involving the loss of electrons by a substance
physical changes changes in which no new chemical substances are formed; a physical change may be a change in shape, size or state, and many of these changes are easy to reverse
physical properties properties that you can either observe using your five senses — seeing, hearing, touching, smelling and tasting — or measure directly
polyester a synthetic fibre; the monomers are joined together by the elimination of water molecules at the joins
precipitate the new, solid product produced when reactants are mixed together; a precipitate is insoluble in water
products new chemical substances that result from a chemical reaction; new chemical bonds are formed to make the products during a chemical reaction
reactants chemical substances used in a chemical reaction; chemical bonds of the reactants are broken during a chemical reaction
reactivity a measure of how likely a particular substance reacts to make new substances
recycle to reuse an unwanted substance or object for another purpose
rust a red-brown substance formed when iron reacts with oxygen and water
rusting the corrosion of iron
spinneret a nozzle with small holes through which a plastic material passes, forming threads; also the organ used by spiders to create their webs
state the condition or phase of a substance; the three main states of matter are solid, liquid and gas
surface protection refers to when a protective coating is applied over a metal surface to prevent corrosion
sustainable describes the concept of using Earth's resources so that the needs of the world's present population can be met, without damaging the ability of future populations to meet their needs
toxicity the danger to your health caused when poisonous substances combine with chemicals in your body to produce new substances with damaging effects

on Resources

eWorkbooks	Study checklist (ewbk-12337) Literacy builder (ewbk-12338) Crossword (ewbk-12340) Word search (ewbk-12342) Reflection (ewbk-12344)
 **Solutions**	Topic 6 Solutions (sol-1118)
 **Practical investigation eLogbook**	Topic 6 Practical investigation eLogbook (elog-2269)
 **Digital document**	Key terms glossary (doc-40131)

6.12 Activities

learn**on**

6.12 Review questions

Select your pathway

■ LEVEL 1	■ LEVEL 2	■ LEVEL 3
1, 5, 7, 13, 15, 17	2, 6, 10, 11, 12, 14, 19	3, 4, 8, 9, 16, 18, 20

These questions are even better in jacPLUS!

- Receive immediate feedback
- Access sample responses
- Track results and progress

Find all this and MORE in jacPLUS

1. Match the substances on the left to the list of properties on the right.

Substance	Properties
a. Glass	**A.** Flexible, biodegradable
b. Metal	**B.** Transparent, unreactive, strong
c. Plastics	**C.** Malleable, ductile, good electrical conductor
d. Paper	**D.** Mouldable, light, strong

2. **a.** True or false? Physical properties are those that you can observe or measure using your senses or measuring instruments. Chemical properties are those that describe how a substance combines with other substances.
   b. Explain your response to part **a**.
3. Match the property on the left to its meaning on the right.

Property	Meaning
a. Ductile	**A.** Able to be rolled or beaten into sheets
b. Reactive	**B.** Substances that can damage living things when taken into the body
c. Malleable	**C.** Readily combines chemically with other substances
d. Lustrous	**D.** Lets light through without scattering
e. Toxic	**E.** Temperature at which a solid changes into its liquid form
f. Transparent	**F.** Able to be drawn into wires
g. Melting point	**G.** Shiny

4. Determine whether each of the following scenarios is a chemical change or physical change. Then write a word equation describing the change.
   a. The wax on a burning candle melts.
   This is a ____________ change.
   ____________ ⟶ ____________
   b. The wax vapour at the top of a candle wick burns with oxygen to produce carbon dioxide, water vapour and heat.
   This is a ____________ change.
   Wax vapour + ____________ ⟶ ____________ + water
   c. Calcium carbonate is dissolved by hydrochloric acid to form calcium chloride, water and carbon dioxide gas.
   This is a ____________ change.
   Solid calcium carbonate + ____________ ⟶ ____________ + carbon dioxide gas + water
   d. Hydrogen gas explodes with oxygen gas to form water.
   This is a ____________ change.
   ____________ ⟶ ____________

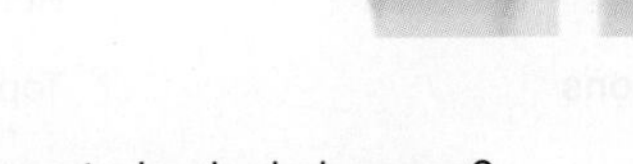

5. **MC** How do you know that toasting bread and the rusting of a nail are not physical changes?
   A. A new colour is produced.
   B. A new chemical has been formed.
   C. A gas is produced.
   D. Heat is produced.

6. When a lead nitrate solution is added to a potassium iodide solution, a chemical reaction takes place. A bright yellow solid appears. It is the compound lead iodide. Another compound, potassium nitrate, remains in the solution and is not visible.
   a. **MC** Which of the following are the reactants in the reaction? Select all possible answers from the options given.
      A. Lead nitrate
      B. Lead iodide
      C. Potassium iodide
      D. Potassium nitrate
      E. Water
   b. **MC** Which of the following are the products in the reaction? Select all possible answers from the options given.
      A. Lead nitrate
      B. Lead iodide
      C. Potassium iodide
      D. Potassium nitrate
      E. Water

   c. The yellow lead iodide will eventually settle to the bottom of the flask. What 11-letter word beginning with 'p' is given to a substance that behaves like the lead iodide?
   d. **MC** Which of the following is the chemical word equation for the reaction?
      A. Lead iodide + potassium iodide ⟶ lead nitrate + potassium nitrate
      B. Lead nitrate + lead iodide ⟶ potassium iodide + potassium nitrate
      C. Lead iodide + potassium nitrate ⟶ lead nitrate + potassium iodide
      D. Lead nitrate + potassium iodide ⟶ lead iodide + potassium nitrate
7. a. True or false? The chemical word equation for rusting is:

   Iron + water + carbon dioxide ⟶ rust (iron oxide)

   b. Explain your response to part **a**.
8. **SIS** For each of the reactions provided, match it to the way that the reaction could be made to happen more quickly.

Reaction	How to speed up the reaction
a. Burning a pile of dry leaves	**A.** Leave it in a warm place
b. Cooking potatoes	**B.** Place it in a salty environment
c. Dissolving marble chips in acid	**C.** Fan the fire to increase the amount of oxygen
d. Making an iron nail go rusty	**D.** Cut them into smaller pieces
e. Milk going sour	**E.** Heat the acid

9. Some chemical reactions can be destructive. Write down three examples of harmful chemical reactions.

Apply and analyse

10. Children's steel swing sets in beachside towns and suburbs rust much faster than those further from the coast.
    a. Explain why this happens.
    b. Suggest two methods of slowing down or preventing the rusting of steel swing sets.

11. The oxyacetylene torch shown is used to melt metals to allow them to be joined together.
   a. **MC** What type of chemical reaction takes place in the oxyacetylene torch?
      A. Corrosion reaction
      B. Combustion reaction
      C. Addition reaction
      D. Rusting reaction
   b. **MC** What is the evidence in the photo that suggests a chemical reaction has taken place?
      A. A bright light and small sparks are being produced.
      B. The photo is mostly dark.
      C. The welder is wearing gloves.
      D. The welder is wearing a face shield.

12. Just as chemicals can be grouped or classified, so can chemical reactions. Match the chemical reactions on the left to the type of reaction on the right.

Chemical reaction	Type of reaction
a. The corrosion of iron	**A.** Combustion
b. The reaction of substances with oxygen	**B.** Rusting
c. Burning	**C.** Oxidation

13. This illustration shows a camper boiling water in a billy over a camp fire.

   a. State the three physical changes that are shown in the image.
   b. Identify which chemical change is shown taking place.
14. Which two properties of the plastic used to make light switches and power points make it right for the job?
15. **MC** What is the difference (other than their properties) between natural and synthetic fibres?
   A. Natural fibres are made from plants, whereas synthetic fibres are made from animals.
   B. Natural fibres are made from plants or animals, whereas synthetic fibres are made from crude oil.
   C. Natural fibres are made from animals, whereas synthetic fibres are made from plants.
   D. Natural fibres are made from crude oil, whereas synthetic fibres are made from plants or animals.

16. Classify whether each of the fibres listed are natural, synthetic or a combination of both.

Fibre	Synthetic	Natural	Combination
a. Nylon			
b. Cotton			
c. Rayon			
d. Lycra			
e. Wool			
f. Polyester			

17. Explain how synthetic fibres, such as nylon, are made.

18. Match the substances on the left to the properties that would be essential for their packaging on the right.

Substances	Properties of packaging
a. Pool chemicals	**A.** Gas-tight, strong, chemically resistant to the drink
b. Eggs	**B.** Airtight, opaque to light
c. Soft drink	**C.** Resistant to the chemicals inside, opaque to light
d. Peanuts	**D.** Lightweight, strong, rigid

Evaluate and create

19. **SIS** Some plastic containers are marked with this symbol.

1

   a. **MC** Which two of the following substances would you expect to find in bottles made from this type of plastic?
   A. Wine
   B. Soft drink
   C. Eggs
   D. Milk

   b. **MC** What two things should you do before placing bottles made from this type of plastic in a recycling bin?
   A. Fill them up with water.
   B. Empty the contents.
   C. Remove the lids.
   D. Screw on the lids.

   c. **MC** Which two of the following are uses for this type of plastic after it has been recycled?
   A. Carpet fibres
   B. This plastic cannot be recycled
   C. Paper
   D. Glass bottles
   E. Flower tubes

20. **SIS** Describe the 'three-bin system' used by many cities and shires in Australia, and explain how it helps the environment.

Fully worked solutions and sample responses are available in your digital formats.

Online Resources

Below is a full list of **rich resources** available online for this topic. These resources are designed to bring ideas to life, to promote deep and lasting learning and to support the different learning needs of each individual.

6.1 Overview

 **eWorkbooks**
- Topic 6 eWorkbook (ewbk-12312)
- Starter activity (ewbk-12314)
- Student learning matrix (ewbk-12188)

 **Solutions**
- Topic 6 Solutions (sol-1118)

 **Practical investigation eLogbooks**
- Topic 6 Practical investigation eLogbook (elog-2269)
- Investigation 6.1: Investigating chemical reactions (elog-2267)

 **Video eLesson**
- A strong acid is poured into a solution containing glucose (eles-2584)

6.2 Physical and chemical properties

 **eWorkbook**
- Properties of materials (ewbk-12316)

 **Practical investigation eLogbook**
- Investigation 6.2: Describing properties (elog-2271)

6.3 Chemical and physical changes

 **eWorkbook**
- Changing states (ewbk-12318)

 **Practical investigation eLogbook**
- Investigation 6.3: A burning candle (elog-2273)

6.4 Chemical reactions

 **eWorkbooks**
- Physical and chemical changes (ewbk-12320)
- Describing chemical changes (ewbk-12322)

 **Practical investigation eLogbooks**
- Investigation 6.4: Heating copper carbonate (elog-2275)
- Investigation 6.5: Magnesium metal in hydrochloric acid (elog-2277)
- Investigation 6.6: Steel wool in copper sulfate solution (elog-2279)

Video eLessons
- Precipitation (eles-2058)
- Magnesium metal burning (eles-2303)
- Baking and carbon dioxide (eles-2059)
- Magnesium and hydrochloric acid (eles-2294)

6.5 Chemical reactions and energy

 **eWorkbook**
- Exothermic and endothermic reactions (ewbk-12324)

 **Practical investigation eLogbooks**
- Investigation 6.7: Exothermic and endothermic processes (elog-2281)
- Investigation 6.8: Instant ice pack (elog-2283)

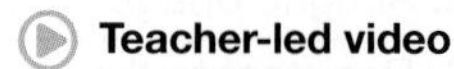 **Teacher-led video**
- Investigation 6.7: Exothermic and endothermic processes (tlvd-10789)

Video eLesson
- An explosion in a quarry (eles-2587)

6.6 Corrosion

 **eWorkbook**
- Rusting (ewbk-12326)

 **Practical investigation eLogbooks**
- Investigation 6.9: Observing rusting (elog-2285)
- Investigation 6.10: Investigating the corrosion of different metals (elog-2287)
- Investigation 6.11: Rusting and salt water (elog-2289)

6.7 Combustion

 **eWorkbook**
- Combustion (ewbk-12328)

 **Practical investigation eLogbook**
- Investigation 6.12: Burning paper (elog-2291)

6.8 Indicators of chemical change in our everyday world

 **Practical investigation eLogbooks**
- Investigation 6.13: Examining flower colour under different pH conditions (elog-2293)
- Investigation 6.14: Water testing (elog-2295)

6.9 Plastics and fibres

 **Practical investigation eLogbook**
- Investigation 6.15: Testing fibres (elog-2297)

 **Video eLesson**
- The future of clothing (eles-0859)

6.10 Recycling

Weblinks

- Waste sorting and recycling
- Recycling

6.12 Review

eWorkbooks

- Topic review Level 1 (ewbk-12331)
- Topic review Level 2 (ewbk-12333)
- Topic review Level 3 (ewbk-12335)
- Study checklist (ewbk-12337)
- Literacy builder (ewbk-12338)
- Crossword (ewbk-12340)
- Word search (ewbk-12342)
- Reflection (ewbk-12344)

Digital document

- Key terms glossary (doc-40131)

To access these online resources, log on to **www.jacplus.com.au**

7 Sedimentary, igneous and metamorphic rocks

CONTENT DESCRIPTION

Describe the key processes of the rock cycle, including the timescales over which they occur, and examine how the properties of sedimentary, igneous and metamorphic rocks reflect their formation and influence their use (AC9S8U04)

LESSON SEQUENCE

SCIENCE INQUIRY AND INVESTIGATIONS

Science inquiry is a central component of the Science curriculum. Investigations, supported by a **Practical investigation eLogbook** and **teacher-led videos**, are included in this topic to provide opportunities to build Science inquiry skills through undertaking investigations and communicating findings.

LESSON
7.1 Overview

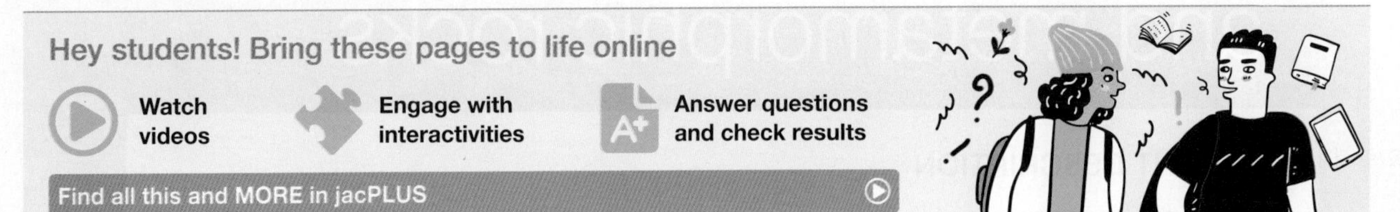

7.1.1 Introduction

Earth's surface is constantly changing. Volcanoes and earthquakes can cause quick changes, but most of the changes to Earth's surface happen slowly. Rocks on and below the surface of Earth are the records of these natural events. A geologist is a scientist who learns to read each type of rock, much like reading a chapter in a book, to discover the chapters of Earth's history. An important record held in the rocks includes the evolution of life.

FIGURE 7.1 Fossil of *Crocodilaemus robustus* that lived 160 million years ago

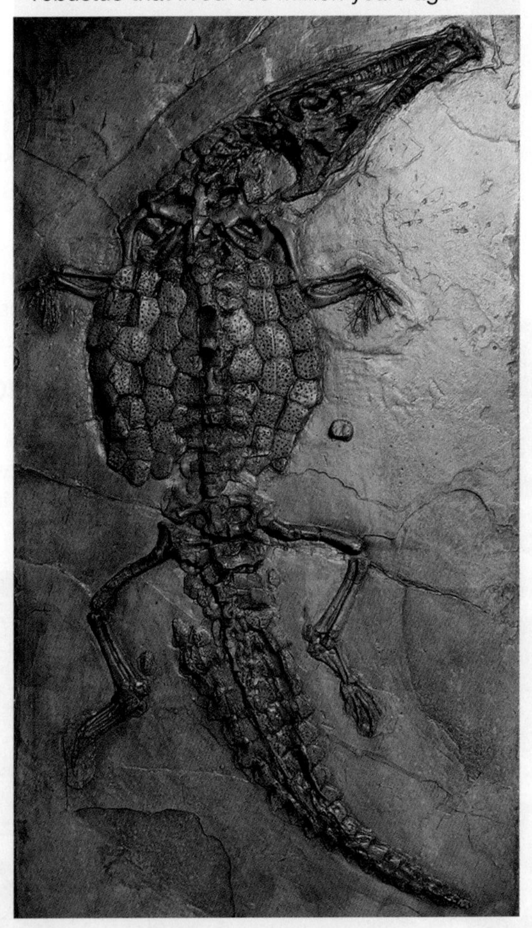

Locked in the limestone rocks of Cerin, France, is a fossil of the *Crocodilaemus robustus*, an extinct reptile similar to crocodiles. The rocks suggest that it lived in a near-tropical lagoon environment about 160 million years ago, long before humans were around. It was a small reptile, only 60 centimetres long, with long back legs that probably made walking on land easier. The strong plated armour on its abdomen and tail protected it, but also made it a slower swimmer. If you visit Cerin today, you won't see a lagoon or a living *Crocodileimus robustus* because Earth's surface is continually changing and life has evolved with it. The best way to learn about the past is to look at the rocks, and know the past is a tool to predicting the future.

7.1.2 Think about rocks

1. Which rock is light enough to float on water?
2. Which rocks are formed from the remains of living things?
3. What do butterflies, frogs, werewolves and metamorphic rocks have in common?
4. How do we know what living things that have not existed for millions of years looked like, how they walked and what they ate?
5. How can whole skeletons of animals be fully preserved for millions of years?
6. What can you learn from a dinosaur footprint?
7. Why did the dinosaurs vanish from Earth 65 million years ago?

7.1.3 Science inquiry

Rock types

Rocks are classified as either igneous, sedimentary or metamorphic. Each of these names provides a clue to how they formed. For example, the word 'igneous' comes from the Latin word *ignis*, meaning 'fire'. That is, these rocks have formed from the cooling and hardening of fiery hot, melted rock. As a class, discuss the following questions.

1. The word 'sediment' comes from the Latin word *sedere*, meaning 'settle', or 'sit'. What could this imply about how sedimentary rocks form?
2. The word *meta* is Greek for 'change' and *morpho* means 'form'. What could this imply about how metamorphic rocks form?

FIGURE 7.2 The distribution of rocks in the crust and across the surface of Earth

a. Crust Sedimentary rock 5% Igneous rock 95%

b. Surface

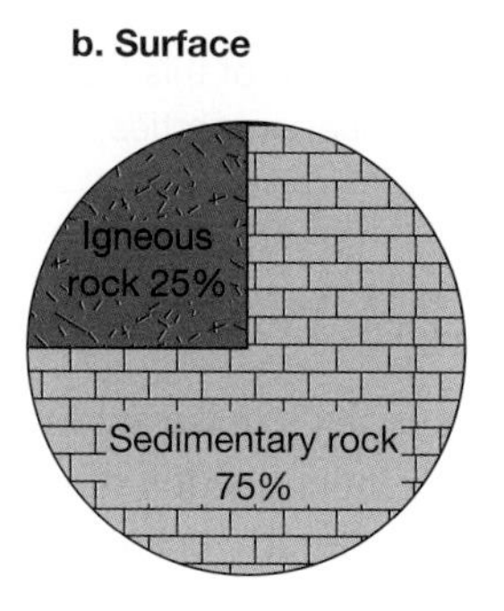

Further to your answers above, consider figure 7.2, which illustrates the distribution of rocks in the crust (the top layer that averages 6–35 km thick) and the surface of Earth. In these pie charts, metamorphic rocks are included within the rocks that they were formed from. That is, metamorphic rocks derived from sedimentary rocks are included in the sedimentary rock total.

3. Igneous rocks, or metamorphic rocks derived from igneous rocks, account for 95 per cent of all rocks in Earth's crust; sedimentary rocks account for 5 per cent. However, when we look at just Earth's surface, sedimentary rocks make up 75 per cent. What does this tell you about the nature of igneous and sedimentary rocks?

Bathroom rocks

When you last used the bathroom, you probably weren't thinking about rocks. After all, what does a bathroom have to do with rocks? But where did the materials to make the shower recess come from? What about the taps and pipes that deliver the water? Where do the materials to make tiles come from? And what about the toothpaste? The answers to all of these questions lead back to rocks. For example, metals are extracted from rocks and are used to make the steel taps.

FIGURE 7.3 Taps, tiles, glass and even mirrors. Where do the materials needed to produce these come from?

Work in small groups to research and answer the following questions. You may wish to use the Mining makes your Smart Home **weblink** in the resources tab.

4. What materials are mirrors made from?
5. What metal is primarily used to make bathroom taps? Where do we get the metal?
6. What are bathroom tiles and the toilet basin made from?
7. List some building materials that are:
   a. made directly from rocks
   b. not made directly from rocks but can be traced back to rocks.

Resources

eWorkbooks	Topic 7 eWorkbook (ewbk-11937) Starter activity (ewbk-11939) Student learning matrix (ewbk-11941)
 **Solutions**	Topic 7 Solutions (sol-1119)
 **Practical investigation eLogbook**	Topic 7 Practical investigation eLogbook (elog-2173)
 **Weblink**	Mining makes your Smart Home

LESSON
7.2 Rocks and minerals

LEARNING INTENTION

At the end of this lesson you will be able to define a mineral and recognise that they have unique chemical and physical properties, and understand how they differ from rocks.

7.2.1 What's in a rock?

Firstly, let's make sure we know what a rock is. To call something a rock, it needs to be a naturally occurring, coherent collection of minerals, organic material and/or glass.

- Naturally occurring: must be formed by natural process; it is not manufactured
- Coherent: holds together; for example, form cliffs
- Collection of minerals, organic matter and/or glass: some rocks only contain a collection of one type of mineral, some contain several different minerals, some are entirely made of glass and some are composed of large volumes of organic matter (e.g. coal).

Never confuse rocks and minerals. Minerals are to rocks as letters are to words.

DISCUSSION

Discuss whether these items are considered to be a rock or not. Be sure you can explain your answer.

- Footpath cement
- Benchtop granite
- Bricks
- Bathroom marble
- Garden soil

How rocks are formed

The rocks we can see have formed in Earth's **lithosphere** (see figure 7.4), which includes Earth's crust and the top part of its mantle. Rocks are classified into one of three groups based on how they formed: **igneous**, **sedimentary** or **metamorphic rocks** (table 7.1).

FIGURE 7.4 The rocks we can see have formed in Earth's lithosphere.

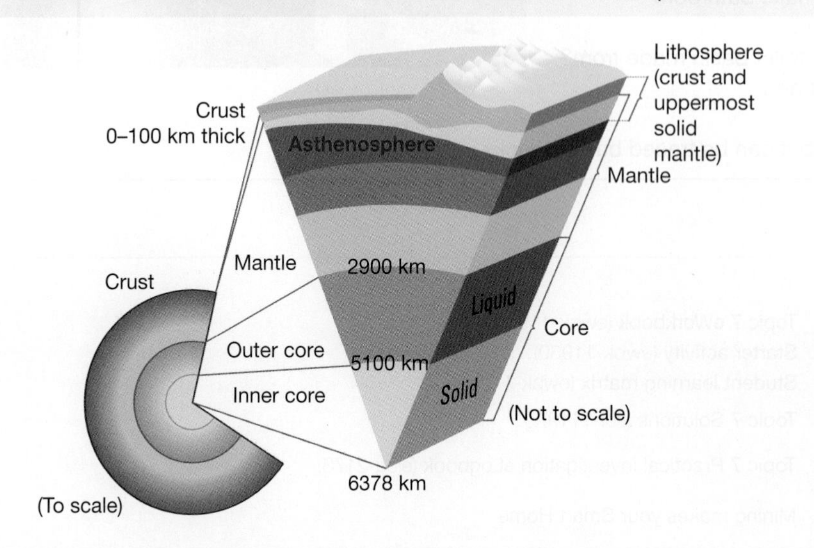

lithosphere the outermost layer of Earth; includes the crust and uppermost part of the mantle

igneous rocks rocks formed when hot, molten rock cools and hardens (solidifies)

sedimentary rocks rocks formed through the deposition and compaction of layered sediment

metamorphic rocks rocks formed from the change (alteration) of pre-existing rocks in response to increasing temperature and/or pressure conditions

asthenosphere the zone of mantle beneath the lithosphere, thought to be much hotter and more fluid than the lithosphere mantle

TABLE 7.1 The three types of rocks and how they formed

Igneous rocks	Sedimentary rocks	Metamorphic rocks
If the right temperature and pressure conditions are met, rocks can melt into **magma**. New rocks are formed when magma gets close to the surface and slowly *cools and solidifies*. Some of that red-hot magma breaks through Earth's crust to form fiery volcanoes, releasing **lava** to cool quickly on the surface or even underwater.	When rocks are exposed on the surface, the presence of water, air and life help to both physically and chemically break them down. This is called **weathering**. New rocks can form as **erosion** and **deposition** create layers of **sediments**; this can include the remains of living things that are hardened by *compaction* as more and more layers of sediment are added.	Rocks can be buried to great depths, where the higher temperatures (greater than 200 °C) and pressures can cause the rock to change form. Both the mineral type and appearance can change. The change happens in the solid state, meaning there is no melting. We can see these rocks because they get brought back up to the surface.

7.2.2 Minerals

Most rocks are made up of substances called **minerals**. A mineral is any naturally occurring solid substance with a definite chemical composition and crystal structure.

Chemical composition

Elements found naturally in their uncombined, pure form are also minerals. These elements, called **native elements**, include diamonds (pure carbon) and gold.

Most minerals in rocks are **compounds**, in which one or more elements bond together. For example, the mineral calcite is the combination of one calcium (Ca), one carbon (C) and three oxygen (O) atoms to make calcium carbonate ($CaCO_3$). Calcite is the primary mineral found in the rocks limestone and marble.

The most common group of minerals is the **silicates**, in which elements bond to oxygen (O) and silicon (Si). This is because oxygen and silicon are the most abundant elements in Earth's crust. The mineral quartz (figure 7.5) is a simple silicate (SiO_2), whereas clay is a complex silicate ($Al_4Si_4O_{10}(OH)_8$).

Wherever you go, a specific type of mineral will have the same chemical composition. The colours and shapes may change a little, which tells us more about how they were formed and provides clues about the past.

Crystal structure

The elemental atoms that join to form minerals create regular geometric shapes called **crystals**. The crystal shape reflects the organisation of the atoms inside. The physical environment around them can also impact the crystal structure, where additional pressure or compaction will force the structure to be more closely packed.

magma a very hot mixture of molten rock and gases, just below Earth's surface, that forms from melting of the mantle and occasionally the crust

lava an extremely hot liquid or semi-liquid rock from the mantle that reaches and flows or erupts on Earth's surface

weathering the physical or chemical breakdown of rocks on the surface

erosion the wearing away and removal of soil and rock by natural elements, such as wind, waves, rivers and ice, and by human activity

deposition the settling of transported sediments

sediment material broken down by weathering and erosion that is moved by wind or water and collects in layers

mineral a naturally occurring, inorganic and solid substance with a defined chemical formula and an ordered arrangement of atoms

native elements elements found uncombined in Earth's crust

compound a substance made up of two or more different types of atoms that are chemically bonded (covalent or ionic) together

silicates a group of minerals consisting primarily of SiO_4^{2-} combined with metal ions, forming a major component of the rocks in Earth's crust

crystal a geometrically shaped substance made up of atoms and molecules arranged in one of seven different shapes; the elements and the conditions present during the crystal's growth determine the arrangement of atoms and molecules and the shape of the crystals

FIGURE 7.5 A close-up of granite shows how the quartz crystals grow due to overcrowding.

The size of crystals depends on how fast they form and how much space is available. If a crystal forms quickly, do you think it would be smaller or bigger? Smaller, because it had less time to grow.

The quartz crystals in figure 7.6 have had a lot of time and space to grow. Quartz, one of the most common minerals, consists of hexagonal crystals of silicon dioxide (SiO_2) that make it look like a six-sided column with a six-sided pyramid at both ends.

FIGURE 7.6 Quartz is one of the most common minerals in Earth's crust.

ACTIVITY

Look at samples of different materials like salt and sugar under a microscope to observe crystal shapes.

EXTENSION: Are diamonds forever?

Although graphite (which can be found in some of your pencils) and diamonds are both minerals with pure carbon as their chemical composition, they are physically very different. Graphite is soft enough to leave behind traces when rubbed on paper, and diamonds are the hardest minerals on Earth. In fact, diamonds are so hard they are used to drill into rocks.

If they are made of the same element, why do they display such different properties? This is because they formed under different pressure conditions. To get the crystal structure of a natural diamond, carbon crystals have to form under pressures that are only reached deeper than 150 km into Earth. This is one of the reasons they are so rare here on the surface. The bad news is that at the surface where the pressure is much less, diamonds are slowly changing back to graphite — very, very slowly.

FIGURE 7.7 Natural graphite form

FIGURE 7.8 Faceted diamond, created by grinding a diamond on a spinning lap

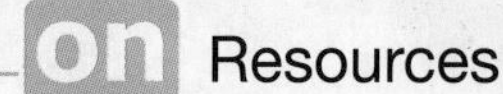

Resources

Interactivity Crystals (int-5338)

Weblinks Mexico giant crystal cave

How do crystals work?

DISCUSSION

Is table salt a mineral? Think carefully about your answer and suggest reasons for and against classifying it as a mineral. How about ice that forms on car windows during a freezing winter night?

ACTIVITY

Grow your own crystals by first making a warm, saturated solution with alum, Epsom salts or non-ionised table salt. Once the solution cools, pour it into either a shallow bowl or a small jar and place a pipe cleaner suspended from the top. Set aside and wait for your crystals to grow! This may take as little as a few hours, or up to a week. Examine your crystals. What shape are they? How big are your largest and smallest crystals? How long did it take for the first crystals to form?

Identifying minerals

Although colour might seem to be the quickest way to identify a mineral, it is not reliable. Many different minerals have similar colours. Some samples of the same mineral can have different colours due to small impurities. For example, quartz can be colourless like glass, or may be pink, violet, brown, black, yellow, white or green. Therefore, we must use a combination of physical properties to identify minerals.

Additional mineral properties include the following:

- The **lustre** of a mineral describes the way that it reflects light. Minerals could be described as metallic or non-metallic. Non-metallic characteristics include dull, pearly, waxy, silky or glassy.
- The **streak** is the colour of a powdery mark left by a mineral when it is scraped across a hard surface, such as an unglazed white ceramic tile. It's a more reliable property to distinguish between two minerals than colour is.
- The **hardness** of a mineral can be determined by trying to scratch one mineral with another. The harder mineral leaves a scratch on the softer mineral.

lustre the high shine and sheen of a substance caused by the way it reflects light

streak the colour of a mineral as a fine powder, found by rubbing it onto an unglazed white ceramic tile

hardness a measure of how difficult it is to scratch the surface of a solid material; hardness can be ranked using Mohs' scale

FIGURE 7.9 Streaks from minerals with a metallic lustre (left) and an earthy, dull lustre (right)

Minerals generally have no 'fingerprint' or single property that sets them apart from others; therefore, we tell them apart by identifying a combination of physical properties.

A mineral's physical properties are the result of its internal atomic arrangement. This means the outward properties of the mineral are the result of its inward atomic structure — the way in which the atoms are bonded to one another results in properties like lustre, streak and hardness.

Friedrich Mohs' scale of hardness is a numbered list of ten minerals ranked in order of hardness. Higher numbers correspond to harder minerals. The hardness of a mineral is determined by comparing it with the minerals or common materials in Mohs' scale. For example, a mineral that can be scratched by quartz but not by orthoclase has a hardness between 6 and 7.

Figure 7.10 shows that some more common materials can also be used to determine the hardness of a mineral.

FIGURE 7.10 Mohs' scale for testing the hardness of minerals

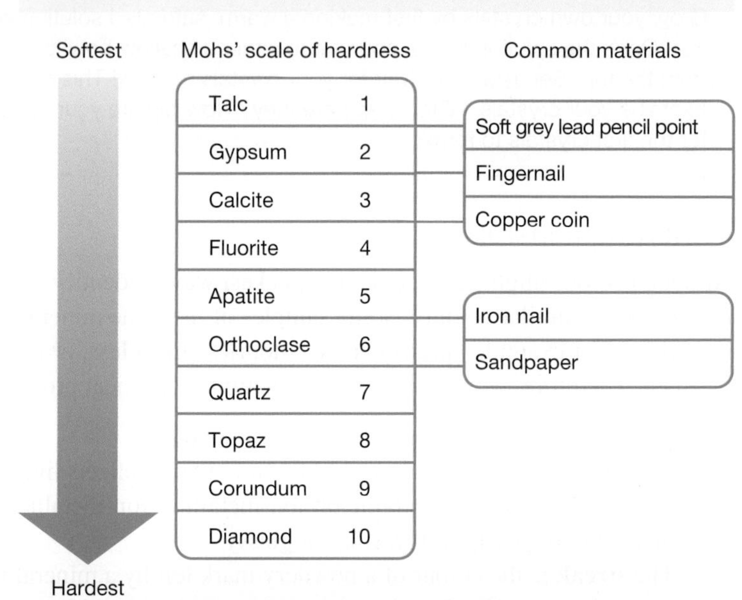

INVESTIGATION 7.1

Identifying mineral properties

Aim

To observe the properties of a range of minerals

Materials

- mineral kit
- common materials to substitute for unavailable Mohs' scale minerals
- hand lens
- white ceramic tile

Method

1. Construct a table like the one in the results section to record your observations as you work through the following steps for each mineral.
2. Write down the mineral name and describe the colour and lustre.
3. Use the magnifying glass to look closely at the mineral and describe the shape and size of its crystal(s).
4. Scrape the mineral across the unglazed side of a white ceramic tile. Record the colour of the streak.
5. Use Mohs' scale minerals or the common materials (figure 7.10) to estimate the hardness of the mineral by trying to scratch it. An approximate range, such as 5–6, is sufficiently accurate.

Results

Complete your table for each mineral to present your results. Remember to add a title to your table.

Mineral	Colour	Lustre	Crystal shape and size	Streak	Hardness

Discussion

1. How similar were some of your minerals? Note two minerals that were close but had one or two different properties.
2. Other than those already described, what additional properties of minerals could be used to identify them?
3. If two unlabelled mineral samples have the same colour and lustre, can you be sure that they are the same mineral? Explain how you would find out.

Conclusion

What can you conclude about the properties across a range of minerals?

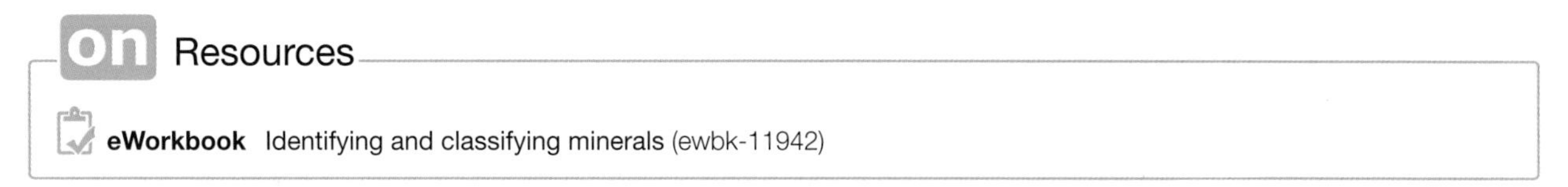

7.2 Activities

learnon

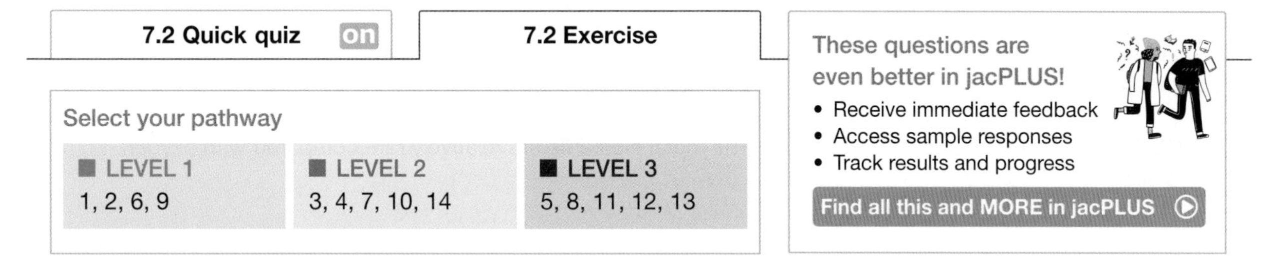

Remember and understand

1. Rocks are naturally forming, coherent combinations of what potential three substances?
2. In which part of Earth are most of the rocks we see formed?
3. List the three ways in which rocks can form.
4. What is a mineral?
5. What is a native element? List two examples.
6. What is the largest group of minerals called and what does their name say about their chemistry?
7. What is the approximate hardness on Mohs' scale (to the nearest whole number) of a mineral that can be scratched by sandpaper but not by an iron nail?
8. List at least five properties that you could observe to help you identify an unknown mineral.

Apply and analyse

9. Explain the difference between a rock and a mineral.
10. A mineral can be scratched by a copper coin but not by a fingernail. Is the mineral quartz, fluorite or calcite?
11. You have two samples, each of a different mineral, but no other equipment to test them for hardness. How could you tell which mineral is harder?
12. You have found a rock with tiny minerals in it and you would like to identify them. How could you go about testing the physical properties to help you identify the minerals?

Evaluate and create

13. **SIS** A geologist has been hired to find some haematite iron ore. In the field, they find lots of rocks. To determine if there is haematite present, they look for rocks with a dark colour and metallic lustre. They then pick it up to see how heavy it is. If it is dark coloured with a steel grey metallic lustre, and is very heavy, they call it haematite.
    a. Reflect on the method the geologist used to identify the mineral. Is anything missing?
    b. How reliable is their claim that the sample contains haematite?
    c. What could you do to improve the conclusion?
14. **SIS** Find out how crystals can be artificially grown and then grow a crystal garden.

Fully worked solutions and sample responses are available in your digital formats.

LESSON
7.3 Mining for metals

LEARNING INTENTION

At the end of this lesson you will be able to describe that minerals containing metals can be mined from Earth's surface but that this process relies on many steps that involve knowledge of the geology, the environmental impact of the mining activity, processing of the ore and rehabilitation once the mining is complete.

SCIENCE AS A HUMAN ENDEAVOUR: Mining metals in Australia

Metals play an important part in our lives every day. The phone you use to chat with your friends has several metal components, including wires and lithium batteries. You use metal cutlery to eat food. The bus, car, bike or scooter you rode to school is made from metal.

The metal elements used to make these things are found in minerals within rocks in Earth's crust. The pie chart in figure 7.11 shows that almost three-quarters of Earth's crust (by weight) is made up of the non-metals oxygen and silicon. Most of the metal elements combine with other elements to form compounds. They commonly combine with oxygen, silicon or other non-metals, like sulfur.

FIGURE 7.11 The elements in Earth's crust. The metal elements are relatively rare compared with oxygen and silicon.

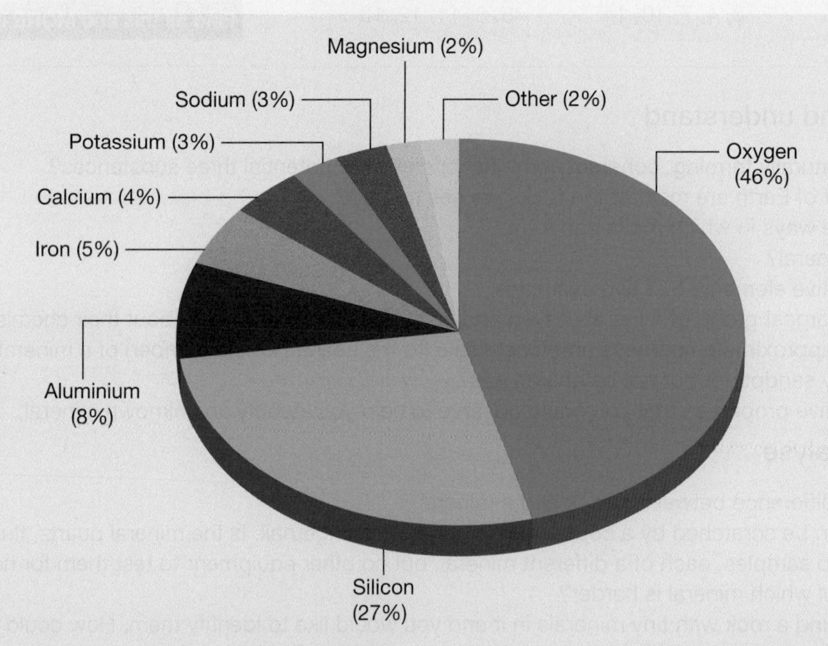

Minerals containing metals of value that can be extracted for profit are called **ore minerals**. It takes a lot of time, effort and money to find the rocks that contain the ore minerals, get them out of the ground, separate the ore from the waste rock and extract the metal element from the ore mineral. The **mining** of an ore can take place only if enough of it is found concentrated at a single location. This makes the potential of finding ore very different from one place to another.

The mining industry makes a major contribution to Australia's economy. Apart from the profits that go to shareholders in mining companies and to the government in taxes, the mining industry employs many thousands of Australians. A report commissioned by the Minerals Council of Australia estimates that between 2010 and 2020, mining (including coal) contributed $132.8 billion in tax and $106 billion in royalties to Commonwealth, state and territory governments! This helps to fund better services for the community, including health care, improved roads, hospitals and other infrastructure.

ore mineral a mineral from which a valuable metal can be removed for profit

mining the process of removing natural resources from Earth

Mineral exploration

Finding minerals below Earth's surface, where you can't see them, is an expensive business. Geologists use their knowledge of soils, rocks and minerals, and the clues they provide, to help them predict where precious ores are likely to be found.

Geologists make use of satellites equipped with cameras, radar and other sensors to search for geological features that are likely to contain high concentrations of ores. The magnetic properties of large bodies of rocks containing some minerals, like haematite iron ore, can be detected by surveys on the ground or conducted by aircraft. Geophysicists are specialists who study the physical properties of Earth, typically using equipment, surveys and modelling.

Minerals in the crust breakdown in rain and running water, and get washed into creeks and rivers. A chemical analysis of the soils, sediments and surface water of lakes and streams can provide evidence of the presence of minerals in the area. Samples of soil and rocks are taken using portable equipment.

On average, only one in one thousand sites explored are eventually mined.

If there is sufficient evidence of useful mineral deposits that might be worth mining, a licence must be obtained before any clearing is done or heavy drilling equipment is brought in. Helicopters are sometimes used to bring in heavy equipment to protect sensitive ecosystems. Drilling allows mining companies to have a very detailed examination of what lies beneath the surface. Mining companies are required by law to clean up exploration drill sites and ensure they are left to an acceptable standard. Landholders and First Nations Australian representatives must also be consulted regarding access to land for exploration and mining.

You can't start until ...

In the past, mining was often carried out without considering its long-term effect on the environment and the people who lived and worked in the area. Today, however, an **environmental impact statement (EIS)** must be prepared before a mining operation can commence. An EIS outlines how the mining company intends to manage all environmental aspects of the proposed mine. It also outlines how the land will be **rehabilitated** or reconstructed, so that it can be used again after the mining is completed.

The EIS — along with any other relevant information, such as a heritage report — is studied by the government before permission to proceed is granted.

The EIS reports on:

- existing flora, fauna and soils
- existing towns and roads in the area
- proposed new towns, roads and other developments
- how the new development might affect the local community and environment
- alternative plans to complete the development that might have less impact on the environment
- measures that will be put in place to monitor and control air, water and noise pollution during the project and while rehabilitation is undertaken
- rehabilitation proposals for the area.

Taking out the ore mineral

To obtain ore from the ground, it is often necessary to remove large amounts of rocks and soil. The way this is done depends on how close the ore deposit is to the surface. If it is close to the surface, the vegetation and topsoil are removed first. Then waste rock from beneath the topsoil, called **overburden**, is removed. The removed topsoil and overburden are used to fill areas that have already been mined, or are left in a pile to restore the newly mined area when mining is completed. This method of mining is called **open-cut mining**.

If the ore deposits are deep below the surface, miners use **underground mining**. This mining method is more dangerous and expensive than open-cut mining. Shafts and tunnels are dug up to four kilometres into the ground to reach the rocks containing the ore. The development of open-cut and underground mining is overseen by mining engineers.

environmental impact statement (EIS) a report on the possible effects of a planned project on the environment

rehabilitated restored to its previous condition or an acceptable, agreed alternative

overburden waste rock removed from below the topsoil; this rock is replaced when the area is restored

open-cut mining mining that removes soil and rocks on the surface of the land

underground mining mining that uses shafts and tunnels to remove rock from deep below the surface

Getting the metal

Obtaining the metal element takes place in two stages:

1. Mineral extraction separates the ore mineral from the rock taken from the ground. This involves crushing, grinding and washing the rock to separate the valuable minerals from the unwanted waste rock.
2. Metal extraction separates the desired metal element from the ore mineral. This always involves chemical reactions. The nature of these reactions depends on a number of factors, including the chemical composition of the ore mineral. Chemical engineers and metallurgists are involved in the design of this process.

Rehabilitation

Before mining of a new site begins, seeds of the natural vegetation of the area are collected so that seedlings can be cultivated at a later stage. The seedlings are grown in special nurseries until they are mature enough to return to the site of the mine.

During open-cut mining, the overburden is used to fill holes left from earlier stages of the mining operation. Fresh topsoil is used to cover the overburden to ensure that new vegetation will grow. The soil surface is shaped to fit in with the surroundings, fertilised and sown with seeds or planted with seedlings. Care is taken to shape the new surface to prevent the newly sown soil from being eroded or washed away by wind or rain. Mines can be rehabilitated for crops, grazing, native restoration or wildlife conservation areas.

FIGURE 7.12 Resurfacing and replanting a former open-cut iron mine on Koolan Island, Western Australia

Resources

Weblinks What's in a mobile phone?

Mt Rawdon mine rehabilitation

NSW Mining rehabilitation site and videos

DISCUSSION

Metals acquired from mining are needed to facilitate the use of greener energies.

In a small group, discuss and list:
a. the factors a mining company should consider when it decides whether or not to start a mining project
b. the different tasks that scientists and engineers might perform from the beginning of mining exploration until mining rehabilitation is complete
c. some advantages to mining
d. some disadvantages to mining.

Compare the lists of your group with those of others in your class. Finally, debate as a class whether or not Australia should move to green energy, considering the impacts of the mining required, and suggest how this could be done more responsibly.

elog-2177

INVESTIGATION 7.2

Searching without disturbing

Aim

To model the search for minerals below the ground

Materials

- a tray of sand
- 10 paperclips
- blindfold (optional)
- compass
- paper and clipboard-ruler

Method

1. Find a partner. Each of you should then draw identical maps of the sand tray. Use a ruler to construct a grid on each map. Label the grids across the top and down the side (e.g. A–J across the top, 1–15 down the side). Each grid should consist of at least 100 equal-sized rectangles or squares.
2. Without showing your partner, hide the paperclips in the tray of sand and mark the location of the 10 clips on your map.
3. Your partner's task is to locate the 10 paperclips and mark them on the map without disturbing the sand. You might wish to set a time limit.
4. Swap roles and repeat the steps above.

Results

1. What property of the paperclips allowed them to be located?
2. Record where and how many you found onto your grid map.

Discussion

1. How could your predictions of the location be checked with a pencil?
2. What was your success rate?
3. After checking, can the sand be restored to its initial condition?

Conclusion

Summarise your findings from this investigation about searching for hidden metals.

Resources

Weblink Mineral exploration interactive

7.3 Activities

learnon

7.3 Quick quiz on	7.3 Exercise

Select your pathway

■ LEVEL 1	■ LEVEL 2	■ LEVEL 3
1, 2, 5, 8	3, 6, 9, 10	4, 7, 11

Remember and understand

1. Where are ore minerals found?
2. Where in Earth's crust are the metal elements found?
3. Describe the method of open-cut mining for removing mineral ores from the ground.
4. Outline the two stages involved in obtaining a metal element from rock.
5. What is an EIS?
6. Outline the information that is included in an EIS.
7. How do mining companies rehabilitate the land used for mining?
8. Explain why it is important to recycle metals as much as possible.

Apply and analyse

9. The most common element in Earth's crust is oxygen. This element is a gas except at extremely low temperatures. In what form is oxygen found in Earth's crust?

Evaluate and create

10. In a table like the one provided, make a list of the benefits and disadvantages of mining.

TABLE Benefits and disadvantages of mining

Benefits	Disadvantages

11. Discuss reasons for and against allowing mining to take place in Australia's national parks.

Fully worked solutions and sample responses are available in your digital formats.

LESSON
7.4 Igneous — the 'hot' rocks

LEARNING INTENTION

At the end of this lesson you will be able to describe the type of environments that igneous rocks form in and how they can be classified according to their composition and texture.

7.4.1 Melting rock

There are places around Earth where physical conditions allow rocks to melt or partially melt deep underground. The molten rock underground is called magma, and it rises slowly towards the surface. If the magma breaks through and flows onto the surface it is then called lava. Rocks that form from the cooling of magma below the surface or lava on the surface are called igneous rocks.

The appearance of all igneous rocks depends on two major factors:

1. How quickly the lava or magma cooled
2. What substances it is made of.

WHAT DOES IT MEAN?

The word 'igneous' comes from the Latin word *ignis*, meaning 'fire'. The words 'ignite' and 'ignition' also come from the same Latin word.

EXTENSION: Is the interior of Earth all liquid?

At Earth's surface, rock begins to melt when heated between 800 and 1000 °C, and will be completely melted at about 1200 °C. However, if you put a rock under pressure, it becomes stronger and requires a higher temperature to melt. This is why the interior of Earth is mostly hot solid material and not all molten rock, despite the fact that temperatures of greater than 1000 °C exist.

The physical conditions required to melt rock would be:

1. adding so much heat that it overcomes the pressure
2. releasing pressure from a hot rock
3. adding fluids, like water.

Releasing pressure and adding fluids lower the melting temperature. Magma is generated only where one or more of these conditions are met.

7.4.2 Extrusive rocks

Lava is released from erupting volcanoes at temperatures of 1000 °C or more. At that temperature, flowing lava could take hours to weeks to cool down and become solid rock. However, if lava is ejected into the air from explosive volcanoes it cools almost instantly. The lava erupting from underwater volcanoes on the ocean floor also cools quickly.

Igneous rocks that form from the cooling of red-hot lava on Earth's surface or lava spilling from underwater volcanoes are classified as **extrusive** or volcanic. Features of extrusive rocks are summarised in table 7.2.

extrusive describes igneous rock that forms when lava cools on Earth's surface

FIGURE 7.13 Red-hot lava flowing on Earth's surface. The cooling of lava forms a crust that can shift with the continued movement of underlying lava, creating a ropey look.

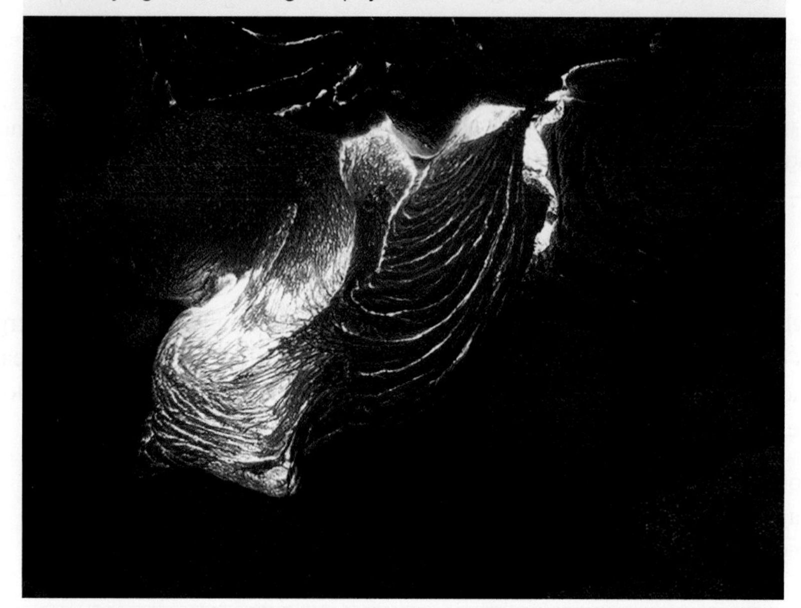

basalt a dark, igneous rock with small crystals formed by fast cooling of hot lava; it sometimes has holes that once contained volcanic gases

rhyolite a light-coloured, extrusive igneous rock with a similar mineral composition to granite, but with smaller crystals

viscosity a measure of a fluid's resistance to flow

TABLE 7.2 Features of extrusive igneous rocks

Crystal size	Rock colour
The size of crystals in extrusive igneous rocks is generally very small because of how fast the lava cools. When it cools quickly, there is not enough time for large crystals to form.	Colours range from black to grey, white or even red. The colour reflects the types of minerals that have formed. Generally, the dark rocks are rich in iron (Fe) and magnesium (Mg) minerals. The lighter coloured rocks contain more minerals that are richer in silicon (Si).

Basalt and rhyolite

Basalt is a common extrusive rock that is dark coloured with small mineral crystals. You may be able to see some of the small crystals, but most require a magnifying tool. If basalt forms from lava cooling in cold ocean water, the crystals will be even smaller and only visible under a microscope. Why do you think that is so?

When rocks are heated up they expand, and when they cool down they contract (shrink). The basalt in figure 7.14 (and in the image opening this topic) formed from a cooling basalt lava flow. During cooling, the new rock contracts and this can form vertical columns of basalt. Beware of these columns on a cliff, as they can topple over.

FIGURE 7.14 When basalt flows cool, they can form hexagonal columns.

Rhyolite is another common extrusive rock. It also has generally small crystals, but, unlike basalt, it is light coloured due to having more silica-rich minerals. More silica-rich minerals make the lava sticky and harder to flow — a term called **viscosity**. A good example of different viscosities is honey versus water. Water flows over a table easily (low viscosity), but honey poured over the same table will move a lot slower (high viscosity). Because the rhyolite lava is viscous, it does not travel far from the volcano. Basalt has a lower viscosity and can flow further from a volcano.

FIGURE 7.15 Basalt and rhyolite are both extrusive rocks. Basalt is dark coloured (left) and rhyolite is light coloured (right).

Scoria, pumice and obsidian

Some violent volcanic eruptions shoot out lava filled with gas. The lava cools very quickly while it is still in the air and traps the gas inside. Rocks that form this way are full of holes from where the gas was trapped. Two examples of this type of rock are **scoria** and **pumice**.

TABLE 7.3 Features of explosive igneous rocks

Scoria	Pumice
Scoria is a dark (black, reddish-brown or grey) volcanic rock full of holes. It has a darker colour because it contains more iron. It is usually found closer to a volcano's crater.	Pumice is a pale-coloured volcanic rock. It is very light because it is mostly made of glass and full of holes. Pumice floats on water and sometimes washes up on beaches thousands of kilometres from where it erupted!

Obsidian is a smooth, black rock that looks like glass because it is a natural volcanic glass. It is formed when silica-rich lava cools almost instantly. Glass is not a mineral because, as it cools so quickly, it does not have a crystal structure.

The unique, curved way in which obsidian fractures when struck, combined with its hardness, makes it a good material to manufacture sharp, blade-like edges. For this reason, obsidian has been used as cutting tools and arrowheads throughout history. Rock technology is explored further in lesson 7.7.

Although obsidian is usually dark in colour like basalt, it is extremely rich in silica and is mostly glass, like pumice. Its dark colour is due to the high amount of impurities caught in the glass.

scoria a dark, igneous rock formed from frothy basalt lava that cools quickly and is full of holes that once contained gas

pumice a glassy, pale igneous rock that forms when frothy rhyolite lava cools in the air; it often floats on water as it is very light and full of holes that once contained gas

obsidian a black, glassy rock that breaks into pieces with smooth shell-like surfaces

FIGURE 7.16 The glassy extrusive igneous rock known as obsidian

FIGURE 7.17 An obsidian arrowhead

on Resources

Video eLesson Volcanoes (eles-0130)

7.4.3 Intrusive rocks

Igneous rocks can also form as magma cools 5–30 kilometres below the surface of Earth. Those that form below the surface are called **intrusive**. They cool very slowly (thousands of years or more) and become visible only when the rocks and soil above them are removed by erosion, or if we drill down into them. Intrusive rocks (sometimes called plutonic rocks) have larger crystals than extrusive rocks because the crystals had more time to grow. Large bodies of intrusive rock are called **batholiths**; they cover an area of over 100 square kilometres.

intrusive describes igneous rock that forms when magma cools below Earth's surface

batholith an intrusive rock mass that covers an area of over 100 square kilometres

int-5337

FIGURE 7.18 If a batholith is exposed to the environment, it will start to wear away along the cracks, which can leave large, rounded boulders called tors balancing on the surface. Over time, the batholith may break down completely. The breakdown of rocks is called weathering.

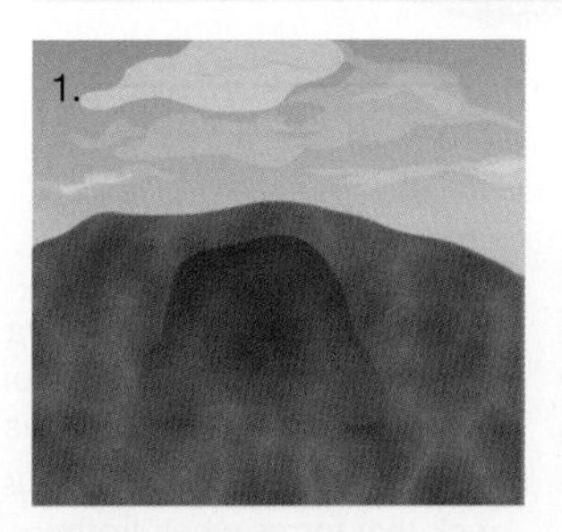

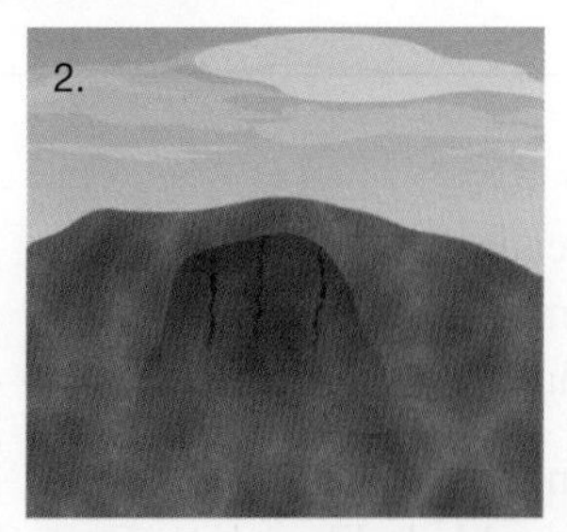

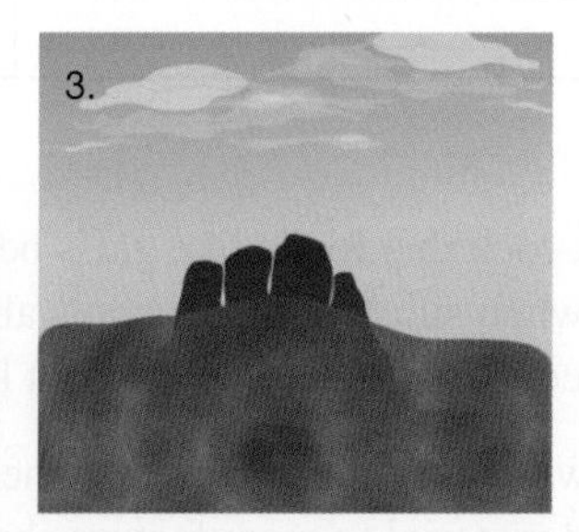

FIGURE 7.19 Igneous rocks can form below or above Earth's surface. Where they form will determine the speed of cooling and thus the crystal size.

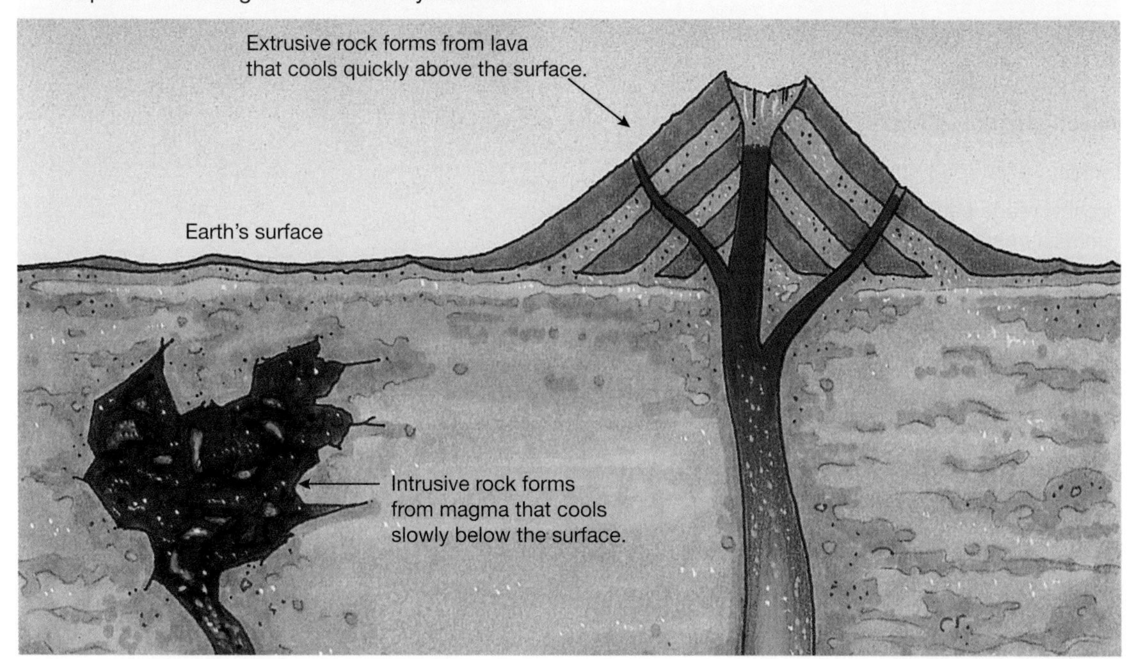

Granite and gabbro

Two common intrusive igneous rocks are **granite** and **gabbro**. The crystals in both form over long periods of time and grow large enough to be easily seen without magnification. Being able to see the individual crystals makes it easier to identify the type of minerals present.

granite a light-coloured, intrusive igneous rock with mineral crystals large enough to see

gabbro a dark-coloured, intrusive igneous rock with a similar mineral composition to basalt, but with larger crystals

TABLE 7.4 Features of intrusive igneous rocks

Granite	Gabbro
Granite is a light-coloured intrusive rock with silica-rich minerals. The crystals found in granite are a mixture of white, pink, clear to grey, and black minerals. These minerals are (in order of most abundant to least): • *feldspar* (white and pink) • *quartz* (clear to grey) • *mica* (black). Granite is the most common igneous rock found in the continental crust. 	Gabbro is a dark-coloured intrusive rock with minerals rich in iron (Fe) and magnesium (Mg). It looks mostly black, but if you look close enough, you will see some white and green. These minerals are (in order of most abundant to least): • *pyroxene* (black) • *feldspar* (white) • *olivine* (green). Gabbro is the most common igneous rock found in the oceanic crust. 

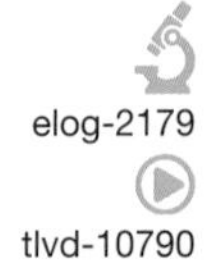

elog-2179

tlvd-10790

INVESTIGATION 7.3

Does fast cooling make a difference?

Aim

To investigate the effect of the cooling rate on the size of crystals

Materials

- freshly made saturated solution of potassium nitrate
- potassium nitrate
- spatula
- 250 mL beaker
- 3 test tubes and test-tube rack
- test-tube holder
- Bunsen burner, heatproof mat and matches
- crushed ice
- foil
- safety glasses
- hand lens

CAUTION

Safety glasses must be worn during this experiment.

Method

1. Half-fill a beaker with crushed ice.
2. Quarter-fill a clean test tube with saturated potassium nitrate solution. Add a spatula of potassium nitrate.
3. Gently heat the solution over a Bunsen burner flame until the added potassium nitrate has dissolved or until the solution starts to boil.
4. Pour half the warm solution into one clean test tube, and then the remaining half into another.
5. Place one test tube in the beaker of crushed ice. Wrap the bottom of the other test tube in foil and place it in the rack to cool.
6. When crystals have formed in each test tube, examine them with a hand lens.
7. Cool one solution quickly and the other one slowly.

Crushed ice

Potassium nitrate solution

Foil

Results

1. Draw a labelled diagram of some crystals in each test tube, concentrating on their shape and size.
2. Which test tube contained the larger crystals: the one that cooled quickly or the one that cooled slowly?

Discussion

1. Which types of igneous rock would you expect to have the larger crystals: those that cool slowly underground or those that cool quickly on the surface?
2. Which types of rock are represented by the two different test tubes?
3. Why do safety glasses need to be worn during this experiment?

Conclusion

Summarise your findings from this investigation, commenting on the relationship between cooling rates and crystal size.

Rhyolite, pumice, obsidian and granite all have the same chemical composition, yet appear very different due to the conditions in which they form.

7.4.4 Useful igneous rocks

Igneous rocks can sometimes host valuable ore minerals but they are also used in several other ways, as summarised in table 7.5.

TABLE 7.5 Uses of igneous rocks

Igneous rocks	Example of modern uses
Basalt	Basalt blocks have been used as a decorative building material. It is also commonly crushed and used for road base, asphalt and concrete.
Scoria	A reddish-brown or grey rock that can be crushed and used in garden paths or as a drainage material around pipes. It is also used in high-temperature insulation.
Pumice	Powdered pumice is used in some **abrasive** cleaning products. It is also used in chemical spill containment, water filtration, horticulture and cement manufacturing.
Granite and gabbro	Commonly used in building due to their strength and beauty. Granite or gabbro that has been polished to give it a glossy finish is also used for grave headstones, benchtops and statues or other monuments.

DISCUSSION

Locate a building, statue or memorial in your area that is made from igneous rock. Describe the rock in the structure, and suggest why it was the chosen material.

abrasive a property of a material or substance that easily scratches another

geothermal energy refers to using heat from Earth as an energy source

SCIENCE AS A HUMAN ENDEAVOUR: Renewable geothermal energy in Australia

Geothermal energy is heat contained within Earth. Australia has great potential for geothermal energy to be used for generating electricity (figure 7.20). Geoscience Australia has calculated that there is sufficient energy contained within the Australian crust around hot rock systems, that if only one per cent of the resource were used, it would provide 26 000 years worth of electricity.

FIGURE 7.20 How heat from Earth can be used to generate electricity

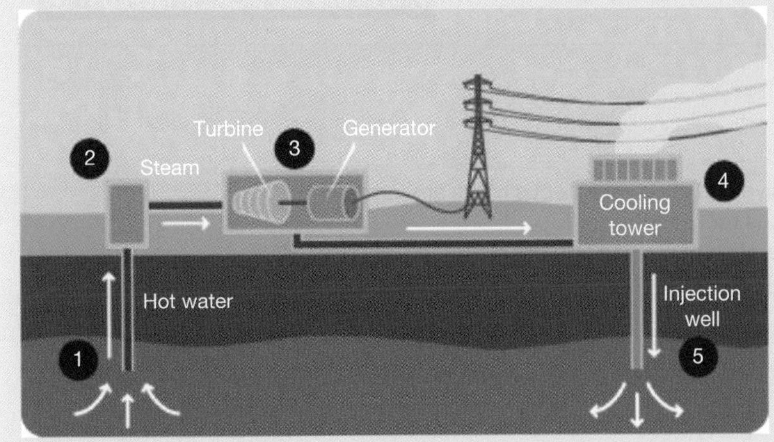

1. Hot water is pumped from deep underground through a well under high pressure.
2. When the water reaches the surface, the pressure is dropped, which causes the water to turn into steam.
3. The steam spins a turbine, which is connected to a generator that produces electricity.
4. The steam cools off in a cooling tower and condenses back to water.
5. The cooled water is pumped back into the Earth to begin the process again.

The hot rock systems in Australia are normally associated with bodies of granite rock 3–5 kilometres deep that contain unusually high concentrations of the naturally radioactive elements uranium (U), thorium (Th) and potassium (K). The radioactive decay of these elements generates heat that is insulated by the rocks above them. Figure 7.21 is a model of the temperature of the crust at 5 kilometres depth. The thicker the insulating layer, the hotter the temperature. For the system to be complete as a geothermal energy source, there also needs to be a fluid circulating through the rock above to transport heat to the surface.

FIGURE 7.21 Modelled crustal temperature at 5 kilometres depth

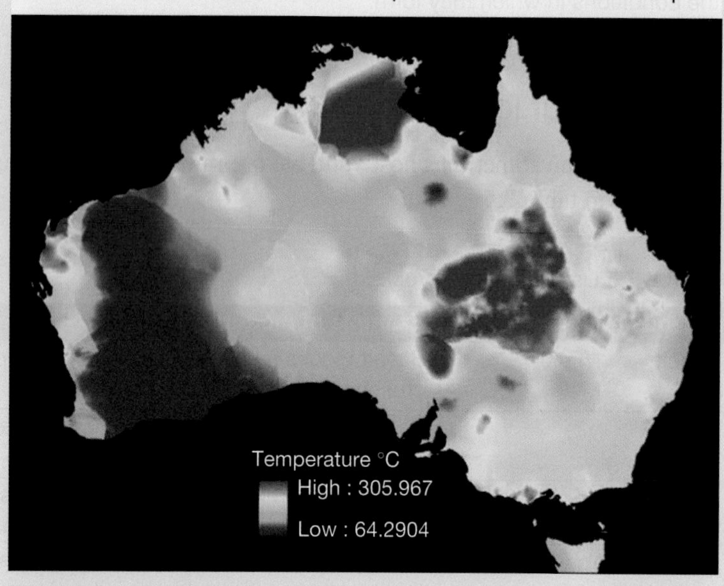

While significant hot rock systems have been identified, there is no present commercial production of geothermal energy in Australia.

- How would exploration be conducted to find these hot rock systems?
- According to the modelled crustal temperatures at 5 kilometres depth, where are the potential hot rock systems?
- What sort of challenges are limiting access to this substantial renewable energy source?

7.4 Activities

learnon

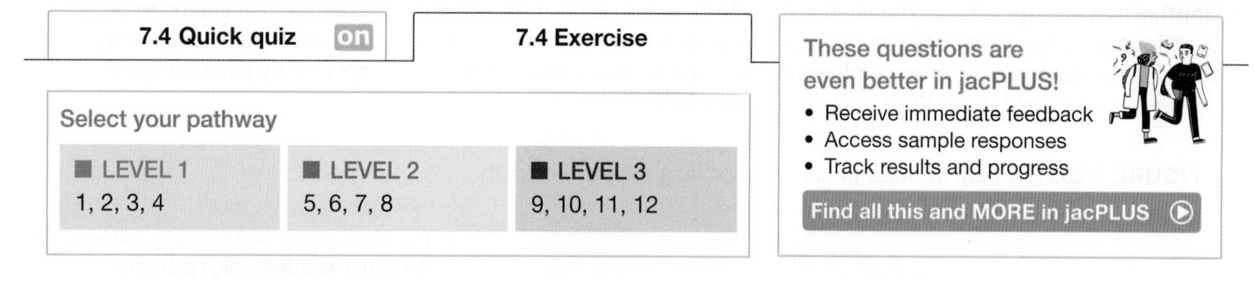

Remember and understand

1. How do igneous rocks form?
2. Distinguish between the ways extrusive and intrusive igneous rocks are formed.
3. What do the varying colours of igneous rocks represent?
4. What causes the frothy (holey) appearance of pumice and scoria?

5. Label the three minerals found in granite.

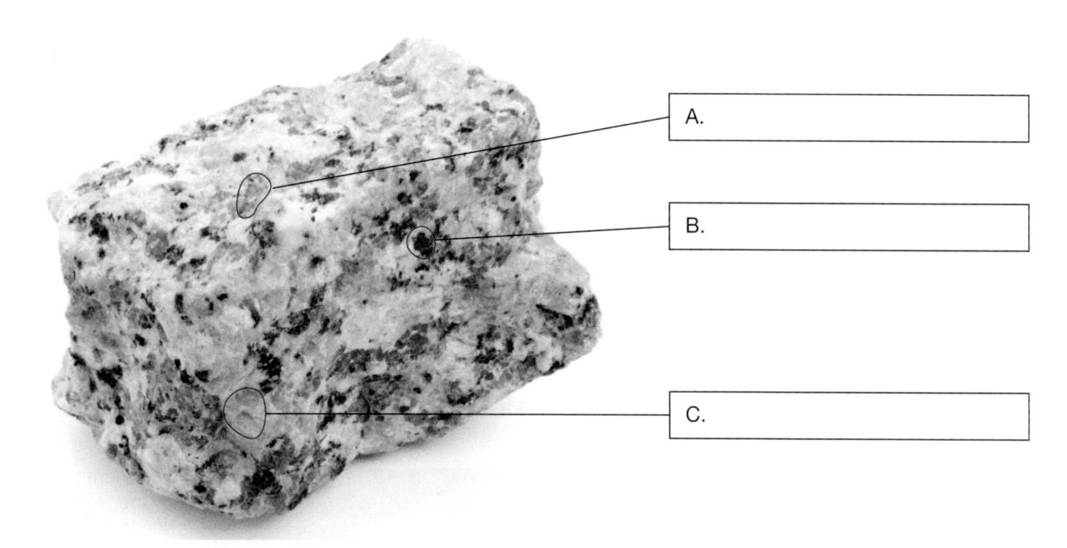

6. Describe two major differences between the appearance of granite and basalt.

Apply and analyse

7. Why are the crystals in basalt that formed under water smaller than those in basalt that formed on the ground?
8. Batholiths form well below the ground. Explain how they become visible on Earth's surface.
9. Explain how you would decide that an igneous rock formed from a volcanic eruption.
10. Rhyolite is an extrusive rock that contains the same minerals as granite. In what way would you expect it to be different from granite?

Evaluate and create

11. **SIS** Geologists like to use classification tables like the one shown to identify relationships between different rock types and their properties.

TABLE Igneous rock classification table

	Silica rich	Iron and magnesium rich
Extrusive		
Intrusive		

a. Complete this igneous-rock classification table by adding the names basalt, granite, rhyolite and gabbro into their proper locations.
b. Where would scoria, pumice and obsidian go?
c. What could you add to your table to include these rocks and identify what makes them different?

12. If you came across an igneous rock that had a mixture of large crystals surrounded by small crystals, suggest how it may have formed.

Fully worked solutions and sample responses are available in your digital formats.

LESSON
7.5 Sedimentary — the 'deposited' rocks

LEARNING INTENTION

At the end of this lesson you will be able to describe how sedimentary rocks are formed and classified, and that they form in layers that record time and changes to Earth's surface.

7.5.1 Weathered, eroded, deposited and lithified

Rocks that are formed from weathered, eroded, deposited and lithified sediments are called sedimentary rocks. Each of these processes can be described as follows:

- Rocks exposed on the surface are physically or chemically broken down by weathering as the rocks are exposed to the atmosphere, water and living things.
- The weathered particles are then transported by wind, running water, waves or flowing glacial ice as sediment. This process is called erosion.
- When the agents of erosion slow down or stop moving, their capacity to transport sediments reduces and the sediments settle onto the surface. This settling is called deposition. Deposits of dead plants and animals are also sediments.
- Sediments will deposit one on top of another, which creates layers, or beds. As beds continue to deposit, the individual sediments are packed closer together by compaction. Water with dissolved minerals can also seep around the sediment. As the water between the fragments gets squeezed out due to increased compaction, the minerals that are left behind act like cement and stick the sediments together to form a sedimentary rock. Compaction and cementation help to **lithify** the beds into rock (figure 7.22).

FIGURE 7.22 Lithification — turning sediments into rock

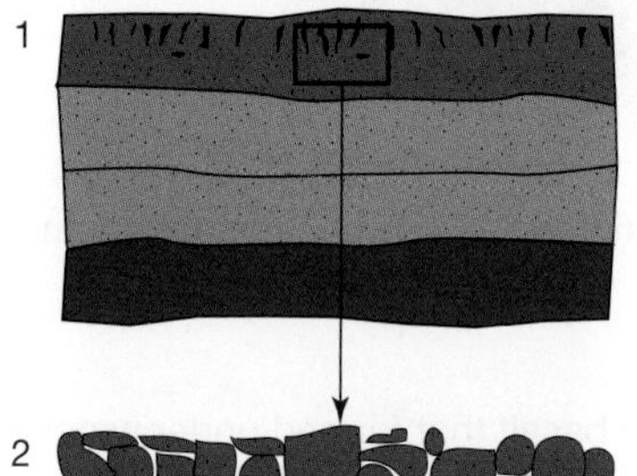

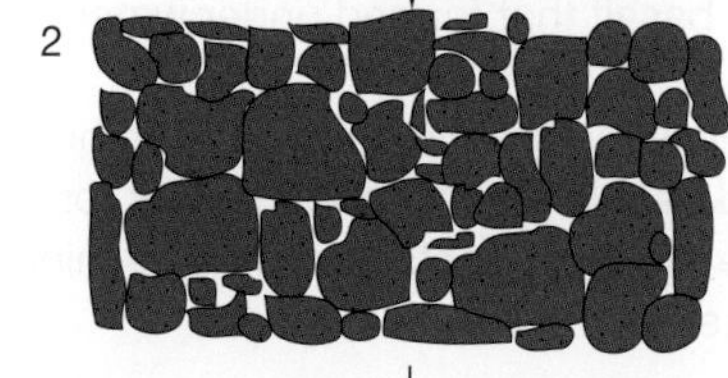

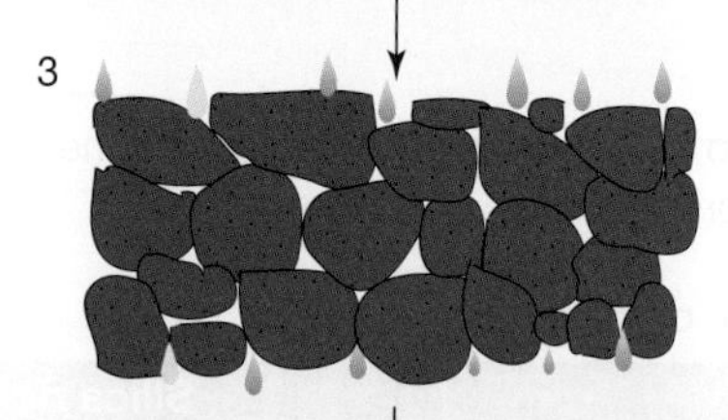

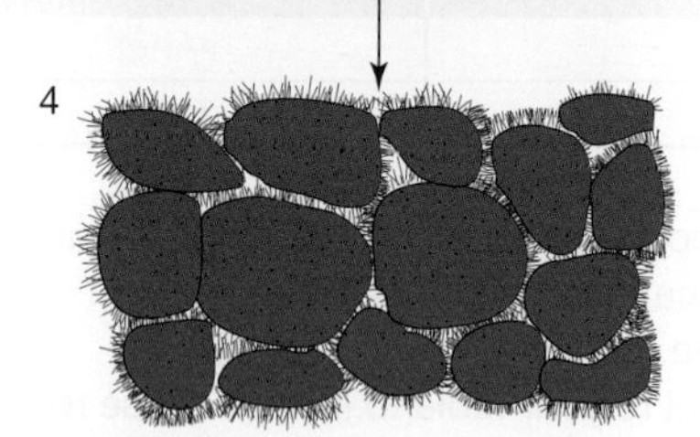

Deposition environments

Sand deposited by the wind forms sand dunes, especially in coastal areas where sand is picked up and blown inland until it is stopped by obstacles such as rock or vegetation.

A fast-moving river is likely to carry sand, gravel and smaller particles. As it slows down on its path to the sea, the river loses energy and will deposit along the river channel. The larger particles, such as gravel and sand, settle first. By the time the river reaches the sea, it is usually travelling so slowly that only the very fine silt and mud particles remain to settle and help form **deltas**.

lithify to transform sediment into rock

delta a landform created by the deposition of sediment at the end of a river as it enters a body of water

During floods, when rivers break out of their channels, sediments are deposited on flat, open land alongside the river. These plains are called **floodplains**.

In the coldest regions of Earth, especially at high altitudes, bodies of ice called **glaciers** slowly make their way down slopes. They generally move between several centimetres and several metres each day. Being solid, glaciers can push boulders, rocks, gravel and smaller particles down the slope. As the glacier melts it can deposit these sediments along the margins of the glacier. The ridges formed are called **moraines**.

floodplain flat, open land beside a river where sediments are deposited during floods

glaciers large bodies of ice that move down slopes and push boulders, rocks and gravel

moraine a ridge made out of sediments deposited by a glacier

int-5339

FIGURE 7.23 Most sedimentary rocks are formed from weathered rock that has been transported and deposited by moving water (rivers and ocean).

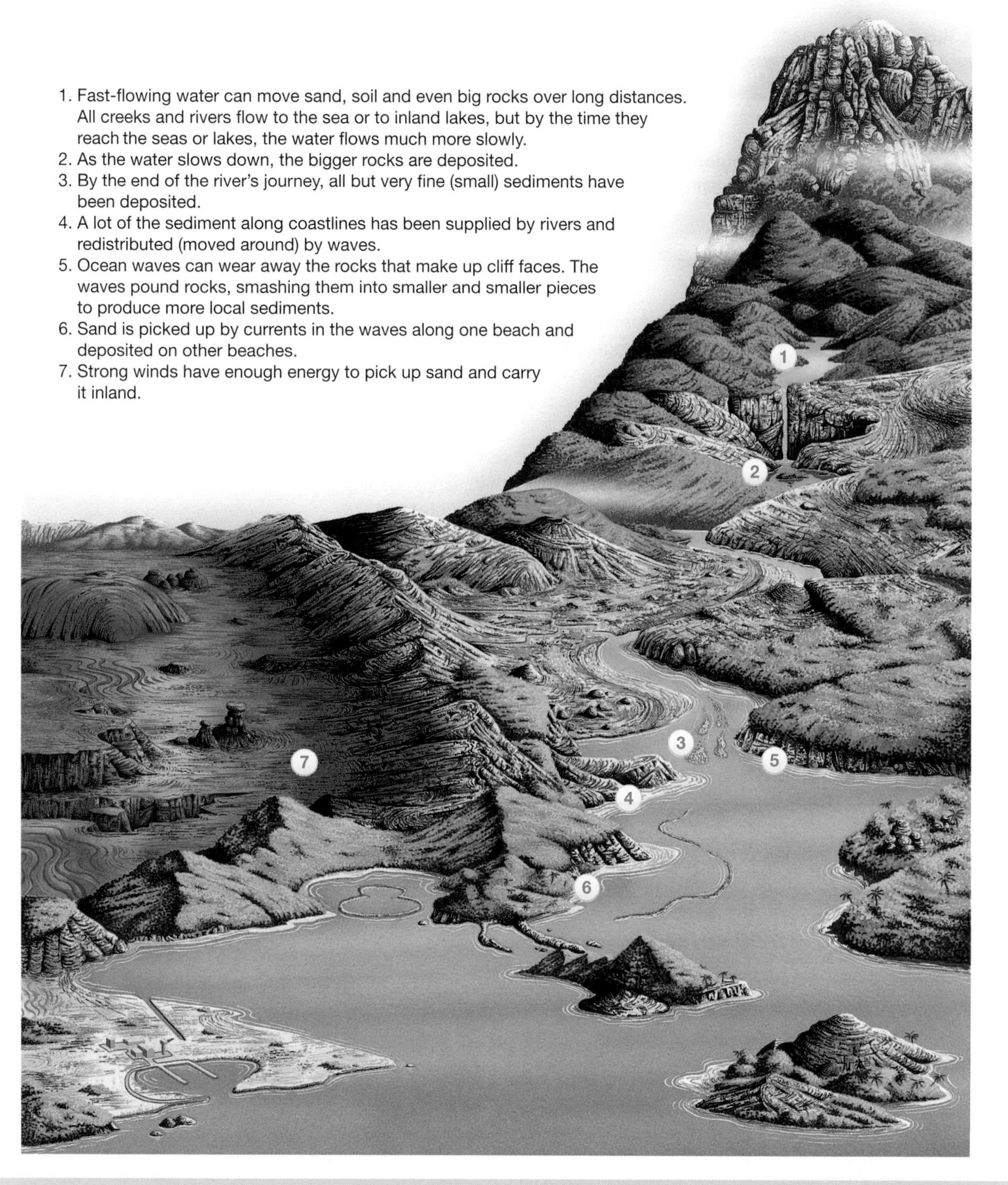

DISCUSSION

A common and fun activity on a warm summer day or weekend is to go to the beach, but where does all that sand come from? Is it a deposition environment? As a class, or in small groups, discuss where you think the sediment on beaches comes from. Consider whether all beaches have the same size and type of sediment.

7.5.2 Clastic sedimentary rocks

Sedimentary rocks are classified by their grain size. The most common type is clastic sedimentary rocks, which are made of fragments/sediments of weathered and eroded pre-existing rocks. These fragments are known as clasts.

TABLE 7.6 The names of sedimentary rocks are based on their grain size.

Sediment clast size	Clastic sedimentary rock names
	Conglomerate contains large clasts surrounded by sediments of different sizes, all cemented together.
	Sandstone is formed from grains of sand that have been cemented together.
	Siltstone particles are smaller than sand, but slightly larger and not as soft as those in mudstone.
	Mudstone and **shale** are formed from muddy particles (clay and silt) deposited by calm water. Shale shows tiny layers of clay (represented by short, horizontal lines), whereas mudstone is a thicker bed of clay.

conglomerate a sedimentary rock containing large fragments of various sizes cemented together

sandstone a sedimentary rock with medium-sized grains; the sand grains are cemented together by silica, lime, mud or salts

siltstone a sedimentary rock with a particle size between that of sandstone and mudstone

mudstone a fine-grained sedimentary rock made of mud (clay and silt)

shale a fine-grained sedimentary rock formed from thinly layered mud

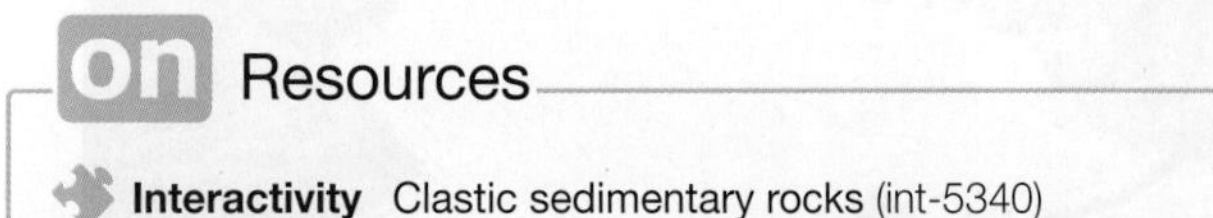

on Resources

Interactivity Clastic sedimentary rocks (int-5340)

WHAT DOES IT MEAN?

Conglomerate is formed from sediments that might be deposited by a fast-flowing or flooded river. The word 'conglomerate' comes from the Latin word *conglomerare*, meaning to 'roll together'.

FIGURE 7.24 A conglomerate

elog-2181

INVESTIGATION 7.4

Sediments and water

Aim

To investigate the order in which different sediments are deposited

Materials

- mixture of garden soil, gravel, sand and clay
- large jar with lid
- watch or clock

Method

1. Before commencing this experiment, form your own hypothesis about the order in which the different types of particles will settle. Give reasons for your hypothesis.
2. Draw a diagram to illustrate your hypothesis.
3. Place enough of a mixture of garden soil, gravel, sand and clay in a large jar to quarter-fill it.
4. Add enough water to three-quarter fill the jar and place the lid on firmly. Shake the jar vigorously.
5. Put the jar down and watch carefully as particles begin to settle. Note the time taken for each layer of sediment to settle completely.
6. Leave the jar for a day or two. Then compare your observations of the jar with your diagram.

Results

Record your answers to the following tasks to present your results:

1. Draw a labelled diagram showing clearly any layers that form. Identify the layers if you can.
2. Which type of sediment settled first?
3. Where are the other particles of sediment while the first layers are settling?
4. Which sediments settled after a day or two?

Discussion

1. Why did the last sediments take so long to settle?
2. Was your hypothesis supported by your observations?
3. What is the relationship between the size of sediment particles and the time taken to settle?

Conclusion

Summarise your findings from this investigation about the order that different sediments are deposited.

7.5.3 Sedimentary rocks from living things

Limestone is a sedimentary rock that is formed from deposits of the remains of sea organisms such as algae, brachiopods and corals. The remains of some of these organisms are still visible as fossils in limestone, while others are microscopic. The skeletal hard parts of these dead animals contain calcium carbonate (the mineral calcite). When the organisms die, fragments of their skeletons deposit as sediments and are cemented together over a period of time.

Coal is sedimentary rock formed from the remains of dead plants that are buried by other sediments. In dense swamps, layers of dead trees and other plants build up. If these layers are covered with water before rotting is completed, they can be buried by other sediments. The weight of the sediments above compacts the partially decayed plant material. Over millions of years the compaction and heating squeezes out the water, forming coal.

limestone a sedimentary rock formed from the remains of sea organisms; it consists mainly of calcium carbonate (calcite)

coal a sedimentary rock formed from dead plants and animals that were buried before rotting completely, followed by compaction and some heating

rock salt a sedimentary deposit formed when a salt lake or seabed dries up; the sediments are made of sodium chloride (halite)

FIGURE 7.25 Limestone is commonly made of marine fossils. Colours can range from white and tan to red and dark grey.

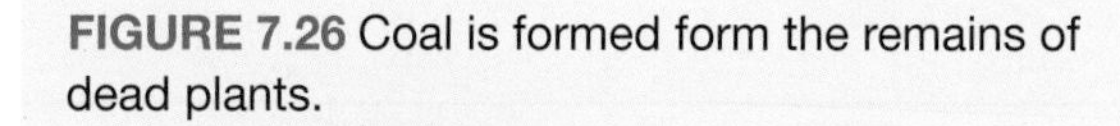

FIGURE 7.26 Coal is formed form the remains of dead plants.

CASE STUDY: Chalk

Chalk is a type of limestone that is not very hard. Chalk is formed from microscopic plankton made of calcium carbonate that separate from sea water and settle to become a white, muddy sediment on the sea floor. The sediment hardens and compacts over time to form chalk. This process takes millions of years. The remains of other sea creatures are also found in the sediment that forms chalk.

FIGURE 7.27 The white cliffs of Dover that overlook the English Channel are composed of chalk.

7.5.4 Chemical sedimentary rocks

Some sedimentary rocks form when water evaporates and leaves behind precipitated mineral crystals that can be compressed and buried by other sediments. **Rock salt** is an example of a rock formed in this way. It forms from residues of salt that remain after the evaporation of water from salt lakes or dried-up seabeds, and can form beds that are hundreds of metres thick.

7.5.5 Rocks in layers

Layers of sedimentary rock are often clearly visible in road cuttings and cliffs, as seen in the spectacular cliffs of the Grand Canyon (figure 7.28). Not only do the layers help you identify them as sedimentary rock, but they are also records of time, with the bottom layers older than the top layers.

When fossils are found in sedimentary rock, the layer they are found in can be used to work out how old the fossils may be.

 Resources

 **Interactivity** Sedimentary rock layers (int-5341)

FIGURE 7.28 The Grand Canyon is a spectacular example of exposed sedimentary rock layers that have been cut into by erosion of a fast-flowing river.

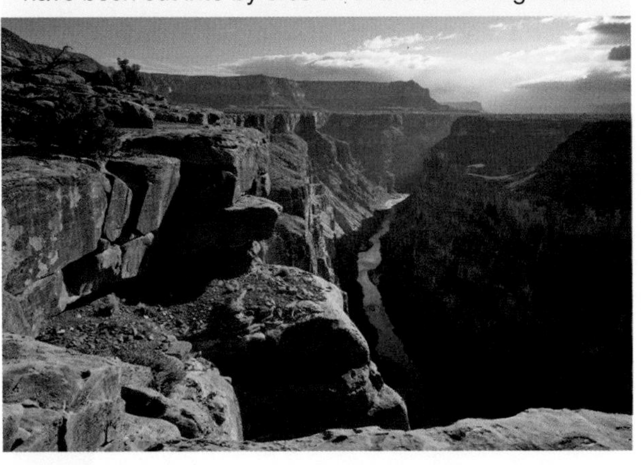

As seen in the Grand Canyon, sedimentary rock layers are originally deposited flat (horizontal). However, layers of sedimentary rocks can be affected by the same forces below Earth's surface that form mountains. Those forces can bend and tilt the rock layers into incredible folds (figure 7.29).

FIGURE 7.29 These layers of limestone formed on the ocean floor and were originally horizontal, but have since been folded by large mountain-building forces.

7.5.6 Useful sedimentary rocks

TABLE 7.7 Uses of sedimentary rocks

Sedimentary rock	Examples of modern uses
Sandstone, limestone and shale	Sandstone and limestone are often used as external walls of buildings. These sedimentary rocks are well suited to carving into bricks of any shape. Shale can be broken up and crushed to make bricks.
Limestone	Limestone is broken up to produce a chemical called lime. Lime is used to make mortar, cement and plaster, and is used in construction for walls and paving. It is also used in the treatment of sewage and in agriculture to neutralise acid in the soil.
Rock salt	Rock salt is used on roads and driveways in very cold areas to combat ice. It is also used as a type of seasoning for food.
Coal	Coal is used in steel-making and cement manufacturing, and in electricity generation. It is burned in electric power stations to boil water. The steam is then used to drive the turbines that produce electricity. In some countries, coal is burned in home heaters, although this can cause air-quality problems. Coal is a non-renewable energy source because it is not replenished within our lifetime; in fact, it has taken millions of years to form a layer of coal.

elog-2183

INVESTIGATION 7.5

Identifying sedimentary rocks

Aim

To use a key to identify a variety of sedimentary rocks

Materials

- several examples of unlabelled sedimentary rocks, including limestone
- dropping bottle of dilute hydrochloric acid

Method

1. Use the following key to identify the samples of sedimentary rocks you have been given.
2. To do the acid test, just add one drop of dilute hydrochloric acid onto the sample and wipe off with a clean paper towel.

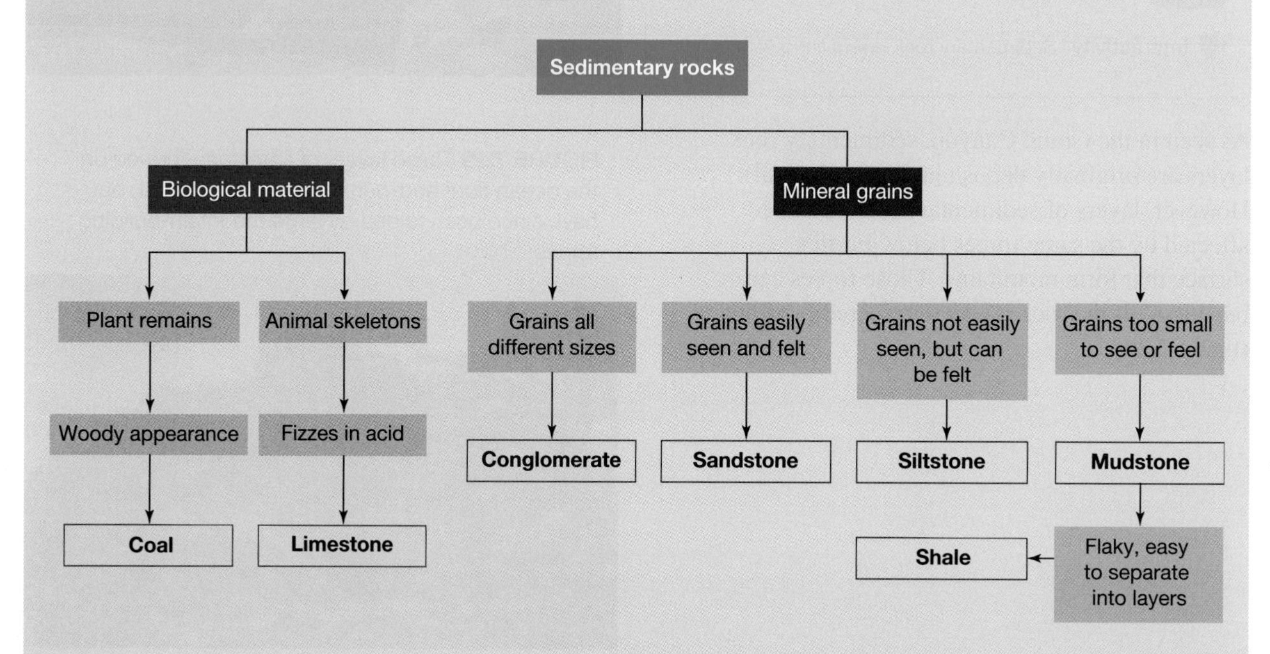

Results

Design a table to record your answers for each step in identifying each sample, particularly the name at the end.

Discussion

1. How many of the unlabelled rocks did you confidently identify?
2. Which of the rock samples were the most difficult to identify, or which are you least confident about?
3. Discuss why it was difficult and how the key might be improved.

Conclusion

What can you conclude about identifying sedimentary rocks?

 Resources

 **eWorkbook** Sedimentary rocks (ewbk-11946)

7.5 Activities

7.5 Quick quiz	7.5 Exercise

Select your pathway

■ LEVEL 1	■ LEVEL 2	■ LEVEL 3
1, 2, 3, 8, 11, 12	4, 5, 9, 13, 16	6, 7, 10, 14, 15

These questions are even better in jacPLUS!

- Receive immediate feedback
- Access sample responses
- Track results and progress

Find all this and MORE in jacPLUS

Remember and understand

1. What are all sedimentary rocks formed from?
2. List in order the process of forming a sedimentary rock.
3. What are clastic sediments before they eventually form a clastic sedimentary rock?
4. As a flooded river slows down, which particles are likely to settle first: gravel, sand or fine clay?
5. Explain, with the aid of a diagram, how sediments lithify and become a sedimentary rock.
6. Clastic sedimentary rocks formed from weathered pieces of other rock are classified based on what characteristic?
7. Explain how a floodplain is created.
8. In which type of sedimentary rock would you most likely find embedded seashells?
9. How is coal formed?

Apply and analyse

10. Explain why sedimentary rocks are found in layers.
11. Explain why limestone and coal are sometimes referred to as 'biological rocks'.
12. A road cutting reveals a layer of sandstone beneath a layer of mudstone. Between them is a much thinner layer of conglomerate.
    a. Which layer would have formed from sediments deep beneath the sea?
    b. Which layer would have formed while the area was flooded by a swollen, fast-flowing river?
    c. Which layer would have formed while the area was near a delta and coastline?
    d. Which layer was formed most recently?

Evaluate and create

13. What type of sediment would you expect to find on the bed of the Yarra River in Melbourne?
14. **SIS** A geologist collects a sedimentary rock sample, shown in the figure. They classify the rock by using the key provided in investigation 7.5. First, they decide it is a sedimentary rock because it came from a layered rock outcrop. Next, they observed that the sample is missing fossils, so they claim that it must contain mineral grains, not biological material. Then, they observe that the grain sizes are the same and small, but still easily seen and felt. In conclusion, they call this rock a sandstone.
    a. The rock was correctly classified as sandstone. Identify two correct applications in the method used to identify the rock.
    b. Is there a step the geologist could have used to gather more information?

15. **SIS** Create a model illustrating how the clast size for sediment deposits would change in a classic path from mountain to coastline to just offshore.

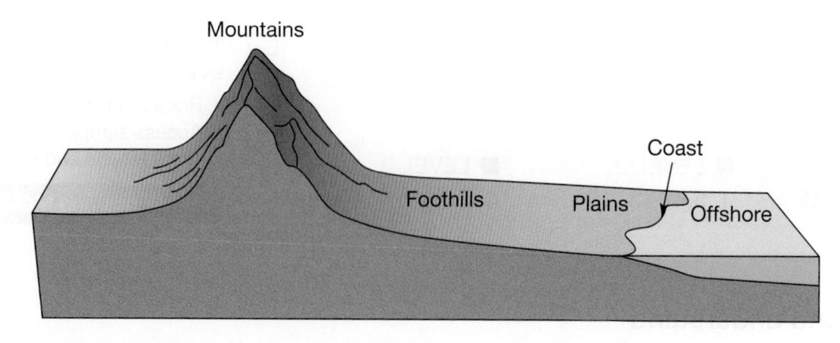

16. What do peat, brown coal and black coal have in common? How are they different from each other?

Fully worked solutions and sample responses are available in your digital formats.

LESSON
7.6 Metamorphic — the 'changed' rocks

LEARNING INTENTION

At the end of this lesson you will be able to describe how some common types of metamorphic rocks form, and what clues they provide to past environments.

7.6.1 Stability and change

So far, we have had rocks melt and solidify, with some materials even blasted out of a volcano to solidify on the surface. On Earth's surface, rocks weather and erode, and sediment ends up depositing and lithifying over time. What else can happen?

Earth never stops changing. As rocks are put under new conditions (like increasing temperature and/or pressure with deep burial or during a mountain-building event) they can 'morph' into another kind of rock — a metamorphic rock.

WHAT DOES IT MEAN?

The word 'metamorphic' comes from the Greek words *meta*, meaning 'change', and *morph,* meaning 'form'.

This change occurs because every mineral forms in a specific set of physical and chemical conditions. When those conditions change, the mineral changes physically and/or chemically to be stable under the new set of conditions. All of this change can happen without melting and is called **metamorphism**.

- A physical shift can occur when the mineral rotates into a new orientation.
- A chemical shift can occur when either the mineral breaks down to form new minerals or crystal structures realign — such as we saw in the formation of diamonds in section 7.2.2.

7.6.2 Metamorphic rocks

Rocks pushed deep below Earth's surface are buried under the weight of the rocks, sediments and soil above them. They are also subjected to higher temperatures with increasing depth. On average, the temperature increases by about 25 °C for every kilometre below the surface. Added heat and pressure can change the type and appearance of the minerals in rocks.

metamorphism the process that changes rocks by extreme pressure or heat (or both)

Any kind of rock (igneous, sedimentary or metamorphic) can undergo metamorphic changes due to heat and/or pressure. The changes that take place during the formation of metamorphic rocks depend on:

- the type of original rock, sometimes called the 'parent' rock
- the amount of heat to which the original rock is exposed
- the type and amount of pressure added to the rock
- how quickly the changes take place.

Heat and pressure cause the minerals to rearrange within the rock, often due to density, and as a result they form bands. We say these rocks have become **foliated**.

Metamorphic rocks do not melt — if they do then they will be classified as igneous rocks.

The higher the amount of heat and/or pressure, or the longer a rock is exposed to metamorphism, the greater the change will be. This is called a metamorphic grade. A low-grade rock has experienced less change than a high-grade rock.

Rocks do not always need to be buried to great depths to experience metamorphism. Figure 7.30 shows how rocks can be changed by the high temperatures that result from contact with hot magma. The metamorphic rocks around the body of magma are baked by the heat escaping the cooling magma body. Where would you expect to find the high-grade metamorphic rocks?

FIGURE 7.30 The formation of metamorphic rock by contact with hot magma

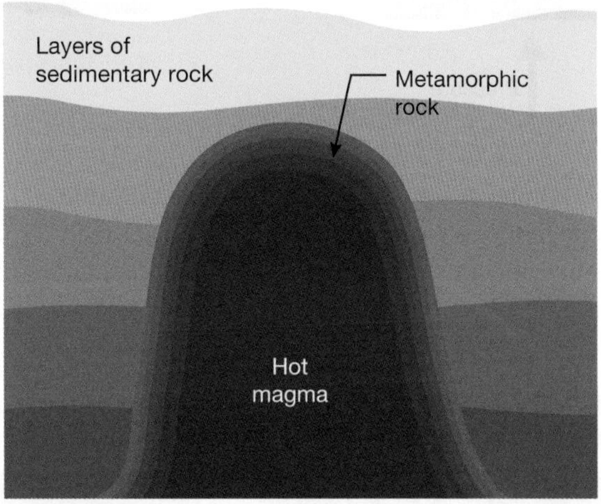

Layers of sedimentary rock

Metamorphic rock

Hot magma

Types of metamorphic rocks

Shale is a common type of sedimentary rock. It is made of tiny clay particles that can be scratched with your fingernail and it is arranged in tiny layers that crumble easily. However, when shale is exposed to heat and pressure, the minerals begin to change, and the rock hardens to form the low-grade metamorphic rock called **slate**. Slate doesn't look much different from shale, in that it can still split into thin layers, but the rock is much harder. Given more time and higher conditions, it will continue to morph into a high-grade rock called *schist* (figure 7.31).

Metamorphic rocks that are mainly the result of great pressure can often be identified by bands of light and dark colours. These bands are evident in the sample of **gneiss** (pronounced 'nice') pictured in figure 7.32. Gneiss is formed mainly as a result of great pressure applied to the igneous rock granite.

foliated consisting of an arrangement of certain mineral grains into distinct bands, which gives the rock a striped appearance

slate a fine-grained metamorphic rock formed as a result of moderate heat and pressure on shale

gneiss a coarse-grained metamorphic rock with light and dark bands formed mainly as a result of great pressure on granite

FIGURE 7.31 When shale (left) is exposed to heat and pressure it becomes slate (centre). With further heat, pressure and time, slate can become schist (right).

FIGURE 7.32 When granite is exposed to high pressure it becomes gneiss.

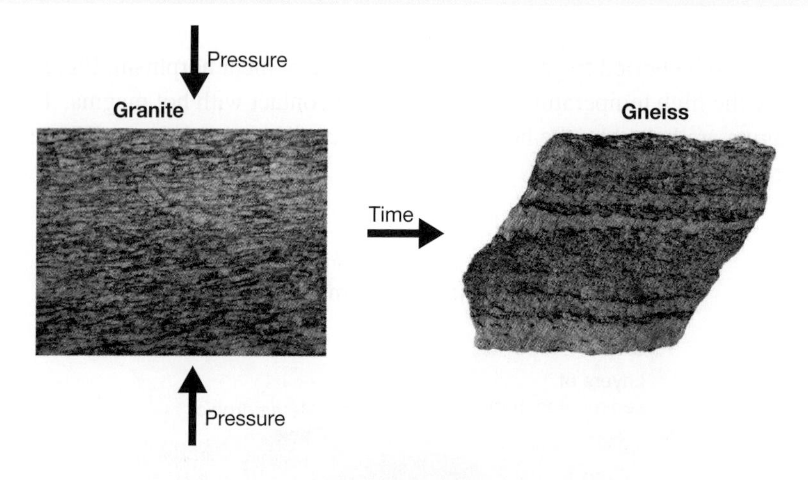

Marble forms when limestone is put under heat and pressure. It contains the same calcite minerals as limestone, although they generally grow larger with metamorphism. If the limestone has minerals other than calcite in it, the marble will have a swirling colour effect.

marble a metamorphic rock formed as a result of great heat or pressure on limestone

FIGURE 7.33 A marble quarry in Italy. Marble forms from the metamorphism of limestone.

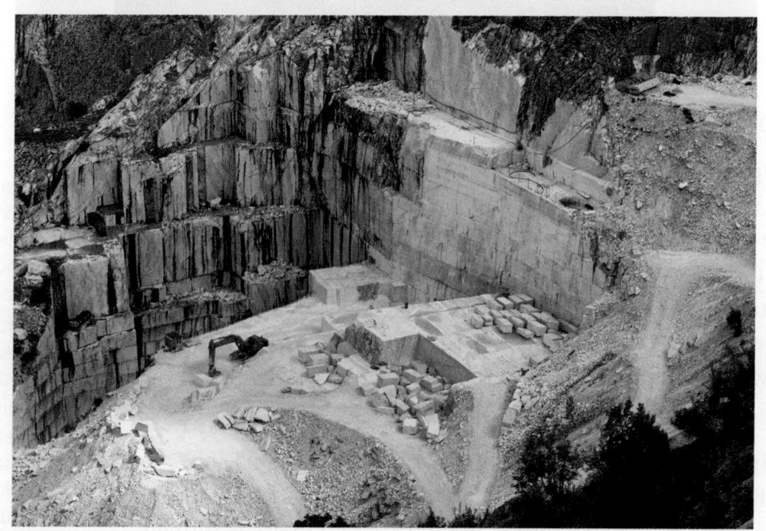

Common examples of the formation of metamorphic rocks are summarised in table 7.8.

TABLE 7.8 How some common metamorphic rocks are formed

'Parent' rock	Condition of metamorphism	Metamorphic rock
Shale (sedimentary)	Mainly low pressure ⇒	Slate
Sandstone (sedimentary)	Mainly heat ⇒	Quartzite
Limestone (sedimentary)	Mainly heat ⇒	Marble
Granite (igneous)	Mainly high pressure ⇒	Gneiss

CASE STUDY: Rocks in your pool table

Have you ever tried to lift one end of a pool table and noticed how incredibly heavy it is? Pool tables are very heavy and difficult to move because the flat surface under the felt is not wood as you may have thought — it's actually made of slate. Because of its natural hardness and flat face, slate makes an ideal even surface!

Resources

 **Interactivity** Metamorphic rocks (int-5343)

Clues from metamorphic rocks

We do not actually see metamorphism, because it takes place entirely underground. This makes metamorphic rocks the most mysterious of the three rock groups. However, the nature of metamorphic rocks above and below the ground can provide clues about the history of an area.

Think about why the presence of **quartzite** or marble high in a mountain range would suggest that the area was once below a shallow sea. The presence of slate might suggest that the area was once the floor of a still lake or deep ocean. These original rocks were either deeply buried, or exposed to magma, or pushed and pulled during a mountain-building event to transform them into new rock.

7.6.3 Useful metamorphic rocks

The unique hardness and ability to split into thin layers has historically made slate useful as roofing or flooring material. Quartzite is also very hard and has been used for building materials.

Marble's softness, colour range and beautiful appearance make it suitable for sculpting. It is also used in building, for tiles, columns, walls and floors (inside and outside). It is usually highly polished. Ground-up marble can also be used in toothpaste, pharmaceuticals, agriculture mixtures, cosmetics, paper, paint and aggregate for construction.

The sedimentary rocks from which marble and slate are formed could not be used for many of these purposes as they are not as durable.

quartzite an extremely compact and hard metamorphic rock consisting essentially of quartz

elog-2185

INVESTIGATION 7.6

Rocks — the new generation

Aim

To examine and compare a selection of metamorphic rocks and their corresponding 'parent' rocks

Materials

- labelled samples of granite, gneiss, limestone, marble, sandstone, quartzite, shale and slate
- hand lens

Method

1. Try to sort the rocks into pairs of 'parent' rock and corresponding metamorphic rock. Use the descriptions and examples in this lesson if you have trouble pairing the rocks.
2. Examine each pair of rocks with a hand lens. Take particular note of grain or crystal size and banding.
3. If necessary, re-sort the rocks into different pairs.

Results

Complete the table provided by noting the similarities and differences between the 'parent' and metamorphic rock of each pair.

TABLE Comparing 'parent' and metamorphic rocks

'Parent' rock	Metamorphic rock	Similarities	Differences	Main cause of metamorphism
Shale				
	Gneiss			
Sandstone				
	Marble			

Discussion

1. Why is the term 'parent' used to describe the original rock before metamorphosis?
2. Use the last column of your table to suggest whether the main cause of metamorphism was heat or pressure.
3. Is there a pattern to the rock's appearance that could help you determine that pressure was a main cause of metamorphism?
4. Suggest an idea or two about why or how the metamorphic layering and banding could form.

Conclusion

What can you conclude about metamorphic rocks and their 'parent' rocks?

DISCUSSION

Some have argued that black coal is actually more of a metamorphic rock than a sedimentary rock. Discuss why this may be, and what you would call it.

7.6.4 The rock cycle

The **rock cycle** in figure 7.34 describes how rocks can change from one type to another. Weathering, erosion, deposition, uplift, heat, pressure, melting and crystallisation are processes that help change rocks. The rock cycle is different from other cycles because there is no particular order in which the changes happen, and it generally takes a long time to make the complete change.

rock cycle a cycle of processes that rocks experience in Earth's crust as they constantly change from one type to another

FIGURE 7.34 The processes of melting, metamorphism by increased heat and pressure, as well as weathering and erosion over time, will change rocks from one type to another. This is called the rock cycle.

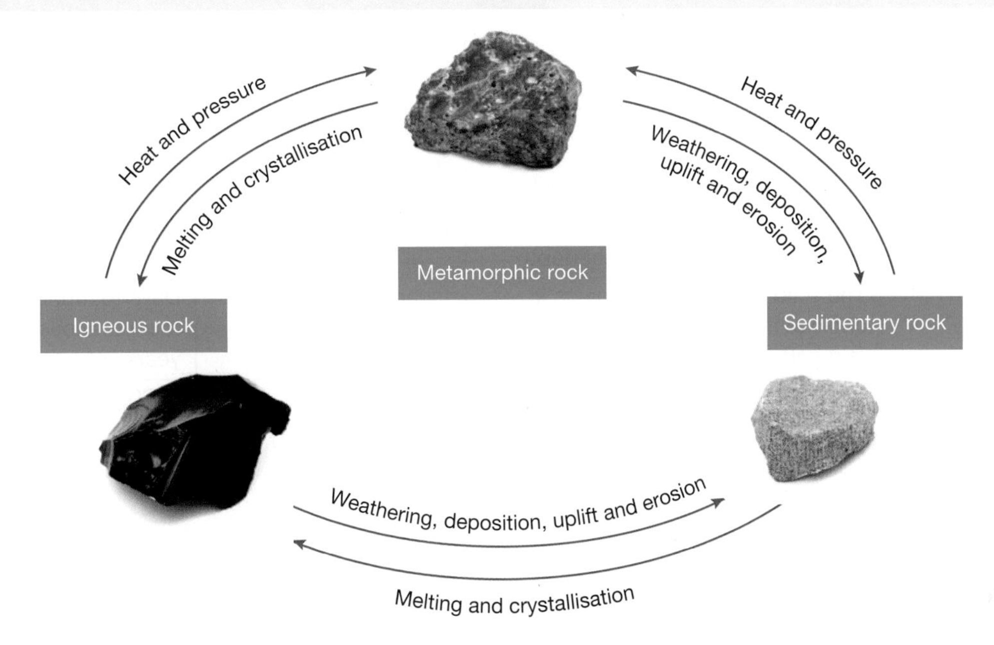

Some rocks have been unchanged on Earth for millions of years and may not change for millions more. Some rocks change a bit quicker (but still slowly), especially near regions of Earth's crust that are pushing, pulling or twisting.

The formation, movement and transformation of rocks results from Earth's internal heat, tectonic processes, and the effects of water, wind, gravity and biological (including human) activities. These processes occur at various rates, over different scales and varying periods of time, as shown in table 7.9.

TABLE 7.9 Processes within the rock cycle

Process	Description	Rock type formed	Timescale (approx.)
Heat and pressure	Earth's internal heat and pressure can transform a rock into a metamorphic rock or melt it completely and form igneous rock. Meteorite impact can cause metamorphism at very high pressure and low temperature.	Igneous Metamorphic	Hundreds of thousands to millions of years (or minutes for a meteorite impact)
Weathering	The action of weather conditions on the breakdown of rocks; it is influenced by precipitation, temperature, topography and vegetation cover	Sedimentary	Days to millions of years
Erosion	The wearing down of rock due to water, wind, ice or gravity, atmospheric and ocean circulation patterns, and regional topography	Sedimentary	Days to millions of years
Uplift	Weathering and erosion of tectonically uplifted, newly exposed rock creates sediments.	Sedimentary	Thousands to millions of years

(continued)

TABLE 7.9 Processes within the rock cycle *(continued)*

Process	Description	Rock type formed	Timescale (approx.)
Transportation and deposition	The movement of particles to a resting place (e.g. river deposits, delta, volcanic ash eruption)	Sedimentary Volcanic	Days to thousands of years
Lithification	Compaction and cementing of sediments during burial, under low pressure conditions, to form rock	Sedimentary Volcanic	Thousands to millions of years
Melting	The melting of a rock, such as in a subduction zone or at great depth	Igneous	Millions of years
Crystallisation	The solidification of magma as it cools at the surface (extrusive rock) and below the ground (intrusive rock)	Igneous	Days to months (extrusive rock) to thousands of years (intrusive rock)
Human activity	Factors such as urbanisation, deforestation, agriculture and mining can result in increased erosion.	Sedimentary	<300 years

FIGURE 7.35 Typical settings and processes that operate in the rock cycle

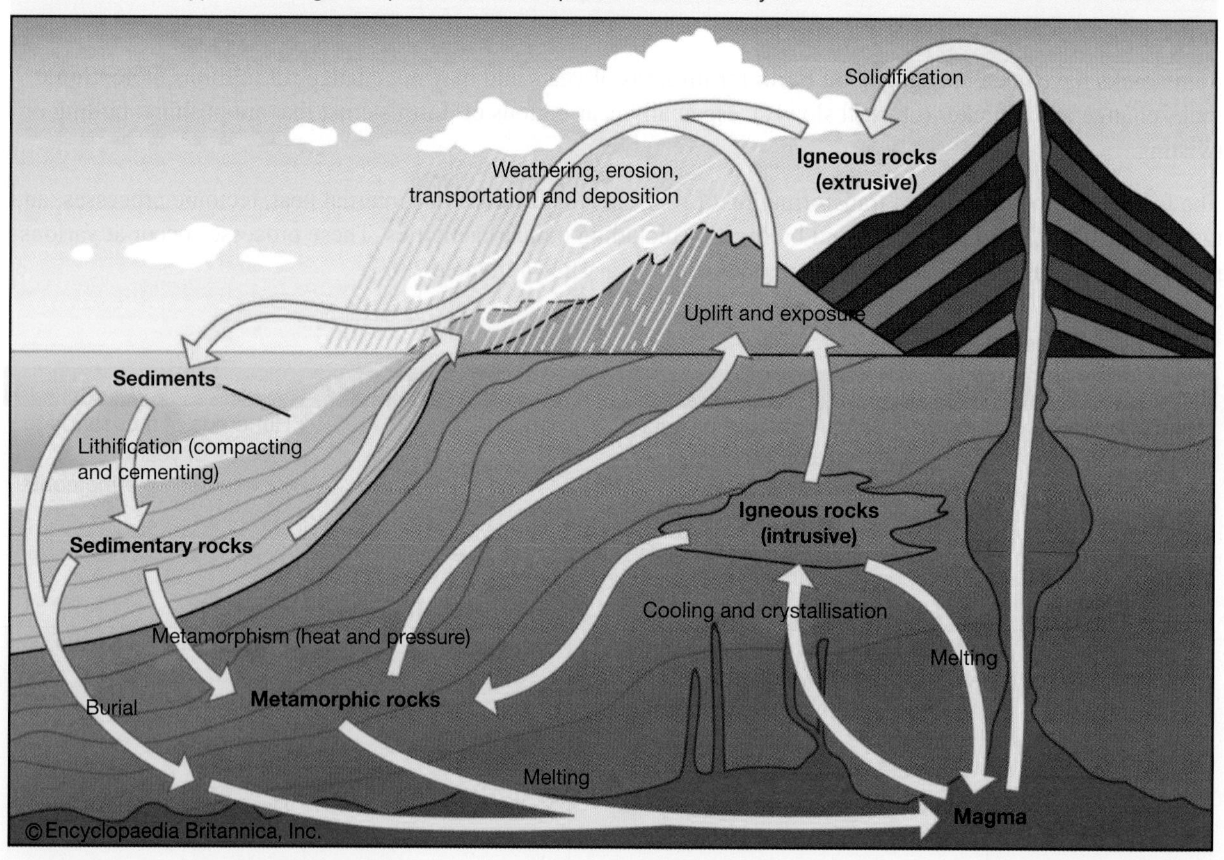

DISCUSSION

A tadpole grows into a frog, female frogs lay eggs, and eventually more tadpoles emerge from the eggs. That's a life cycle.

Some of the changes in rocks can be described as cycles too. Weathered rock is moved by erosion and the particles form sediments, which can be cemented together to form sedimentary rocks, which in turn may eventually change into metamorphic rocks. Once those rocks are exposed at the surface the weathering starts all over again. A complete cycle normally takes millions of years, but sometimes never takes place at all. Why?

FIGURE 7.36 There are many cycles in nature. Some happen faster than others.

7.6.5 Identifying major rock types

The key in figure 7.37 can be used to identify some of the more common rocks. Start in the middle and choose a path based on your observations. First, identify whether the rock consists mainly of crystals (that sparkle) or organic matter (that don't). A magnifier might help you.

FIGURE 7.37 A key for identifying major rock types

What rock is it?

Igneous

Granite ← Black, white and pink intergrown, large crystals

Diorite ← Black and white intergrown crystals

Rhyolite ← Light-coloured with scattered, pale crystals

Basalt ← Dark with small crystals; may have holes

Obsidian ← Glassy, dark conchoidal fracture

Pumice ← Small and light

Scoria ← Large and dark

Scattered or intergrown crystals of different types

Very fine-grained or glassy

Intergrown crystals of similar size and colour

Crystals react with HCl → Marble

Quartzite

Non-foliated (not layered)

Crystals

Foliated (layered)

Black and white layers → Gneiss

Thin layers that may have other crystals present → Schist

Flattened, fine-grained and dull → Slate

Metamorphic

Start here

Particles

Holes ← No layers

Layered

Made of sediment

Lightweight and made of dark, organic material → Coal

Marine fossils, fizzes with HCl → Limestone

Cemented rounded gravel → Conglomerate

Cemented angular pebbles → Breccia

Sand → Sandstone

Mud → Shale

Sedimentary

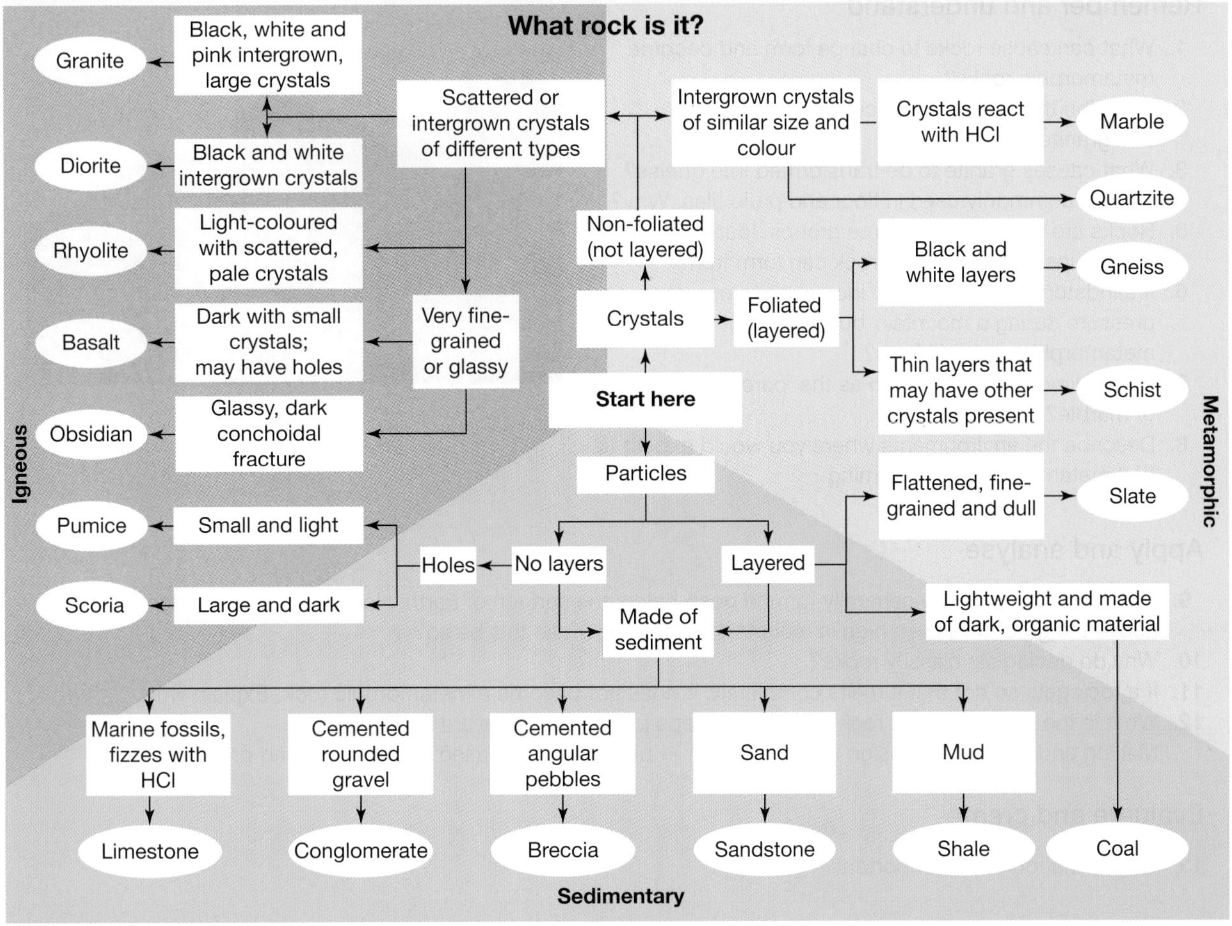

Note: Not all rock types are included in this key.

ACTIVITY

Find a rock outside or in your neighbourhood and see if you can identify it!

Resources

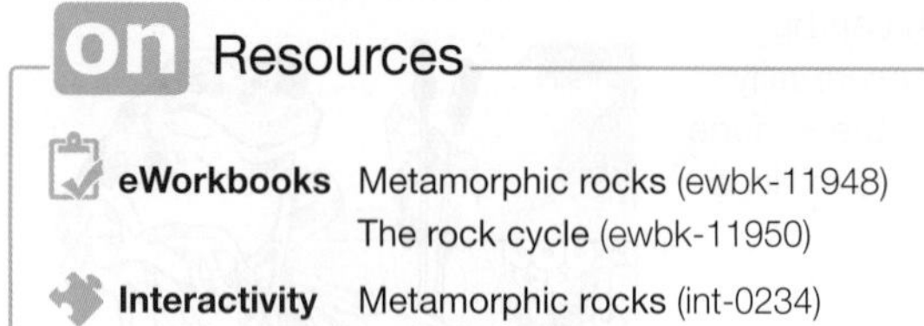

eWorkbooks Metamorphic rocks (ewbk-11948)
The rock cycle (ewbk-11950)

Interactivity Metamorphic rocks (int-0234)

7.6 Activities

learnon

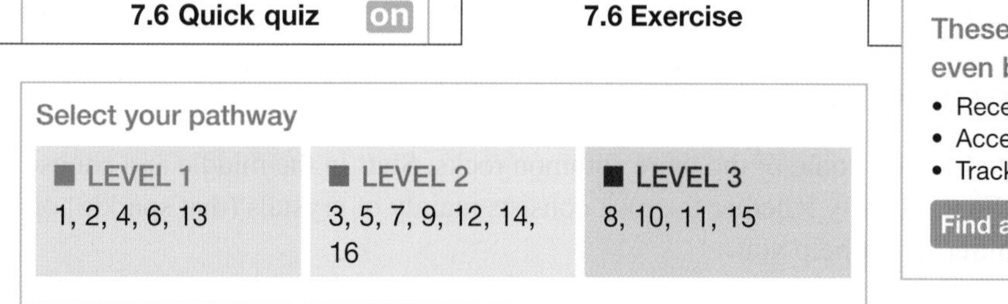

7.6 Quick quiz on | **7.6 Exercise**

Select your pathway

LEVEL 1	LEVEL 2	LEVEL 3
1, 2, 4, 6, 13	3, 5, 7, 9, 12, 14, 16	8, 10, 11, 15

These questions are even better in jacPLUS!

- Receive immediate feedback
- Access sample responses
- Track results and progress

Find all this and MORE in jacPLUS

Remember and understand

1. What can cause rocks to change form and become metamorphic rocks?
2. Describe the visual differences between gneiss and granite.
3. What causes granite to be transformed into gneiss?
4. Slate is commonly used in floor and patio tiles. Why?
5. Rocks are classified into three groups. Identify which of the groups a metamorphic rock can form from.
6. If sandstone is subjected to increased temperature and pressure during a mountain-building event, what metamorphic rock will form?
7. Why is limestone referred to as the 'parent' rock of marble?
8. Describe the environments where you would expect to find metamorphic rocks forming.

Apply and analyse

9. Metamorphic rocks are generally formed deep below the surface of Earth. However, they are often found above the ground — even high in mountain ranges. How can this be so?
10. Why do geologists classify rocks?
11. If a rock gets so hot that it melts completely, it does not become a metamorphic rock. Explain why.
12. What is the progression of rock types if the steps in the rock cycle are as follows?
    Melting and cooling → erosion and deposition → burial with increased temperature and pressure

Evaluate and create

13. Why is the rock cycle important?

14. **SIS** Consider the provided figure, which shows the different grades of metamorphism in rocks. As the pressure and/or temperature changes, the minerals in the rock become unstable, break down and form new minerals. The growth of particular minerals indicates that a grade boundary on this graph has been crossed.
    a. What are the relationships between temperature, pressure, depth and metamorphic grade?
    b. What is the temperature range for a low-grade rock found at a depth of 10 km?
    c. What are geologists using to recognise the difference between low- and high-grade metamorphic rocks?
    d. Suggest a definition for *diagenesis* and find out how it differs from lithification.

Grades of metamorphism in rocks

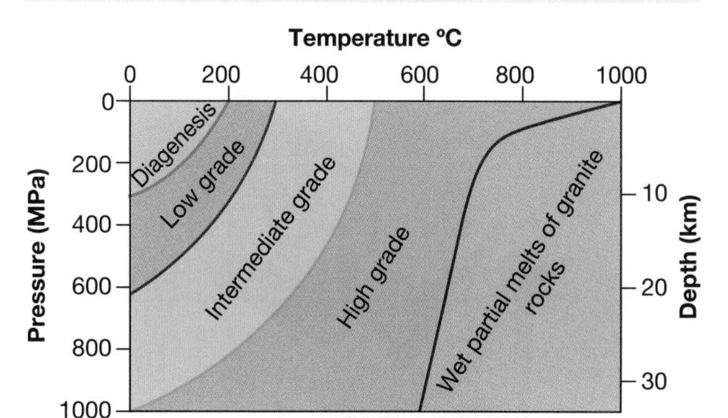

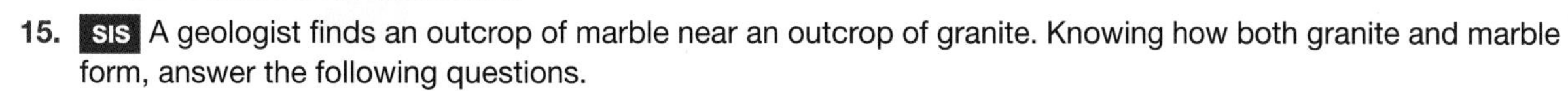

15. **SIS** A geologist finds an outcrop of marble near an outcrop of granite. Knowing how both granite and marble form, answer the following questions.
    a. Suggest an idea that explains this relationship.
    b. How could you test your idea?
16. Devise a 'buildings trail' in your city or town to locate buildings made of different kinds of rock. Draw a map to show the location of the buildings and the type of rock used in constructing them.

Fully worked solutions and sample responses are available in your digital formats.

LESSON
7.7 Rock technology

First Nations Australian readers are advised that this lesson and relevant resources may contain images of and references to people who have died.

LEARNING INTENTION

At the end of this lesson you will be able to describe how development of Stone Age tools required knowledge of rock types and how First Nations Australians created tools for many specific purposes using their knowledge of different rock types.

7.7.1 Stone Age tools

SCIENCE AS A HUMAN ENDEAVOUR: Rocks as specialised tools

Rock technology began about two million years ago when early humans started using rocks to make simple chopping tools. This was the beginning of the period known as the **Stone Age**. For the great civilisations of Asia, Europe and North Africa, the Stone Age ended around 3000 BC with the discovery of bronze, an **alloy** of copper and tin.

The most commonly used resource in the Stone Age was a fine-grained sedimentary rock called **flint**. When flint breaks, it leaves a razor-sharp edge, so it was ideal for making sharp tools like knives, axes and spearheads.

Stone Age a prehistoric time when weapons and tools were made of stone, bone or wood

alloy a mixture of a metal with a non-metal or another metal

flint a fine-grained sedimentary rock that leaves a very sharp edge when broken

FIGURE 7.38 The use of chipping one rock with another to make the desired shape

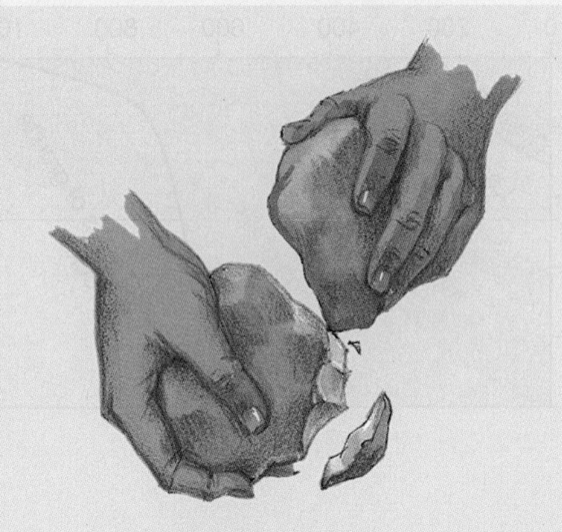

FIGURE 7.39 Flint arrowheads were attached to wooden shafts with twine or animal sinews.

Small tools were made by striking tool stones like flint or the glass-like igneous rock obsidian with harder stones (hammerstones) such as quartzite, a metamorphic rock. To remove large flakes from the tool stone, a sharp blow was delivered by the harder rock. If the tool stone was struck correctly, a flake sheared from it. This process is called **percussion flaking**. The toolmaker continued to remove flakes from the stone until the desired shape was obtained. The flakes were then used to make tools such as knife blades, scrapers and engravers.

Larger items such as axeheads and spearheads were made with a combination of techniques, such as percussion flaking, grinding stones against each other and chiselling against the edge of a stone with tools made of bone or wood.

percussion flaking a process in which tool stones, such as flint or obsidian, were struck with harder stones, such as quartzite, to shear large flakes off until they were a desired shape

Indigenous ingenuity

First Nations Australians were still using Stone Age tools when Europeans began to colonise Australia in 1788. First Nations Australians were highly skilled at working with stone, being the first people to use ground edges on cutting tools and to grind seed.

Their stone axes and other sharp tools were used to cut wood, shape canoes, chop plants for food, skin animals and make other tools out of stone or wood. The sharpened stones were often attached to wooden handles with twine from trees, resin from plants or with animal sinews.

FIGURE 7.40 Hand axes made and used by the Ngadjon-Jii Peoples of the tropical rainforests of northern Queensland

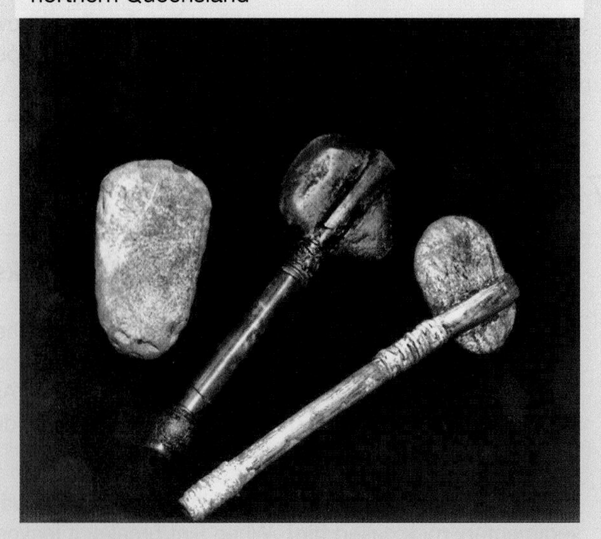

Grinding stones (figure 7.41) are slabs of stone used with a smaller, harder top stone to grind grains and seeds such as wheat, as well as corn, berries, roots, insects and many other things to prepare food for cooking. Leaves and bark were sometimes ground to make medicines. First Nations Australians also used grinding stones to grind various types of soil and rock like ochre to make the powders used to paint shields and other wooden implements with distinct markings specific to the nation group.

The tools and the type of stone used to make them varied from group to group, depending on the location. First Nations Australians were skilled at making good use of the available resources. Apart from grinding stones, axes and other cutting tools, they made items such as bowls, cups and food graters out of stone.

TABLE 7.10 Rock uses and their functional properties

Tool	Rock type	Rock properties
Ground-edge axe head	Volcanic rocks (e.g. diorite, basalt), slate	Durable and hard; can be shaped and take a sharpened edge
Stone knives and spear points	Slivers of quartzite, silcrete, obsidian	Hard with sharp edges; can be worked to an elongated shape
Hand axes and choppers	Volcanic rocks, silcrete	Hard with sharp edges; durable
Cutting tools and flakes	Quartzite, silcrete, obsidian	Durable with sharp edges; fits in hand
Anvil	Quartzite	Resistant; forms flat blocks 30–50 cm wide
Hammerstone	River pebbles of quartz or quartzite	Dense, round and hard
Grinding stone	Sandstone, basalt, quartzite (lower stone)	Rough/abrasive; slab-forming (15–70 cm across) (lower stone)
	River pebble (top stone)	Dense, round and smooth; fits in hand (top stone)
Sharpening stone	Sandstone	Rough; forms broad, flat horizontal surfaces
Paint	Ochre (weathered rock containing iron oxide)	Has a range of colours; can be mixed with fat
Fuel	Coal	Burns slowly and retains heat; light to carry
Sacred stones	River pebbles	Waterworn and naturally shaped
Magic stones	Quartz crystals and unusually coloured or shaped stones	Attractive and unusual

FIGURE 7.41 A First Nations Australian grinding stone made from sandstone. The top stone is a hard, smooth river pebble.

FIGURE 7.42 A food grater made from stone by the Ngadjon-Jii Peoples of northern Queensland

7.7.2 Quarrying

First Nations Australian quarries are places where First Nations Australians removed stone for many different tools and purposes. Stone was often traded too, with some axes traded up to 800 km away! Types of rock mined by First Nations Australians included amphibolite, andesite, basalt, chalcedony, chert, dolerite, granite, greenstone, greywacke, ironstone, limestone, mudstone, obsidian, porphyry, quartzite, sandstone, silcrete, siltstone and trachyte. Soils and weathered rock containing ochre were also quarried.

Quarries were a valuable resource and ranged in size from one boulder to outcrops, riverbeds and large areas of scattered stone (gibber plains). The higher the quality or value of the stone, the larger the quarrying operation and the greater the distance over which the stone was traded. Different clans would gather at a meeting place near a quarry site to trade for the stone or ochre, and to hold ceremonies, initiations and other important cultural events. Quarry ownership rested with the local custodians of that land, who could provide permission for access. Quarrying was often done by people with special knowledge of how to efficiently extract the resource.

First Nations Australians used at least two methods of stone quarrying. One method was to hit the surface of the outcrop at an angle with a hammerstone to break off small pieces. The other method involved digging around and under outcrops to find buried stone that was unweathered. Some operations even extended underground. In South Australia there is evidence that flint mining extended about 75 metres below the surface and up to 300 metres from the entrance of a cave.

Work at a quarry would also have involved the trimming of 'blanks' — pieces of a convenient size and shape for making into axes. Final trimming of the axe and grinding of the blade was often done elsewhere. Hammerstones, anvils and grinding stones were used at the quarry and were often left there because they are heavy.

Quarries are important to First Nations Australians today, providing a link with their culture and their past.

CASE STUDY: The Wilgie Mia red ochre mine

Wilgie Mia, or Thuwarri Thaa (the place of red ochre), in Western Australia is the largest and deepest ochre mine in Australia. Believed to be 27 000 years old, it is thought to be the oldest continually worked mine site in human history. The ochre is very high quality and was traded around Australia.

Initiated Wajarri Yamatji men mined the ochre using heavy stone hammers and fire-hardened wooden wedges to pry away rock, which was then broken up to separate the ochre. Pole scaffolding and wooden platforms were used so they could simultaneously mine at different heights, increasing output.

FIGURE 7.43 Mining for red ochre at Wilgie Mia, Weld Range, Murchison, Western Australia

7.7 Activities

7.7 Quick quiz on	7.7 Exercise

Select your pathway

LEVEL 1	LEVEL 2	LEVEL 3
1, 2, 3	4, 5, 8	6, 7

These questions are even better in jacPLUS!
- Receive immediate feedback
- Access sample responses
- Track results and progress

Find all this and MORE in jacPLUS

Remember and understand

1. List one example of each of the following types of rock that were used in the Stone Age to make tools.
   a. Igneous
   b. Sedimentary
   c. Metamorphic
2. Which alloy replaced stone to make tools when the Stone Age ended?
3. What role did animal sinews play in tool-making by First Nations Australians?
4. List three different uses of grinding stones.

Apply and analyse

5. What properties of flint made it so useful during the Stone Age?
6. Suggest how the process of percussion flaking got its name.
7. List some properties that you would look for when selecting a suitable top stone for a grinding stone.

Evaluate and create

8. Research and report on a range of tools and other devices made from rocks or other natural materials that First Nations Australians used in their daily lives.

Fully worked solutions and sample responses are available in your digital formats.

LESSON
7.8 Geologic history

LEARNING INTENTION

At the end of this lesson you will be able to identify the relationships between rocks, and the clues they contain as to how and when they formed, including the fossil record.

7.8.1 Clues in rocks

If only rocks could talk! They would have so much to say. They would tell us about Earth's history — about prehistoric creatures whose fossils lie within them; about explosive volcanoes, earthquakes and flooded rivers that washed them away; and about what it is like inside Earth.

Although rocks can't talk, geologists are able to read them to answer questions such as:
- How has Earth's climate changed over millions of years?
- When did the Himalayas form?
- What were the first signs of life?
- What caused the extinction of the dinosaurs?

The clues lie in the appearance of the rocks, the minerals they contain, and how they are layered or located relative to one another. There are also clues in fossils. A **fossil** is evidence of living things preserved in rocks.

fossil any remains, impression or trace of a life form of a former geological age; evidence of life in the past

relative age the age of a rock compared with the age of another rock

7.8.2 Geologic history

Over very long periods of time, rivers change their course, mountains form where seas once existed and the climate changes. As these changes take place, different layers of sediment can be deposited at the same location. Some layers will be thicker than others, but the layer below will have been deposited before the one on top. Sudden events that occurred in the area of deposition, such as erupting volcanoes or landslides, are also recorded in the sediment layers.

Sedimentary rocks, which are formed by the hardening of the different layers of sediments, provide many clues about the order in which events took place. Slow movements caused by the forces beneath the surface can tilt, break, curve and push up the layers. This will disturb the pattern of layers and mark their occurrence in the record. Reading sedimentary rocks is particularly important to interpreting a geologic history of Earth.

A geologic history uses the concept of **relative age** to interpret an order to events. The relative age of a rock simply indicates whether it was formed before or after another rock.

FIGURE 7.44 Rocks tell the history of the Earth.

Principles of relative age

- A layer of sedimentary rock is older than the rocks above it and younger than those below it.
- Layers of sedimentary rock were originally deposited and lithified as horizontal layers. Thus, if you see bent or broken layers, the bending or breaking event must have happened after deposition.
- A time of weathering and erosion can remove layers, disturbing the surface. This results in a time gap in the record but can also be a fresh start for a younger sequence to deposit on top.

The relative ages of some igneous rocks and metamorphic rocks can be determined in the same way.

DISCUSSION

The present is a key to the past. As a class, or in a small group, discuss how much time you think it took to complete the sequence of events in figure 7.45. What observations of the modern world can we use to judge how quickly geologic events occur?

int-0233

FIGURE 7.45 Illustrated side-view of a portion of the crust. It highlights the relationships of rock layers relative to one another, which can be used to propose a geologic history.

7. These layers were deposited last. They have started to weather and erode.
6. A long period of weathering and erosion left the layer of limestone with a flat surface. When a volcano then erupted nearby, lava from the volcano cooled to form basalt on the flat surface.
5. A sudden event such as an earthquake has occurred to break the layers of rocks like this. This event took place after the lower layers were folded. A break like this is called a fault.
4. A slow event has caused the lower levels to buckle. This is called folding. Folding can occur when rock layers are under pressure from both sides.
3. The third event to occur was deposition of limestone. It tells us that there were probably marine organisms present in the area during this time.
2. This is the second layer deposited. Shale is a fine-grained rock that is deposited in a quiet environment — such as a swamp, lake or slow-flowing part of a river.
1. Conglomerate was deposited first in this rock sample. This layer was deposited by a glacier or an active environment — such as a very fast-flowing river.

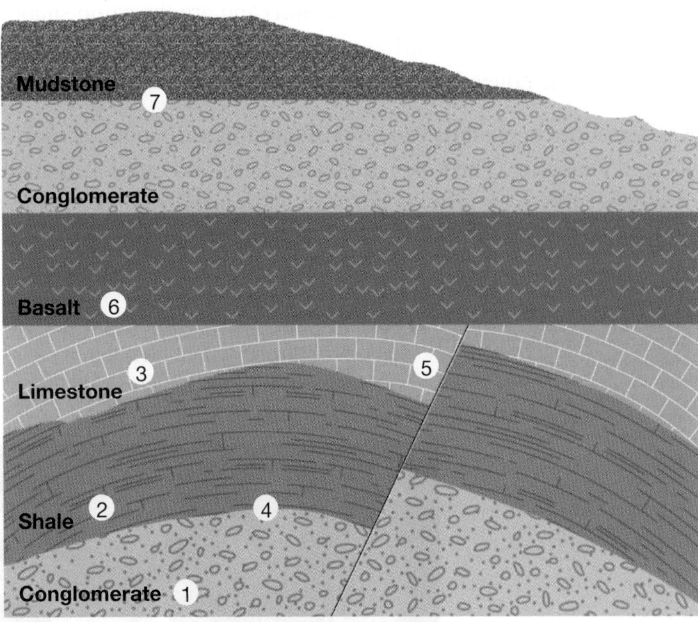

It's all relative

Fossils provide a way of finding out how living things have changed over time. Evidence of the very oldest living things is buried within the deepest and oldest layers of rock. Scientists who study fossils are called **palaeontologists**.

Since it is almost certain that a layer of sedimentary rock is older than the rocks above it and younger than those below it, it can be assumed that the fossils in lower layers are older than those in the layers above. By comparing fossils found in rocks in different areas, including different continents, it is possible to compare the relative age of rocks throughout the world.

FIGURE 7.46 Fossils provide clues about life in the past. This is a fossil of an ancient fish.

7.8.3 How fossils form

The remains of most animals and plants decay or are eaten by other organisms, leaving no trace behind. However, if the remains are buried in sediments before they disappear, they can be preserved, or fossilised. Fossils can form in several ways.

palaeontologist a scientist who studies fossils

The hard parts of plants and animals are more likely to be preserved than the softer parts. Wood, shells, bones and teeth can be replaced or chemically changed by minerals dissolved in the water that seeps into them. Fossils are most commonly formed in this way (**permineralisation**) and are the same shape as the original remains but are made of different chemicals. Petrified wood, opalised fossils and fossil dinosaur bones are examples of fossils formed by permineralisation.

Animal bones and shells can be preserved in sediments or rock for many years without changing. The types of bones, shells and other remains found in the layers of sedimentary rock provide clues about the environment, behaviour and diets of ancient animals.

permineralisation the most common method of fossilisation, in which minerals fill the cellular spaces and crystallise; the shape of the original plant or animal is preserved in great detail

carnivore an animal that eats other animals

scavenger an animal that eats dead plant and animal material

FIGURE 7.47 A selection of opalised fossils. These are opals that have formed within cavities in sand and clay left by the remains of living things, like shells, teeth, and bones, creating incredible fossil replicas.

int-5342

Dinosaurs preserved in rock

1. After the death of a dinosaur, its body would usually be eaten by meat-eating animals (**carnivores** or **scavengers**). Its bones would be crushed or weathered, leaving no remains. If, however, the remains of a dinosaur were buried in sediment, the bones could be preserved.
2. If a dinosaur died near a muddy swamp, shallow lake or riverbed, its remains sank in the mud or were washed into a river in a flood. The bones were quickly buried in sediment.

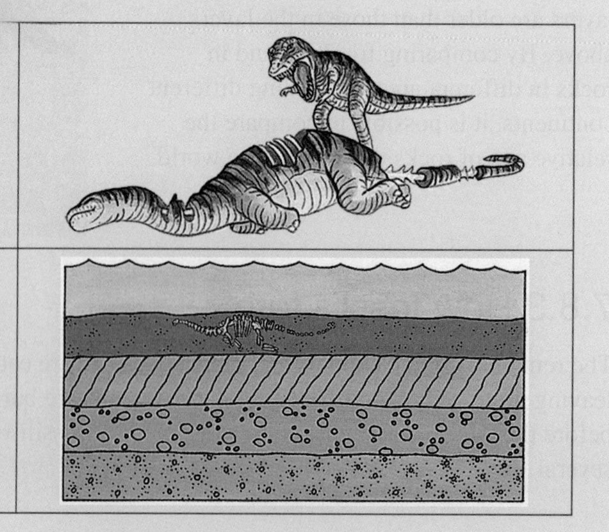

3. Over millions of years, more layers of sediment were deposited on top of the buried remains. Chemicals that dissolved in the water that seeped into the remains changed their colour and chemical composition. The shape, however, was preserved. The sediments were gradually transformed into sedimentary rock.	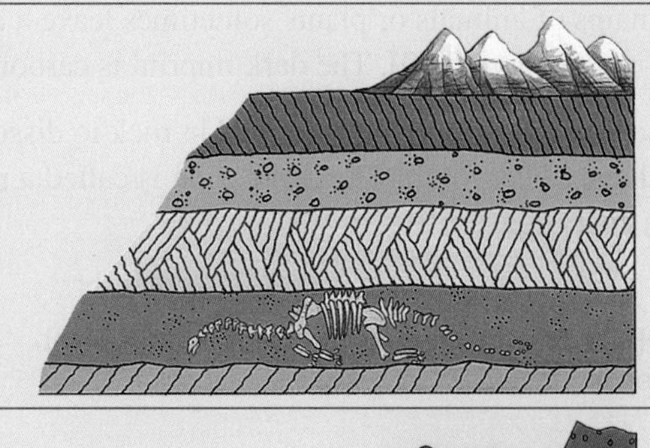
4. The layers of rock containing the fossilised remains were pushed upwards, bent and tilted by forces beneath Earth's surface. Weathering and erosion by the wind, sea, rivers or glaciers might expose one or more of the bones or teeth. If the exposed fossils were discovered before being buried again, palaeontologists might be able to recover them.	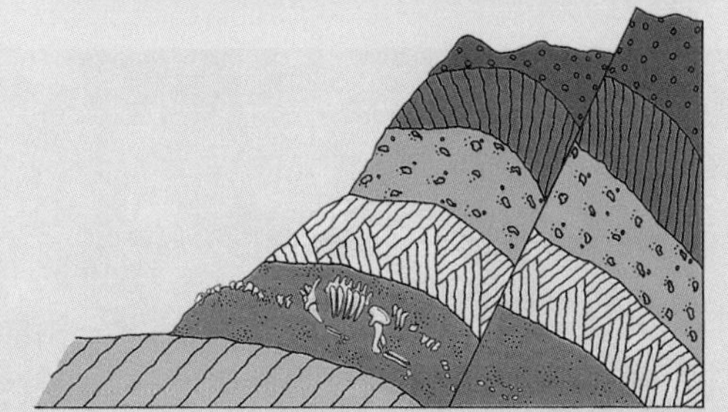

Whole bodies

Sometimes, fossils of whole organisms, including the soft parts, are preserved. Such fossils are rare and valuable. Insects that became trapped in the resin of ancient trees (the fossilised resin is called amber) have sometimes been wholly preserved (figure 7.48). Similarly, if the remains of animals or plants are frozen and buried in ice, they can be fully preserved.

Whole bodies of ancient woolly mammoths (including skin, hair and internal organs) have been found trapped in the ice of Siberia and Alaska (figure 7.49). Whole bodies and preserved skulls of animals can even reveal evidence of their last meal before death. Scientists collect DNA from these remains and compare the DNA sequences to those of modern elephants. Could we one day clone a woolly mammoth?

FIGURE 7.48 These insects were trapped in the resin of a tree millions of years ago.

FIGURE 7.49 Whole bodies of ancient woolly mammoths have been discovered in the ice of Siberia and Alaska.

Making an impression

The remains of animals or plants sometimes leave a dark impression, or imprint, in hardened sediments or newly formed rock (figure 7.50). The dark imprint is carbon, a reminder that the imprint came from a once-living thing.

It is also possible for remains trapped in rock to dissolve and be broken down; this leaves behind an empty space in the shape of the fossil. The depression is called a **mould**. If the mould is filled with minerals over time it forms a **cast** (figure 7.51).

FIGURE 7.50 The imprint of a leaf from an ancient fern left in stone

FIGURE 7.51 The coiled external shell of an ammonite has created a mould (left), which filled with minerals to create the cast (right).

Leaving just a trace

Some fossils only provide signs of the presence of animals, not the animal itself, and are called **trace fossils**. A trace fossil can be a footprint, trail or burrow. Footprints preserved in rock can provide clues about ancient animals, including dinosaurs, and how they lived. By studying the shape, size and depth of footprints, hypotheses can be made about the size and weight of extinct animals, as well as how they walked or ran.

mould a cavity in a rock that shows the shape of the hard parts of an organism

cast a fossil cavity that has been filled with minerals or other matter

trace fossils fossils that provide evidence, such as footprints, that an organism was present when the rock was formed

DISCUSSION

How complete is the fossil record?

Having just learned about the various ways a fossil is formed, discuss what parts of organisms are most likely to be preserved in the fossil record, and what external conditions are required for those parts to be preserved.

With your responses in mind, consider then, what percentage of life today would actually become part of the modern fossil record? Are there any plants or animals that you think are very unlikely to create a fossil? What does this tell you about how complete our fossil record is?

7.8.4 Earth's living history

Despite the fossil record being incomplete, every fossil found helps to piece together the history of life on Earth. The oldest undisputed fossils are from rocks dated around 3.5 billion years ago in Western Australia. The fossils are photosynthetic single-celled organisms that formed features called *stromatolites*. They would have helped to introduce free oxygen into the atmosphere!

Fossils from the Ediacaran Period (636 to 541 million years ago) were the first multicellular forms, with soft bodies like jellyfish or worms. Because the Ediacaran fauna did not have hard parts, they were not well preserved. However, after the Ediacaran Period, the rock record explodes with an abundance of fossils because organisms evolved with hard body parts like claws, scales, shells and bones. Early life was dominated by marine organisms, but both animals and plants slowly grew in abundance and complexity on land.

mass extinction a widespread and rapid decrease in the biodiversity and abundance of life

FIGURE 7.52 A summary of Earth's living history for the last 600 million years, marking the first fossil records and major mass extinction events (stars)

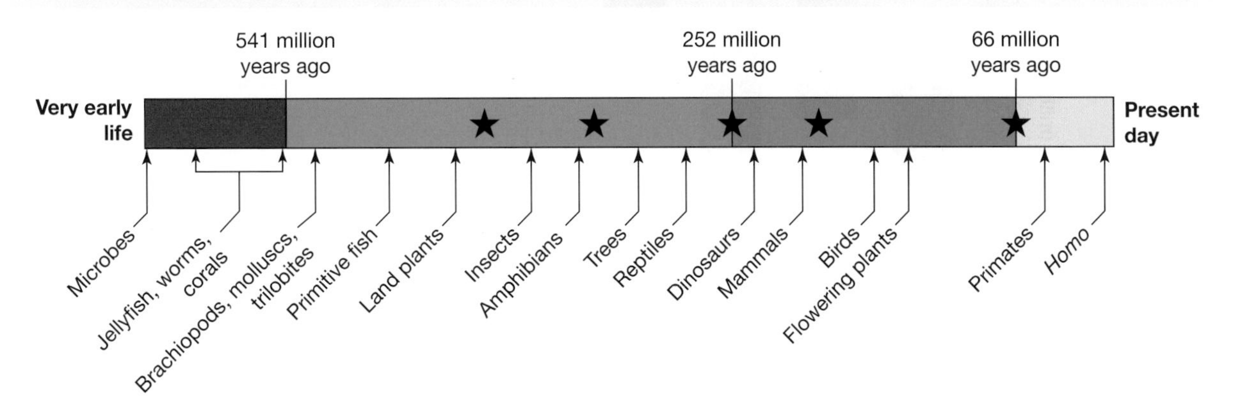

The last 541 million years has seen five major **mass extinctions**, where large volumes of life disappear from the rock record. The most significant of these was around 252 million years ago, which saw the extinction of over 80 per cent of all species, including the trilobites (figure 7.53). Palaeontologists have placed most of the divisions of geologic time at points in the fossil record where there are major changes in the type of organisms observed in the rocks.

FIGURE 7.53 Trilobites were some of the first organisms with hard parts preserved in the fossil record, but are not found in any rocks younger than 252 million years.

SCIENCE AS A HUMAN ENDEAVOUR: Fossils help to date rocks

By studying the fossil record, geologists over the years have built up a record of fossil distribution through time and across Earth. This is used to date and relate rocks over vast distances. An interval of rocks may contain a useful fossil of a known age, called an 'index species', that is relatively abundant, easy to recognise, only lived for a limited period of geologic time and was geographically widespread. Fossils that are commonly used for dating include brachiopods, conodonts, dinoflagellate cysts, foraminifera, graptolites, spores and pollen, and trilobites.

Conodonts are extinct microfossils found in marine limestones of eastern Australia and China, indicating that these regions were located together about 450 million years ago!

If the rock we are studying has several types of fossils in it, and we can assign time ranges to those fossils, we might be able to narrow the age range of the rock. First, we evaluate the age range for each fossil, then determine the overlap in age between fossils. The overlap is the age, or time period, of the rock sample.

FIGURE 7.54 When a rock is found containing multiple fossils, the overlap of the fossil age ranges can narrow down the age of the rock.

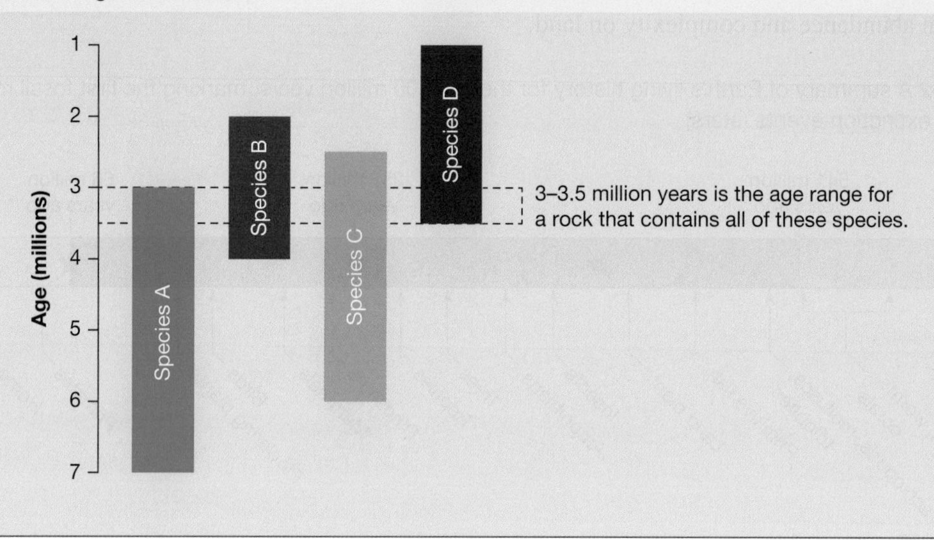

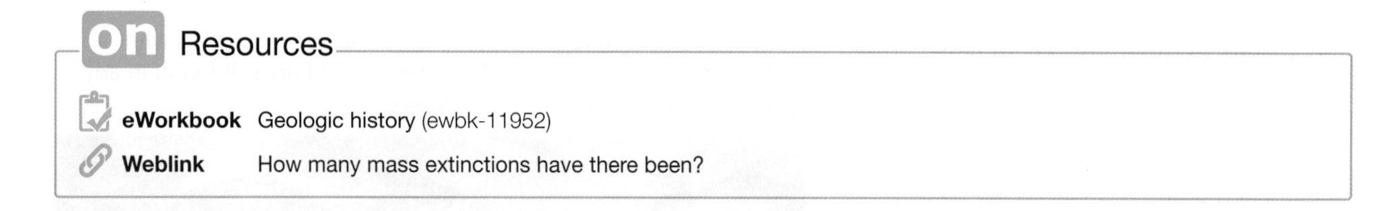

7.8 Activities

learnon

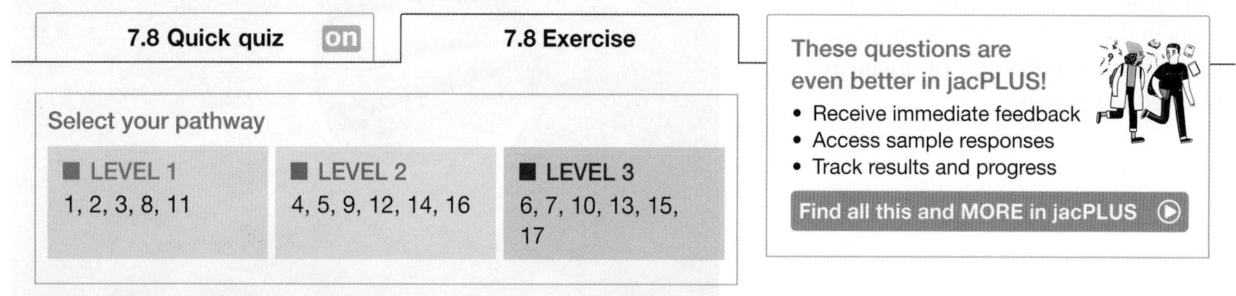

Remember and understand

1. Explain why some layers of sedimentary rock are tilted, even though the sediments that formed them were originally laid in horizontal beds.
2. What does a palaeontologist study?
3. What rock group is most likely to have fossils?
4. What clues about life in the past do fossils provide?
5. Under what circumstances can whole ancient living things be preserved as fossils?
6. Describe trace fossils and how are they useful.
7. What is the difference between a cast and a mould?
8. List the information about dinosaurs that can be obtained from fossils.
9. Fossils of dinosaurs form when their remains are buried deep under many layers of sediment that eventually turns into rock. Explain why fossils are often discovered on Earth's surface.
10. Describe four ways an animal or plant can be fossilised.

Apply and analyse

Use the following figure to answer questions **11–13**.

Layers of rock exposed by a road cutting

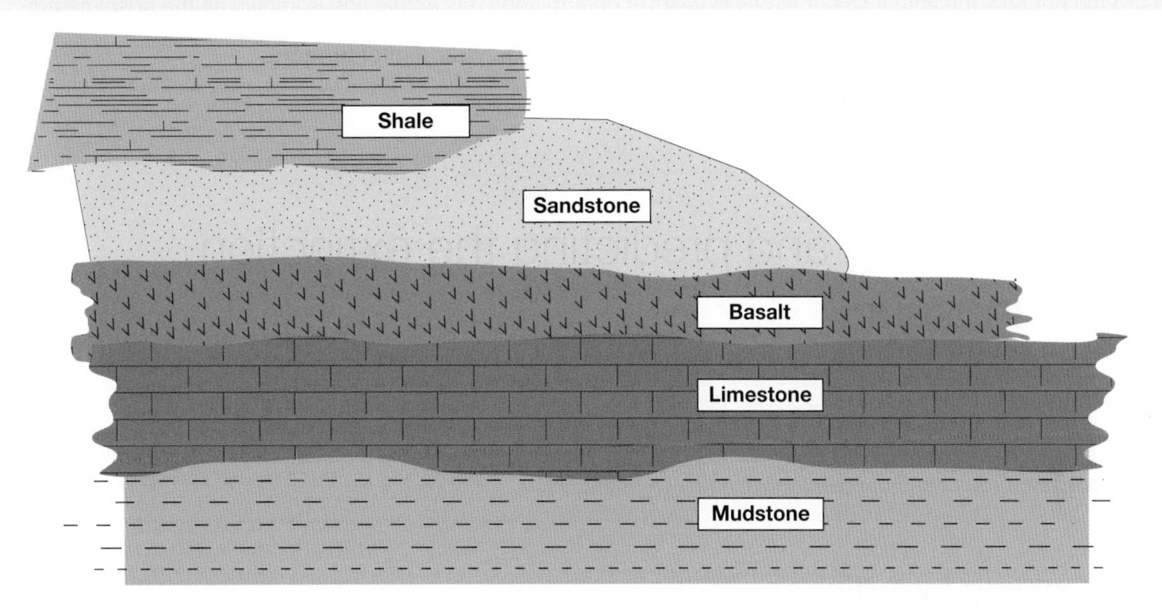

11. a. A road cutting reveals the layers of rock shown in the figure. Which of the rocks in the cutting is:
   i. the oldest rock
   ii. the youngest rock
   iii. evidence of volcanic activity?
   b. Why are some layers in the diagram thicker than others?

12. In which rocks in the figure would you most likely find the fossil of:
   a. a seashell
   b. the leaf of a fern usually found in swamps?

13. If a fault cuts and displaces the mudstone, limestone and basalt layers in the figure:
   a. what is the relative age of the basalt and the fault
   b. what is the relative age of the sandstone and the fault?

14. **SIS** Explain why the hard parts of plants and animals are more likely to be preserved than the softer parts.

15. **SIS** Using the fossil age ranges (arrows) in the following figure, determine the time period of the three rocks shown.

Fossil age ranges 541 million years ago to present

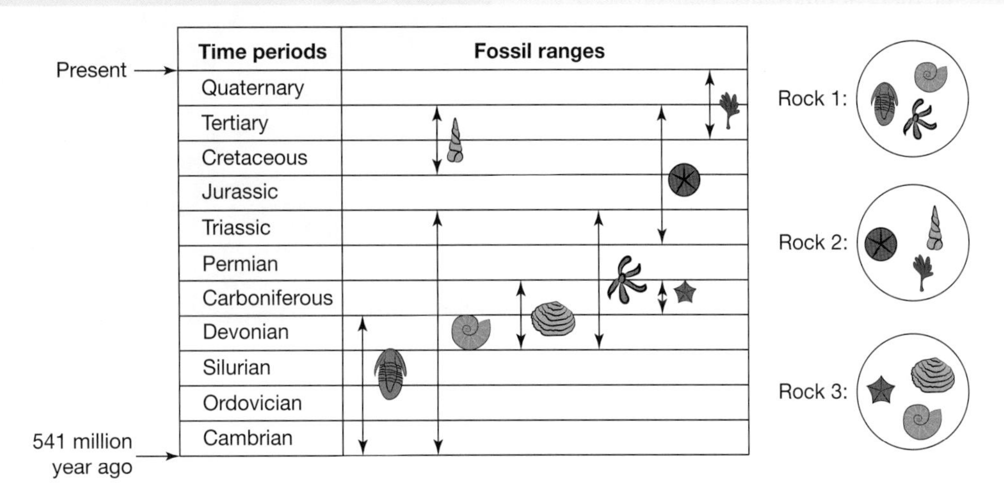

16. Even an animal's droppings can become fossilised. Use the internet or books to research and report on the following.
    a. Which animal was responsible for a huge fossilised dropping found in Canada in 1998?
    b. How long was the dropping?
    c. What can palaeontologists find out from it?
17. Find out how the actual age of a rock in years is determined. This actual age is known as the absolute age.

Fully worked solutions and sample responses are available in your digital formats.

LESSON
7.9 Questioning and predicting the extinction of dinosaurs

LEARNING INTENTION

At the end of this lesson you will understand what fossils teach us about the dinosaurs, and the scientific questions and predictions concerning why they went extinct.

7.9.1 Delving into dinosaurs

It is about 66 million years since the last non-flying dinosaurs existed on Earth.

- What did they look like?
- What colour were they?
- How fast could they move?
- How did they behave?
- What did they eat?

Palaeontologists use fossils to try to answer all of these questions and more!

Not just a pile of bones

Dinosaur fossils are not all bones. They may include the following:

- *Fossilised teeth.* The shape of the teeth and the way they are arranged provide vital clues about the diets of dinosaurs. Flat-surfaced grinding teeth would have belonged to a dinosaur with a plant diet. Sharp-pointed teeth suited to tearing flesh would have belonged to a meat-eating dinosaur.
- *Footprints.* Dinosaur footprints have been preserved in rock. Footprints from a single dinosaur provide clues about its size and weight. They also indicate whether the dinosaur walked on two legs or four, and how its weight was spread. The distance between footprints enables palaeontologists to estimate how fast the dinosaur moved. Footprints also provide clues about the behaviour of dinosaurs and whether they lived in herds or alone.
- *Impressions.* Skin impressions may be left in mud that has hardened.

FIGURE 7.55 A human hand compared to a three-toed dinosaur footprint

on Resources

Video eLesson A palaeontologist (eles-2054)

Close-up of an Edmontosaurus femur being excavated by a paleontologist.
Notice the flakes and layers of sedimentary rock being removed.

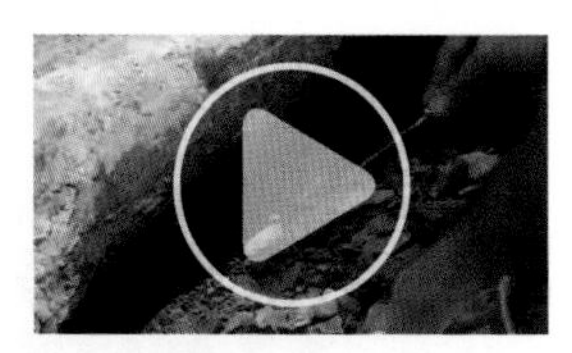

7.9.2 What happened to the dinosaurs?

Between about 66 and 252 million years ago, dinosaurs were the most successful animals on Earth. In fact, those years are known as 'the age of the dinosaurs'. Dinosaurs thrived and dominated the land while mammals lived in their shadow. Fossil evidence indicates that the last of the dinosaurs died about 66 million years ago.

Two of the inquiry skills that geologists and other scientists use are questioning and predicting. The question of how the dinosaurs died out has intrigued scientists for many years. In answering this question, scientists use scientific knowledge to make 'predictions' about what happened many millions of years ago.

FIGURE 7.56 Dinosaurs dominated the earth 66 million years ago.

7.9.3 Solving the dinosaur riddle

There are several lines of thought about the extinction of the dinosaurs. Scientists and others argue about whether the end of the dinosaurs was sudden or gradual. Scientists do generally agree that the riddle of the dinosaur extinction remains unsolved. Palaeontologists and other scientists continue to look for clues that might provide the final solution.

The asteroid theory

The most widely accepted solution to the dinosaur riddle is that an asteroid collided with Earth around 66 million years ago.

The asteroid's impact threw billions of tonnes of dust into the air, blocking out sunlight and plunging Earth into darkness for two or three years.

- Plants stopped growing but their seeds remained intact.
- The temperature dropped.
- The large plant-eating dinosaurs would have died quickly of starvation.
- The meat-eating dinosaurs would probably have died next, having lost their main food supply but surviving for a while by eating smaller animals.
- Many smaller animals would have survived by eating seeds, nuts and rotting plants.

FIGURE 7.57 Dust from a colliding asteroid could have blocked out sunlight for two or three years.

As the debris began to settle and sunlight filtered through the thinning dust clouds, many of the plants began to grow again. The surviving animals continued to live as they did before the impact. The surviving mammals were no longer competing with dinosaurs for food. It was the beginning of the age of mammals.

The volcano theory

The June 1991 eruption of Mount Pinatubo in the Philippines showed that ash and gases from volcanoes could reduce average temperatures all over the world. The average global temperatures during 1992 and 1993 were almost 0.2 °C less than expected. While this is not a large drop in temperature, the size of the eruption of Mount Pinatubo was much smaller than those of many ancient volcanoes.

The ash from a large volcano could have the same effect on sunlight and Earth's temperature as an asteroid impact. If there was an unusually large amount of volcanic activity about 66 million years ago, the extinction of the dinosaurs could be explained. The largest known volcanic eruption occurred about 252 million years ago in what is now Siberia. It is believed that many types of marine animals became extinct at about the same time.

FIGURE 7.58 Huge volcanic eruptions could have caused a global cooling.

The cooling climate theory

The gradual cooling of Earth's climate due to changes in the Sun's activity or Earth's orbit around the Sun could have caused the extinction of the dinosaurs. Dinosaurs, with no fur or feathers, had less protection from cold weather than mammals and birds. The larger dinosaurs would have found it very difficult to shelter from the cold conditions. Many smaller animals could burrow below the ground, or shelter in the hollow trunks of trees or in caves. Many mammals and birds would have been able to migrate to warmer regions closer to the equator.

ectothermic describes an animal whose body temperature is determined by its environment

endothermic describes an animal that can internally generate heat to maintain its body temperature

FIGURE 7.59 Could global climate change have killed the dinosaurs?

EXTENSION: Cold-blooded or warm-blooded?

Until recently, it was believed that dinosaurs were **ectothermic**. Ectothermic animals have body temperatures that depend on the temperature of their surroundings. As the surrounding temperature decreases, their body temperature decreases, and they become less active.

Mammals are **endothermic**. Endothermic animals are able to maintain a constant body temperature that is usually above that of their surroundings. They are able to remain warm and active in lower surrounding temperatures.

If dinosaurs were in fact ectothermic, a cooler climate would have made it more difficult for them to compete with other animals for food. However, many scientists now believe that dinosaurs may have been endothermic. The question of whether dinosaurs were cold-blooded or warm-blooded needs to be answered before the riddle of the dinosaurs can be solved.

The emerging plants theory

During the Cretaceous period (145 million to 66 million years ago), new types of plants began to appear. Flowering plants evolved, competing with the more primitive plants such as ferns for nutrients, water and sunlight. The plant-eating dinosaurs did not eat flowering plants. According to this theory, as their traditional food supply became more scarce, the plant-eating dinosaurs could not survive, and the meat-eating dinosaurs that preyed on them starved as well.

FIGURE 7.60 The evolution of flowering plants may have wiped out much of the plant-eating dinosaurs' main food source.

The declining oxygen theory

During the 'age of the dinosaurs' the amount of oxygen in the atmosphere was higher than it is today — 10 to 14 per cent more. This meant that despite the dinosaurs having large bodies and muscles, they could get away with having small lungs.

Around 95 million years ago, the oxygen levels appear to drop rapidly to the 21 per cent we have today. It is possible that the decline in available oxygen was quicker than evolution could take place, and the dinosaur's biology did not cope with the lower levels of oxygen.

CASE STUDY: Australian megafauna

The name *megafauna* means 'big animals' and is a general term used to describe a group of large land animals that appeared millions of years after the dinosaurs became extinct. The megafauna was at its largest and most widespread during the last 2.5 million years. In Australia, the megafauna was unique, including giant marsupials such as *Diprotodon*. *Diprotodon* is often referred to as a giant wombat with a nose like a koala. It was about the size of a two-tonne white rhinoceros.

Most of the Australian megafauna became extinct around 40 000 years ago. Scientists have been debating the causes of the extinction for decades. Some claim the animals could not have survived changes in climate, which would have changed landscapes from wooded eucalypts to arid and sparsely vegetated. Others have suggested the animals were hunted to extinction by Australia's earliest immigrants who had colonised most of the continent by 50 000 years ago. Perhaps it is a combination of the two.

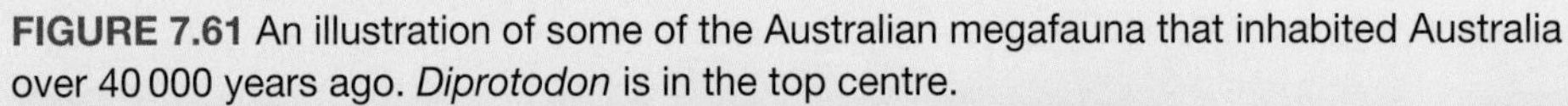

FIGURE 7.61 An illustration of some of the Australian megafauna that inhabited Australia over 40 000 years ago. *Diprotodon* is in the top centre.

7.9 Activities

learnon

7.9 Quick quiz on	7.9 Exercise

Select your pathway

■ LEVEL 1	■ LEVEL 2	■ LEVEL 3
1, 2, 3, 6, 7, 11	4, 8, 9, 12, 14	5, 10, 13, 15

Remember and understand

1. When did a mass extinction take place that included the loss of most dinosaurs?
2. How is the shape of a fossil tooth helpful for understanding how a dinosaur lived?
3. What is the most widely accepted theory to explain the extinction of the dinosaurs?
4. Why would smaller animals be more likely to survive the effects of an asteroid impact or large volcanic eruption than larger animals?
5. What is the difference between an ectothermic animal and an endothermic animal?

Apply and analyse

6. In what ways were the dinosaurs different from mammals?
7. How could volcanic eruptions affect life throughout the whole world?
8. How could meat-eating dinosaurs be endangered by the evolution of new types of plants?
9. **SIS** The boundary that marks the extinction of the dinosaurs is known as the K-T boundary. When scientists sampled this boundary across the world, they found that it had a much higher concentration of iridium than normal. Iridium is rare on Earth's surface.
   a. Of all the theories presented, which theory does this finding best support?
   b. Are there any patterns to the iridium concentrations that would help convince you even further?
10. **SIS** Explain how it is possible to use preserved animal footprints to form hypotheses about:
    a. whether the animal lived alone or in herds
    b. the way that the animal walked
    c. the weight of the animal
    d. the walking or running speed of the animal.

Evaluate and create

11. Which group of animals benefited the most as a result of the extinction of the dinosaurs?
12. List as many weaknesses as you can in each of the five theories about the dinosaur extinction presented.
13. Which theory of the extinction of the dinosaurs do you think is most likely to be correct? Explain your answer.
14. Imagine what it would have been like 66 million years ago if an asteroid plunged into Earth. Write a story about the first 24 hours after the impact.
15. Which animals and plants do you think would be most likely to survive if an asteroid struck central Australia now? Explain your answer.

Fully worked solutions and sample responses are available in your digital formats.

LESSON
7.10 Thinking tools — Fishbone diagrams and tree maps

7.10.1 Tell me

A fishbone diagram is a useful thinking tool when you are trying to determine or understand what could have caused something to happen. They show visually the reasons for what is happening or has happened. They are sometimes called cause-and-effect analysis or Ishikawa diagrams.

FIGURE 7.62 A fishbone diagram

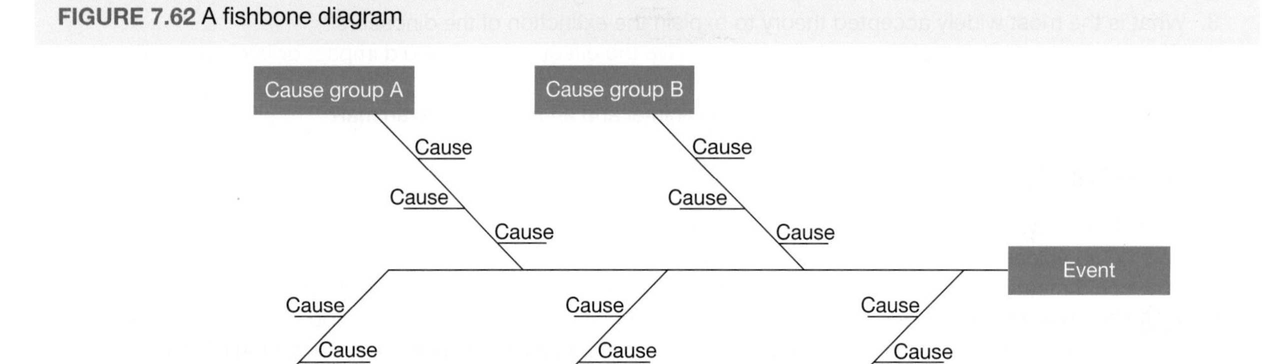

Similar to fishbone diagrams, a tree map divides ideas into groups or categories. However, fishbone diagrams always categorise causes, but tree maps do not.

FIGURE 7.63 A tree map

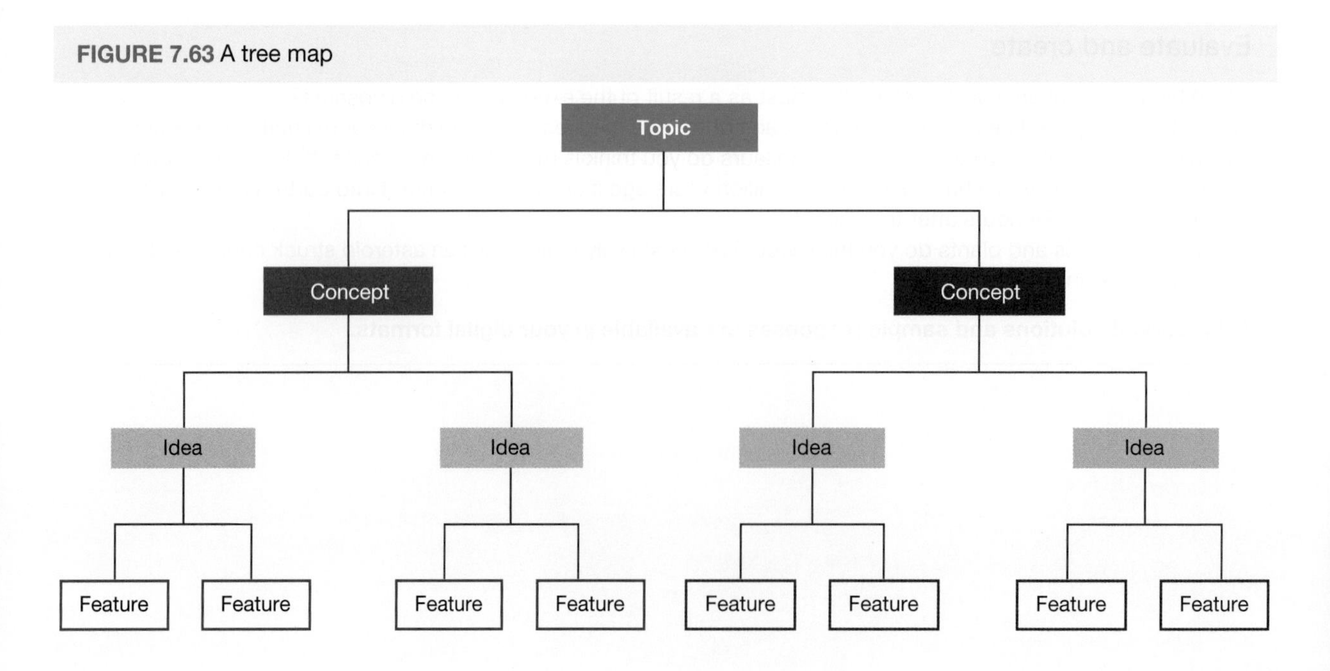

7.10.2 Show me

1. Think of an event of which you do not know the causes.
2. Brainstorm as many possible causes as you can for this event.
3. In pairs, or teams of four, organise your list of causes into groups.
4. Write the event that you are analysing as the 'fish's head' of a fishbone diagram. Your groups of causes then become the main 'bones' of the diagram, one bone for each group.
5. Write the title for each of your groups of causes on its relevant 'fishbone'.
6. Write the causes on the smaller 'fishbones' that are joined to the sides of the main bones. (You can attach causes to more than one bone or group of causes.)

7.10.3 Let me do it

7.10 Activity

These questions are even better in jacPLUS!

- Receive immediate feedback
- Access sample responses
- Track results and progress

Find all this and MORE in jacPLUS

1. Create a fishbone diagram that shows the possible causes of the extinction of the dinosaurs. Use the one provided as a template, adding the smaller 'bones' and causes yourself. Four of the theories discussed in this lesson are already on the template. Include a separate theory of your own on the diagram. It could be a combination of the four other theories, or something completely different.

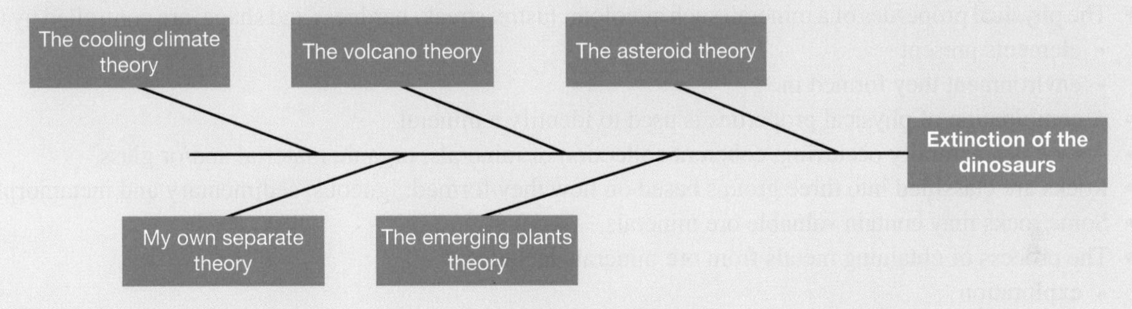

2. Working with a partner, complete the tree map given to represent the three main classes of rock found in the crust. Add further branches below the existing ones if you can.

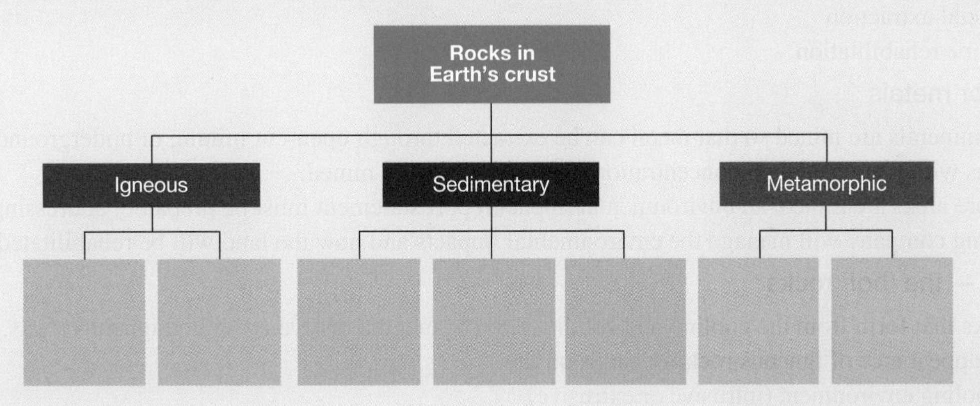

3. Create a fishbone diagram to represent the formation of rocks. Use the cause categories listed, and add any others that you think should be included.
   Cause categories:
   - Weathering
   - Erosion
   - Volcanoes
   - Pressure
   - Heat.

Fully worked solutions and sample responses are available in your digital formats.

LESSON
7.11 Review

Hey students! Now that it's time to revise this topic, go online to:

Review your results

Watch teacher-led videos

Practise questions with immediate feedback

Find all this and MORE in jacPLUS

Access your topic review eWorkbooks

on Resources

Topic review Level 1	Topic review Level 2	Topic review Level 3
ewbk-11954	ewbk-11956	ewbk-11958

7.11.1 Summary

Rocks and minerals

- A mineral is any naturally occurring solid substance with a definite chemical composition and crystal structure.
- The physical properties of a mineral, such as colour, lustre, streak, hardness and shape, are controlled by the:
  - elements present
  - environment they formed in.
- A combination of physical properties is used to identify a mineral.
- A rock is a naturally occurring, coherent collection of minerals, organic material and/or glass.
- Rocks are classified into three groups based on how they formed: igneous, sedimentary and metamorphic.
- Some rocks may contain valuable ore minerals.
- The process of obtaining metals from ore minerals includes:
  - exploration
  - environmental impact studies
  - mining
  - metal extraction
  - mine rehabilitation.

Mining for metals

- Ore minerals are mined so that metal can be extracted through open-cut mining or underground mining.
- Areas with a large enough concentration of ore minerals are mined.
- Before areas are mined, an environmental impact report statement must be prepared, addressing how the mining company will manage the environmental impacts and how the land will be rehabilitated.

Igneous — the 'hot' rocks

- Rocks that form from the cooling and solidification of magma or lava are called igneous rocks.
- The appearance of igneous rocks depends on the:
  - cooling environment (intrusive or extrusive)
  - types of minerals or glass substances present.
- Igneous rocks that formed in an extrusive environment will have small mineral crystals or volcanic glass because they solidified quickly (days to weeks) on or near the surface.
  - Basalt is dark coloured, with minerals rich in iron and magnesium.
  - Rhyolite is light coloured, with more silica-rich minerals.
  - Scoria is like basalt but with lots of holes from trapped gas.
  - Pumice also has a lot of holes, but is made of glass and can float on water.
  - Obsidian is a dark and dense volcanic glass.

- Igneous rocks that formed in an intrusive environment will have large mineral crystals because they solidified slowly (thousands of years) underground.
  - Granite is light coloured, like rhyolite, but with visible crystals.
  - Gabbro is dark coloured, like basalt, but with visible crystals.
- Igneous rocks can be useful because they may contain ore minerals or they can be used as building material.
- Granite bodies underground can be a source of renewable geothermal energy.

Sedimentary — the 'deposited' rocks

- Sedimentary rocks form by weathering, erosion, deposition and lithification of sediments on the surface.
- Lithification can occur by:
  - long periods of compaction
  - quick cementing.
- Clastic sedimentary rocks are made of weathered fragments of pre-exiting rocks and are named based on clast size.
  - From largest to smallest: conglomerate, sandstone, siltstone, mudstone/shale
  - Mudstone and shale are soft because they are mostly clay minerals.
- Biological sedimentary rocks are made of fragments of living things.
  - Limestone is the most common and is made from the calcite skeletal remains of warm shallow sea animals.
  - Coal is made from the sediment remains of swamp plants that have been buried and heated for millions of years.
- Rock salt and gypsum are examples of chemical sedimentary rocks, which form when water evaporates, and minerals grow from the solution.
- Sedimentary rocks come in layers, as records of changing surface conditions. The lower layers are older than the top layers.
- There are many modern uses for sedimentary rocks, such as gravels and lime in concrete, and coal as a non-renewable fossil fuel.

Metamorphic — the 'changed' rocks

- Metamorphic rocks form when pre-existing rocks are put under increased temperature and pressure.
  - The rock changes to become more stable to the new conditions.
- The degree and type of metamorphism is determined by the:
  - original rock (parent rock)
  - amount of pressure or temperature change
  - length of time.
- Some of the common metamorphic rocks are slate, quartzite, marble and gneiss.
  - Apply low pressure to shale to form slate that splits into thin sheets.
  - Apply high pressure to granite to form gneiss with light and dark bands.
  - Apply higher temperatures to limestone to form marble.
  - Apply higher pressure or temperature to sandstone to form quartzite.
- Each type of metamorphic rock is a clue to the original rock as well as the amount of temperature or pressure change that caused the metamorphism.
- Metamorphic rocks like slate, quartzite and gneiss are hard and can be used as building material or percussion tools.
- Marble is softer and easier to carve into sculptures.
- Geologists refer to the constant and slow change of rocks from one type to another over time as the 'rock cycle'.

Rock technology

- First Nations Australians created specialised tools using different rocks based on their properties, including flint arrowheads, axes and knives.
- First Nations Australians removed stones from quarries to make tools, as well as trade.
- Quarries were also a meeting place for First Nations Australians to hold ceremonies, initiations and other important cultural events.

Geologic history

- Rock types and their relationships to one another are used to interpret past environments and events.
- Principles of relative age help to interpret these rocks or events as either older or younger than another.
- Fossils in sedimentary rocks are recorded and compared across the world to help with determining rock age.
- Fossils of plant or animal bodies can form by:
  - permineralisation
  - whole-body preservation
  - imprints.
- Hard parts of plants and animals are likely to be better preserved than soft parts.
- The fossil record is incomplete, making the finding and addition of any new fossil records important to building the history of life on Earth.
  - Complexity of life on Earth exploded 541 million years ago and continues today.
  - There have been five major mass extinction events in the past 600 million years.

Questioning and predicting the extinction of dinosaurs

- Dinosaurs dominated Earth from 66 to 252 million years ago.
- Palaeontologists use dinosaur bones as well as fossilised teeth, footprints and skin impressions to question and predict dinosaur:
  - size and weight
  - walking style
  - diet
  - social behaviour.
- Fossil evidence indicates that the last of the dinosaurs died about 66 million years ago.
- Scientific questioning and predicting skills have helped to develop several theories as to why the dinosaurs went extinct:
  - Asteroid impact
  - Large volcanic eruptions
  - Global climate cooling
  - Evolution of flowering plants
  - Decreasing oxygen levels.
- The most widely accepted theory is that an asteroid collided with Earth.

7.11.2 Key terms

abrasive a property of a material or substance that easily scratches another
alloy a mixture of a metal with a non-metal or another metal
asthenosphere the zone of mantle beneath the lithosphere, thought to be much hotter and more fluid than the lithosphere mantle
basalt a dark, igneous rock with small crystals formed by fast cooling of hot lava; it sometimes has holes that once contained volcanic gases
batholith an intrusive rock mass that covers an area of over 100 square kilometres
carnivore an animal that eats other animals
cast a fossil cavity that has been filled with minerals or other matter
coal a sedimentary rock formed from dead plants and animals that were buried before rotting completely, followed by compaction and some heating
compound a substance made up of two or more different types of atoms that are chemically bonded (covalent or ionic) together
conglomerate a sedimentary rock containing large fragments of various sizes cemented together
crystal a geometrically shaped substance made up of atoms and molecules arranged in one of seven different shapes; the elements and the conditions present during the crystal's growth determine the arrangement of atoms and molecules and the shape of the crystals
delta a landform created by the deposition of sediment at the end of a river as it enters a body of water
deposition the settling of transported sediments

ectothermic describes an animal whose body temperature is determined by its environment
endothermic describes an animal that can internally generate heat to maintain its body temperature
environmental impact statement (EIS) a report on the possible effects of a planned project on the environment
erosion the wearing away and removal of soil and rock by natural elements, such as wind, waves, rivers and ice, and by human activity
extrusive describes igneous rock that forms when lava cools on Earth's surface
flint a fine-grained sedimentary rock that leaves a very sharp edge when broken
floodplain flat, open land beside a river where sediments are deposited during floods
foliated consisting of an arrangement of certain mineral grains into distinct bands, which gives the rock a striped appearance
fossil any remains, impression or trace of a life form of a former geological age; evidence of life in the past
gabbro a dark-coloured, intrusive igneous rock with a similar mineral composition to basalt, but with larger crystals
geothermal energy refers to using heat from Earth as an energy source
glaciers large bodies of ice that move down slopes and push boulders, rocks and gravel
gneiss a coarse-grained metamorphic rock with light and dark bands formed mainly as a result of great pressure on granite
granite a light-coloured, intrusive igneous rock with mineral crystals large enough to see
hardness a measure of how difficult it is to scratch the surface of a solid material; hardness can be ranked using Mohs' scale
igneous rocks rocks formed when hot, molten rock cools and hardens (solidifies)
intrusive describes igneous rock that forms when magma cools below Earth's surface
lava an extremely hot liquid or semi-liquid rock from the mantle that reaches and flows or erupts on Earth's surface
limestone a sedimentary rock formed from the remains of sea organisms; it consists mainly of calcium carbonate (calcite)
lithify to transform sediment into rock
lithosphere the outermost layer of Earth; includes the crust and uppermost part of the mantle
lustre the high shine and sheen of a substance caused by the way it reflects light
magma a very hot mixture of molten rock and gases, just below Earth's surface, that forms from melting of the mantle and occasionally the crust
marble a metamorphic rock formed as a result of great heat or pressure on limestone
mass extinction a widespread and rapid decrease in the biodiversity and abundance of life
metamorphic rocks rocks formed from the change (alteration) of pre-existing rocks in response to increasing temperature and/or pressure conditions
metamorphism the process that changes rocks by extreme pressure or heat (or both)
mineral a naturally occurring, inorganic and solid substance with a defined chemical formula and an ordered arrangement of atoms
mining the process of removing natural resources from Earth
moraine a ridge made out of sediments deposited by a glacier
mould a cavity in a rock that shows the shape of the hard parts of an organism
mudstone a fine-grained sedimentary rock made of mud (clay and silt)
native elements elements found uncombined in Earth's crust
obsidian a black, glassy rock that breaks into pieces with smooth shell-like surfaces
open-cut mining mining that removes soil and rocks on the surface of the land
ore mineral a mineral from which a valuable metal can be removed for profit
overburden waste rock removed from below the topsoil; this rock is replaced when the area is restored
palaeontologist a scientist who studies fossils
percussion flaking a process in which tool stones, such as flint or obsidian, were struck with harder stones, such as quartzite, to shear large flakes off until they were a desired shape
permineralisation the most common method of fossilisation, in which minerals fill the cellular spaces and crystallise; the shape of the original plant or animal is preserved in great detail
pumice a glassy, pale igneous rock that forms when frothy rhyolite lava cools in the air; it often floats on water as it is very light and full of holes that once contained gas
quartzite an extremely compact and hard metamorphic rock consisting essentially of quartz
rehabilitated restored to its previous condition or an acceptable, agreed alternative
relative age the age of a rock compared with the age of another rock

rhyolite a light-coloured, extrusive igneous rock with a similar mineral composition to granite, but with smaller crystals
rock cycle a cycle of processes that rocks experience in Earth's crust as they constantly change from one type to another
rock salt a sedimentary deposit formed when a salt lake or seabed dries up; the sediments are made of sodium chloride (halite)
sandstone a sedimentary rock with medium-sized grains; the sand grains are cemented together by silica, lime, mud or salts
scavenger an animal that eats dead plant and animal material
scoria a dark, igneous rock formed from frothy basalt lava that cools quickly and is full of holes that once contained gas
sediment material broken down by weathering and erosion that is moved by wind or water and collects in layers
sedimentary rocks rocks formed through the deposition and compaction of layered sediment
shale a fine-grained sedimentary rock formed from thinly layered mud
silicates a group of minerals consisting primarily of SiO_4^{2-} combined with metal ions, forming a major component of the rocks in Earth's crust
siltstone a sedimentary rock with a particle size between that of sandstone and mudstone
slate a fine-grained metamorphic rock formed as a result of moderate heat and pressure on shale
Stone Age a prehistoric time when weapons and tools were made of stone, bone or wood
streak the colour of a mineral as a fine powder, found by rubbing it onto an unglazed white ceramic tile
trace fossils fossils that provide evidence, such as footprints, that an organism was present when the rock was formed
underground mining mining that uses shafts and tunnels to remove rock from deep below the surface
viscosity a measure of a fluid's resistance to flow
weathering the physical or chemical breakdown of rocks on the surface

on Resources

eWorkbooks	Study checklist (ewbk-11960) Literacy builder (ewbk-11961) Crossword (ewbk-11963) Word search (ewbk-11965) Reflection (ewbk-11967)
Solutions	Topic 7 Solutions (sol-1119)
Practical investigation eLogbook	Topic 7 Practical investigation eLogbook (elog-2173)
Digital document	Key terms glossary (doc-39988)

7.11 Review questions

learnon

7.11 Review questions

Select your pathway

■ LEVEL 1	■ LEVEL 2	■ LEVEL 3
1, 2, 3, 4, 5, 6, 7, 8, 9, 11, 19, 26, 27	10, 12, 13, 14, 15, 20, 21, 23, 28	16, 17, 18, 22, 24, 25, 29, 30, 31, 32, 33

Remember and understand

1. What are the three requirements needed to call something a rock?
2. The rocks formed in Earth's lithosphere are grouped into which three groups based on how the rock was formed?
3. How are all igneous rocks formed?
4. Explain the difference between the ways in which extrusive igneous rocks and intrusive rocks are formed.
5. List three examples of extrusive igneous rocks.
6. What is the difference between basalt and rhyolite?
7. What clues does the size of the mineral crystals in an igneous rock provide about how the rock was formed?
8. What are sediments?
9. Explain why some layers of sedimentary rocks are tilted or bent.
10. Describe three ways that sedimentary rocks can form.
11. What is a 'parent' rock?
12. Describe two ways in which igneous and sedimentary rocks can be transformed into metamorphic rocks.
13. Complete the table to summarise what you know about igneous, sedimentary and metamorphic rocks.

TABLE Summary of rock types

Class of rock	How it is formed	Special features	Example	Uses
Igneous				
Sedimentary				
Metamorphic				

14. Complete the diagram of parent rocks and their rock type, on the left, and the common metamorphic rocks they form, on the right.

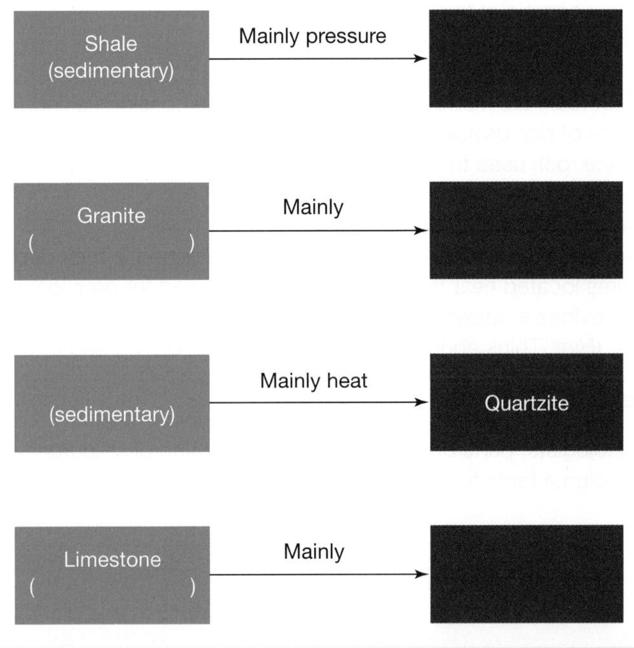

15. Explain why the mineral crystals in granite are larger than those in basalt.
16. How does applying pressure change the appearance of a rock?
17. The changes that lead to the formation of the three main groups of rocks can be drawn as a cycle, as shown in the figure.

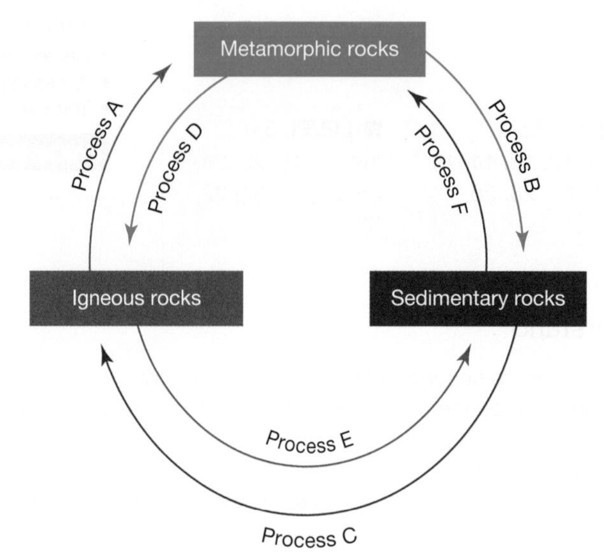

Which of processes A–F involve:

a. weathering and erosion **b.** heat and pressure **c.** melting and crystallisation?

18. What characteristic of minerals do the following terms describe?

a. Lustre **b.** Streak **c.** Hardness

19. What mineral property does Mohs' scale provide an approximate measure of?
20. Which metamorphic rock is used by First Nations Australians to make anvils? What properties make it useful?
21. Not all fossils are the actual remains of living things. Name and describe two types of fossils that are not preserved remains.
22. Of the several ways a fossil can form, which is the most common?

Apply and analyse

23. **SIS** Suggest a way of testing that a rock sample is limestone.
24. **SIS** While studying sedimentary rocks in a railway cutting, a geologist discovers a layer of graded bedding, which features larger-sized clasts at the bottom and the gradual decrease in size towards the top. How could the graded bedding have been made?
25. **SIS** If you were given a sample of two different minerals:
    a. how could you tell which one had the greater hardness
    b. which one would you prefer to make a road out of?
26. First Nations Australian tools are made by a process called percussion flaking.
    a. Describe the process of percussion flaking.
    b. Which property of the rock used to make the tool must be different from those of the stone from which the tool is formed?
27. What is the most common element in Earth's crust?
28. One factor that determines the way in which ore minerals are mined is their depth. Compare the mining processes used for ores located near the surface with those used for ores located deeper in Earth's crust.
29. The mining industry provides employment for many Australians. Make a list of occupations that are involved in the mining industry. (*Hint*: Think about what happens before, during and after mining is undertaken.)

Evaluate and create

30. According to many geologists, parts of Antarctica are rich in mineral resources, similar to those found in Australia. Use a two-column table to list reasons these mineral resources should be mined and why they should not be mined.

31. **SIS** Imagine that the set of fossilised dinosaur footprints as shown in the figure were found in a layer of sedimentary rock.
   a. Use the footprints to write a description of what might have happened millions of years ago.
   b. Compare your interpretation of the footprints with others.
   c. Does each person interpret the evidence in the same way?
   d. If there are differences of opinion about what happened, is there any way of knowing who is right?
   e. List as many differences as you can between the two types of dinosaurs making these footprints.

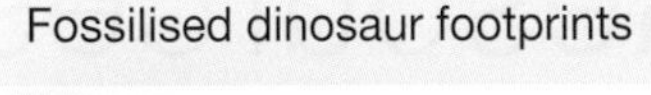

Fossilised dinosaur footprints

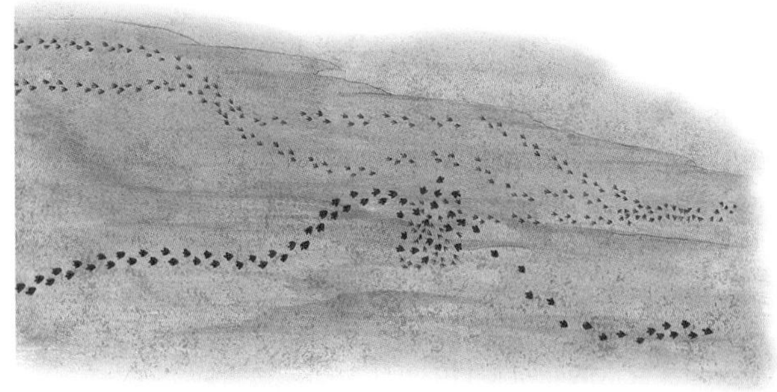

32. This figure is a sketch of some rock layers and their relationships as seen from a cliff face of a canyon.
   a. What is the oldest rock?
   b. What is the youngest rock?
   c. What sort of event does the youngest rock suggest?
   d. Which feature represents a period of weathering and erosion?
   e. Is the period of weathering and erosion younger or older than faulting?
   f. **SIS** Which layer in the diagram would most likely have fossils?
   g. Normally, old layers of rock are found below younger layers. Sometimes, however, younger layers are found beneath older layers. Can you identify a spot on the diagram where this has happened?

Cross-section through a rock sequence

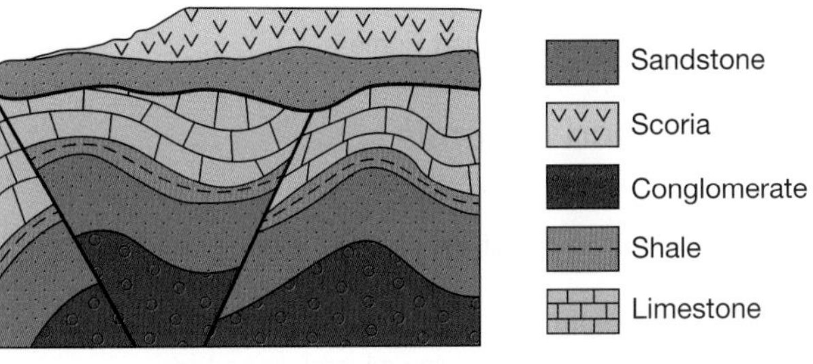

33. **SIS** The photograph provided shows dinosaur footprints that have been preserved in rock at Gantheaume Point near Broome.
   a. What type of fossil is it?
   b. Why is it classified as a fossil even though it could be described as a dent in a rock?
   c. Have all dinosaur footprints been preserved? Why have these been preserved for hundreds of millions of years?
   d. What can be learned about the features of the dinosaur that left these footprints?
   e. What forms of evidence, apart from preserved footprints, can be used to gather knowledge about dinosaurs?

Preserved footprints, Gantheaume Point

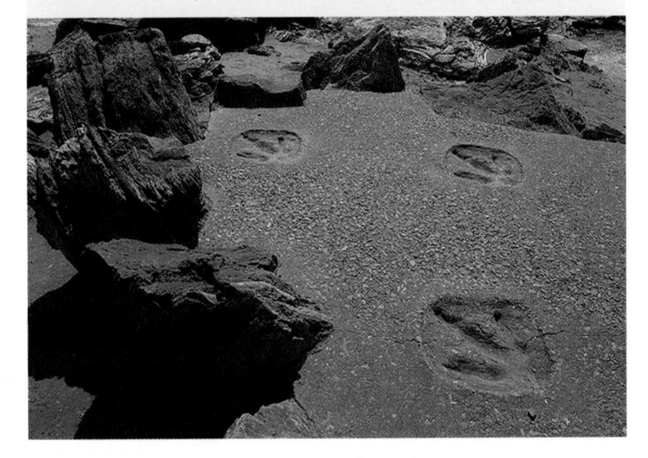

Fully worked solutions and sample responses are available in your digital formats.

Online Resources

Below is a full list of **rich resources** available online for this topic. These resources are designed to bring ideas to life, to promote deep and lasting learning and to support the different learning needs of each individual.

7.1 Overview

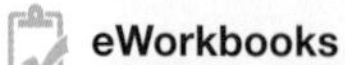

eWorkbooks
- Topic 7 eWorkbook (ewbk-11937)
- Starter activity (ewbk-11939)
- Student learning matrix (ewbk-11941)

Solutions
- Topic 7 Solutions (sol-1119)

Practical investigation eLogbook
- Topic 7 Practical investigation eLogbook (elog-2173)

Weblink
- Mining makes your Smart Home

7.2 Rocks and minerals

eWorkbook
- Identifying and classifying minerals (ewbk-11942)

Practical investigation eLogbook
- Investigation 7.1: Identifying mineral properties (elog-2175)

Interactivity
- Crystals (int-5338)

Weblinks
- Mexico giant crystal cave
- How do crystals work?

7.3 Mining for metals

Practical investigation eLogbook
- Investigation 7.2: Searching without disturbing (elog-2177)

Weblinks
- What's in a mobile phone?
- Mt Rawdon mine rehabilitation
- NSW Mining rehabilitation site and videos
- Mineral exploration interactive

7.4 Igneous — the 'hot' rocks

eWorkbook
- Igneous rocks (ewbk-11944)

Practical investigation eLogbook
- Investigation 7.3: Does fast cooling make a difference? (elog-2179)

Teacher-led video
- Investigation 7.3: Does fast cooling make a difference? (tlvd-10790)

Video eLesson
- Volcanoes (eles-0130)

Interactivity
- The weathering of a batholith (int-5337)

7.5 Sedimentary — the 'deposited' rocks

eWorkbook
- Sedimentary rocks (ewbk-11946)

Practical investigation eLogbooks
- Investigation 7.4: Sediments and water (elog-2181)
- Investigation 7.5: Identifying sedimentary rocks (elog-2183)

Interactivities
- Weathering (int-5339)
- Clastic sedimentary rocks (int-5340)
- Sedimentary rock layers (int-5341)

7.6 Metamorphic — the 'changed' rocks

eWorkbooks
- Metamorphic rocks (ewbk-11948)
- The rock cycle (ewbk-11950)

Practical investigation eLogbook
- Investigation 7.6: Rocks — the new generation (elog-2185)

Interactivities
- Metamorphic rocks (int-5343)
- Metamorphic rocks (int-0234)

7.8 Geologic history

eWorkbook
- Geologic history (ewbk-11952)

Interactivities
- Relative age of rocks (int-0233)
- Formation of a fossil (int-5342)

Weblink
- How many mass extinctions have there been?

7.9 Questioning and predicting the extinction of dinosaurs

Video eLesson

- A palaeontologist (eles-2054)

7.11 Review

eWorkbooks

- Topic review Level 1 (ewbk-11954)
- Topic review Level 2 (ewbk-11956)
- Topic review Level 3 (ewbk-11958)
- Study checklist (ewbk-11960)
- Literacy builder (ewbk-11961)
- Crossword (ewbk-11963)
- Word search (ewbk-11965)
- Reflection (ewbk-11967)

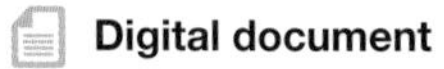

Digital document

- Key terms glossary (doc-39988)

To access these online resources, log on to **www.jacplus.com.au**

8 Dynamic Earth

CONTENT DESCRIPTION

Investigate tectonic activity including the formation of geological features at divergent, convergent and transform plate boundaries and describe the scientific evidence for the theory of plate tectonics (AC9S8U03)

LESSON SEQUENCE

SCIENCE INQUIRY AND INVESTIGATIONS

Science inquiry is a central component of the Science curriculum. Investigations, supported by a **Practical investigation eLogbook** and **teacher-led videos**, are included in this topic to provide opportunities to build Science inquiry skills through undertaking investigations and communicating findings.

LESSON
8.1 Overview

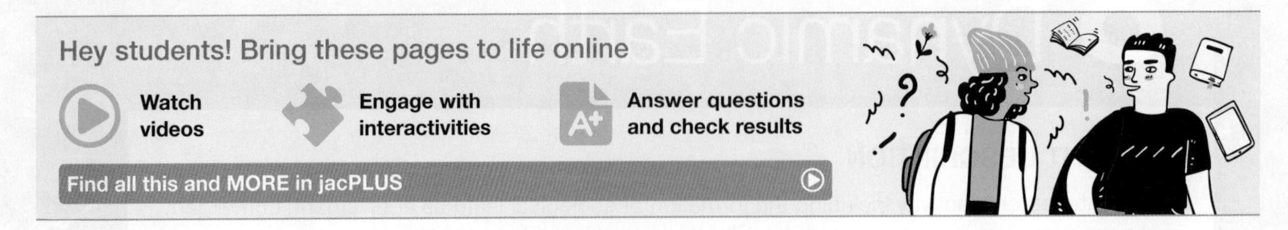

8.1.1 Introduction

The ground beneath you seems still. It might even seem dull. But first appearances can be deceiving. In fact, Earth's crust is not still — it is constantly moving and changing as seen by earthquakes. Nor is it dull — there are locations where red-hot molten rock is created, bursts through and creates a volcano like the one shown in figure 8.1 (Eyjafjallajökull, Iceland, which erupted in 2010).

FIGURE 8.1 Molten lava erupts from Eyjafjallajökull, Iceland, 2010

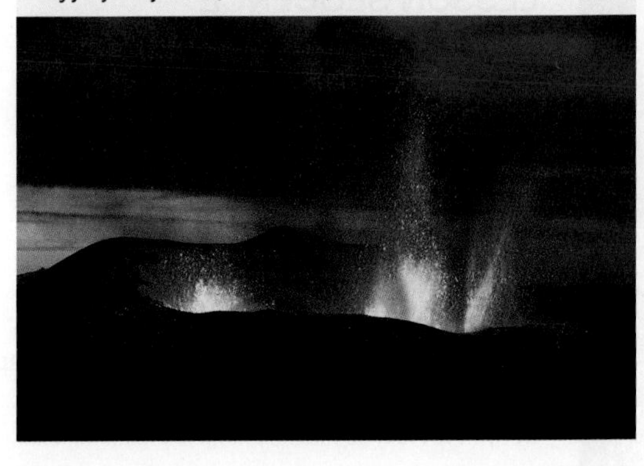

Volcanoes and earthquakes provide spectacular evidence that Earth is a dynamic, ever-changing planet. This activity can be both amazing and scary at the same time. However, there is a pattern to this activity, as the location of most volcanoes and earthquakes are controlled by the movements of Earth's tectonic plates. Tectonic plates are pieces of Earth's crust that shift and move around relative to each other; more than 80 per cent of all volcanoes and earthquakes are formed along the boundaries of the tectonic plates. Take Iceland as an example. Here, two tectonic plates are moving away from each other, forming large fissures in the crust and allowing hot material to rise up from the middle, melt and erupt to form inspiring volcanoes.

on Resources

Video eLesson Volcanic eruption in Iceland 2010, Eyjafjallajökull (eles-2661)

This short video of the eruption of Eyjafjallajökull was taken less than a kilometre from the crater. A lava flow, creating new rocks, can be seen in the bottom right of the screen.

8.1.2 Think about our dynamic Earth

1. How can something as large as a continent move?
2. Why do volcanoes make a 'ring of fire' around the Pacific Ocean?
3. What did the world map look like 250 million years ago?
4. Why do the Himalayas have many of the highest mountains on Earth?
5. What causes tsunamis?
6. How can a volcano suddenly appear from nowhere?
7. Where is the largest volcano in the solar system?

8.1.3 Science inquiry

Journey to the Centre of the Earth

'Descend into the crater of Yokul of Sneffels, which the shade of Scataris caresses before the Kalends of July, audacious traveller, and you will reach the centre of the Earth. I did it.'

So wrote Jules Verne in his science fiction novel *Journey to the Center of the Earth*, which was published in 1864. The novel describes a fascinating journey by the adventurous Professor Lidenbrock, his nephew Axel and their guide Hans to the centre of Earth. Their quest begins with a descent into the crater of the extinct volcano Snæfellsjökull in Iceland.

Although no one has ever been able to drill a hole to, much less visit, the centre of Earth, geologists have made some scientific discoveries about what is deep inside Earth. For starters, they found that Earth is layered. The surface layer is called the crust, followed by the mantle and then the core at the very centre. As you move towards the centre, each layer is made of denser material — the heavy material sank to the bottom, while the lighter material stayed on top.

FIGURE 8.2 The layers of Earth

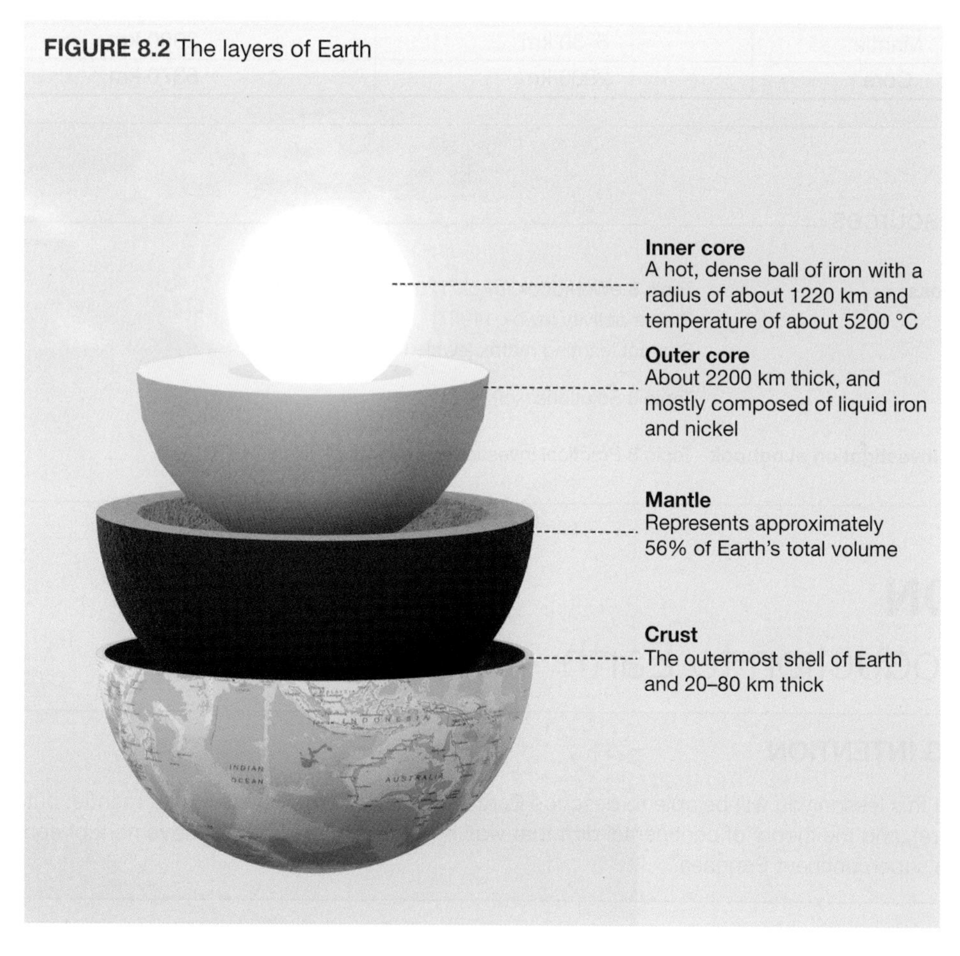

Think about the crust

When you look at an image of Earth's surface the largest features are continents and ocean basins.

1. Are there any patterns to the size or locations of the continents and oceans?
2. Geologists consider the bulk of continental crust to be made of different rocks types than the oceanic crust.
   a. How might the wide range of crustal thickness be related to this?
   b. Which do you think is thinner — oceanic or continental crust?
3. Consider the materials found close to the surface of Earth — close enough to be able to reach with drills and tunnels. Make a list of materials that are:
   a. used to provide energy for heating, transport and industry
   b. used for building and other construction
   c. exceptionally valuable.

Think about the mantle and core

As we travel towards the centre of Earth, the weight of the rock above gets heavier, thus pressure ever increases as we make the journey. Temperature also increases, because rocks can contain radioactive material that decays and releases heat, keeping the interior of Earth nice and toasty.

4. What do you imagine the state of the mantle to be like?
5. Jules Verne described Earth's interior to be full of interconnected caverns, a bit like swiss cheese. How valid is his description?
6. What would you expect to find at the very centre of Earth?
7. Earth's surface is constantly experiencing change. Make a list of events that cause change to the surface and evaluate which ones may be the result of a dynamic interior.
8. Scientists have estimated Earth to be approximately a sphere with an average radius of 6370 kilometres. Create a scaled-down model for the interior of Earth to illustrate the three layers.

TABLE 8.1 Depth to each layer of Earth

Earth layer	Average depth to top of layer	Average depth to bottom of layer
Crust	0 km	6–30 km
Mantle	6–30 km	2900 km
Core	2900 km	6370 km

on Resources

eWorkbooks	Topic 8 eWorkbook (ewbk-12016) Starter activity (ewbk-11967) Student learning matrix (ewbk-11969)
Solutions	Topic 8 Solutions (sol-1120)
Practical investigation eLogbook	Topic 8 Practical investigation eLogbook (elog-2187)

LESSON
8.2 Introduction to Earth

LEARNING INTENTION

At the end of this lesson you will be able to describe Earth's distinct internal layers (crust, mantle, outer core and inner core), and the theory of continental drift that was proposed to explain the movement of continents, including the supercontinent Pangaea.

8.2.1 Structure of Earth

The interior of Earth has three basic layers: **crust**, **mantle** and **core**. The crust is the very thin, hard, outer layer of our planet. To get an idea of how thin Earth's crust is, imagine that Earth is an apple. The crust would be as thin as the skin of the apple. Two questions have intrigued geologists for more than a hundred years:

1. What lies beneath the crust?
2. Is everything stationary?

crust the hard and thin outer rock layer of Earth

mantle the solid but soft middle rock layer of Earth

core the hot centre of Earth made of iron and nickel

ewbk-11970
eles-4148
int-8163

FIGURE 8.3 Layers of Earth

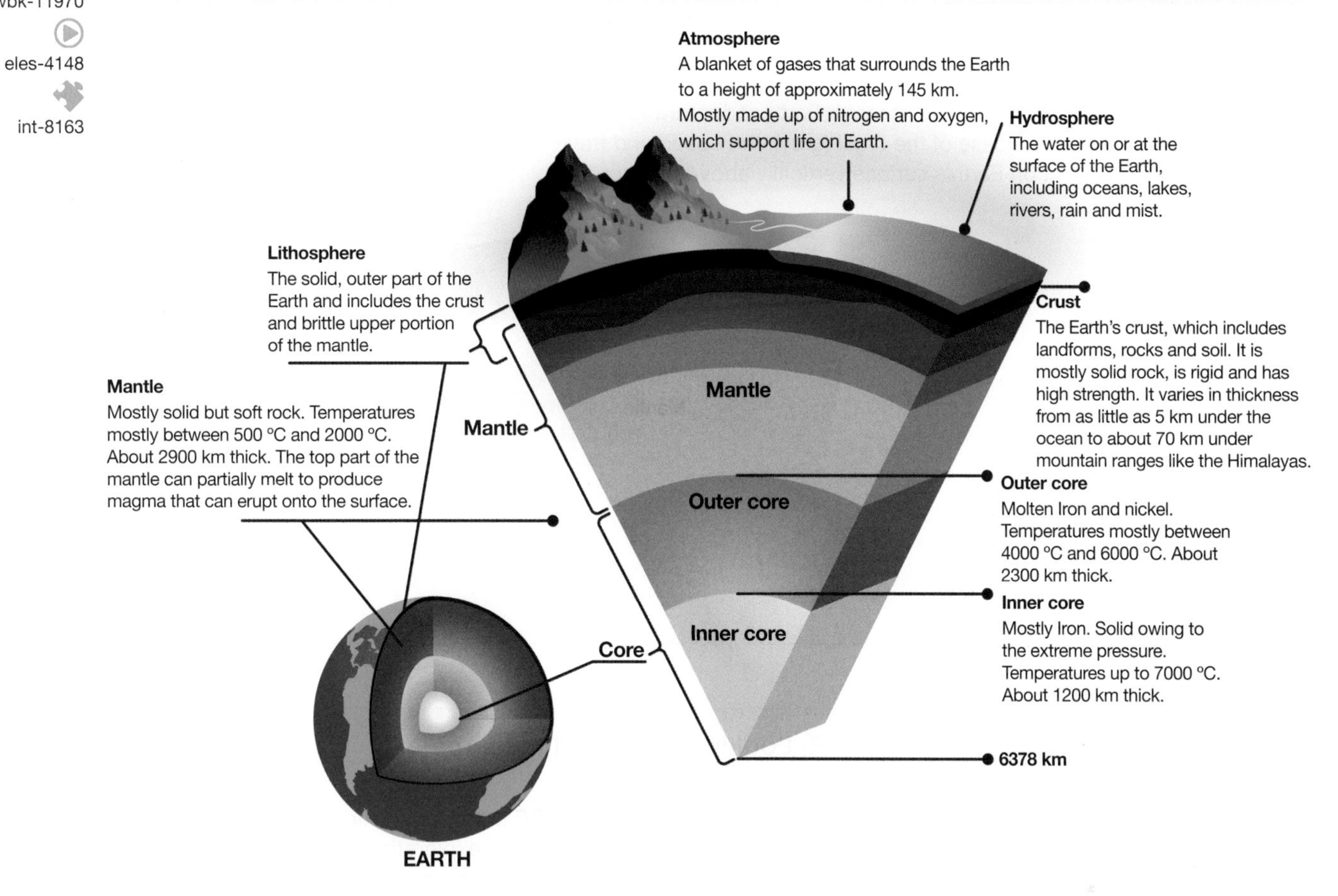

Questions about what is beneath Earth's surface have inspired curiosity and imaginative writing — such as Jules Verne's novels. The idea of drilling through to or even travelling to the centre of Earth is appealing. There could be no better way to find out what is down there. But the deepest man-made holes in Earth have been drilled to only around 12 km of the 6370 km distance to the centre. So, how do geologists know that a mantle and core exist, and how do they investigate what they are like? Some methods include:

- the study of meteorites from space
- laboratory studies to determine temperature and pressure stability conditions for minerals and rocks
- measuring and interpreting seismic wave signals that have travelled through Earth
- looking at features of the crust, which can provide clues to what happens beneath.

seismic waves waves released when rock breaks or is rapidly moved

outer core the liquid outer layer of the core, which is about 2300 km thick

inner core the solid inner-most layer of the core under extreme pressure conditions, with an approximate 1200 km radius

SCIENCE AS A HUMAN ENDEAVOUR: Imaging the interior of Earth

Geophysicists use data from earthquakes to find out what lies inside Earth. Earthquakes produce **seismic waves** that transfer energy from the site of the earthquake (the focus) through the crust. It is the energy of these waves that causes destruction at the surface.

Seismic waves travel differently (speed and behaviour) as they pass through different substances below the crust. By analysing seismic waves, scientists have been able to identify the state and chemical composition of the substances inside Earth. For example, Earth's core is likely made of iron and nickel, but is divided into two layers: a liquid **outer core** and a solid **inner core**. Flow of the liquid outer core plays an important role for life on Earth, as it generates Earth's magnetic field, which protects the surface from some of the most harmful solar radiation.

Consider the following questions.

- What else is the magnetic field useful for?
- Is the outer core the only Earth layer that moves?

FIGURE 8.4 Seismic waves travel through Earth and return to the surface. When they interact with a new medium, like at the base of the crust (grey), the speed and travel path of the waves is altered. The epicentre is the point on Earth's surface vertically above the focus.

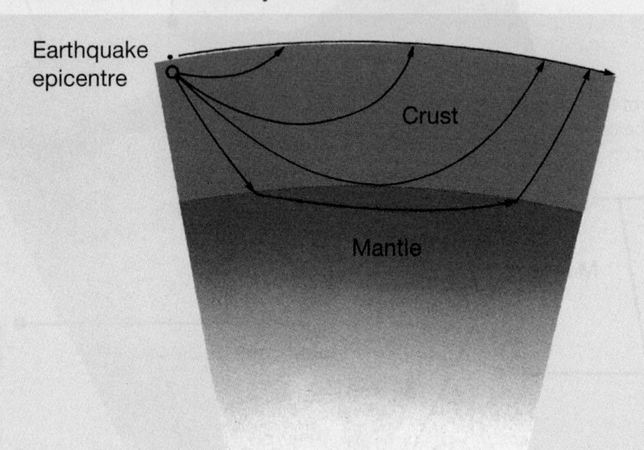

SCIENCE AS A HUMAN ENDEAVOUR: Looking for buried resources

Most of Australia's mineral, energy and water resources have been found in the exposed or shallow rocks of the crust that make up about 20 per cent of its area. Poorly exposed rocks under sedimentary basins or weathering profiles, or even the sea, may also contain resources. These concealed areas will be the focus of future exploration.

Government and industry scientists are adapting new technologies to explore for buried resources. Penetrative geophysical surveys collect data that is used in 3D modelling of Earth's sub-surface. Geophysical investigations include seismic, magnetic, electromagnetic and gravity surveys. These measure different rock properties, which are used in computer modelling. Geological, geochemical and groundwater data are also fed into models of the sub-surface.

FIGURE 8.5 Seismic survey technique used in resource exploration

The Australian Academy of Science is leading the government program UNCOVER, which involves many organisations. This program will use smart analytics and algorithms to simulate geological models and rock properties. New analytical software tools will use predictive technology, machine learning, geological uncertainty analysis and geoscience modelling to improve deep-Earth imaging, develop new exploration technology and support the future search for resources.

8.2.2 The crust

There are two kinds of crust: continental and oceanic as described in table 8.2:

TABLE 8.2 Thickness and density of continental crust and oceanic crust

Continental crust	Oceanic crust
Ranges from 25 to 70 kilometres thick	Ranges from 4 to 10 kilometres thick
Average density is 2.7 g/cm^3, similar to the rock granite	Average density is 3.0 g/cm^3, similar to the rock basalt

- The continents are made of continental crust, which you will recognise as the land.
- The ocean basins are made of oceanic crust, but in a few locations on Earth, slices of the oceanic crust have been observed on land. These unusual rocks are called **ophiolites**. Macquarie Island is an exposed ophiolite in the Pacific Ocean, halfway between New Zealand and Antarctica.
  It is a part of Tasmania and is a UNESCO World Heritage Site.

ophiolites pieces of oceanic crust observed on continental crust (land)

Their shapes, nature and features can help tell geologists what may be happening beneath.

DISCUSSION

Where do you think the thickest continental crust would be? How could a geologist test your hypothesis?

8.2.3 Moving continents

Most geologists of the 1800s believed that Earth started off as a hot, molten ball of rock material. As it cooled, a crust formed, and Earth began to shrink. The shrinking size would cause the solid crust to wrinkle, in the same way that the skin of an apple wrinkles when it begins to rot. Geologists hypothesised that the continents, particularly mountain ranges, were the high parts of the 'wrinkles' and that oceans covered the lower parts. Accordingly, mountains would appear randomly all over Earth's surface and constantly grow; volcanoes and earthquakes would also occur randomly.

However, all scientific theories can be challenged and evolve when new information is gathered. During the late 1800s and early 1900s, evidence was found that showed that the continents were moving.

SCIENCE AS A HUMAN ENDEAVOUR: Early speculation about moving continents

Ever since the coastlines of the continents around the Atlantic Ocean were first mapped, people have been intrigued by the similarity of the coastlines of the Americas and of Europe and Africa. Possibly the first to speculate that the land masses were once joined was Abraham Ortelius in 1596. The idea gained more attention in the eighteenth and nineteenth centuries.

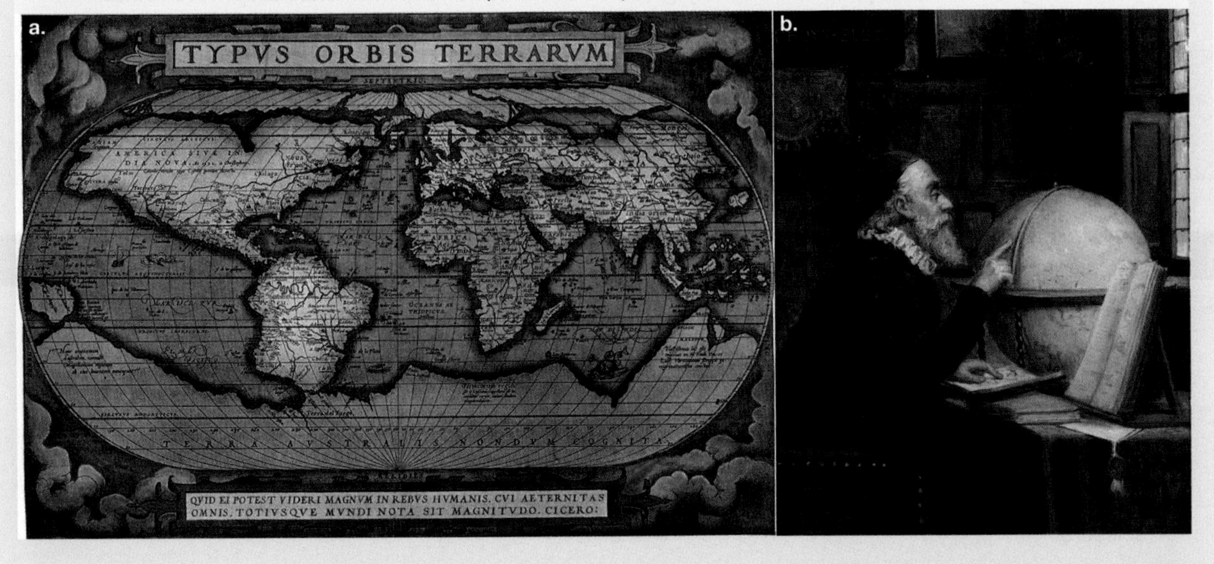

FIGURE 8.6 a. Abraham Ortelius' World Map, 1910 **b.** A portrait of Ortelius in his study

The continental drift theory

In 1912, a German meteorologist and polar explorer named Alfred Wegener proposed a new theory. He suggested that Earth wasn't shrinking, but that continents were slowly drifting across Earth over a weaker mantle, sometimes pushing through ocean crust and colliding with another continent. This process became known as **continental drift**.

Wegener also proposed that, at one time, all the continents were joined like pieces of a giant jigsaw puzzle into a single 'supercontinent' that he called **Pangaea**. Pangaea was surrounded by a vast ocean called **Panthalassa** and a smaller inlying ocean called Tethys. By about 200 million years ago, Pangaea began to break into separate continents that have slowly drifted apart to their present positions.

FIGURE 8.7 The supercontinent of Pangaea as it would have appeared 200 million years ago, surrounded by the Panthalassa Ocean

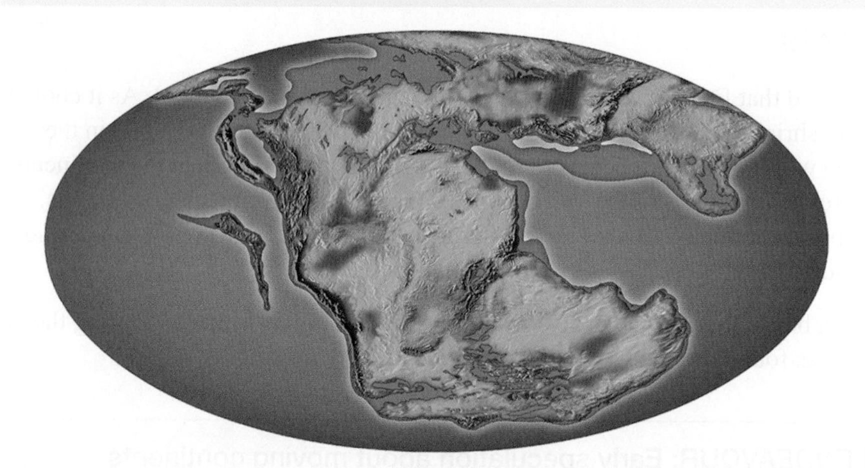

continental drift the movement of Earth's continents relative to each other over geologic time

Pangaea a supercontinent that existed about 299 to 200 million years ago; all landmasses were joined together to form it

Panthalassa the vast ocean surrounding the supercontinent of Pangaea

fossil any remains, impression or trace of a life form of a former geological age; evidence of life in the past

Wegener's claims were based on several lines of evidence, including:

1. the present-day continents looked as though they would fit together, very much like a jigsaw puzzle
2. the discovery of **fossils** of the same land plants and animals on different continents now separated by large oceans
3. the distribution of unique rock deposits and features of the same age across continents
4. the discovery of fossils of plants that clearly grew in a different climate to where they were found.

DISCUSSION

In what ways did continental drift affect the evolution of animals and plants living on Earth at the time?

elog-2189

INVESTIGATION 8.1

Continental drift

Aim

To create a simple model to demonstrate continental drift

Materials

- enlarged copy of the map
- scissors

Method

1. Cut out the continents from the enlarged copy of the map provided.
2. Examine the distribution of fossils on each continent.
3. Rearrange the continents into one supercontinent by matching the distribution of fossils. For example, you want the pink trend of the *Glossopteris* (fern) fossil on one continent to align with another trend on a different continent.

Results

1. Once you have rearranged the continents for your model, glue them into your logbook.
   a. What continent aligns with the east side (right side) of South America?
   b. Is there a continent along the southern margin of Australia?
2. Note which continents must be rotated from their modern-day positions.

Discussion

1. What is the reason for using the distribution of land-based fossils as evidence of Wegener's theory of continental drift?
2. What part of Pangaea does your landmass represent?
3. What latitude and climate conditions during the time of Pangaea would you predict for Australia? How could you investigate your hypothesis?
4. How valid do you think your results are? (*Hint:* Compare your result with others around you; did everyone come up with the same configuration? Is there more than one possible configuration?)
5. Suggest at least one other line of evidence that you could look for that would strengthen your results.

Distribution of a selection of fossils of ancient organisms

INDIA
AFRICA
SOUTH AMERICA
AUSTRALIA
ANTARCTICA
Lystrosaurus – a land reptile
Cynognathus – a land reptile
Mesosaurus – a freshwater reptile
Glossopteris – a fern

Conclusion

Summarise the findings from this investigation about continental drift.

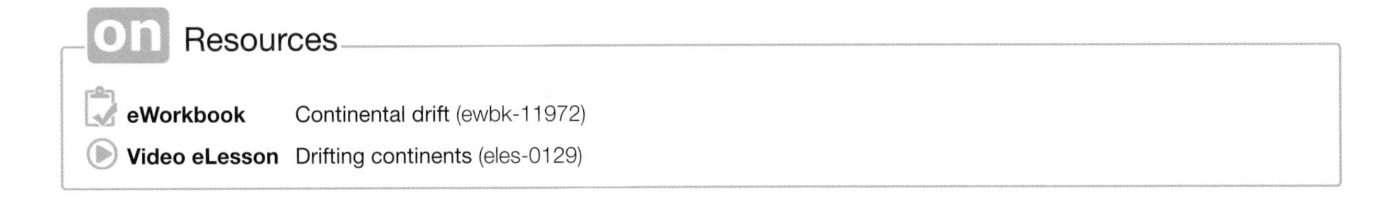

on Resources

eWorkbook Continental drift (ewbk-11972)

Video eLesson Drifting continents (eles-0129)

8.2 Activities

learnon

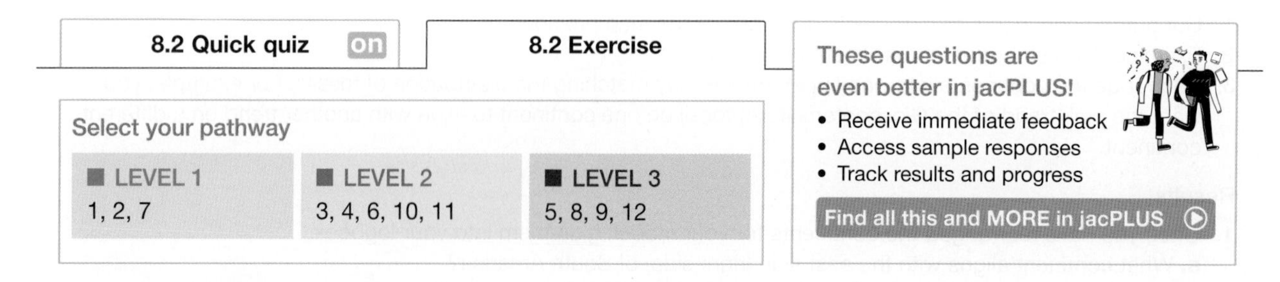

Remember and understand

1. Complete the table by adding descriptions for each of Earth's layers.

Layer	Description
Atmosphere	
Hydrosphere	
Crust	
Mantle	
Outer core	
Inner core	

2. Provide the appropriate labels for the model of Earth shown (*don't forget about the two different types of crust*).

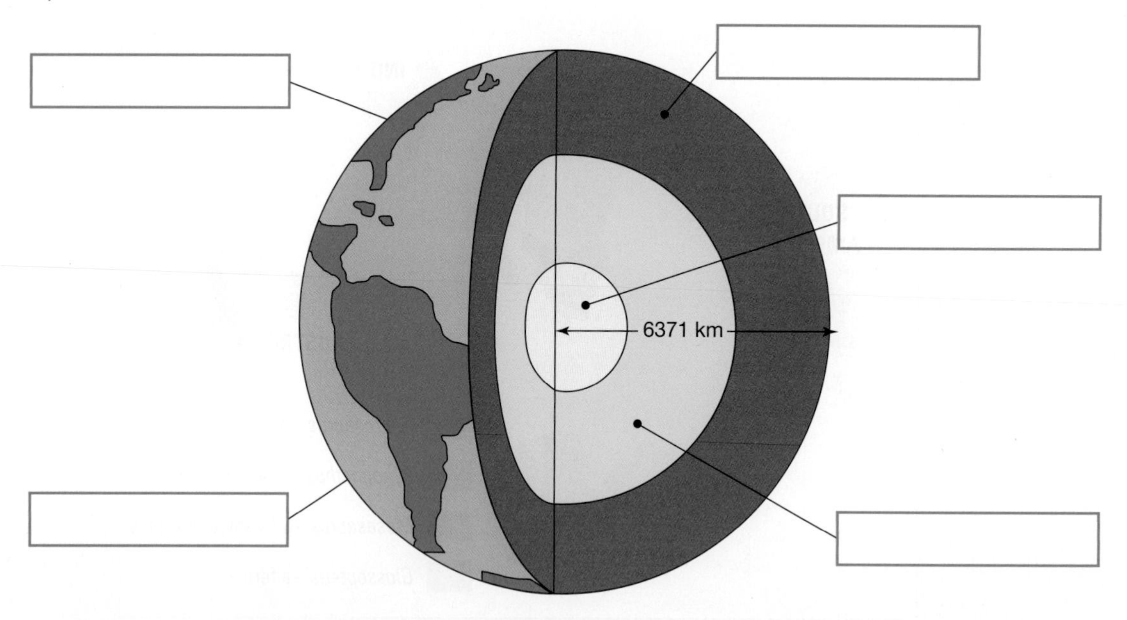

3. Even though the inner core is hotter than the molten outer core, it is solid. Explain why this is the case.
4. How is oceanic crust different from continental crust?
5. Describe two observations that provided evidence for Wegener's theory of continental drift.
6. According to Wegener's theory of continental drift, upon which layer of Earth are the continents floating?
7. What were Pangaea and Panthalassa?

Apply and analyse

8. How might the study of meteorites improve our understanding of Earth's interior?
9. **SIS** Seismic waves travel fast, measured in kilometres per second (km/s). The velocity changes of a seismic wave for the depths around the mantle–core boundary within Earth are shown in the graph.
   a. Mark the boundary between mantle and core.
   b. What variables can influence the speed of the seismic wave?
10. Explain why geologists have not been able to drill a hole deeper than 12 km, considering the distance to the centre of the Earth is over 6000 km deep.

Velocity of seismic waves around mantle–core boundary

P-wave velocity (km/s): 0 2 4 6 8 10 12 14

Depth (km): 660, 2900, 5140

Evaluate and create

11. **SIS** The Shrinking Earth theory was popular during the 1800s to explain the existence of continents and mountain ranges.
    a. Outline how it explained the existence of continents, particularly mountain ranges.
    b. Suggest an investigation that could help test the theory.
12. **SIS** Alfred Wegener's theory of continental drift was not widely accepted, despite all the evidence put forward. Review the theory and suggest a reason why the scientific world may have had a hard time accepting continental drift, as he proposed it.

Fully worked solutions and sample responses are available in your digital formats.

8.3 The theory of plate tectonics

LEARNING INTENTION

At the end of this lesson you will be able to explain how sea-floor spreading works and its importance to the theory of plate tectonics.

8.3.1 Mapping the sea floor

Alfred Wegener's theory of continental drift was not widely accepted in the beginning because it failed to explain how the continents moved. It wasn't until the 1940s, when additional information from the sea floor was gathered, that the story started to piece together.

During World War II, submarines were used as a defensive tool. They used echo-sounding to avoid collisions and search for other submarines. Harry Hess, a United States naval officer and marine geologist, took advantage of the echo-sounding to also survey and map the ocean floor. He discovered some unexpected features: **abyssal plains**, **ocean ridges** and **deep-ocean trenches**. He only understood his ocean-floor profiles after detailed mapping in the North Atlantic Ocean by Marie Tharp and Bruce Heezen revealed the character of a major rift running along the Mid-Atlantic Ridge.

abyssal plains relatively flat underwater deep-ocean floor, around 4000 metres depth

ocean ridges submarine mountains that tower 2000 metres above the abyssal plains

deep-ocean trenches narrow and deep troughs in the ocean floor, generally greater than 5000 metres depth

The Mariana Trench of the western Pacific is the deepest place on Earth, with a depth of 10.9 kilometres — deep enough to swallow Mount Everest without a trace.

FIGURE 8.8 This topographic model of the Pacific Ocean highlights sea-floor features such as abyssal plains, ocean ridges and deep-ocean trenches.

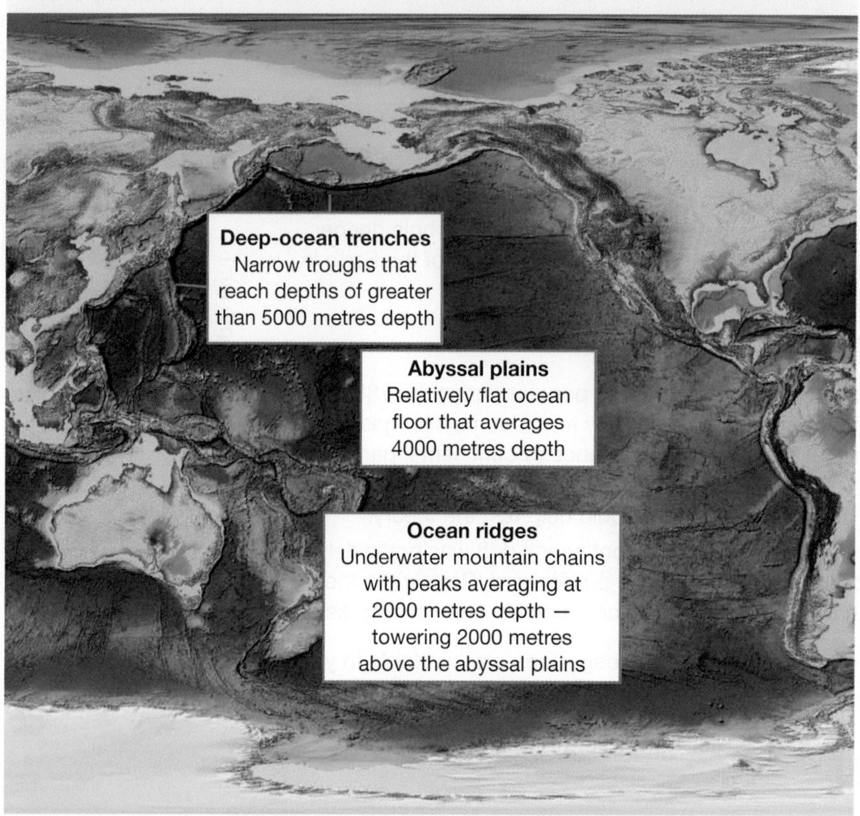

By the 1960s, much of the ocean floor had been mapped, and geologists had gathered more information about the oceanic crust. For example, they discovered that *the rocks further away from the ocean ridges are older and colder than those closer to the centre*. The new data from the ocean floor led to the hypothesis of **sea-floor spreading**.

sea-floor spreading the formation of oceanic crust, which occurs by the rising and melting mantle at ocean ridges that push older crust away from the ridge

The hypothesis of sea-floor spreading states that new oceanic crust forms at the centre of ocean ridges as mantle rises, melts, and erupts through underwater volcanoes and then cools again. The new crust then splits in half to allow even younger crust to form in the middle. This pushes the older crust away from the ridges. With time, sea-floor spreading will make an ocean basin wider as more crust is made.

FIGURE 8.9 The process of sea-floor spreading can move continents apart.

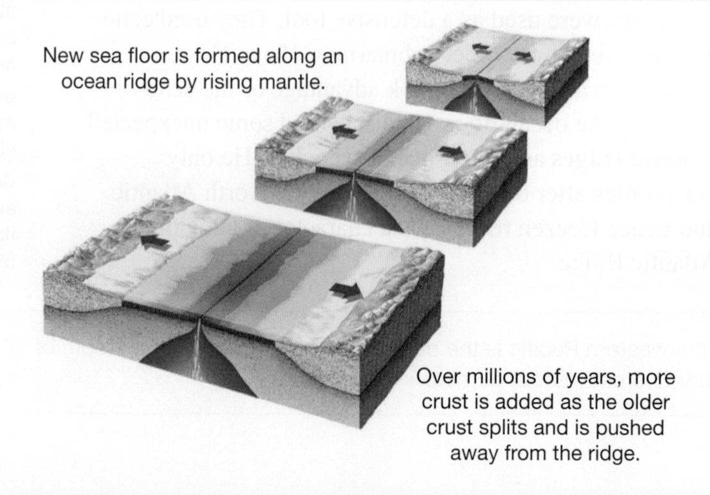

SCIENCE AS A HUMAN ENDEAVOUR: The woman who mapped the sea floor

Marie Tharp was a geologist and cartographer whose sharp mind and quiet determination allowed her to make a huge impact on our understanding of Earth. A pioneer in a field dominated by men, she worked mostly in the background. Today she is openly acknowledged for major contributions to oceanic exploration and helping to prove the theory of plate tectonics.

Science was a male-dominated subject in 1942, but Tharp was given the rare chance to study geology at university because many men were fighting in World War II. In 1948, she began work at Columbia University and collaborated with Bruce Heezen on a project mapping the sea floor, which was thought to be mostly flat and featureless. They wanted to understand its geology and how it was connected to the continents. Heezen went out on research vessels to sea and collected sonar measurements of the ocean depths. At the time, it was considered bad luck to allow women on ships, so Tharp stayed home and used the data collected to draw a detailed map of the ocean floor using only pens, ink, and rulers. Her drawings revealed that the ocean floor was not flat, but had canyons, ridges and mountains!

FIGURE 8.10 Marie Tharp working on the map of the Atlantic Ocean sea floor in the early 1950s

In 1953, Tharp made the amazing discovery of a 16 000-kilometre ridge in the middle of the Atlantic Ocean. The ridge surrounded a huge tear in the ocean floor (a rift), caused by the gradual separation of two massive tectonic plates. She suggested this was evidence of continental drift but others disagreed, as this theory was unpopular at the time. Heezen dismissed her hypothesis as 'girl talk', but once he discovered clustered earthquakes along Tharp's ridge he began to agree with her. An expedition by ocean explorer Jacques Cousteau, who also did not believe Tharp at first, produced video evidence of a lava-filled valley surrounded by two sharp ridges, exactly where Tharp predicted it would be!

Tharp and Heezen completed several maps of the ocean floor; the North Atlantic map came out in 1957, followed by maps of the South Atlantic and Indian Ocean in the early 1960s. In 1977, Tharp published the first complete world map of the ocean floors. Tharp and Heezen's work helped to prove the theory of plate tectonics — the idea that the continents move over time — and revolutionised our understanding of earth science.

Tharp's name does not appear on any of the major papers on plate tectonics that Heezen and others published between 1959 and 1963. Although later recognised and attributed for her work on the Mid-Atlantic Ridge, it was Heezen who received credit in 1956 for the discovery that was made. This is an example of the Matilda effect.

The Matilda effect

The 'Matilda effect' is a bias against acknowledging the achievements of female scientists, whose work is instead attributed to their male colleagues. Coined by science historian Margaret W. Rossiter in 1993, the term is named in honour of US activist Matilda Gage, who first described the phenomenon in an essay in 1870. The 'Matilda effect' contrasts with the 'Matthew effect', in which scientists who are already famous are over-credited with new discoveries. We've come a long way towards providing all scientists with equal opportunities, but there's still some work to do.

FIGURE 8.11 World Ocean Floor Panorama, 1978. This map by Tharp was painted by Heinrich Berann.

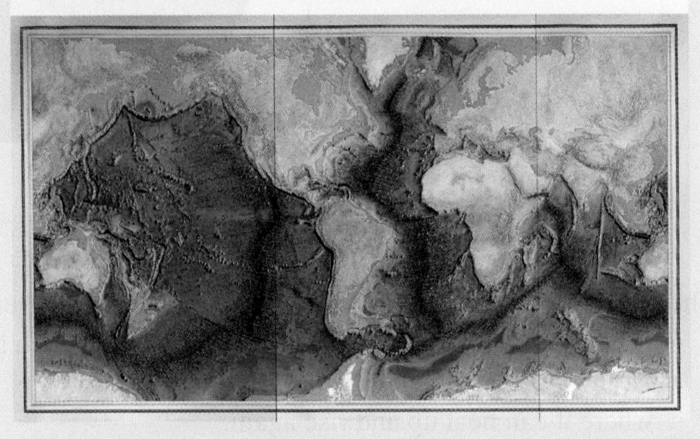

DISCUSSION

If Earth's size and volume do not change, is there a problem with new crust continually being made at ocean ridges? Formulate an idea or hypothesis that could solve this problem.

8.3.2 How plate tectonics works

The theories of continental drift and sea-floor spreading paved the way for the more recent theory of **plate tectonics**. This theory proposes that Earth's crust is broken into pieces, called plates, that move around and interact with one another (tectonics). The theory of plate tectonics explains much more than the movement of continents; for example, it explains why the Himalayas have grown to such great heights, why Iceland is slowly splitting in two and why new rock is being formed along ocean ridges.

plate tectonics a scientific theory that describes the relative movements and interaction of plates of Earth's crust over the underlying mantle

magma a very hot mixture of molten rock and gases, just below Earth's surface, that forms from melting of the mantle and occasionally the crust

radioactive refers to when atoms are unstable and emit a particle to remove excess energy; these particles are capable of ionising other atoms upon collision, which can cause harm to living tissue

convection currents the movement of particles in a liquid or gas resulting from a temperature or density difference

The centre of Earth is hot. When the world formed about 4.5 billion years ago, the whole planet would have been one big ball of **magma**. After all this time you might expect it to have cooled to become a solid ball of rock. However, the Earth is **radioactive**. Heavier atoms like uranium break apart, releasing particles that are absorbed by the rock surrounding them. This heats the rock so that the core of Earth is hot enough to keep the rock in a liquid state. As the centre is hotter than the outside, heat will flow by **convection currents** upwards (figure 8.12). You experience convection currents daily. Boiling water or feeling the wind blow are examples of convection currents, where heated material is moving from warmer or higher-pressure conditions to cooler or lower-pressure conditions.

FIGURE 8.12 Convection currents in the mantle

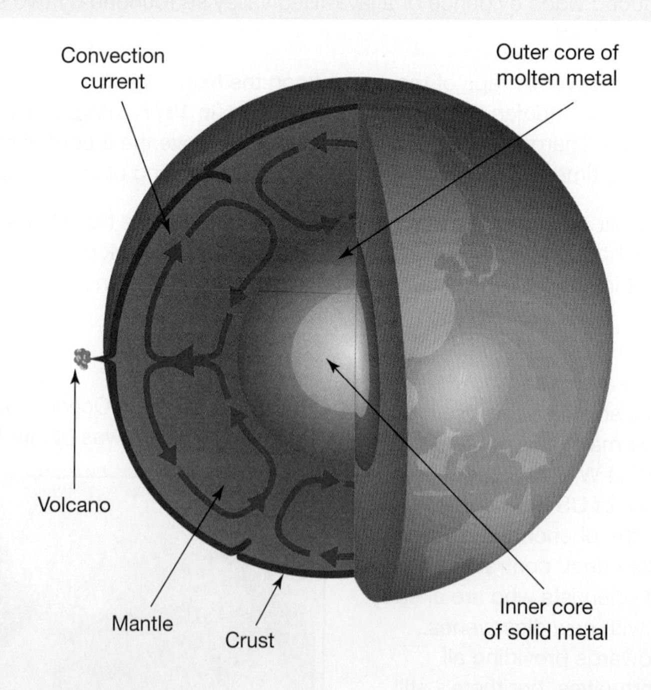

The convection currents within the mantle form due to thermal expansion and contraction of rock.

- Heat causes the rock in the mantle to expand, which helps it to rise towards the surface.
- At the top of the mantle the rock cools and contracts, encouraging it to sink under the force of gravity, where it can heat up and rise again.

This process of the plates sliding over a weak layer of slow-flowing rock in the mantle is called plate tectonics.

The plates can consist of only continental crust or oceanic crust, or be a combination of both. In any case they move slowly (usually just a few centimetres in a year), and can move away from each other, push against each other, or slide past each other.

TABLE 8.3 Milestones in the development of the theory of plate tectonics

Year	Idea or observation	Who was responsible?
1596	Speculation about Earth's continents having moved	Abraham Ortelius
1858	Maps produced fitting continents together, based on correlation of rock type and fossils	Antonio Snider-Pellegrini
1872–1876	Suggestion of a vast mountain range in the Atlantic Ocean and deep troughs elsewhere, based on 'sounding', a technique that used lead-weighted rope over the side of a ship	*Challenger* expedition, Sir Charles Thomson
1896	Discovery of radioactivity	Henri Becquerel
1912	Theory of continental drift proposed	Alfred Wegner
1925–1927	Sonar surveys identified a mid-ocean ridge in the Atlantic Ocean, continuing to the Indian Ocean	*Meteor* expedition
1927	Convection in mantle proposed to move continents	Arthur Holmes
1928	Zone of earthquakes discovered dipping down from ocean trenches, later recognised to be caused by subducting slabs	Kiyoo Wadati; independently discovered by Hugo Benioff in 1949
1953	Detailed map of the Mid-Atlantic Ridge completed and a central rift proposed	Marie Tharp
1950s–1960s	Theory of the expansion of Earth (later disproved) stimulated development of the idea of the sea floor spreading	Sam W. Carey
1960s	Discovery of the apparent movement of magnetic poles over geological time; this 'polar wander' varied between continents, but came together if the continents were grouped into the supercontinents of Gondwana and Pangaea	Stanley Runcorn, Kenneth Creer and Ted Irving
1961–1962	Proposal that oceanic crust forms along mid-ocean ridges and spreads out laterally away from them; 'spreading ridges' named	Harry Hess and Robert Dietz
1963	Magnetic striping in ocean-floor rocks symmetrical about mid-ocean ridges was used to calculate rates of plate movement	Frederick Vine, Drummond Matthews and Lawrence Morley
1963	Hotspots proposed to explain volcanoes a long way from a plate boundary; transform faults recognised as a third type of plate boundary	John Tuzo-Wilson
1960s	Geophysical evidence gathered, helping define the driving forces of plate tectonics	
1966	The Wilson Cycle was proposed, stating that oceans opened and closed throughout Earth's history	John Tuzo-Wilson
1967	Spherical geometry and Euler's theory of motion on a sphere were used to determine plate motion across divergent boundaries	Jason Morgan and Dan McKenzie
1968	The term 'plate tectonics' was introduced.	Frederick Vine and Harry Hess
1968	Computer model produced of the motion of six plates that form Earth's crust; it showed the total crust created at the ocean ridges equalled the amount lost due to subduction	Xavier Le Pichon

(continued)

TABLE 8.3 Milestones in the development of the theory of plate tectonics *(continued)*

Year	Idea or observation	Who was responsible?
1968	Dating of deep-ocean drill cores showed that the ages of the ocean-floor rocks increase away from mid-ocean ridges	*Glomar Challenger* expedition
1970s	Seismic tomography revealed more of Earth's interior, producing 3D images by combining information from many earthquakes	
1975	Modelling showed that of all the forces likely to be driving plate motion, slab pull is the strongest	Don Forsyth and Seiya Uyeda
1977	First worldwide map of the ocean floor produced	Marie Tharp and Bruce Heezen

WHAT DOES IT MEAN?

The word 'tectonic' is derived from the Greek word *tektonikos*, meaning 'builder'.

elog-2191

INVESTIGATION 8.2

Convection currents

Aim

To observe convection currents

Materials

- 250 mL beaker
- water
- 1 g of potassium permanganate
- Bunsen burner
- tripod
- gauze mat
- matches

Method

1. Set up the Bunsen burner, tripod and gauze mat.
2. Add 200 mL of water into the beaker and heat it until it is boiling.
3. Add 1 g of potassium permanganate and record your observations.

Results

1. Draw a labelled diagram showing your observations.

Discussion

1. What is the purpose of the potassium permanganate in this investigation?
2. Using appropriate scientific terminology, explain your observations.
3. Describe similarities and differences between the convection currents in water and in Earth's mantle.

Conclusion

Summarise your findings from this investigation about convection currents.

Extension

Use a digital thermometer to explore if a temperature difference can be observed in the water at the top of the beaker compared to the bottom of the beaker.

Note: The thermometer should not touch the bottom of the beaker.

What drives plate movement?

Three forces have been proposed as the main drivers of tectonic plate movement:

- *Mantle convection currents.* Currents in the upper mantle carry Earth's outer rigid tectonic plates along on top due to friction, like shopping on a supermarket conveyor belt. Intruding magma at spreading ridges pushes the plates apart. The upper mantle along with the crust are part of the **lithosphere**.
- *Slab pull.* The weight of the older, colder and denser parts of a plate sink into the mantle at the subduction zone (or trench) and pulls the rest of the plate along behind it. The cold slab sinking into the mantle likely contributes to mantle convection.
- *Ridge push.* Newly formed plate material at oceanic ridges is warm, and so has a higher elevation at the oceanic ridge than the colder, more dense plate material further away. Gravity causes the higher plate at the ridge to push away the material further from the ridge.

lithosphere the outermost layer of Earth; includes the crust and uppermost part of the mantle

FIGURE 8.13 Drivers of tectonic activity

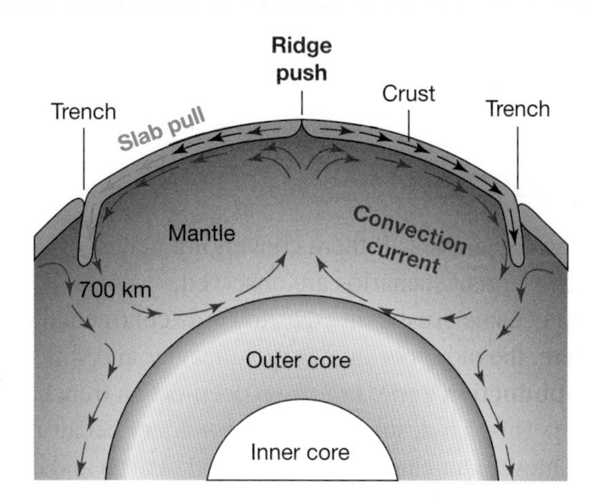

For many years mantle convection currents were believed to be the major driver of plate movement. However, recent research has shown that the major driving force for most plate movement is slab pull, because the plates with more of their edges being subducted are the faster-moving ones. Ridge push only seems to be effective where there are no slab-pull forces. There is little geophysical evidence for mantle convection being an important contributor. It seems likely that gravity drives the motion of plates, which help stir the mantle, rather than convection of the mantle driving the plates.

8.3 Activities

learnon

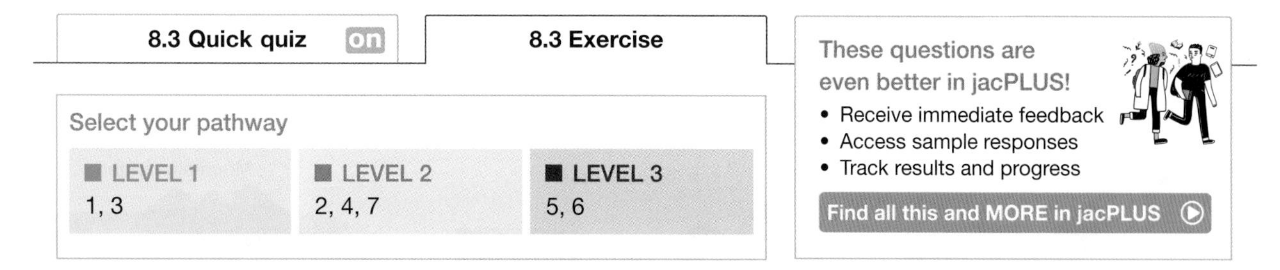

Remember and understand

1. What tools were used to survey and map the ocean floor during World War II?
2. What is the theory of plate tectonics?
3. If Earth's surface consists of moving plates, what are the plates moving on?
4. Why does oceanic crust subduct and continental crust does not?

Apply and analyse

5. List, in point form, at least three pieces of evidence that support the theory of plate tectonics.
6. The theory of continental drift was first proposed in 1912, over 50 years before the theory of plate tectonics evolved. The evidence for the theory of continental drift also supports the theory of plate tectonics. Explain the difference between the two theories.
7. Explain why the older ocean floor is found furthest away from the ridge.

Fully worked solutions and sample responses are available in your digital formats.

LESSON
8.4 Plate boundaries

LEARNING INTENTION

At the end of this lesson you will be able to recognise different plate boundaries and their features.

8.4.1 Plates coming together

When two plates move towards each other we call them converging plates, or a **convergent boundary**, and two convergent scenarios are observed.

1. **Subduction** occurs when old oceanic crust converges towards continental crust or younger oceanic crust. Here, the older oceanic crust is heavier and sinks beneath the other crust at a **subduction zone**, to form a deep-ocean trench (figure 8.14, left-hand image). This is a destructive plate boundary because ocean crust enters the mantle. This movement causes powerful earthquakes and creates arcs of **volcanoes** that are parallel with the deep-ocean trenches. The Ring of Fire is a circle of subduction convergent boundaries.

   As time progresses, this subduction can lead to a collision with another continent as they are brought closer together (figure 8.14, centre image).

2. **Collision** occurs when subduction brings two continents together. Here, both continents are light and thick, which prevents them from entering the mantle. Instead, huge **mountain ranges**, like the Himalayas, are formed as the continents crumple together (figure 8.14, right-hand image). The Himalayas are the result of the collision of the Indian continental plate and the Eurasian continental plate.

convergent boundary where two tectonic plates move towards each other

subduction refers to a convergent plate boundary where one plate moves under another

subduction zone where old oceanic crust enters the mantle

volcano a landscape feature through which melted rock is erupted onto Earth's surface

collision refers to when two continents crumple together to form a mountain range

mountain range a group of high-ground features, commonly the result of tectonic collision

FIGURE 8.14 Convergent boundaries from subduction of oceanic crust beneath continental crust through to continent–continent collision, which forms mountain ranges

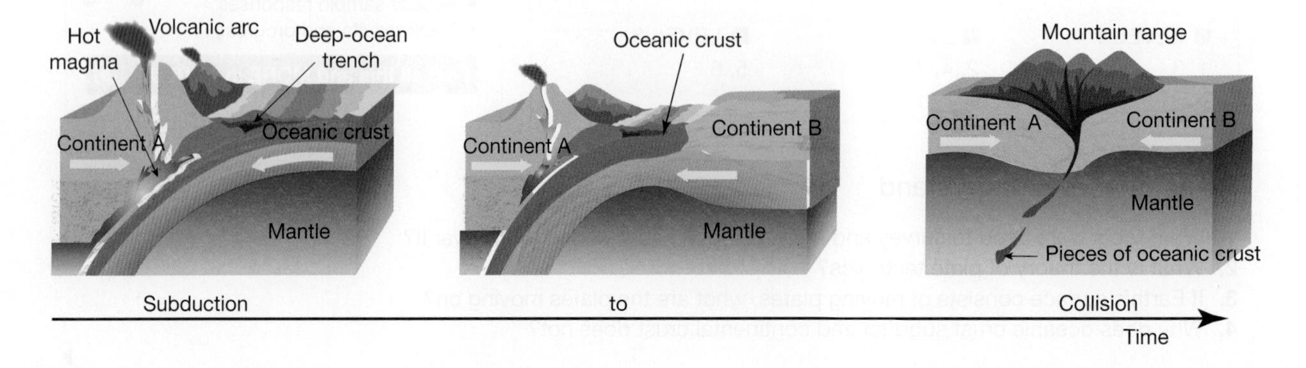

CASE STUDY: The Pacific Ring of Fire

The majority of the world's active volcanoes are not random; they lie along a circle around the Pacific Ocean that is known as the Ring of Fire. Why do you think volcanoes are distributed like this?

FIGURE 8.15 The Ring of Fire

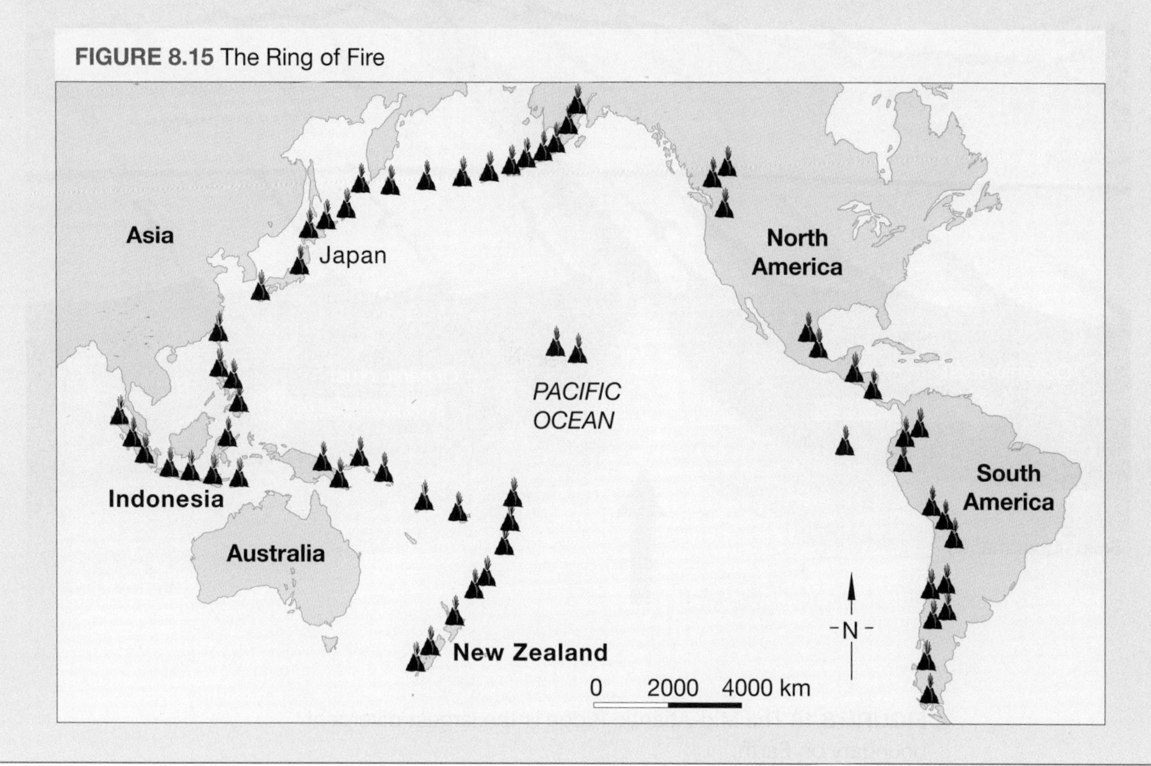

8.4.2 Plates moving apart

Divergent boundaries are found where plates are diverging or moving apart (figures 8.16 and 8.17). As the plates move apart, hot mantle rises to fill in the middle and partially melts because of a decrease in pressure. The melt continues to rise and forms small volcanoes along the divergent boundary; as it cools and solidifies, new oceanic crust is created. Because of the emergence of new crust, these boundaries are known as **constructive plate boundaries**.

Sea-floor spreading is an example of diverging plates and the centre of an ocean ridge is the divergent boundary between plates. The most famous divergent boundary is the Mid-Atlantic Ridge (figure 8.18), which is spreading apart at 2.5 cm per year or 25 km in one million years! Sea-floor spreading over the past 100–200 million years has caused the Atlantic Ocean to grow from a small body of water between Europe, Africa and the Americas, to the enormous ocean it is today.

FIGURE 8.16 Silfra divergent tectonic drift in Thingvellir, Sudurland, Iceland caused by the North American and Eurasian plates

divergent boundary where two tectonic plates move apart

constructive plate boundary where new crust is formed

FIGURE 8.17 Divergent boundary: the spreading sea floor

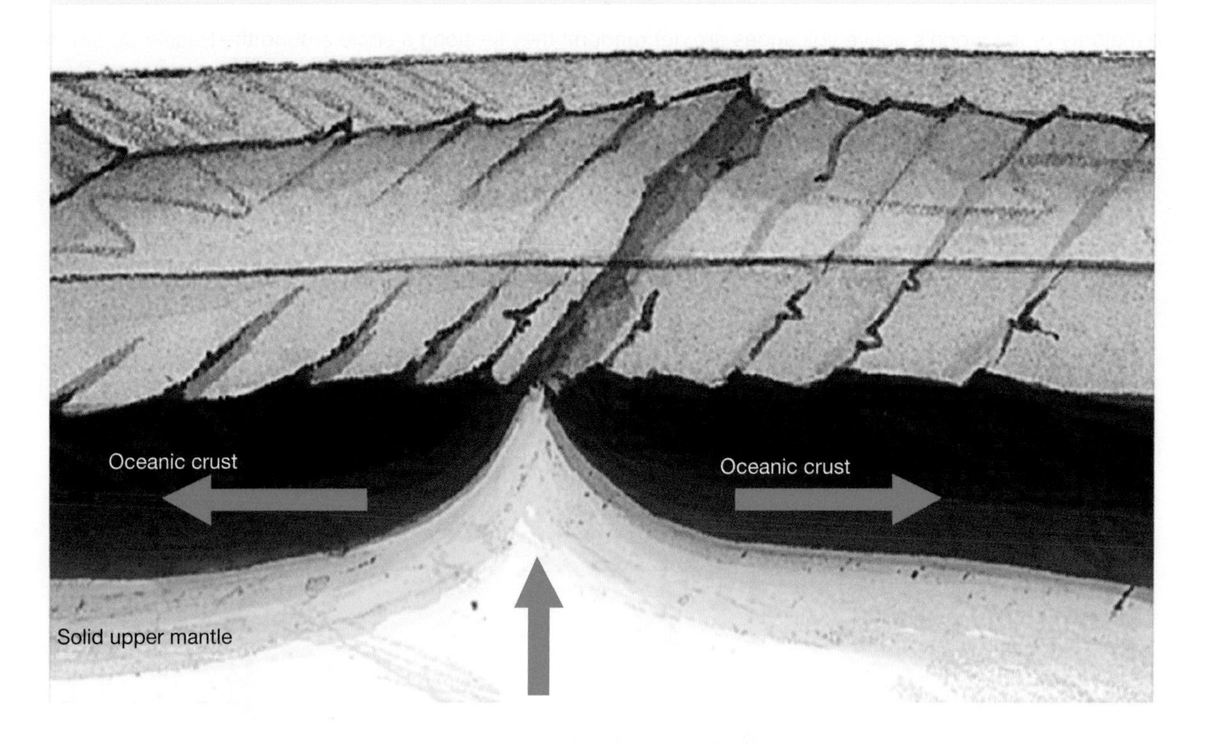

FIGURE 8.18 The Mid-Atlantic Ridge is the largest divergent boundary on Earth.

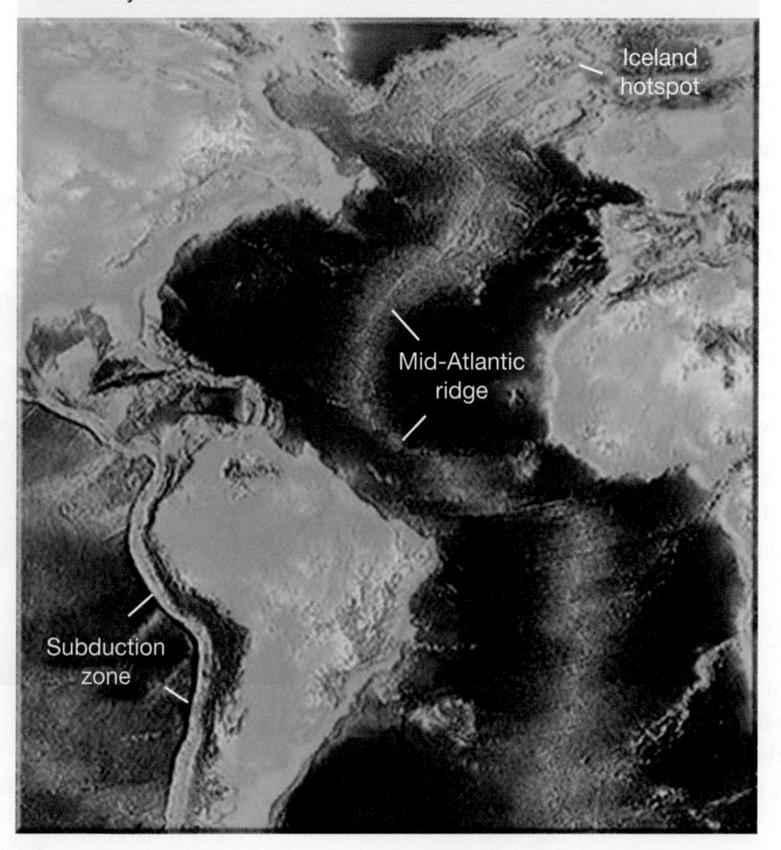

DISCUSSION

Look at the image of the Mid-Atlantic Ridge (figure 8.18) and locate Iceland. Can you predict why Iceland has so much geological activity? What is causing the volcanic eruptions?

8.4.3 Plates sliding side by side

When two neighbouring plates slide past each other, we call it a **transform boundary**, and earthquakes commonly occur. Large earthquakes occur when something prevents the plates from sliding. Pressure builds up until there is enough force to restart the sliding with a jolt.

transform boundary where two tectonic plates slide past one another

conservative plate boundary where crust is neither created nor destroyed

The boundaries between sliding plates are known as **conservative plate boundaries**. This is because the crust is conserved (it is neither created nor destroyed). The San Andreas Fault in California, United States, is perhaps the best-known example of a transform boundary.

FIGURE 8.19 Transform boundary: plates sliding side by side

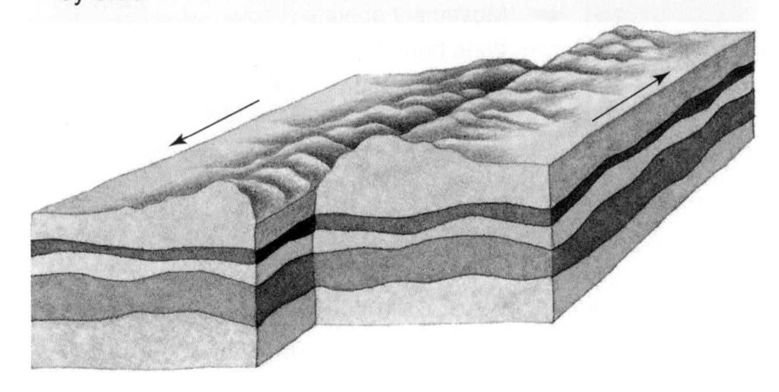

FIGURE 8.20 A view of the San Andreas Fault in California

Resources

Video eLesson San Andreas Fault (eles-4149)

8.4.4 Identifying the current plate boundaries

The major plate boundaries of today can be identified by observing the pattern of volcanoes and earthquakes, and the mapping of spreading ocean ridges and growing mountain ranges. Geologists have directly measured movement of the continents to confirm the plate boundaries.

Geologists have been able to demonstrate that Earth's crust is divided into over 20 plates, not just the separate continents. Some of the plates are very large, while others are quite small. Figure 8.21 shows the location of some of the major plates and the direction of plate movement. The location of some of the boundaries is still not certain; these are shown on figure 8.21 by the red lines.

eles-4150

FIGURE 8.21 A simplified map showing the major tectonic plates that make up Earth's crust. The arrows show the direction of plate movement.

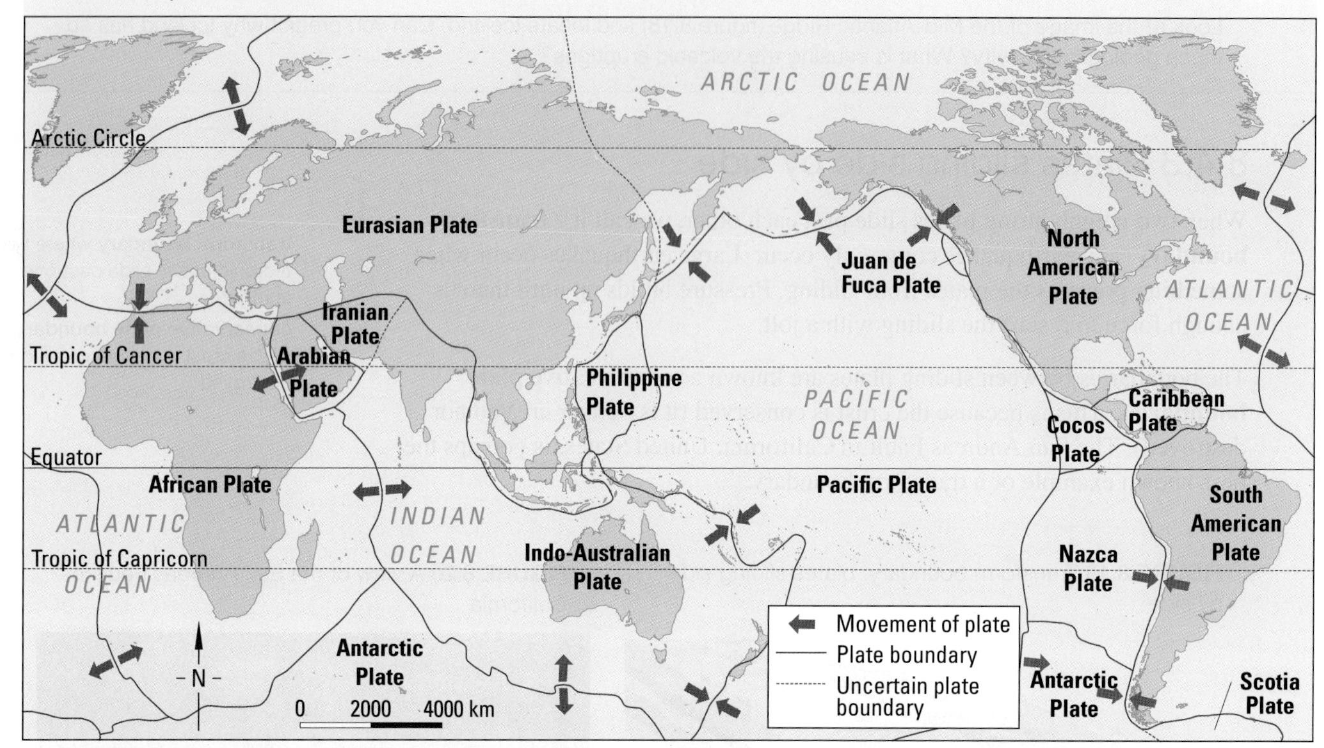

Earth recycles itself

While mantle rises and partially melts to form new oceanic crust at ocean ridges, old oceanic crust pushed away from the ridges sinks back down into the mantle at subduction zones. This slow and continuing natural process of 'recycling' old crust and producing new crust takes place over millions of years. But in the process, ocean basins can open and close, and continents get to go along for the ride. The result is that the continents continually shift and reorganise themselves. The configuration of continents we see today is not what it was in the past, and is not what it will be in the future.

8.4.5 The continental jigsaw

The theory of plate tectonics enabled a more complete reconstruction of the movement of continents proposed by the continental drift theory. Geologists now accept that about 200 million years ago the supercontinent Pangaea broke up into two smaller continents called **Laurasia** and **Gondwana** (or Gondwanaland). The continents of Africa, South America, Antarctica and Australia were all part of Gondwana.

One of the most famous fossils to support the theory of plate tectonics is from a seed fern called *Glossopteris*, which has been found across all of the now detached southern continents of South America, Africa, India, Australia, New Zealand and Antarctica. It existed for nearly 50 million years as the dominant plant of Gondwana. Gondwana was named after the region of India where *Glossopteris* was found.

Laurasia the northern part of the broken-up supercontinent of Pangaea, which included the continents of North America, Europe and Asia

Gondwana the southern part of the broken-up supercontinent of Pangaea, which included the continents of Africa, South America, Antarctica and Australia; also known as Gondwanaland

FIGURE 8.22 Two hundred million years ago Pangaea began to break apart, first into two large masses called Laurasia and Gondwana. Which one was Australia a part of?

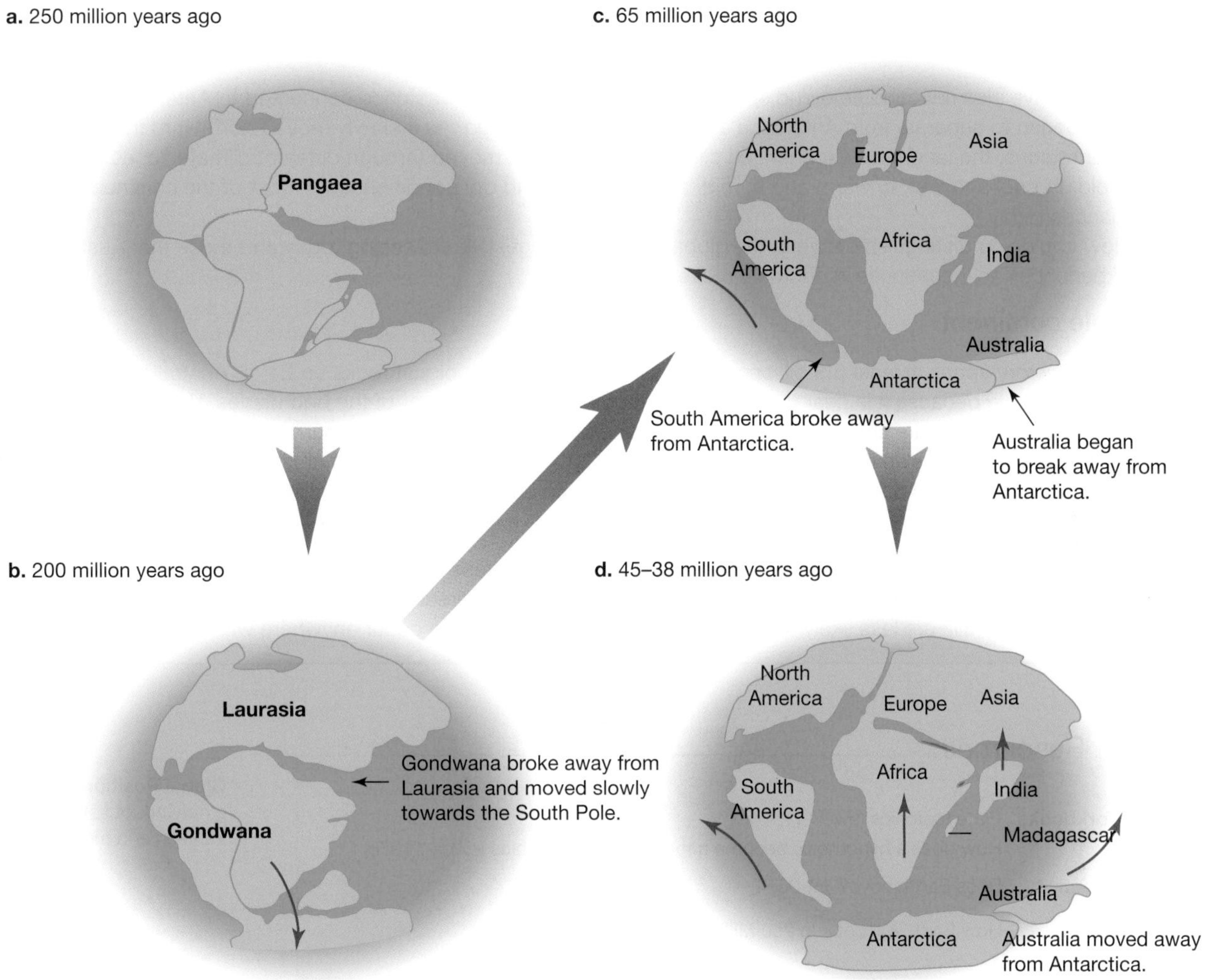

a. 250 million years ago

Pangaea

b. 200 million years ago

Laurasia

Gondwana

Gondwana broke away from Laurasia and moved slowly towards the South Pole.

c. 65 million years ago

North America

Europe

Asia

South America

Africa

India

Australia

Antarctica

South America broke away from Antarctica.

Australia began to break away from Antarctica.

d. 45–38 million years ago

North America

Europe

Asia

Africa

India

South America

Madagascar

Australia

Antarctica

Australia moved away from Antarctica.

FIGURE 8.23 *Glossopteris* fossils are evidence of the theory of plate tectonics.

SCIENCE AS A HUMAN ENDEAVOUR: Australia on the move

Using the high-resolution **Global Positioning System (GPS)**, geologists have measured that Australia is moving approximately north at 7 centimetres per year. Scientists predict that Australia will join with Asia and Europe in about 60 million years!

Geologists have also rewound time and space to identify that Australia was once connected to Antarctica. Australia began to separate from Antarctica about 65 million years ago. As it slowly moved northward from the polar regions, the landmass experienced climate changes as it moved through different climate zones — from cold, to cool and wet, to warm and humid, and to the hot and dry conditions that most of the continent experiences today.

- How could you distinguish between the global climate changes we are seeing today and those that Australia experiences as it continues its tectonic movement?

A stable continent

Australia's history spans at least 3.8 billion years, so it has some of the oldest rocks on Earth. Australia is geologically stable because it is near the centre of a tectonic plate, well clear of the boundaries. Volcanic activity and severe earthquakes are unlikely, but may occur; for example, residents of Melbourne felt an earthquake that originated from the town of Mansfield (approximately 53 km SSE of Melbourne) on 22 September 2021, measuring a magnitude of 5.9. Some very old rocks are preserved due to Australia's distance from tectonic boundaries, where rocks are formed or destroyed.

The Australian plate is slowly colliding with plates to the north, so stress builds up in the interior of the continent. A sudden release of this stress occurs when rocks deep underground break and move along a fault line. While some parts of Australia are less likely to experience earthquakes, they can occur anywhere without warning.

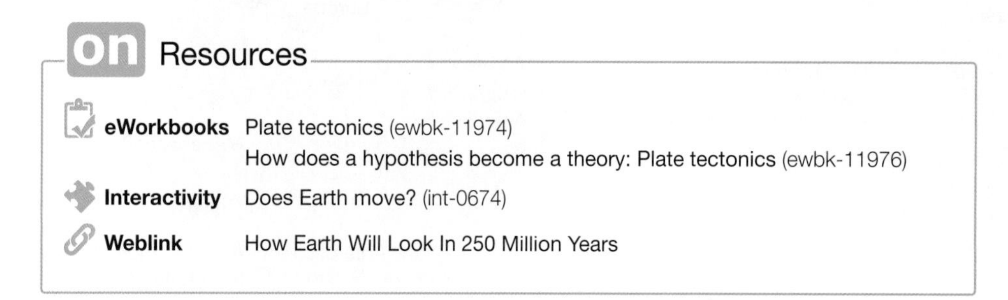

Global Positioning System (GPS) a network of satellites that tracks location and movement

8.4 Activities

learnon

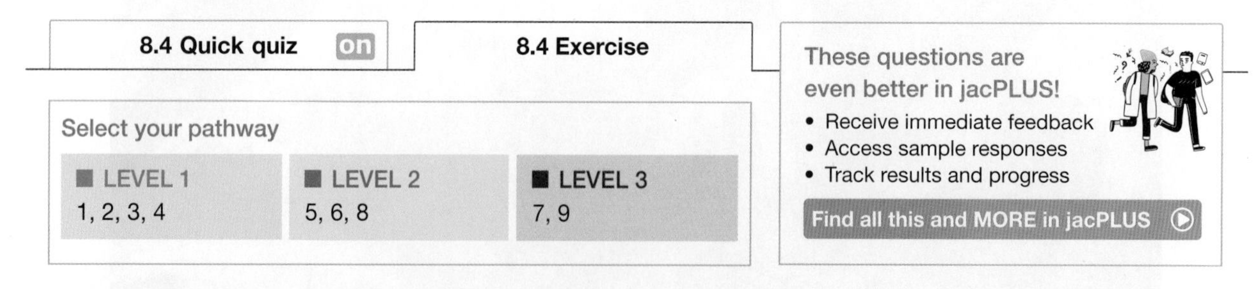

Remember and understand

1. **MC** Where does sea-floor spreading occur?
   A. At subduction zones
   B. At convergent plate margins
   C. Along ocean ridges
   D. From undersea volcanoes
2. Which plate boundary are ocean ridges associated with?

3. Describe what happens between plates at the following boundaries.
   a. Transform boundary
   b. Divergent boundary
   c. Convergent boundary — subduction
   d. Convergent boundary — collision
4. What is Gondwana?
5. Explain why earthquakes are common in the regions surrounding the Himalayas.

Apply and analyse

6. **SIS** What is the Ring of Fire and why, according to the theory of plate tectonics, does it exist?
7. The illustration represents part of a plate boundary.
   a. Identify the type of boundary present.
   b. Describe the movement of the plates on either side of the plate boundary.
   c. Should this boundary be described as a constructive or a destructive boundary? Explain your answer.

Evaluate and create

8. **SIS** Examine this topographic map of the Indonesian islands and surrounding countries. Focus on the southwestern islands of Sumatra, Java, Bali and Lombok to complete the following.

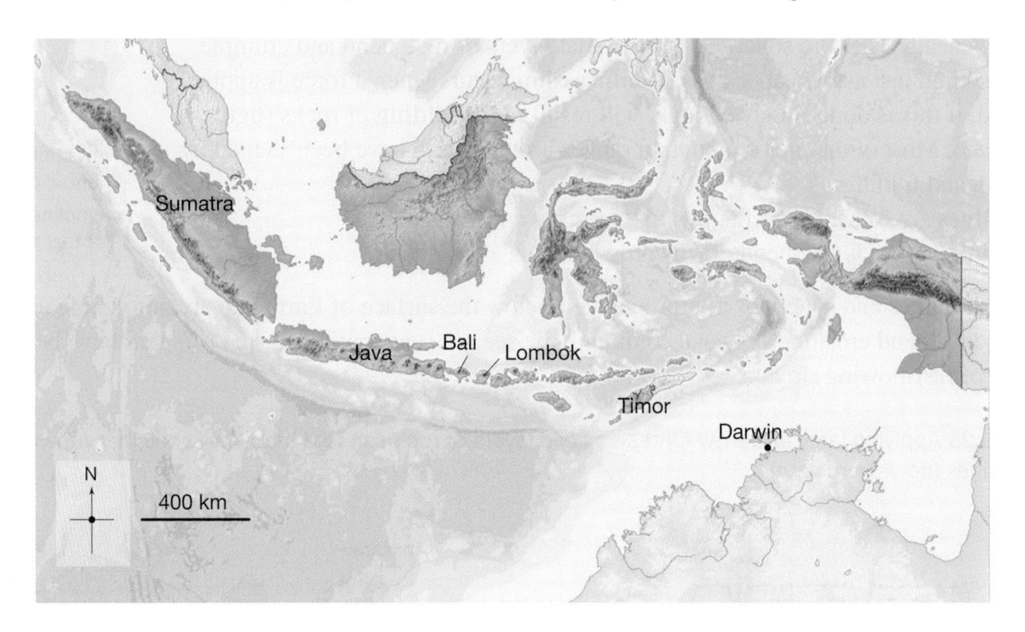

   a. Trace the arc of volcanoes that form on Sumatra, Java, Bali and Lombok.
   b. Trace the deep-ocean trench in that same area.
   c. There have been several earthquakes recently in this region. Identify the type of boundary present that is causing all these earthquakes.
   d. Should this boundary be described as a constructive or a destructive boundary? Explain your answer.
   e. Present a hypothesis about the tectonics east of Lombok (around Timor, north of Darwin).
9. Explain why the climate of most of the Australian continent has changed from cold to hot and dry during the past 65 million years.

Fully worked solutions and sample responses are available in your digital formats.

LESSON
8.5 Folding and faulting

LEARNING INTENTION

At the end of this lesson you will be able to explain why rocks bend into folds or break into faults when put under tectonic force, and you will be able to identify different types of folds and faults resulting from different tectonic forces.

8.5.1 Rocks under pressure

As the plates that make up Earth's crust slowly move, solid rock is pushed, pulled, bent and twisted. The tectonic forces on the rocks are huge — large enough to break them, but also large and slow enough to bend them. The forces are concentrated along the plate boundaries but can extend beyond the boundaries.

FIGURE 8.24 Folded layers of limestone in Greece that were formed by tectonic forces

8.5.2 Bending without breaking

If you hold a sheet of paper with one hand on each end and move the ends toward each other, the paper bends upwards or downwards.

The forces beneath Earth are so large and slow that layers of rock bend and crumple without breaking, just as the paper does. **Compression** occurs when a force is applied to rocks and, if this is done slowly enough, will result in the **folding** of rocks (figures 8.24 and 8.25). Most of the major mountain ranges around Earth have been shaped by compression and folding.

- **Anticlines** are folds that bend upwards, forming an 'A' shape.
- **Synclines** are folds that bend downwards, forming a 'U' shape.

compression a squeezing force
folding refers to when rocks bend into anticlines or synclines
anticline a fold in a rock with the narrow point facing upwards
syncline a fold in a rock with the narrow point facing downwards

Generally, anticlines and synclines are formed well below the surface of Earth and are not visible unless they are exposed by uplift and erosion. They can commonly be seen in road cuttings or in cliffs that have been formed by the erosion of fast-flowing streams.

FIGURE 8.25 Applying large and slow forces on solid layers of rock can fold them into anticlines (upward arch) and synclines (downward arch).

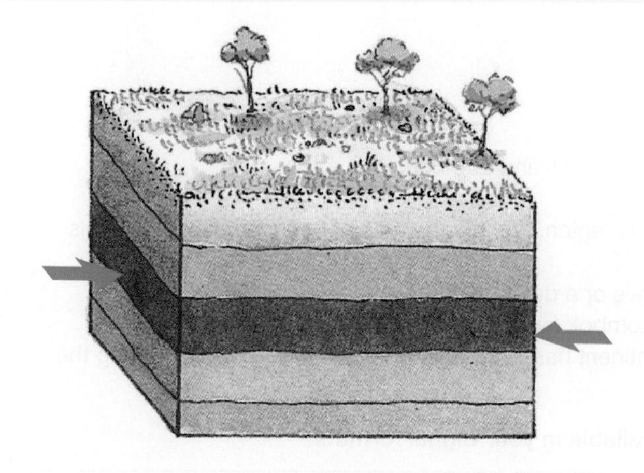

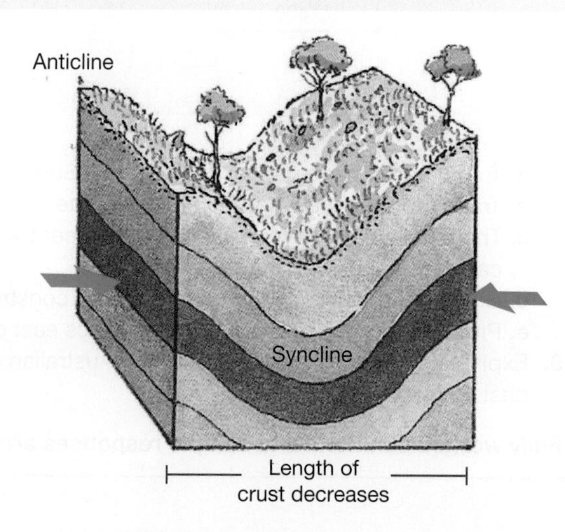

CASE STUDY: Forming the world's largest mountains

Did you know fossils of ancient sea creatures can be found at the top of the Himalayas, thousands of metres above sea level? How did they get there? We now know that the Himalayas are a convergent plate margin between two continental plates — the Indian Plate and the Eurasian Plate. Geologists estimate this collision began around 40 to 50 million years ago. As these plates are both made of continental crust, one plate will not easily slide under the other. Instead, the two are crumpling against each other, forming the mountains. Sediments that once lay at the bottom of the sea between the two landmasses have been forced upward and can be found at the peaks of the mountain range.

The Himalayas are estimated to be rising by more than 1 cm per year as India continues to collide with Asia to the north. This is why the region still experiences shallow earthquakes. However, Earth's gravitational forces, as well as the processes of weathering and erosion, lower the Himalayas at about the same rate. Mountains on Earth can't grow much higher than Mount Everest, which is about 8840 metres above sea level.

FIGURE 8.26 The folding of rocks is important in the creation of the Himalayas, as two parts of Earth's crust collide with each other.

elog-2193

INVESTIGATION 8.3

Modelling folds

Rocks are usually folded well below Earth's surface. The anticlines and synclines can be seen only along road cuttings or where erosion has exposed the layers of rock. A model is a useful way to describe how folded rocks would appear under the surface.

Aim

To model the folding of rocks

Materials

- 3 or 4 pieces of differently coloured plasticine
- ruler
- knife or blade or dental floss
- rolling pin
- board

Method

1. Using the rolling pin, roll the individual colours of plasticine into 0.5–1 cm thick layers.
2. Stack the layers of coloured plasticine on top of each other. Press down lightly on the layers, so that they stick together, but not too much as to cause the plasticine to stick to the table.
3. Measure the length and thickness of your model.
4. With the palms of your hands or books on opposite ends, very gently compress the layers from the side by bringing your hands (or books) closer to each other.
5. Measure the new length and thickness of your model.

Results

1. Describe the appearance of the plasticine when the layers are compressed. Include the measure of length and thickness change.
2. Draw a diagram of the plasticine after compression, labelling anticlines and synclines (don't forget a scale).

Discussion

1. Discuss the relationship between the change in length and the change in thickness. Include a link to building a mountain range.
2. Consider why rocks need to be compressed slowly (or gently) to form folds.

Conclusion

Summarise the findings from this investigation about modelling folds.

Extension

Imagine that the rock layers are eroded at Earth's surface. With the tools provided, model erosion and draw a set of new diagrams of the eroded model, as viewed from above and when viewed from the side.

Where are the oldest and youngest rocks? (Recall the relative age of rock layers with older deposited first.) Is there any relationship between the geometry of anticlines or synclines and the age of rocks?

8.5.3 Breaking under stress

We have learnt that slow compression forces produce folds. But if rock layers break, rather than bend, they produce a **fault**. A number of different forces can produce a variety of faults. These are:

- compression — the force of pushing something together
- **tension** — the force of pulling something apart
- **shearing** — the force of smearing or moving something along the side of something else.

These different forces result in different kinds of movement, which produce different faults: **reverse**, **normal** and **strike-slip faults**. Reverse and normal faults move rock vertically, whereas strike-slip faults move rock horizontally. Sometimes there is both horizontal and vertical movement, which produces more complicated faults.

In all cases, the movement happens very rapidly and produces earthquakes.

fault a break in the crust where one side moves relative to the other

tension a stretching force

shearing a smearing force

reverse fault a break where the rock above the fault moves 'up' due to compression

normal fault a break where the rock above the fault moves 'down' due to tension

strike-slip fault a break where the rocks on either side of the fault move horizontally due to shearing

FIGURE 8.27 Compression, tension and shear are different forces. Each produce different types of faults.

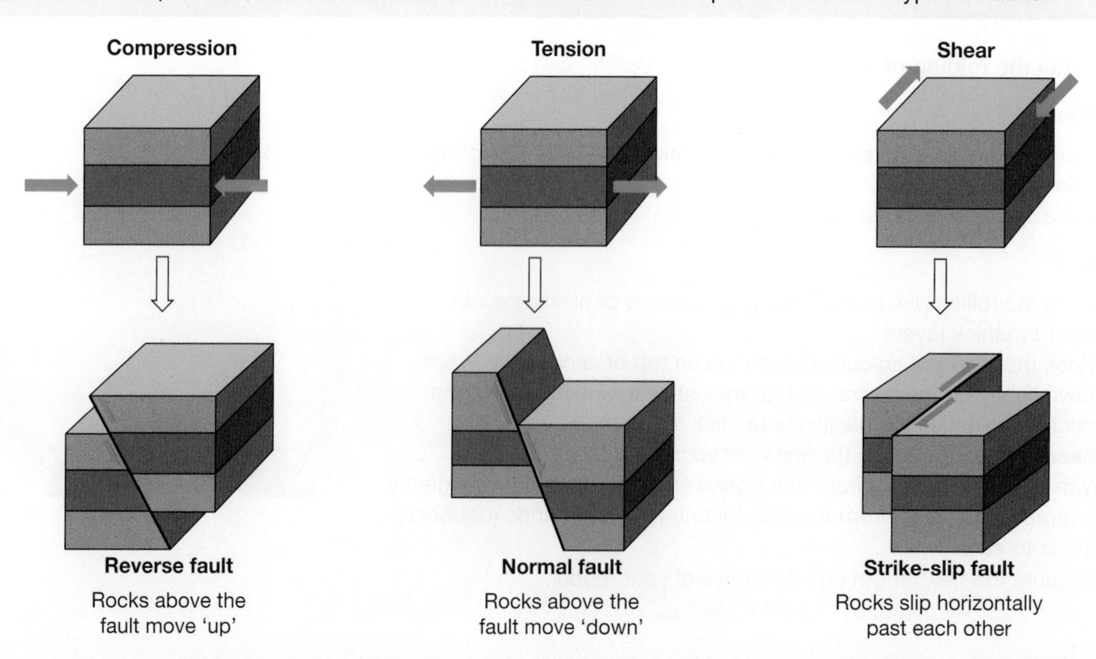

CASE STUDY: Forming valleys and mountains in South Australia

FIGURE 8.28 Faulting has shaped the Gulf region of South Australia.

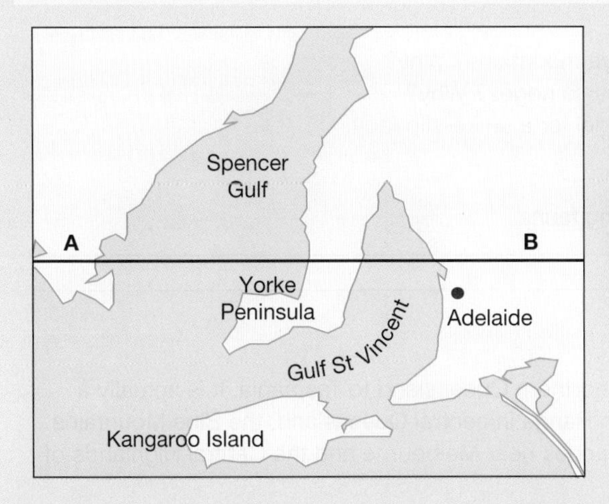

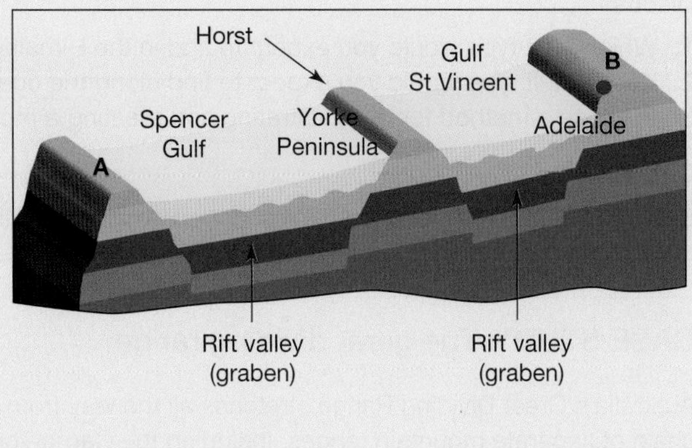

The Gulf region of South Australia has been shaped by a series of normal faults. Two blocks of crust have dropped down between faults to form Spencer Gulf and Gulf St Vincent. These sunken blocks are called **rift valleys** or grabens. Between them is a block that is kept at a higher elevation than the rift valleys. This block, called a **horst**, has formed the Yorke Peninsula.

What type of force do you think causes horsts and grabens to form?

Occasionally earthquakes are felt in the Adelaide area from movement along these faults, but the movement has changed from when the current landscape was formed. Recent earthquakes are the result of compression, which has changed the normal faults into reverse faults!

rift valley a sunken lowland between two normal faults; a graben

horst a highland between two normal faults

elog-2195

INVESTIGATION 8.4

Modelling faults

Aim

To model normal and reverse faults

Materials

- 3 or 4 pieces of differently coloured plasticine
- a thin sheet of polystyrene
- knife or blade

Method

1. Place the first piece of plasticine on the bench and flatten it into a rectangular shape. Do not make it too thin. Cut a piece of polystyrene the same size and fit it over the plasticine rectangle.
2. Add two or three more layers of plasticine with a layer of polystyrene between each layer.
3. Cut through the layers at an angle as shown in the diagram. Use the two parts to model each of the two types of faults shown.

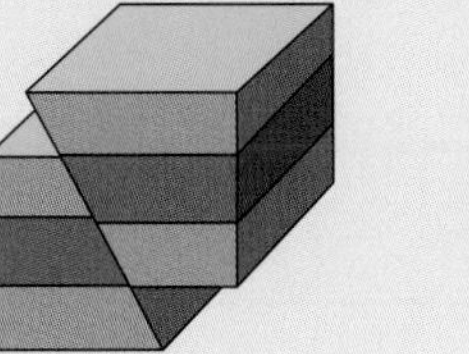

Results

Photograph or draw a diagram of each fault. Label it with arrows to show the direction in which each block moved to create the fault.

Discuss

1. Which fault type would you expect to find in the Himalaya mountains? Why?
2. Which fault type would you expect to find along the oceanic ridges? Why?
3. Propose a method for demonstrating and creating a model for a strike-slip fault.

Conclusion

Summarise the findings for this investigation about modelling faults.

CASE STUDY: The great dividing range

Australia's Great Dividing Range stretches all the way from northern Queensland to Tasmania. It is actually a chain of separate mountain ranges, including the Carnarvon Range in central Queensland, the Blue Mountains of New South Wales, the Australian Alps, the Dandenong Ranges near Melbourne and the Central Highlands of Tasmania.

About 80 million years ago, the Tasman Sea between Australia and New Zealand began to open by sea-floor spreading as Gondwana was splitting apart. The western edge of this rift basin was uplifted to form the Great Dividing Range. Several volcanoes were active along it, erupting large volumes of material that included diamonds and sapphires!

FIGURE 8.29 Australia's Great Dividing Range is the longest mountain range in Australia.

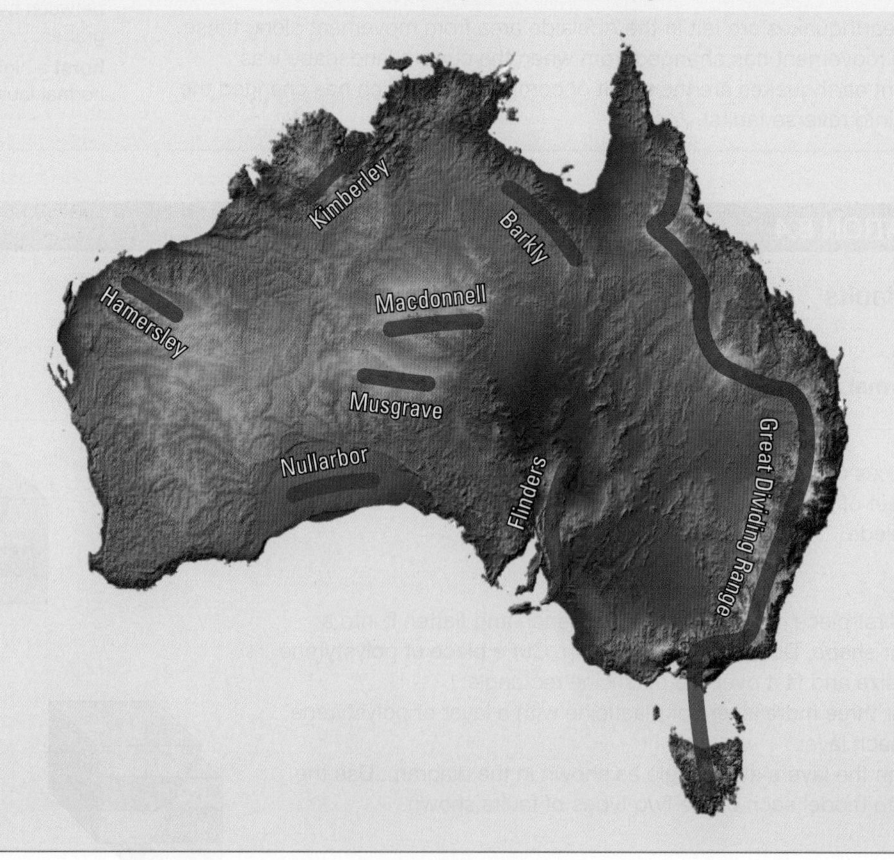

Resources

eWorkbook Folding and faulting (ewbk-11978)

8.5 Activities

8.5 Quick quiz on	8.5 Exercise

These questions are even better in jacPLUS!
- Receive immediate feedback
- Access sample responses
- Track results and progress

Find all this and MORE in jacPLUS

Select your pathway

LEVEL 1	LEVEL 2	LEVEL 3
1, 2, 4, 6	3, 5, 7, 10	8, 9, 11

Remember and understand

1. Why do rocks bend or break?
2. MC When referring to layers of rock, what is *folding*?
   A. The bending and crumpling of rock without breaking
   B. The breaking and crumpling of rock
   C. The uplifting of super-cooled magma to create rock
   D. The uplifting of rock along a fault
3. What is the cause of folding?
4. Explain the difference between a syncline and an anticline.
5. What are the three different types of forces responsible for developing different fault movements?

Apply and analyse

6. a. Explain the difference between a reverse fault and a normal fault.
   b. Sketch a reverse fault and a normal fault.
7. What causes earthquakes along the San Andreas Fault?
8. SIS There is a lot of faulting as well as folding in the Himalayas. Explain how it is possible for both folding and reverse faulting to develop during mountain building.
9. When the Tasman Sea started forming around 80 million years ago, the Great Dividing Range experienced uplift and faulting. What type of faults would you predict dominated this event?

Evaluate and create

10. SIS Explain, with the aid of some labelled diagrams, how mountains could be formed by faulting.
11. SIS Why don't mountains grow forever? Use Mount Everest as an example.

Fully worked solutions and sample responses are available in your digital formats.

LESSON
8.6 Earthquakes

LEARNING INTENTION

At the end of this lesson you will be able to describe how and where earthquakes form, how they are measured, their potential hazards and their relationship to plate tectonics.

8.6.1 Shake, rattle and roll

Earthquakes result from movements on faults in Earth's crust. Rocks can only stretch or bend so far before they 'snap', like stretching a rubber band too far. When the rocks 'snap' and move, we get episodes of ground shaking, where the vibrations travel outwards in all directions. Fortunately, most of the vibrations are too weak to be felt. These are called **tremors**. However, when they are strong enough to be felt, they are called **earthquakes**.

FIGURE 8.30 The focus is where the earthquake begins. The epicentre is the surface directly above the focus.

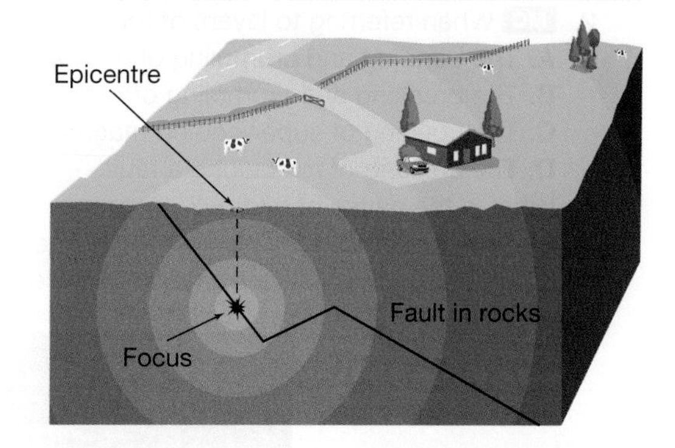

The point at which the earthquake begins is called the **focus**. The **epicentre** of an earthquake is the point on the surface directly above the focus. The epicentre generally experiences the largest vibrations on the surface. If the focus is deeper in the crust, the epicentre will experience less vibration.

tremors minor vibrations of the ground that are commonly not felt

earthquake a sudden and violent shaking of the ground

focus the location underground of the fault movement causing an earthquake

epicentre the surface point directly above the earthquake focus

Tremors and minor earthquakes can take place wherever there is a fault or weakness in Earth's crust. Major earthquakes generally occur at or near the plate boundaries where plates are:

- pushing against each other in subduction zones or collisions
- spreading apart at rift valleys or ocean ridges
- slipping and sliding against each other.

SCIENCE AS A HUMAN ENDEAVOUR: Locating earthquakes

Scientists record earthquake vibrations at seismic stations. The records help to identify how far away the earthquake took place. Records from a single station cannot tell what direction the earthquake came from, because the vibrations travel outwards in all directions. Scientists use a method known as triangulation to determine the position of the epicentre. Triangulation uses, as the name suggests, at least three points to locate the epicentre, as shown in figure 8.31.

FIGURE 8.31 Locating the epicentre using triangulation

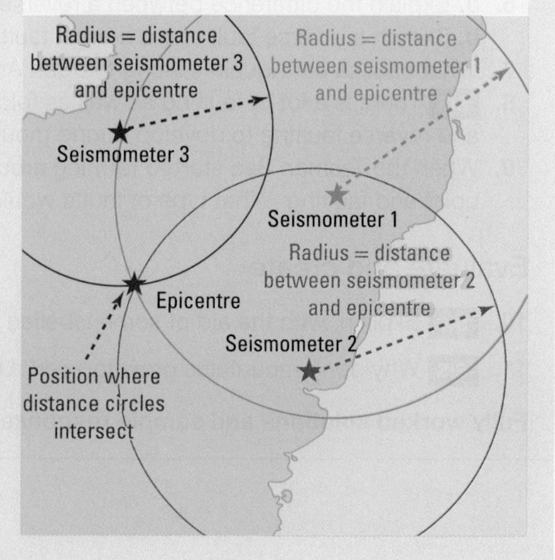

8.6.2 Seismic waves

Energy released during an earthquake travels in the form of waves. There are two basic groups of waves that are generated by earthquakes: body waves and surface waves.

- **Body waves** radiate outward and travel through the interior of Earth.
- **Surface waves** tend to travel only along Earth's surface.

Body waves

Body waves have two kinds of motions:

- **P-waves or primary waves** are compression waves (push-and-pull motion) that move through Earth in the same way that sound waves move through air. They are the fastest of the seismic waves. They can travel through all of Earth's interior layers.
- **S-waves or secondary waves** are the second set of waves, which travel in the form of transverse waves (up-and-down motion). They are second because they are slower than P-waves. S-waves cannot travel through fluids. Because of this characteristic, the outer core was determined to be liquid, because no S-waves have been observed to travel through it.

body waves seismic waves that quickly travel through the interior of Earth

surface waves seismic waves that travel slower than body waves and only along the surface of Earth; their energy is lost with depth and distance

P-waves or primary waves body seismic waves with a compressional (push-and-pull) motion; they are the fastest and first to arrive

S-waves or secondary waves body seismic waves with a transverse (up-and-down) motion; they are slower than P-waves and cannot travel through fluids

Love waves surface seismic waves that have a side-to-side motion

Rayleigh waves surface seismic waves that have a rolling motion

FIGURE 8.32 P-waves travel through Earth as compression waves, while S-waves are transverse waves.

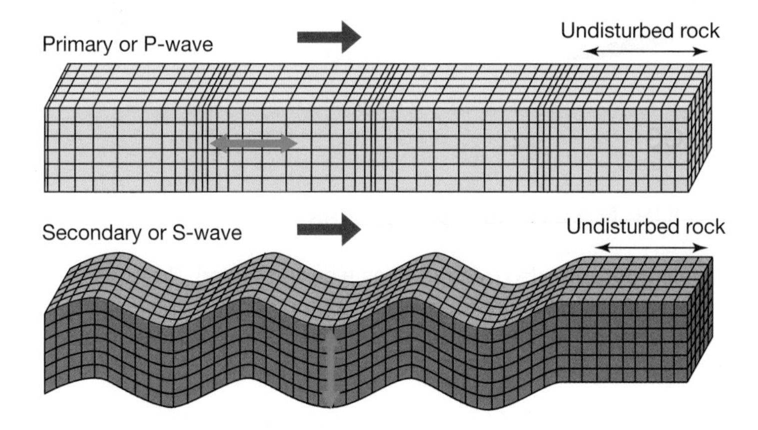

Surface waves

The surface waves are the *slowest* seismic waves, and lose energy with depth and distance travelled. The motions can be complicated, but two types of motions identified are:

- **Love waves** (or L-waves) — waves that move in a side-to-side motion, like a snake
- **Rayleigh waves** — waves that move with a rolling motion, like an ocean wave.

These surface waves are responsible for most of an earthquake's destructive power. This is because all of the wave energy is distributed across the surface of Earth, rather than being spread out through Earth's interior like P- and S-waves.

FIGURE 8.33 L-waves travel along the surface with a side-to-side motion, while Rayleigh waves have a rolling motion.

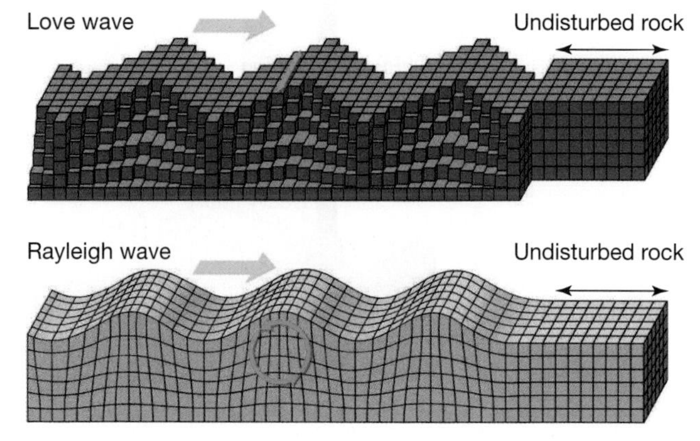

FIGURE 8.34 P-waves are able to travel through all of Earth's interior; S-waves cannot travel through liquid, and are thus not observed in the outer core; L-waves and Rayleigh waves will only travel along Earth's surface.

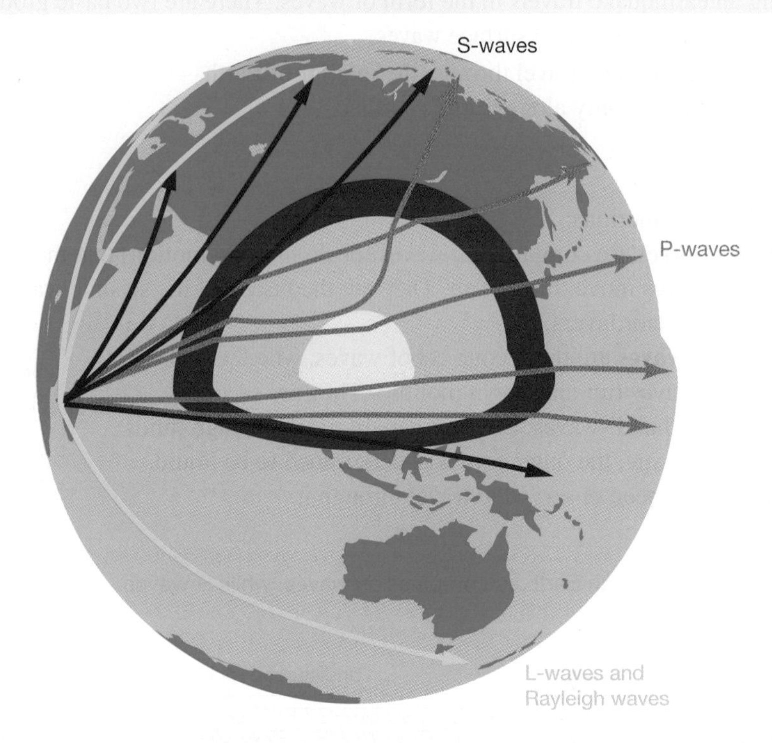

8.6.3 Measuring earthquakes

SCIENCE AS A HUMAN ENDEAVOUR: Measuring earthquakes

Movements in Earth are recorded with a **seismograph** (also referred to as a seismometer). Previously, earthquakes were measured using a device as shown in figure 8.35, where a strip of paper moves past a stationary pen. The record of an earthquake is shown in figure 8.36. Today, earthquakes are measured using electrical currents to create digital graphs.

The strength of an earthquake can be measured in a number of ways, the most well-known of which is the **Richter scale**. As the different types of seismic waves each travel at different speeds, they are recorded as separate groups on a seismograph. The further apart they are, the further away the seismograph is from the epicentre. Also, the waves record as shorter peaks the further away the seismograph is from the epicentre.

FIGURE 8.35 An earthquake recorded on a seismograph

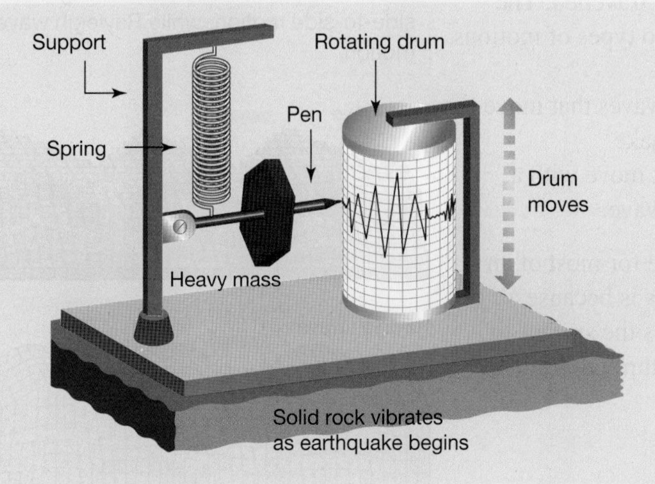

seismograph an instrument used to detect and measure the intensity of an earthquake; also called a seismometer

Richter scale a logarithmic scale that measures the amount of energy released during an earthquake, thus allowing one earthquake to easily be compared to another

FIGURE 8.36 Different types of waves travel at different speeds, which allows scientists to tell how far away the earthquake is.

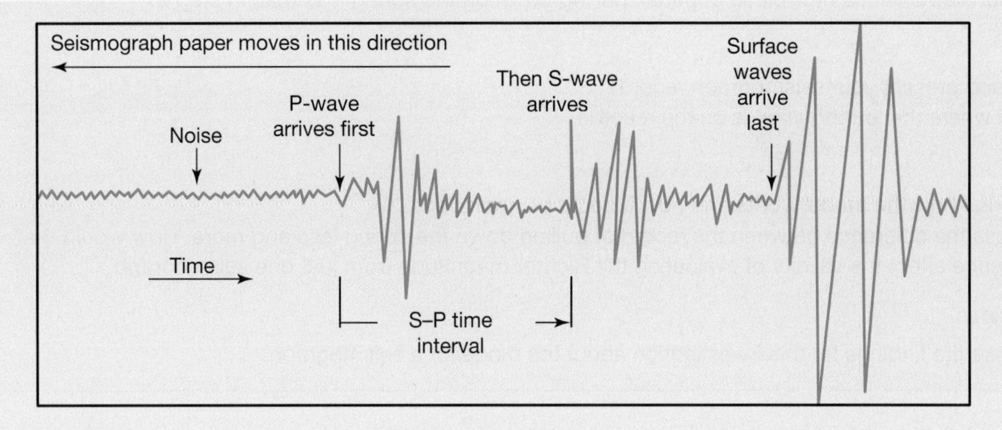

The Richter scale

The Richter scale is a measure of the amount of energy released by an earthquake, and is used to calculate the magnitude (or size) of the earthquake. Scientists determine the magnitude of the earthquake from the amplitude (height) of the surface waves recorded by seismographs.

The Richter scale ranges from 0 to 10, with each increase of 1.0 on the scale representing a *30-fold increase in the amount of energy released*. So, an earthquake of magnitude 6.0 releases 30 times as much energy as one of magnitude 5.0, and an earthquake of magnitude 8.0 releases 900 (30 × 30) times as much energy as one of magnitude 5.0. How much larger is a magnitude 8.0 relative to a 5.0?

Microquakes measure less than 2.0 on the Richter scale and are rarely felt. Earthquakes of magnitude 4.0 on the Richter scale are felt and may even cause objects on shelves or in cupboards to rattle. The largest recorded earthquake was in southern Chile, May 1960, with a magnitude of 9.5.

The Richter scale is not always a good indication of the destructive power of an earthquake. In a crowded city, small earthquakes can cause many deaths, injuries and a great deal of damage, including cutting off water, gas and electricity supplies. Larger earthquakes in remote areas cause few injuries and little damage.

elog-2197

tlvd-10791

INVESTIGATION 8.5

Making a seismograph

Aim

To construct a working model of a seismograph

Materials

- retort stand, bosshead and rod
- spring
- cardboard
- 500 g or 1 kg weight (or a can full of sand)
- sticky tape
- felt pen
- A4 paper

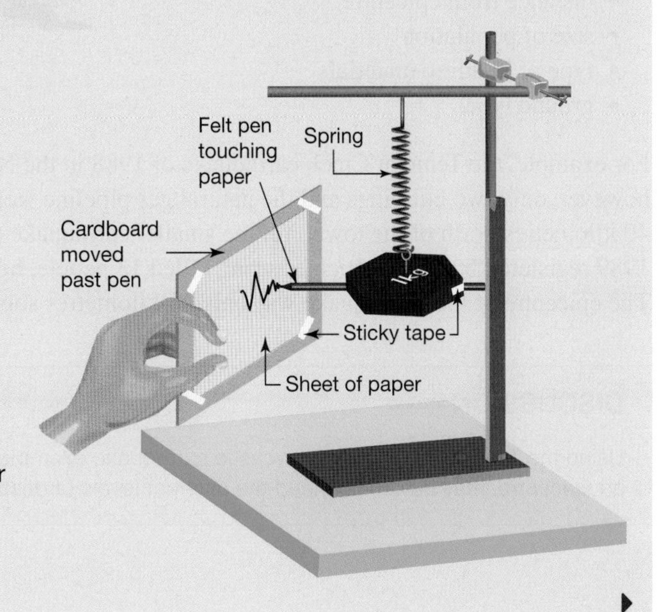

Method

1. Set up the equipment as shown in the diagram. *Note:* a cardboard guide can be made to sit either side of the pen so there is no sideway motion.

2. Pull down on the spring, then release it. Have your partner slide the cardboard past the pen (keeping the board in contact with the table the entire time).
3. Repeat step 2 with a new piece of paper, pulling down on the spring less than in step 2.

Results

1. Title and present your seismograph records.
2. Label where the 'earthquake' is on the records.

Discuss

1. Describe how the model works and how it could be improved.
2. Discuss the difference between the record of pulling down the spring less and more. How would that difference affect the validity of evaluating the Richter magnitude from just one seismograph?

Conclusion

Summarise the findings for this investigation about the model of a seismograph.

8.6.4 Destructive power

Australia does experience earthquakes, even though it is not on the edge of a tectonic plate boundary. They are called shallow intraplate earthquakes. They are caused mostly by compression coming from the northern and eastern convergent boundaries of the Indo-Australian Plate on which Australia sits. This compression can build up within the interior of the plate and be released as earthquakes. There are, on average, 100 earthquakes of magnitude 3.0 or more in Australia each year.

FIGURE 8.37 In February 2011, a magnitude 6.3 earthquake struck Christchurch, New Zealand. The earthquake destroyed many buildings and homes, injured thousands, and killed 185 people.

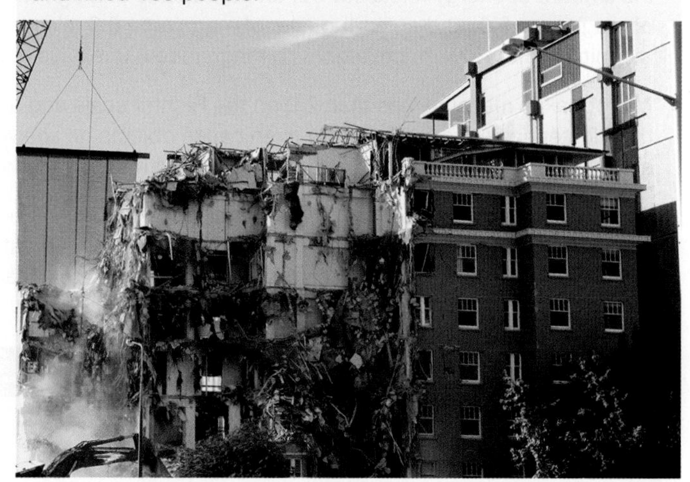

Ground shaking or fracturing by surface waves can cause destruction. The destructive power of an earthquake in any location depends on factors such as:

- earthquake magnitude
- distance from epicentre
- size of population
- type of building materials
- ground type.

For example, the Tennant Creek earthquake of 1988 in the Northern Territory had a Richter magnitude of 6.6; however, only two buildings and the natural gas pipeline were damaged. The epicentre of the earthquake was 40 kilometres north of the town. Yet the smaller earthquake that devastated Newcastle in New South Wales in 1989 registered 5.6 on the Richter scale, killed 13 people, hospitalised 160 others and demolished 300 buildings. The epicentre of that earthquake was only 15 kilometres southwest of the CBD.

DISCUSSION

Using the Tennant Creek and Newcastle earthquake examples, discuss what other factors, besides distance from an epicentre, may have influenced the different levels of destruction.

CASE STUDY: Waves of destruction

Earthquakes occurring under the water or near the coast can cause giant waves called **tsunamis**. To form a tsunami, movement on a fault underwater must push up the sea bed, which also lifts up the water above it. This causes huge water waves to form, which travel through the ocean at speeds of up to 900 kilometres per hour. When the waves approach land the water gets shallower. This causes the waves to slow down and build to heights of up to 30 metres.

FIGURE 8.38 Illustration of how a tsunami forms

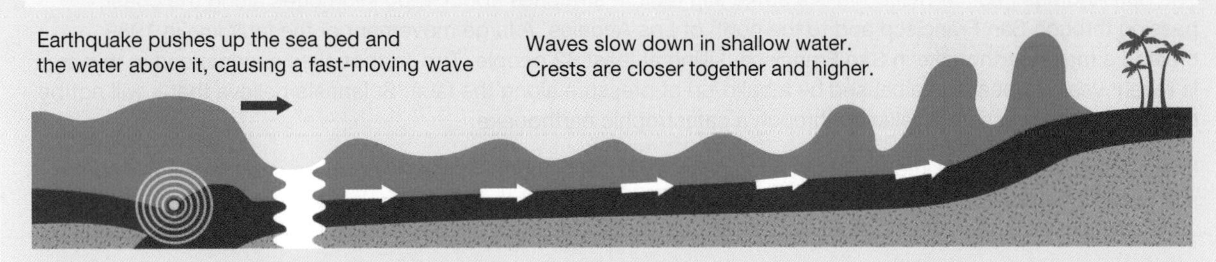

The destructive power of tsunamis became very clear on 26 December 2004, when about 300 000 people died across South-East Asia, southern Asia and eastern Africa. Millions more lost their homes. The tsunami, known as the Sumatra–Andaman tsunami, was caused by a huge earthquake under the ocean floor about 250 kilometres off the coast of the Indonesian island of Sumatra. The earthquake measured 9.0 magnitude on the Richter scale. It pushed a 1000-kilometre-long strip of the ocean floor about 30 metres upwards.

The tsunami flooded 10 kilometres inland, near the Sumatran city of Banda Aceh, with a 3-metre-high wall of water, mud and debris. Thousands were killed in Sri Lanka, India and Thailand as well. Death and destruction also occurred in Malaysia, Myanmar, Bangladesh and the Maldives. More than eight hours after the earthquake, the tsunami arrived at the east coast of Africa, more than 5000 kilometres from the epicentre of the earthquake. Even at that distance from the earthquake, the tsunami caused flooding that killed more than 160 people on the coasts of Somalia, Kenya and Tanzania.

FIGURE 8.39 This map shows the huge area affected by the Sumatra–Andaman tsunami on 26 December 2004.

tsunami a powerful ocean wave triggered by an undersea earth movement

on Resources

▶ **Video eLesson** Tsunami wave propagation during the 2004 Sumatra–Andaman tsunami (eles-4151)

DISCUSSION

The San Andreas Fault in California is a strike-slip fault. It stretches about 1200 kilometres along the coast, passing through San Francisco and to the north of Los Angeles. A large movement of the fault line in 1989 created a major earthquake in San Francisco, killing at least 62 people. The earthquakes experienced in this area in recent years appear to be caused by a build-up of pressure along the fault. Scientists believe that it will not be long before the pressure is relieved through a catastrophic earthquake.

Imagine that you were offered the chance to spend a year at a school in a leafy northern suburb of Los Angeles, just two kilometres from the San Andreas Fault. Would you accept the offer? Explain your response.

SCIENCE AS A HUMAN ENDEAVOUR: Living on the edge

For the people living near the plate boundaries, particularly on the edges of the Pacific Ocean, the ability of scientists to predict earthquakes and tsunamis is critical. The scientists who study earthquakes are called **seismologists**.

Although it is difficult to predict the time, location and size of earthquakes, seismologists use:

- patterns of past earthquake events to identify the probability of earthquakes of different sizes
- sensors to monitor movement and pressure build-up along plate boundaries and fault lines.

Early warning systems

Tsunami early warning systems rely on the early detection of earthquakes and a system of buoys placed around the Pacific and Atlantic Oceans. This system is called DART (Deep-ocean Assessment and Reporting of Tsunamis). Sudden rises in sea level are detected by the buoys and alerts are sent to tsunami warning centres.

seismologist a scientist who studies earthquakes to both understand how they work and how to better predict them

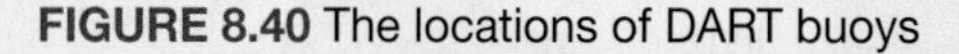

FIGURE 8.40 The locations of DART buoys

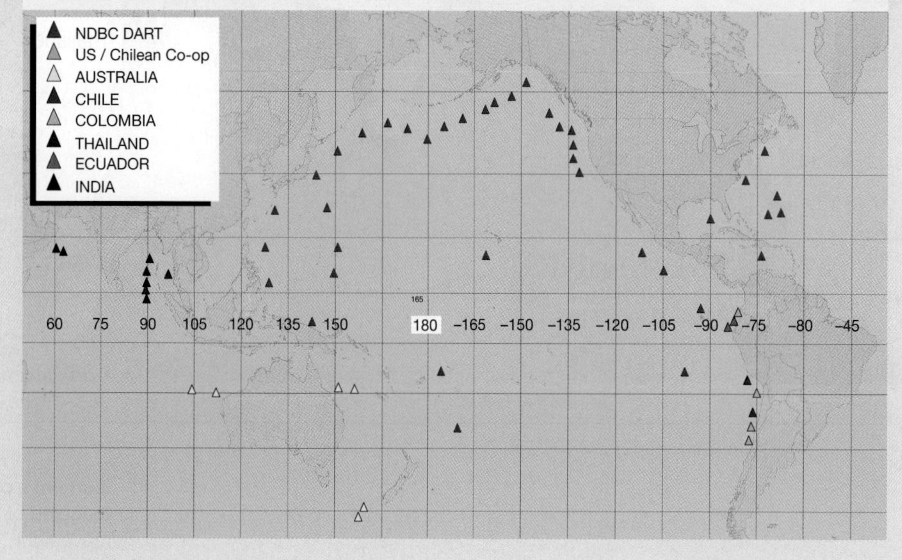

SCIENCE AS A HUMAN ENDEAVOUR: Sharing seismic data to issue alerts

Geoscience Australia, a Commonwealth Government agency, monitors seismic data from more than 60 stations on the Australian National Seismograph Network and over 300 stations worldwide 24 hours a day, seven days a week. Within 30 seconds of being recorded at a seismometer, most data arrive at Geoscience Australia's central processing facility in Canberra through various digital satellite and broadband communication systems.

Overseas governments that have national seismic networks also provide data. Geoscience Australia uses data provided by the governments of New Zealand, Indonesia, Malaysia, Singapore and China, and has access to data from global seismic networks provided by the USA, Japan, Germany and France. The Comprehensive Nuclear-Test-Ban Treaty Organization's International Monitoring System also provides seismic data for tsunami warning purposes.

The seismic data are collected and analysed automatically and immediately reviewed by Geoscience Australia's Duty Seismologist. As part of the Joint Australian Tsunami Warning Centre, Duty Seismologists must also analyse and report within 10 minutes of the origin time on earthquakes that have the potential to generate a tsunami. An earthquake alert is then sent to the Australian Bureau of Meteorology, to determine tsunami advice and publish tsunami bulletins.

The parameters of all other earthquakes with a magnitude greater than 3.5 are generally computed within 20 minutes. The analysis includes its magnitude, origin time and date, and the location of its hypocentre (or focus).

FIGURE 8.41 Geoscience Australia's National Earthquake Alerts Centre

Source: © Commonwealth of Australia (Geoscience Australia) 2021

CASE STUDY: The 2011 Japanese earthquake and nuclear meltdown

The world was reminded of the destructive power of earthquakes and tsunamis in March 2011 when an earthquake struck Japan that was of the same magnitude as the 2004 Sumatra–Andaman earthquake. The epicentre of this earthquake was only 70 kilometres off the coast of the Japanese island of Honshu.

Residents of Tokyo received a one *minute warning* before the strong shaking hit the city. The alerts were received over television but also by text message to mobile phones. The early warning, even though only a minute, prevented many deaths from the earthquake by stopping high-speed trains and factory assembly lines.

The nearest major city to the epicentre was Sendai, where the port and airport were almost totally destroyed by a tsunami produced by the earthquake. In that city, at least 670 people were killed and about 2200 were injured. Around 6900 houses were destroyed, with many more partially destroyed. Waves of up to 40 metres in height were recorded on the coast and some caused damage as far as 10 kilometres inland.

Several nuclear reactors were shut down immediately following the earthquake that caused the tsunami. However, that wasn't enough to prevent meltdowns in three reactors at the Fukushima Daiichi Power Plant, resulting in explosions and the leakage of radiation into the atmosphere, water and soil.

FIGURE 8.42 Destruction left behind after the 2011 Japan earthquake and tsunami

DISCUSSION

As a class, consider which other countries the 2011 Japan earthquake and tsunami may have affected. Remember to consider the effects of damaged nuclear power stations.

on Resources

eWorkbooks Earthquakes (ewbk-11980)
Plotting earthquakes activity (ewbk-11982)

8.6 Activities

learnon

8.6 Quick quiz on | **8.6 Exercise**

Select your pathway

■ LEVEL 1	■ LEVEL 2	■ LEVEL 3
1, 2, 3, 4, 7	5, 6, 8, 9	10, 11, 12

These questions are even better in jacPLUS!

- Receive immediate feedback
- Access sample responses
- Track results and progress

Find all this and MORE in jacPLUS

Remember and understand

1. What causes earthquakes?
2. Distinguish between an Earth tremor and an earthquake.
3. What name is given to the point at which an earthquake begins?
4. Where is the epicentre of an earthquake relative to the focus?
5. What does the Richter scale measure?

Apply and analyse

6. **SIS** A seismograph record is shown. Use this image to complete the following.
   a. Label the P- and S-waves.
   b. Label the surface waves.
   c. How would the record look if the seismograph was further away from the epicentre?

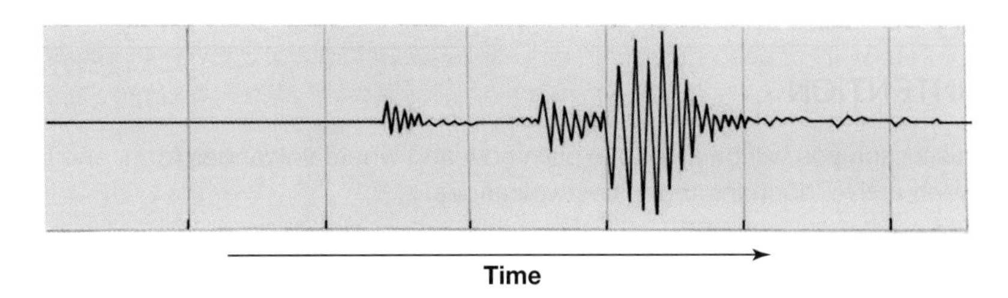

7. Explain how seismologists are able to make predictions about the likelihood of an earthquake.
8. Explain why a tsunami only a few metres high in open ocean can reach heights of up to 30 metres by the time it reaches land.
9. **SIS** The table shows the number of people killed in some of the major earthquakes in recent years.

TABLE Earthquake year, location, fatalaties and magnitude

Year	Location	Number of deaths (approx.)	Richter scale magnitude
1994	Los Angeles, USA	57	6.6
1995	Kobe, Japan	6400	8.2
1999	Iznit, Turkey	17 000	8.4
2001	Gujarat, India	20 000	8.9
2003	Bam, Iran	26 000	6.6
2004	Sumatra, Indonesia	230 000	9.0
2008	East Sichuan, China	90 000	8.9
2010	Haiti (Caribbean Sea)	316 000	8.0
2011	Sendai, Japan	21 000	9.0
2015	Nepal	8964	8.8
2021	Haiti	2250	8.2

   a. List a pair of earthquakes that provide evidence that the Richter scale does not indicate the loss of life in earthquakes.
   b. What factors, apart from the magnitude, affect the number of deaths in an earthquake?
   c. How much more energy was released by the 2004 Sumatra earthquake than the 2010 Haiti earthquake?
   d. Suggest why there may have been more fatalities during the Haiti earthquake.

Evaluate and create

10. **SIS** You are requested to measure the magnitude and location of an earthquake. How would you go about each?
11. Explain why Indonesia is more likely to experience major earthquakes compared to Australia.
12. **SIS** Earthquakes are mostly generated at depths of 5 to 20 kilometres, where rocks are relatively cool and easier to break; however, they have been measured at depths of up to 670 kilometres.
   a. Which plate boundary would you associate with the deepest earthquakes? Explain your reasoning.
   b. Would you predict these deeper earthquakes to be as destructive as the shallow earthquakes?

Fully worked solutions and sample responses are available in your digital formats.

LESSON
8.7 Volcanoes

First Nations Australian readers are advised that this lesson and relevant resources may contain images of and references to people who have died.

LEARNING INTENTION

At the end of this lesson you will be able to explain how and where volcanoes form, and be able to identify the difference between active, dormant and extinct volcanoes.

8.7.1 Mountains of fire

Although most changes in Earth's crust are slow and not readily observable, the eruption of volcanoes provides evidence that the changes can also be explosive, fiery and spectacular.

Volcanoes are formed when molten rock from below Earth's surface, called magma, bursts through a weakness in Earth's crust. The eruption of a volcano ejects the magma as red-hot **lava** (molten rock flowing on the surface), ash and gas. Most visualise a volcanic eruption as highly explosive with tall clouds of ash. However, a volcanic eruption can vary wildly from highly explosive to calm. A scientist who studies volcanoes is called a volcanologist.

WHAT DOES IT MEAN?

The word 'volcano' comes from the name of the ancient Roman god, Vulcan, who was the god of fire.

What comes out?

Commonly deep below a volcano is a magma chamber. When the pressure builds up in the magma chamber, steam is the first to emerge from the vents of a volcano. When the volcano erupts, lava flows from the vents, and red-hot fragments of rock, dust and ash, steam and other gases shoot out of the crater. The larger fragments of rock blown out of the crater are called **volcanic bombs** as seen in figure 8.43. The gases include carbon monoxide and hydrogen sulfide ('rotten egg' gas).

FIGURE 8.43 Volcanic bombs from Arintica in Chile

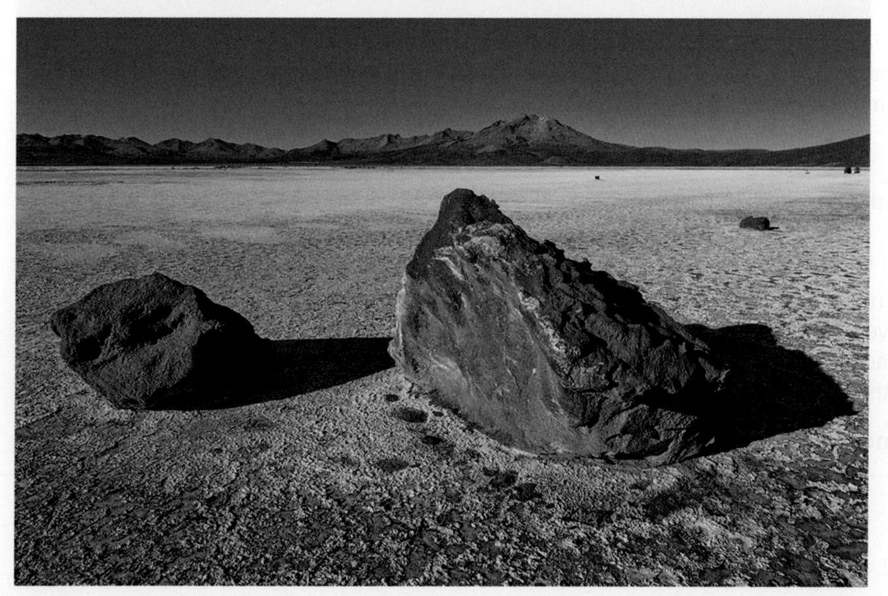

lava an extremely hot liquid or semi-liquid rock from the mantle that reaches and flows or erupts on Earth's surface

volcanic bomb refers to when a large rock fragment that falls from an eruption, formed as lava, is blown out of a volcano and is rapidly cooled in the air

The lava flowing from a volcano can be runny or sticky.

- If it is runny, gas escapes easily, which generally results in a 'calm' or fountaining eruption. It can also flood large areas, cooling to form large basalt plains like those in Victoria's western district as well as in Melbourne, and to the city's north and south.
- If the lava is sticky, it can build up within and on the sides of steep volcanoes, and can also block the vents as it cools. When this happens, gases build up in the magma below. As the pressure increases, the volcano can bulge and 'blow its top', thrusting rocks, gases and hot lava high into the air. Exploding gases often destroy part of the volcano.

FIGURE 8.44 Parts of an erupting steep volcano formed with sticky lava

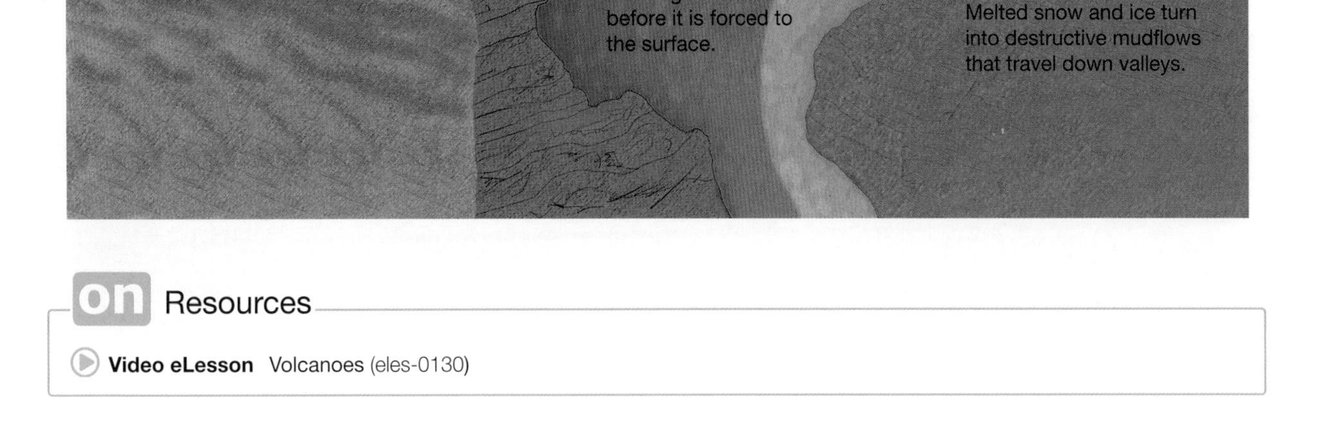

on Resources

Video eLesson Volcanoes (eles-0130)

CASE STUDY: Forming a volcano

On a cool winter's day in 1943, a small crack opened in a field of corn on a quiet, peaceful Mexican farm. When red-hot cinders shot out of the crack, the shocked farmer tried to fill it with dirt. The next day, the crack had opened into a hole over 2 metres in diameter. A week later, the dust, ash and rocks erupting from the hole had formed a cone-shaped mound 150 metres high! Explosions roared through the peaceful countryside and molten lava began spewing from the crater, destroying the village of Paricutin. The eruptions continued and, when the eruptions stopped in 1952, the new mountain named Paricutin was 410 metres high.

This volcano of Paricutin is one of several volcanoes that string down the western side of Mexico. Why are they there? They are a part of the Ring of Fire that circles the Pacific Ocean, and are formed from the subduction of oceanic crust under continental crust.

FIGURE 8.45 The volcano of Paricutin formed in just nine years.

8.7.2 Where do volcanoes occur?

Most volcanic activity on Earth occurs along two types of plate boundaries: mid-ocean ridges and subduction zones. At mid-ocean ridges, basalt lava erupts to produce new oceanic crust. At subduction zones, volcanoes are created on the overriding plate as melt from the subducting plate rises up through the mantle and crust.

A third setting for volcanoes is within plates.

FIGURE 8.46 Volcanism resulting from two different plate boundaries: convergent and divergent plate boundaries. Can you identify a volcano in the diagram that is not related to a tectonic plate boundary?

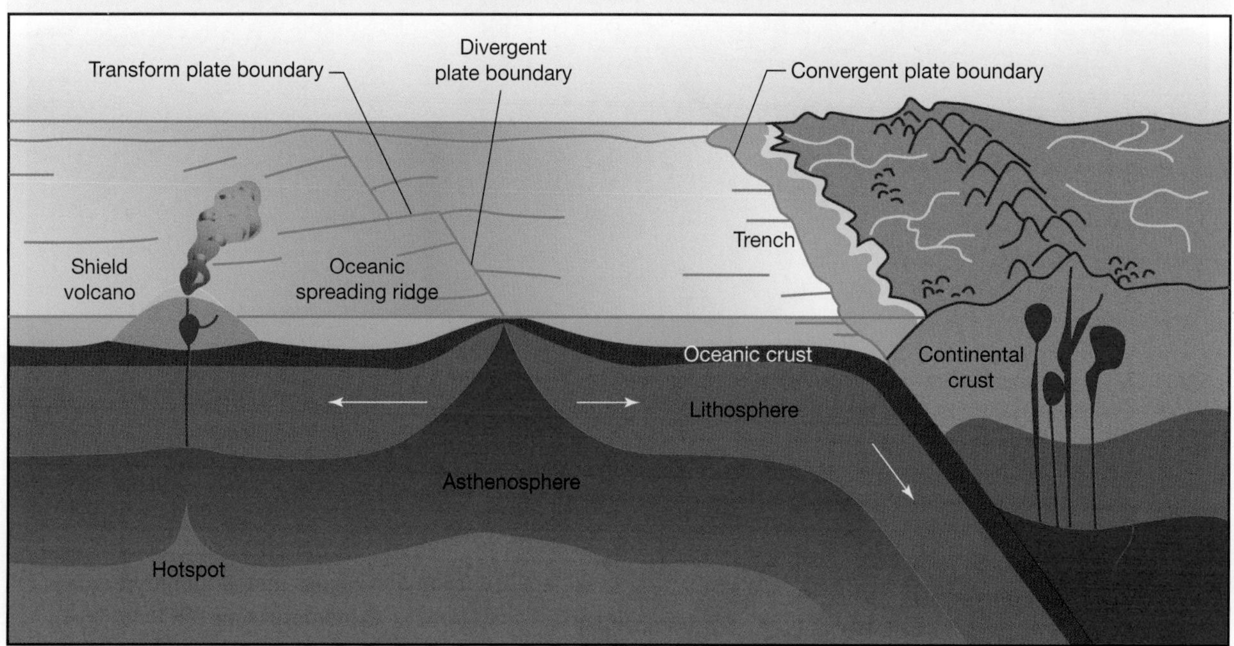

Hotspots

What about the Hawaiian Islands in the middle of the Pacific Ocean? They are clearly volcanic, but don't align with any tectonic boundaries. Although most of the world's volcanoes are found at the edges of the plates of Earth's crust, some lie over features we call **hotspots**. These hotspots are regions of the crust where the mantle below is extremely hot and magma surges upwards into the crust.

Hotspots create chains of volcanic features that are older the further away they are from the active site of the hotspot. Therefore, *the hotspot is believed to be stationary as the crust moves over it.* While the hotspot theory explains relationships in some volcanic chains, it doesn't explain others. Some scientists think that hotspots lie at shallower depths and migrate slowly over time, rather than stay in the same spot. Others don't think they exist. Such is the nature of science and why we continue to explore!

FIGURE 8.47 The formation of a hotspot volcanic chain in the ocean, where the oldest volcanic feature is the furthest away from the hotspot. As the older volcanoes become further away from the source of magma and heat, they erode and sink below the ocean.

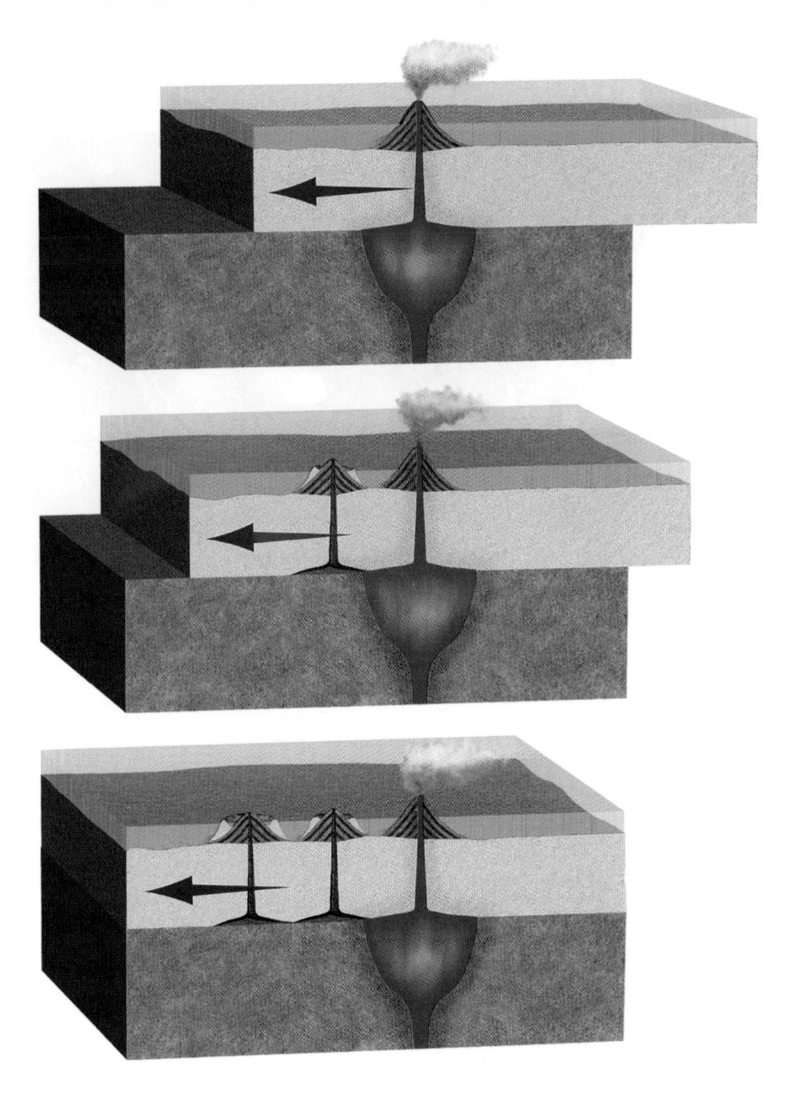

Underwater volcanoes

Active volcanoes also erupt under the sea. In fact, an ocean ridge is a continuous belt of volcanoes with **black smokers**. Black smokers are thermal vents on the sea floor, which eject superheated water rich in elements that were dissolved from the ocean crust. Theories suggest that the beginnings of life could have started around black smokers under the ocean.

hotspot a volcanic region directly above an area of extremely hot mantle

black smoker a geothermal vent on the sea floor that ejects superheated, mineral-rich water

An active volcano below the sea is generally not visible. However, if layers of lava build up, they may eventually emerge from the sea as a volcanic island. Lord Howe Island, off the coast of New South Wales, was formed in this way about 6.5 million years ago. A more recent example is the island of Surtsey, which emerged from the sea off the coast of Iceland in 1963.

CASE STUDY: A violent submarine eruption in Tonga

On 15 January 2022, a powerful explosive eruption from an underwater volcano in Tonga threw volcanic ash and gas high into the stratosphere, and sent atmospheric shockwaves and tsunami waves around the world. Hot magma full of volcanic gas was blasted out of the sea floor, rapidly hitting cold ocean water and causing an explosive eruption.

Thick ash blanketed the nearby islands, and tsunami waves severely damaged coastal communities and undersea communications cables. Tonga airports were shut down for several days due to ashfall. The umbrella cloud of ash reached 500 kilometres in diameter and could be seen by satellite and the International Space Station!

The volcano is part of a chain of volcanic islands that forms part of the Pacific Ring of Fire.

FIGURE 8.48 Underwater volcano eruption, Tonga, 2022

15 Jan 2022 05:00Z NOAA/NESDIS/STAR GOES-West ABI GEOCOLOR

Resources

Weblink Tonga volcano eruption January 2022
Power of Tonga volcano eruption

8.7.3 Active, extinct or dormant

Active volcanoes are erupting or have recently erupted. Mount Pinatubo in the Philippines, which erupted in June 1991, killing 300 people, is an active volcano. There was so much smoke and ash coming from Mount Pinatubo that scientists believe that Earth's weather was cooler for over a year.

Volcanoes that have erupted in the last 10 000 years but are not currently erupting are called **dormant volcanoes**. Dormant means 'asleep', and these volcanoes could 'wake up' at any time and erupt. Mount Pinatubo was a dormant volcano before its eruption in 1991.

Extinct volcanoes are those that have not erupted for at least 10 000 years. They are effectively dead and are very unlikely to erupt again.

active volcano a volcano that is erupting or has recently erupted
dormant volcano a volcano that has erupted in the last 10 000 years but is not currently erupting; they are considered likely to erupt again
extinct volcano a volcano that *has not* erupted in the last 10 000 years; they are considered dead and not to erupt again

DISCUSSION

As a class, or in small groups, list some physical observations you could make about the appearance of a volcano that may help to label it as active, dormant or extinct.

There are many extinct volcanoes in Australia. The Glasshouse Mountains of Queensland are the remains of lava that cooled in volcanic vents. In Victoria, Tower Hill — near Warrnambool — is another example, and there are many others. Mount Gambier in South Australia could be dormant or extinct, as estimates of its last eruption range from over 28 000 to a little less than 5000 years ago. In either case, the craters collapsed and have filled with water to form beautiful clear lakes. Some radiocarbon dating of plant fibres in the main crater indicates the last eruption was a little before 6000 years ago, suggesting more of a dormant status.

FIGURE 8.49 A small crater on the extinct Tower Hill volcano in Victoria

Australia is close to the centre of the Indo-Australian Plate so it is very stable geologically. This is why there are few recent volcanic eruptions and only a small number of substantial earthquakes.

8.7.4 Volcanoes, climate and life

Volcanoes have a significant influence on climate and life on Earth. They produced the warm and acidic environments in the deep ocean where life first evolved. They contributed gases to build our atmosphere and continue to play a vital role in climate change by increasing or decreasing the amount of carbon dioxide in the atmosphere. Volcanic rocks contain mineral deposits that society relies upon to support modern lifestyle and technology. They also provide mineral nutrients that lifeforms need.

The long-term effects of volcanic eruptions are significant and can be devastating. Destruction of farmland could lead to famine, and poisoning of water sources could lead to drought. Poor air conditions could result in widespread health problems. Volcanoes have historically been known to wipe out entire civilisations, much like Pompeii, which fell victim to the eruption of Mount Vesuvius in 79 AD.

Volcanic activity is also thought to be an important cause of several mass extinctions. For example, the vast eruptions associated with the end-Permian extinction about 252 million years ago left behind lava that covers an area the size of Western Europe and is more than a kilometre thick. The lava reacted with rocks and produced toxic gases (a combination of high hydrogen sulfide levels and low oxygen levels) that affected the chemistry of the atmosphere, causing global cooling and then warming. This was Earth's most severe extinction event, wiping out many lifeforms on land and in the sea.

The deep-sea vents, or black smokers, shown in figure 8.50 release mineral-rich fluid at over 300 °C! The minerals react on contact with the cold surrounding water, producing 'smoke'. They provide an unusual habitat for specialised lifeforms, including bacteria. Deep-sea crabs, fish and giant tube worms are visible here, adapted to the high temperature, pressure and sulphur conditions. Some mineral deposits once formed in such environments.

FIGURE 8.50 Black-smoker volcanic vents in the Pacific Ocean

SCIENCE AS A HUMAN ENDEAVOUR: First Nations Australians' knowledge of volcanic activity

First Nations Australians have witnessed volcanic activity over thousands of years. The oral history of the Boandik Peoples of south-eastern South Australia includes a story that suggests their ancestors witnessed volcanic activity in the Mount Gambier area, the site of Australia's most recent eruption around 5000 years ago. Four volcanic craters can be seen there today.

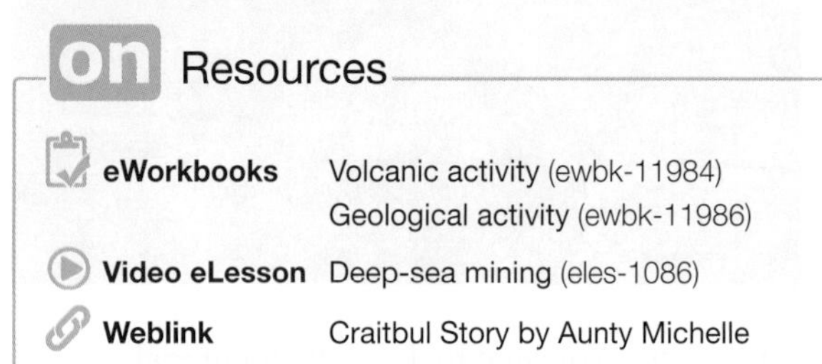

on Resources

eWorkbooks Volcanic activity (ewbk-11984)
Geological activity (ewbk-11986)

Video eLesson Deep-sea mining (eles-1086)

Weblink Craitbul Story by Aunty Michelle

8.7 Activities

learnon

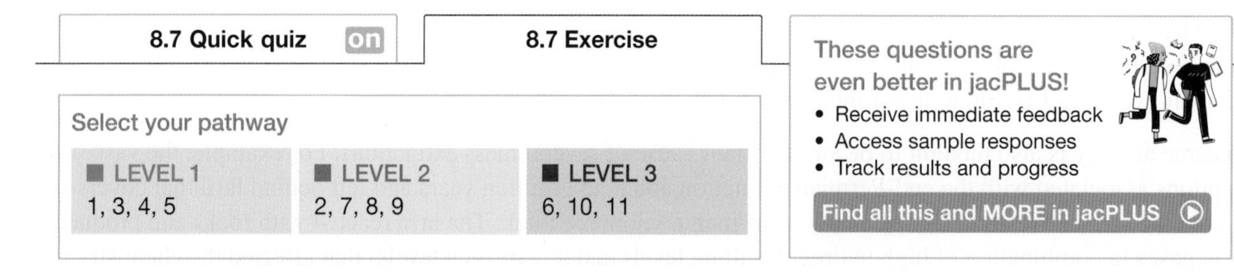

8.7 Quick quiz on | **8.7 Exercise**

Select your pathway

LEVEL 1	LEVEL 2	LEVEL 3
1, 3, 4, 5	2, 7, 8, 9	6, 10, 11

These questions are even better in jacPLUS!
- Receive immediate feedback
- Access sample responses
- Track results and progress

Find all this and MORE in jacPLUS

Remember and understand

1. What can cause a volcano to erupt?
2. List the substances that emerge from a volcanic crater during an eruption.
3. Explain the difference between a dormant volcano and an extinct volcano.

4. What is a hotspot?
5. Explain in terms of the plates that form Earth's crust why Australia experiences little volcanic or earthquake activity.
6. Use a Venn diagram to show the differences and similarities between magma and lava.
7. How do you know that many of the volcanoes in the western district of Victoria had runny lava?

Apply and analyse

8. Explain how a volcano can affect Earth's weather.
9. A photograph of the crater at Mount Gambier is shown. Should Mount Gambier be described as an extinct or dormant volcano? Explain your answer.

10. Explain how the islands of Hawaii were formed.

Evaluate and create

11. **SIS** A volcanologist working for the government is assigned to assess the probability of a volcanic eruption from a local volcanic feature. The volcanologist decides to first map the rocks around the volcano and obtain the ages of these rocks to piece together the eruption history.
    a. Explain why it is important for the geologist to map the rocks and obtain their ages.
    b. Suggest one additional investigation that may improve the final conclusion.

Fully worked solutions and sample responses are available in your digital formats.

LESSON
8.8 Human response to tectonic events

LEARNING INTENTION

At the end of this lesson you will understand the impacts of tectonic events on humans, and engineering solutions to mitigate them.

8.8.1 Natural disasters

Tectonic forces produce natural hazards such as volcanoes, earthquakes and associated tsunamis and avalanches. If these hazards cause significant loss of life or major damage to property or the economy, they are called natural disasters.

For example, on 14 August 2021 a 7.2 magnitude earthquake struck Haiti, which sits on a tectonic plate boundary. It killed at least 2248 people, possibly many more. It is estimated that 1.2 million people were affected and at least 137 500 buildings were damaged or destroyed.

To help manage the impact of such events, geologists strive to understand, monitor and predict geological processes. Engineers also investigate the forces involved, and design monitoring devices, technology to collect data, and equipment to assist in rescues. They design structures such as buildings and bridges to withstand earthquakes, build channels to deflect lava flows from volcanoes, and raise buildings on stilts to limit the damage from high water.

SCIENCE AS A HUMAN ENDEAVOUR: First Nations Australians' knowledge of natural events

First Nations Australians have witnessed many geological changes and extreme weather events over the last 60 000 years. Their oral traditions are highly detailed and enable knowledge from many thousands of years ago to be passed down through generations. These narratives include evidence of geological events such as volcanic eruptions, earthquakes and tsunamis that have shaped Australia's landscape.

Keeping the knowledge intact is done by encoding the information in story, song, dance and art. Details are also kept safe across thousands of years by minimising the number of people who are allowed to be the custodians of any particular knowledge. Some knowledge is only for women, some only for men, and is passed down only when the recipient is ready.

FIGURE 8.51 First Nations Australians' hand stencil art at Carnarvon Gorge, Queensland

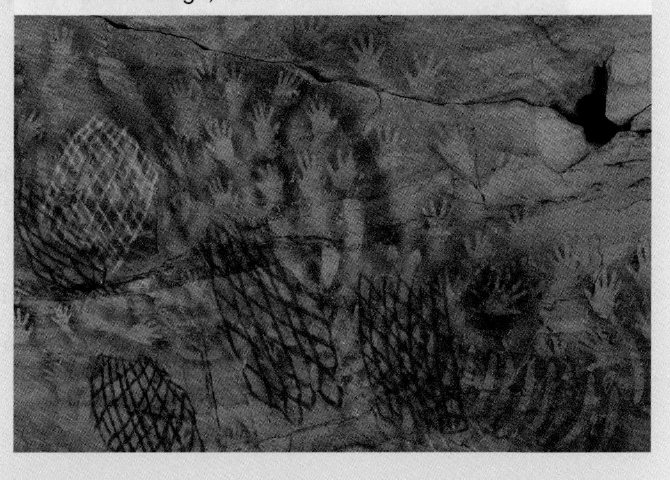

Stories of the Awabakal Peoples of the mid-north coast of New South Wales feature knowledge of earthquake activity. Similarly, tsunami events are preserved in the knowledge of the Gundungarra Peoples of south-eastern New South Wales and the Kambure Peoples of the Kimberley region in Western Australia. Volcanic activity in Queensland over 17 000 years ago features in the knowledge of the Ngadjon-Jii Peoples in Queensland, explaining the origin of the region's volcanic crater lakes, and the environmental and landscape changes that resulted from the volcanic eruption.

The knowledge held by First Nations Australians precedes European investigations, so provides a deeper understanding of historical geological events. Recent scientific research endorses the importance of this historical information.

8.8.2 Monitoring and predicting geohazards

The best way to avoid or limit damage to life and property is prevention, by monitoring and predicting destructive tectonic events (or geohazards). Geological maps and geophysical images show the position of faults and volcanoes, where geohazards are likely. It is important for governments to undertake disaster recovery planning to deal with tectonic events and communicate them to the population. Planning should include education and training, emergency assistance, communication, evacuation, physical and mental health services, restoration of services (such as power, telecommunications and transport) and rebuilding of damaged communities.

One of the best predictors of geological hazards is historical records, which include those documented or contained in the knowledge of First Nations Australians. If a disaster has happened in the last few hundred to thousands of years, it is likely to occur again.

Technology

Today's technology allows us to monitor tectonic activity and attempt to predict it. Engineers, computer programmers, physicists, chemists and geoscientists are continually working on improvements, and developing new tools and ideas.

Satellite images, especially radar images, can reveal changes in the shape of volcanoes, ash clouds, glaciers, fault lines and ground prone to landslides. The Global Positioning System (GPS) is useful for monitoring ground movements that may occur near faults, volcanoes and landslides. GPS consists of a group of satellites that orbit Earth twice per day at an altitude of about 20 000 km and continuously transmit information to receivers on the ground. The data can be used to calculate the exact position of the receiver on the ground at a particular time.

Networks of seismometers on the ground detect the location and magnitude of seismic activity that may warn of a larger earthquake or a volcanic eruption. A change in shape of a volcano can indicate an imminent eruption. A **tiltmeter** measures tiny changes in the slope angle or 'tilt' of the ground and a **strainmeter** measures extremely minute strain (change in shape) of the crust.

FIGURE 8.52 A small seismometer, about 20 cm across

Gas and water monitoring is important for prediction of volcanic eruptions. An increase in gas output, the appearance of new vents or a change in the chemistry of the gas and water can signal increasing volcanic activity. Emissions also become hotter before an eruption, so temperature is monitored. Some satellites can also detect temperature changes.

Computer software is also a useful tool. Computer simulations can present a disaster event scenario to test the plans and procedures in place to manage a natural hazard. Computer modelling of the effects of a natural disaster are also useful to identify potential problems and areas to be impacted. For example, the 2011 meltdown of a nuclear reactor in Fukushima, Japan that resulted from an earthquake and tsunami released a lot of radiation. Scientists immediately produced a detailed model of the emission, transport and deposition of radioactive material from the site to estimate the health impacts.

satellite images images taken of Earth's surface

tiltmeter equipment that measures changes in the angle of the ground

strainmeter equipment that measures any change in the shape of Earth's crust

SCIENCE AS A HUMAN ENDEAVOUR: An early Chinese earthquake detector

In the year 132 AD, a Chinese scholar by the name of Zhang Heng invented the first seismoscope to measure earthquakes. Long before the theory of plate tectonics, he believed that observations of the direction, force and timing of winds could indicate and predict events on Earth and in space.

A seismoscope records the motions of Earth's shaking, but unlike a seismometer, it does not retain a time record of those motions. To indicate the direction of a distant earthquake, Zhang's device dropped a bronze ball from one of eight tubed projections shaped as dragon heads. The ball fell into the mouth of one in a circle of metal toads, each representing a compass direction.

One day Zhang's device was triggered but no seismic disturbance was felt. Several days later a messenger arrived from the west and reported that an earthquake had occurred 500 km away in the direction that Zhang's device had indicated!

FIGURE 8.53 An antique Chinese seismoscope

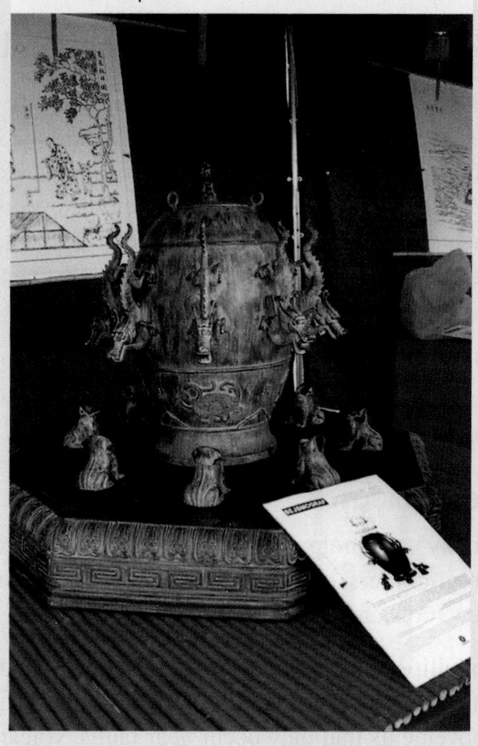

8.8.3 Engineering solutions

Earthquakes

The violent shaking and fracturing of the ground that accompanies an earthquake causes a great deal of damage to buildings and can destroy cities, with falling debris leading to much loss of life and injury. **Liquefaction** can make the ground soft and unstable, causing buildings to sink and destroy roads, pipes and cables. Damage to services and loss of housing has a huge impact on the population and local economy, which can last for years.

liquefaction a phenomenon in which soil loses its strength and stiffness due to strong ground shaking

New Zealand is a country with a long history of earthquakes, and has been developing and improving building codes to deal with this over many years. The University of Canterbury has done research on how concrete behaves during earthquakes and contributed to building design codes used around the world. Many buildings and bridges are protected with lead dampers and lead-and-rubber bearings invented in New Zealand. These devices in building foundations can reduce the motion caused by ground shaking. Flexible joints or ductile pipes are used for water pipelines across unstable ground to prevent rupture. Similarly, gas pipelines have been welded to prevent breakage, or replaced by polythene pipes.

FIGURE 8.54 Damage from a magnitude 6.3 earthquake that hit Christchurch, New Zealand on 22 February 2011

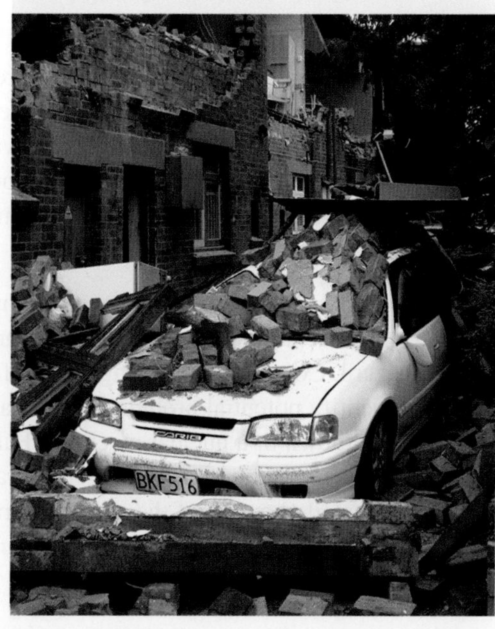

FIGURE 8.55 Earthquake solutions: a. A lead-and-rubber bearing b. A damper under a building

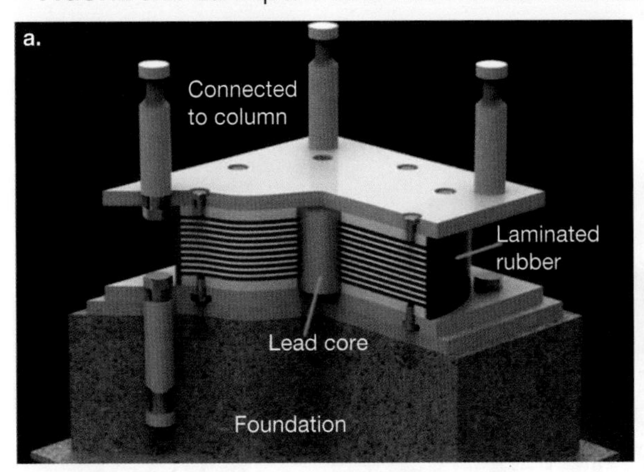

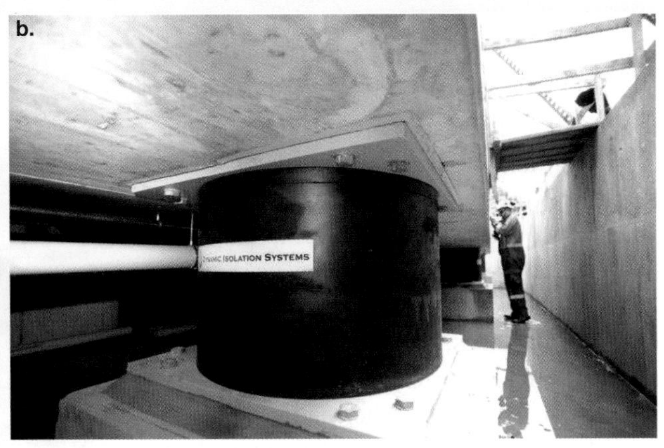

CASE STUDY: A dam designed to shake

The Clyde Dam in New Zealand is designed to withstand intense shaking. It is built across a fault and has been constructed with a specially designed slip joint. If the land on either side of the fault moves during an earthquake, the joint will allow sections of the dam to shift up to 2 metres horizontally and 1 metre vertically without the dam failing.

FIGURE 8.56 The Clyde Dam in New Zealand is designed to sustain earthquakes

SCIENCE AS A HUMAN ENDEAVOUR: Bamboo houses

In 2018 an earthquake reduced much of the island of Lombok in Indonesia to rubble, and killed 560 people. A visiting engineer noticed that bamboo houses withstood the shaking better than concrete ones. This is because bamboo houses move during earthquakes, allowing the energy to be dispersed. Bamboo is lightweight, strong, cheap and readily available in many earthquake-prone, developing areas of the world, so is an ideal and sustainable building material. Engineers and University College London have since been working closely with the people of Lombok to design strong and safe bamboo houses.

FIGURE 8.57 A bamboo house

Volcanoes

Volcanoes can have devastating impacts on people and landscapes. Lives are lost, lava and mud flows can destroy towns and natural resources, and landscapes or habitats can be drastically altered. For example, in 79 AD the Roman city of Pompeii was destroyed by the volcano Mount Vesuvius. Unlike earthquakes, however, volcanoes usually show warning signs. Geologists investigate the history and types of volcanoes, map volcanic deposits, and use instruments to predict what type of eruption may occur, when and where. Pyroclastic eruptions are much more explosive and faster than lava flows.

FIGURE 8.58 Mount Vesuvius and the remains of Pompeii

To help predict eruptions, engineers design and build many different devices to detect subtle changes in a volcano that occur before it erupts. They build specialised instruments that can detect gases and changes in the shape of the volcano, as well as monitor earthquakes that may signal a possible eruption.

Building design solutions can prevent death and injury in volcanic areas, particularly in low-income areas that rely on cheap materials. Volcanic ash is much like snowfall; however, it is more than twice as heavy, it is corrosive, it can make the air potentially toxic, and the eruption of it may be associated with severe wind. Buildings typically require clear evacuation routes; windows that allow good views of the volcano; and smooth and steep roofs to allow ash to slide off, but with good roof support, short spans and lateral bracing to withstand winds. Windows and doors should be able to be sealed completely, to keep out toxic gases. Emergency supplies in each building should include gas masks and oxygen.

Tsunamis

Destructive tsunamis can result from earthquakes and volcanic activity, producing similar social, economic and environmental impacts. Damage is caused by the impact of fast-flowing water and waves carrying debris (which may be burning), then erosion of the land, accompanied by strong winds.

FIGURE 8.59 Damage at Banda Aceh, Indonesia, after an earthquake and tsunami in December 2005

When building a house in an area at risk of tsunamis, there are four main ways to mitigate the destructive power of tsunami waves:

- Build structures with reinforced concrete instead of wood, designed to let water flow through, with deep foundations and vertical evacuation routes that allow people to climb quickly above the water level.
- Elevate buildings using stilts or a podium.
- Create friction; forest, slopes, ditches, reefs and coastal mangroves between the coast and the house lessen the wave power.
- Guide the tsunami wave by strategically positioning angled walls and ditches.
- Stop the force of the waves with hardened structures, like seawalls, rigid walls and embankments. This may, however, increase the wave height or direct the waves to other features.

FIGURE 8.60 A tsunami evacuation structure at a beach in Japan. The platform is 12.5 m above sea level.

8.8 Activities

learnon

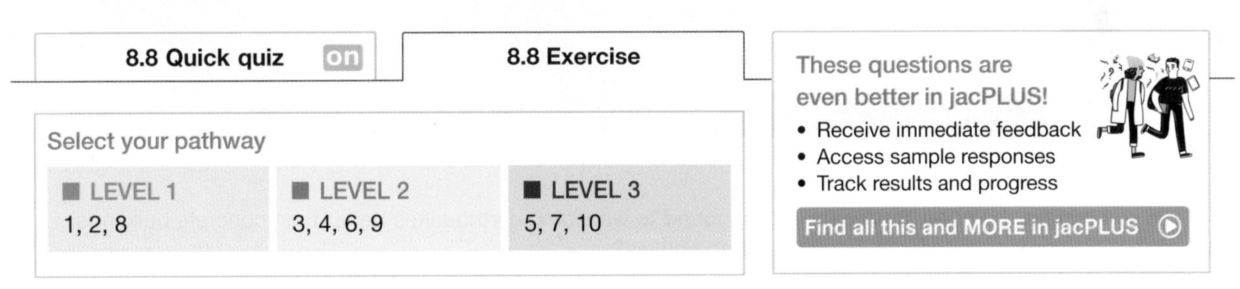

Remember and understand

1. What is the difference between a natural hazard and a natural disaster?
2. What are some hazards related to tectonic events?
3. What devices are used to monitor tectonic activity?
4. How do First Nations Australians pass on knowledge about geological events?
5. How do engineers help society deal with natural hazards? Give some examples.

Apply and analyse

6. The following image shows the locations of earthquakes that occurred in Australia between the years 1800 and 2000. Based on this, which two capital cities are likely to have the greatest risk of earthquakes in the future?

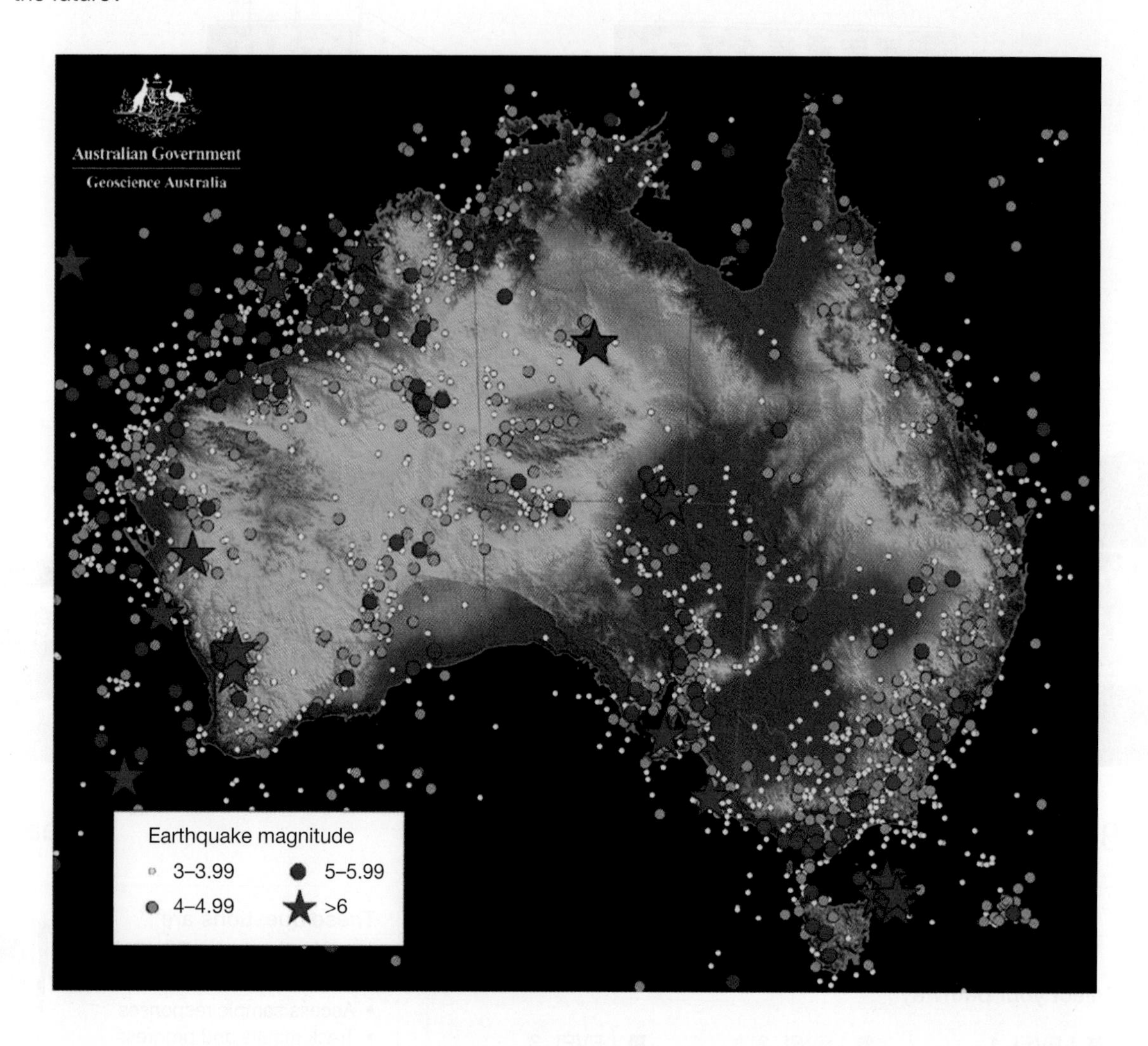

7. In some areas, bamboo houses have been found to withstand earthquakes better than concrete buildings. What are some other advantages of bamboo houses?

Evaluate and create

8. A geologist mapping a volcano in Queensland finds First Nations Australian art on the rocks near the summit. What could they do, apart from making geological observations, to find out more about the volcanic history?
9. You have a 200-metre square, forested block of land that extends from the beach to a low ridge at 10 metres above sea level. Locate and design a simple house to best withstand a tsunami.
10. What type of app might be useful during and following a natural disaster?

Fully worked solutions and sample responses are available in your digital formats.

LESSON
8.9 Thinking tools — Double bubble maps

8.9.1 Tell me

What is a double bubble map?

A double bubble map is a diagram that breaks up two ideas or topics into different categories and allows you to compare and contrast them. It shows features that are similar — these are joined to both topics — and features that are different — these are joined to just one topic.

A double bubble map helps you understand how topics can be broken into different categories and the relationships that exist between them. They allow you to make connections between two ideas that have both similarities and differences.

FIGURE 8.61 Double bubble map

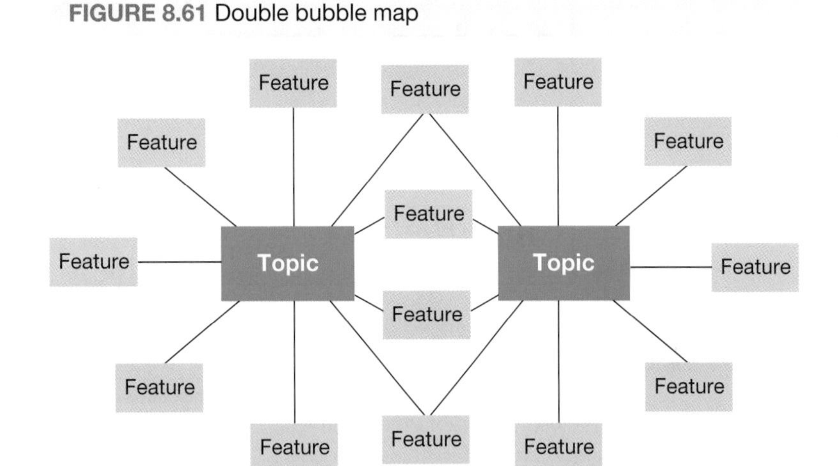

For example, you would use a double bubble map to:
- compare two concepts
- break your ideas down into smaller features
- brainstorm ideas for a project or essay.

Comparing a double bubble map to an affinity diagram

An affinity diagram lets you explore one idea and see how it can be broken into smaller groups. These smaller groups each contain ideas or features that are similar. It can be helpful to think of your topic and then write down all the ideas you have about it on small cards or pieces of paper. You can then move these cards to place them in groups that are similar.

FIGURE 8.62 Affinity diagram

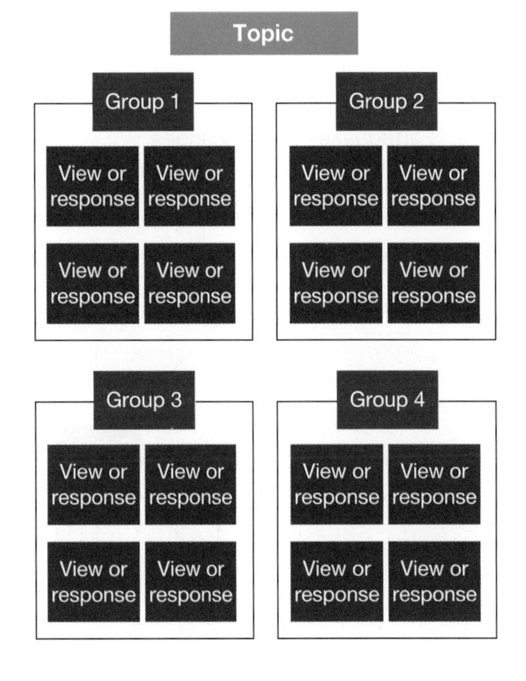

8.9.2 Show me

To create a double bubble map:
1. Choose two topics that are related to each other and write any ideas you may have onto small pieces of paper. For example, you might choose to compare volcanoes and plate tectonics.
2. Examine your pieces of paper and put ideas that only relate to one topic into one group, and any ideas that relate to the second idea in another.
3. Any ideas you have that are related to both topics will now form the middle of your double bubble map.

4. Using these words, you might create a double-bubble map that looks like the example given.

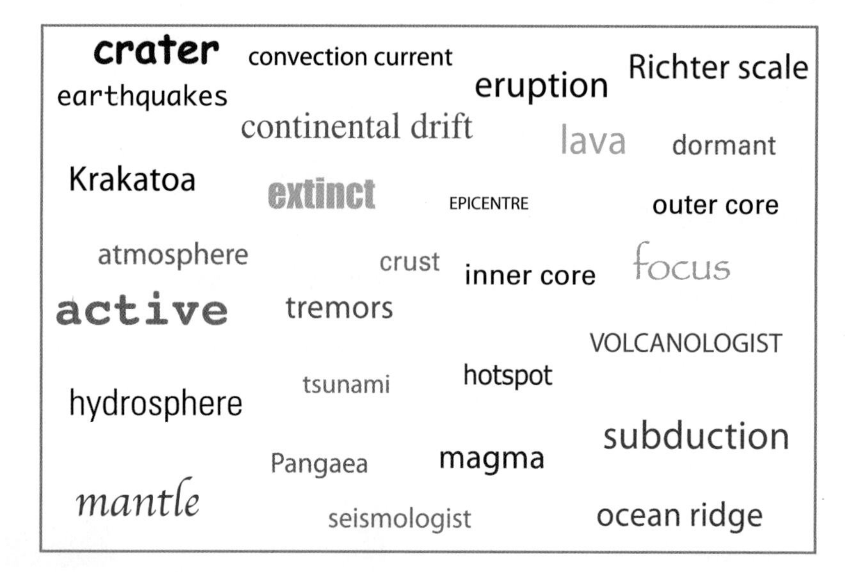

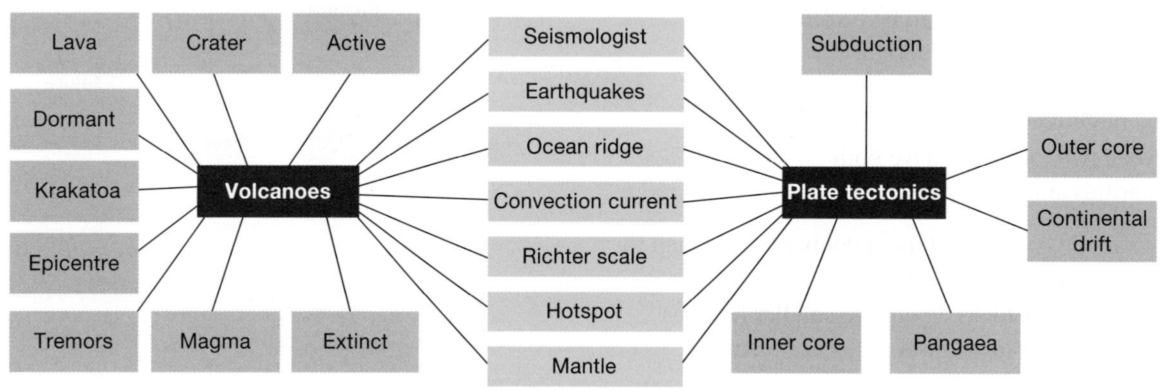

8.9.3 Let me do it

8.9 Activity

Create double bubble maps that illustrate the similar and different features of the following pairs of topics:

a. Folding and faulting
b. Earthquakes and volcanoes
c. Continental drift and plate tectonics.

These questions are even better in jacPLUS!
- Receive immediate feedback
- Access sample responses
- Track results and progress

Find all this and MORE in jacPLUS

Use this figure to help you get started on a double bubble map for folding and faulting.

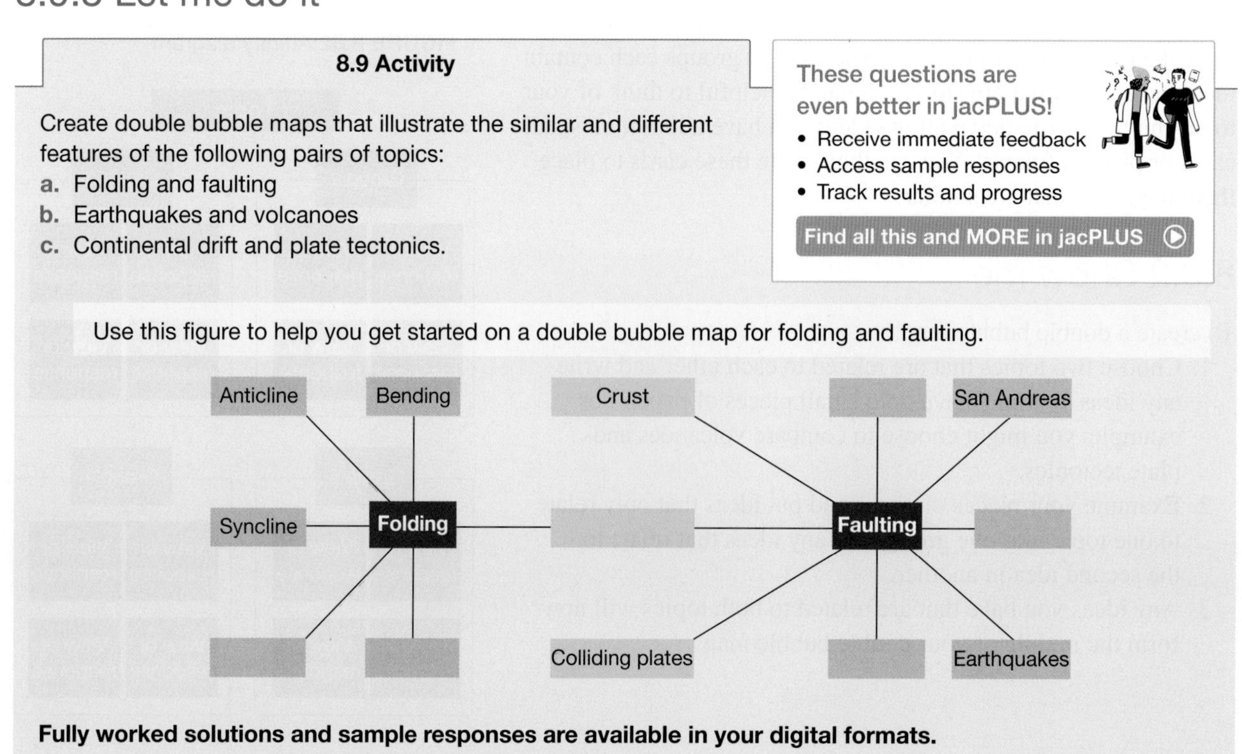

Fully worked solutions and sample responses are available in your digital formats.

LESSON
8.10 Project — Disaster-proof

Scenario

Earthquakes occur when pressure built up between adjacent sections of rock in Earth's crust is suddenly released. The bigger the earthquake's magnitude, the greater the amount of energy that shakes Earth. However, the magnitude of the earthquake is not necessarily a good indication of how deadly it will be. The May 2006 earthquake in Java had a magnitude of 6.2 and caused the deaths of nearly 6000 people, yet the 2004 Guadalupe earthquake was the same size but killed only 1 person. In some cases, magnitude 5.3 earthquakes have killed more people than those with magnitude 8.1. In fact, the key predictors other than magnitude of how deadly an earthquake will be are how heavily populated the area is and what type of buildings are there. Sadly, the majority of people who die in earthquakes do so because the buildings around them fail.

Unlike the more earthquake-prone regions of the world, Australia is not near a plate boundary, but we are not out of danger. The 1989 Newcastle earthquake had a magnitude of 5.6 and resulted in 13 deaths, 160 injuries and damage to over 60 000 buildings. With this in mind, your company — Shakeless Seismic Solutions — has been approached by a wealthy client who wishes to build an earthquake-proof five-storey office block in Perth. However, yours is not the only company that she has approached. In order to determine which business she will award the contract to, she is asking each company not only to come up with a design, but also to have a scale model of their design tested on a shake-table earthquake simulator.

FIGURE 8.63 Damage caused by the 1989 Newcastle earthquake

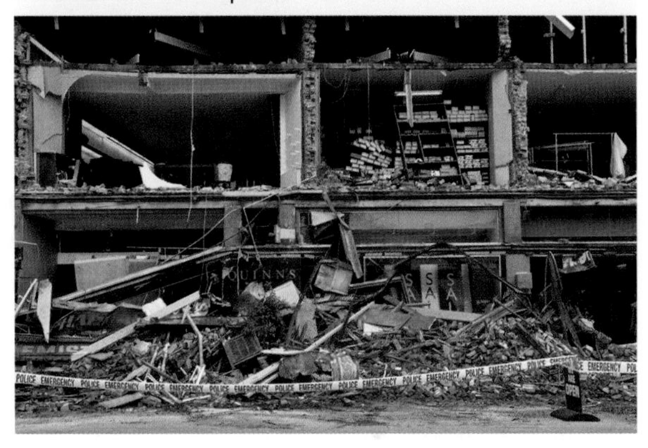

Your task

Your group will use research, ingenuity and online simulators to design a five-storey office block that will survive an earthquake. You will build a scale model of your design and compete with other groups to determine which model/design is able to withstand the most energetic shaking on the simulator. Your model will need to fulfil the following criteria:

- It should have a total mass of no more than 1.5 kg.
- It should have a base area no bigger than 20 cm × 20 cm and should have a height of at least 50 cm.
- No glue, staples, nails or pins are allowed; however, you may use interlocking pieces.
- It must be freestanding (it may not be stuck to the table in any way).

Before testing, you will be required to explain the main aspects of your design to the client (your teacher) and describe what makes the model and the real building earthquake-proof.

Resources

ProjectsPLUS Disaster-proof (proj-0108)

LESSON
8.11 Review

Hey students! Now that it's time to revise this topic, go online to:

Review your results

Watch teacher-led videos

Practise questions with immediate feedback

Find all this and MORE in jacPLUS

Access your topic review eWorkbooks

on Resources

Topic review Level 1	Topic review Level 2	Topic review Level 3
ewbk-11988	ewbk-11990	ewbk-11992

8.11.1 Summary

Introduction to Earth

- Earth has three basic interior layers: crust, mantle and core.
- There are two types of crust: continental crust and oceanic crust. The oceanic crust is thinner and heavier.
- The mantle is mostly solid but soft rock.
- The core is made of iron and nickel, with a liquid outer core and solid inner core.
- The layers are not stationary; they move or flow.
- Alfred Wegener proposed that Pangaea was a supercontinent that broke apart by continental drift millions of years ago.
- Many continental shapes fit together like a jigsaw.
- Rock types and land-based fossils match across continents.

The theory of plate tectonics

- The theory of sea-floor spreading explains how ocean basins form and continents move apart.
- At a spreading centre, crust splits apart, and melted mantle rises to fill the gap and forms new crust.
- Continental drift and sea-floor spreading theories evolved into the theory of plate tectonics.
- Earth's surface is broken into fragments called tectonic plates that move relative to one another.
- Plates move under the influence of slab pull, ridge push and convection currents.
- There are three types of plate boundaries: convergent, divergent and transform.
  - Convergent subduction is destructive, as old oceanic crust sinks back into the mantle, forming deep-ocean trenches and volcanoes.
  - Convergent collision forms mountain ranges as continental crusts collide.
  - Divergent boundaries are constructive because new oceanic crust forms by volcanoes along the spreading centre of ocean ridges.
  - Transform boundaries are conservative as crust is neither created nor destroyed.
- All types of tectonic boundaries have earthquakes.
- The modern tectonic boundaries are identified and mapped by:
  - volcano and earthquake patterns
  - growing mountain ranges and ocean ridges
  - direct GPS measurements.
- Tectonic boundaries shift with time, constantly changing how the surface of Earth appears.
- Australia is an old continent. It sits within the middle of the Indo-Australian Plate and was once a part of a larger continent called Gondwana.

Plate boundaries

- When there are converging plates, two scenarios can be observed:
  - subduction is when old oceanic crust converges to continental or younger oceanic crust
  - collision is when two continents come together.
- Divergent boundary is when two plates move apart. This results in sea-floor spreading.
- Two types of boundaries occur as plates slide side by side:
  - transform boundary where earthquakes occur
  - conservative plate boundaries where the crust is conserved.

Folding and faulting

- When huge tectonic forces are applied, rocks can bend into folds or break as faults.
- Folding occurs when there is a slow compression force, which shortens and thickens the crust.
- Folds can be anticlines (shaped like an 'A') and synclines (shaped like a 'U').
- Faults form when rocks in the upper crust (cool) are exposed to tectonic forces.
- There are three major fault types:
  - Normal faults — the rock above the fault moves down, related to tension force like those at divergent boundaries
  - Reverse faults — the rock above the fault moves up, related to compression force like those at convergent boundaries
  - Strike-slip faults — rocks move horizontally side by side, related to shearing force like those at transform boundaries.
- Australia has a long, complex geologic history with many events of both folding and faulting.

Earthquakes

- Earthquakes are caused by breaking crust, releasing the built-up pressure.
- The vibrations (seismic waves) of an earthquake travel from the focus as both body and surface waves.
- P- and S-waves are body waves. P-waves are the fastest.
- Love and Rayleigh waves are slower surface waves.
- Seismic waves are recorded with a seismograph.
  - The size and spacing of the waves recorded relate to the amount of energy released and the station's distance from the epicentre.
- Earthquake magnitude is measured on the Richter scale, which measures the amount of energy released by an earthquake.
- Large earthquakes are concentrated along the current plate boundaries.
- Australia's earthquake potential is lower than other countries near plate boundaries.
- Tsunamis can form from underwater earthquakes that have lifted the sea bed and the water column above it.
- The extent of earthquake damage is not only related to the earthquake magnitude, but also the:
  - distance from the epicentre
  - population
  - building materials.

Volcanoes

- Volcanoes form when magma erupts onto the surface from over-pressured magma chambers underground.
- The erupted material can include red-hot lava, gas, rock, ash and volcanic bombs.
- Runny lava produces a calmer eruption with long-distance lava flows.
- Sticky lava produces an explosive eruption with steep-sided volcanoes.
- Volcanoes can be active, dormant or extinct.
- Most active volcanoes form along divergent and subduction boundaries.
- Most volcanoes in Australia are classified as extinct.
- Volcanoes can also be related to hotspots, which occur in areas with extremely hot mantle close to the surface.
- Oceanic spreading ridges contain chains of underwater active volcanoes with black smokers.

Human response to tectonic events

- To determine when a natural disaster may strike, sites are monitored using a variety of technologies including GPS, seismometers, tiltmeters, strainmeters and gas emissions. Scientists look for any changes to an imminent event so people can be evacuated if needed.
- Events can be predicted using historical records, to determine if and when it is likely to occur.
- Engineers are also helping to minimise the damage that can occur during an event by designing the foundations so buildings and bridges do not move as much.

8.11.2 Key terms

abyssal plains relatively flat underwater deep-ocean floor, around 4000 metres depth
active volcano a volcano that is erupting or has recently erupted
anticline a fold in a rock with the narrow point facing upwards
black smoker a geothermal vent on the sea floor that ejects superheated, mineral-rich water
body waves seismic waves that quickly travel through the interior of Earth
collision refers to when two continents crumple together to form a mountain range
compression a squeezing force
conservative plate boundary where crust is neither created nor destroyed
constructive plate boundary where new crust is formed
continental drift the movement of Earth's continents relative to each other over geologic time
convection currents the movement of particles in a liquid or gas resulting from a temperature or density difference
convergent boundary where two tectonic plates move towards each other
core the hot centre of Earth made of iron and nickel
crust the hard and thin outer rock layer of Earth
deep-ocean trenches narrow and deep troughs in the ocean floor, generally greater than 5000 metres depth
divergent boundary where two tectonic plates move apart
dormant volcano a volcano that has erupted in the last 10 000 years but is not currently erupting; they are considered likely to erupt again
earthquake a sudden and violent shaking of the ground
epicentre the surface point directly above the earthquake focus
extinct volcano a volcano that *has not* erupted in the last 10 000 years; they are considered dead and not to erupt again
fault a break in the crust where one side moves relative to the other
focus the location underground of the fault movement causing an earthquake
folding refers to when rocks bend into anticlines or synclines
fossil any remains, impression or trace of a life form of a former geological age; evidence of life in the past
Global Positioning System (GPS) a network of satellites that tracks location and movement
Gondwana the southern part of the broken-up supercontinent of Pangaea, which included the continents of Africa, South America, Antarctica and Australia; also known as Gondwanaland
horst a highland between two normal faults
hotspot a volcanic region directly above an area of extremely hot mantle
inner core the solid inner-most layer of the core under extreme pressure conditions, with an approximate 1200 km radius
Laurasia the northern part of the broken-up supercontinent of Pangaea, which included the continents of North America, Europe and Asia
lava an extremely hot liquid or semi-liquid rock from the mantle that reaches and flows or erupts on Earth's surface
liquefaction a phenomenon in which soil loses its strength and stiffness due to strong ground shaking
lithosphere the outermost layer of Earth; includes the crust and uppermost part of the mantle
Love waves surface seismic waves that have a side-to-side motion
magma a very hot mixture of molten rock and gases, just below Earth's surface, that forms from melting of the mantle and occasionally the crust
mantle the solid but soft middle rock layer of Earth
mountain range a group of high-ground features, commonly the result of tectonic collision
normal fault a break where the rock above the fault moves 'down' due to tension

ocean ridges submarine mountains that tower 2000 metres above the abyssal plains
ophiolites pieces of oceanic crust observed on continental crust (land)
outer core the liquid outer layer of the core, which is about 2300 km thick
P-waves or primary waves body seismic waves with a compressional (push-and-pull) motion; they are the fastest and first to arrive
Pangaea a supercontinent that existed about 299 to 200 million years ago; all landmasses were joined together to form it
Panthalassa the vast ocean surrounding the supercontinent of Pangaea
plate tectonics a scientific theory that describes the relative movements and interaction of plates of Earth's crust over the underlying mantle
radioactive refers to when atoms are unstable and emit a particle to remove excess energy; these particles are capable of ionising other atoms upon collision, which can cause harm to living tissue
Rayleigh waves surface seismic waves that have a rolling motion
reverse fault a break where the rock above the fault moves 'up' due to compression
Richter scale a logarithmic scale that measures the amount of energy released during an earthquake, thus allowing one earthquake to easily be compared to another
rift valley a sunken lowland between two normal faults; a graben
S-waves or secondary waves body seismic waves with a transverse (up-and-down) motion; they are slower than P-waves and cannot travel through fluids
satellite images images taken of Earth's surface
sea-floor spreading the formation of oceanic crust, which occurs by the rising and melting mantle at ocean ridges that push older crust away from the ridge
seismic waves waves released when rock breaks or is rapidly moved
seismograph an instrument used to detect and measure the intensity of an earthquake; also called a seismometer
seismologist a scientist who studies earthquakes to both understand how they work and how to better predict them
shearing a smearing force
strainmeter equipment that measures any change in the shape of Earth's crust
strike-slip fault a break where the rocks on either side of the fault move horizontally due to shearing
subduction refers to a convergent plate boundary where one plate moves under another
subduction zone where old oceanic crust enters the mantle
surface waves seismic waves that travel slower than body waves and only along the surface of Earth; their energy is lost with depth and distance
syncline a fold in a rock with the narrow point facing downwards
tension a stretching force
tiltmeter equipment that measures changes in the angle of the ground
transform boundary where two tectonic plates slide past one another
tremors minor vibrations of the ground that are commonly not felt
tsunami a powerful ocean wave triggered by an undersea earth movement
volcanic bomb refers to when a large rock fragment that falls from an eruption, formed as lava, is blown out of a volcano and is rapidly cooled in the air
volcano a landscape feature through which melted rock is erupted onto Earth's surface

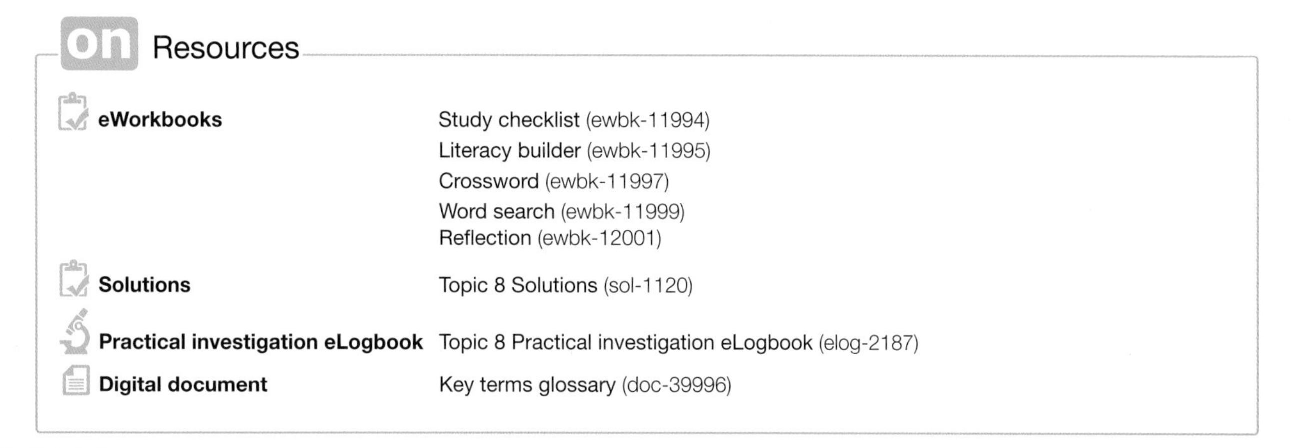

on Resources

eWorkbooks	Study checklist (ewbk-11994) Literacy builder (ewbk-11995) Crossword (ewbk-11997) Word search (ewbk-11999) Reflection (ewbk-12001)
Solutions	Topic 8 Solutions (sol-1120)
Practical investigation eLogbook	Topic 8 Practical investigation eLogbook (elog-2187)
Digital document	Key terms glossary (doc-39996)

8.11 Activities

learn on

8.11 Review questions

Select your pathway

■ LEVEL 1	■ LEVEL 2	■ LEVEL 3
1, 2, 4, 5, 9, 11, 13, 15	3, 6, 10, 12, 16, 18, 19	7, 8, 14, 17, 20, 21, 22

These questions are even better in jacPLUS!

- Receive immediate feedback
- Access sample responses
- Track results and progress

Find all this and MORE in jacPLUS

Remember and understand

1. Identify the layers of Earth that have the following characteristics.
   a. Completely molten
   b. Solid but soft
   c. Solid and hard, and includes rock, soil and landforms
   d. Solid and mostly made of iron
   e. Lies above the surface
2. SIS Describe two pieces of evidence that were used to develop Wegener's theory of continental drift.
3. Explain how scientists know about what lies deep below the surface of Earth without going there.

The Earth and what lies beneath

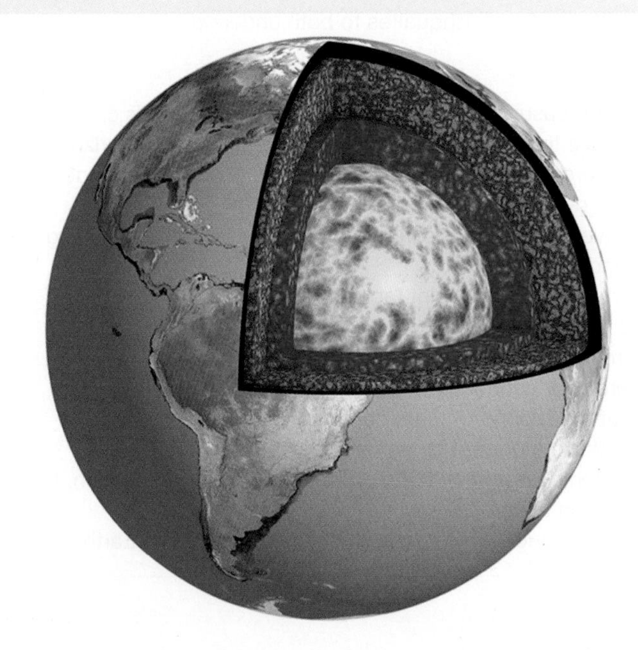

4. Where would you find the youngest oceanic crust according to the theory of sea-floor spreading?
5. How is an ocean ridge different from a subduction zone?
6. When oceanic crust pushes against continental crust, why does the oceanic crust slide underneath the continental crust?
7. What is the major difference between the continental drift theory and the theory of plate tectonics?
8. Describe the movements in Earth's crust that cause the folding of rock and have shaped many of Earth's mountains.
9. Explain how faults are created.

10. Examine the following diagrams and label the features A–H using the following words: anticline, continental crust, magma, normal fault, oceanic crust, reverse fault, upper mantle, syncline.

A subduction zone

A B C D

Two types of faulting

E F

Folding upwards and downwards

G H

11. Distinguish between the epicentre of an earthquake and its focus.
12. What is a seismograph used to measure?
13. Name three gases that are released from a volcano.

Apply and analyse

14. The San Andreas Fault runs along much of coastal California, including the cities of Los Angeles and San Francisco, and is susceptible to earthquakes.
    a. Explain why the San Andreas Fault is called a strike-slip fault.
    b. What causes major earthquakes along this fault?
15. Using the image provided, identify where on Earth the Ring of Fire is and why it exists? Explain why it is called the Ring of Fire.

A map of the Earth

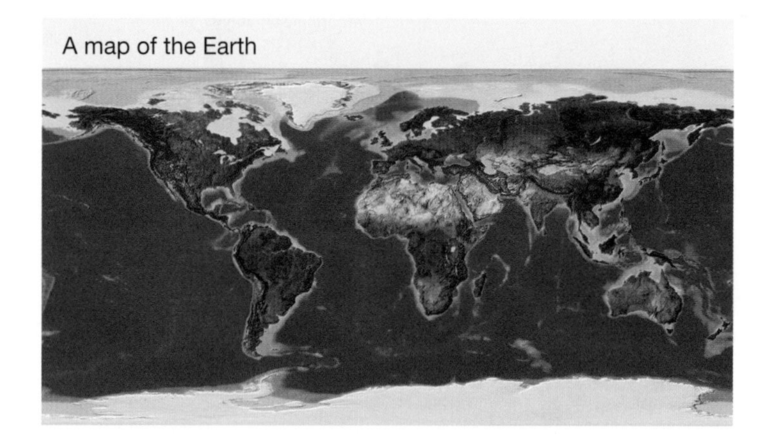

16. Suggest two reasons why an earthquake that registers 6.6 on the Richter scale can cause more deaths and devastation than an earthquake that registers 8.9.
17. How much energy is released by an earthquake that registers 6.0 on the Richter scale relative to one that registers 8.0?
18. Explain why Australia is less likely to experience volcanic activity and major earthquakes than New Zealand.
19. Tsunamis can form due to fault movements on the ocean floor, but there are other geologic events that can trigger a tsunami, such as a huge coastal or underwater landslip. How can a large landslip trigger a tsunami?
20. Before an explosive volcano erupts, its vents are blocked with thick, sticky lava.
    a. What change takes place to cause the volcano to erupt?
    b. How would runny lava change the eruption style?

Evaluate and create

21. **SIS** Observe in the following figure the age of the volcanic shield volcanoes along the Great Dividing Range. Recall that the eastern margin of Australia has not been located near a plate boundary for the last 50 million years.

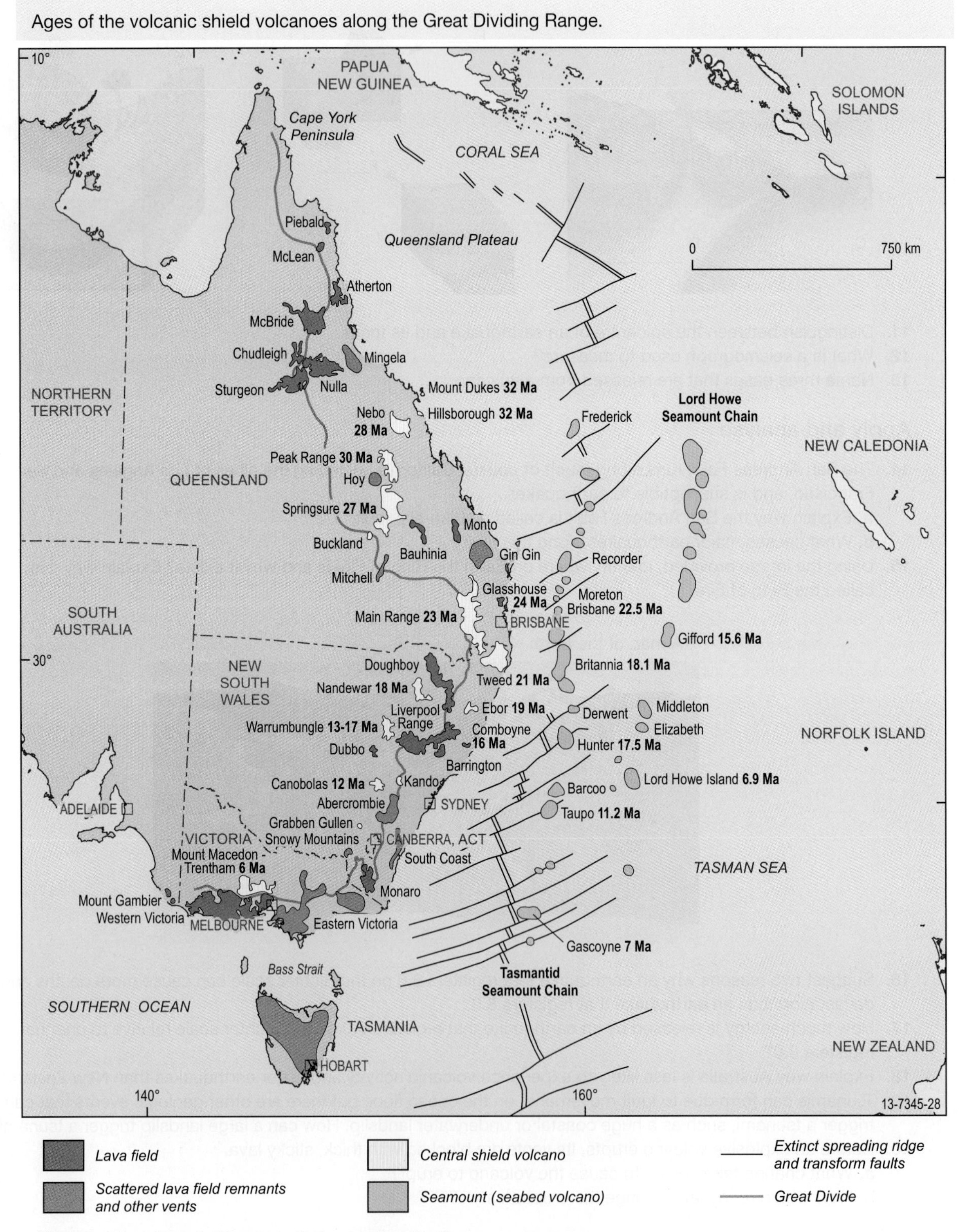

Ages of the volcanic shield volcanoes along the Great Dividing Range.

a. What is the pattern to the age relative to direction?

b. Based on your knowledge and observations, how would you explain the origin of the volcanoes?

c. Are there any inconsistencies to that pattern or other reasons as to why you may question the conclusion?

22. According to the theory of plate tectonics, Earth's crust is divided into slowly moving plates.
   a. What makes the plates move?
   b. What can happen when two plates slide past each other?
   c. How does the theory of plate tectonics explain the growth of the Himalayas? Your answer should include reference to relevant plate boundaries, rock density and any limiting factors on the height of mountain chains.

Fully worked solutions and sample responses are available in your digital formats.

Online Resources

Below is a full list of **rich resources** available online for this topic. These resources are designed to bring ideas to life, to promote deep and lasting learning and to support the different learning needs of each individual.

8.1 Overview

eWorkbooks
- Topic 8 eWorkbook (ewbk-12016)
- Starter activity (ewbk-11967)
- Student learning matrix (ewbk-11969)

Solutions
- Topic 8 Solutions (sol-1120)

Practical investigation eLogbook
- Topic 8 Practical investigation eLogbook (elog-2187)

Video eLesson
- Volcanic eruption in Iceland 2010, Eyjafjallajökull (eles-2661)

8.2 Introduction to Earth

eWorkbooks
- Labelling the layers of Earth (ewbk-11970)
- Continental drift (ewbk-11972)

Video eLessons
- Interior of Earth (eles-4148)
- Drifting continents (eles-0129)

Practical investigation eLogbook
- Investigation 8.1: Continental drift (elog-2189)

Interactivity
- Labelling the layers of Earth (int-8163)

8.3 The theory of plate tectonics

Practical investigation eLogbook
- Investigation 8.2: Convection currents (elog-2191)

8.4 Plate boundaries

Video eLessons
- San Andreas Fault (eles-4149)
- Plate margins of the world (eles-4150)

eWorkbooks
- Plate tectonics (ewbk-11974)
- How does a hypothesis become a theory: Plate tectonics (ewbk-11976)

Interactivity
- Does Earth move? (int-0674)

Weblink
- How Earth Will Look In 250 million Years

8.5 Folding and faulting

eWorkbook
- Folding and faulting (ewbk-11978)

Practical investigation eLogbooks
- Investigation 8.3: Modelling folds (elog-2193)
- Investigation 8.4: Modelling faults (elog-2195)

Teacher-led video
- Investigation 8.5: Making a seismograph (tlvd-10791)

8.6 Earthquakes

Video eLesson
- Tsunami wave propagation during the 2004 Sumatra–Andaman tsunami (eles-4151)

eWorkbooks
- Earthquakes (ewbk-11980)
- Plotting earthquakes activity (ewbk-11982)

Practical investigation eLogbook
- Investigation 8.5: Making a seismograph (elog-2197)

8.7 Volcanoes

Video eLessons
- Volcanoes (eles-0130)
- Deep-sea mining (eles-1086)

eWorkbooks
- Volcanic activity (ewbk-11984)
- Geological activity (ewbk-11986)

Weblinks
- Tonga volcano eruption January 2022
- Power of Tonga volcano eruption
- Craitbul Story by Aunty Michelle

8.10 Project — Disaster-proof

ProjectsPLUS
- Disaster-proof (proj-0108)

8.11 Review

eWorkbooks
- Topic review Level 1 (ewbk-11988)
- Topic review Level 2 (ewbk-11990)
- Topic review Level 3 (ewbk-11992)
- Study checklist (ewbk-11994)
- Literacy builder (ewbk-11995)
- Crossword (ewbk-11997)
- Word search (ewbk-11999)
- Reflection (ewbk-12001)

Digital document
- Key terms glossary (doc-39996)

To access these online resources, log on to **www.jacplus.com.au**

9 Energy

CONTENT DESCRIPTION

Classify different types of energy as kinetic or potential and investigate energy transfer and transformations in simple systems (AC9S8U05)

LESSON SEQUENCE

SCIENCE INQUIRY AND INVESTIGATIONS

Science inquiry is a central component of the Science curriculum. Investigations, supported by a **Practical investigation eLogbook** and **teacher-led videos**, are included in this topic to provide opportunities to build Science inquiry skills through undertaking investigations and communicating findings.

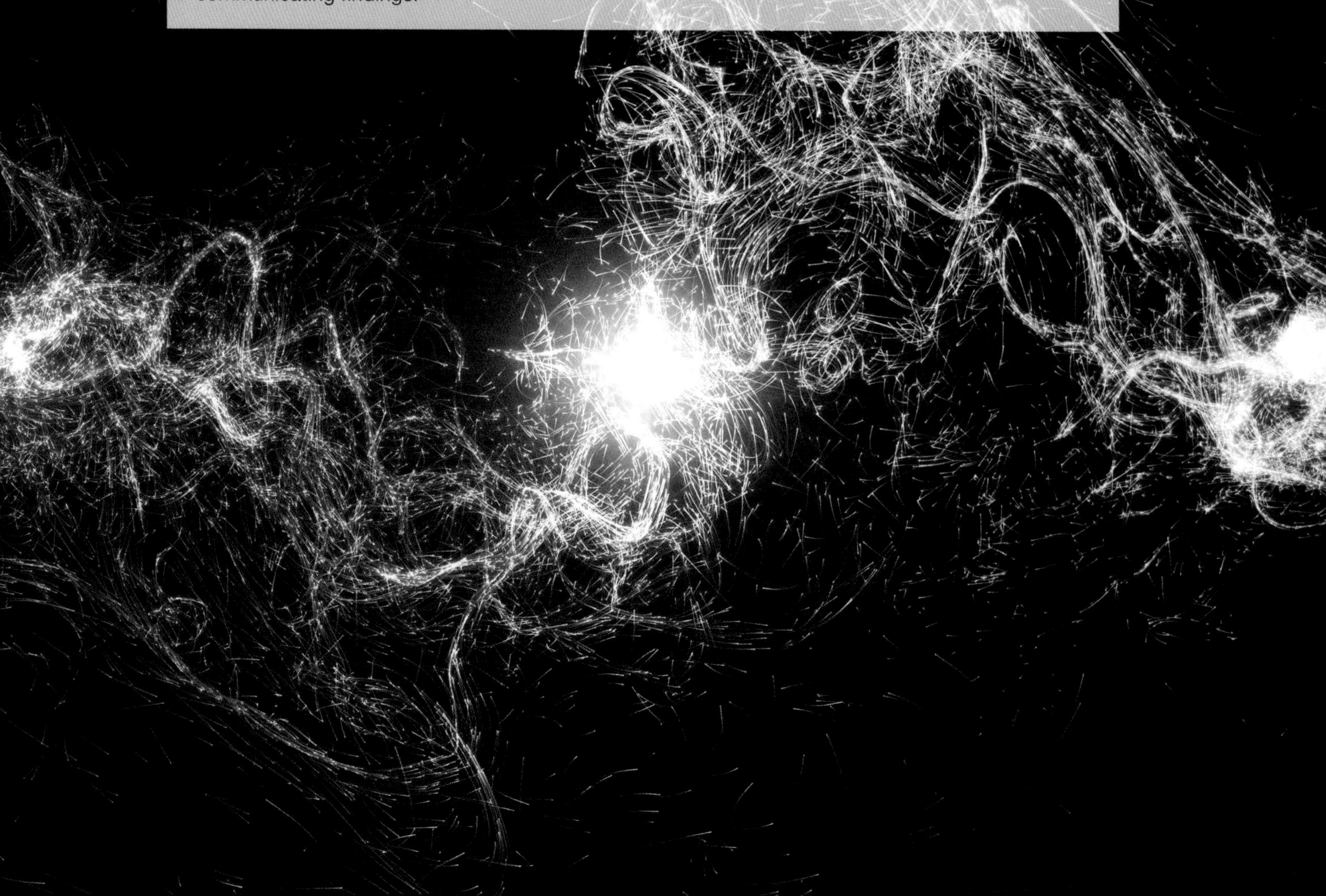

LESSON
9.1 Overview

Hey students! Bring these pages to life online

 Watch videos

 Engage with interactivities

 Answer questions and check results

Find all this and MORE in jacPLUS

9.1.1 Introduction

A fireworks display is one of the most spectacular energy transformations; you can not only see it, but also hear, feel and smell it. When fireworks are ignited, the energy stored in the substances inside them (potential energy) is quickly transformed into movement (kinetic energy), light energy, sound energy and thermal energy (more commonly called heat). Energy that is stored is known as potential energy.

FIGURE 9.1 Fireworks are a spectacular display of energy transformation.

 Resources

Video eLesson Energy transformations (eles-2677)

Watch the video of energy transformations to visualise the conversion between potential energy and kinetic energy as an object moves along a rollercoaster.

9.1.2 Think about energy

1. Which type of energy do you find in chocolate?
2. When you drop a tennis ball to the ground, why doesn't it return to its initial height?
3. How much electrical energy is wasted as heat by an incandescent light globe?
4. How do you get a swing started without someone to push you?
5. How do glow-in-the-dark stickers work?
6. Why do professional tennis players have racquets with different string tensions?

9.1.3 Science inquiry

Potential energy and kinetic energy

All substances and objects possess potential energy, and you will not be able to see it until it transforms into other types of energy. For example, the energy stored in fireworks only becomes apparent when they explode, transferring the stored energy into light, heat and sound. When divers dive from a platform or diving board, the energy stored in them because of their height above the ground is transformed into kinetic energy they gain on the way down. The energy stored in the stretched string of a bow is transformed into the kinetic energy of the arrow when it is released.

FIGURE 9.2 Potential energy is all around us. Examples include elastic potential energy in bows and the gravitational potential energy we transform into kinetic energy when we dive off a diving board.

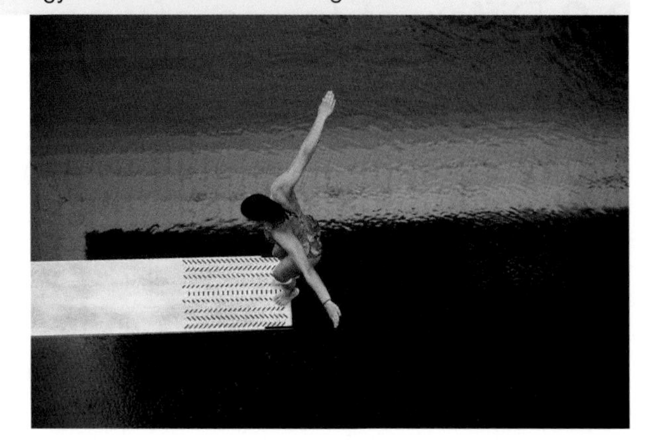

1. Complete the table given. One example has been completed for you.

TABLE Releasing and transforming energy of different objects

Object	What to do to release the stored energy	Potential energy is transformed into ...
Torch battery	Switch it on	electrical energy and light energy
Chocolate		
Petrol		
Dynamite		
Olympic diver on platform		
Match		
Stretched elastic band		

2. Answer the following questions about the wind-up toy shown.
   a. Where is the energy stored when it is wound up?
   b. What do you have to do to allow the stored energy to be transformed into different forms?
   c. Name two forms of energy into which the potential energy is transformed.
   d. From where does the energy come that allows the user to wind up the toy?

on Resources

eWorkbooks	Topic 9 eWorkbook (ewbk-12018) Starter activity (ewbk-12020) Student learning matrix (ewbk-12022)
Solutions	Topic 9 Solutions (sol-1121)
Practical investigation eLogbook	Topic 9 Practical investigation eLogbook (elog-2199)

LESSON
9.2 Different forms of energy

LEARNING INTENTION

At the end of this lesson you will be able to describe the different forms that energy can take, and be able to classify them as potential energy or kinetic energy.

9.2.1 What is energy?

Energy is a word that you sometimes use to describe how active you feel. However, in Science, we can define energy as the ability to make something happen.

We know that:
- all things possess energy — even if they are not moving
- energy cannot be created or destroyed. This statement is known as the **Law of Conservation of Energy**. It means that the amount of energy in the universe is always the same.
- energy can be transferred to another object (e.g. from a cricket bat to a ball) or **transform** into a different form (e.g. from electrical into sound)
- energy can be stored.

Law of Conservation of Energy a law that states that energy cannot be created or destroyed

transform in energy terms, this refers to the changing of one form of energy into another

The SI unit of energy is the joule (J).

SCIENCE AS A HUMAN ENDEAVOUR: Energy through the ages

Humans have always used energy. The earliest humans mainly used their own 'muscle power' to get jobs done, with the required energy coming from the food they ate. Fire was also important, being used for warmth, light and cooking. Fire was no doubt discovered through natural events such as lightning strikes, but methods to generate it were eventually discovered, mainly through utilising friction to produce heat. The methods used by First Nations Australians to do this are discussed later in section 9.3.3.

FIGURE 9.3 A replica of an ancient trireme. The sail and oars are clearly visible.

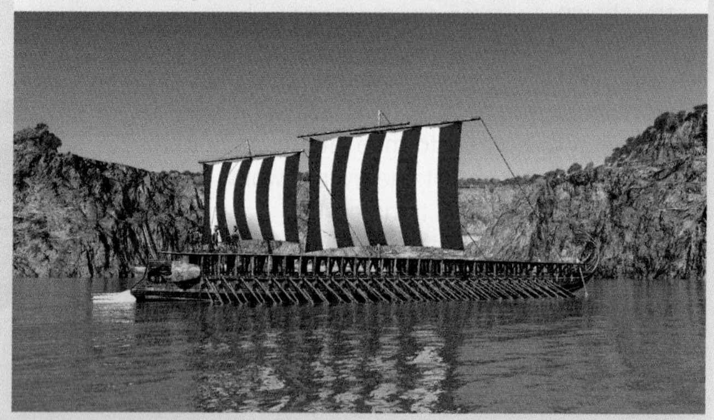

Solar energy was also used in a passive way for warmth, and for drying food and clothes. In some parts of the world, people undertook annual migrations to follow prey animals as they moved to warmer climates during the winter months. With the domestication of animals such as horses, increased muscle power became available for farming and transport as well as warfare. As a result, people had more time to develop and spread new ideas rather than just hunting for prey and collecting firewood. This led to the development of specialised trades and an overall advance in civilisation. In one form or another, this occurred in many different geographical regions around the world.

The natural environment was also harnessed as an energy source. Waterwheels were one of the first inventions that allowed mechanical energy to replace the muscle power of animals and humans. One of their most well-known tasks was to grind wheat to make flour (for making bread). Other uses included:
- the operation of pumps to lift water
- driving sawmills
- operating bellows in forges
- powering heavy hammers.

However, because waterwheels required running water, their use was restricted to areas where this was available. Added to this was the problem that in some locations, the water sources were privately owned and therefore under the control of a few individuals.

Windmills were also an important invention for humans in our endeavours to harness energy from the environment. It is thought that the windmill was first used in the Middle East, from where its use then spread to India, China and Europe. Mainly used for grinding grain and pumping water, windmills are still used in many parts of the world today. Australia has many remote locations where these are used to lift water to the surface from under the ground, thus permitting livestock to survive in otherwise marginal locations.

The other great need for harnessing energy was for transport and this followed a similar path. On water, muscle power in the form of rowing with oars was used, then supplemented and eventually replaced by wind energy captured via sails. Entire civilisations rose and fell based on their ability to trade, travel and conduct warfare over the seas adjoining their lands. In the Mediterranean, vessels called triremes were a common sight. These were propelled by three banks of rowers on each side of the boat, and a sail. The eventual progression to ships powered entirely by sails is well known, as are the resulting discoveries that followed.

The dawn of the Industrial Revolution in the 1700s and the discoveries that followed completely changed the way in which humans harnessed and used energy. As a result of this:

- traditional energy sources were able to be used in new ways. For example, heat energy was able to be obtained from fuels and used in a steam engine to produce movement (kinetic energy). This could then be used to replace windmills and waterwheels, as well as the sails on ships.
- steam engines could be used as a reliable source of energy to operate the machinery in new factories, producing goods that were cheaper and more plentiful than ever before
- new fuels were discovered — some for use in older technologies and some to power new technologies. Examples include the use of gas in homes for heating and lighting (replacing wood and coal), the use of petrol and diesel in cars, and the use of oil in ships and railway locomotives (replacing coal).
- the amount of energy 'used' has increased astronomically. As explained previously, this energy still exists after it has been 'used'. This is the *Law of Conservation of Energy*. Unfortunately, in nearly all cases, the form in which the energy exists after it is used is heat energy. This is now one of the factors that is leading to a small but significant warming of Earth's atmosphere.

Today, our demand for energy continues to increase. Most of this energy comes from fossil fuels, which are burnt to produce heat energy and carbon dioxide. Together with the 'leftover' heat energy mentioned previously, carbon dioxide is another major cause of Earth warming. This cannot be allowed to continue and, because of this, there is now much work and research being done to supply and use our energy in different ways. Examples of this include:

- the use of hydrogen as a fuel. This produces water rather than carbon dioxide when it is used. It can be produced in a number of ways, such as by solar or wind-generated electricity. This is known as 'green' hydrogen.
- the use of technologies such as solar cells to convert sunlight directly into electrical energy
- the use of new and better technologies to produce electrical energy from wind and moving water
- new transport technologies such as electric vehicles (EVs). Currently these have batteries that need recharging. If this can be done using solar-powered electricity, the vehicles will be almost pollution-free. In the future, EVs may contain fuel cells that use the green hydrogen mentioned previously. These will also be almost pollution-free. Further, in heavy transport and shipping, ammonia (which can be made from hydrogen) may prove itself to be a 'fuel of the future'.
- other new technologies such as artificial photosynthesis. Scientists are currently attempting to copy nature and use versions of photosynthesis to produce new, more environmentally friendly fuels.

9.2.2 Types of energy

All forms of energy can be classified as either **potential energy** or **kinetic energy**. Potential energy is the stored energy related to position. Kinetic energy is the energy of a moving object.

potential energy energy stored within an object that depends on its position within a system, such as gravitational energy, elastic energy and chemical energy

kinetic energy energy due to the motion of an object

Many forms of kinetic energy, such as light, sound and thermal energy, are very easily observed. Potential energy has the 'potential' to make something happen, so is not easily observed until it is transformed into another type of energy.

9.2.3 Potential energy

Potential energy is the stored energy associated with the position of an object. Potential energy can be a result of stretching or squashing an object, lifting an object above the ground, or keeping unlike charges apart. Figure 9.4 lists the six different types of potential energy that will be explored.

FIGURE 9.4 Types of potential energy

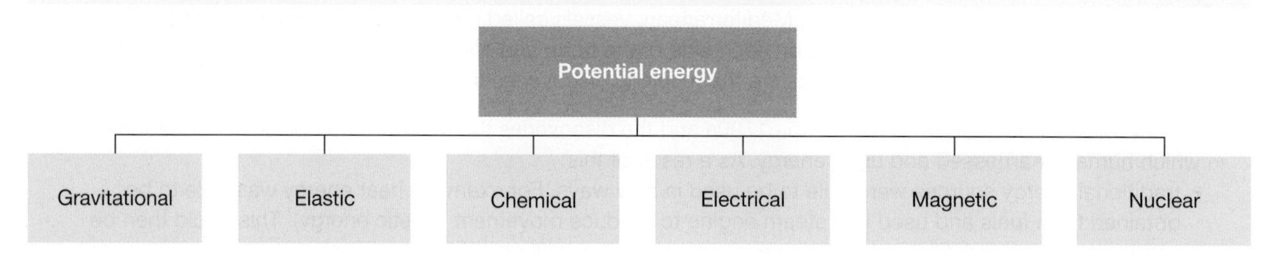

Gravitational potential energy

Gravitational potential energy is the result of gravity. When an object is lifted above the ground or moved away from Earth's surface, it has the potential to fall back to Earth's surface as soon as it is released; hence, it is said to have gravitational potential energy. The heavier the object and the further it is from Earth, the more gravitational potential energy it has (figure 9.5).

Elastic potential energy

When the shape of an object is changed, it gains or loses elastic potential energy. A good example of this is a slingshot (figure 9.6). When the rubber band is pulled back, it gains elastic potential energy. The more the band is stretched, the more elastic potential energy it gains. While the rubber band is held stretched, it maintains its elastic potential energy that has the potential to make something happen as soon as it is released.

FIGURE 9.5 A diver diving off the 10 m board will have more gravitational potential energy than a diver on the 5 m board, where the pool surface is the 'ground'.

FIGURE 9.6 The stretched rubber in a slingshot stores elastic potential energy.

Chemical potential energy

An object is said to have chemical potential energy when the chemicals inside it have the potential to react and make something happen. One example of this is a battery. When the two terminals of a battery are connected, a chemical reaction takes place that results in the flow of electricity. Another example is the chemical energy in food and drinks as seen in figure 9.7. When eaten or drunk, food and drink release their stored chemical energy to our body so that we have energy to do things.

FIGURE 9.7 The chemical potential energy in foods and drinks is converted into the energy that we use to power our muscles.

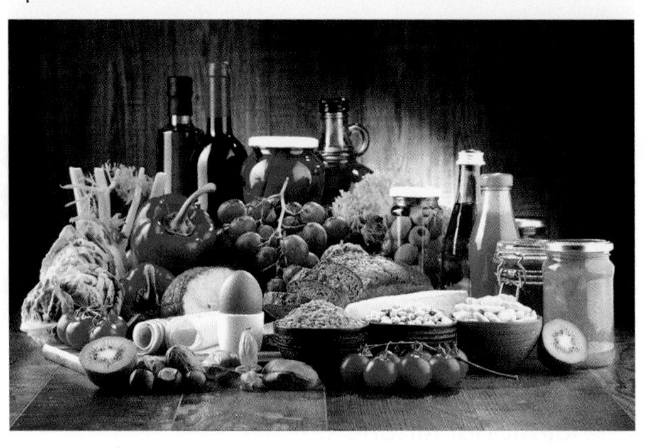

Magnetic potential energy

Magnetic fields provide another form of potential energy, called magnetic potential energy. It is easy to understand magnetic potential by playing around with some magnets. If you hold two magnets so that they are attracted to each other you will feel the pull of the magnets attempting to reach each other. If released, the magnetic potential energy will cause them to accelerate together.

FIGURE 9.8 Electrical potential energy causes electricity to flow through the circuit shown, providing power to the light.

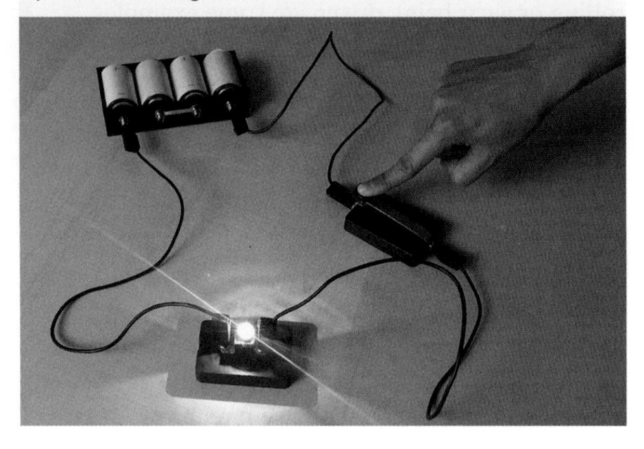

Electrical potential energy

All substances are made up of positive and negative charges. When opposite charges are separated, they are said to have electrical potential energy, because as soon as they are released they are attracted together again. As the charges come together again, they release their electrical potential energy into other forms of energy. If this takes place in an electrical circuit, the stored electrical potential energy is released to the connecting wires and components of the circuit, such as a light globe (figure 9.8).

FIGURE 9.9 The amount of nuclear potential energy that is released in nuclear reactions is so large it causes enormous explosions like this one.

Nuclear potential energy

The energy stored in the nuclei of atoms is called nuclear potential energy, because if the nuclei can be made to split or combine, a huge amount of energy has the potential to be released. An atomic bomb is an example of the energy released as a result of nuclear fission as seen in figure 9.9 (whereby a large nucleus splits into smaller fragments), while our Sun is an example of nuclear fusion (whereby two small nuclei combine to form a large nucleus, releasing large amounts of energy).

9.2.4 Kinetic energy

Kinetic energy exists in many different forms. All kinetic energy involves movement, whether it be movement of objects that we see every day, the vibration of particles as thermal energy, or other forms such as sound, light or electrical energy.

FIGURE 9.10 Types of kinetic energy

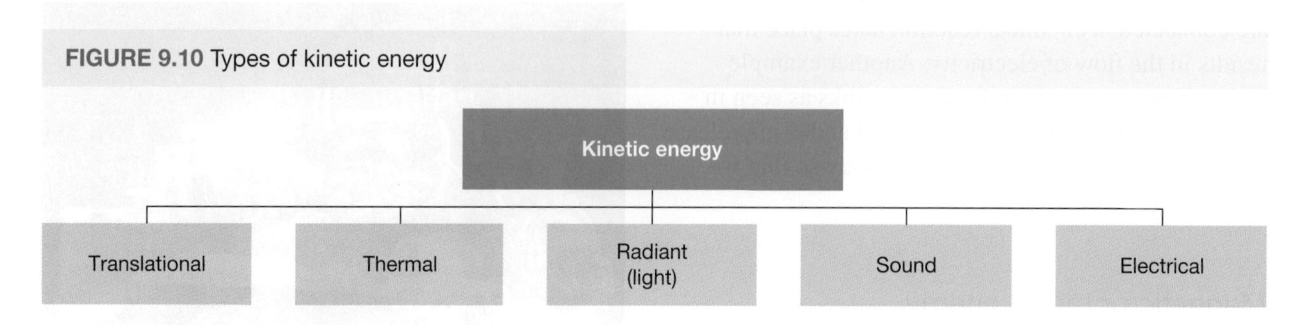

Translational kinetic energy

Any object that is moving has translational kinetic energy. The heavier an object and the faster it travels, the more translational kinetic energy it possesses. A person walking, a ball rolling or a car being driven are some examples of translational kinetic energy.

FIGURE 9.11 The blades in wind turbines contain large amounts of translational kinetic energy when they rotate.

Thermal energy

Thermal energy is more commonly known as heat, although more specifically, heat is defined as the transfer of thermal energy, which flows from hotter objects to cooler objects. Thermal energy transfer can occur as a result of the movement of atoms, molecules or ions within a solid, liquid or gas. This type of transfer is a form of kinetic energy because it requires the movement of particles. However, thermal energy can also occur as a result of radiation, even without the presence of particles, such as between the Sun and Earth, and in outer space. We experience heating when thermal energy is transferred into our body from an object or substance with a higher temperature. Cooling is experienced when thermal energy is transferred out of your body to an object or substance with a lower temperature. If you touch a cold object with your hand, such as an ice block, thermal energy moves from your hand to the ice block. If you touch a hot object, such as a pan on a stove, thermal energy moves from the pan to your hand.

FIGURE 9.12 When you touch a cold object you transfer thermal energy to it, heating it slightly.

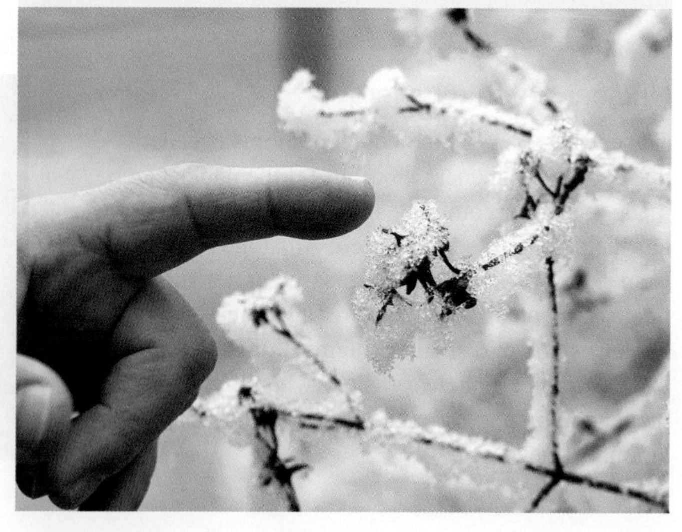

Radiant energy

Radiant energy is the energy of electromagnetic waves. The light that we can see with our eyes is electromagnetic waves with particular frequencies in the **visible spectrum**, which is only one part of a broader **electromagnetic spectrum**. Not all light can be seen with our eyes. Examples of radiant energy include light from the Sun, light bulbs, lamps, torches and flames, x-rays, radio waves, infrared, gamma rays and microwaves. Radiant energy is classified as a form of kinetic energy because waves (and/or particles) carry radiation from one source to another. Light (**electromagnetic radiation**) is the fastest thing known. It travels through space at 300 000 kilometres per second.

visible spectrum different colours that combine to make up white light; they are separated in rainbows

electromagnetic spectrum the complete range of wavelengths of energy radiated as electric and magnetic fields

electromagnetic radiation the radiant energy such as radio waves, infrared, visible light, x-rays and gamma rays released by magnetic or electric fields

FIGURE 9.13 Hot objects release radiant energy in the form of light.

Sound energy

Sound involves the vibration of particles in the air or another medium. It is therefore a form of kinetic energy. A sound source — such as an instrument or our voice — vibrates, causing the nearby particles in the air to vibrate. We are able to hear some sounds because our ear can detect the vibration of particles and send a message to our brain, which tells us the type of sound we are hearing.

FIGURE 9.14 The strings on a guitar vibrate, sending sound waves through the air.

Electrical energy

Electrical energy can be a form of both potential energy and kinetic energy. When electric charges are moving through a circuit, it is a form of kinetic energy called electricity. Electricity is used to power most of your favourite devices, such as your television, your smart phone and your computer. When electric charges are separated they possess potential energy that will be converted into other forms once a critical amount is reached. An example of this is in lightning, whereby a charge imbalance builds up to the point where it cannot be sustained, leading to a lightning strike that converts the electrical potential energy into thermal energy, light and sound.

FIGURE 9.15 Electrical energy powers all electrical devices, such as this tablet.

TABLE 9.1 A summary of the types of energy

Potential energy (stored energy that, when released, is converted to forms of kinetic energy)	**Kinetic energy (often converted from potential energy, these forms are more easily observed by our senses)**
Gravitational — potential energy of an object elevated above the ground 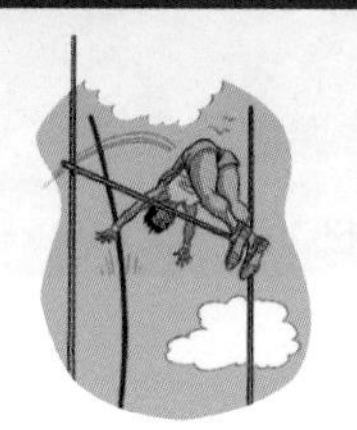	Translational kinetic — energy possessed by objects that are moving 
Elastic — energy stored by an elastic object that is stretched, such as a spring or rubber band 	Thermal — energy that causes objects to gain temperature 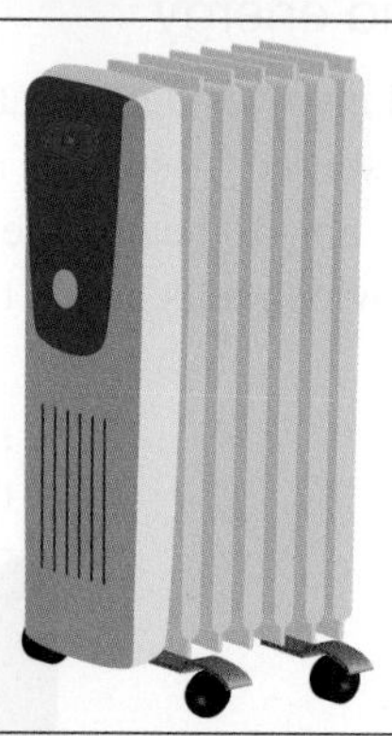
Chemical — energy stored in chemicals that is released as heat, sound, light or other forms of energy in a chemical reaction 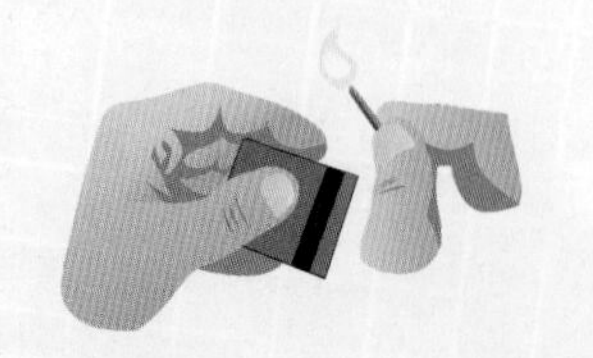	Radiant — energy that may be released, for example, when an object is hot or by a nuclear reaction in a star 

Nuclear — energy stored in the nucleus of atoms that can be released slowly, such as in a nuclear reactor, or quickly, such as in a nuclear explosion		Sound — energy carried by vibrating particles and detected by the ear	
Electrical — energy stored by the build-up of charge		Electrical — energy provided by the movement of electrons	
Magnetic — energy stored in magnets or metals placed in a magnetic field	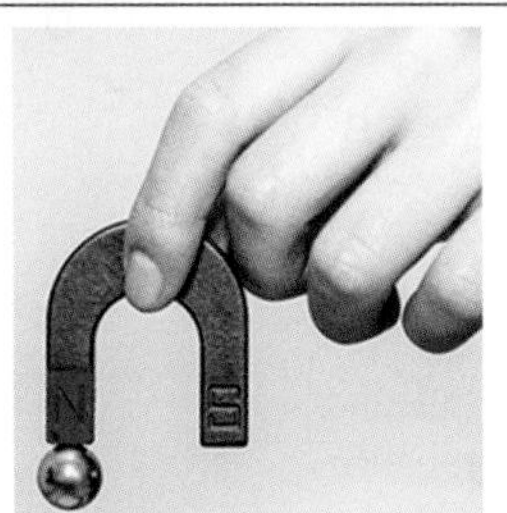		

on Resources

Video eLesson Energy in disguise (eles-0063)

9.2 Activities

learnon

9.2 Quick quiz on | **9.2 Exercise**

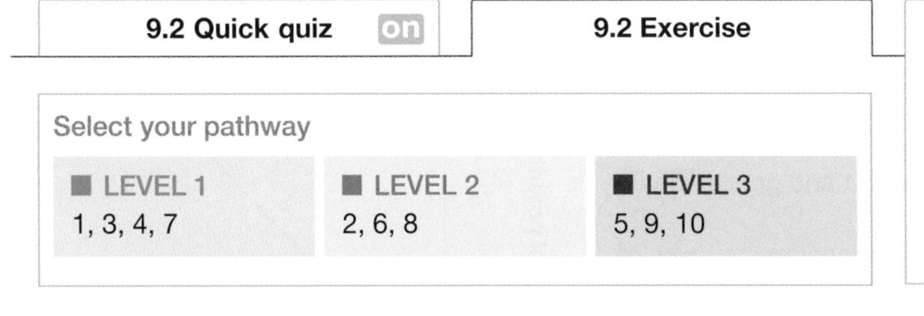

These questions are even better in jacPLUS!
- Receive immediate feedback
- Access sample responses
- Track results and progress

Find all this and MORE in jacPLUS

Remember and understand

1. State the Law of Conservation of Energy.
2. Classify the following as examples of potential energy or kinetic energy:
   - An athlete running
   - A spring being squashed
   - Sound coming from a speaker
   - A skydiver about to jump from an aeroplane
   - The light emitted from a globe.
3. List five types of potential energy.

Apply and analyse

4. For each of the following statements:
   i. determine whether it is true or false
   ii. justify your response.
      a. As a ball is thrown up into the air, it gains more gravitational potential energy the higher it moves.
      b. Elastic energy is a type of kinetic energy.
      c. Only springs and rubber bands can have elastic potential energy.
      d. Sound is a type of kinetic energy.
      e. Fusion is the process of splitting the nucleus.
5. Identify four types of energy that are present during a lightning strike.
6. How can you tell that a high diver has gravitational potential energy?
7. **MC** A student throws a paper aeroplane and it follows the path shown by the dotted line. Consider the points in its flight labelled A, B, C, D and E.
   At which point does the paper aeroplane have the greatest gravitational potential energy?
   A. Point A
   B. Point B
   C. Point C
   D. Point D
   E. Point E

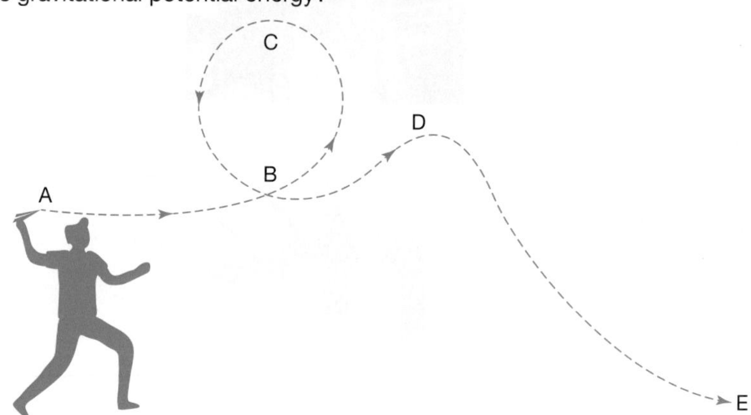

Evaluate and create

8. **SIS** Create a poster that illustrates the different forms of energy found in a moving car. Include a diagram of a car with arrows and labels indicating where different forms of energy exist.
9. **SIS** For one whole day, keep a tally of the number of times you come across each of the different forms of energy. Present your results in a bar chart.
10. **SIS** The graph shows the relationship between gravitational potential energy (J) against height above the ground (m) of a 1 kg ball thrown into the air.
    a. How much gravitational potential energy does the ball have when it is 10 m above the ground?
    b. At what height above the ground is the ball when it has 75 J of gravitational potential energy?
    c. Describe the relationship between height and gravitational potential energy shown in this graph.
    d. If gravitational energy is directly proportional to the mass of an object, sketch a graph showing the gravitational potential energy versus height for a ball with mass 2 kg thrown into the air.

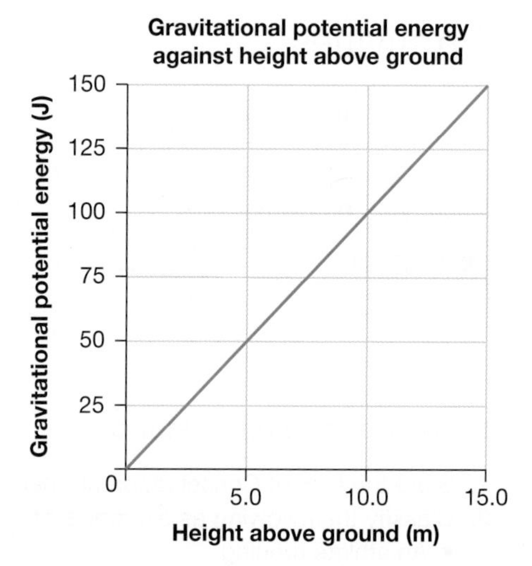

Fully worked solutions and sample responses are available in your digital formats.

LESSON
9.3 Transforming energy

First Nations Australian readers are advised that this lesson and relevant resources may contain images of and references to people who have died.

LEARNING INTENTION

At the end of this lesson you will be able to describe energy transformations using flow diagrams, explain why sometimes energy appears to be lost, and calculate the efficiency of an energy-converting device.

9.3.1 Energy can be transformed

Energy can change from one form to another; we call this an energy transformation or an energy conversion. Sometimes energy transformations result in something happening that we can see with our eyes, while other times the result of an energy transformation may not be so obvious.

Examples illustrating some everyday energy transformations are shown in figure 9.16.

- The chemical energy stored in food is transformed into kinetic energy in the body when you move.
- Electrical energy is transformed into light when a lamp is plugged into a power point.
- Light from the Sun is transformed into chemical energy by plants via a process called photosynthesis.
- Chemical energy stored in batteries is transformed into light when a torch is switched on.
- Chemical energy stored in petrol is transformed into kinetic energy when a car is moving.
- Electrical energy is transformed into thermal energy when a toaster is switched on.

FIGURE 9.16 Examples of different energy transformations, or energy conversions

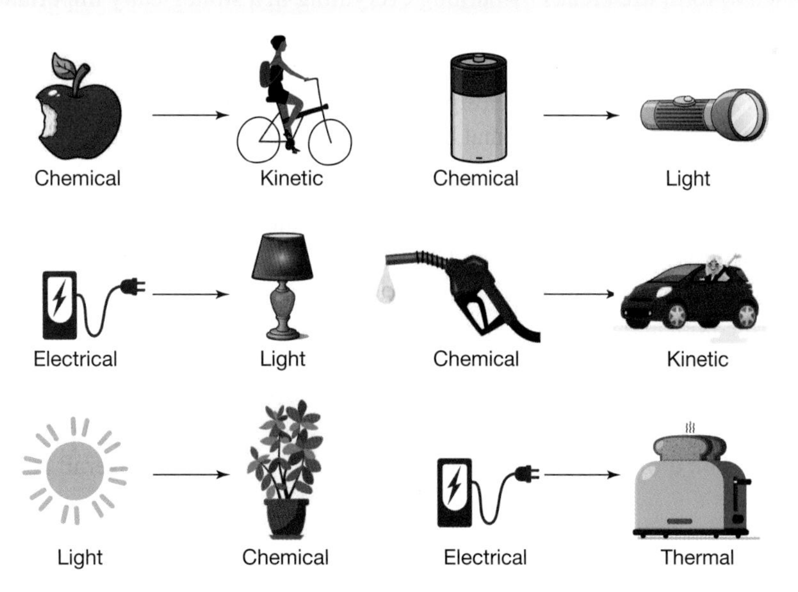

9.3.2 Energy flow diagrams

Energy flow diagrams are a visual way to show the energy transformations occurring in a system. In an energy flow diagram, an arrow is drawn from the energy input to the **useful energy** output (see figure 9.17).

It is usual for there to be more than one energy output, but only the useful forms of energy are listed in energy flow diagrams. In the example given in figure 9.17, heat would not usually be listed as it is not a useful form of energy for a mobile phone. Minor energy outputs that are not useful are known as **by-products**.

useful energy the energy that produces the desired output

by-products in energy terms, energy produced that is not useful

FIGURE 9.17 An example of the energy transformations occurring in a mobile phone are shown in the energy flow diagram.

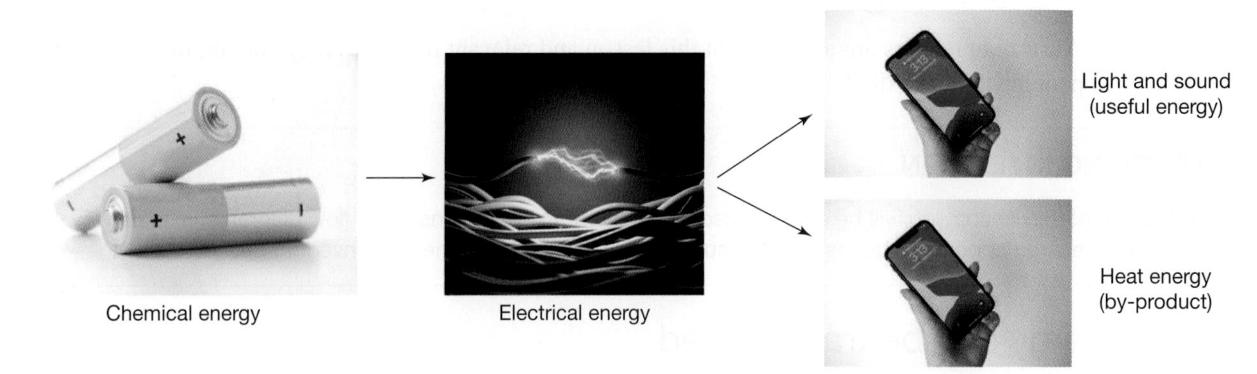

9.3.3 First Nations Australians and fire

Fire was, and still is, a very important tool used by First Nations Australians. It is used for cooking, for warmth, for hunting, for smoking ceremonies, and, very importantly, to manage the landscape. It also holds a deep spiritual meaning to First Nations Australians. Many stories are told and passed on around campfires, either verbally or through various songs and dances. Fire is regarded as a friend and is a very important part of their culture. Fire-starting is an important feature in many First Nations Australian ceremonies.

It is well known that Australia is one of the most bushfire-prone countries on Earth. Not only do these wildfires destroy habitats, but they also add enormous amounts of carbon dioxide to the atmosphere, thus adding to the enhanced greenhouse effect. Since European colonisation, one of the main methods used in attempts to control these bushfires has been hazard reduction burning. These are small but controlled burns of relatively high intensity, designed to form fire breaks by burning everything in a strategically important area. If a bushfire hits one of these areas there will be nothing left to burn, and it will therefore be easier to manage and even extinguish. However, such fire breaks need to be maintained on a regular basis due to regrowth of unsuitable plant types. Also, because everything within a designated area is burnt, unnecessary amounts of carbon dioxide are added to the atmosphere, and plant and animal habitats are destroyed.

The necessity for managing fire, together with the problems mentioned previously, have led many people in recent years to investigate and learn from the practices of First Nations Australians in this area. For many thousands of years prior to European colonisation, First Peoples of Australia practiced *cool burning* as a means of controlling fire and managing their land. This is also called *cultural burning*. This practice requires a detailed knowledge of climate, the flora and fauna in an area, soil types and fire temperature. It also involves 'patchwork' burning rather than the burning of whole, continuous sections. Because the fires are cool, only the necessary plants are burnt and a relatively small amount of carbon dioxide is added to the atmosphere. When regrowth occurs it happens quickly, and more suitable plant types often establish themselves. This regrowth also removes a portion of the added carbon dioxide from the atmosphere, further reducing the overall amount that was added.

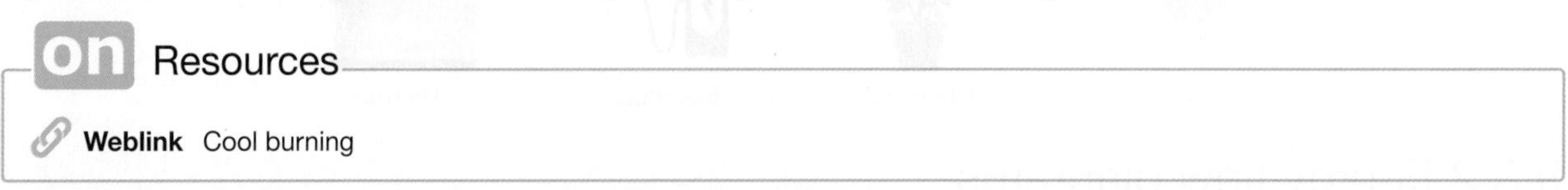

DISCUSSION

Australia has experienced many serious bushfires recently.

- Do you think current methods for preventing bushfires are working?
- Should we be looking at alternative methods for preventing bushfires that are more environmentally friendly?
- Where could you find more information on current fire prevention methods and alternative fire prevention methods?
- Should First Nations Australians fire management practices be incorporated more widely across Australia?

Transforming energy to start a fire

The most common way that fire is started anywhere in the world today is through friction matches. Striking a match along the side of its box produces **friction** and thermal (heat) energy. This then ignites a small phosphorus head. When this and the wood that it is attached to ignite, chemical energy in the wood and phosphorus are transformed into thermal energy. The energy transformations involved may be summarised as follows:

Kinetic energy (by striking)	→	Thermal energy (by friction)	→	Chemical energy (in phosphorus and wood)	→	Thermal energy (heat from burning match)

This sequence is really just an adaptation of methods that were traditionally used by many First Peoples right across the world. Most methods utilised involved friction of wood on wood to raise the temperature high enough to start burning. It was just the phosphorus component that was missing.

Across the length and breadth of Australia, First Nations Australians use two main methods to start a fire. The first is friction of wood on wood and the second, but less widely used, is striking stones such as flint or ironstone together to make sparks.

In the most common method, wood is rubbed against wood to generate heat by friction. There are a number of ways that this is done, but one of the most common is the *fire drill* method.

This involves a round stick (the drill), and a flat piece of wood (the hearth). A small hole is made in the hearth to locate the drill. A small notch is also often cut into the side of this hole so that the hot sawdust produced can fall out onto some type of flammable dried grass or similar material (this is called **tinder**). The drill is then rubbed back and forth between the palms to generate the required friction. Also, small amounts of sand are often added to increase this friction. Once the hot sawdust falls onto the tinder and starts smouldering, it is blown gently to produce flame.

FIGURE 9.18 Fire-starting is a part of many First Nations Australian ceremonies.

This process involves a number of energy **transfers** and transformations. Kinetic energy from the hands is transferred to kinetic energy in the rotating stick. This energy is then transformed to thermal energy through friction, raising the temperature of the sawdust produced. This sawdust eventually reaches a temperature at which it starts to smoulder. At this point, the chemical energy in the sawdust begins to be transformed into more thermal energy. Finally, this thermal energy is transferred to the tinder material. The chemical energy within this material is then transformed into even more thermal energy when it begins to burn. The key transformations involved are:

friction the rubbing together of surfaces; all friction produces thermal energy (heat)

tinder light, dry material that is easy to ignite

transfer in energy terms, the movement of energy from one place to another

Kinetic energy (in rotating drill)	→	Thermal energy (by friction)	→	Chemical energy (in sawdust)	→	Thermal energy (heat from burning tinder)

9.3.4 Falling objects

As an object falls from a height, its gravitational potential energy is converted to kinetic energy. When the object reaches the ground, all its gravitational potential energy has been transformed to kinetic energy.

In figure 9.19, notice that at position A, where the ball has just been kicked, the ball has no potential energy because it is on the ground — all its energy is kinetic.

When the ball reaches position B, some of its kinetic energy has been converted to potential energy as it rises above the ground.

FIGURE 9.19 The energy transformations of a soccer ball in flight

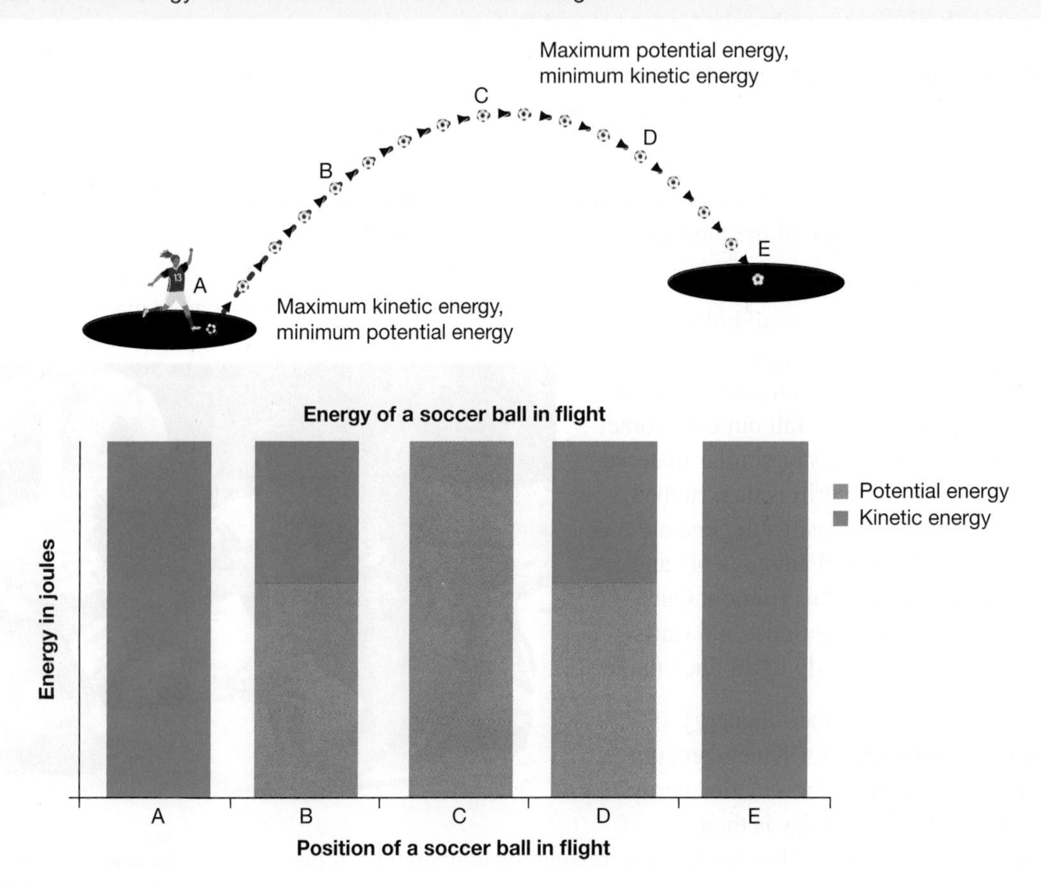

At position C, the ball reaches its highest point, meaning its gravitational potential energy reaches its maximum and its kinetic energy its minimum.

At position D, some of the ball's potential energy is converted back to kinetic energy as it falls.

When the ball reaches position E, all of its energy has been converted back into kinetic energy, and the ball hits the ground at its maximum speed.

Notice that although the ball has different amounts of potential and kinetic energy at each position, the total amount of energy is the same at each position. The ball never gains or loses energy; it simply transforms kinetic energy to potential energy and then back again. Thus, we say that the energy is conserved.

elog-1966

INVESTIGATION 9.1

Energy transformations in a pendulum

Aim

To investigate the behaviour and energy changes in a pendulum

Materials

Laboratory version

- retort stand (as tall as possible)
- boss head and clamp
- string
- object to attach to the string
- book or similar object to mark release height
- timing device

Outdoors 'mega' version

- rope
- heavy object (e.g. concrete block)
- tree branch or beam to tie rope over
- object to mark the release height
- timing device

Method

1. Tie one end of the string or rope to the object that will be the weight at the end of your pendulum.
2. Tie the other end to the clamp or beam that will be the top of your pendulum. Make sure that both ends are tied securely.
3. Pull the weight back to a certain height whilst keeping the string or rope taut.
4. You now need a marker for this height. If inside, you could place a book just behind the pendulum at this point. If outside, you could just use your hand to mark this point, being careful to keep it still.
5. Release the pendulum (do not push it) and make sure it swings freely without hitting anything (including yourself!).
6. Observe what happens each time the pendulum returns to its starting position. Time how long it takes for each complete swing (this is called the **period**).
7. Repeat steps 3–6 using different starting heights.

Results

1. Describe the pattern that occurs each time the pendulum swings back to its starting position.
2. Do different starting heights affect this pattern?
3. Draw up a table to record the different starting heights you used, and the period obtained in each case.

Discussion

1. Do different starting heights affect the period?
2. The energy contained in your pendulum is distributed between gravitational potential energy (GPE) and kinetic energy (KE). Describe the energy changes that occur during each complete swing of your pendulum.
3. Does your answer to question **2** account for all the energy transformations involved? Support your answer with observations from your experiment.
4. A child's swing is an example of a pendulum. In energy terms, explain why a child can start from rest and make the swing go higher with each swing.
5. Years ago, many clocks had a pendulum mechanism to help them keep accurate time. The pendulum was able to keep swinging backwards and forwards for a considerable period of time without dying out. Find out why this was possible and explain your answer in terms of energy.

Conclusion

Summarise your findings from the investigation about the total energy in a pendulum and the forms in which it is presented.

period the time taken for one oscillation of a pendulum

Next, consider the energy transformations involved in bouncing on a trampoline (figure 9.20). As the person jumping falls towards the trampoline, some of their gravitational potential is converted to kinetic energy. When they reach the trampoline mat, their gravitational potential and kinetic energy are converted to elastic potential energy stored in the springs. Then, as the person jumps back up into the air, the elastic potential energy is converted back into kinetic and gravitational potential energy.

FIGURE 9.20 Energy conversions that occur when bouncing on a trampoline

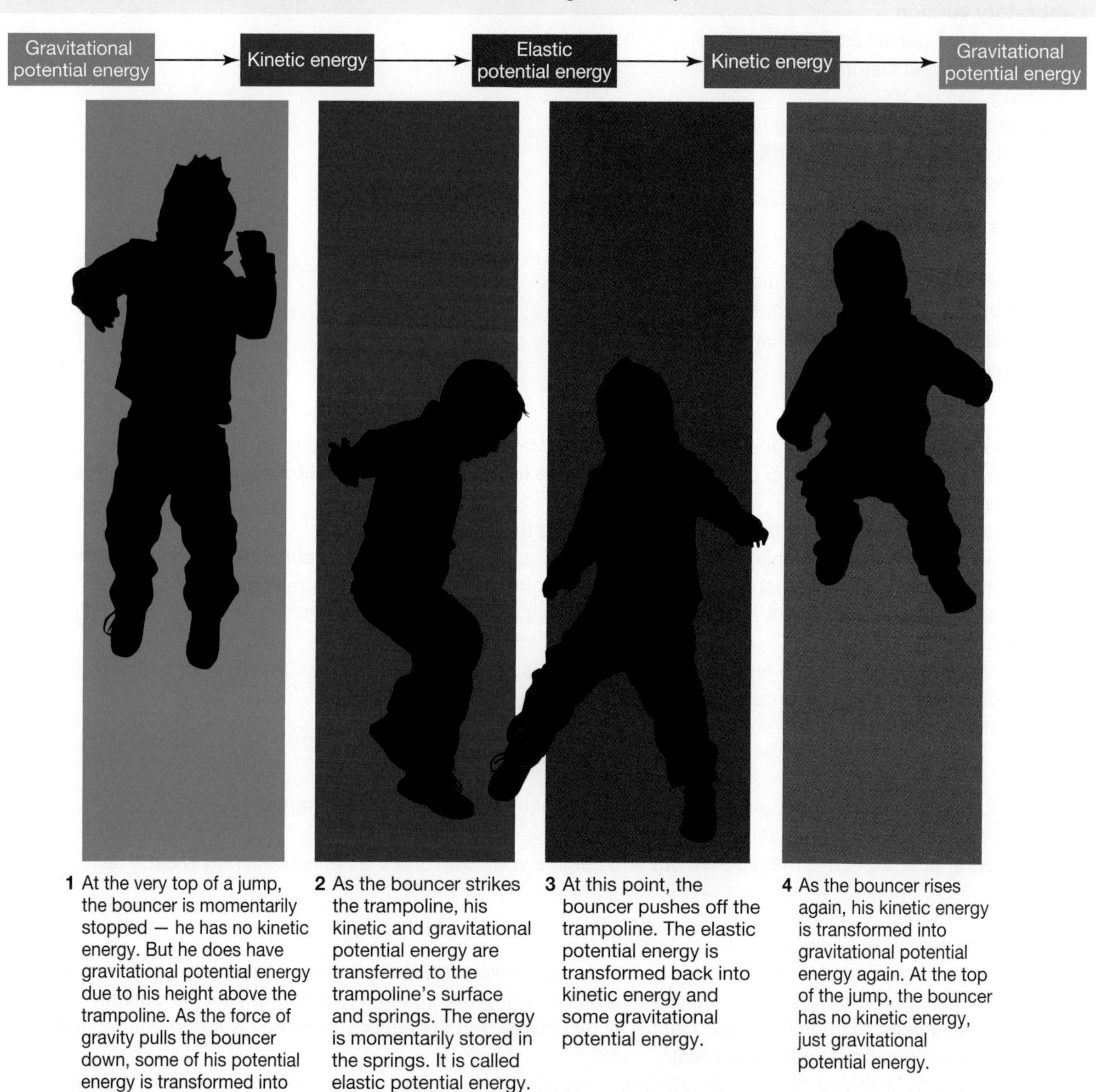

1 At the very top of a jump, the bouncer is momentarily stopped — he has no kinetic energy. But he does have gravitational potential energy due to his height above the trampoline. As the force of gravity pulls the bouncer down, some of his potential energy is transformed into kinetic energy.

2 As the bouncer strikes the trampoline, his kinetic and gravitational potential energy are transferred to the trampoline's surface and springs. The energy is momentarily stored in the springs. It is called elastic potential energy.

3 At this point, the bouncer pushes off the trampoline. The elastic potential energy is transformed back into kinetic energy and some gravitational potential energy.

4 As the bouncer rises again, his kinetic energy is transformed into gravitational potential energy again. At the top of the jump, the bouncer has no kinetic energy, just gravitational potential energy.

elog-1967

tlvd-10796

INVESTIGATION 9.2

Ice-cream stick energy

Elastic potential energy is energy stored in an object due to a change in its shape.
In this activity you will make a basketweave pattern from some ice-cream sticks. When you make this, each time you add a new stick to the pattern a small amount of bending is required. This stores energy in the arrangement that you are producing.
You will then observe what happens when this energy is released.

Aim

To demonstrate the storage and subsequent release of a form of potential energy

Materials

- a bundle of ice-cream sticks (If you can get tongue depressors these are even better, as they are the same shape but more flexible.)
- clamp(s)
- two pieces of ruler (a 30 cm wooden ruler cut in half)
- video recorder (preferably one with a slow-motion option)

Method

1. Make a starting piece as follows.
   Place the ends of three sticks between the two ruler halves to make a sandwich. Make sure the sticks poke out to the *right*, as shown in the following diagram.
   Next, clamp the two ruler halves tightly together. You may need to cut off the bits that poke out to the *left* of sticks 1 and 2 so that you can apply the clamps.

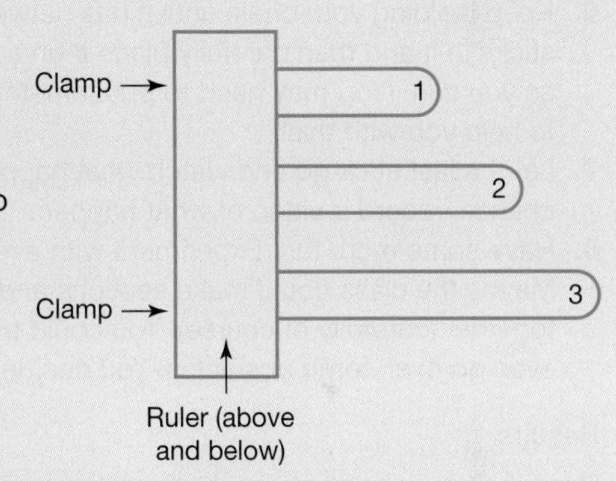

2. Now take two more sticks (4 and 5) and add them as shown in the following diagram. Make sure that they go above and below sticks 1, 2 and 3 as shown.

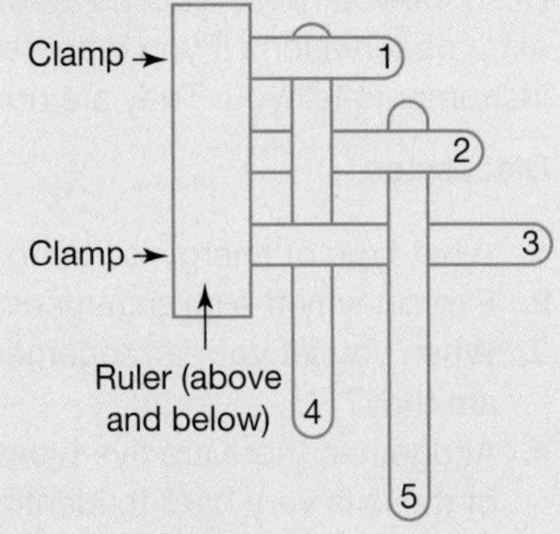

3. Add one more stick (6) as shown in the following diagram, once again making sure that it goes above and below as shown.

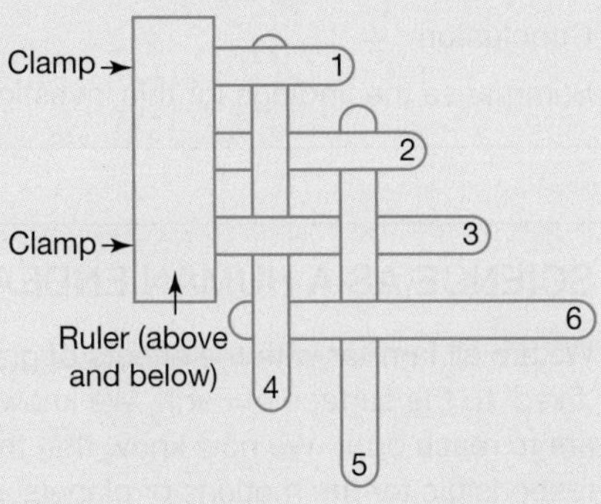

4. Add your next stick (7) as shown in the following diagram.
   At this stage you should notice that a 'basketweave' pattern is starting to form. You may also notice that your construction is starting to feel 'springy'.

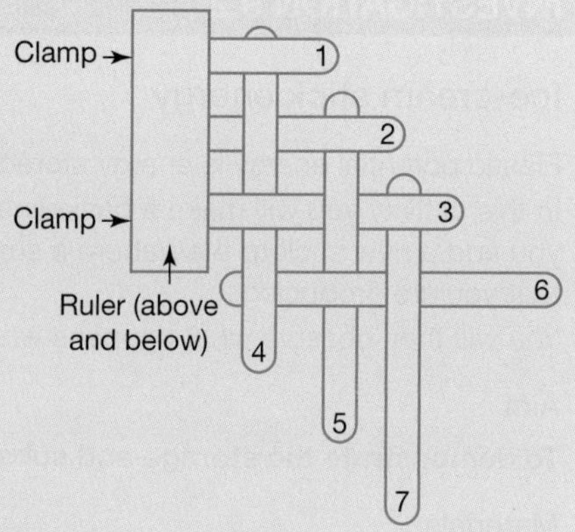

5. Continue adding sticks as in steps 3 and 4.
   Tip: The 'springyness' in your arrangement will probably become a problem for you. To overcome this, temporarily swing each new stick sideways, as shown in the following diagram, whilst you organise yourself to add the next stick. Make certain you return it to its proper position though before you add the next stick.

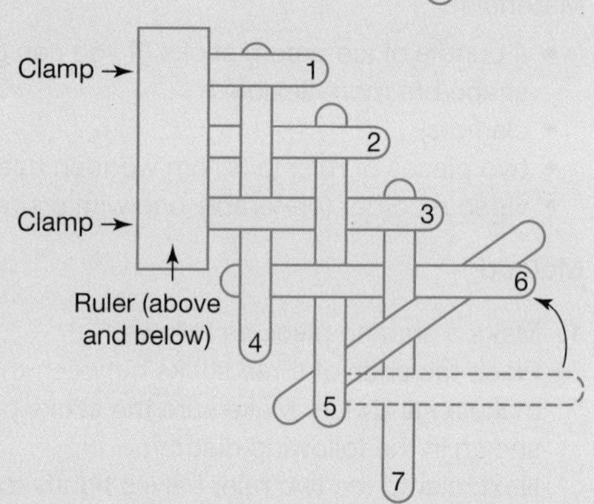

6. Keep building your chain until it has between twenty to thirty sticks in it and then carefully place it on a hard floor. Lay it as flat as you can. (You may need to put something under the clamps to help you with this!)
7. Let the last stick go and watch what happens. If you have a chance, record a video of what happens.
8. Have some more fun! Experiment with even longer chains. Maybe the class could make sections and you could add these together (carefully of course). You could try different surfaces or even go over some obstacles. You decide!

Results

Describe your observations, both while you were building your chain and when you let it off. Be sure to add any extra observations if you repeated this experiment. Remember that your observations are what your senses or instruments tell you. They are not explanations.

Discussion

1. What type of energy is stored in the ice-cream sticks when you bend each one to fit it into the pattern?
2. Explain why the arrangement feels springy, especially as you add more and more sticks.
3. When you let your arrangement off, there are three main energy transformations that take place. What are they?
4. Altogether, there are five types of energy involved in this experiment. Try to name them. (You may find that one of these is very hard to identify!)

Conclusion

Summarise the findings for this investigation about the types of energy that were demonstrated.

SCIENCE AS A HUMAN ENDEAVOUR: Newton, Einstein and gravity

We are all familiar with the effects of gravity. We know that it causes objects to fall. We know that it keeps us 'fixed' to the surface of Earth. We know that spacecraft have to travel fast enough to overcome its pull if they are to reach orbit. We now know that the same force that we are familiar with here on Earth's surface is also responsible for the motions of planets, stars and entire galaxies.

Although we take it for granted, gravity is one of the most important forces in nature. Although its effects have long been observed, it took the work of two men — Sir Isaac Newton and Albert Einstein — to describe it and unravel its consequences. In science, theories have to not only explain current observations, but also need to predict new observations that can be tested by experiments. Newton's and Einstein's theories enabled this to be done. People's knowledge and understanding of gravity was therefore moved from just a descriptive basis to a true scientific basis.

Sir Isaac Newton

Sir Isaac Newton is regarded as probably the most influential figure in western science.

Newton was born in England in 1642 and died in 1727. During his lifetime he was responsible for many discoveries, inventions and theories. These included:

- the invention of the first reflecting telescope. Reflecting telescopes use mirrors and a relatively small lens. Prior to this, telescopes relied on arrangements of lenses to gather and focus light. Today, virtually all astronomical telescopes are reflectors, whereas nearly all terrestrial telescopes use lenses.
- the discovery that white light could be broken into the colours of the rainbow by a prism. From this and further work, he put forward a new theory that light was composed of tiny particles.
- the invention of a simple form of calculus — an entirely new branch of mathematics
- the discovery that motion could be explained and described by the action of forces. The laws that he put forward have since come to be known as Newton's first, second and third laws of motion.
- his theory of universal gravitation.

A popular story is that Newton 'discovered' gravity when he saw an apple fall from a tree. We don't know if this is true, but we do know that he was forced into long-term isolation by the Great Plague of London in 1665. (A bit like our lockdowns in recent times due to the COVID-19 pandemic!) This gave him plenty of uninterrupted time to imagine and think, which allowed him to develop his theory that the gravity experienced on Earth was the same force that operated throughout the universe.

The essential features of this theory are as follows:

- Gravity is an attractive force between objects and is due to their mass. (All objects will therefore attract each other.)
- Gravity is a very weak force, but is due to the amount of mass an object has. (This means that its effects are really only noticed when objects have a very large mass; for example, planets stars and moons.)
- The strength of gravitational attraction increases as objects get closer together, and vice versa. The amount of this change is predicted by the 'inverse square law'. For example, if two objects halve the distance between them, gravity will be four times stronger; if they double their distance apart, it will be four times weaker.

After Newton, much of the physical world could now be described with great mathematical accuracy. In 1846, for example, the planet Neptune was discovered. It had been earlier observed that the orbit of the planet Uranus seemed to be affected by something further out in the solar system. Newton's laws were used to predict where this new planet would be, thus leading to Neptune's discovery.

Newton's laws are still used today and give accurate results in most situations. However, science is an everchanging subject and is always being influenced by new observations and ideas. In situations where gravity is very strong it has been observed that these laws do not always apply as expected. It took until the early twentieth century for another great scientist, Albert Einstein, to figure out why.

Albert Einstein

Born in Germany in 1879, Albert Einstein is also regarded as one of the most influential scientists of all time.

In 1905, Einstein published three papers that immediately established his reputation as a brilliant scientist.

- The first of these provided mathematical proof for the existence of molecules.
- The second concerned itself with something called the 'photoelectric effect'. This led to the invention of many devices that are important and familiar today, including transistors, photelectric cells, computers, electron microscopes and LEDs, to name just a few. Einstein was awarded a Nobel Prize for this paper in 1921.
- The third introduced Einstein's famous 'special theory of relativity'. This paper dealt with the motion of objects relative to each other, especially when they were travelling at close to the speed of light. It also said that nothing can travel faster than the speed of light (300 000 kilometres per second).

Following this, Einstein spent a number of years working on a broader theory of relativity by incorporating gravity into it. By 1916 he was ready to publish this work, which has since become known as the 'general theory of relativity'. This theory was not only able to incorporate Newton's ideas about gravity, but was also able to correct certain problems that had been discovered in them. A new way of looking at gravity was born.

One of the predictions from the general theory of relativity was that light could be bent in a certain predictable amount when it passes a massive object. Einstein predicted that the position of a star just behind the Sun would appear to shift a little as its light passed the massive Sun. Normally, of course, this would not be noticeable because of the Sun's brightness. However, during a complete solar eclipse, it should be noticed. In 1919, such an eclipse occurred and scientists were able to verify Einstein's predictions.

FIGURE 9.21 Einstein correctly predicted how light would be bent as it passes the Sun.

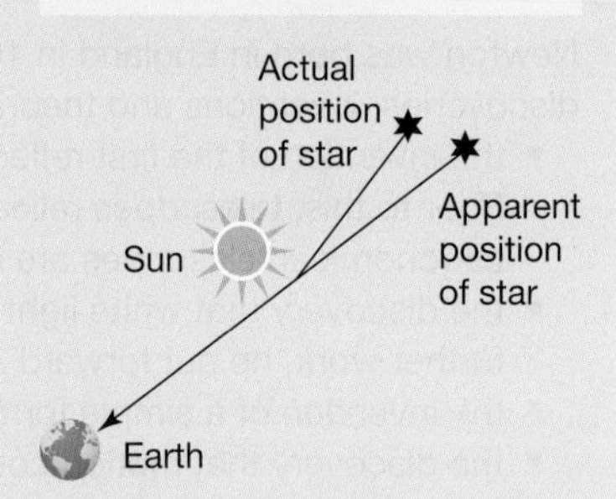

In 1933, Einstein moved to the USA to avoid being in Germany during the rise of Adolf Hitler. In his lifetime he held German, Swiss and American citizenships. At the time of his death in 1955, he was working on a way to bring together many important scientific theories into one 'supertheory', a quest that continues to this day.

What inspired Einstein to think about gravity?

Scientists were aware of a number of things that could not be explained by Newton's ideas of gravity. One example was the irregularities in the orbit of the planet Mercury. Mercury is the smallest planet in our solar system and is the closest to the Sun. It was even thought for a while that there must be another planet even closer to the Sun pulling on Mercury and affecting its orbit. This proposed planet would have been almost invisible due to the Sun's brightness. Einstein questioned this and his general theory of relativity was able to explain the irregularities in Mercury's orbit.

Einstein also questioned why light must travel in straight lines. He was able to explain why light might sometimes appear to travel in a curved path, especially when viewed from different viewpoints, and also when passing close to massive objects with strong gravity.

Another question that worried Einstein was how gravity actually worked. He wanted to explain its workings, rather than just describe them as Newton had done. This led him to predict the existence of gravity waves. These were only discovered recently in 2015.

Resources

Weblink What led Einstein to think of gravity in terms of warps and curves in space time?

9.3.5 Energy and electrical circuits

One of the most useful forms of energy in today's society is electrical energy. It is easily transmitted and so many devices exist to convert it into other forms of energy. Its use is seen just about everywhere, and we are so accustomed to using electrical energy that it can often be taken for granted. How many times have you turned a light switch on during a blackout and then been surprised that the light has not come on?

To make use of electricity in our devices, an **electrical circuit** is required. You will recall from topic 5 that atoms are made up of even smaller sub-atomic particles. One of these particles is the negatively charged electron. Electrons are the most common type of particles that flow around an electrical circuit and they carry energy in the form of electrical kinetic energy. This flow is called an **electric current**.

electrical circuit consists of a power supply, a conducting path for charge to flow and one or more loads

electric current a measure of the number of electrons flowing through a circuit every second

As the name suggests, a circuit is a path that ends up back at its beginning and can be travelled upon without any breaks or gaps.

What is an electric circuit?

All electric circuits consist of three essential items:

- a **power supply** to provide the electrical energy
- a **load** (or loads) in which electrical energy is converted into other useful forms of energy
- a **conducting path** that allows electric charge to flow around the circuit.

FIGURE 9.22 This electric circuit is a model of an electric kettle. The wire coil, or the loading coil, will heat up when electrical energy is passed through it.

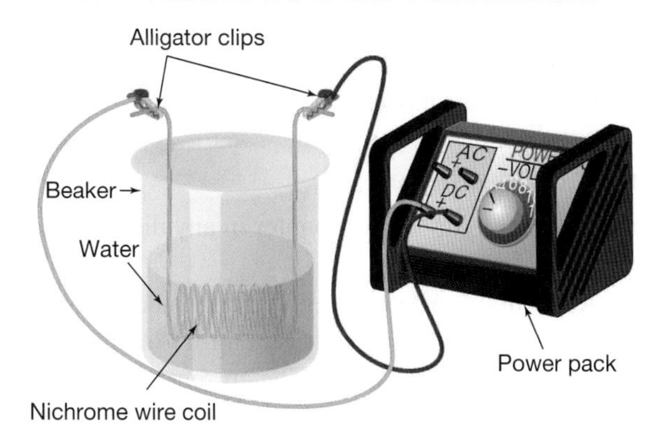

Power supplies

The job of a power supply is to provide a supply of electrons and push them through a wire.

Different types of power supplies include:

- batteries (one on its own is called a **cell**), which are used in torches and many other devices. These *store chemical energy* in the substances inside them. The chemical energy is transformed into electrical energy when a chemical reaction takes place inside the cell. Many battery-operated devices use more than one battery connected in **series**. They are connected end-to-end. It is important to ensure that the positive end of one battery is connected to the negative end of the other.
- generators at power stations. The electrical energy that is used when you turn on a light switch or an appliance connected to a power point comes from a power station. This is where a generator pushes electrons.
- a solar panel, which can take the energy of sunlight and convert it directly into a flow of electrons in a circuit.

DISCUSSION

Do we need wires to transfer electrical energy? If not, why do we use them?

Loads

The load in an electric circuit is an energy converter (**transducer**). It is here that most of the electrical energy carried by electric charge is transformed into useful forms of energy such as light, heat, sound and movement. In specialist light globes the load is the **filament** (figure 9.23), a coiled tungsten wire inside the globe. The filament glows brightly when it gets hot; here, the load converts electrical energy into light (and thermal) energy. In a hairdryer there are two loads: a heater and a fan motor. Loads 'push back' against the power supply. We call this **resistance**. The greater the resistance, the less current can flow in the circuit.

Conducting path

The electrical energy provided by batteries and power outlets is transformed into other forms of energy *only* when the conducting path is complete, which allows electric charge to flow through the circuit.

Conducting paths have the following features:

- In an efficient electric circuit most of the electrical energy provided by the power supply is transformed in the load.
- Some of the electrical energy is transformed in the conducting path, heating the path and its surroundings. The more this occurs, the less efficient the circuit.

power supply a device that can provide an electric current

load a device that uses electrical energy and converts it into other forms of energy

conducting path a connected series of materials along which an electric current can flow

cell a single battery

series a formation of electricity-generating or electricity-using devices whereby the electricity passes from one device to the next in a single conducting loop, one after the other

transducer a device that converts energy from one form into another form

filament a coil of wire made from a metal that glows brightly when it gets hot

resistance a measure of the opposition to an electric current passing through an object, measured in ohms (Ω)

- The conducting paths in the electric circuits in appliances are usually made of metals such as copper so that they have little resistance to the flow of electric charge. The conducting path in a torch consists of copper wires covered with an insulating layer of plastic.
- Any kind of break in the path, such as a broken wire or burned-out component, will stop the current flowing.
- A deliberate break in a circuit can be made using a **switch**. This allows you to have control over whether or not the conducting path is complete.

FIGURE 9.23 The filament in this light globe is an example of a load in an electric circuit. The word 'filament' comes from the Latin term *filamentum*, meaning 'spin'.

Current and voltage

Two terms that you will encounter in any discussion on electric circuits are current and voltage.

- As mentioned previously, current is the flow of electric charge (most often electrons) in a circuit. (It is helpful to think of it in similar terms to the flow of water in a river.) The current in an electrical circuit is a measure of the amount of electrons that pass a particular point in the circuit each second. Current is measured in units called *amperes* (or *amps* for short). A current of 10 amps in a circuit will therefore have ten times the amount of electrons passing through it each second compared to a circuit with a current of 1 amp.
- Electrons do not move around a circuit by themselves. They need to be 'pushed'. **Voltage** measures the energy that is given to the electrons to push them around a circuit.

Current and voltage are linked to one another by a third term — resistance. As mentioned previously, resistance measures how difficult it is for electrons to flow around a circuit. If resistance is high, more energy (voltage) will be required to push electrons around a circuit. Typically, metals have low resistance and electrons flow through them easily. Because of this, metals and other substances with low resistance are called **conductors**. However, there are other substances, such as plastics, that have such a high resistance that no amount of voltage can push electrons through them. These substances are called **insulators**. The wires that are used to construct electrical circuits are good examples of both these points. The core of these wires is usually a metal, such as copper or aluminium. This offers low resistance, and electrons travel easily through this. Surrounding this is a plastic wrapping that insulates the metal core. This stops the electrons from accidentally leaving the wire and travelling in unintended paths if the wire accidentally touches another conductor.

Watch that load!

In an electrical circuit, it is the load that determines how the electrical energy is converted. Some common situations are:

- *converting electrical energy into light energy.* This is done by using a globe of some sort. There are many different types of globes that do this. Examples include older-type incandescent globes, fluorescent globes and halogen globes. A relatively new type of globe that is rapidly becoming popular is the LED globe. LED stands for 'light-emitting diode'. These globes have a much lower resistance, and so much less electrical energy is wasted when the energy from the electrons is converted into light energy.
- *converting electrical energy into sound energy.* Devices that do this include speakers, bells and buzzers. These devices have an internal structure that features a number of energy conversions. For example, a simple bell involves conversions between electrical, magnetic, kinetic, elastic and sound energy. The overall effect, however, is that when placed in an electrical circuit, electrical energy is converted into sound energy.

switch a device that opens and closes the conducting path through which a current flows

voltage the amount of energy that is pushing electrons around a circuit, per coulomb of charge that flows between two points

conductors materials that have a very low resistance, allowing current to flow through them with ease

insulators materials that do not allow electricity to flow easily

- *converting electrical energy into heat energy.* This is done by placing a load that has a high resistance into the circuit. As the electrons travel through this load, they lose a lot of the energy that they originally had to overcome this resistance. However, this energy is not totally lost. It is converted into thermal (heat) energy. This is the Law of Conservation of Energy at work.

The following activities and investigations all illustrate how electrical circuits can be used. You will notice that in each case they demonstrate how the electrical energy is *transferred* from the power supply to the load (through the conductors made up of wires and other metallic objects), and how the electrical energy is *transformed* (which depends on the type of load).

elog-2203

INVESTIGATION 9.3

Making the right connections

Aim

To connect a battery to a light globe so that it lights up

Materials

- 2.5-volt torch light globe
- 1.5-volt battery
- two connecting leads

Method

1. Connect one or two connecting leads, a 2.5-volt light globe and a 1.5-volt battery to make the globe light up.
2. Try the different arrangements as shown in figures A, B, C and D to see whether there is more than one way to make the globe light up.

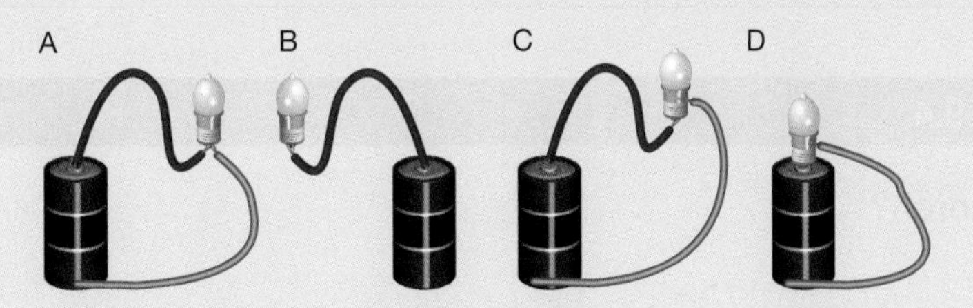

3. Try different arrangements to see whether there are other ways to make the globe light up.

Results

1. In which of the electric circuits shown are the components correctly arranged so that the light globe will work?
2. Describe, with the aid of a diagram, any other arrangements that cause the globe to light up.

Discussion

1. Draw a flow diagram to show the energy transformations that take place when the globe lights up.
2. Are all the energy transformations that take place useful? Explain your answer.

Conclusion

Summarise the findings for this investigation about creating electric circuits.

CASE STUDY: The torch circuit

Features of the torch circuit:

- The power supply of a torch usually consists of two or more 1.5-volt batteries connected in series. When two 1.5-volt batteries are connected in series, the total voltage is 3.0 volts. Twice as much electrical energy is available to move the electric charge around the circuit.
- The load in a torch circuit is the globe.
- When the switch is closed, electric current flows around the circuit.
- As electric charge passes through the globe, its electrical energy is released as heat in the filament. The filament is the coiled wire inside the globe. It is made of the metal tungsten and glows brightly when it gets hot.
- The conducting path consists of the spring that pushes the battery against the base of the globe (or a metal globe holder) and the metal strip that includes the switch. When the switch is open, the metal strip does not make contact with the globe and the circuit is not complete.

FIGURE 9.24 a. Components of a torch
b. Circuit diagram for a torch (see 9.3 Exercise, question 10)

a.

b.

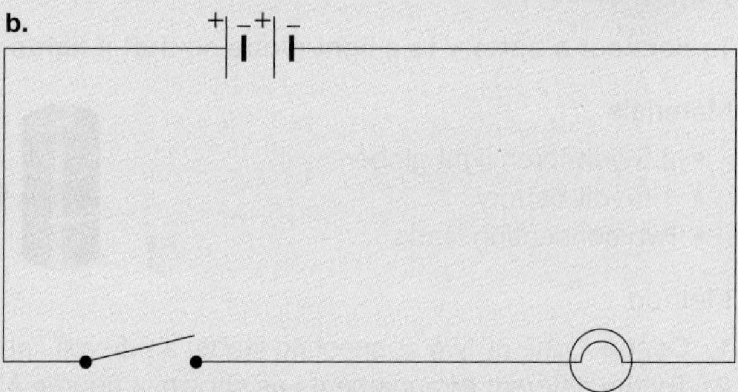

elog-2205

INVESTIGATION 9.4

What's inside a torch?

Aim

To investigate the electric circuit in a torch

Materials

- torch fitted with two 1.5-volt batteries
- hand lens

Method

1. Check that closing the switch makes the globe light up.
2. Unscrew the end of the torch and remove the batteries. Look closely at the batteries.
3. Look at the globe.
4. Carefully remove the globe and examine it with a hand lens.
5. Look inside the case of the torch and locate the spring and metal strip.
6. Close the switch.

Results

1. How were the batteries connected together inside the torch?
2. Draw a diagram to show what is inside the globe.

Discussion

1. What is the voltage of each battery?
2. What does the bottom of the globe touch when it is inside the torch?
3. What does the side of the globe touch when it is inside the torch?
4. Which two parts of a working torch does the spring make contact with?

5. What happens to the metal strip while the switch is being closed?
6. What does the metal strip in front of the switch touch when the switch is closed?
7. What other forms of energy is the electrical energy changed into when the circuit is closed?

Conclusion

Summarise the findings for this investigation.

ACTIVITIES

Constructing a model torch

Construct your own model torch circuit using the following items: a torch globe and holder; two 1.5-volt batteries and holders; connecting leads with alligator clips or banana plugs; and a switch. Use other available materials to make your model torch circuit more realistic.

Steady-hand tester

Make a steady-hand tester. You will need: an old wire coathanger or similar thickness wire that can hold its own shape; a loop of thin wire; wire cutters; a battery; an electric bell or light globe; connecting wires; and a shirt box, shoe box or cereal packet for the base.

The 'alarm' can be a bell hidden in the base or a globe attached to the base. Hide as much of the connecting wires as you can.

FIGURE 9.25 A steady-hand tester

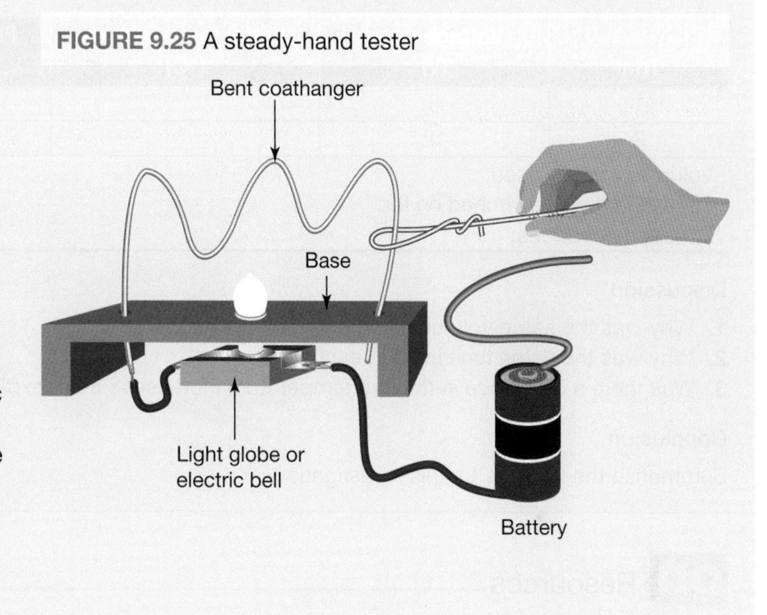

elog-1968

INVESTIGATION 9.5

Making a model kettle

Aim

To make a model kettle and investigate the effect of voltage on its function

Materials

- 100 mL beaker
- wires with alligator clips attached
- power supply with different voltage options
- length of nichrome wire
- thermometer
- glass stirring rod
- measuring cylinder
- pencil

Method

1. Use figure 9.22 to make your electric kettle model.
   Tip: Wind the nichrome wire around the pencil to make your coil. Make sure individual coils do not touch one another.

2. Ask a member from another group to check your circuit. Then ask your teacher to check it.
3. Use the measuring cylinder to half-fill your beaker with water. Make sure that the coils in the nichrome wire are well covered. Also make sure that you have accurately measured how much water you used.
4. Place the stirring rod in the beaker and gently stir the water. Be careful not to damage the wire coil.
5. Use the thermometer to record the temperature of the water.
6. Set the power supply to 6 volts and turn it on. Let it run for about 10–15 minutes. Gently stir the water during this time.
7. When the time is up, switch off the power supply and record the new temperature.
8. Empty out the water and then repeat steps 3–7, but this time set the power supply to 12 volts. Make sure that you use the same volume of water and that you let it run for the same time.

Results

Record your results in a table like the one shown.

Trial	Voltage used (V)	Initial temperature (°C)	Final temperature (°C)	Temperature increase (°C)
1	6			
2	12			

Volume of water used:
Time circuit was switched on for:
Other observations:

Discussion

1. Why was the same volume of water used in each trial?
2. Why was the same time used in each trial?
3. Was there a difference in the two temperature increases? If so, explain why.

Conclusion

Summarise the findings for this investigation.

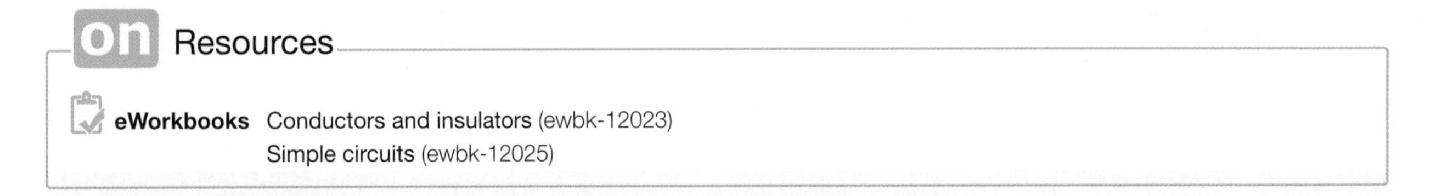

on Resources

eWorkbooks Conductors and insulators (ewbk-12023)
Simple circuits (ewbk-12025)

9.3.6 Solar cells

A solar cell, or photovoltaic cell, is a device that converts light energy from the Sun into electrical energy. When light from the Sun strikes the thin semiconductor layer in the solar cell, electrons are knocked free from their atoms. If the solar cell is connected to an electrical circuit, the free electrons flow through the circuit, creating electricity that can be used to power devices. Energy can also be stored in batteries for later use; for example, at night when there is little light. The most efficient solar cells designed for home use convert around 20 per cent of the energy arriving from the Sun into useful energy.

Several solar cells can be connected together to form a photovoltaic module, more commonly known as a solar panel. Multiple modules can then be wired to form an array. You may have seen an array of solar panels on the roof of a house (figures 9.26 and 9.27).

FIGURE 9.26 Solar arrays are made up of modules, which are made up of cells.

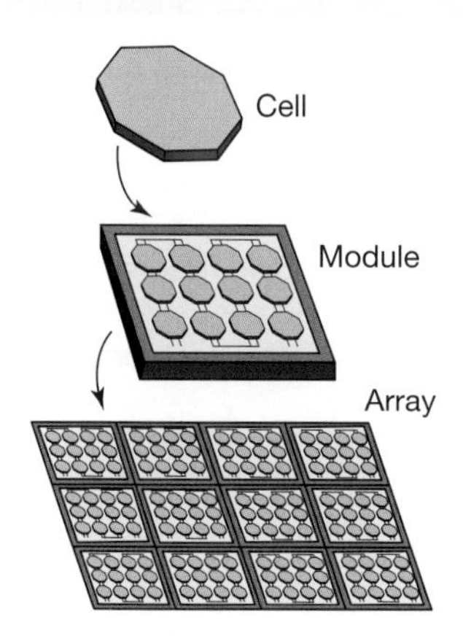

FIGURE 9.27 Solar arrays are often placed on roofs to provide cheap, sustainable energy.

9.3.7 Energy 'loss'

Every electrical appliance you use, whether powered by batteries or plugged into a power point, converts electrical energy into other forms of energy. Most of that energy is usually converted into useful energy, but some is converted into forms of energy that are wasted or not so useful. Nevertheless, all of the electrical energy is converted — that's the Law of Conservation of Energy in action. None of the wasted energy is actually lost; it is just transformed into less useful forms of energy. Table 9.2 shows some examples of energy conversion by electrical appliances.

TABLE 9.2 Energy conversion by electrical appliances

Appliance	Electrical energy usefully converted to ...	Electrical energy wasted ...
Microwave oven	thermal energy of food	heating air in the oven, plates and cups, etc.
Television	light and sound	heating the television and the surrounding air
Hair dryer	thermal energy and kinetic energy of air	as sound
Electric cooktop	thermal energy of food	as light and heating the surrounding air

This loss of useful energy is also apparent when you step on the brake pedal in a car — not all the energy you transfer to the pedal is used to stop the car. Much of it is lost in the brakes; it is converted to thermal energy and is released to the surrounding air as heat. The same applies to using the brakes of a bicycle. Also, when you drop a tennis or cricket ball it never bounces back to its original height because some energy is lost as heat. On a larger scale it is seen in power stations, where the fuel, falling water, solar energy or any other energy source is used to produce electricity. Some of the energy of the source is transformed to heat, warming the power equipment, the surrounding air and the water used as coolant. The 'loss' of useful energy is unavoidable.

FIGURE 9.28 Heat is a by-product in a number of energy transformations. a. When recharging electronic devices, not all the electrical energy is converted into chemical energy. b. When you exercise you get hot as your muscles convert the chemical energy into kinetic energy. This is why you perspire — to help cool you down.

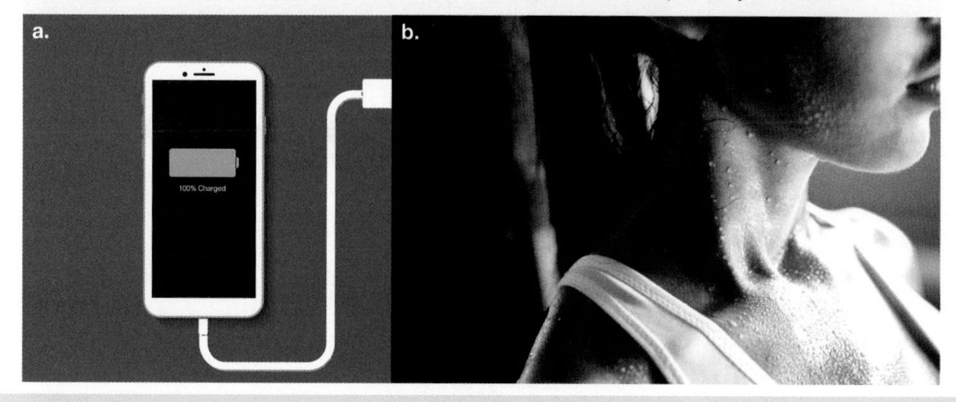

Some types of lighting waste more energy than others. Old-fashioned incandescent light bulbs convert more energy to wasted heat than to light. They emit light only when the filament inside gets white hot. Fluorescent lights and LEDs waste substantially less energy. Almost all of the electrical energy is converted to light, so you use much less energy to produce the same amount of light than you would using an incandescent bulb.

CASE STUDY: Comparing the energy efficiency of light bulbs

In old-fashioned incandescent light bulbs, electricity passes through a thin filament in the bulb filled with nitrogen or argon gas, causing it to glow white hot. The light is a useful form of energy, but about 90 per cent of the electrical energy is wasted as heat. Compact fluorescent lights (CFLs) offer a more energy-efficient form of lighting, but light-emitting diodes (LEDs) are even more efficient.

FIGURE 9.29 LEDs are much more efficient than the alternatives. Note that the figures quoted are approximate.

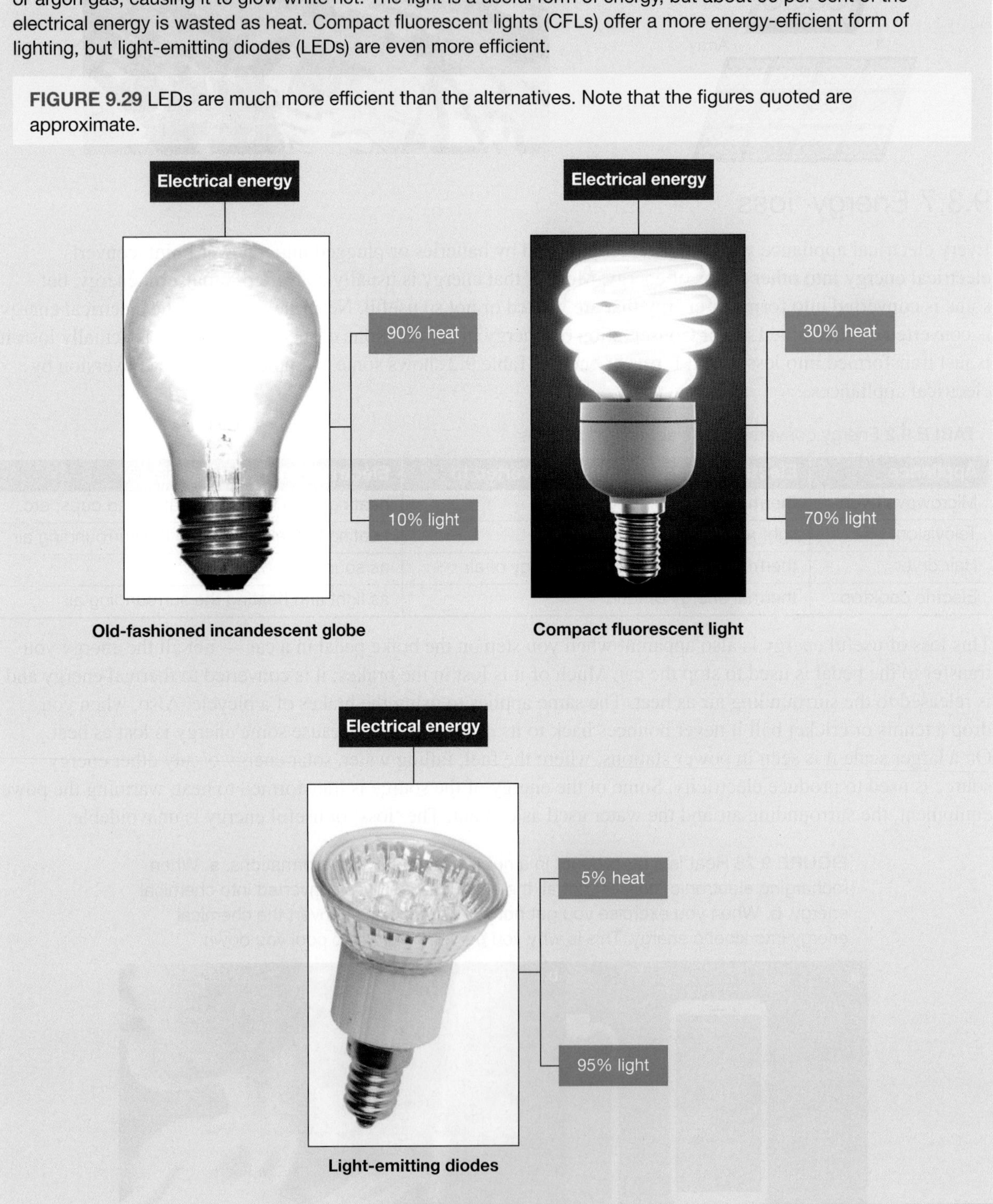

9.3.8 Efficiency

The **efficiency** of a car, light bulb, gas heater, power station, solar cell or any other energy converter is a measure of its ability to provide useful energy.

Efficiency is usually expressed as a percentage, and can be calculated using the following formula:

$$\text{Efficiency} = \frac{\text{useful energy output}}{\text{energy input}} \times 100\%$$

The efficiency of the incandescent light globe in figure 9.29 is 10 per cent because 10 per cent of the total electrical energy input is usefully transformed into light. The efficiency of the compact fluorescent light is 70 per cent, and the LED light is 95 per cent efficient.

The efficiency of every device that uses fossil fuels is very important for the environment and life on Earth. Scientists and automotive engineers are constantly working on methods of reducing fuel consumption by:

- increasing the efficiency of burning petrol and other fossil fuels such as diesel by reducing the amount of energy wasted as heat
- changing the external design of cars to reduce the amount of energy needed to overcome air resistance
- searching for alternative fuels such as ethanol that can be produced from sugar cane and grain crops.

Efficiency is discussed in more detail in Year 9.

efficiency in energy terms, the fraction of energy supplied to a device as useful energy

DISCUSSION

- Should it be mandatory to use energy-efficient devices?
- Outline at least one reason efficiency is important for devices that use fossil fuels.
- Are solar-powered cars a realistic alternative to cars that run on fossil fuels or biofuels such as ethanol? What criteria would you use to evaluate this?

Resources

Video eLesson The Australian–International Model Solar Challenge (eles-0068)

eWorkbooks Skateboard flick cards (ewbk-12027)
Types of energy (ewbk-12029)

9.3 Activities

learnon

9.3 Quick quiz on	9.3 Exercise

These questions are even better in jacPLUS!
- Receive immediate feedback
- Access sample responses
- Track results and progress

Find all this and MORE in jacPLUS

Select your pathway

LEVEL 1	LEVEL 2	LEVEL 3
1, 2, 4, 5, 8	3, 6, 9, 10, 11, 13, 16	7, 12, 14, 15

Remember and understand

1. Complete the table, listing the useful energy and the wasted energy converted by each of the devices.

TABLE The useful energy and the wasted energy converted by the devices

Device	Source of energy	Energy usefully converted to ...	Forms of energy wasted
A torch			
A wind-up toy			
A pop-up toaster			
A gas cooktop			
A car engine			

2. Outline at least three reasons efficiency is important for devices that use fossil fuels.
3. If a stretched rubber band has 12 J of elastic potential energy, and 9 J of kinetic energy is produced when the band is released:
   a. what is its percentage efficiency
   b. where has the 'lost' 3 J of energy gone?
4. A friend tells you that a light globe transforms 60 J of electric potential energy into 100 J of light. Are they correct? Why or why not?
5. Give definitions for the following terms.
   a. Conducting path **b.** Load **c.** Electrical circuit **d.** Electric current

Apply and analyse

6. An object is dropped from a height of 20 m. At a point during its fall towards the ground, it has 15 J of gravitational potential energy and 5 J of kinetic energy.
   a. What is the total amount of energy of the ball at any time during its flight?
   b. How much gravitational energy did the ball have before it was dropped?
   c. How much kinetic energy will the ball have just before it hits the ground (assuming it is 100 per cent efficient)?
7. When a tennis ball is bounced on the ground, it never returns to its original height.
   a. Does this break the Law of Conservation of Energy? Explain your answer.
   b. Why has the ball not reached its original height after the bounce? Explain with the aid of an energy flow diagram.
   c. Will the ball have the same amount of gravitational potential energy when it reaches the maximum height of its path after the bounce, compared to when it was originally dropped? Explain.
8. **SIS** A student investigating the energy of a skateboarder in a half pipe records the data shown in the table. Fill in the missing values.

TABLE The change in energy of a skateboarder in a half pipe

Position of skateboarder	Gravitational potential energy (J)	Kinetic energy (J)	Total energy (J)
Top of half pipe	600	0	
Part way down		200	600
Bottom of half pipe		600	
Part way up	300		
Top of half pipe			

9. Suggest some methods that drivers could use to increase the fuel efficiency of their vehicles.
10. Circuit diagrams are a way that circuits can be drawn easily and accurately. The diagram shows the circuit for the torch in figure 9.24.

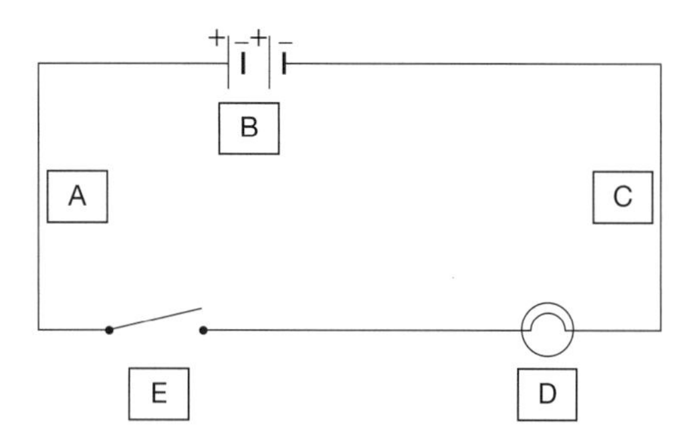

What do you think each of the components labelled by the letters A to E are?
11. When a toaster is plugged into a wall socket and switched on, a circuit is completed that can be traced all the way back to a power station.
    a. What is the power supply in this circuit?
    b. What are some of the conductors in this circuit?
    c. Describe how the switch stops the supply of electricity to the toaster when it is switched off.
    d. State two energy transformations that occur in a working toaster.

Evaluate and create

12. The energy we get from eating a piece of fruit starts from the Sun! Describe the energy transformations involved in this process using a flow diagram.
13. A catapult like the one in figure 9.30 was used by the Romans more than 2000 years ago to attack castles, cities and invading armies. The long arm was held in its usual vertical position with rope twisted around its base in what is known as a torsion bundle (figure 9.31). The arm was pulled back towards the ground using a second rope so that the bucket could be loaded with a missile. This caused the torsion bundle to twist more tightly. When the arm was released, the torsion bundle quickly untwisted and it returned to its vertical position, releasing the missile from the bucket at high speed towards the target. The missiles fired included rocks, burning tar and even human corpses. Use flow diagrams to show:
    a. the energy transfers that took place during the loading and firing of the missile
    b. the energy transformations that took place from the time that the missile was loaded until the time that the missile found its target.

FIGURE 9.30 A Roman catapult

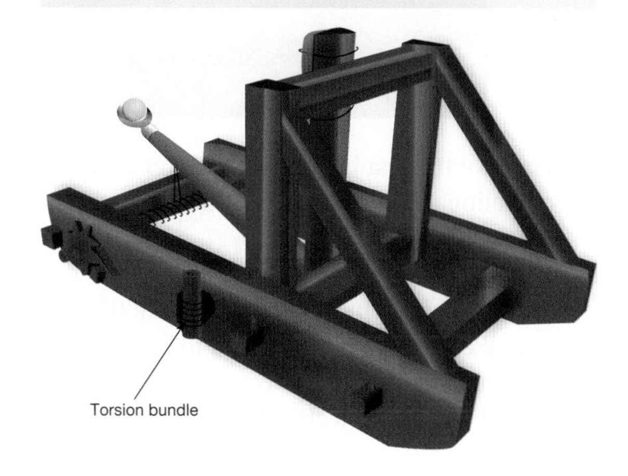

FIGURE 9.31 The torsion bundle

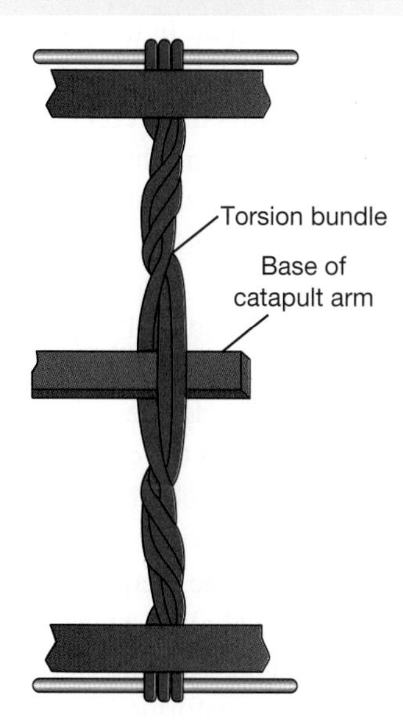

14. **SIS** Create a poster-sized flow diagram to show the energy transformations that take place to produce lightning and thunder. (Think first about how the clouds become electrically charged during an electrical storm.)
15. **SIS** Waterwheels have been used in the past (and are still being used) to convert the energy of moving water to other useful forms of energy. Research and report on one example of the use of a waterwheel. In your report, use flow diagrams to illustrate the transformations and transfers of energy that take place.
16. **SIS** Are solar-powered cars a realistic alternative to cars that run on fossil fuels or biofuels such as ethanol? Find out what scientists, engineers and members of the public have contributed to the design of solar-powered vehicles.

Fully worked solutions and sample responses are available in your digital formats.

LESSON
9.4 Having fun with energy

LEARNING INTENTION

At the end of this lesson you will be able to describe how energy can be transferred from one object or place to another during common leisure activities.

9.4.1 Playing sport

Energy transfer from one object to another object is usually easy to observe because one or both objects slow down, speed up or change direction. A transfer of energy to or from an object can also cause it to start or stop spinning. Some examples of energy transfer from object to object are explained here.

FIGURE 9.32 Energy transfers from your muscles into the club and finally into the ball in a golf swing.

- When the golfer in figure 9.32 swings his club, energy is transferred from his body to the club. When the club strikes the ball, most of its energy is transferred to the ball to make it move. The ball gains both kinetic energy and gravitational potential energy as a result of the transfer. It might also spin. In this case the force on the ball is supplied by the club.
- When a weightlifter lifts the weights into the air, they are transferring energy from their body to the weights in the form of additional gravitational potential energy. When we throw a javelin, we transfer energy from our body to the javelin in the form of kinetic energy and gravitational potential energy. In both cases the force is supplied by your body.
- The guernseys of AFL and AFLW players contain pockets into which tracking devices are placed. These record a player's movements during a game. The devices are powered by small batteries. In the electrical circuit that powers these devices, chemical energy in the cell or battery is transferred to electric charge, causing it to move around the circuit. The force causing the movement is the electrical force of attraction between opposite electric charges. The electrical energy of the moving electric charge is then transformed to operate the tracking device.
- When a tennis ball hits the racquet strings, its kinetic energy is temporarily stored as elastic potential energy in the squashed ball and the racquet strings. Most of this is transferred back to kinetic energy as the ball rebounds, along with the extra energy that has been added from the player's moving arm. A small amount of energy is transformed into heat.

- In billiards and snooker, a moving cue ball strikes a stationary target ball. After they collide the target ball will move off with a certain speed, while the cue ball will be moving slower or may even stop. Exactly what happens depends on the angle at which they collide. All the energy involved here is kinetic energy.
- In tenpin bowling, a heavy bowling ball strikes pins, which are themselves quite heavy. These pins then strike other pins, resulting in the pins being knocked over. In energy terms, the object of the game is to transfer kinetic energy from the bowling ball to the first pins that are struck. As these fly off, their kinetic energy and the remaining kinetic energy in the ball itself are transferred to other pins as they collide. This results in the pins being knocked down.

FIGURE 9.33 Tenpin bowling involves transferring energy from a bowling ball to the target pins.

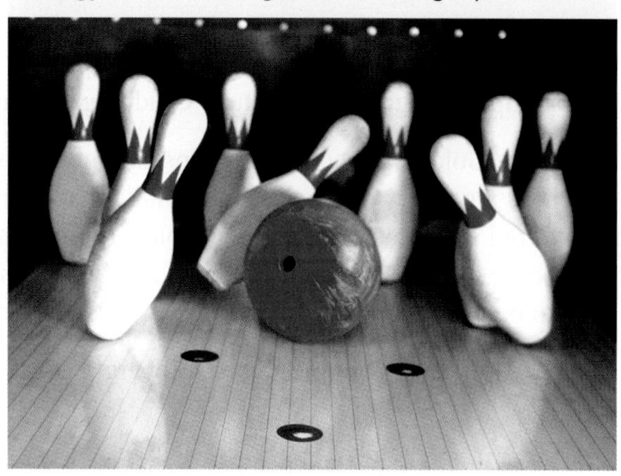

9.4.2 Riding a rollercoaster

We all know what fun it is to ride a rollercoaster — at least those of us who do not scare easily! But did you know that the operation of a rollercoaster is a very good example of energy being converted backwards and forwards between gravitational potential energy and kinetic energy, with some thermal energy also being produced through friction? All rollercoasters begin by lifting a car to the top of the track using a chain that is usually operated by an electric motor. The car then goes through a number of ups and downs, with the ride finishing at the lowest point of the track.

So what are the energy changes involved? To understand what happens, look at figure 9.34.

- As the chain pulls the car up from the start (S) to point A, gravitational potential energy is being added to the car. When the car reaches point A it is at its highest point and therefore has the most gravitational potential energy that it will ever have during the ride.
- As the car descends from point A to point B, its potential energy due to gravity is changed to kinetic energy as it picks up speed.
- As the car rises from point B to point C, the process is reversed as it gains height and slows down.
- This changing backwards and forwards between gravitational potential energy and kinetic energy is repeated throughout the rest of the ride.
- Throughout the ride the total energy of the car is the sum of its gravitational potential energy and its kinetic energy. However, this amount is slowly decreasing due to friction with the rails that the car runs on. This causes some of this energy to be converted into heat (thermal energy) and sound energy. Useful energy is therefore lost — there is less energy for climbing and speeding. The car will never reach its starting height and it gradually slows down.
- At the end of the ride, the car is usually still moving a bit. It still has some kinetic energy. To stop it, and for safety reasons, there is usually a braking device in the track. This converts most of this remaining kinetic energy into thermal energy through friction and brings the car to a stop.

FIGURE 9.34 A simple rollercoaster. Notice that the cars never return to their starting height.

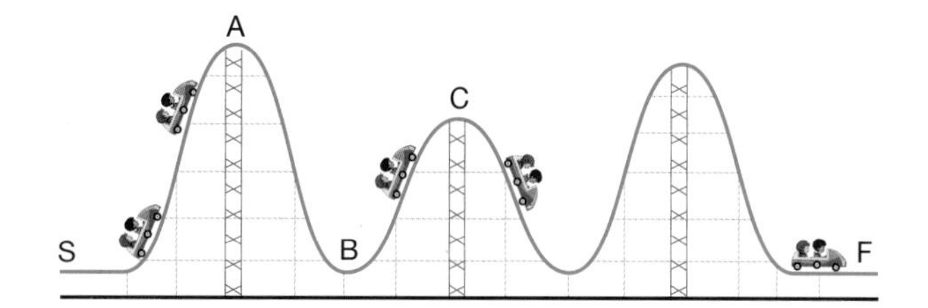

 Resources

 **Interactivity** Coaster (int-0226)

9.4.3 Rube Goldberg machines

A **Rube Goldberg machine** is a machine that is designed to perform a simple task in a complicated way. It is named after an American engineer and cartoonist who was famous for drawing such devices. You may have seen these machines in animations, video clips or at science displays. You may even have played the game Mousetrap®, designed after a Rube Goldberg machine, or seen them in advertisements on television.

Any Rube Goldberg machine contains a number of sections linked together. Each section performs a task, which, as we know, requires energy. These machines are therefore good examples of how energy is used, transferred and transformed. Rube Goldberg machines may be simple and only have a few steps, or they be impressive and very complicated with a very large number of steps. When you build these machines, you often have to be patient and modify each section a number of times before everything works properly.

Figure 9.35 shows a simple Rube Goldberg machine designed to burst a balloon using water.

FIGURE 9.35 A Rube Goldberg machine designed to burst a balloon with water

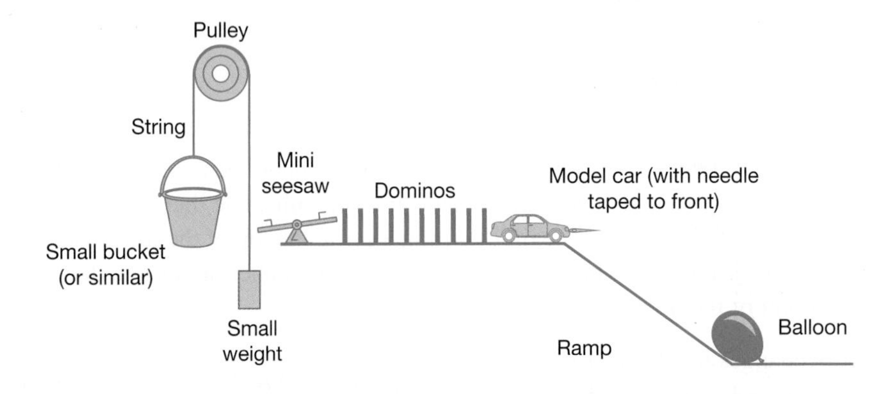

An explanation of how this machine would work is as follows. The energy changes involved are shown in brackets.

- Water is poured into the small bucket. It starts to move down and the small weight starts to move up. (Gravitational potential energy in the small bucket is transformed into kinetic energy in the small weight).
- As the weight passes the mini seesaw it just grazes it, causing the opposite end to tip down. (Some of the kinetic energy in the weight is transferred to kinetic energy in the seesaw.)
- The end of the seesaw makes the first domino fall. This sets off a chain reaction that causes the remaining dominoes to fall. (Kinetic energy from the seesaw is transferred to kinetic energy in the falling dominos.)
- The model car is positioned so that it is just about to slide down the ramp. When the last domino falls, it hits the car and pushes it over the edge. It then slides down the ramp. (The car has gravitational potential energy due to its height before it starts to move. When the last domino falls, it hits the car and transfers a little bit of kinetic energy to the car, causing it to move. As the car slides down the ramp, it gains extra kinetic energy as it loses height by changing its gravitational potential energy into kinetic energy.)
- At the bottom of the ramp the needle poking out the front of the car bursts the balloon. The car then rolls across the floor until it comes to a stop. (The kinetic energy in the car is converted into sound energy and heat (thermal energy) as friction brings the car to a stop.)

Rube Goldberg machine a machine that is designed to perform a simple task in a complicated way

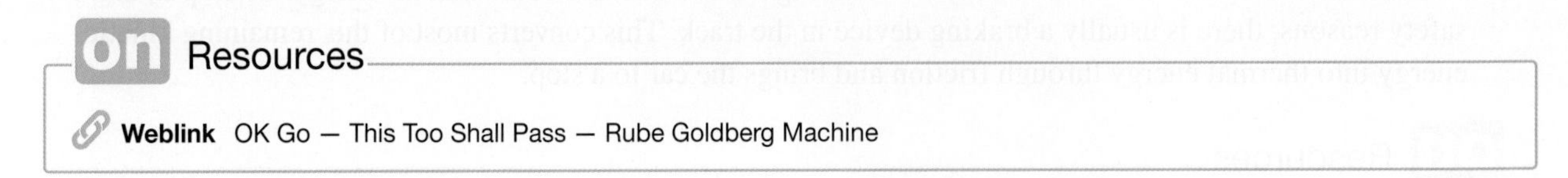

elog-1969

INVESTIGATION 9.6

Building a Rube Goldberg machine

Aim

To build a Rube Goldberg machine and analyse the energy changes taking place

Materials

These will depend on exactly what you are going to build. See method step 2.

Method

1. Make a design of your machine on paper. Remember that you need to use materials and objects that are easily available.
2. Make a list of what you will need. Go and get these.
3. Construct your machine.
   Tip 1: Make your machine in sections, ensuring each section works before going on to the next one.
   Tip 2: Be patient. It may take a bit of trial and error before you get each stage and each connection right.
4. Set your machine off. You might like to do this as a class demonstration.

Results

Draw a neatly labelled diagram of your machine and describe how it worked.

Discussion

1. What modifications to your original design did you have to make to get your machine to work properly?
2. Identify and describe the energy transfers and transformations in your machine.
3. By definition, a Rube Goldberg machine performs a simple task in a complicated way. However, many machines perform a necessary task in a way that *needs* to be complicated.
   Give two examples of such machines.

Conclusion

Summarise your findings for this investigation, referencing the energy aspects involved.

9.4.4 Newton's cradle.

A **Newton's cradle** (see figure 9.36) is a desk toy that is often seen in offices. It consists of a number of metal balls (usually five) that are suspended by metal threads so that they just touch each other and form a perfectly straight line. When the ball on one end is pulled back and released, it swings down and hits the ball it was next to. The ball on the other end then flies up before coming back down and repeating the process from the other end. All the while, the balls in the middle remain nearly motionless. This sequence repeats itself before gradually dying out. Each ball strike is accompanied by a loud 'clacking' sound.

FIGURE 9.36 In a Newton's cradle the outside balls swing backwards and forwards while the middle ones stay almost still.

When it is working, Newton's cradle demonstrates both the transformation and conservation of energy as follows:

- When everything is still, there is no kinetic energy in the system because nothing is moving. Also, because the balls are at their lowest points there is no gravitational potential energy.
- When the first ball is lifted up to its release height it now has gravitational potential energy due to this height.
- When the ball is released it swings down, loses height and picks up speed. It loses gravitational potential energy and gains kinetic energy.

Newton's cradlea decorative device consisting of suspended metal balls that that can be made to swing backwards and forwards in a particular way

- At the bottom of its swing, just before it hits the next ball, all its gravitational potential energy has been converted into kinetic energy.
- When it hits the next ball it stops and therefore has no kinetic energy. Where does this energy go? It goes into the next ball as elastic potential energy because this ball is compressed slightly by the collision. When this ball returns to its original shape it converts this energy back into kinetic energy, which then causes it to compress the next ball. This is repeated down the chain. The balls are acting like a spring!
- When this chain reaction reaches the last ball, the kinetic energy has no further ball to be transferred to. This ball then swings up with the same speed as the first ball and reaches the same height.

In theory, this should continue for a very long time. However, as we learned in section 9.3.7, energy conversions are never one hundred percent efficient. There will always be some loss of useful energy. Frequently this is as heat (thermal energy), which is often generated through friction. In this case, sound energy is also produced when the balls collide with each other. All this means that the amount of energy available for each transformation gets less and less until everything stops because there is no energy left.

Resources

Weblink Amazing Demonstration Of A Giant Newton's Cradle!

SCIENCE AS A HUMAN ENDEAVOUR: Nikola Tesla and Swami Vivekananda

How could two people, with very different world views, collaborate to explore the relationship between mass and energy?

In 1896, a meeting between Nikola Telsa — an engineer and physicist — and Swami Vivekananda — a monk, religious teacher, author and philosopher — saw them exploring the relationship between mass, energy and the elementary concepts of the Vedantic doctrine, which include:

- Prana (energy)
- Akasha (matter)
- Dyuloka (electric sphere), where Prana and Akasha existed as one.

Based on their conversation, Telsa set about trying to prove mathematically the equivalence of mass and energy; however, he was unsuccessful because he believed that mass might be converted into energy.

It wasn't until 1905 that Albert Einstein published the formula $E = mc^2$, where he proved that energy and mass are the same thing, just in different forms.

FIGURE 9.37 Nikola Tesla

FIGURE 9.38 Swami Vivekananda

9.4 Activities

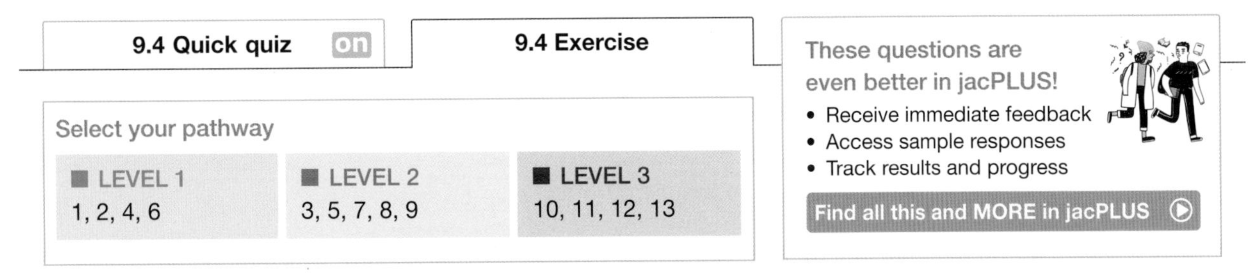

Remember and understand

1. What are the two methods by which energy can be transferred?
2. When a person kicks a soccer ball, energy from their body is transferred to the soccer ball. Which two useful forms of energy does the soccer ball gain as a result?
3. When playing squash, a very compressible ball that does not bounce very much gets warm. What types of energy are involved just before and just after the ball is hit?
4. During exercise, kinetic energy is generated by your muscles. What other type of energy is also generated?
5. Identify four types of energy that are present during an explosion.
6. A billiard ball strikes another ball exactly head on and stops completely. Where has all its energy gone?

Apply and analyse

7. What are two ways that you could hit a ball further?
8. Explain why it is safe to stand right behind a swing that is swinging if you do not push it, but why it might be dangerous to you if you do push it.
9. **SIS** Rubber bands store energy as elastic potential energy when they are stretched. State two hypotheses that you could test experimentally concerning the amount of energy stored in a stretched rubber band.
10. Using oil is very important in a car engine, even though it is not used as a fuel. Explain why this is so.
11. **SIS** The following bar graph shows not only the total energy involved, but also the proportions of kinetic energy and gravitational potential energy when a golf ball is hit into the air before rolling down the fairway and coming to rest. Match each bar to the correct position of the ball during its flight.

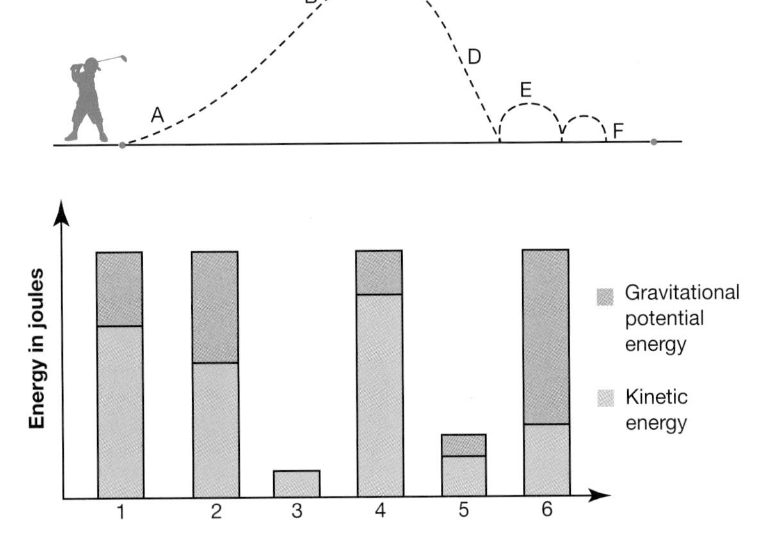

Evaluate and create

12. Prepare a poster that demonstrates the energy changes that take place during a bungee jump. Start from the moment the person leaves the platform until they are lowered back to the ground.
13. **SIS** Design an experiment to investigate what happens in a Newton's cradle if:
    a. two balls are released to start it instead of one
    b. an even number of balls are present.
    Write up your experiment in the usual way. Make sure you report on the energy changes involved in your discussion. Are you able to make any conclusions about the amount of energy involved in the collisions?

Fully worked solutions and sample responses are available in your digital formats.

LESSON
9.5 Thinking tools — Matrices and Venn diagrams

9.5.1 Tell me

What is a matrix?

A matrix is a very useful thinking tool for comparing topics and identifying ways in which these topics are similar and different. A matrix shows similarities and differences between topics. It is sometimes called a table, grid or decision chart.

FIGURE 9.39 A matrix

Topic	Feature A	Feature B	Feature C	Feature D	Feature E
1	✓		✓	✓	✓
2		✓			✓
3		✓		✓	✓
4			✓	✓	✓

Why use a matrix over a Venn diagram?

Similar to a matrix, a Venn diagram identifies common points between two separate topics. However, it uses a different graphic format to show the common features. A matrix uses a grid or table, whereas a Venn diagram uses overlapping circles.

FIGURE 9.40 A Venn diagram

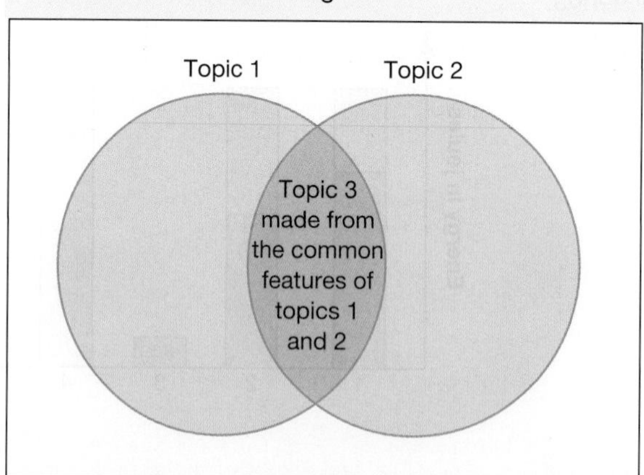

9.5.2 Show me

To create a matrix

1. Write the topics in the left-hand column of the matrix.
2. Write the characteristics to be compared along the top row of the matrix.
3. If a characteristic applies to a topic, put a tick in the appropriate cell of the matrix.
4. The matrix now shows how the various topics are related.

Table 9.3 shows a matrix indicating the forms of energy that electrical energy is transformed into by each of the electrical devices listed.

TABLE 9.3 Conversion of electrical energy by different electrical devices

Device	Electrical energy is converted into ...				
	light energy	sound energy	thermal energy	kinetic energy	potential energy
Hair dryer		✓	✓	✓	✓
Television	✓	✓	✓	✓	✓
Desk lamp	✓		✓	✓	✓
Vacuum cleaner		✓	✓	✓	✓
Home computer	✓	✓	✓	✓	✓
Incandescent light bulb	✓		✓	✓	✓
Air conditioner		✓	✓	✓	✓
Elevator going up		✓	✓	✓	✓

Note: All substances and objects possess potential energy, but you can't tell unless something happens to transform the potential energy into a different energy type.

9.5.3 Let me do it

9.5 Activity

1. **a.** Complete the following matrix. Use ticks to show which statements refer to light and which refer to sound. Some of the statements refer to both light and sound.

Statement	Radiant energy	Sound energy
Travels through air at 300 000 kilometres per second		
Travels faster than a speeding bicycle		
Is a form of kinetic energy		
Includes light energy		
Is always caused by vibrating objects or substances		
Is observed in an electrical storm		
Can travel through empty space		
Can be produced from another form of energy		
Is produced in a fireworks display		

b. The information in the matrix provided can be represented in a Venn diagram. Convert the information in the matrix from part a into a large Venn diagram based on the example provided.

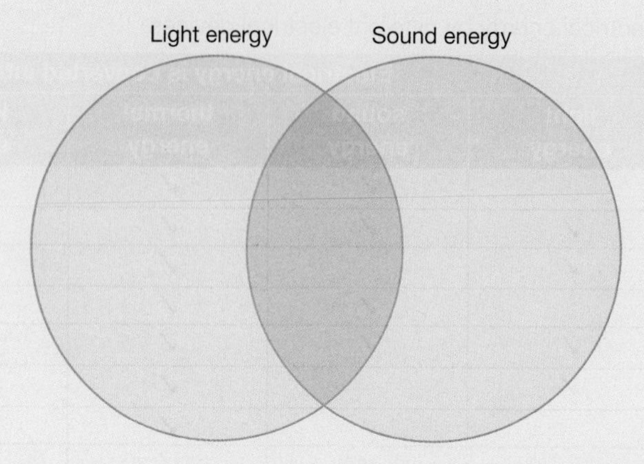

Fully worked solutions and sample responses are available in your digital formats.

LESSON
9.6 Project — Going green

Scenario

As the supply of fossil fuels dwindles, cities become more crowded and human-caused global warming becomes an unavoidable reality, an increasing number of people are opting for a more self-sufficient lifestyle. To meet this need, an increasing number of architecture and building firms specialise in the design and construction of energy-efficient houses that are able to exist off the electricity grid indefinitely, because they use electricity generation systems that meet all of the household's needs using renewable energy sources.

You and your team at Sustainable Housing Solutions have been approached by a potential client who wants to build a series of sustainable eco-tourist cottages in remote locations across the country. To see whether your company should be awarded the lucrative contract to oversee the work on the whole chain of cottages, the client has asked you to make a presentation detailing how you would make one of these cottages as energy efficient and self-sustaining as possible. You can place this trial cottage anywhere in the country for your presentation purposes, provided that it is at least 100 km away from any town with a population greater than 10 000 people. Other criteria must also be met as follows:

- All of the cottages will have the same layout and will be constructed of mud bricks and have tiled roofs (you will be given a copy of the plan). While you can change the orientation and location of the cottage, you cannot change the design or the construction materials.
- Each cottage must have the following appliances: refrigerator, washing machine, stove, microwave, TV set, DVD player and stereo system. Smaller appliances such as toasters, shavers, hair dryers and computers may occasionally be used by guests as well.

FIGURE 9.41 Sustainable eco-tourist cottages

- The cottages must be cool in summer and warm in winter; the client is not opposed to the idea of a reverse-cycle air conditioner or fans.
- There must be sufficient lighting to be able to read in every room.
- The cottages will not be attached to the national electricity grid — all of the electricity needs of each cottage must be met using a renewable energy source in its area. (Water will be provided from rainwater tanks, and septic tanks will take care of the sewage.)

Your task

Your team will prepare and deliver a report for the client that provides the following information.

- The best location to place the trial cottage (keeping in mind that it can be placed somewhere close to a source of renewable energy)
- Suggestions as to how the cottage can be made as energy efficient as possible
- A detailed estimate of how much electricity will need to be generated to power the cottage and run appliances
- A justified recommendation as to which renewable energy system should be used to generate that amount of electricity and how it would be supplied to the trial cottage
- An estimate of how much the energy system will cost, using costs for similar systems available on the internet as a guideline

Resources

ProjectsPLUS Going green (pro-0093)

LESSON
9.7 Review

Hey students! Now that it's time to revise this topic, go online to:

Review your results

Watch teacher-led videos

Practise questions with immediate feedback

Find all this and MORE in jacPLUS

Access your topic review eWorkbooks

Topic review Level 1	Topic review Level 2	Topic review Level 3
ewbk-12031	ewbk-12033	ewbk-12035

9.7.1 Summary

Different forms of energy

- Energy cannot be created or destroyed. This statement is known as the Law of Conservation of Energy.
- There are many different types of energy, all of which fall under one of two categories: potential or kinetic.
- Potential energy is energy that is stored in objects. Types of potential energy include gravitational, elastic, chemical, nuclear, electrical and magnetic.
- Kinetic energy is energy that involves movement. Types of kinetic energy include translational, thermal, radiant, sound and electrical.
- Visible light is only a small segment of the larger electromagnetic spectrum.
- Other forms of electromagnetic radiation are radio waves, infrared, ultraviolet, x-rays, gamma rays and microwaves.
- Sound is the vibration of particles through substances.

Transforming energy

- Energy can be transformed from one type to another type.
- Potential energy can be transformed into kinetic energy in many ways. An example is gravitational potential energy transforming into kinetic energy when an object is dropped.
- Most energy transformations convert energy from one form into multiple other forms. Some of the forms that it is converted into are not very 'useful' and are sometimes classified as wasted energy.
- The efficiency of an energy transformation is the percentage of the initial energy that is converted into the desired form.

$$\text{Efficiency} = \frac{\text{useful energy output}}{\text{energy input}} \times 100\%$$

Having fun with energy

- Energy can be transferred from an object to another object, or from one form into another form by forces or by heating.
- For example, when an object falls the gravitational force converts gravitational potential energy into kinetic energy. When the object hits the ground the upwards force on it transforms the kinetic energy into sound and heat energy, along with elastic potential energy, causing it to bounce.
- Energy transfers and transformations are readily observed in many sports and in devices such as rollercoasters, Rube Goldberg machines and a Newton's cradle.

9.7.2 Key terms

by-products in energy terms, energy produced that is not useful
cell a single battery
conducting path a connected series of materials along which an electric current can flow
conductors materials that have a very low resistance, allowing current to flow through them with ease
efficiency in energy terms, the fraction of energy supplied to a device as useful energy
electric current a measure of the number of electrons flowing through a circuit every second
electrical circuit consists of a power supply, a conducting path for charge to flow and one or more loads
electromagnetic radiation the radiant energy such as radio waves, infrared, visible light, x-rays and gamma rays released by magnetic or electric fields
electromagnetic spectrum the complete range of wavelengths of energy radiated as electric and magnetic fields
electromagnetic waves waves of electromagnetic radiation, light being just one example
filament a coil of wire made from a metal that glows brightly when it gets hot
friction the rubbing together of surfaces; all friction produces thermal energy (heat)
insulators materials that do not allow electricity to flow easily
kinetic energy energy due to the motion of an object
Law of Conservation of Energy a law that states that energy cannot be created or destroyed
transform in energy terms, this refers to the changing of one form of energy into another
load a device that uses electrical energy and converts it into other forms of energy
Newton's cradlea decorative device consisting of suspended metal balls that that can be made to swing backwards and forwards in a particular way
period the time taken for one oscillation of a pendulum
potential energy energy stored within an object that depends on its position within a system, such as gravitational energy, elastic energy and chemical energy
power supply a device that can provide an electric current
resistance a measure of the opposition to an electric current passing through an object, measured in ohms (Ω)
Rube Goldberg machine a machine that is designed to perform a simple task in a complicated way
series a formation of electricity-generating or electricity-using devices whereby the electricity passes from one device to the next in a single conducting loop, one after the other
switch a device that opens and closes the conducting path through which a current flows
tinder light, dry material that is easy to ignite
transducer a device that converts energy from one form into another form
transfer in energy terms, the movement of energy from one place to another
transform in energy terms, this refers to the changing of one form of energy into another
transmitted passed through something, such as light or sound passing through air
useful energy the energy that produces the desired output
visible spectrum different colours that combine to make up white light; they are separated in rainbows
voltage the amount of energy that is pushing electrons around a circuit, per coulomb of charge that flows between two points

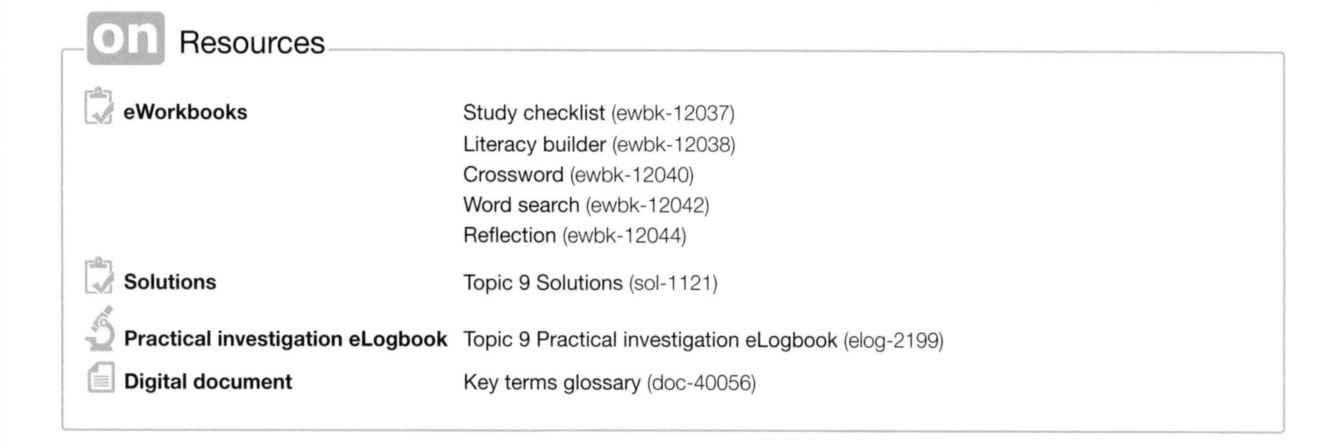

on Resources

eWorkbooks	Study checklist (ewbk-12037) Literacy builder (ewbk-12038) Crossword (ewbk-12040) Word search (ewbk-12042) Reflection (ewbk-12044)
Solutions	Topic 9 Solutions (sol-1121)
Practical investigation eLogbook	Topic 9 Practical investigation eLogbook (elog-2199)
Digital document	Key terms glossary (doc-40056)

9.7 Activities

9.7 Review questions

Select your pathway

■ LEVEL 1	■ LEVEL 2	■ LEVEL 3
1, 2, 3, 7, 8, 9, 12	4, 6, 10, 13, 14, 17, 21	5, 11, 15, 16, 18, 19, 20

Remember and understand.

1. Replace each of the following descriptions with a single word.
   a. Energy associated with all moving objects
   b. Energy associated with the position of an object
   c. The form of energy that causes an object to have a high temperature
   d. The form of energy stored in a battery that is not connected to anything
   e. The source of most of Earth's light
2. Explain why the amount of energy in the universe never changes.
3. Describe an example of an object that has:
   a. elastic potential energy
   b. gravitational potential energy.
4. Explain why elastic potential energy is present in a working Newton's cradle.
5. Explain why thermal energy is a form of kinetic energy.
6. Describe one example of evidence that white light is made up of many different colours.
7. To which form of electromagnetic radiation do microwaves belong?
8. Consider points A, B, C and D on the rollercoaster in the following diagram.

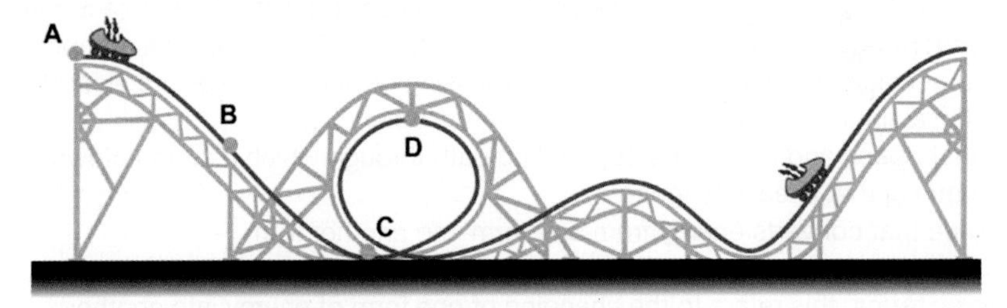

   At which point does the rollercoaster have the greatest gravitational potential energy?
9. When First Nations Australians start a fire using traditional methods,
   A. friction converts kinetic energy into thermal energy.
   B. chemical energy in tinder is converted into thermal energy.
   C. chemical energy in tinder is converted into light energy.
   D. all of the above statements are true.

Apply and analyse

10. Calculate the efficiency of the light bulbs shown in figures a and b.

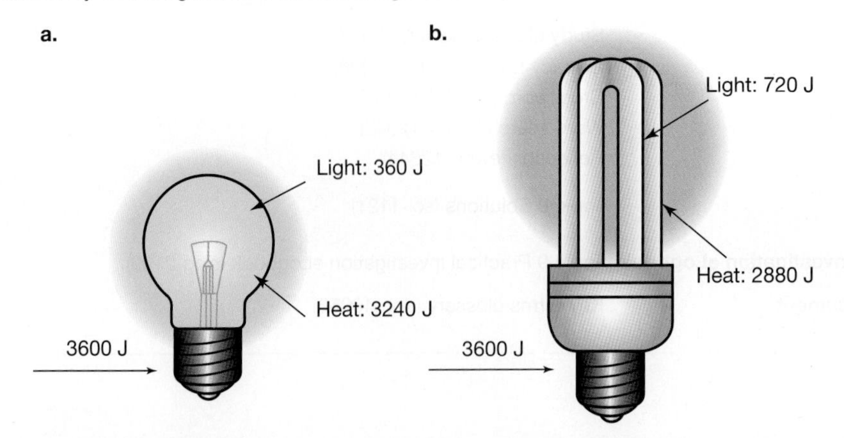

11. When a kettle of water is boiled on a gas cooktop, not all of the energy stored in the gas is used to heat the water. Where does the rest of the energy go?
12. Explain why it is not possible for an energy converter like a battery or car to have an efficiency of 100 per cent.
13. Give two ways that a golfer could impart more energy to a golf ball, enabling them to hit it further.
14. Explain why 'in space, no one can hear you scream'.
15. How would the efficiency of a Rube Goldberg machine compare to a simpler machine designed to do the same job?
16. Explain why electrical energy can be both a form of kinetic energy and a form of potential energy.
17. High jumpers train to develop both their muscles and their technique. Explain why, in terms of energy, this enables them to jump greater heights.
18. A student set up four different circuits, labelled A, B, C and D.
    Using suitable instruments, the voltage and current in each circuit was measured. The results are shown in the following table.

Circuit	Voltage (V)	Current (A)
A	2.5	1.4
B	2.5	0.6
C	3.8	0.8
D	3.1	0.8

    a. Compare circuits A and B. Which circuit has the higher resistance? Explain.
    b. Compare circuits C and D. Which circuit has the higher resistance? Explain.

Evaluate and create

19. Draw a flow diagram to illustrate the energy transformations that take place:
    a. after you switch on a torch
    b. when a firecracker is lit
    c. when a ball rolls down a hill and then up another hill.
20. Plants use energy from the Sun in a process called photosynthesis that enables them to grow. Produce a poster that summarises all the energy changes that would take place if some wood was burnt in a fireplace for warmth.
21. The graph shows the fuel consumed versus distance travelled for two cars. These cars are identical except for the fact that one is fuelled by ethanol and the other by petrol. What does this graph tell you about the amount of chemical potential energy stored in each litre of each fuel?

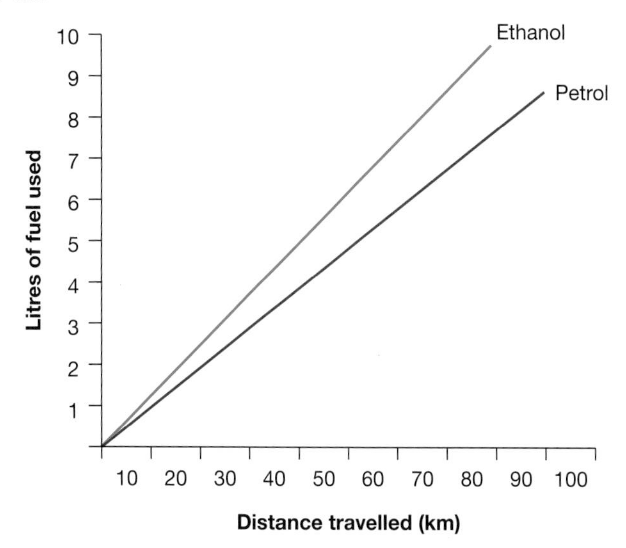

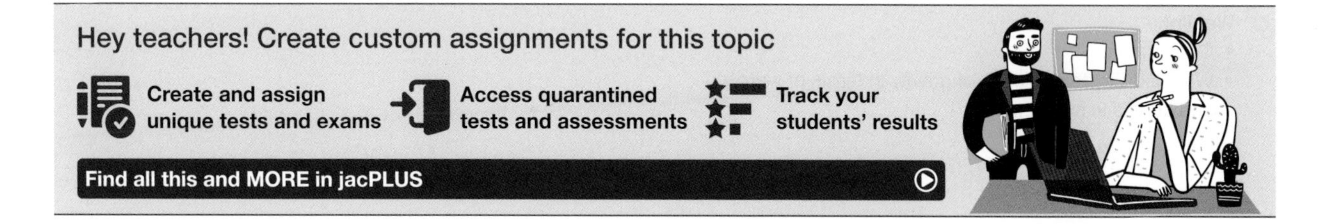

Online Resources

Below is a full list of **rich resources** available online for this topic. These resources are designed to bring ideas to life, to promote deep and lasting learning and to support the different learning needs of each individual.

9.1 Overview

eWorkbooks
- Topic 9 eWorkbook (ewbk-12018)
- Starter activity (ewbk-12020)
- Student learning matrix (ewbk-12022)

Solutions
- Topic 9 Solutions (sol-1121)

Practical investigation eLogbook
- Topic 9 Practical investigation eLogbook (elog-2199)

Video eLesson
- Energy transformations (eles-2677)

9.2 Different forms of energy

Video eLesson
- Energy in disguise (eles-0063)

9.3 Transforming energy

eWorkbooks
- Conductors and insulators (ewbk-12023)
- Simple circuits (ewbk-12025)
- Skateboard flick cards (ewbk-12027)
- Types of energy (ewbk-12029)

Practical investigation eLogbooks
- Investigation 9.1 Energy transformations in a pendulum (elog-1966)
- Investigation 9.2 Ice-cream stick energy (elog-1967)
- Investigation 9.3 Making the right connections (elog-2203)
- Investigation 9.4 What's inside a torch? (elog-2205)
- Investigation 9.5 Making a model kettle (elog-1968)

Teacher-led video
- Investigation 9.2: Ice-cream stick energy (tlvd-10796)

Video eLesson
- The Australian–International Model Solar Challenge (eles-0068)

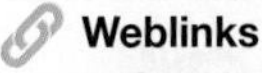

Weblinks
- Cool burning
- What led Einstein to think of gravity in terms of warps and curves in space time?

9.4 Having fun with energy

Practical investigation eLogbook
- Investigation 9.6 Building a Rube Goldberg machine (elog-1969)

Interactivity
- Coaster (int-0226)

Weblinks
- OK Go — This Too Shall Pass — Rube Goldberg Machine
- Amazing Demonstration of a Giant Newton's Cradle!

9.6 Project — Going green

ProjectsPLUS
- Going green (pro-0093)

9.7 Review

eWorkbooks
- Topic review Level 1 (ewbk-12031)
- Topic review Level 2 (ewbk-12033)
- Topic review Level 3 (ewbk-12035)
- Study checklist (ewbk-12037)
- Literacy builder (ewbk-12038)
- Crossword (ewbk-12040)
- Word search (ewbk-12042)
- Reflection (ewbk-12044)

Digital document
- Key terms glossary (doc-40056)

To access these online resources, log on to **www.jacplus.com.au**

PERIODIC TABLE OF THE ELEMENTS

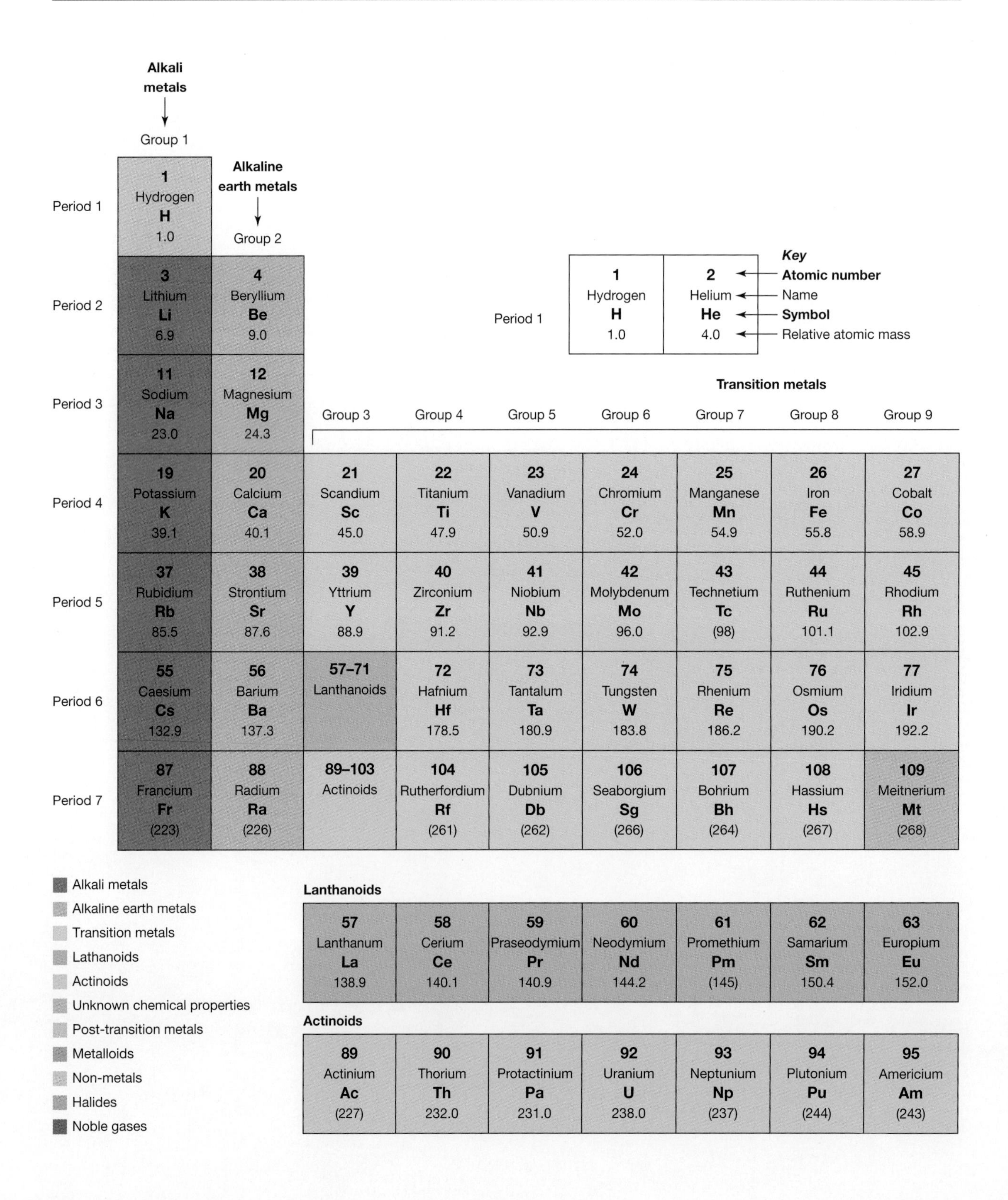

Key

Period 1	1 Hydrogen **H** 1.0	2 Helium **He** 4.0	**Atomic number** Name **Symbol** Relative atomic mass

Alkali metals → Group 1; **Alkaline earth metals** → Group 2; **Transition metals** → Groups 3–9

	Group 1	Group 2	Group 3	Group 4	Group 5	Group 6	Group 7	Group 8	Group 9
Period 1	1 Hydrogen **H** 1.0								
Period 2	3 Lithium **Li** 6.9	4 Beryllium **Be** 9.0							
Period 3	11 Sodium **Na** 23.0	12 Magnesium **Mg** 24.3							
Period 4	19 Potassium **K** 39.1	20 Calcium **Ca** 40.1	21 Scandium **Sc** 45.0	22 Titanium **Ti** 47.9	23 Vanadium **V** 50.9	24 Chromium **Cr** 52.0	25 Manganese **Mn** 54.9	26 Iron **Fe** 55.8	27 Cobalt **Co** 58.9
Period 5	37 Rubidium **Rb** 85.5	38 Strontium **Sr** 87.6	39 Yttrium **Y** 88.9	40 Zirconium **Zr** 91.2	41 Niobium **Nb** 92.9	42 Molybdenum **Mo** 96.0	43 Technetium **Tc** (98)	44 Ruthenium **Ru** 101.1	45 Rhodium **Rh** 102.9
Period 6	55 Caesium **Cs** 132.9	56 Barium **Ba** 137.3	57–71 Lanthanoids	72 Hafnium **Hf** 178.5	73 Tantalum **Ta** 180.9	74 Tungsten **W** 183.8	75 Rhenium **Re** 186.2	76 Osmium **Os** 190.2	77 Iridium **Ir** 192.2
Period 7	87 Francium **Fr** (223)	88 Radium **Ra** (226)	89–103 Actinoids	104 Rutherfordium **Rf** (261)	105 Dubnium **Db** (262)	106 Seaborgium **Sg** (266)	107 Bohrium **Bh** (264)	108 Hassium **Hs** (267)	109 Meitnerium **Mt** (268)

- Alkali metals
- Alkaline earth metals
- Transition metals
- Lathanoids
- Actinoids
- Unknown chemical properties
- Post-transition metals
- Metalloids
- Non-metals
- Halides
- Noble gases

Lanthanoids

57 Lanthanum **La** 138.9	58 Cerium **Ce** 140.1	59 Praseodymium **Pr** 140.9	60 Neodymium **Nd** 144.2	61 Promethium **Pm** (145)	62 Samarium **Sm** 150.4	63 Europium **Eu** 152.0

Actinoids

89 Actinium **Ac** (227)	90 Thorium **Th** 232.0	91 Protactinium **Pa** 231.0	92 Uranium **U** 238.0	93 Neptunium **Np** (237)	94 Plutonium **Pu** (244)	95 Americium **Am** (243)

Noble gases → Group 18

Non-metals →

Group 10	Group 11	Group 12	Group 13	Group 14	Group 15	Group 16	Group 17	Group 18
								2 Helium **He** 4.0
			5 Boron **B** 10.8	6 Carbon **C** 12.0	7 Nitrogen **N** 14.0	8 Oxygen **O** 16.0	9 Fluorine **F** 19.0	10 Neon **Ne** 20.2
			13 Aluminium **Al** 27.0	14 Silicon **Si** 28.1	15 Phosphorus **P** 31.0	16 Sulfur **S** 32.1	17 Chlorine **Cl** 35.5	18 Argon **Ar** 39.9
28 Nickel **Ni** 58.7	29 Copper **Cu** 63.5	30 Zinc **Zn** 65.4	31 Gallium **Ga** 69.7	32 Germanium **Ge** 72.6	33 Arsenic **As** 74.9	34 Selenium **Se** 79.0	35 Bromine **Br** 79.9	36 Krypton **Kr** 83.8
46 Palladium **Pd** 106.4	47 Silver **Ag** 107.9	48 Cadmium **Cd** 112.4	49 Indium **In** 114.8	50 Tin **Sn** 118.7	51 Antimony **Sb** 121.8	52 Tellurium **Te** 127.6	53 Iodine **I** 126.9	54 Xenon **Xe** 131.3
78 Platinum **Pt** 195.1	79 Gold **Au** 197.0	80 Mercury **Hg** 200.6	81 Thallium **Tl** 204.4	82 Lead **Pb** 207.2	83 Bismuth **Bi** 209.0	84 Polonium **Po** (210)	85 Astatine **At** (210)	86 Radon **Rn** (222)
110 Darmstadtium **Ds** (271)	111 Roentgenium **Rg** (272)	112 Copernicium **Cn** (285)	113 Nihonium **Nh** (280)	114 Flerovium **Fl** (289)	115 Moscovium **Mc** (289)	116 Livermorium **Lv** (292)	117 Tennessine **Ts** (294)	118 Oganesson **Og** (294)

↓ Metals

64 Gadolinium **Gd** 157.3	65 Terbium **Tb** 158.9	66 Dysprosium **Dy** 162.5	67 Holmium **Ho** 164.9	68 Erbium **Er** 167.3	69 Thulium **Tm** 168.9	70 Ytterbium **Yb** 173.1	71 Lutetium **Lu** 175.0
96 Curium **Cm** (247)	97 Berkelium **Bk** (247)	98 Californium **Cf** (251)	99 Einsteinium **Es** (252)	100 Fermium **Fm** (257)	101 Mendelevium **Md** (258)	102 Nobelium **No** (259)	103 Lawrencium **Lr** (262)

GLOSSARY

abrasive a property of a material or substance that easily scratches another
absorption the taking in of a substance; for example, from the intestine to the surrounding capillaries
abyssal plains relatively flat underwater deep-ocean floor, around 4000 metres depth
accuracy how close an experimental measurement is to a known value
active volcano a volcano that is erupting or has recently erupted
aim a statement outlining the purpose of an investigation
alchemist an olden-day 'chemist' who mixed chemicals and tried to change ordinary metals into gold; alchemists also tried to predict the future
alimentary canal see **gastrointestinal tract**
allergen an antigen that elicits an allergic response
allergy an abnormal immune response to a substance that is harmless for most people
alloy a mixture of a metal with a non-metal or another metal
alveoli tiny air sacs in the lungs at the ends of the narrowest tubes; oxygen moves from alveoli into the surrounding blood vessels, in exchange for carbon dioxide; singular = alveolus
ammonia a nitrogenous waste product of protein break down
amylases enzymes found in saliva that break starch down into sugar
anaphylaxis an acute and potentially lethal allergic reaction to an allergen to which a person has become hypersensitive
Animalia the kingdom of organisms that have cells with a membrane-bound nucleus, but no cell wall, large vacuole or chloroplasts (e.g. animals)
antibiotic a substance derived from a microorganism and used to kill bacteria in the body
anticline a fold in a rock with the narrow point facing upwards
antiseptic a mild disinfectant used on body tissue to kill microbes
anus the end of the digestive system, through which faeces are passed as waste
aorta a large artery through which oxygenated blood is pumped at high pressure from the left ventricle of the heart to the body
aqueous a solution with water as the solvent
arteries hollow tubes (vessels) with thick walls carrying blood pumped from the heart to other body parts
arterioles vessels that transport oxygenated blood from the arteries to the capillaries
arthritis a condition in which inflammation of the joints causes them to swell and become painful
asthenosphere the zone of mantle beneath the lithosphere, thought to be much hotter and more fluid than the lithosphere mantle
asthma narrowing of the air pipes that join the mouth and nose to the lungs
atomic number the number of protons in the nucleus of an atom, which identifies the element to which the atom belongs
atoms very small particles that make up all things; atoms have the same properties as the objects they make up
bactericidal describes an antiseptic that kills bacteria
bacteriostatic describes an antiseptic that stops bacteria from growing or dividing but doesn't kill them
ball-and-socket joints joints in which the rounded end of one bone fits into the hollow end of another
basalt a dark, igneous rock with small crystals formed by fast cooling of hot lava; it sometimes has holes that once contained volcanic gases
batholith an intrusive rock mass that covers an area of over 100 square kilometres
beaker a container for mixing or heating substances
bibliography a list of references and sources at the end of a scientific report
bile a substance produced by the liver that helps digest fats and oils
binocular microscope a microscope with two eyepieces through which the specimen is seen using both eyes
biodegradable describes a substance that breaks down or decomposes easily in the environment
black smoker a geothermal vent on the sea floor that ejects superheated, mineral-rich water

bladder a sac that stores urine
blood cells living cells in the blood
blood pressure measures how strongly the blood is pumped through the body's main arteries
blood vessels the veins, arteries and capillaries through which the blood flows around the body
body systems groups of organs within organisms that carry out specific functions
body waves seismic waves that quickly travel through the interior of Earth
bolus a round, chewed-up ball of food made in the mouth that makes swallowing easier
bonded joined by a force that holds particles of matter, such as atoms, together
bone marrow a substance inside bones in which blood cells are made
bones the pieces of hard tissue that make up the skeleton of a vertebrate
Bowman's capsule a cup-like structure at one end of a nephron within the kidney, surrounding the glomerulus; it serves as a filter to remove wastes and excess water
breathing the movement of muscles in the chest causing air to enter the lungs and the altered air in the lungs to leave; the air entering the lungs contains more oxygen and less carbon dioxide than the air leaving the lungs
brittle can easily break if hit; the opposite of malleable
bronchi the narrow tubes through which air passes from the trachea to the smaller bronchioles and alveoli in the respiratory system; singular = bronchus
bronchioles small, branching tubes in the lungs leading from the two larger bronchi to the alveoli
burning combining a substance with oxygen in a flame
burping the release of swallowed gas through the mouth
by-products in energy terms, energy produced that is not useful
calcium an element occurring in limestone and chalk, and also present in vertebrates and other animals as a component of bone and shell; it is necessary for nerve conduction, heartbeat, muscle contraction and many other physiological functions
calibrate to check or adjust a measuring instrument to ensure accurate measurements
capillaries numerous tiny blood vessels that are only a single cell thick to allow exchange of materials to and from body cells; every cell of the body is supplied with blood through capillaries
carbon dioxide a colourless gas (CO_2) made up of one carbon and two oxygen atoms; it is essential for photosynthesis and is a waste product of cellular respiration; the burning of fossil fuels also releases carbon dioxide
carcinogens chemicals that cause cancer
cardiac muscle a special kind of muscle in the heart that never tires; it is involved in pumping blood through the heart
carnivore an animal that eats other animals
cartilage a waxy, whitish, flexible substance that lines or connects bone joints or, in some animals such as sharks, replaces bone as the supporting skeletal tissue; the ears and tips of noses of humans are shaped by cartilage
cast a fossil cavity that has been filled with minerals or other matter
cell (in biology terms) the smallest unit of life; cells are the building blocks of living things and can be many different shapes and sizes; (in energy terms) a single battery
cell membrane the structure that encloses the contents of a cell and allows the movement of some materials in and out
cell theory the theory that states that all living things are made up of cells and that all cells come from pre-existing cells
cellular respiration a series of chemical reactions in which the chemical energy in molecules such as glucose is transferred into ATP molecules, which is a form of energy that the cells can use
cellulose a natural substance that keeps the cell walls of plants rigid
chemical change a change that results in at least one new substance being formed due to the breaking and forming of chemical bonds and rearrangement of atoms in a reaction
chemical digestion the chemical reactions that change food into simpler substances that are absorbed into the bloodstream for use in other parts of the body
chemical energy energy stored in chemical bonds that is released during chemical reactions

chemical formula shows the ratio of the atoms of each element present in a molecule or compound
chemical properties properties that describe how a substance combines with other substances to form new chemicals, or how a substance breaks up into two or more different substances
chemical reaction a chemical change between two or more substances in which one or more new chemical substances are produced
chemical symbol the standard way that scientists write the names of the elements, using either a capital letter or a capital followed by a lower-case letter; for example, carbon is C and copper is Cu
chlorophyll the green-coloured chemical in plants, located in chloroplasts, that absorbs light energy so that it can be used in the process of photosynthesis
chloroplasts oval-shaped organelles that are involved in the process of photosynthesis, which results in the conversion of light energy into chemical energy
circulatory system the heart, blood and blood vessels, which are responsible for circulating oxygen and nutrients to body cells, and carbon dioxide and other wastes away from them
coal a sedimentary rock formed from dead plants and animals that were buried before rotting completely, followed by compaction and some heating
collision refers to when two continents crumple together to form a mountain range
colloid a mixture in which a microscopically insoluble substance is dispersed and suspended throughout another substance
colon the part of the large intestine where food mass passes from the small intestine, and where water and other remaining essential nutrients are absorbed into the body
combustion the process of combining with oxygen, most commonly burning with a flame
compound a substance made up of two or more different types of atoms that are chemically bonded (covalent or ionic) together
compression a squeezing force
conducting path a connected series of materials along which an electric current can flow
conductors materials that have a very low resistance, allowing current to flow through them with ease
conglomerate a sedimentary rock containing large fragments of various sizes cemented together
conservative plate boundary where crust is neither created nor destroyed
constructive plate boundary where new crust is formed
continental drift the movement of Earth's continents relative to each other over geologic time
control an experimental set-up in which the independent variable is not applied; a control is used to ensure that the result is due to the variable and nothing else
controlled variables the conditions that must be kept the same throughout an experiment
convection currents the movement of particles in a liquid or gas resulting from a temperature or density difference
convergent boundary where two tectonic plates move towards each other
core the hot centre of Earth made of iron and nickel
corrosion a chemical reaction between air, water or chemicals in the air or water with a metal, which causes the metal to wear away
corrosive describes a chemical that wears away the surface of substances, especially metals
crust the hard and thin outer rock layer of Earth
crystal a geometrically shaped substance made up of atoms and molecules arranged in one of seven different shapes; the elements and the conditions present during the crystal's growth determine the arrangement of atoms and molecules and the shape of the crystals
cullet used glass
cytoplasm the jelly-like material inside a cell; it contains many organelles, such as the nucleus and vacuoles
cytosol the fluid found inside cells
data information collected that can be used for studying or analysing
decomposition the breaking up of a substance into smaller parts
deep-ocean trenches narrow and deep troughs in the ocean floor, generally greater than 5000 metres depth
delta a landform created by the deposition of sediment at the end of a river as it enters a body of water
denatured describes the condition of proteins after they have been overheated

deoxygenated blood blood from which some oxygen has been removed
dependent variable a variable that is expected to change when the independent variable is changed; the dependent variable is observed or measured during the experiment
deposition the settling of transported sediments
diaphragm a flexible, dome-shaped, muscular layer separating the chest and the abdomen; it is involved in breathing
diarrhoea excessive discharge of watery faeces
diastolic pressure the lower blood pressure reading during relaxation of the heart muscles
diffusion movement of molecules through the cell membrane
digestion the breakdown of food into a form that can be used by an animal; it includes both mechanical digestion and chemical digestion
digestive system a complex series of organs and glands that processes food to supply the body with the nutrients it needs to function effectively
discussion a detailed area of a scientific report that explains the results and how they link back to the relevant concepts; it also includes suggestions for improvements to the experiment
disinfectant a chemical used to kill bacteria on surfaces and non-living objects
dissolved refers to when a solid substance integrates into the liquid solvent
divergent boundary where two tectonic plates move apart
dormant volcano a volcano that has erupted in the last 10 000 years but is not currently erupting; they are considered likely to erupt again
ductile capable of being drawn into wires or threads; a property of most metals
durability the quality of lasting; not easily being worn out
dynamite a relatively stable explosive invented by Alfred Nobel in 1866; it is created by mixing nitroglycerine with an absorbent substance such as silica, forming a paste that can be shaped into rods
earthquake a sudden and violent shaking of the ground
ectothermic describes an animal whose body temperature is determined by its environment
efficiency in energy terms, the fraction of energy supplied to a device as useful energy
elastic describes a material that is able to return to its original size after being stretched
elasticity the property that allows a material to return to its original size after being stretched
electric current a measure of the number of electrons flowing through a circuit every second
electrical circuit consists of a power supply, a conducting path for charge to flow and one or more loads
electrocardiogram (ECG) a graph made using the tiny electrical impulses generated in the heart muscle, giving information about the health of the heart
electromagnetic radiation the radiant energy such as radio waves, infrared, visible light, x-rays and gamma rays released by magnetic or electric fields
electromagnetic spectrum the complete range of wavelengths of energy radiated as electric and magnetic fields
electromagnetic waves waves of electromagnetic radiation, light being just one example
electron microscope an instrument used for viewing very small objects; an electron microscope is much more powerful than a light microscope and can magnify things up to a million times
electrons very light, negatively charged particles inside an atom; electrons orbit around the atom's nucleus
elements pure substances made up of only one type of atom
emphysema a condition in which the air sacs in the lungs break open and join together, reducing the amount of oxygen taken in and carbon dioxide removed
emulsify combine two liquids that do not normally mix easily
endocrine system the body system of glands that produce and secrete hormones into the bloodstream to regulate processes in various organs
endoskeleton a skeleton that lies inside the body
endothermic (in chemistry terms) refers to chemical reactions that absorb heat energy from the surroundings; (in biology terms) describes an animal that can internally generate heat to maintain its body temperature
environmental impact statement (EIS) a report on the possible effects of a planned project on the environment
enzymes special chemicals that speed up reactions but are themselves not used up in the reaction

epicentre the surface point directly above the earthquake focus
epiglottis a leaf-like flap of cartilage behind the tongue that closes the air passage during swallowing
erosion the wearing away and removal of soil and rock by natural elements, such as wind, waves, rivers and ice, and by human activity
erythrocytes red blood cells
eukaryote any cell or organism with a membrane-bound nucleus (e.g. plants, animals, fungi and protists)
evaporates changes state from a liquid to a gas
excretion the removal of wastes from the body
excretory system the body system that removes waste substances from the body
exoskeleton a skeleton or shell that lies outside the body
exothermic refers to chemical reactions that give out heat energy to the surroundings
extinct volcano a volcano that *has not* erupted in the last 10 000 years; they are considered dead and not to erupt again
extrusive describes igneous rock that forms when lava cools on Earth's surface
fair test a test that changes only one variable and controls all other variables when attempting to answer a scientific question
falsifiable can be proven false
fault a break in the crust where one side moves relative to the other
filament a coil of wire made from a metal that glows brightly when it gets hot
filter funnel a funnel used with filter paper to separate solids from liquids
flaccid refers to cells that are not firm due to loss of water
flammability an indicator of how easily a substance catches fire
flammable describes substances, such as methylated spirits, that burn easily
flatulence the release of gas through the anus; this gas is produced by bacteria in the large intestine
flint a fine-grained sedimentary rock that leaves a very sharp edge when broken
floodplain flat, open land beside a river where sediments are deposited during floods
focus the location underground of the fault movement causing an earthquake
folding refers to when rocks bend into anticlines or synclines
foliated consisting of an arrangement of certain mineral grains into distinct bands, which gives the rock a striped appearance
fossil fuel a substance, such as coal, oil or natural gas, that has formed from the remains of ancient organisms; coal, oil and natural gas are often used as fuels — that is, they are burnt in order to produce heat
fossil any remains, impression or trace of a life form of a former geological age; evidence of life in the past
fracture a break in a bone
friction the rubbing together of surfaces; all friction produces thermal energy (heat)
Fungi the kingdom of organisms made up of cells that possess a membrane-bound nucleus and cell wall, but no chloroplasts (e.g. mushrooms); some fungi can help to decompose dead and decaying matter
gabbro a dark-coloured, intrusive igneous rock with a similar mineral composition to basalt, but with larger crystals
gall bladder a small organ that stores and concentrates bile within the body
galvanising protecting a metal by covering it with a more reactive metal that will corrode first
gastrointestinal tract also called the digestive tract or the alimentary canal, it is a tubular passage that starts with the mouth and ends with the anus; it intakes and digests food (absorbing energy and nutrients) and expels waste
geothermal energy refers to using heat from Earth as an energy source
glaciers large bodies of ice that move down slopes and push boulders, rocks and gravel
Global Positioning System (GPS) a network of satellites that tracks location and movement
glomerulus a cluster of capillaries in the kidney that acts as a filter to remove wastes and excess water
glucose a six-carbon sugar (monosaccharide) that acts as a primary energy supply for many organisms
gneiss a coarse-grained metamorphic rock with light and dark bands formed mainly as a result of great pressure on granite

Gondwana the southern part of the broken-up supercontinent of Pangaea, which included the continents of Africa, South America, Antarctica and Australia; also known as Gondwanaland

granite a light-coloured, intrusive igneous rock with mineral crystals large enough to see

greenstick fracture a break that is not completely through the bone, often seen in children

group in the periodic table of elements, a single vertical column of elements with a similar nature

guard cells cells on either side of a stoma that work together to control the opening and closing of the stoma

haemodialysis the process of passing blood through a machine to remove wastes

haemoglobin the red pigment in red blood cells that carries oxygen

hardness a measure of how difficult it is to scratch the surface of a solid material; hardness can be ranked using Mohs' scale

heart a muscular organ that pumps deoxygenated blood to the lungs to be oxygenated and then pumps the oxygenated blood to the body

heartbeat a contraction of the heart muscle occurring about 60–100 times per minute

heartburn a burning sensation caused by stomach acid rising into the oesophagus

herbivore an animal that eats only plants

heterogeneous has a non-uniform composition throughout

hinge joints joints in which two bones are connected so that movement occurs in one plane only

homogeneous has a uniform composition throughout

horst a highland between two normal faults

hotspot a volcanic region directly above an area of extremely hot mantle

hydrogen the element with the smallest atom and the most common element in living things; by itself, it is a colourless gas and combines with other elements to form a large number of substances, including water

hypothesis a suggested, testable explanation for observations or experimental results; it acts as a prediction for the investigation

igneous rocks rocks formed when hot, molten rock cools and hardens (solidifies)

immovable joints joints that allow no movement except when absorbing a hard blow

independent variable the variable that the scientist chooses to change to observe its effect on another variable

inert not reactive

inner core the solid inner-most layer of the core under extreme pressure conditions, with an approximate 1200 km radius

insectivore a carnivore that eats only insects

insulators materials that do not allow electricity to flow easily

intrusive describes igneous rock that forms when magma cools below Earth's surface

investigations activities aimed at finding information

involuntary muscles muscles not under the control of the will; they contract slowly and rhythmically, and are at work in the heart, intestines and lungs

joint a region where two bones meet

kidneys body organs that filter the blood, removing urea and other wastes

kinetic energy energy due to the motion of an object

large intestine the penultimate part of the digestive system, where water is absorbed from the waste before it is transported out of the body

Laurasia the northern part of the broken-up supercontinent of Pangaea, which included the continents of North America, Europe and Asia

lava an extremely hot liquid or semi-liquid rock from the mantle that reaches and flows or erupts on Earth's surface

Law of Conservation of Energy a law that states that energy cannot be created or destroyed

left atrium the upper-left section of the heart where oxygenated blood from the lungs enters the heart

left ventricle the lower-left section of the heart, which pumps oxygenated blood to all parts of the body

leucocytes white blood cells

ligament a band of tough tissue that connects the ends of bones or keeps an organ in place

light microscope an instrument used for viewing very small objects; a light microscope can magnify things up to 1500 times

lignin a hard substance in the walls of dead xylem cells that make up the tubes carrying water up plant stems; lignin forms up to 30 per cent of the wood of trees
limestone a sedimentary rock formed from the remains of sea organisms; it consists mainly of calcium carbonate (calcite)
line of best fit a smooth curve or line that passes as close as possible to all plotted points on a graph
lipases enzymes that break fats and oils down into fatty acids and glycerol
lipids a class of nutrients that include fats and oils
liquefaction a phenomenon in which soil loses its strength and stiffness due to strong ground shaking
lithify to transform sediment into rock
lithosphere the outermost layer of Earth; includes the crust and uppermost part of the mantle
liver the largest gland in the body; it secretes bile for digestion of fats, builds proteins from amino acids, breaks down many substances harmful to the body and has many other essential functions
load a device that uses electrical energy and converts it into other forms of energy
logbook a complete record of an investigation from the time a search for a topic is started
Love waves surface seismic waves that have a side-to-side motion
'lub dub' the sound made by the heart valves as they close
lungs the organ for breathing air; gas exchange occurs in the lungs
lustre the high shine and sheen of a substance caused by the way it reflects light
magma a very hot mixture of molten rock and gases, just below Earth's surface, that forms from melting of the mantle and occasionally the crust
magnification the number of times the image of an object has been enlarged using a lens or lens system; for example, a magnification of two means the object has been enlarged to twice its actual size
malleable able to be beaten, bent or flattened into shape
mantle the solid but soft middle rock layer of Earth
marble a metamorphic rock formed as a result of great heat or pressure on limestone
mass extinction a widespread and rapid decrease in the biodiversity and abundance of life
measuring cylinder a cylinder used to measure volumes of liquids accurately
mechanical digestion digestion that uses physical factors such as chewing with the teeth
metabolism the chemical reactions occurring within an organism that enable the organism to use energy and grow and repair cells
metalloids elements that have the appearance of metals but not all the other properties of metals
metals elements that conduct heat and electricity; shiny solids that can be made into thin wires and sheets that bend easily; mercury is the only liquid metal at room temperature
metamorphic rocks rocks formed from the change (alteration) of pre-existing rocks in response to increasing temperature and/or pressure conditions
metamorphism the process that changes rocks by extreme pressure or heat (or both)
micrometre one millionth of a metre
microscope an instrument used for viewing small objects
mineral (in biology terms) any of the inorganic elements that are essential to the functioning of the human body and are obtained from foods; (in geology terms) a naturally occurring, inorganic and solid substance with a defined chemical formula and an ordered arrangement of atoms
mining the process of removing natural resources from Earth
mitochondria small, rod-shaped organelles that are involved in the process of cellular respiration, which results in the conversion of energy into a form that the cells can use
mixture a combination of substances in which each keeps its own properties (i.e. not chemically bonded)
molecule two or more atoms joined (bonded) covalently together
monocular microscope a microscope with a single eyepiece through which the specimen is seen using only one eye
moraine a ridge made out of sediments deposited by a glacier
mould a cavity in a rock that shows the shape of the hard parts of an organism
mountain range a group of high-ground features, commonly the result of tectonic collision
mouth the opening of the gastrointestinal tract through which food is taken into the body

mudstone a fine-grained sedimentary rock made of mud (clay and silt)
multicellular made up of many cells
multicellular organisms living things comprised of specialised cells that perform specific functions
muscles tissue consisting of cells that can shorten
musculoskeletal system consists of the skeletal system (bones and joints) and the skeletal muscle system (voluntary or striated muscle); working together, these two systems protect the internal organs, maintain posture, produce blood cells, store minerals and enable the body to move
nanometre one billionth of a metre
nanotechnology a science and technology that focuses on manipulating the structure of matter at an atomic and molecular level
native elements elements found uncombined in Earth's crust
natural fibres fibres that form naturally — that is, they have not been made by humans; they include wool and silk from animals, and cotton from plants
nephrons the filtration and excretory units of the kidney
nervous system consists of neurons, nerves and the brain, which are responsible for detecting and responding to both internal and external stimuli
neutrons tiny, but heavy, particles found in the nucleus of an atom; they have no electrical charge
Newton's cradle a decorative device consisting of suspended metal balls that that can be made to swing backwards and forwards in a particular way
nitrogenous wastes waste products from protein breakdown, including ammonia, urea and uric acid
noble gases elements in the last column of the periodic table; they are extremely inert gases
non-metals elements that do not conduct electricity or heat; they melt and turn into gases easily, and are brittle and often coloured
normal fault a break where the rock above the fault moves 'down' due to tension
nucleus (in biology terms) a roundish structure inside a cell that acts as its control centre; (in chemistry terms) the central part of an atom, made up of protons and neutrons; plural = nuclei
nutrients substances that provide the energy and chemicals that living things need to stay alive, grow and reproduce
nylon a synthetic fibre; the monomers are joined together by the elimination of water molecules at the joins
observations information obtained by the use of our senses or measuring instruments
obsidian a black, glassy rock that breaks into pieces with smooth shell-like surfaces
ocean ridges submarine mountains that tower 2000 metres above the abyssal plains
oesophagus part of the digestive system, composed of a tube connecting the mouth with the stomach
omnivore an animal that eats plants and other animals
open-cut mining mining that removes soil and rocks on the surface of the land
ophiolites pieces of oceanic crust observed on continental crust (land)
ore mineral a mineral from which a valuable metal can be removed for profit
organelle any specialised structure in a cell that performs a specific function
organisms living things
organs structures, composed of tissue, that perform specific functions
oscillation one complete swing of a pendulum
ossification hardening of bones
osteoporosis loss of bone mass that causes bones to become lighter, more fragile and more easily broken
outer core the liquid outer layer of the core, which is about 2300 km thick
overburden waste rock removed from below the topsoil; this rock is replaced when the area is restored
oxidation a chemical reaction involving the loss of electrons by a substance
oxygen an atom that forms molecules (O_2) of tasteless and colourless gas; it is essential for cellular respiration for most organisms and is a product of photosynthesis
oxygenated blood the bright red blood that has been supplied with oxygen in the lungs
P-waves or primary waves body seismic waves with a compressional (push-and-pull) motion; they are the fastest and first to arrive

pacemaker an electronic device inserted in the chest to keep the heart beating regularly at the correct rate; it works by stimulating the heart with tiny electrical impulses

palaeontologist a scientist who studies fossils

pancreas a large gland in the body that produces and secretes the hormone insulin and an important digestive fluid containing enzymes

Pangaea a supercontinent that existed about 299 to 200 million years ago; all landmasses were joined together to form it

Panthalassa the vast ocean surrounding the supercontinent of Pangaea

pathogen a microorganism such as bacteria or a virus that can cause disease

pendulum an object swinging on the end of a string, chain or rod

percussion flaking a process in which tool stones, such as flint or obsidian, were struck with harder stones, such as quartzite, to shear large flakes off until they were a desired shape

period the time taken for one oscillation of a pendulum

periodic table a table listing all known elements; the elements are grouped according to their properties and in order of the number of protons in their nucleus

peristalsis the process of pushing food along the oesophagus or small intestine by the action of muscles

permineralisation the most common method of fossilisation, in which minerals fill the cellular spaces and crystallise; the shape of the original plant or animal is preserved in great detail

phloem a type of tissue that transports sugars made in the leaves to other parts of a plant

phosphorus a substance that plays an important role in almost every chemical reaction in the body; together with calcium, it is required by the body to maintain healthy bones and teeth

photosynthesis a series of chemical reactions that occur within chloroplasts in which the light energy is converted into chemical energy; the process also requires carbon dioxide and water, and produces oxygen and sugars, which the plant can use as 'food'

physical changes changes in which no new chemical substances are formed; a physical change may be a change in shape, size or state, and many of these changes are easy to reverse

physical properties properties that you can either observe using your five senses — seeing, hearing, touching, smelling and tasting — or measure directly

pivot joints joints that allows a twisting movement

Plantae the kingdom of organisms that have cells with a membrane-bound nucleus, cell wall, large vacuole and chloroplasts (e.g. plants)

plasma the yellowish liquid part of blood that contains water, minerals, food and wastes from cells

plastic a synthetic substance capable of being moulded

platelets small bodies involved in blood clotting; they are responsible for healing by clumping together around a wound

plate tectonics a scientific theory that describes the relative movements and interaction of plates of Earth's crust over the underlying mantle

polyester a synthetic fibre; the monomers are joined together by the elimination of water molecules at the joins

polymer a substance made by joining smaller identical units; all plastics are polymers

pop test a test that uses a flame to test for the presence of hydrogen; a 'pop' sound will be heard on ignition if the gas has been produced

potential energy energy stored within an object that depends on its position within a system, such as gravitational energy, elastic energy and chemical energy

power supply a device that can provide an electric current

precipitate the new, solid product produced when reactants are mixed together; a precipitate is insoluble in water

precision how close multiple measurements of the same investigation are to each other

products new chemical substances that result from a chemical reaction; new chemical bonds are formed to make the products during a chemical reaction

prokaryote any cell or organism without a membrane-bound nucleus (e.g. bacteria)

proteases enzymes that break proteins down into amino acids

protein a chemical made up of amino acids needed for growth and repair of cells in living things
Protista the kingdom of organisms made up of cells that possess a membrane-bound nucleus but vary in other features and do not fit into other groups (e.g. protozoans); also called Protoctista
protons tiny, but heavy, positively charged particles found in the nucleus of an atom
pulmonary artery the vessel through which deoxygenated blood, carrying wastes from respiration, travels from the heart to the lungs
pulmonary vein the vessel through which oxygenated blood travels from your lungs to the heart
pulse the alternating contraction and expansion of arteries due to the pumping of blood by the heart
pumice a glassy, pale igneous rock that forms when frothy rhyolite lava cools in the air; it often floats on water as it is very light and full of holes that once contained gas
qualitative data categorical data that examines the quality of something (e.g. colour or gender) rather than a measurement or quantity
quantitative data numerical data that examines the quantity of something (e.g. length or time)
quartzite an extremely compact and hard metamorphic rock consisting essentially of quartz
radioactive refers to when atoms are unstable and emit a particle to remove excess energy; these particles are capable of ionising other atoms upon collision, which can cause harm to living tissue
Rayleigh wave surface seismic waves that have a rolling motion
reactants chemical substances used in a chemical reaction; chemical bonds of the reactants are broken during a chemical reaction
reactivity a measure of how likely a particular substance reacts to make new substances
rectum the final section of the digestive system, where waste food matter is stored as faeces before being excreted through the anus
recycle to reuse an unwanted substance or object for another purpose
red blood cells living cells in the blood that transport oxygen to all other living cells in the body
rehabilitated restored to its previous condition or an acceptable, agreed alternative
relative age the age of a rock compared with the age of another rock
reliable data data that is able to be replicated in different circumstances but the same conditions
reproductive system the different reproductive organs required by many organisms to reproduce and create offspring
research question a question that is the focus of the investigation, which the experiment aims to answer
resistance a measure of the opposition to an electric current passing through an object, measured in ohms (Ω)
respiratory system the lungs and associated structures that are responsible for getting oxygen into the organism and carbon dioxide out
reverse fault a break where the rock above the fault moves 'up' due to compression
rhyolite a light-coloured, extrusive igneous rock with a similar mineral composition to granite, but with smaller crystals
ribosomes small structures within a cell in which proteins such as enzymes are made
Richter scale a logarithmic scale that measures the amount of energy released during an earthquake, thus allowing one earthquake to easily be compared to another
rift valley a sunken lowland between two normal faults; a graben
right atrium the upper-right section of the heart where deoxygenated blood from the body enters
right ventricle the lower-right section of the heart, which pumps deoxygenated blood to the lungs
rock cycle a cycle of processes that rocks experience in Earth's crust as they constantly change from one type to another
rock salt a sedimentary deposit formed when a salt lake or seabed dries up; the sediments are made of sodium chloride (halite)
Rube Goldberg machine a machine that is designed to perform a simple task in a complicated way
rust a red-brown substance formed when iron reacts with oxygen and water
rusting the corrosion of iron
S-waves or secondary waves body seismic waves with a transverse (up-and-down) motion; they are slower than P-waves and cannot travel through fluids
safety glasses plastic glasses used to protect the eyes during experiments

saliva a watery substance in the mouth that contains enzymes involved in the digestion of food
salivary glands glands in the mouth that produce saliva
sandstone a sedimentary rock with medium-sized grains; the sand grains are cemented together by silica, lime, mud or salts
satellite images images taken of Earth's surface
scavenger an animal that eats dead plant and animal material
scientific method a systematic and logical process of investigation to test hypotheses and answer questions based on data or experimental observations
scientists people skilled in or working in the fields of science; scientists use experiments to find out about the material world around them
scoria a dark, igneous rock formed from frothy basalt lava that cools quickly and is full of holes that once contained gas
sea-floor spreading the formation of oceanic crust, which occurs by the rising and melting mantle at ocean ridges that push older crust away from the ridge
sediment material broken down by weathering and erosion that is moved by wind or water and collects in layers
sedimentary rocks rocks formed through the deposition and compaction of layered sediment
seismic waves waves released when rock breaks or is rapidly moved
seismograph an instrument used to detect and measure the intensity of an earthquake; also called a seismometer
seismologist a scientist who studies earthquakes to both understand how they work and how to better predict them
sensor a device connected to an instrument, such as a data logger, that measures and sends information
series a formation of electricity-generating or electricity-using devices whereby the electricity passes from one device to the next in a single conducting loop, one after the other
shale a fine-grained sedimentary rock formed from thinly layered mud
shearing a smearing force
silicates a group of minerals consisting primarily of SiO_4^{2-} combined with metal ions, forming a major component of the rocks in Earth's crust
siltstone a sedimentary rock with a particle size between that of sandstone and mudstone
skeletal muscle system voluntary or striated muscle
skeletal system consists of the bones and joints
skeleton the bones or shell of an animal that support and protect it as well as allowing movement
skin the external covering of a vertebrate's body
slate a fine-grained metamorphic rock formed as a result of moderate heat and pressure on shale
small intestine the part of the digestive system between the stomach and large intestine, where much of the digestion of food and absorption of nutrients takes place
solute a dissolved substance in a solution
solvent the substance in which the solute is dissolved
spinneret a nozzle with small holes through which a plastic material passes, forming threads; also the organ used by spiders to create their webs
sprain an injury caused by tearing a ligament
state the condition or phase of a substance; the three main states of matter are solid, liquid and gas
stem cells undeveloped cells found in blood and bone marrow that can reproduce themselves indefinitely
stereo microscope a type of binocular microscope through which the detail of larger specimens can be observed
stomach a large muscular organ that churns and mixes food with gastric juice to start to break down protein
stomata openings mainly on the lower surface of leaves; these pores are opened and closed by guard cells; singular = stoma
Stone Age a prehistoric time when weapons and tools were made of stone, bone or wood
strainmeter equipment that measures any change in the shape of Earth's crust
streak the colour of a mineral as a fine powder, found by rubbing it onto an unglazed white ceramic tile
strike-slip fault a break where the rocks on either side of the fault move horizontally due to shearing
subduction refers to a convergent plate boundary where one plate moves under another

subduction zone where old oceanic crust enters the mantle
surface protection refers to when a protective coating is applied over a metal surface to prevent corrosion
surface waves seismic waves that travel slower than body waves and only along the surface of Earth; their energy is lost with depth and distance
suspension a mixture in which solid substances do not dissolve and are dispersed throughout the volume of the liquid
sustainable describes the concept of using Earth's resources so that the needs of the world's present population can be met, without damaging the ability of future populations to meet their needs
switch a device that opens and closes the conducting path through which a current flows
syncline a fold in a rock with the narrow point facing downwards
synovial fluid the liquid inside the cavity surrounding a joint that helps bones to slide freely over each other
systolic pressure the higher blood pressure reading during contraction of the heart muscles
teeth hard structures within the mouth that allow chewing
tendon tough, rope-like tissue connecting a muscle to a bone
tennis elbow occurs when repeated grasping and bending back of your wrist leads to the inflammation of the tendon that connects the muscles of your forearm to the bone in your upper arm, causing pain
tension a stretching force
testable able to be supported or proven false through the use of observations and investigation
test tube a thin glass container for holding, heating or mixing small amounts of substances
tiltmeter equipment that measures changes in the angle of the ground
tinder light, dry material that is easy to ignite
tissue a group of cells of similar structure that perform a specific function
torn hamstring a common sporting injury caused by overstretching the hamstring muscle, which joins the pelvis to the knee joint
toxic describes chemicals that are dangerous to touch, inhale or swallow
toxicity the danger to your health caused when poisonous substances combine with chemicals in your body to produce new substances with damaging effects
trace fossils fossils that provide evidence, such as footprints, that an organism was present when the rock was formed
trachea the narrow tube from the mouth to the lungs through which air moves
transducer a device that converts energy from one form into another form
transfer in energy terms, the movement of energy from one place to another
transform in energy terms, this refers to the changing of one form of energy into another
transform boundary where two tectonic plates slide past one another
translocation the process in which sugars and amino acids are transported within a plant by phloem tissue
transmitted passed through something, such as light or sound passing through air
transpiration the loss of water from plant leaves through their stomata
transpiration stream the movement of water through a plant as a result of loss of water from the leaves
tremors minor vibrations of the ground that are commonly not felt
tsunami a powerful ocean wave triggered by an undersea earth movement
turgid refers to cells that are firm
underground mining mining that uses shafts and tunnels to remove rock from deep below the surface
unicellular made up of only one cell
urea a nitrogen-containing substance produced by the breakdown of proteins and removed from the blood by the kidneys
ureters tubes from each kidney that carry urine to the bladder
urethra the tube through which urine is emptied from the bladder to the outside of the body
uric acid a nitrogenous waste product of protein breakdown
urination the passing of urine from the bladder to the outside of the body
urine a yellowish liquid, produced in the kidneys; it is mostly water and contains waste products from the blood such as urea, ammonia and uric acid
useful energy the energy that produces the desired output

vacuoles sacs within a cell used to store food and wastes; plant cells usually have one large vacuole, while animal cells have several small vacuoles or none at all

valid sound or true; a valid conclusion can be supported by other scientific investigations

valves flap-like folds in the lining of a blood vessel or other hollow organ that allow a liquid, such as blood, to flow in one direction only

variables quantities or conditions in an experiment that can change

varicose veins expanded or knotted blood vessels close to the skin, usually in the legs; they are caused by weak valves that do not prevent blood from flowing backwards

vascular bundles groups of xylem and phloem vessels within plant stems

veins blood vessels that carry blood back to the heart; they have valves and thinner walls than arteries

vena cava the large vein leading into the top-right chamber of the heart

venules small veins

villi tiny, finger-like projections from the wall of the intestine that maximise the surface area of the structure to increase the efficiency of nutrient absorption; singular = villus

viscosity a measure of a fluid's resistance to flow

visible spectrum different colours that combine to make up white light; they are separated in rainbows

vital capacity the largest volume of air that can be breathed in or out at one time

vitamin D a nutrient that regulates the concentration of calcium and phosphate in the bloodstream and promotes the healthy growth and remodelling of bone

volcanic bomb refers to when a large rock fragment that falls from an eruption, formed as lava, is blown out of a volcano and is rapidly cooled in the air

volcano a landscape feature through which melted rock is erupted onto Earth's surface

voltage the amount of energy that is pushing electrons around a circuit, per coulomb of charge that flows between two points

voluntary muscles muscles attached to bones; they move the bones by contracting and are controlled by an animal's thoughts

vomiting the forceful ejection of matter from the stomach through the mouth

weathering the physical or chemical breakdown of rocks on the surface

white blood cells living cells that fight bacteria and viruses as part of the human body's immune system

wilt refers to when plant stems and leaves droop due to insufficient water in their cells

xylem vessels pipelines for the flow of water up plants, made up of the remains of dead xylem cells fitted end to end with the joining walls broken down; lignin in the cell walls gives them strength

INDEX

D

E

Q

R

S